2020 IEEE 26th International Symposium for Design and Technology in Electronic Packaging (SIITME 2020)

Pitesti, Romania
21 – 24 October 2020

IEEE Catalog Number: CFP2007I-POD
ISBN: 978-1-7281-7507-2

**Copyright © 2020 by the Institute of Electrical and Electronics Engineers, Inc.
All Rights Reserved**

Copyright and Reprint Permissions: Abstracting is permitted with credit to the source. Libraries are permitted to photocopy beyond the limit of U.S. copyright law for private use of patrons those articles in this volume that carry a code at the bottom of the first page, provided the per-copy fee indicated in the code is paid through Copyright Clearance Center, 222 Rosewood Drive, Danvers, MA 01923.

For other copying, reprint or republication permission, write to IEEE Copyrights Manager, IEEE Service Center, 445 Hoes Lane, Piscataway, NJ 08854. All rights reserved.

*** This is a print representation of what appears in the IEEE Digital Library. Some format issues inherent in the e-media version may also appear in this print version.

IEEE Catalog Number: CFP2007I-POD
ISBN (Print-On-Demand): 978-1-7281-7507-2
ISBN (Online):   978-1-7281-7506-5
ISSN: 2641-287X

Additional Copies of This Publication Are Available From:

Curran Associates, Inc
57 Morehouse Lane
Red Hook, NY 12571 USA
Phone: (845) 758-0400
Fax: (845) 758-2633
E-mail: curran@proceedings.com
Web: www.proceedings.com

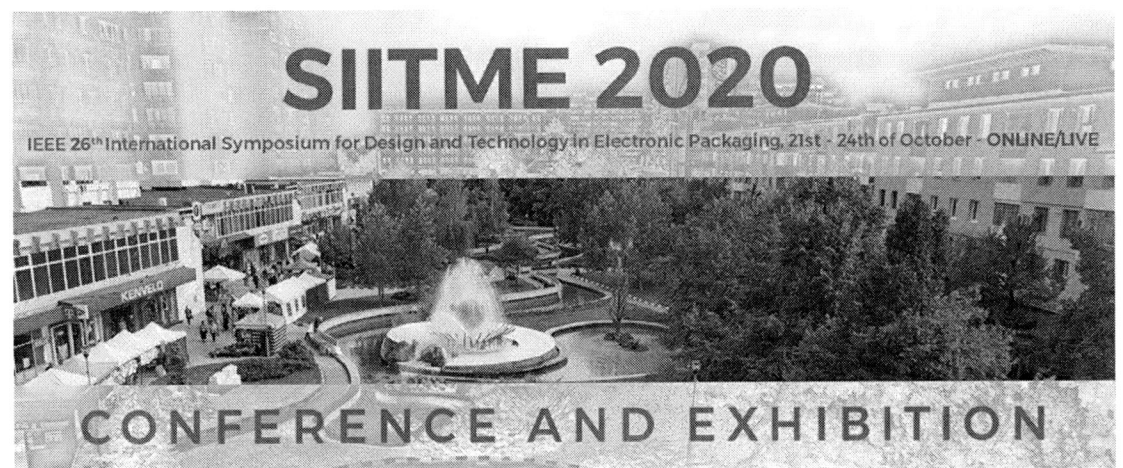

2020 IEEE 26th International Symposium for Design and Technology in Electronic Packaging (SIITME)

21st– 24th of October 2020, Pitesti, Romania

Conference Proceedings

Contact: Gabriel Chindris, Publication Chair
Tel: +40 264-401469 Fax: +40 264-594806
E-mail: gabriel.chindris@ael.utcluj.ro

Foreword

SIITME has become, during its long existence, a meaningful event focused on highlighting topics that are relevant for both industry and education, in the field of electronics technology. This statement is supported by the fact that, this year, an entire week is dedicated to electronics technology, bringing together students, valuable teachers, researchers and prominent representatives of the electronics industry in what that is called "Electronic Industry Week in Central and South Eastern Europe". In this way, the main goal of SIITME conference, to ensure a strong connection between industry and academia, is achieved.

I would like to congratulate the SIITME organizers, especially Prof. Paul Svasta and his team, for their work to raise, year by year, the quality of this event, making this conference one of the best European conference in the field of Electronic Packaging. In fact, based on these achievements, due to the involvement of Prof. Paul Svasta in promoting electronic technology, Romania was chosen to host in 2022 the ESTC conference, which represents the IEEE-EPS flagship conference in Europe.

I would like to express my gratitude to be part of SIITME 2020 conference and I wish to all the participants of "Electronic Industry Week in Central and South Eastern Europe" event a fruitful experience.

Prof. Ovidiu A. Pop, PhD
Technical University of Cluj-Napoca

CALL FOR PAPERS - SIITME 2020

The organising committee of SIITME 2020 kindly invites you to submit an abstract/paper to the 2020 IEEE 26th International Symposium for Design and Technology in Electronic Packaging (SIITME). The scientific event will be held as a virtual, live conference, on October 21st–24th, 2020.

This will ensure a high-quality conference to be held independent of what restrictions may apply in October 2020. The live format will allow interaction similar to an in-person conference. The digital format will give new possibilities, such as recording of presentations allowing to catch up presentations in parallel sessions.

TOPICS

A. Emerging Topics in Advanced Packaging;
B. New Components and Manufacturing Technologies;
C. Printed Electronics, Smart Textiles and Healthcare;
D. Sensors, Actuators and Microsystems;
E. Nanomaterials, Nanoelectronics and Nanotechnology;
F. Embedded Systems, Robotics and Artificial Intelligence;
G. Power Electronics and Thermal Management;
H. Smart Grid and Renewable Energy;
I. Virtual Prototyping and System Validation;
J. Quality Management, Applied Reliability, Characterization and Testing Failure Diagnosis;
K. Corrosion in Electronics;
L. Challenges in Digitalisation and Global Education for Electronics.

CHAIRS

General Chair:
Paul SVASTA, "Politehnica" University of Bucharest, Romania
General Academic Co-Chair:
Dan PITICĂ, Technical University of Cluj-Napoca, Romania
General Industrial Co-Chair:
Marian PETRESCU, Continental Automotive, Iaşi, Romania

Conference Chair:
Ioan LIȚĂ, University of Piteşti, Romania
Conference Co-Chair:
Cosmin MOISA, Continental Automotive, Timişoara, Romania

Technical Program Chair:
Detlef BONFERT, Fraunhofer EMFT, Münich, Germany
Technical Program Co-Chair:
Norocel CODREANU, "Politehnica" University of Bucharest, Romania

Awards Committee Chair:
Heinz WOHLRABE, Dresden University of Technology, Dresden, Germany

Scientific Committee Chair:
Balázs ILLÉS, Budapest University of Technology and Economics, Hungary
Scientific Co-Chairs:
Heinz WOHLRABE, Dresden University of Technology, Germany
Ciprian IONESCU, "Politehnica" University of Bucharest, Romania

Human Resource Education and Training Committee Chair:
Aurelia FLOREA, MIELE Tehnica, Braşov, Romania

Human Resource Education and Training Committee Co-Chair:
Maria MARCOVICI, Continental Automotive, Timisoara, Romania

Publication Chair:
Gabriel CHINDRIŞ, Technical University of Cluj- Napoca, Romania
Publication Co-Chair:
Bogdan MIHĂILESCU, "Politehnica" University of Bucharest, Romania

Professional Development Courses and International Publication Advisor:
Zsolt ILLYEFALVI-VITÉZ, Budapest University of Technology and Economics, Hungary

Virtual Conference Management Committee
Chair: Ioana MANEA, Cisco Romania
Co-Chairs: Alin DAVID, Cisco Romania
Robert DOBRE, University Politehnica of Bucharest, Romania

*The originality of papers will be checked by the IEEE CrossCheck plagiarism detection and prevention software.

Contents

O1	*Simulation Model of a GMR Based Current Sensor*	Elena-Mirela Stetco, Ovidiu Aurel Pop, Alin Grama	17
O2	*Analysis of single-cell force-spectroscopy data of Vero cells recorded by FluidFM BOT*	Ágoston G. Nagy, Attila Bonyár, Inna Székács, Robert Horvath	21
O3	*Enhanced X-Ray Inspection of Solder Joints in SMT Electronics Production using Convolutional Neural Networks*	Konstantin Schmidt, Nils Thielen, Christian Voigt, Reinhardt Seidel, Jörg Franke, Yannik Milde, Jochen Bönig, Gunter Beitinger	26
O4	*Surface-Enhanced Raman Spectroscopy Investigation of DNA Molecules on Gold/Epoxy Nanocomposite Substrates*	Shireen Zangana, Tomáš Lednický, István Rigó, István Csarnovics, Miklós Veres, Attila Bonyár	32
O5	*Small-Signal Modelling of the Three Switch 1L2C Boost Converter*	Septimiu Lica, Alex Molcuţ, Ioan Lie, Dan Lascu	37
O6	*Simulating the Evolution of Infectious Agents Through Human Interaction*	P.V. Vezeteu, D.I. Năstac	43
O7	*Study of Ceramic Capacitor technology link to Electro Chemical Migration in Automotive Electronics*	Szasz Francisc	47
O8	*Electromigration in lead-free solder joints on ceramic PCB substrates*	Dániel Straubinger, Dániel Rigler, Attila Géczy, Beata Synkiewicz-Musialska	52
O9	*Change Detection in the Complexity of Time Series with Information-based Criteria*	Dorel Aiordachioaie, Sorin Marius Pavel	57
O10	*On the Performance of LMS-Based Algorithms for the Identification of Low-Rank Systems*	Roxana-Elena Mihaescu, Constantin Paleologu, Silviu Ciochina, Jacob Benesty	63
P1.1	*Investigations at the Interface of a Multilayer Structure Made of Non-conductive and Conductive Resins*	Mihai Branzei, Gaudentiu Varzaru, Razvan Ungurelu, Ciprian Ionescu, Bogdan Mihailescu, Paul Svasta, Marin Gheorghe	67
P1.2	*Realization and Testing of a Supercapacitor, Pouch Type Cell*	R. Negroiu, P. Svasta, Al. Vasile, C. Ionescu, M. R. Buga	71
P1.3	*Data Mining System Architecture for Industrial Internet of Things in Electronics Production*	Reinhardt Seidel, Hassan Amada, Jonathan Fuchs, Nils Thielen, Konstantin Schmidt, Christian Voigt, Jörg Franke	75
P1.4	*Theoretical and Practical Aspects in the Design and Construction of Active Electrodes for EEG*	Daniela Andreea Coman, Silviu Ionita, Ioan Lita	81
P1.5	*Wearable Smart Prototype for Personal Air Quality Monitoring*	Attila Géczy, Lajos Kuglics, László Jakab, Gábor Harsányi	84

P1.6	Decision support platform for intelligent and sustainable farming	Mihaela Bălănescu, Andreea Bădicu, George Suciu, Carmen Poenaru, Adrian Pasat, Alexandru Vulpe, Marius Vochin	89
P1.7	Study on Unmanned Surface Vehicles used for Environmental Monitoring in Fragile Ecosystems	Mihaela Bălănescu, George Suciu, Andreea Bădicu, Andrei Bîrdici, Adrian Pasat, Carmen Poenaru, Ionel Zătreanu	94
P1.8	A Pupil Detection Algorithm Based on Contour Fourier Descriptors Analysis	Petronela Bonteanu, Radu Gabriel Bozomitu, Arcadie Cracan, Gabriel Bonteanu	98
P1.9	Intelligent Warning System for Drivers	Loredana-Maria Burciu, Rodica Constantinescu, Radu-Petru Fotescu, Paul Svasta	102
P1.10	Algorithm to Design Conductive Mesh for Tamperproof Envelope	Sorin Chițu, Daniel Ciprian Vasile, Tudor Ioan Honceriu, Paul Svasta	106
P1.11	Machine Learning algorithms for air pollutants forecasting	Marius Dobrea, Andreea Bădicu, Marina Barbu, Oana Șubea, Mihaela Bălănescu, Geroge Suciu, Andrei Bîrdici, Oana Orza, Ciprian Dobre	109
P1.12	Intelligent System for Vehicle Recognition	Radu-Petru Fotescu, Loredana-Maria Burciu, Paul Svasta, Rodica Constantinescu	114
P1.13	Investigation on modified SRR for accurate dielectric measurements	R. Gavrilă, I.A. Mocanu	118
P1.15	IoT Based Automatic Electronic System for Monitoring and Control of Street Lighting	Seher Kadirova, Teodor Nenov, Daniel Kajtsanov	122
P1.16	Dynamic adaptation of power emissivity for mobile microstrip antennas in a variable impedance environment with microcontroller	L. Baicu, B. Dumitrascu, M.Culea, N. Nistor	127
P1.17	Software Controlled Radio Receiver for Versatile Wireless Communications	Daniel Alexandru Visan, Mariana Jurian, Ioan Lita, Laurentiu Mihai Ionescu, Alin Gheorghita Mazare	132
P1.18	Automation Module for Precision Irrigation Systems	Ioan Lita, Daniel Alexandru Visan, Alin Gheorghita, Mazare, Laurentiu Mihai Ionescu, Adrian Ioan Lita	136
P1.19	LoRa and Bluetooth-based IoT alarm clock device for hearing-impaired people	Cătălina Mărculescu, Alina Machedon, Ana-Maria Claudia Drăgulinescu, Ioana Marcu, Ciprian Zamfirescu	140
P1.20	Android Application for Data Processing from a Gas Detection Sensor in Atmosphere	A. Alexandrescu, D.I. Năstac	144
P1.22	Application of Ultrasonic Sensors in Mapping Vineyard Parameters	E. Szilagyi, S. Meza, D. Petreus, T. Patarau, R. Etz	150
P1.23	Electronic system for measuring frequency in GHz range	F. Vasile, A. Craciun, M. Vladescu, P. Schiopu, V. Feies, N. D. Codreanu	155
P1.26	An Approach for Calculating the Temperature at a Point in the Cross Section Formed by Temperature	Snezhinka Lubomirova Zaharieva, Adriana Naydenova Borodzhieva, Iordan	159

	Sensors	Ivanov Stoev, Svilen Ivanov Stoyanov	
P1.27	*Analyzing the RFID Failure Impact on Availability of IoT Services*	C. Corches, I. C. Donca, O. Stan, L. Miclea, M. Daraban	163
P1.28	*A Metamodel Residual-based Stopping Criterion for Adaptive Verification of Integrated Circuits*	Ingrid Kovacs, Marina Ţopa, Monica Ene, Andi Buzo, Georg Pelz	169
P1.29	*Prediction algorithms using specialized software tools for steel industry equipment*	E. Raducan, V. Nicolau, M. Andrei, G. Petrea, G.M. Vlej	174
P1.31	*Numerical Models of the Electrochemical Migration: a short review*	A. Gharaibeh, B. Illés, A. Géczy, B. Medgyes	178
P2.1	*SoC based IoT sensor network hub for activity recognition using ML.net framework*	Alexandru Alexan, Anca Alexan, Oniga Ştefan	184
P2.2	*Machine learning activity detection using ML.Net*	Anca Alexan, Alexandru Alexan, Oniga Ştefan	188
P2.3	*Embedded System for Smart Controlling Electronic Devices*	D. G. Bălan, A. Drumea, A. E. Marcu	192
P2.5	*Impedance matching for UHF band antennas on ceramic substrate*	Călin Mircea, Svasta Paul	196
P2.6	*Key Expansion in Cryptographic Systems*	Sorin Chiţu, Daniel Ciprian Vasile, Ionuţ Daniel Trămândan, Paul Svasta	202
P2.7	*Blockchain-Based Image Copyright Protection System using JPEG Resistant Digital Signature*	Robert Alexandru Dobre, Radu Ovidiu Preda, Radu Alexandru Badea, Mihai Stanciu	206
P2.8	*High performance interconnecting technique using power line communication*	Elena Valentina Dumitraşcu, Paul Mugur Svasta, Madalin Vasile Moise, Aurelian Kotlar	211
P2.9	*Image Compression and Noise Reduction through Algorithms in Wavelet Domain*	Cătălin Dumitrescu, Maria Simona Răboacă, Ioana Manta	215
P2.10	*Usage of ZigBee and LoRa wireless technologies in IoT systems*	Vlad-Dacian Gavra , Ovidiu Aurel Pop	221
P2.11	*Towards real-time and real-life image classification and detection using CNN: a review of practical applications requirements, algorithms, hardware and current trends*	Mariana Eugenia Ilas, Constantin Ilas	225
P2.13	*Investigating the performance of MicroPython and C on ESP32 and STM32 microcontrollers*	Valeriu Manuel Ionescu, Florentina Magda Enescu	234
P2.14	*A Neuro Model for Weather Forecasting*	T.G. Predună, V.A. Rusu, D.I. Năstac	238
P2.15	*Intelligent Control for Dual-Boiler System with Digital Communication for Smart Buildings*	V. Nicolau, M. Andrei, G. Petrea, E. Raducan	242

P2.16	On Image Processing System for Robot Control using DSK 6713 DSP Kit	G. Petrea, V. Nicolau, M. Andrei	246
P2.17	Development and Test of a Data Framework for Prediction of Soldering Quality in Selective Wave Soldering Applying K-Nearest Neighbors	Reinhardt Seidel, Nils Thielen, Konstantin Schmidt, Christian Voigt, Jörg Franke	250
P2.18	Resource Utilization Comparison between Plain FPGA and SoC Combined with FPGA for Image Processing Applications Used by Robotic Arms	Roland Szabo, Aurel Gontean	256
P2.19	Smart System for Incubating Eggs	L.A. Szolga, A. Bondric	260
P2.20	Phosphor Based White LED Driver by Taking Advantage on the Remanence Effect	L.A. Szolga, R.G. Groza	265
P2.21	Integration of Internet of Things technology into a pill dispenser	Madalin Vasile Moise, Ana-Maria Niculescu, Andreea Dumitraşcu	270
P2.22	Design of a command and control system for an automatic pill dispenser	Madalin Vasile Moise, Daniela-Mihaela Pavel, Nicolae Elisei	274
P2.23	Design of touch ECG detection system based on STM32 and Android mobile phone	Junzhuo Zhou, Ang Li, Zeying Tian	278
P2.24	Sensorless BLDC Control Method	A. Zîrnea, G. Bărbulescu, M. Păunoiu, C. Pop, N. Codreanu, M. Enăchescu	283
P2.25	Increasing Students' Motivation Using Project-Based Learning on the Topic of Electrical Filters	Adriana N. Borodzhieva	287
P2.26	Computer-Based Education for Teaching the Topic "Galois Linear Feedback Shift Registers"	Adriana N. Borodzhieva	291
P2.27	Computer-Aided Tools for Synthesis and Analysis of Pseudorandom Number Generators	Adriana Borodzhieva, Iordan Stoev, Snezhinka Zaharieva , Valentin Mutkov	294
P2.28	Education 4.0: An Adaptive Framework with Artificial Intelligence, Raspberry Pi and Wearables - Innovation for Creating Value	Monica Ionita Ciolacu , Ali Fallah Tehrani, Paul Svasta, Ioan Tache , Dan Stoichescu	298
P2.29	Adaptation of Electrical Engineering Education to the COVID-19 Situation: Method and Results	B. I. Evstatiev, T. V. Hristova	304
P2.30	Higher Education with Distance Learning during COVID-19 Pandemic – a Transitional Semester from the Viewpoint of Teachers	Attila Géczy, Olivér Krammer, László Sujbert	309
P2.31	Integrated topics approach for teamwork students projects	Madalin Vasile Moise, Paul Mugur Svasta, Elena Valentina Dumitrascu	314
P3.1	Optimization of Silver-PDMS and Gold-PDMS Surface Nanocomposite Fabrication Technologies Considering	Alexandra Borók, Zsanett Izsold, Shireen Zangana, Attila Bonyár	318

	LSPR and SERS applications		
P3.2	*Solar Cell Types and Technologies with Applications in Energy Harvesting*	Andrei Drăgulinescu, Ana-Maria Claudia Drăgulinescu	323
P3.3	*Characteristics of a Dilute Nitride InGaAsN Double Quantum Well Laser at 1047 nm*	Andrei Drăgulinescu, Mihail Dumitrescu	327
P3.4	*Dependence of Shear Strength of Adhesive Conductive Joints on Adhesive Modification with the Silver Nanoparticles and Climatic Aging*	Pavel Mach	331
P3.5	*Effect of different analyte solutions on the SERS process examined on gold nanoisland samples*	Petra Pál, István Csarnovics, Miklós Veres, Attila Bonyár	335
P3.6	*Optical properties of core-shell Ag@Au and Au@Ag nanoparticles*	Géza Szántó, István Csarnovics, Attila Bonyár	338
P3.7	*Modelling Thermally-Induced Mechanical Faults in Power Integrated Circuits Assemblies*	A. Bojiţă, M. Purcar, V. Ţopa, R. Oneţ, M. Neag	342
P3.9	*DC/DC Converter Output Capacitor Bank's Reliability Comparison using Prediction Standard MIL-HDBK-217F and IEC 61709*	Dan Butnicu, Luminiţa-Camelia Lazăr	346
P3.10	*An Efficiency Comparative Workbench Study of eGaN and Silicon Discrete Transistor based Buck Converters*	Dan Butnicu, Luminiţa-Camelia Lazăr	350
P3.11	*Modelling of the Thermal Conditions of a LED Driver*	N. L. Evstatieva, B. I. Evstatiev	354
P3.12	*Spent Battery Classification by Electrical Characterization*	A. Fazakas, M. Purcar, A. C. Vonsza	358
P3.13	*Converter topologies for MVDC traction transformers*	Izsák Ferencz, Dorin Petreuş, Pietro Tricoli	362
P3.14	*Electro-Thermal Simulation of Power DMOS Devices Operating under Fast Thermal Cycling*	Ciprian Florea, Vasile Ţopa, Dan Simon	368
P3.15	*Estimating Power Dissipation through Thermal Measurements in Power Circuits*	A. Fodor, G. Chindris	372
P3.16	*A Comparison between State of Charge Estimation Methods: Extended Kalman Filter and Unscented Kalman Filter*	Adelina Ioana Ilieş, Gabriel Chindriş, Dan Pitică	376
P3.18	*Cooling Techniques for M.2 to PCI(e) Adapters*	Rajmond Jánó, Alexandra Fodor	382
P3.19	*A Generalized Model for Stacked Boost Single-Switch Converters*	Septimiu Lica, Vlad Vătău, Dan Lascu, Mircea Tomoroga	386

P3.20	Vector Control Of Permanent Magnet Synchronous Machine	Ana-Maria Petri, Dorin Petreuş	390
P3.22	Comparison between LLC and Phase-Shift converter with synchronous rectification for high power, high current applications	T. M. Patarau, D. M. Petreus, I. Ferencz, Z. Orban	398
P3.23	Impact Protection of Vehicles by Automatic Cutting of General Power Supply with GTO	Alexandru Vasile, Irina Bristena Bacis, Ciprian Ionescu	404
P3.24	Two-Stage Converter for Piezoelectric Energy Harvesting using Buck Configuration	C. Covaci, A. Gontean	408
P3.25	The Energy Efficiency of a Prosumer in a Photovoltaic System	Marius-Alexandru Dobrea, Mihaela Vasluianu, Stefan Bichiu, Ioana Opris	412
P3.26	3D tracking system at maximum solar emissivity with microcontroller	B. Dumitrascu, L. Baicu, A. Culea Florescu, N. Nistor	417
P3.27	Virtual Investigations of a Stand-Alone Photovoltaic System with Supercapacitor Bank Used to Power an Irrigation System	B. I. Evstatiev, N. D. Codreanu, K. G. Gabrovska-Evstatieva	421
P3.28	Islanded Microgrid Simulation and Cost Optimisation	A. Ignat, E. Szilagyi, D. Petreuş	426
P3.29	Theoretical and Numerical Aspects Concerning the Stress in a Superconducting Solenoid	Radu Jubleanu, Dumitru Cazacu, Nicu Bizon	430
P3.31	Comparative Analysis of Pad Geometries Used for Multi-Layer Ceramic Capacitors in Power Distribution Networks	Adrian-Razvan Petre, A.Drumea, M.Pantazica, C.I.Marghescu	434
P3.32	EMC Simulation of Conducted Emissions Produced by a DC-DC converter	Andrei-Marius Silaghi, Florin Berinde, Ciprian Bleoju, Aldo De Sabata	440

Authors

Author	Page
Aiordachioaie, D.	57
Alexan, A.	184, 188
Alexan, Al.	184, 188
Alexandrescu, A.	144
Amanda, H.	75
Andrei, M.	174, 242, 246
Bacis, I. B.	404
Badea, R. A.	206
Bădicu, A.	89, 94, 109
Baicu, L.	127, 417
Bălan, D. G.	192
Bălănescu, M.	89, 94, 109
Barbu, M.	109
Bărbulescu, G.	283
Beitinger, G.	26
Benesty, J.	63
Berinde, F.	440
Bichiu, St.	412
Bîrdici, A.	94, 109
Bizon, N.	430
Bleoju, C.	440
Bojiță, A.	342
Bondric, A.	260
Bönig, J.	26
Bonteanu, G.	98
Bonteanu, P.	98
Bonyár, A.	21, 32, 318, 335, 338
Borodzhieva, A. N.	159, 287, 291, 294
Borók, A.	318
Bozomitu, R. G.	98
Branzei, M.	67
Buga, M. R.	71
Burciu, L. M.	102, 114
Butnicu, D.	346, 350
Buzo, A.	169
Cazacu, D.	430
Chindris, G.	372, 376
Chițu, S.	106, 202

Ciochina, S.	63
Ciolacu, M. I.	298
Codreanu, N.	283
Codreanu, N. D.	155, 421
Coman, D. A.	81
Constantinescu, R.	102, 114
Corches, C.	163
Covaci, C.	408
Cracan, A.	98
Craciun, A.	155
Csarnovics, I.	32, 335
Csarnovics, I.	335
Culea Florescu, A.	417
Culea, M.	127
Daraban, M.	163
De Sabata, A.	440
Dobre, C.	109
Dobre, R. A.	206
Dobrea, M.	109
Dobrea, M. Al.	412
Donca, I. C.	163
Drăgulinescu, A.	323, 327
Drăgulinescu, A. M. C.	140, 323
Drumea, A.	192, 434
Dumitraşcu, A.	270
Dumitrascu, B.	127, 417
Dumitrascu, E. V.	211, 314
Dumitrescu, C.	215
Dumitrescu, M.	327
Elisei, N.	274
Enăchescu, M.	283
Ene, M.	169
Enescu, F. M.	234
Etz, R.	150
Evstatiev, B. I.	304, 354, 421
Evstatieva, N. L.	354
Fazakas, A.	358
Feies, V.	155
Ferencz, I.	362, 398
Florea, C.	368

Fodor, A.	372, 382
Fotescu, R. P.	102, 114
Franke, J.	26, 75, 250
Fuchs, J.	75
Gabrovska-Evstatieva, K. G.	421
Gavra, V. D.	221
Gavrilă, R.	118
Géczy, A.	52, 84, 178, 309
Gharaibeh, A.	178
Gheorghe, M.	67
Gontean, A.	256, 408
Grama, A.	17
Groza, R. A.	265
Harsányi, G.	84
Honceriu, T. I.	106
Horvath, R.	21
Hristova, T. V.	304
Ignat, A.	426
Ilas, C.	225
Ilas, M. E.	225
Ilieş, A. I.	376
Illés, B.	178
Ionescu, C.	67, 71, 404
Ionescu, M. I.	132, 136
Ionescu, V. M.	234
Ionita, S.	81
Izsold, Z.	318
Jakab, L.	84
Jánó, R.	382
Jubleanu, R.	430
Jurian, M.	132
Kadirova, S.	122
Kajtsanov, D.	122
Kotlar, A.	211
Kovacs, I.	169
Krammer, O.	309
Kuglics, L.	84
Lascu, D.	37, 386
Lazăr, L. C.	346, 350
Lednický, T.	32

Li, A.	278
Lica, S.	37, 386
Lie, I.	37
Lita, A. I.	136
Lita, I.	81, 132, 136
Mach, P.	331
Machedon, A.	140
Manta, I.	215
Marcu, A. E.	192
Marcu, I.	140
Mărculescu, C.	140
Marghescu, C. I.	434
Mazare, A. G.	132, 136
Medgyes, B.	178
Meza, S.	150
Miclea, L.	163
Mihaescu, R. E.	63
Mihailescu, B.	67
Milde, Y.	26
Mircea, C.	196
Mocanu, I. A.	118
Moise, M. V.	211, 270, 274, 314
Molcuț, A.	37
Mutkov, V.	294
Nagy, G. A.	21
Năstac, D. I.	43, 144, 238
Neag, M.	342
Negroiu, R.	71
Nenov, T.	122
Nicolau, V.	174, 241, 246
Niculescu, A. M.	270
Nistor, N.	127, 417
Oneț, R.	342
Oniga, St.	184, 188
Opris, I.	412
Orban, Z.	398
Orza, O.	109
Pál, P.	335
Paleologu, C.	63
Pantazica, M.	434

Pasat, A.	89, 94
Patarau, T.	150, 398
Pavel, D. M.	274
Pavel, S. M.	57
Pelz, G.	169
Petre, A. R.	434
Petrea, G.	174, 241, 246
Petreus, D.	150, 362, 390, 398, 426
Petri, A. M.	390
Pitică, D.	376
Poenaru, C.	89, 94
Pop, C.	283
Pop, O. A.	17, 221
Preda, R. O.	206
Predună, T. G.	238
Purcar, M.	342, 358
Răboacă, M. S.	215
Raducan, E.	174, 241
Rigler, D.	52
Rigó, I.	32
Rusu, V. A.	238
Schiopu, P.	155
Schmidt, K.	26, 75, 250
Seidel, R.	26, 75, 250
Silaghi, A. M.	440
Simon, D.	368
Stan, O.	163
Stanciu, M.	206
Stetco, E. M.	17
Stoev, I. I.	159, 294
Stoichescu, D.	298
Stoyanov, S. I.	159
Straubinger, D.	52
Şubea, O.	109
Suciu, G.	89, 94, 109
Sujbert, L.	309
Svasta, P. M.	67, 71, 102, 106, 114, 196, 202, 211, 298, 314
Synkiewicz-Musialska, B.	52
Szabo, R.	256

Szántó, G.	338
Szasz, F.	47
Székács, I.	21
Szilagyi, E.	150, 426
Szolga, L. A.	260, 265
Tache, I.	298
Tehrani, A. F.	298
Thielen, N.	26, 75, 250
Tian, Z.	278
Tomoroga, M.	386
Țopa, M.	169, 342
Țopa, V.	368
Trămândan, I. D.	202
Tricoli, P.	362
Ungurelu, R.	67
Varzaru, G.	67
Vasile, Al.	71, 404
Vasile, D. C.	106, 202
Vasile, F.	155
Vasluianu, M.	412
Vătău, V.	386
Veres, M.	32, 335
Vezeteu, P.V.	43
Visan, D. A.	132, 136
Vladescu, M.	155
Vlej, G. M.	174
Vochin, M.	89
Voigt, C.	26, 75, 250
Vonsza, A. C.	358
Vulpe, A.	89
Zaharieva, S. L.	159, 294
Zamfirescu, C.	140
Zangana, S.	32, 318
Zătreanu, I.	94
Zhou, J.	278
Zîrnea, A.	283

Simulation Model of a GMR Based Current Sensor

Elena-Mirela Stetco, Ovidiu Aurel Pop, Alin Grama
Applied Electronics Department
Technical University of Cluj-Napoca
Cluj-Napoca, Romania
elena.stetco@ael.utcluj.ro

Abstract— **The paper presents a Pspice electrical model of a current sensor based on the giant magnetoresistance effect. The novelty of the proposed model consists in the addition of non-electrical parameters that influence the correct measurement of current. By using these parameters, it is possible to perform tests by simulating the behaviour of the sensor at different distances from the current trace and different positions of the current trace (current path placed on the top layer or on the bottom layer of the PCB). Also, the simulation of the current sensor model behaviour take into consideration different standard FR4 PCB thicknesses used in the manufacture process.**

Keywords—Giant MagnetoResistance; PSpice model; Spin-Valve; PCB.

I. Introduction

Starting from previous results [1] based on the PSpice model developed for a GMR spin-valve, this paper proposes a simulation model of a commercial current sensor, also build on a GMR structure. The model was developed based on the transfer characteristic of the current sensor, provided by the manufacturer [2-6], which shows the dependence of the output voltage of the sensor as a function of the measured magnetic field. The proposed model is designed depending on the magnetic field created by the current passing through a conductor located in the vicinity of the sensor, as well as on the distance at which the conductor is placed. The design method is suitable to describe the behaviour of different types of magnetoresistive sensors. In the present paper the behaviour of a commercial sensor NVE AA002 is modeled using the methodology validated in [1]. The principle of operation of the described current sensor is shown in Fig.1.

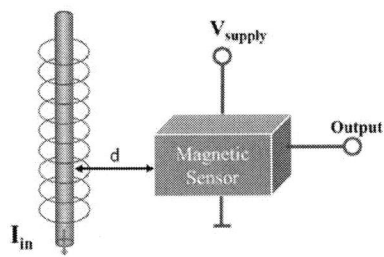

Fig. 1. Operating principle of the current sensor

The sensor model is implemented using the Look-Up Table method, based on the parameters taken from the datasheet, which is a table-based approach to empirical modelling [1][7].

The main steps of SPICE modelling are:

- Data collection – the data are taken from the experimental characteristic of the sensor [7]
- Model specification – methodology [1]
- SPICE implementation
- Model testing and evaluation of fit
- Optimization
- Results interpretation and validation

Existing PSpice models of GMR magnetic sensors take into account only the electrical parameters of the device. This paper proposes a PSpice model that can be used to study the behaviour of the device according with the position of the current line on the top or bottom side of the PCB and, in case in which the current trace is located on the bottom side, for different types of PCB's thickness. In order to do that, another parameter was included in the PSpice model to allow the study of the influence of the distance at which the sensor is positioned from the current trace.

II. Design and Implementation

A. Principle of operation

The purpose of this research is to model a commercial GMR current sensor NVE AA002 using the methodology proposed in [1]. The NVE AA series sensors are designed to be placed over a circuit board current trace, at a certain distance d, according to the block diagram from Fig.2 and Fig. 3. The magnetic field generated by the current passing through the path is determined with Ampére's Law [2-6].

$$B = \frac{\mu_0 \cdot I}{2\pi d} \ [T] \tag{1}$$

where μ_0 represents the magnetic permeability of the vacuum, I is the current passing through the current trace and d represents the distance between the trace and the GMR current sensor [1].

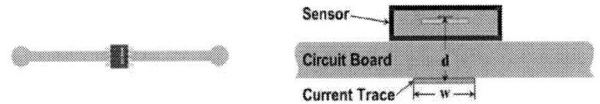

Fig. 2. The geometry of current - sensing over a circuit board trace [4].

978-1-7281-7507-2/20 $31.00 © 2020 IEEE

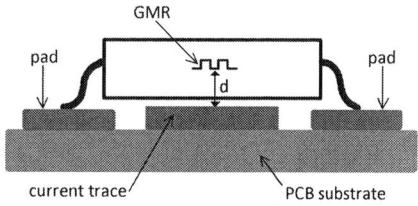

Fig. 3. GMR sensor chip placed on PCB.

The sensor operates as a Wheatstone bridge with four GMR elements, from which, two are magnetically shielded and two are active elements, Fig.4. This means that the bridge behaves as a half bridge, whose output voltage can be determined mathematically according to equation (1) [8-9].

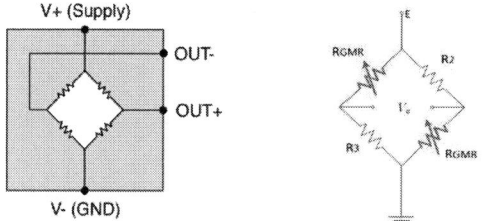

Fig.4. NVE AA002 GMR sensor functional block diagram [4].

$$V_o = \frac{R_{GMR}^2 - R_2 \cdot R_3}{(R_2 + R_{GMR}) \cdot (R_3 + R_{GMR})} \cdot E \qquad (2)$$

If $R_2 = R_3 = R$; $R_{GMR} = R \pm \Delta R$, where ΔR is the variation of the electrical resistance of the sensor when applying an external magnetic field, results

$$V_{out} = E \cdot \frac{\frac{\Delta R}{R}}{2 \cdot \left(2 + \frac{\Delta R}{R}\right)} \qquad (3)$$

For small variations in sensor resistance, $\Delta R / R \ll 1$, it results:

$$V_{out} = E \cdot \frac{\frac{\Delta R}{R}}{2} \qquad (4)$$

The sensitivity of the sensor can be determined according to the equation (5)

$$S = \frac{\partial V_{out}}{\partial \Delta R} = E \cdot \frac{\frac{\Delta R}{R}}{2} \qquad (5)$$

B. PSpice design

The electrical model is implemented in Cadence, using the ABM (Analogue Behavioral Model) library of the SPICE software package. ABM blocks are used to describe the behavior of electronic devices by modeling the transfer function or by means of a correspondence table - Look-Up Table, which allows a table-based approach to empirical modeling [1] [11-14].

The model is implemented based on the current sensor transfer characteristic, provided by the manufacturer in the

catalog sheets [2-6], which shows the dependence of the sensor output voltage depending on the measured magnetic field. In Fig.5 it can be observed the typical AA002 output.

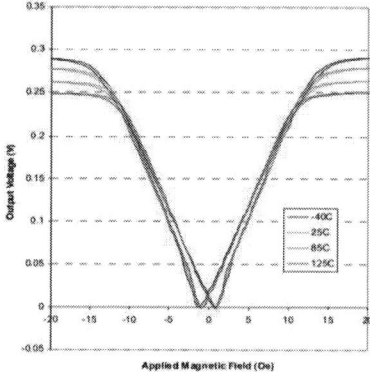

Fig. 5. Transfer characteristic of AA002 current sensor from datasheet [4].

Taking into account the characteristic at a temperature of 25°C, with the help of a specialized tool the pairs of H-V points describing the curve were extracted and then introduced in the ETABLE component of the model. In general, in Orcad, the magnetic field can be modeled using a current source, but because the ETABLE component only accepts voltages as input quantities, a current-controlled voltage source was used.

The model created allows the setting of the distance between the GMR NVE AA002 sensor and the current path and is represented in Fig. 6.

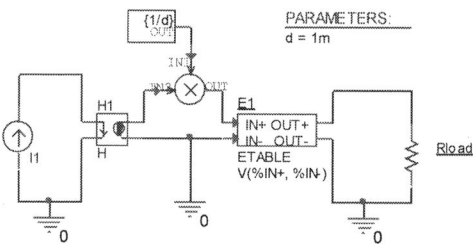

Fig. 6. PSpice model of the current sensor.

III. RESULTS

The results returned by the PSpice model simulation are compared with the measurements performed on an evaluation board. The output voltage versus field characteristic of GMR NVE AA002 sensor is shown in Fig. 7.

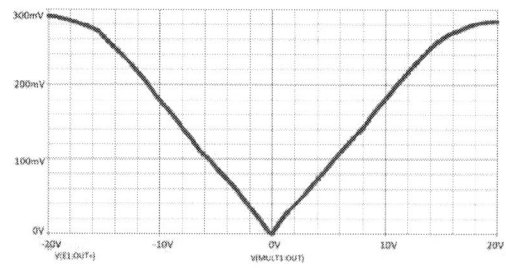

Fig. 7. Output voltage - field characteristic of GMR NVE AA002 sensor.

On the x-axis are represented the voltage values that correspond to the magnetic field values from Fig.5, because the simulation environments of the electronic circuits do not contain PSpice models for the magnetic field source, and the ABM blocks operates with voltages.

All tests were performed on the most used dielectric PCB - FR4. Also, an important aspect is the placement of the current path on the top or bottom layer of the PCB, according with the schematic representation from Fig.8, taken from [5].

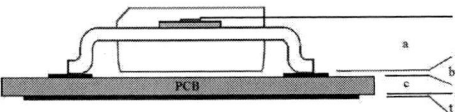

Fig. 8. Schematic representation of placement of the current path on PCB [5].

where a = the distance between bottom of leads and the sensor element, b = the distance between bottom of package leads and top of PCB, c = PCB thickness, t = current trace thickness [5].

In a first stage, the configuration in which the current line was placed on the top-side and the integrated circuit at a distance d above the current line was tested. The commercial sensors NVE AA002 are delivered in two types of package: MSOP and SOIC. The producer recommends the larger SOIC package to reduce the influence of mechanical tolerances that can cause errors in distance calculus. The datasheet delivered by NVE company specifies a distance between the sensing element and the bottom side of leads about a=1.15mm [5].

Another dimension that must be added to the total distance is the distance between leads and top of PCB. Knowing that the copper thickness for a standard PCB is 0.035mm per ounce, for a 3 oz copper with approximately 0.020 mm plated we obtain a complete distance between package leads and top of PCB about b=0.125 mm [5][15].

After summing the two distances calculated previously, the distance between the sensor and the PCB is obtained. From this value, half of trace thickness (t/2=0.035/2mm=0.0175mm) is extracted to obtain the total distance between the sensor and the center of current trace, d [5].

$$d = a + b - t/2 = 1.15mm + 0.125mm - 0.0175mm = 1.2575mm \quad (6)$$

Running the simulation model of the sensor in which, the distance was set at 1.2575mm, the output voltage of the sensor versus the measured current is represented in Fig. 9.

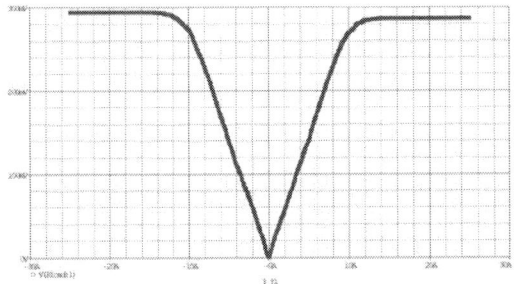

Fig. 9. Output voltage depending by measured current.

Using d =1.2575 mm in equation (1) and considering maximum value of applied magnetic field that can be measured Bmax=12Oe, it can be determined the maximum current value that can be measured $I_{max} = 7.545A$.

In the following stage the current line was placed on the bottom-side of the PCB. This means that it should be taken in account the thickness of the PCB (c). In this way the total distance between the sensor and the current path can be determined according to equation (7).

$$d = a + b + c + t/2 \quad (7)$$

The standard thicknesses for a FR4 PCB considered and the distance determined for each situation are shown in Table 1 [15].

TABLE I.   FR4 PCB STANDARD THICKNESS.

c [mils]	35	39	47	59	93
c [mm]	0.8890	0.9906	1.1938	1.4986	2.3622
d [mm]	2.1815	2.2831	2.4863	2.7911	3.6547

For a specific value of the measured current, the dependency between the output voltage and the distance between the GMR sensor and the current trace is plotted in Fig. 10. As it can be noted, if the distance increases, the output voltage decreases.

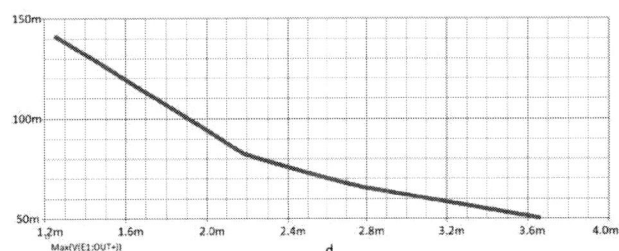

Fig. 10. Output voltage versus distance between the sensor and current trace.

Table 2 gives correspondence values between the distance at which the sensor must be placed and the maximum current that can be measured.

This allows the designer to set the correct distance at which a GMR sensor must be placed so that it can measure a certain current range. The calculus was performed considering that the sensor has a linear behavior at values of the magnetic field of up to ± 12 Oe, according to Fig. 5.

TABLE II.   CORRESPONDENCE DISTANCE – MEASURED CURRENT.

d [mm]	1.2575	2.1815	2.2831	2.4863	2.7911	3.6547
Imax [A]	7.5450	13.089	13.6986	14.9178	16.7466	21.9282

From Fig.11 a linear dependence can be observed between the maximum value of the current that can be measured as a function of the distance at which the sensor must be placed to the current path.

The marked points correspond to the calculated ones from Table 2.

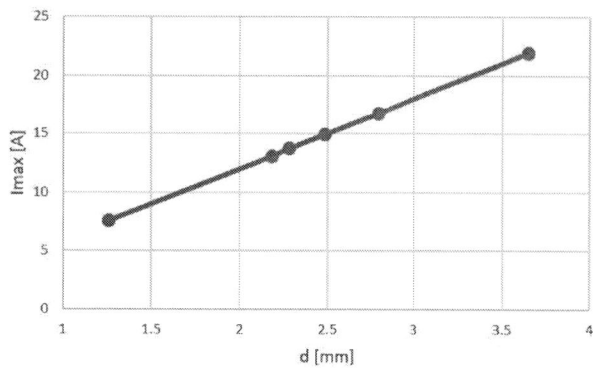

Fig. 11. Maximum current versus distance.

The results certify that the PSpice model and the methodology proposed are validated and, also, can be used to model other types of GMR magnetoresistive sensors.

IV. CONCLUSIONS

The primary objective of this paper is to create a simulation model for a commercial GMR current sensor - NVE AA002 validating a methodology proposed in a previous paper [1].

The novelty of the proposed model consists in the addition of non-electrical parameters that influence the correct measurement of current (the distance between the sensor and the current trace; the position of the current trace on the top layer or on the bottom layer of the PCB).

All tests were performed on the most used dielectric PCB - FR4 and situations in which the current path was placed on the top or bottom layer of the PCB were analyzed. Also, placing the current path on the bottom layer, considering standard values of PCB thicknesses, the maximum values of the current that can be measured were estimated.

From the results obtained, it was demonstrated that using the methodology based on Look-Up Table the characteristic obtained after testing the model closely match the characteristic given by the manufacturer in the data sheet.

In this way the proposed method was validated and can be used to model the behavior of other types of magnetoresistive sensors.

ACKNOWLEDGMENT

This work was supposed by a grand of the Romanian Ministry of Research and Innovation, CCCDI − UEFISCDI, project number 52/2020, PN-III-P2-2.1-PTE-2019- 0867, within PNCDI III.

REFERENCES

[1] E. M. Stetco, O. A. Pop, M. S. Gabor, A. Grama, and D. Pitica, "PSpice Model for a Current Sensor Based on Spin-Valve Magnetoresistive Microstructure", Electronics System-Integration Technology Conference, 2020, Vestfold, Norway.

[2] NVE Corporation, AA and AB-Series Analog Sensors, https://www.nve.com/.

[3] NVE Corporation, GMR Sensor Catalog, SB-00-014, February 6, 2012, https://www.nve.com/ .

[4] NVE Corporation, AA/AB-Series Analog Magnetic Sensors. https://www.nve.com/ datasheet, July 2019.

[5] NVE Corporation, Current Measurement Using GMR Sensors, Sensor Application, SB-SA-001; rev. November 2008, https://www.nve.com/ .

[6] NVE Corporation, Application Notes for GMR Sensors, https://www.nve.com/.

[7] E.M. Stetco, O. A. Pop, C. A. Davidas, T. Petrisor Jr., and M. S. Gabor, "Design and Characterization of a Micrometric Magnetoresitive sensor", IEEE 25th International Symposium for Design and Technology in Electronic Packaging (SIITME), 2019, Cluj-Napoca, România, ISSN: 2642-7036, DOI: 10.1109/SIITME47687.2019.8990875 .

[8] C. Muşuroi, M. Oproiu, M. Volmer, I. Firastrau, "High Sensitivity Differential Giant Magnetoresistance (GMR) Based Sensor for Non-Contacting DC/AC Current Measurement" Sensors 2020, vol. 20, no. 1, pp 323-340, DOI: 10.3390/s20010323.

[9] C. Reig, S. C de Feritas, S. C. Mukhopadhyay, Giant Magnetoresistance (GMR) Sensors, Smart Sensors, Measurement and Instrumentation, vol. 6, Springer- Verlag Berlin Heidelberg, 2013, ISBN 978-3-642-37171-4.

[10] C. Reig, M.-D. Cubells-Beltran, and D. R. Munoz, "Magnetic Field Sensors Based on Giant Magnetoresistance (GMR) Technology: Applications in Electrical Current Sensing", Sensors, Vol.9, pp.:7919-7942, 2009, DOI: 10.3390/s91007919.

[11] B. M. Nikolova, G. T. Nikolov, M. H. Hristov, "Analogue Behavioural Modelling of Integrated Sensors," International Conference on Computer Engineering and Systems, Cairo, 2006, pp. 107-112, DOI: 10.1109/ICCES.2006.320433.

[12] H. Fardi, G. Alaghband, "Analog behavioral modeling of magnetoresistive sensors," 53rd IEEE International Midwest Symposium on Circuits and Systems, Seattle, WA, 2010, pp. 408-411, DOI: 10.1109/MWSCAS.2010.5548883.

[13] B. Nikolova, G. T. Nikolov, M. Todorov, "SPICE modelling of magnetoresisitive sensors", Int. Journal of Reasoning-based Intelligent Systems, vol. 6, pp. 12-18, 2014, DOI: 10.1504/IJRIS.2014.063948.

[14] B. Nikolova, M. Todorov, T. Brusev, "Curve Fitting for Sensors' Analog Behavioural Modelling", International Scientific Conference on Information, Communication and Energy Systems and Technologies (iCEST), vol. 3, Niš, Serbia, 2011.

[15] Association Connecting Electronics Industries (IPC), "IPC-4101E, Specification for Base Materials for Rigid And Multilayer Printed Boards", revision E, March 2017, published April 2020.

[16] M. D. Cubells-Beltran, C. Reig, D. R. Munoz, S. I. P. C. de Freitas and P. J. P. de Freitas, "Full Wheatstone Bridge Spin-Valve Based Sensors for IC Currents Monitoring," IEEE Sensors Journal, vol. 9, no. 12, pp. 1756-1762, 2009, DOI: 10.1109/JSEN.2009.2030880.

[17] A. Roldán, C. Reig, M.D.Cubells-Beltrán, J.B.Roldán, D.Ramírez, S.Cardoso, P.P.Freitas, "Analytical compact modeling of GMR based current sensors: Application to power measurement at the IC level", Solid-State Electronics, vol. 54, pp. 1606-1612, 2019, DOI: 10.1016/j.sse.2010.07.012.

[18] P.P. Feritas, S. Cardoso, et., "Optimization and Integration of Magnetoresistive Sensors" SPIN, vol.1, no.1, pp. 71-91, 2011, DOI: 10.1142/S2010324711000070.

Analysis of single-cell force-spectroscopy data of Vero cells recorded by FluidFM BOT

Ágoston G. Nagy[1,2], Attila Bonyár[1]
[1]Department of Electronics Technology
Budapest University of Technology and Economics
Budapest, Hungary
ag.nagy@ett.bme.hu

Inna Székács[2], Robert Horvath[2]
[2]Nanobiosensorics Momentum Group
Institute of Technical Physics and Material Science, Energy
Research Centre, Eötvös Loránd Research Network
Budapest, Hungary

The robotized fluidic-force microscopy (FluidFM BOT) technology enables label-free, high-throughput, and flexible data acquisition, and have found a wide range of applications in single-cell biology. One of its features is to measure the characteristic force-distance (FD) curves produced by living cells during single-cell force-spectroscopy (SCFS). In the present work, African green monkey kidney epithelial (Vero) cells were seeded on a gelatin-coated substrate to mimic their attachment to the extracellular matrix. Vero cells were detached from the substrate with 8 μm aperture micropipette cantilever, and their FD-curves were investigated besides the actual cell size. The results on 19 individual cells indicate that cell size and adhesion force may deviate; however, highly adhered cells require a greater distance to detach them compared to weakly adhered cells. Adhesion force and energy of the cell normalized by their cell area also show rising linear tendencies with the detachment distance. The investigation presents how cellular morphology influences the adhesive properties of Vero cells.

Keywords: FluidFM, single-cell force-spectroscopy, nanofluidic manipulation, Vero cell, adhesion

I. INTRODUCTION

The fluidic force microscopy (FluidFM) is a novel and versatile tool for the adhesion measurement of living cells[1]. The technique applies hollow fluidic cantilevers with a micron-scale opening, which can be easily positioned on top of living cells to attach the cell onto the cantilever by employing negative pressure inside the fluidic channel [2]. Standard atomic force microscopy (AFM) protocols for single-cell force-spectroscopy (SCFS), having compact cantilevers, employ time-consuming chemical protocols to attach the cells and require excessive know-how with low volume data output [3]. In contrast, FluidFM offers much higher throughputs, especially the robotized FluidFM BOT platform [4]–[6] having a user-friendly software interface for automatized cell manipulation and data acquisition pipeline. However, the special design of the FluidFM microfabricated cantilevers requires an extensive investigation of the mechanical properties of the cantilever in order to obtain a precise force value from its bending during cell detachment measurements. For example, the in-depth study of the inverse optical lever sensitivity (*InvOLS*) was a must, since the parallel rows of pillars located in the cantilever influence the softness of the cantilever in each laser spot position reflected from the back of the cantilever [7]. We have followed the guidelines set by our previous investigations to minimize the error in our force data [7]. Our research focused on the measurement of living Vero cells seeded on 0.2 mg/ml gelatin-coated surfaces to study the relations of parameters characteristic of SCFS. Vero cells are capable of forming cellular monolayers and, therefore, an optimal candidate to investigate the self-assembly and mechanics of monolayer formation. To understand self-assembled epithelial monolayers formed by Vero cells, the fundamentals of these cellular structures must be studied. Therefore, we focused on single Vero cells, which have no physical intercellular connections with other Vero cells. The experimental surface, gelatin, is packed with various peptide motifs allowing integrin transmembrane anchor proteins expressed by cells to adhere [8]. With this experimental set-up, it is possible to determine the adhesive forces exhibited by individual Vero cells capable of forming confluent monolayers.

II. THE ARCHITECTURE OF FLUIDFM CANTILEVERS AND THE CALIBRATION OF INVOLS

The calculation of the force (F [N/m]) exerted by a living cell onto the underlying surface is based on Hooke's law (1). It uses the spring constant (k [N/m]) and the *InvOLS* [μm/V] of the cantilever, and the output of the two-segmented photodetector (U [V]). The photodetector (PD) receives the reflected laser beam focused on the backside of the cantilever covered with a reflective coating.

$$F = U * k * InvOLS \tag{1}$$

For the precise calculation of the force values, one must acquire the k and *InvOLS* before measuring individual cells set by our guidelines [7]. The microarchitecture of the cantilever differs from the standard AFM cantilever since it incorporates a nanofluidic channel filled with buffer solution and a parallel row of pillars supporting this channel. At the end of the cantilever, an outlet of the fluidic channel is located to apply negative or positive pressure from - 800 to 1000 mBar. The outlet of the cantilever is called the aperture, and it has three main structures: two of them have a pyramidal shape and are used for injection or extraction of cellular fluids, or nanoprinting. The third structure is completely flat, and its aperture is circular with diameters of 2, 4, and 8 μm; this micropipette cantilever is used for microprinting [4], colloidal spectroscopy [6], and SCFS [5].

The FluidFM cantilever is bonded to a microchip glued to a plastic reservoir holder, together they form the probe. The reservoir can be filled with any liquid, which can be moved by the pressure control system of the unit. The probe can be easily mounted onto the Z-stage of the FluidFM platform with the help of automatic features, thus saving time during the experimental set-up.

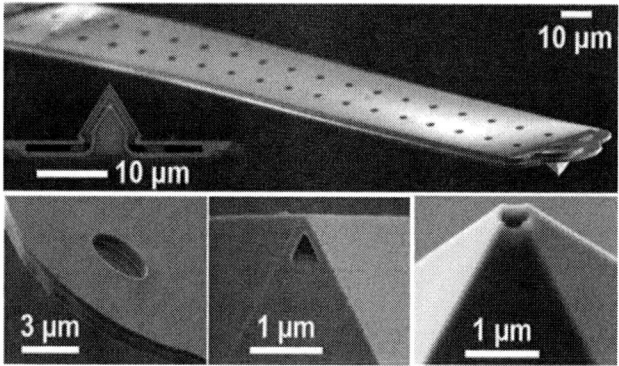

Figure 1. The architecture of FluidFM cantilevers. The upper panel shows the parallel row of pillars supporting the nanofluidic channel located between the silicon nitride wafers with the aperture at the end of the consol. The bottom panel shows from left to right the micropipette, the rapid prototyping, and the nanopipette apertures. The opening of the channel is fabricated with focused ion beam milling. Images are courtesy of Tomaso Zambelli[9].

III. METHODS

A. Cell culture maintenance

Vero cells were maintained in completed culture medium (Dulbecco's modification of Eagle Medium with 10% Fetal Bovine Serum) in a humidified incubator at 37°C and 5% CO_2 for up to 30 passages[10]. For the experiment, non-coated 6-well polystyrene plates were used covered with 0.2 mg/ml gelatin diluted in PBS. One well must be kept intact for the future calibration of *InvOLS*. Gelatin solution was polymerized in a humidified incubator for 20 minutes. Afterward, the surface was rinsed 3-times with PBS, and then cell culture media was added.

Vero cells were rinsed with DPBS and detached from the bottom of the tissue culture petri-dish with trypsin-EDTA. Cells were picked up in 1 ml culture media and transferred in 50, 100, 150, and 200 μl quantities to the gelatin-coated wells. After 24 hours of incubation and cellular adherence, wells are rinsed with DPBS and filled with HPMI buffer (9 mM glucose, 10 mM $NaHCO_3$, 119 mM NaCl, 9 mM HEPES, 5 mM KCl, 0.85 mM $MgCl_2$, 0.053 mM $CaCl_2$, 5 mM $Na_2HPO_4 \times 2H_2O$, pH 7.4). Thus SCFS curves could be measured of independent Vero cells without cell-cell connection to other Vero cells.

B. Experimental set-up and SCFS recording

FluidFM BOT (Cytosurge AG., Zürich, Switzerland) was placed on a vibration-free table, and the 6-well plate with the cultured cells was inserted into the sample holder XY-stage of the BOT. Before the measurement, the probe must be filled with the buffer solution of the experiment and mounted to the z-stage of the instrument. The mounted cantilever can now be used after positioning the laser and measuring the k with the instruments built-in feature using the Sader-method[7]. The BOT can handle two different plates, from which one is used for measurements and the other one for cleaning the probe after each recording. To clean the probe from cellular debris, at least

six wells are needed, from which two are filled with 5% sodium hypochlorite and 4 with MQ water. By dipping the probe first in MQ water followed by few seconds of hypo, the surface of the cantilever will be free of any cellular remedies influencing *InvOLS* calibration and measurement. Following the hypo treatment, the cantilever must be rinsed with MQ in different wells so that the detergent does not reach the cells. After cleaning the probe, it must be transferred to a cell-free buffer solution, so the MQ water does not dilute the buffer in the experimental cell covered wells. In the cell-free buffer, the InvOLS can be calibrated on the gelatin-free polystyrene bottom before each measurement according to our guidelines[7].

SCFS recording was carried out with a FluidFM micropipette cantilever with 8 μm aperture filled with HPMI buffer. The following parameters were used to pick-up individual Vero cells from the surface: set-point 20 mV, approach and retract speed 1 μm/s, pressure -500 mBar, pause 5 s, retraction distance 150 μm. When a cell was already detached from the bottom as it was visible in the output signal and in the microscope, the recording was stopped before the piezo reached the 150 μm retraction distance to save time between recordings. Characteristic Force-Distance (FD) curves (Fig. 2.) and cellular images were exported from the software of the instrument and were evaluated in a custom made software and Origin 8.5, which was also used for correlation analysis using Pearson's r correlation. Cell areas were defined with CellProfiler[11] based on the microscopical images taken before the actual adhesion measurement.

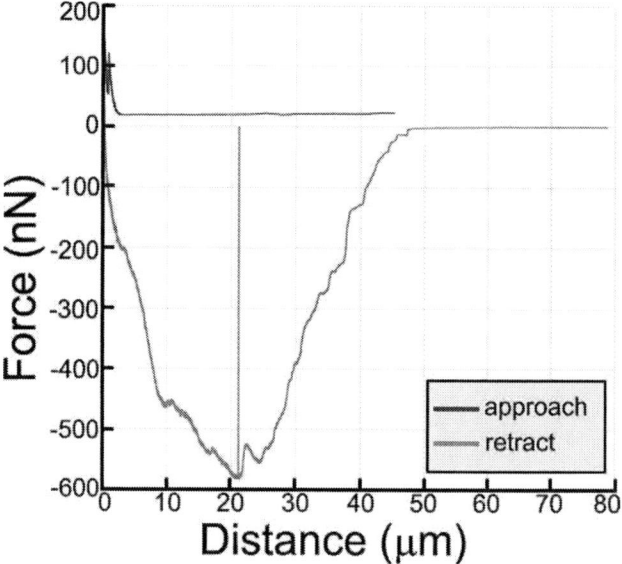

Figure 2. A FD-curve from a single Vero cell. The instrument records the force based on the laser reflection from the back side of the cantilever. The distance is the actual position of the piezo from the Z-stage. The curve represents how a the cantilever is bend due to the adhesion of the cell to the substrate. Its peak value where the force is the greatest is considered F_{max} and its distance from the surface D_{max}. The integrated curve is the energy of the adhesion and is called E_{max}.

IV. RESULTS AND DISCUSSION

During the experiments 19 single VERO cells have been targeted and photographed, and their characteristic FD-curves were measured, which contain valuable information regarding the surface adhesive capabilities of the cells. With the FD-curves' parameters, such as maximal detachment force (F_{max} [nN]), detachment energy (E_{max} [pJ]), and the distance from the surface at which the cell exerts the F_{max} (D_{max} [μm]) could be determined. With image analysis, we were able to determine the area [A_{cell} μm2] of individual Vero cells and investigate the relationship between these four parameters. Our measurements revealed that there is a direct correlation between the parameters F_{max}–E_{max} (Fig. 3.) and F_{max}–D_{max} (Fig. 4.). It is known in the literature that F_{max} correlates with E_{max}[5],[10]–[15], which is also validated by our investigations (see Fig. 3).

Cells which require a larger D_{max} to detach them exert higher F_{max} and E_{max} on the substrate (Fig. 4. and 5.), which possibly means an intensive expression of integrin-receptors involved in the adhesion processes. The role of this information is to elucidate how the formation of epithelial monolayers characteristic of Vero cells is influenced by adhesive properties of individual cells. The correlation between F_{max}-D_{max} and E_{max}-D_{max} may not be linear; however, to understand the nature of this behavior, a larger dataset would be advantageous.

Moreover, from the results, the following observation can be made: There may be a deviation in the same cell population, meaning that cells with similar areas adhere differently to the substrate (Fig. 6.). It is interesting to see, that cells with adhesion forces in the upper range have cell areas equal to those cells, which exerted low adhesion forces, however, the largest cells do not exert the highest force. Increasing the sample size may, however, result in cell F_{max}-A_{cell} correlation. It can also be supposed that the number of ligands in the gelatin coating does not distribute equally in their structure, which is the reason that equally sized cells adhere differently when more or less adhesive bonds are present on the same area [8]. We also can not rule out the possibility that individual Vero cells express different amounts of integrin molecules responsible for binding, which proteins copy number may deviate through the life cycle of a cell [18], [19].

Despite the previously mentioned deviation F_{max}-A_{cell} a correlation of D_{max}-A_{cell} can be seen (Fig. 7.), which means that larger cells' F_{max} is "harder to reach". The mechanism behind this is that larger cells have greater volume compared to smaller cells, and so it is possible to extensively pull-up their elastic membrane during an SCFS recording compared to smaller cells. However, this statement must be handled with care, since adhesion is a complex process where many cellular features are involved, and also smaller cells can exhibit larger forces (Fig. 6.) and energies (Fig. 8.). The F_{max} and E_{max} normalized with the A_{cell} of the measured cell and compared against D_{max} gives an impression on how the adhesion force and energy depend on the elasticity of the cell (Fig. 9. and 10.). For the figures 3, 4, 8 and 9 a total of 3 datapoints have been selected as outliers and excluded from the correlation analysis. This resulted in an increasing relationship of the displayed parameters. The strongest correlation is between D_{max} and E_{max} (Fig. 5. and 10.) followed by D_{max} and F_{max} (Fig. 4. and 9.).

The quantity of expressed integrin molecules forming adhesive bonds represented by F_{max} may not depend on the cell size according to the data measured on individual Vero cells. However, it is possible that subpopulations exist and that the cellular life-cycle plays a role involved in cell growth and integrin expression. Also, the quality of these adhesive bonds formed by integrin-RGD interactions (E_{max}) shows no clear distribution. On the figures, it is visible that F_{max}-E_{max} shows correlation, but also here, the question of outliers from the distribution arises: Why do cells with similar F_{max} values (quantity of integrin-RGD bonds) represent slightly different quality (E_{max}) features, which can be considered a deviation from the correlation? Also, it is important to mention that D_{max} correlates with the A_{cell}, however, the D_{max}-F_{max} graph shows a rather scattered behavior of cells. The D_{max} is an exact representation of a cell's elastic capability, and it is important to mention that elastic cells do not necessarily adhere stronger to the substrate. The elasticity of a cell depends on its size and the structure and amount of intracellular scaffold proteins, such as actin.

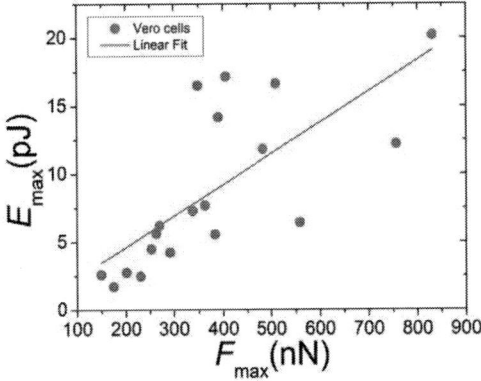

Figure 3. Linear correlation of F_{max}-E_{max} values of single Vero cells. The correlation is 0.724, the linear fit is described by y=0.023x+0.027.

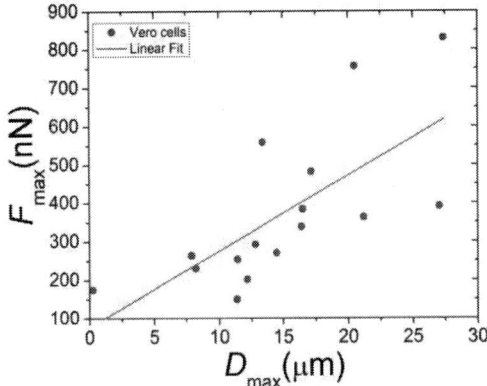

Figure 4 D_{max}-F_{max} correlation of single Vero cells. To calculate the correlation between -D_{max}-F_{max} values, 3 values have been excluded as outliers. The correlation of D_{max}-F_{max} is 0.697.

*2020 IEEE 26th International Symposium for Design and Technology in Electronic Packaging (**SIITME**)*

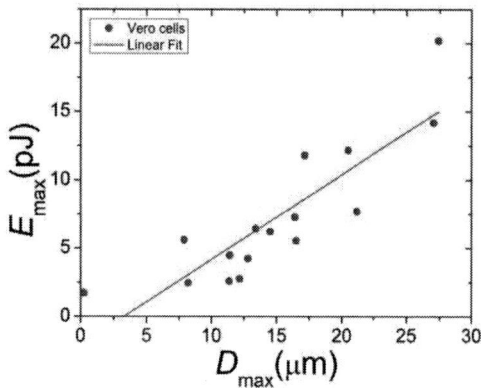

Figure 5. D_{max}-E_{max} correlation of single Vero cells. The higher the E_{max} of the cells the D_{max} will be also higher. To caluculate the correlation 3 data points have been excluded as outliers, which resulted in a correlation of 0.862.

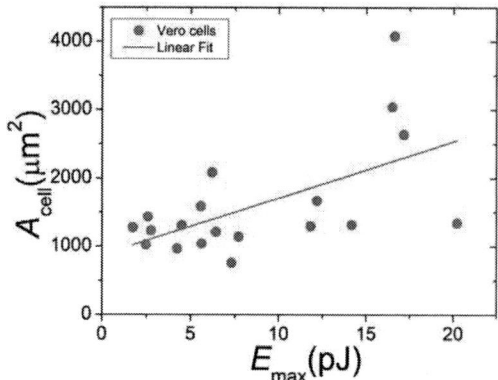

Figure 8. E_{max}-A_{cell} relationship of single Vero cells. Under 2000 µm² the cells' E_{max} is is independent of the cell size. Above this certain value, the E_{max} seems to be linearly rising with the A_{cell} of the cells. The correlation of the data points is 0.586.

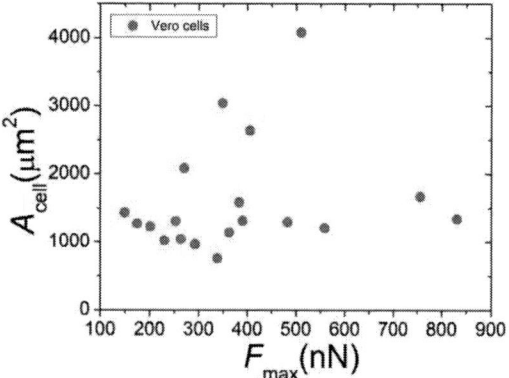

Figure 6. F_{max}-A_{cell} a relationship of single Vero cells. For cells under 2000 µm² the F_{max} seems to be independent from the size, above 2000 µm² this F_{max} seems to rise linearly with A_{cell}.

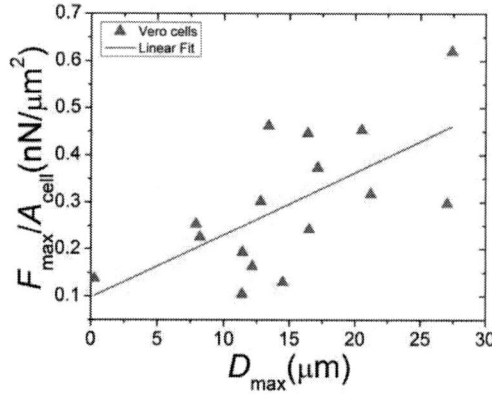

Figure 9. The normalized F_{max} with A_{cell} shows a correlation with D_{max} of 0.64. From the original dataset 3 values have been discarded as outliers.

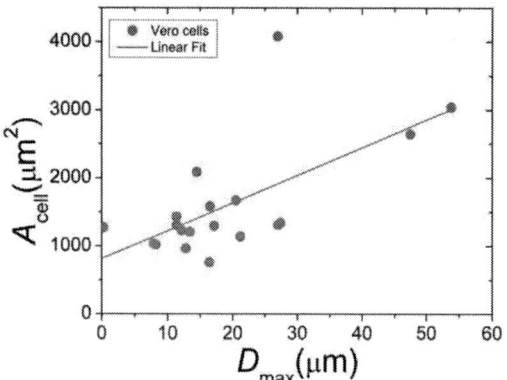

Figure 7. D_{max}-A_{cell} correlation of single Vero cells. As expected, larger cells require a larger D_{max} to detach them, which is understandable due to their increased volume. No outliers have been excluded, correlation is 0.645.

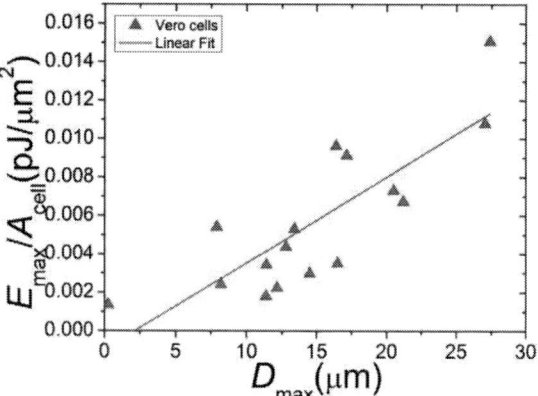

Figure 10. E_{max} normalized with A_{cell} shows a correlation with D_{max} of 0.816. From the original dataset 3 values have been discarded as outliers.

V. CONCLUSION

With our investigation, we have shown that it is possible to gather information on individual epithelial Vero cells incubated for 24 hours. The individual cells are probably in different stages of the cellular life-cycle and therefore express various numbers of anchor proteins, which cause different adhesive capabilities dependent on the cells' actual behavior (e.g. mobility). To further investigate the assembly of Vero cells, the experiments must be expanded to an island-like assembly of Vero cells and to confluent monolayers. The increase of sample-size would also be beneficial; however, the number of measurements is still higher compared to previous investigations on Vero cells. Our results prove that the size and elasticity of the cells influence adhesion forces and energies of Vero cells. From the perspective of monolayer formation, the important question here would be, how different cell morphologies play different roles in the cell-cell and cell-substrate adhesion of Vero cells located in confluent monolayers.

ACKNOWLEDGMENT

Szilvia Bősze kindly provided Vero kidney epithelial cells and HPMI buffer from the Research Group of Peptide Chemistry, Department of Organic Chemistry, Eötvös L. University, Budapest, Hungary. Milán Sztilkovics wrote and shared the evaluation app designed for FluidFM SCFS measurements.

This work was supported by the "Lendület" (HAS) research program, the National Research, Development and Innovation Office of Hungary ("Élvonal" KKP_19 and KH grants). The research reported in this paper was partially supported by the Higher Education Excellence Program of the Ministry of Human Capacities in the frame Biotechnology (BME-FIKP-BIO) research areas of Budapest University of Technology and Economics.

REFERENCES

[1] O. Guillaume-Gentil, E. Potthoff, D. Ossola, C. M. Franz, T. Zambelli, and J. A. Vorholt, "Force-controlled manipulation of single cells: From AFM to FluidFM," *Trends Biotechnol.*, vol. 32, no. 7, pp. 381–388, 2014.

[2] A. Meister *et al.*, "FluidFM: Combining atomic force microscopy and nanofluidics in a universal liquid delivery system for single cell applications and beyond," *Nano Lett.*, vol. 9, no. 6, pp. 2501–2507, 2009.

[3] J. Helenius, C.-P. Heisenberg, H. E. Gaub, and D. J. Muller, "Single-cell force spectroscopy," *J. Cell Sci.*, vol. 121, no. 11, pp. 1785–1791, 2008.

[4] A. Saftics *et al.*, "Biomimetic Dextran-Based Hydrogel Layers for Cell Micropatterning over Large Areas Using the FluidFM BOT Technology," *Langmuir*, vol. 35, no. 6, pp. 2412–2421, 2019.

[5] M. Sztilkovics *et al.*, "Single-cell adhesion force kinetics of cell populations from combined label-free optical biosensor and robotic fluidic force microscopy," *Sci. Rep.*, pp. 1–13, 2020.

[6] T. Gerecsei *et al.*, "Journal of Colloid and Interface Science Adhesion force measurements on functionalized microbeads: An in-depth comparison of computer controlled micropipette and fluidic force microscopy," *J. Colloid Interface Sci.*, vol. 555, no. July, pp. 245–253, 2019.

[7] Á. G. Nagy, J. Kámán, R. Horváth, and A. Bonyár, "Spring constant and sensitivity calibration of FluidFM micropipette cantilevers for force spectroscopy measurements," *Sci. Rep.*, vol. 9, no. 1, Dec. 2019.

[8] N. Davidenko *et al.*, "Evaluation of cell binding to collagen and gelatin: a study of the effect of 2D and 3D architecture and surface chemistry," *J. Mater. Sci. Mater. Med.*, vol. 27, no. 10, 2016.

[9] T. Zambelli, "Electrochemical additive manufacturing of metal microstructures with the FluidFM," *Smart Manufacturing Seminar Series "Micro Additive Manufacturing-Metals,"* 2017. [Online]. Available: https://additivemanufacturingseries.com/wp-content/uploads/2017/10/zambelli.pdf.

[10] N. C. Ammerman, M. Beier-sexton, and A. F. Azad, "Growth and Maintenance of Vero Cell Lines," no. November, pp. 1–7, 2008.

[11] A. E. Carpenter *et al.*, "CellProfiler: image analysis software for identifying and quantifying cell phenotypes," vol. 7, no. 10, 2006.

[12] L. Jaatinen, E. Young, J. Hyttinen, J. Vörös, T. Zambelli, and L. Demkó, "Quantifying the effect of electric current on cell adhesion studied by single-cell force spectroscopy," *Biointerphases*, vol. 11, no. 1, p. 011004, 2016.

[13] S. Sankaran, L. Jaatinen, J. Brinkmann, T. Zambelli, J. Vörös, and P. Jonkheijm, "Cell Adhesion on Dynamic Supramolecular Surfaces Probed by Fluid Force Microscopy-Based Single-Cell Force Spectroscopy," *ACS Nano*, vol. 11, no. 4, pp. 3867–3874, 2017.

[14] A. Sancho, I. Vandersmissen, S. Craps, A. Luttun, and J. Groll, "A new strategy to measure intercellular adhesion forces in mature cell-cell contacts," *Sci. Rep.*, vol. 7, pp. 1–14, 2017.

[15] E. Potthoff *et al.*, "Rapid and Serial Quantification of Adhesion Forces of Yeast and Mammalian Cells," *PLoS One*, vol. 7, no. 12, 2012.

[16] P. Doerig, "Manipulating cells and colloids with FluidFM," 2013.

[17] P. Dörig *et al.*, "Force-controlled spatial manipulation of viable mammalian cells and micro-organisms by means of FluidFM technology," *Appl. Phys. Lett.*, vol. 97, no. 2, p. 023701, 2010.

[18] J. Lin and A. Amir, "Homeostasis of protein and mRNA concentrations in growing cells," *Nat. Commun.*, no. 2018.

[19] Y. Li and K. Burridge, "Cell-Cycle-Dependent Regulation of Cell Adhesions: Adhering to the Schedule," vol. 1800165, pp. 1–5, 2019.

Enhanced X-Ray Inspection of Solder Joints in SMT Electronics Production using Convolutional Neural Networks

Konstantin Schmidt, Nils Thielen, Christian Voigt,
Reinhardt Seidel and Jörg Franke
Institute for Factory Automation and Production Systems
Friedrich-Alexander University Erlangen-Nürnberg
Nürnberg, Germany
Konstantin.Schmidt@faps.fau.de

Yannik Milde, Jochen Bönig
and Gunter Beitinger
Segment of Factory Automation, Manufacturing
Siemens AG
Amberg, Germany

Abstract—The electronics production is prone to a multitude of possible failures along the production process. Therefore, the manufacturing process of surface-mounted electronics devices (SMD) includes visual quality inspection processes for defect detection. The detection of certain error patterns like solder voids and head in pillow defects require radioscopic inspection. These high-end inspection machines, like the X-ray inspection, rely on static checking routines, programmed manually by the expert user of the machine, to verify the quality. The utilization of the implicit knowledge of domain expert(s), based on soldering guidelines, allows the evaluation of the quality. The distinctive dependence on the individual qualification significantly influences false call rates of the inbuilt computer vision routines. In this contribution, we present a novel framework for the automatic solder joint classification based on Convolutional Neural Networks (CNN), flexibly reclassifying insufficient X-ray inspection results. We utilize existing deep learning network architectures for a region of interest detection on 2D grayscale images. The comparison with product-related meta-data ensures the presence of relevant areas and results in a subsequent classification based on a CNN. Subsequent data augmentation ensures sufficient input features. The results indicate a significant reduction of the false call rate compared to commercial X-ray machines, combined with reduced product-related optimization iterations.

Keywords—electronics production; automated X-ray inspection; machine learning; computer vision; Convolutional Neural Networks

I. INTRODUCTION

The process performance in industrial production is commonly specified in occurring defects per million opportunities (DPMO). Trough progressing automation and enhanced process capability, electronics manufacturers nearly achieve single-digit DPMO-rates within their production processes. [1] Insufficient solder connections in the electronics production can result from a multitude of influencing factors so that a direct attribution to a single influencing factor is improbable. [2] The validation of the quality levels is realized with an extensive inspection coverage. Besides electrical test methods, optical inspection processes characterize the outer appearance of the products defined regions of interest (ROI). The increased application of integrated circuits (ICs) with

covered ROIs like ball grid arrays (BGAs) justify the usage of X-ray based methods in addition to inspections using light in the visible spectrum.

The automated X-ray inspection (AXI) typically represents the last optical inspection step in the SMD production. Accurate classification is essential and leads to narrow checking routines in the qualification process. As a direct consequence of the static decision rules, two main problems arise. For the setup of these rule-based test routines, extensive product- and process-specific know-how are required. The high overhead of creating and maintaining product-specific checking routines, leads to increased costs, and severe dependency on expert knowledge, impeding the continuous improvement of existing test procedures. Static checking routines lead to reduced sensitivity with varying conditions, as misclassification rises.

The aforementioned issues justify the investigation of a deep learning approach. A flexible inspection algorithm is expected to increase the classification accuracy, reduce optimization costs, and lower dependency on expert knowledge. A system based on object detection is capable of autonomously detecting relevant ROIs for new inspection tasks derived from historic data. For this study, the solder joints of an electrical connector used in programmable logic controllers are evaluated. A flexible computer vision system is developed from the automated classification of the extracted grayscale images.

II. STATE OF THE ART

A. Industrial X-Ray Inspection of Solder Joints

The quality assessment of a covered solder joint is carried out by the utilization of X-ray inspection based on grayscale images. The intensity of certain areas of the image varies depending on the material being irradiated. This results in darkened image areas for metallic material accumulations. Automated evaluation of predefined image areas is carried out based on these differences in intensity. If a certain amount of pixels within this area exceeds a set threshold, the area is marked as insufficient. A confirmation from the operator must be carried out. These product-specific test routines are manually created and adjusted.

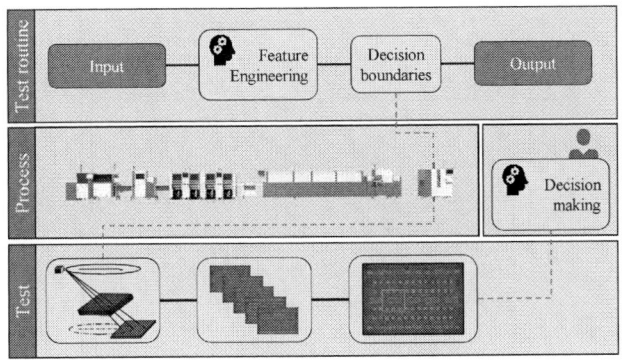

Fig. 1: Schematic overview of the human influence on the test result.

B. Object Detection based on Deep Learning

Machine Learning based Computer Vision applications have gained the highest interest with increasing computing power. Besides the image recognition, the object detection and instance segmentation are in general exercised for knowledge extraction from images. These applications are based on deep learning techniques. They use multiple processing layers to learn representations from the abstracted data. [3] The feature engineering is no longer a manual process, executed by human engineers, but learned in the abstraction layers of the network. With a sufficiently large combination of layers, functions of any complexity can be approximated, using general learning techniques.

One of the most frequently applied types of learning is supervised learning. This is an attempt to learn the relationship between an input vector and an assigned output. It requires a sufficient quantity of labeled inputs to create significant relations. An error function computes the difference between the given and the predicted outcome to indicate the score during the training. With continuous updating of function parameters (weights), the result is being optimized towards a minimal error. This adjustment is based on optimization methods. Stochastic gradient descent is a commonly exploited method, due to its ability to properly generalize on unknown data. It minimizes the loss by computing the gradient of the loss curve for training examples and updating function parameters accordingly. [4]

The general architecture of Neural Networks (NNs) consists of an input layer, directing the information to the output over a generalized linear function (activation function). Intermediate layers, hidden between input and output are applied for the non-linear abstraction of the forwarded information. For updating the weights according to the gradient of the error function, the information has to be propagated back into the network. With the backpropagation algorithm, the loss of every node can be computed and the weights can be optimized subsequently. [5]

A Convolutional Neural Network (CNN) is a specific type of feed-forward network commonly used for grid-structured data (arrays). So that this type of network can be used for image classification or object detection, a mathematical operation called convolution is used instead of the conventional matrix-vector multiplication. The convolution is a cross-correlation between the signal and a given filter (feature map) to detect structural similarities between both functions. [3] With the extension to the two-dimensional space, image data can be covered in height and width. For improved generalization and reduced computational costs, pooling layers are applied. The array is divided into a grid of a set resolution and only certain features are transferred. [6] The conclusive fully connected layer with multiple 1D layers is the successor of the output layer with the predicted classes. [7] Two main reasons, why CNNs show superior performance on image data, are a high local correlation of associated values and the local invariance of the values, both being considered with the specific architecture. [3]

C. Industrial applications of Machine Learning

The growing number of industrial applications strengthen the suitability and even the superiority over rule-based approaches in versatile production processes. Among others, Cia et al. [8] present a sophisticated approach for solder joint inspection using a cascaded CNN. For the region proposal of the ROIs, a sliding window approach was conducted. For the proposed regions, CNN-based quality predictions are executed, surpassing conventional SMT solder joint inspection. [9, 10] Metzner et al. [11] use a sophisticated approach to challenge the false call rate of commercial Automated Optical Inspection (AOI) systems via a Neural Network classifier. Within a supervised learning approach, static ROIs are defined for the relevant inspection areas, by manually labeling the associated areas of the images. In a benchmark, the precision of the proposed model surpassed commercial AOI systems.

D. Need for Action in Research

State of the art X-ray inspection systems demand extensive domain expert knowledge for the setup and optimization of testing routines. The test programs suffer from insufficient classification accuracy, particularly under varying conditions. Subsequent false call reclassification demands high personnel retention. Current research efforts have mainly focused on training Neural Networks with predefined regions of interest.

This contribution is to research if deep learning-based algorithms improve the generation of testing routines regarding flexibility and classification accuracy of SMD solder joints. Grayscale images of the solder joints of an electrical connector used in programmable logic controllers are evaluated. The classification is trained and tested on multiple types of solder joints and error types. In a subsequent approach, an object detection is trained for the autonomous detection of solder joints with no predefined regions of interest. It is evaluated if the developed concept applies to the existing manufacturing infrastructure, regarding prediction accuracy and computation time. The used training images are generated as a by-product in a real production environment.

III. SYSTEM DESIGN

For the realization of the tests, the concept shown in Fig. 2 is proposed. After the inspection process of the X-ray machine, the result images are fed into the inspection algorithm, running on an industrial computer. After the detection, the result is transferred to the test station, where it can be further processed, depending on the outcome. The setup allows a direct inspection of components marked as defective by the test routine. Only marked components are evaluated and considered in the learning

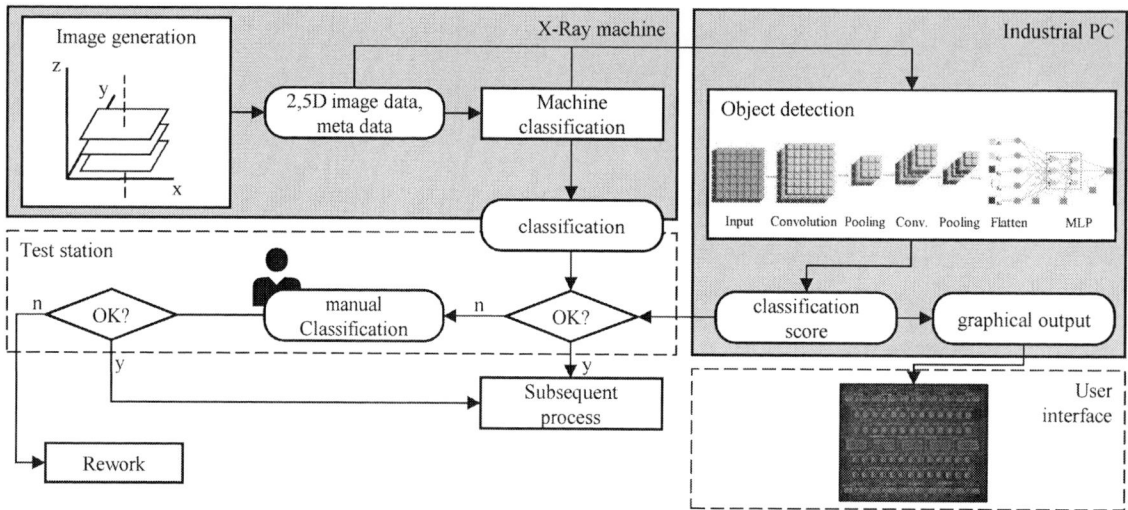

Fig. 2: Data flow in the proposed, CNN-based system design

process so that the algorithm has to distinguish between false calls and true failures.

A. Data acquisition and preparation

On the investigated PCB, two functionally different types of solder connections are defined. These connections are subdivided into six categories for the detection, according to their geometry. Each side of the board contains 79 or 52 solder joints, depending on the current component.

For the X-ray inspection, an Omron VT-750 X-ray machine is conducted. The imaging system is based on a slice imaging method using transversal computer tomography (CT). The assembled PCB is positioned between the X-ray tube and the detector, which are horizontally circulating the PCB. [12] The generated image data consists of 65 8-bit grayscale images vertically varying over the inspected product. One image layer consists of 623x483 pixels for the inspected component.

The image data is extracted as *.rec-files, containing the multilayer pixel information. In combination with the provided meta-data file, containing information about layers and resolution, the information can be transferred into an array structure of the form 623x483x65. In the second step of the data preparation, the array is reduced to one layer, selected as sufficient by the inspection system after the quality assessment. It is ensured that the same database is used for the training as seen and evaluated by the operator. The feature space is reduced to a two-dimensional vector.

Subsequently, the positioning of the solder connections on the board is defined. A self-developed graphical user interface (GUI) was conducted for the position-labeling and type categorization of the solder connections. This step is performed each time a new connector is trained.

For the labeling of the individual solder connections, the database of the in-plant quality system is utilized: The true label, assigned by the operator in the confirmation process, is directed to the system, overruling the X-ray inspection result if required.

The pin-based evaluation result is merged with the 2D image array by a unique identifier. For the training 6387 images are available.

To counteract potential overfitting, data augmentation is executed, to synthetically create a variance within the image data. This includes rotation of the image (90° - 270°), relocation of the ROIs, and insertion of interfering elements. For the training of the image classification, the data is split in the proportions 70 % (training), 20 % (validation), and 10 % (testing). Due to the high yield (> 99.99%), a class imbalance is created. To prevent a model, biased towards one class, class weights are introduced. According to the proportion of a class in the dataset, a weight is added to relativize the effect of this class during training. The minority classes are rated > 1, whereas the majority class is rated < 1. The application is written and executed in Python (v3.7). The used TensorFlow-GPU (v1.15) backend is addressed via Keras (v2.0). [13, 14]

B. Test Setup

In Table I. the selected test setup investigated in this work is shown. A distinction is made between the detection method and the image section used (cut out, full image). Consequently, the required classes and the available data for the training are adapted. The first training and testing (A) are done on cut out areas of the images. The defined ROIs are derived from the labeling process. Further tests (B-D) were carried out on the complete images. The second dataset (B) includes two classes, functional pins, and open solder connections. For the third dataset (C) functional pins and solder bridges are trained. In the last image classification (D) the tests are merged to distinguish functional connections from general defects.

In the training of the object detection, the different pin types occurring on the component are classified. Besides, the error types 'open solder' and 'solder bridge' are included for the detection.

TABLE I. EVALUATED TEST SETUPS

Test Set	Test Setups			
	Detection method	*Image type*	*Classes*	*Class distribution*
A	Image classification	Cut out solder connections	2	5560/827
B	Image classification	Complete image	2	562/728
C	Image classification	Complete image	2	562/326
D	Image classification	Complete image	2	562/1054
--	Object detection	Complete image	8	526/1054

C. Image Classification

As a suitable CNN architecture for the image classification, the VGG-16 is chosen. [7] Compared to AlexNet, the kernel size is reduced to 3x3 over multiple layers. With the increased number of non-linear layers, more complex structures can be learned. [15] The pre-trained convolutional block of the VGG-16 is utilized. As described earlier, the local correlation of associated values within images and the local invariance of the values allow the training of a generic feature extractor. It is suitable for edge and structure detection. [3, 15] After the convolutional block, the compressed data is then transformed into a one-dimensional vector (flatten layer) and transferred to a multi-layer perceptron (MLP) for classification. In three fully connected layers, the output of the convolutional part is converted into a 1x1x4096 vector to be interpreted by the last dense layer for the classification. Two output neurons are defined for the final classification.

In the input layer, the images are resized to a vector of the size 224x224x3. Since X-ray inspection generates grayscale images, which corresponds to a width of one, two channels of the images used are left empty. The MLP is replaced with a two-class soft matrix, which means that the predefined weights are lost. Binary cross-entropy is chosen as the cost function, as a two-class categorization is required. Weight adjustment is carried out by the SGD optimizer, according to the loss calculated by backpropagation. For the training a learning rate of $\eta = 0.0001$ is chosen. The learning rate regulates the convergence of the weights over the epochs. The model is trained over 100 epochs.

For Training A an accuracy of 98 % could be achieved after 4 epochs on the validation set (Fig. 3). The similar learning curve for both sets indicates minimal overfitting in the process. During Training B similar indicators were observed (accuracy = 98 %). In Training C, the increase in accuracy of the training and validation set is significantly reduced in comparison to Training A and B. After 60 epochs the accuracy exceeds 90 %. This is considered to be a result of the higher variance in the occurrence of errors. Training D showed improved performance compared to Training C (40 epochs > 90 % accuracy), indicating an enhanced learning effect due to the higher volume of training data.

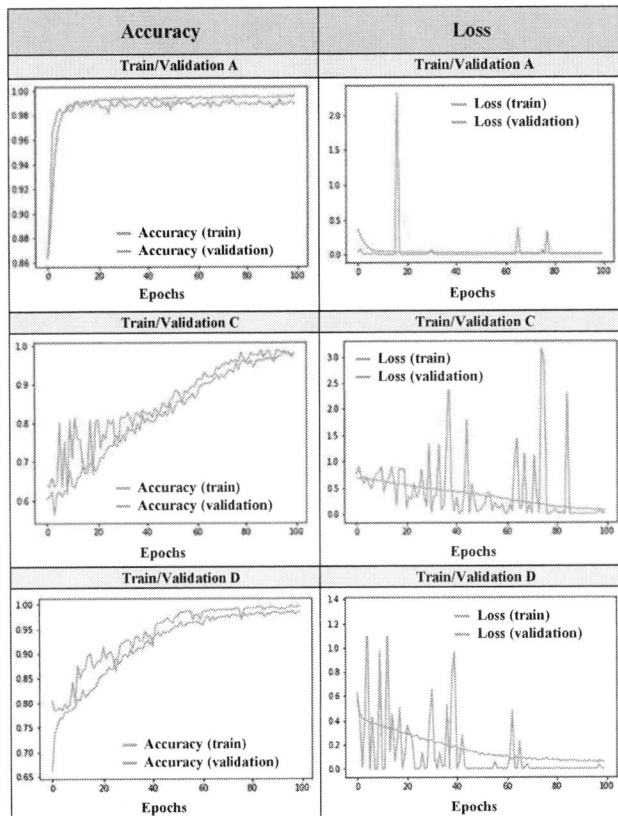

Fig. 3: Accuracy and loss for the training and the validation of the image classification (A, C, D).

D. Object Detection

As a modification to image classification, an object recognition, taking the location under consideration is evaluated. A Mask-RCNN architecture is conducted. [16] With this architecture, the possibility of an instance segmentation is given but is not considered in this work. [17] The according layers are removed. A ResNet101 architecture is used as a backbone for the feature extraction. [18]

As eight different objects are targeted with the detection, categorical cross-entropy is chosen as the cost function. The SGD optimizer is regulated by a learning rate of $\eta = 0.0001$. Due to the high computing complexity, the batch size is reduced to one, and the epochs for the training set to 5. The previous tests with the VGG-16 architecture have shown, that after only a few epochs an accuracy of over 90 % could be achieved. Consequently, the reduced number of epochs is considered sufficient. For the evaluation, the mean average precision (mAP) is used. The loss curves of the training and validation data fall from 1 to 0.5 after the first epoch and converge towards zero with a negative gradient in the following epochs, as shown in Fig. 4.

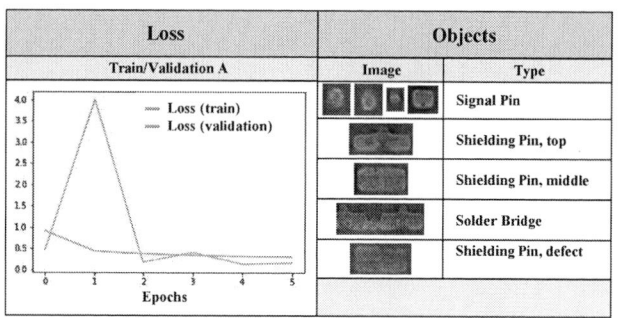

Fig. 4: Loss for the training and the validation of the object detection and the according object classes.

E. Evaluation of the Results

For the evaluation of the trained model, the confusion matrices are shown in Fig. 5. Misclassification by the model occurred in five of 647 cases for Test A (cut out pin images). The misclassification of true failures commonly represents the most cost-intensive decision in production. With the adaption of the discrimination threshold, the False Positive Rate (FPR) can be reduced to zero, resulting in a decreased True Positive Rate (TPR) of 0.95, as shown in the receiver operating characteristic (ROC) curve (see Fig. 6). The classification of Test B to D show a similar proportion but result in a stronger decay of the TPR for FPR of zero (Test C).

The object detection models each achieve an average mAP score of 0.99, with an intersection over union (IoU) threshold of 0.5. For the classification of the functional solder connections, the confidence score varies between $0.98 - 1.00$ for functional solder connections and $0.96 - 0.99$ for bad solder connections. The slight difference in favor of the functional pins could be attributed to the increased variance of errors (open solder vs. solder bridge). To ensure that an unexecuted classification is not resulting in an unclassified solder connection, a product-related quantity of objects must be identified for a positive test result.

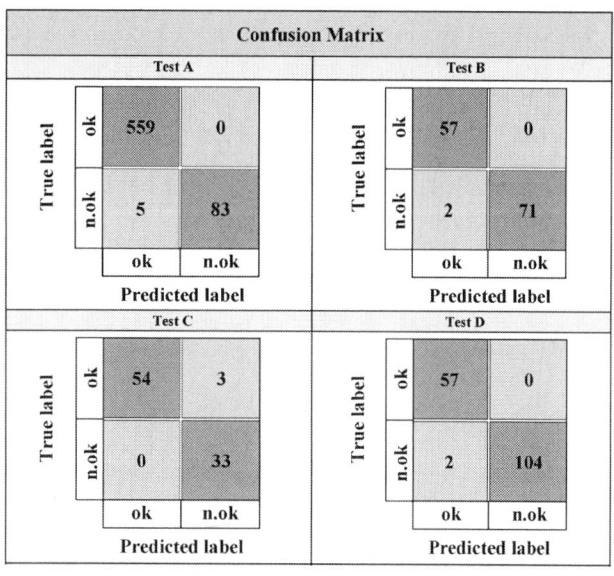

Fig. 5: Confusion matrices of the testing of image classification (Test A-D)

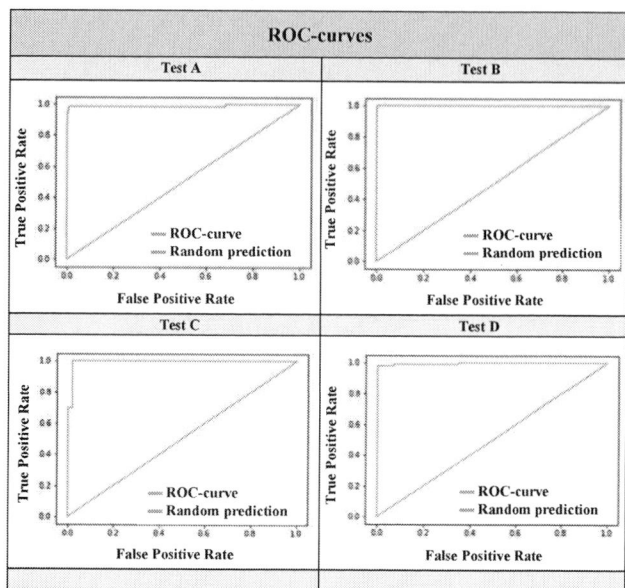

Fig. 6: ROC curves for the testing of the image classification (Test A-D)

F. Transferability of the model

The evaluation of various images showed promising results, regarding the transferability to further components. The model was able to identify objects as solder connections of components, which have not been included in the training set. The fully connected layer of the model can learn a generalized concept of a solder connection, as shown in Fig. 7. On 50 test images of an unknown component, the model reached an average mAP score of 0.98. The application of a generalized model, based on a similar training, evaluating multiple products, could enable a more time-efficient and flexible generation of test programs for X-ray detection.

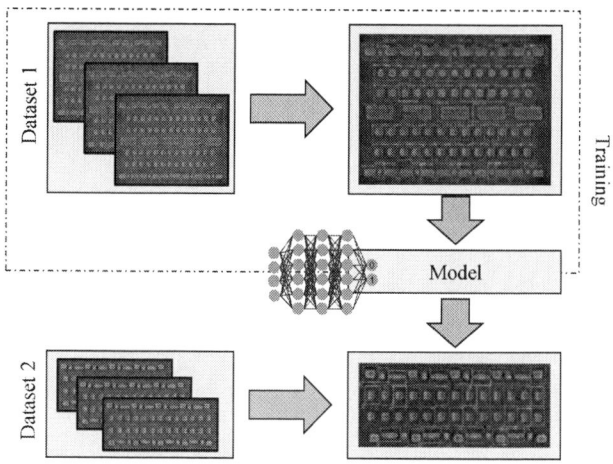

Fig. 7: Concept for the transferability of the evaluated model.

IV. CONCLUSIONS AND FUTURE RESEARCH

The application of different CNN architectures, for the detection and classification of different solder connections and error types in grayscale X-ray images, could successfully be evaluated. The proposed system shows superior precision compared to the commercial X-ray inspection test routines. The dependence on domain-specific know-how for the generation and optimization of the test-routines could be reduced with the CNN-based classification approach. The application of an object detection algorithm shows promising results, regarding the autonomous generation of test routines for the X-ray inspection. The model was able to perform cross-component detections of solder connections. For a reliable integration of the system, improved classification accuracy is still demanded. The sufficiently large dataset required for this purpose is one of the main challenges to be addressed. Further research activities will investigate the application of the proposed supervised techniques in combination with unsupervised techniques, to focus on the recognition of unknown failure types and eliminate the prerequisite of a comprehensive dataset.

Also, there is still a dependency on the selection of the appropriate focus level of the image files made by the machine, which requires further development of the proposed solution. A promising approach could be the integration of a multilayer feature extractor. Moreover, the image data could be extended by a further dimension, increasing the utilized information, as 3D information is provided by the image data generated by the X-ray inspection.

ACKNOWLEDGMENT

The provision of the measurement, process, and quality data used, as well as the IT infrastructure, was made possible by Siemens AG's electronics plant Amberg. The results were obtained within the scope of research activities at the Institute for Factory Automation and Production Systems (FAPS) at the Friedrich-Alexander University Erlangen-Nuremberg (FAU).

REFERENCES

[1] G. Reinhart, Ed., *Handbuch Industrie 4.0: Geschäftsmodelle, Prozesse, Technik*. München: Hanser, 2017.

[2] K. Schmidt, J. Bönig, G. Beitinger, N. Thielen, and J. Frank, "An approach to quality prediction using intelligent SMT solder joint inspection through the use of Machine Learning," in *GMM-Fachbericht*, vol. 94, *EBL 2020 - Elektronische Baugruppen und Leiterplatten*, M. Nowottnick, Ed., Berlin, Offenbach: VDE VERLAG GMBH, 2020.

[3] Y. LeCun, Y. Bengio, and G. Hinton, "Deep learning," *Nature*, vol. 521, no. 7553, pp. 436–444, 2015, doi: 10.1038/nature14539.

[4] Léon Bottou and Olivier Bousquet, "The Tradeoffs of Large Scale Learning," in *Advances in Neural Information Processing Systems*, 2008, pp. 161–168. [Online]. Available: http://leon.bottou.org/publications/pdf/nips-2007.pdf

[5] T. Hastie, R. Tibshirani, and J. Friedman, *The elements of statistical learning, second edition: Data mining, inference, and prediction*, 2nd ed. New York: Springer, 2009.

[6] V. Christlein, L. Spranger, M. Seuret, A. Nicolaou, P. Kral, and A. Maier, "Deep Generalized Max Pooling," in *2019 International Conference on Document Analysis and Recognition (ICDAR)*, Sydney, Australia, Sep. 2019 - Sep. 2019, pp. 1090–1096.

[7] A. Z. K. Simonyan, "Very Deep Convolutional Networks for Large-Scale Image Recognition," [Online]. Available: http://arxiv.org/pdf/1409.1556v6

[8] N. Cai, G. Cen, J. Wu, F. Li, H. Wang, and X. Chen, "SMT Solder Joint Inspection via a Novel Cascaded Convolutional Neural Network," *IEEE Trans. Compon., Packag. Manufact. Technol.*, vol. 8, no. 4, pp. 670–677, 2018, doi: 10.1109/TCPMT.2018.2789453.

[9] G. Acciani, G. Brunetti, and G. Fornarelli, "Application of Neural Networks in Optical Inspection and Classification of Solder Joints in Surface Mount Technology," *IEEE Trans. Ind. Inf.*, vol. 2, no. 3, pp. 200–209, 2006, doi: 10.1109/TII.2006.877265.

[10] W. Hao, Z. Xianmin, K. Yongcong, O. Gaofei, and X. Hongwei, "Solder joint inspection based on neural network combined with genetic algorithm," *Optik*, vol. 124, no. 20, pp. 4110–4116, 2013, doi: 10.1016/j.ijleo.2012.12.030.

[11] Maximilian Metzner, Daniel Fiebag, Andreas Mayr, and Jorg Franke, "Automated Optical Inspection of Soldering Connections in Power Electronics Production Using Convolutional Neural Networks,"

[12] Omron Corporation, *VT-X750: High-speed automated X-ray CT inspection system Handbook*.

[13] Abadi et al., "TensorFlow: Large-Scale Machine Learning on Heterogeneous Distributed Systems," [Online]. Available: https://arxiv.org/pdf/1603.04467.pdf

[14] F. Chollet and others, *Keras*.

[15] Alex Krizhevsky, I. Sutskever, and G. E. Hinton, "ImageNet Classification with Deep Convolutional Neural Networks," in *Advances in Neural Information Processing Systems 25*, F. Pereira, C. J. C. Burges, L. Bottou, and K. Q. Weinberger, Eds.: Curran Associates, Inc, 2012, pp. 1097–1105. [Online]. Available: http://papers.nips.cc/paper/4824-imagenet-classification-with-deep-convolutional-neural-networks.pdf

[16] Waleed Abdulla, *Mask R-CNN for object detection and instance segmentation on Keras and TensorFlow*: Github, *GitHub repository*.

[17] E. Shelhamer, J. Long, and T. Darrell, "Fully Convolutional Networks for Semantic Segmentation," *IEEE transactions on pattern analysis and machine intelligence*, vol. 39, no. 4, pp. 640–651, 2017, doi: 10.1109/TPAMI.2016.2572683.

[18] K. He, X. Zhang, S. Ren, and J. Sun, "Deep Residual Learning for Image Recognition," in *Proceedings of the IEEE Conference on Computer Vision and Pattern Recognition (CVPR)*, 2016.

Surface-Enhanced Raman Spectroscopy Investigation of DNA Molecules on Gold/Epoxy Nanocomposite Substrates

Shireen Zangana[1], Tomáš Lednický[2], István Rigó[3], István Csarnovics[4], Miklós Veres[3], Attila Bonyár[1]

[1]Department of Electronics Technology, Budapest University for Economics and Informatics, Budapest, Hungary.
[2]CEITEC - Central European Institute of Technology, Brno University of Technology, Brno, Czech Republic
[3]Institute for Solid State Physics and Optics, Wigner Research Centre for Physics, Budapest, Hungary
Department of Solid State Physics, University of Debrecen, Debrecen, Hungary
sh.zangna@ett.bme.hu

Abstract— **Gold/epoxy nanocomposites were prepared by template-assisted solid-state dewetting and tested as a Surface-Enhanced Raman Scattering (SERS) substrates for DNA detection. The effect of excitation wavelength on the detected peak intensities was investigated with three different laser sources (532, 633, and 785 nm). The relationship between the plasmon absorbance band of the substrate and the excitation wavelength was studied. It was shown that the shift in the plasmon absorbance band, caused by the binding of the target-DNA molecules could have a major effect on the intensity of the measured peaks, which effect can be more significant than the signal contribution of target molecules on the surface.**

Keywords—Nanocomposite; Gold nanoparticle; DNA; SERS; biosensors.

I. INTRODUCTION

Sensitive and accurate identification and quantification of DNA molecules can influence clinical decisions in many application areas, such as disease diagnostics, pathogen detection, environmental monitoring, etc. Surface-enhanced Raman spectroscopy (SERS) is a promising ultrasensitive method in which the signal intensity of Raman scattering is enhanced in the presence of nanoparticles [1]. The enhancement – which can reach several orders of magnitude – is due to the intense electric near-fields around the particles, present during plasmon excitation. The absorption band of localized surface plasmon resonance (LSPR) – the collective oscillation of electrons incited by the electromagnetic wave – is very sensitive to the properties of the nanostructures (e.g., particle size, shape, inter-particle distance), and many current works are focusing on the optimization of these parameters to maximize the SERS enhancement at different operating wavelengths [1-3].

In our recent paper, we have demonstrated that the excitation wavelength should match the plasmon absorbance of the utilized SERS substrate for high enhancements [2]. In this work, a small target molecule (benzophenone) was measured in a static environment (i.e., the absorbance of the substrate did not change during the detection of the target). In affinity-type nanoplasmonic biosensors, a receptor molecule is attached to the surface of the transducer (i.e., the nanostructures), and the binding of the target molecule causes a measurable plasmonic response, which redshifts the absorbance peak of the sensor. Depending on the sensitivity of the nanoplasmonic sensor, this shift can be in the several 10 nm range, even for small molecules, such as DNA [4,5].

Our current work aims to investigate the effect of this dynamic sensor response of the SERS substrate on the measured signal, and its relation to the used excitation wavelength, through label-free DNA sensing.

Recently different methods using various noble metal NPs, silver (Ag) and gold (Au), have been proposed for the label-free SERS detection and sensing of DNA molecules [6-9]. For our investigations, we used gold/epoxy surface nanocomposites as SERS substrates. These nanocomposites were successfully utilized as LSPR nucleotide sensors to detect the binding of target-DNA molecules in low concentrations [5]. The aim of our work is to test the applicability of these nanocomposites as SERS substrates, besides to investigate and optimize the experimental parameters (e.g., laser excitation wavelength and power) in accordance with the dynamically changing plasmonic absorption of the substrate, in order to get the most distinctive signal from the DNA molecules with high intensity.

II. MATERIALS AND METHODS

A. Gold/epoxy nanocomposite fabrication

In this work, an organized array of gold nanoparticles (Au NPs) on epoxy substrates were fabricated. The detailed experimental procedure regarding the fabrication of the samples is discussed in our previous work [5]. Briefly, a high purity aluminum foil was electrochemically polished then anodized to create a 50 μm thick porous anodic alumina (PAA) layer. The next step was the dissolving of the PAA to create the nano-bowled template structure for the Au nanoparticles. This template was later used as a substrate for gold nanoparticle synthesis. A thin Au film was deposited by RF magnetron sputtering, then the substrate was thermally annealed at 300°C for 5 min. In order to use this Au NP's layer as SERS substrate, two-compound epoxy resins were used to transfer the Au NPs, and the Al layer was dissolved. The samples were cut into cc.

1×2 cm^2 sizes and etched by using O$_2$ plasma prior to any surface functionalization.

An advantage of the fabrication technology is that the hexagonal arrangement of nanoparticles can be controlled on a large surface area. The particle size and the interparticle distance can be set through the anodizing conditions and the deposition/annealing parameters, as discussed in [5]. For our current investigations, two types of substrates were used, denoted as type A and type B. SEM (scanning electron microscopy) images of the arrangements are presented in Fig. 1. The particle size distribution is 102 ± 9 nm for Type A, and 92 ± 6 nm for Type B [5]. As can be seen in the SEM images, the interparticle distance is in the few 10 nm range, the high-density of hot-spots is expected to be ideal for SERS applications.

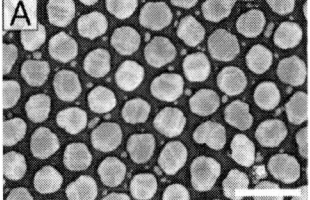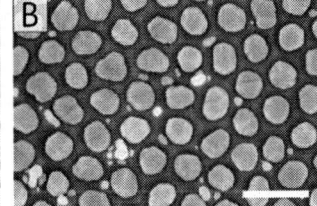

Fig. 1. SEM images of the hexagonal arrangement of Au NPs on a nano-dimpled aluminum substrate, before their transfer to the epoxy substrates. Scale bar: 300 nm.

B. DNA protocols

DNA tests were performed by using 20 bases long probe- and target-DNA molecules that form a specific sequence from the parasite Giardia lamblia (the β-giardin gene) [12], purchased from Sigma-Aldrich. The base sequences of the probe and target ss-DNA chains are listed in Table I. The same surface functionalization protocols that were used to test the gold/epoxy substrates as LSPR sensor elements [5] were applied in this work as well. A mixture of 0.75 M NaCl and 50 µM Na$_2$HPO$_4$ was prepared as the buffer solution. 6-mercapto-1-hexanol (MCH) in 1 mM concentration were also prepared and used to get rid of nonbonding DNA strands on the Au NPs. *Giardia probe* solution *Giardia target* solution of (both 1 µM) was prepared by mixing 10 µl of the probe or target stock with 290 µl of buffer solution.

The experiments were done in steps, and characterizations (optical spectrometry and SERS) were done in each step. First, the samples were cleaned with low-power O$_2$ plasma (20 W at 0.4 mbar for 15 s), then 1 µM of *Giardia probe* solution was drop coated on the surface of the cleaned sample and incubated overnight at RT in a humidified, sealed Petri dish to prevent solvent evaporation. After immobilization of the probe-DNA, the excess of the probe-DNA was rinsed with the same buffer solution and 1 mM of MCH was added to the surface for 30 min at RT in order to get rid of the non-specifically bound probe-DNA strands. After 30 min, the samples were rinsed by using the same buffer. DNA hybridization was done by drop coating the 1 µM *Giardia target* on the surface of the sample for 2 h (at RT). Finally, the surface was rinsed again with buffer, prior to the measurements. It is important to note that only buffer was used for rinsing, and all the Raman-spectroscopy measurements were performed in air. The samples were dried after each step

without rinsing the surface with de-ionized water. This means that there are salt residues on the surface of the samples. These could be removed by rinsing with water, but that would damage the DNA layers. The dried salt on the surface does not affect the Raman-spectroscopy measurements, however, it redshifts the plasmon absorbance peak of the substrates, which should be taken into consideration.

TABLE I. The name and base sequences of the probe and target ss-DNA molecules

Name	Sequences (5'-3')
Giardia probe	CGTACATCTTCTTCCTTTTT(ThiC6)
Giardia target	AGGAAGAAGATGTACGACCA

C. Characterization Methods

Renishaw 1000 micro-Raman spectrometer at Wigner Research Centre for Physics was used to perform the Raman spectroscopy measurements. Three laser wavelengths were used for excitation (532, 633, and 785 nm) between 1 % and 100 % power (depending on the wavelength, as indicated for each experiment separately), 10 s exposure time, and 5 accumulations. The diameter of the excitation spot was around 1 µm, which was monitored with a 50× objective.

Avantes Avaspec 2048-4DT spectrometer and an Avantes Avalight DHS halogen light source between 300 nm and 800 nm have been used to measure the plasmon absorbance bands and sensitivity of the SERS substrates. The sensitivities were measured by changing the medium above the substrates between air and water. Absorbance was also measured after binding DNA or MCH on the surface to evaluate the shift in the substrates' absorbance peak to select the most appropriate laser excitation for the SERS measurements.

Scanning electron microscopy (SEM) images were performed with a high-resolution SEM (FEI Verios 460L) in secondary electron detector mode and an acceleration voltage of 5 keV.

III. RESULT AND DISCUSSION

A. Absorbance peak shift investigation

The absorption spectra of gold/epoxy nanocomposite substrates were measured with the spectrophotometer in each step of the experiments. The results are collected in Fig. 2. Directly after cleaning with O$_2$ plasma, the plasmonic sensitivity was measured by changing the medium above the substrates between air and water. As can be seen in Fig. 2 the bulk refractive index sensitivity of all four tested samples was between 85–100 nm/RIU. Samples A1 and B1 were only treated with MCH as control, while samples A2 and B2 went through the whole described procedure, including probe-DNA binding, MCH treatment and finally, target-DNA binding. Fig. 2 shows their absorbance spectra after target-DNA binding. It has to be noted again, that the "after t-DNA" phases in Fig. 2 were measured in air, after washing the substrates with buffer and drying. The contribution of the plasmonic shift thus also contains the dried salt on the surface of the chip.

It can clearly be seen that the addition of DNA and MCH on the surface of the nanoparticles causes a significant redshift. A more detailed, step-by-step investigation is discussed in [5].

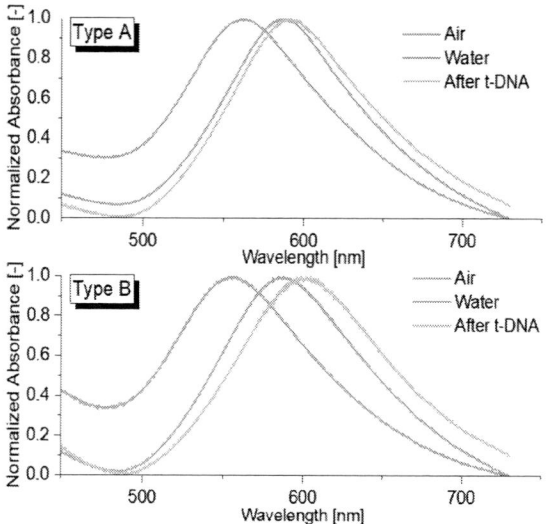

Fig.2. Absorbance spectra of gold/epoxy nanocomposite substrates in the different phases of sample preparation.

B. SERS experiments

As it was shown in Fig. 2, the nanoplasmonic substrate has a significant dynamic response for binding the various molecules in the subsequent phases. Fig. 3 demonstrates that this dynamic sensor response has a significant effect on the measured SERS intensities, depending on the utilized excitation wavelength. The SERS spectra for Type A substrates were measured with 633 nm excitation wavelength while the Type B substrates with 532 nm excitation, both with 1% laser power. The measurements were carried out in each step of the sensor preparation process: 1) clean reference; 2) addition of only MCH (as control); 3) addition of *Giardia probe* (p-DNA) only; 4) p-DNA + MCH; 5) finally p-DNA + MCH and hybridization with *Giardia target* (t-DNA), respectively. The spectra clearly show the effect of the excitation wavelength on the peaks' positions and intensity.

As can be seen in Fig. 2, the 633 nm excitation is on the right side of the substrates' plasmon absorbance peak. The redshift, caused by the increasing amount of molecules bound to the gold surface, brings the resonance peak closer to this excitation wavelength. This causes an observable increase in absolute SERS intensities, from reference to t-DNA in Fig. 3. In the other hand, the 532 nm excitation, which is on the left side of the plasmon peak, gets further away from the redshifting peak in each step, resulting in a significant decrease in absolute SERS intensities. This observed connection between absolute SERS intensities and the relation between the excitation wavelength and plasmon resonance peak is in good accordance with our previous investigations [2].

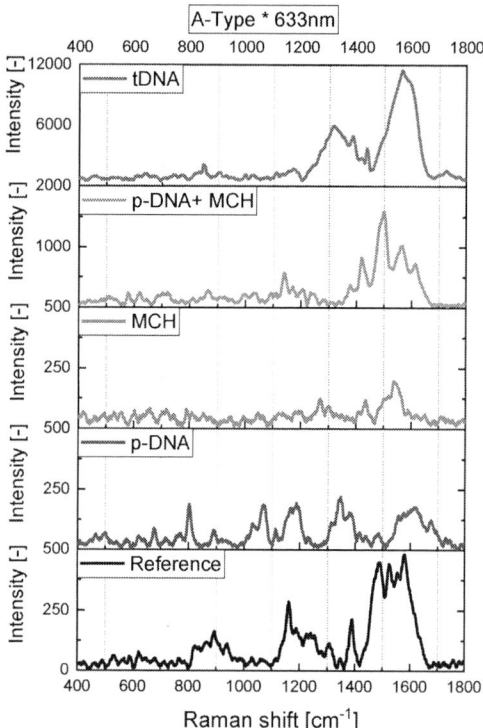

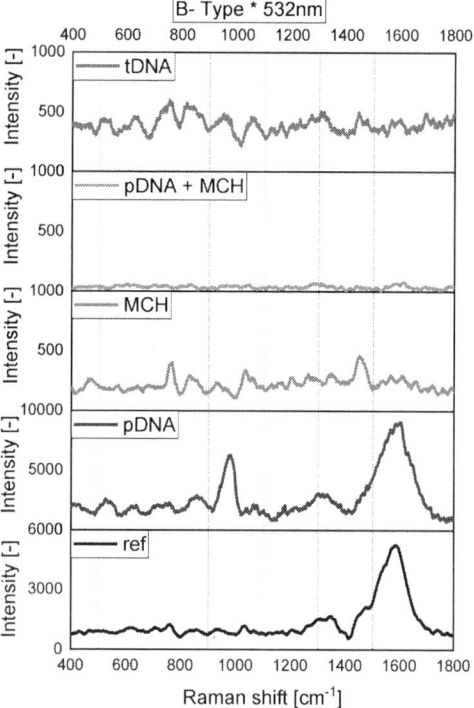

Fig. 3. SERS spectra of gold/epoxy nanocomposites, in different steps of a DNA sensor fabrication. Ref: bare (cleaned) reference. pDNA: *Giardia probe* DNA only. MCH: 6-mercapto-1-hexanol only. pDNA+MCH: *Giardia probe* + MCH treatment. tDNA: *Giardia probe* + MCH treatment + *Giardia target* DNA hybridization. Top: Type A substrate, measured with 633 nm excitation. Bottom: Type B substrate, measured with 532 nm excitation.

However, this dynamic effect was previously not considered for dynamically changing absorption bands. Selecting the excitation wavelength based on the absorption measured on a clean sample in air can be misleading: the 532 nm excitation seems to be a perfect fit, since it has around 70–80% of the absorbance at the peak position (i.e., normalized absorbance is between 0.7–0.8 in Fig.2). However, this drops to only 10–20% after target-DNA binding (with dried salt also on the surface). The 633 nm has around 30–40% absorbance measured on a clean sample, which increases to 70–80% after target binding. Based on our previous work, the binding of target-DNA molecules alone could cause a 6–7 nm shift with this sensor, which can influence the target's SERS signal, depending on the excitation.

Another effect that has to be considered in our experiments is the substrate's thermalization during the SERS experiments. The incited plasmonic particles can heat up their environment with their intensive near fields. This depends on the laser power and focus (energy density), but also on the relation between the excitation wavelength and plasmonic absorption. In our experiments, increasing the laser power to 5% caused observable damage on the surface. With 1% power, the peaks of the DNA cannot be detected clearly, especially with the 532 nm excitation, where the gap between laser wavelength and absorption peak is large. The thermalization of the substrate could also be observed (at 5% laser power) as signal instability: clearly observable peaks disappeared (in a matter of seconds) with time as, presumably, the heated epoxy substrate melted and the nanoparticles moved out of focus. The relatively high intensity of a broad peak, corresponding to carbon (at 1500–1600 cm⁻¹ in Fig. 3), could also be the effect of this thermal damage. It is clear that the laser power should be set according to the relation between the absorbance peak position and the excitation wavelength. Unfortunately, with the used equipment, it was not possible to set a power value between 1% (which was too low) and 5% (too high) for the 532 nm and 633 nm sources.

To test this hypothesis further, a third excitation wavelength was also tested. Based on Fig. 2, the 785 nm excitation is far away from the plasmon peak of the substrates. With this excitation, even 100% laser power did not cause any damage to the substrates; thermalization was thus avoided. Fig. 4 presents SERS spectra obtained on Type B substrates with the three different excitation wavelengths after binding the target-DNA. The 523 and 633 nm excitations were measured with 1% power, while the 785 nm was obtained on maximum, 100% laser power. Target DNA peaks were not detected clearly, and the intensity of the peaks was very low in the case of the 532 nm excitation. Also, with 532 nm and 633 nm excitation, the carbon peaks were observed to be dominant around 1500 cm⁻¹. With 633 nm excitation, the peaks corresponding to the DNA bases started to appear. Here, the 1% power was not enough to increase the relative intensity of these peaks. With 5% intensity, the peaks corresponding to DNA usually 'flashed' with high intensity for a couple of seconds, before the substrate started to degrade due to thermalization.

On the other hand, by using 785 nm excitation wavelength with 100% power intensity, no damage was observed on the surface, and the high-intensity peaks of target-DNA bound to the Au NPs have been observed. The obtained spectra were also very stable in time. The results again confirm

the relationship between excitation and plasmon absorbance and that damaging thermalization can be avoided by selecting an operating wavelength with lower absorbance.

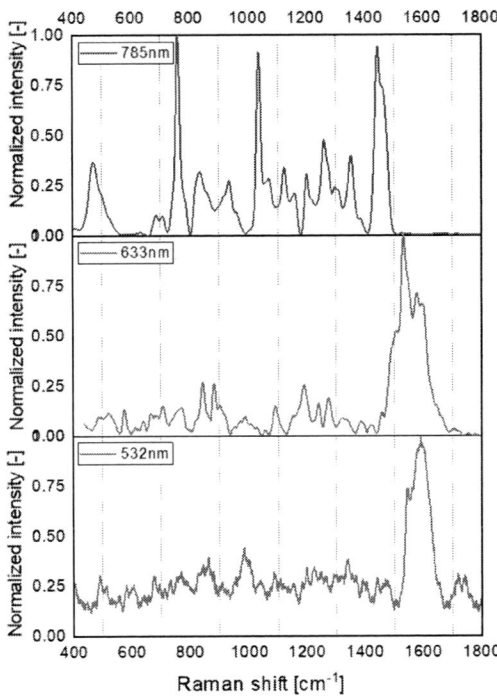

Fig.4. Excitation wavelength-dependent SERS spectra of *Giardia target* measured at 785, 633 and 532 nm.

Based on the spectrum measured with 785 nm excitation, the peak positions can be assigned to specific DNA base resonances in the following way (see Table II and Fig. 5). Ring breathing mode of Adenine peak at 761 cm⁻¹ and Adenine, Thymine and Cytosine related peak at 1032 cm⁻¹.

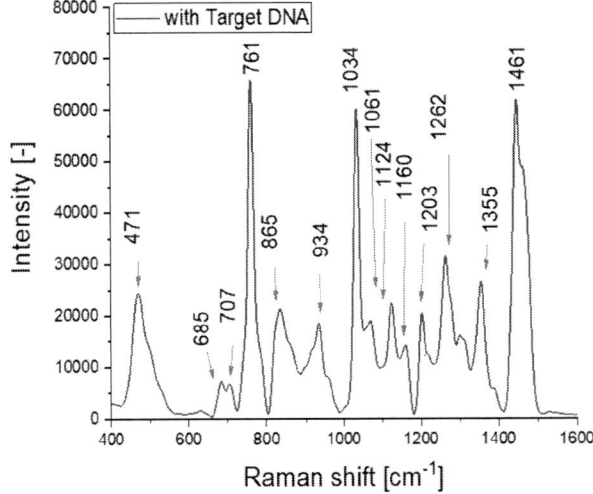

Fig.5. SERS spectrum of *Giardia target* measured with 785 nm excitation, and the assignment of the observed peaks (see Table 2).

The strong peak of Adenine at 1461 cm⁻¹ could be observed only with the excitation of 785 nm. The peak at 1261 cm⁻¹ was detected under both 633 and 785 nm, corresponding to Adenine and Thymine, due to stretching in (C–C) and (C–N). From the SERS spectra, it can be said that the majority of DNA distinctive peaks are related to Adenine, which is the most active base on the Au nanocomposite besides Guanine and deoxyribose. This observation is reasonable since the *Giardia target* consists of Adenine and Guanine more than other bases.

TABLE II. Tentative band assignments

Raman shift (cm^{-1})	Assignment
471	T ,C
685	stretching A, G, deformation (C-H)
707	scissoring A, vibration (C–S)
761	Ring breathing mode A
865	G ring stretching
934	stretching deoxyribose, G
1034	A,G,T
1061	Vibration (C–N)
1124	A, stretching of deoxyribose phosphate backbone
1160	A,G,T, stretching of deoxyribose phosphate backbone
1203	stretching of deoxyribose phosphate backbone
1262	A,T, stretching C–C and C–N
1355	stretching C4–O of T
1461	A,(C–H) deformation of deoxyribose

Peak references from [6-7, 9-11]

IV. CONCLUSION

It was demonstrated that the recently introduced, highly ordered gold/epoxy surface nanocomposite could be effectively used as a SERS substrate to detect the presence of target-DNA molecules. It was shown that the dynamic behavior of the nanoplasmonic sensor, namely the redshift of the absorbance spectra due to the addition of molecules, influences the obtainable absolute Raman signals by shifting the absorbance peak closer or further away from the excitation wavelength. This effect should always be considered when selecting the appropriate excitation wavelength for SERS applications. Also, it was shown that using an excitation wavelength close to the plasmonic resonance of the substrate can damage the nanocomposite by intensive heating, even at low laser powers. For our nanocomposites, using an excitation wavelength further away from the resonance (785 nm compared to the 523 and 633 nm) with increased power is beneficial to avoid thermalization and to detect specific DNA peaks with high intensity. Based on the SERS spectra, it was found that Adenine, Guanine, and deoxyribose yielded the most signal at the operation wavelength of 785 nm.

ACKNOWLEDGMENTS

The research reported in this paper and partially carried out at the Budapest University of Technology and Economics has been supported by the National Research Development and Innovation Fund (TKP2020) based on the charter of bolster issued by the National Research Development and Innovation Office under the auspices of the Ministry for Innovation and Technology.

This work was partially supported by Nanoplasmonic Laser Fusion Research Laboratory project financed by the National Research and Innovation Office (NKFIH) and by the Eotvos Roland Research Network (ELKH), Hungary.

CzechNanoLab project LM2018110 funded by MEYS CR is gratefully acknowledged for the financial support of the measurements/sample fabrication at CEITEC Nano Research Infrastructure.

REFERENCES

[1] Shinki and Subhendu Sarkar, "Au0.5Ag0.5 Alloy Nanolayer Deposited on Pyramidal Si Arrays as Substrates for Surface-Enhanced Raman Spectroscopy" ACS Appl. Nano Mater. 2020, 3, 7, 7088–7095.

[2] P.Pál, A. Bonyár, M. Veres, L. Himics, L. Balázs, L. Juhász, I. Csarnovics, "A generalized exponential relationship between the surface-enhanced Raman scattering (SERS) efficiency of gold/silver nanoisland arrangements and their non-dimensional interparticle distance/particle diameter ratio" Sensors and Actuators A: Physical,

[3] Vignesh Suresh, Lu Ding, Ah Bian Chew, and Fung Ling Yap, "Fabrication of Large-Area Flexible SERS Substrates by Nanoimprint Lithography", ACS Appl. Nano Mater. 2018, 1, 2, 886–893.

[4] Attila Bonyár, "Label-Free Nucleic Acid Biosensing Using Nanomaterial-Based Localized Surface Plasmon Resonance Imaging: A Review", ACS Appl. Nano Mater. 2020, 3, 9, 8506–8521

[5] T. Lednický and Attila Bonyar, "Large Scale Fabrication of Ordered Gold Nanoparticle–Epoxy Surface Nanocomposites and Their Application as Label-FreePlasmonic DNA Biosensors," ACS Applied Materials & Interfaces, 4th ed.,vol. 12, pp.4804-4814, January 2020.

[6] L.Xu, Z. Lei, J. Li, C.Zong, C.Yang and B. Ren, "Label-Free Surface-Enhanced Raman Spectroscopy Detection of DNA with Single-Base Sensitivity," Journal of the American Chemical Society, vol. 137, pp. 5149–5154, April 2015.

[7] S. Mirajkar, A. Dhayagude, N. Maiti, P. Suprasanna and S. Kapoor, "Distinguishing genomic DNA of Brassica juncea and Arabidopsis thaliana using surface-enhanced Raman scattering," J. Raman Spectrosc., 1st ed., vol. 51, pp. 89–103, October 2019.

[8] I. Khalil *et.al.*, "Dual platform based sandwich assay surfaceenhanced Raman scattering DNA biosensor for the sensitive detection of food adulteration," Analyst, vol. 145, pp.1414-1426, December 2020.

[9] T. Chan, *et al.*, "SERS Detection of Biomolecules by Highly Sensitive and Reproducible Raman-Enhancing Nanoparticle Array,". Nanoscale Research Letters 12, vol. 344, May 2017.

[10] G. Szekeres and J. Kneipp, "SERS Probing of Proteins in Gold Nanoparticle Agglomerates," Frontiers in Chemistry, vol. 7, January 2019.

[11] E. Pyrak, A. Jaworska and A. Kudelski, "SERS Studies of Adsorption on Gold Surfaces of Mononucleotides with Attached Hexanethiol Moiety: Comparison with Selected Single-Stranded Thiolated DNA Fragments," Molecules, vol. 24, October 2019.

[12] Guy, R. A.; Xiao, C.; Horgen, P. A. Real-Time PCR Assay for Detection and Genotype Differentiation of Giardia Lamblia in Stool Specimens. J. Clin. Microbiol. 2004, 42 (7), 3317–3320.

*2020 IEEE 26th International Symposium for Design and Technology in Electronic Packaging (**SIITME**)*

Small-Signal Modelling of the Three Switch 1L2C Boost Converter

Septimiu Lica, Alex Molcuț, Ioan Lie, Dan Lascu

Applied Electronics Department
Politehnica University Timisoara,
Timisoara, Romania
Septimiu.Lica@upt.ro, Dan.Lascu@upt.ro

Abstract— The paper derives the small-signal state-space model, the continuous and discrete transfer functions for the three switch 1L2C boost DC-DC converter. A capacitive loop occurs in the first topological state of the converter and thus it is difficult to calculate the small-signal model using usual approaches. The paper presents the analytical and simulation results together with the computations in MATLAB™ and finally the validation provided by measurements on a practical prototype.

Keywords— DC-DC PWM converters, state-space model, small-signal analysis, transfer functions.

I. INTRODUCTION

The three-switch 1L2C DC-DC converters were introduced by dr. D. Zhou in [1] together with their DC analysis.

This converter is useful for high voltage applications, the output voltage doubler being possible to be extended into a multiplier [1]. Also, an application could be in symmetric power sources, because of the negative output voltage. The hard switching may be alleviated by adding a small inductor that transforms the converter into a hybrid one [2]. The converter can also be modified such that to provide isolation [1].

For complete description and practical use as a constant voltage or constant current source in a close loop, the AC model is also needed.

The paper structure is as follows: in Section II the DC model of the converter is presented, in the third section the state-space model is derived, then section IV is used to obtain a discrete-time small-signal model. A mathematical artifice is proposed in section V to directly obtain the continuous-time transfer functions. The results are compared and validated by simulation in section VI and sustained by experiment in section VII. Finally, some conclusions are drawn.

II. THE DC MATHEMATICAL MODEL

The 1L2C converter [1] which is analysed in this paper is depicted in Fig. 1. The converter comprises of one active switch S, two diodes $D_{1,2}$, one inductor L, one internal capacitor C_1 and the output capacitor C_o-. The transistor is driven by a pulse-width modulated (PWM) signal, of switching frequency f_s and duty cycle D, as in the traditional converters [3].

All the electrical magnitudes in this section are average values and thus written with capitals. The analysis is performed in continuous conduction mode (CCM) relative to diodes. All circuit elements are considered ideal.

The converter has two topological states corresponding to the levels in the control signal.

The first topological state occurs when the active switch S conducts and is detailed in Fig. 2. Initially, diode D_1 is also ON and D_2 is OFF. It can be remarked that the two capacitors are connected in parallel through D_1 and S. In this case it seems like only one capacitor exists in the circuit. Therefore, the number of state variables changes from 2 to 3 from one topological state to another. A question that arises is what is the order of the system? Another issue, even for simulation, is that connecting the two capacitors in parallel, a high spike current occurs since they

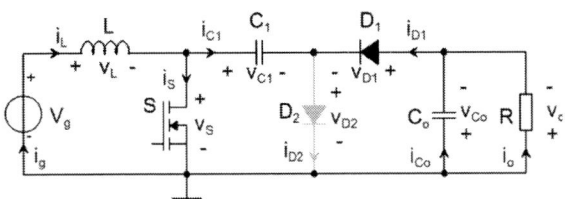

Fig. 2. The first topological state of the three switch 1L2C boost.

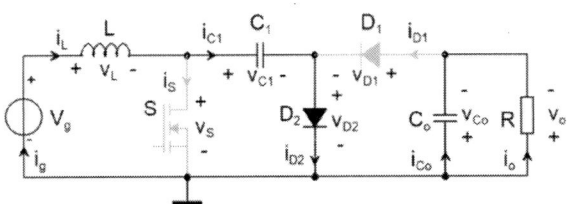

Fig. 3. The second topological state of the three switch 1L2C boost converter.

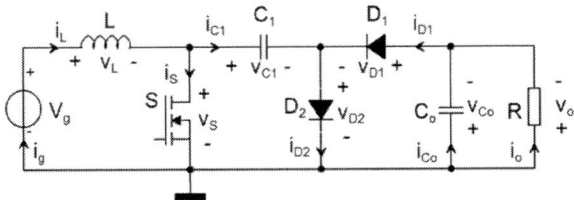

Fig. 1. The 1L2C boost DC-DC converter [1].

exhibit different voltages when connected and they need rapidly to redistribute the charge. These aspects were not studied yet. Finally, as it may be observed, in this state of the 1L2C converter, two current loops exist, as in the classical boost converter. The first one contains the input voltage source V_g, the inductor L and returns to the ground terminal of the voltage source through the transistor. The second one consists of the output load resistor R that discharges the parallel assembly of capacitors.

In the second topological state, depicted in Fig. 3, the active switch S is OFF and only diode D_2 is forward biased. In this case other two loops are formed: one containing the input voltage source V_g, the inductor L, the internal capacitor C_1 and returning to the ground of the source through D_2. The other one is at the output, where capacitor C_o is discharged on the load resistor R.

By using the fundamental conservative laws, the steady-state DC values of the capacitive voltages and inductive current may be derived [3]. For the capacitors, they have equal DC voltage [1], calculated from the volt-second balance equation:

$$V_{C1} = \frac{1}{1-D} \tag{1}$$

$$V_{Co} = \frac{1}{1-D} \ g \tag{2}$$

Also, from charge balance equation, the DC current through the inductor would be:

$$I_L = \frac{1}{(1-D)^2} \frac{V_g}{R} \tag{3}$$

The static conversion ratio, from (2), results in:

$$M = -\frac{1}{1-D} \tag{4}$$

The minus sign emphasizes the fact that the output voltage V_o has opposite sign compared to the input voltage V_g.

Other currents or voltages from the circuit can be derived from these fundamental formulas.

III. THE STATE-SPACE MODEL

As a first step the state-space model in vector form was derived [4]. The notations in lower case represent instantaneous values. The input vector is:

$$\mathbf{u} = [V_g] \tag{5}$$

The state vector is defined as:

$$\mathbf{x} = [i_L \ v_{C1} \ v_{Co}]^t \tag{6}$$

The system matrix for the first topological state would be:

$$\mathbf{A}_1 \begin{vmatrix} 0 & 0 & 0 \\ 0 & -\dfrac{1}{R \cdot (C_1 + C)} & \\ 0 & & -\dfrac{1}{R \cdot (C_1 + C)} \end{vmatrix} \tag{7}$$

The input matrix in the same state:

$$\mathbf{B}_1 = \begin{bmatrix} \dfrac{1}{L} \\ 0 \\ 0 \end{bmatrix} \tag{8}$$

If the output vector is defined as:

$$\mathbf{y} = [v_o] \tag{9}$$

Then the output and the feedforward matrices would be:

$$\mathbf{E}_1 = [0 \ 0 \ 1] \tag{10}$$

$$\mathbf{F}_1 = [0] \tag{11}$$

For the second topological state, the system matrix is:

$$\mathbf{A}_2 = \begin{vmatrix} 0 & -\dfrac{1}{L} & \\ \dfrac{1}{C_1} & 0 & \\ 0 & & -\dfrac{1}{R \cdot C_2} \end{vmatrix} \tag{12}$$

The input matrix in this state would result the same as in (8):

$$\mathbf{B}_2 = \begin{bmatrix} \dfrac{1}{L} \\ 0 \\ 0 \end{bmatrix} \tag{13}$$

The output and the feedforward matrices would also result as in the first, state, like (10) and (11):

$$\mathbf{E}_2 = [0 \ 0 \ 1] \tag{14}$$

$$\mathbf{F}_2 = [0] \tag{15}$$

The first thought was to use the averaged model with:

$$\mathbf{A_D} = D \cdot \mathbf{A}_1 + (1-D) \cdot \mathbf{A}_2 \tag{16}$$

$$\mathbf{B_D} = D \cdot \mathbf{B}_1 + (1-D) \cdot \mathbf{B}_2 \tag{17}$$

$$\mathbf{E_D} = D \cdot \mathbf{E}_1 + (1-D) \cdot \mathbf{E}_2 \tag{18}$$

$$\mathbf{F_D} = D \cdot \mathbf{F}_1 + (1-D) \cdot \mathbf{F}_2 \tag{19}$$

In this approach the first difficulty appears, the matrix $\mathbf{A_D}$ being singular, thus being unable to apply the procedure to obtain the transfer function [4]. It is needed to find out an approach that overcomes the singularity of the system matrix or does not use it, going through the nonlinearity induced by the capacitive Dirac-like current pulse and the voltage step when the charge is redistributed on the parallel coupled capacitors.

IV. THE DISCRETE-TIME MODDELING

A first approach is to use the discrete model [5] to sample the waveforms with the switching frequency f_s and jump over the nonlinearities. As it is known, the small-signal model is a low frequency model, that is exact only below the half of the switching frequency. The precision given by Whittaker–Kotelnikov–Shannon sampling theorem.

With lower case and hat the perturbations for the variables are defined. In the linearized analyses [5] the signals are composed by their average value and a first-order perturbation:

$$x(t) = X + \hat{x}(t) \qquad (20)$$

The recurrence equation for the discretized state signal is:

$$x[k+1] = \mathbf{\Phi} \cdot x[k] + \mathbf{\Psi} \cdot u_r[k] \qquad (21)$$

where k is the no. of the sample and matrices $\mathbf{\Phi}$ and $\mathbf{\Psi}$ will be calculated below.

A. The Discret Model Particularities

As it has previously stated, because of the high capacitive current spike that occurs when the first topological state begins, the usual approach cannot be used. We shall consider the sampling moment, the time moment just before switching ON the transistor, index k. The value of the voltage on both capacitors v_c after they balance their charge will be proportional to their capacitance and would occur after the transistor is switched ON:

$$v_C[k+] = \frac{C_1}{C_1 + C_o} \cdot v_{C1}[k] + \frac{C_o}{C \quad C} \cdot v_{Co}[k] \qquad (22)$$

For the sake of simplicity, let us denote:

$$p_1 = \frac{C_1}{C_1 + C_o} \qquad (23)$$

$$p_2 = \frac{C_o}{C_1 + C_o} \qquad (24)$$

With the sum $p_1 + p_2 = 1$.

The equation that relates the next sampling moment to the moment $k+$ is given by:

$$x[k+1] = \varphi_2 \cdot \varphi_1 \cdot \begin{vmatrix} i_L[k+ \\ v\ [k+] \\ c[k+] \end{vmatrix} + (\varphi_2 \cdot \psi_1 + \psi_2) \cdot u_r[k] \qquad (25)$$

By defining two extractor matrices $\mathbf{M_1}$ and $\mathbf{M_2}$,

$$\mathbf{M_1} = \begin{vmatrix} 1 & 0 & 0 \\ 0 & 1 & 0 \\ 0 & 1 & 0 \end{vmatrix} \qquad (26)$$

$$\mathbf{M_2} = \begin{vmatrix} 1 & 0 & 0 \\ 0 & 0 & 1 \\ 0 & 0 & 1 \end{vmatrix} \qquad (27)$$

the required recurrence equation is obtained:

$$x[k+1] = \varphi_2 \cdot \varphi_1 \cdot (p_1 \cdot \mathbf{M_1} + p_2 \cdot \mathbf{M_2}) \cdot x[k] + (\varphi_2 \cdot \psi_1 + \psi_2) \cdot u_r[k] \qquad (28)$$

Here the matrices of the model are defined as usual:

$$\varphi_1 = e^{\mathbf{A_1} \cdot d_k \cdot T_s} \qquad (29)$$

$$\varphi_2 = e^{\mathbf{A_2} \cdot (1 - d_k) \cdot} \qquad (30)$$

$$\psi_1 = e^{\mathbf{A_1} \cdot d_k \cdot T} \cdot \left(\int_0^{d_k \cdot T} e^{-\mathbf{A_1} \cdot \tau} \cdot d\tau \right) \qquad (31)$$

$$\psi_2 = e^{\mathbf{A_2} \cdot (1 - d) \cdot T_s} \cdot \left(\int_0^{(1 - d) \, T_s} e^{-\mathbf{A_2} \cdot \tau} \cdot d\tau \right) \qquad (32)$$

Comparing to (21), it could be easily identified that:

$$\mathbf{\Phi} = \varphi_2 \cdot \varphi_1 \cdot (p_1 \cdot \mathbf{M_1} + p_2 \cdot \mathbf{M_2}) \qquad (33)$$

$$\mathbf{\Psi} = \varphi_2 \cdot \psi_1 + \psi_2 \qquad (34)$$

Denoting, $\mathbf{M} = (p_1 \cdot \mathbf{M_1} + p_2 \cdot \mathbf{M_2})$, it is easily observed that in comparison to a classic converter are valid the relationships $\mathbf{\Phi} = \mathbf{\Phi}_{classic} \cdot \mathbf{M}$ and $\mathbf{\Psi} = \mathbf{\Psi}_{classic}$.

Continuing the methodology from [5], the small-signal recurrence equation is written as:

$$\hat{x}[k+1] = \mathbf{\Phi} \cdot \hat{x}[k] + \mathbf{\Psi} \cdot \hat{u}_r[k] + \gamma \cdot \hat{d}_k \qquad (35)$$

The third matrix γ that defines the model is derived as:

$$\gamma = (\varphi_2' \cdot \varphi_1 + \varphi_2 \cdot \varphi_1') \cdot \mathbf{M} \cdot X + (\varphi_2' \cdot \psi_1 + \varphi_2 \cdot \psi_1' + \psi_2') \cdot U_r \qquad (36)$$

with $X = (\mathbf{I_3} - \mathbf{\Phi})^{-1} \cdot \mathbf{\Psi} \cdot U_r$, where $\mathbf{I_3}$ is the 3×3 unity matrix.

The derivates φ_1', φ_2', ψ_1', and ψ_2' with respect to d_k are analytically calculated:

$$\varphi_1' = \mathbf{A_1} \cdot T_s \cdot \varphi_1 \qquad (37)$$

$$\varphi_2' = -\mathbf{A_2} \cdot T_s \cdot \varphi_2 \qquad (38)$$

$$\psi_1' = T_s \cdot \left(\mathbf{A_1} \cdot \varphi_1 \cdot \int_0^{d_k T_s} e^{-\mathbf{A_1} \tau} \cdot d\tau + \mathbf{I_3} \right) \cdot \mathbf{B_1} \qquad (39)$$

$$\psi_2' = -T_s \cdot \left(\mathbf{A_2} \cdot \varphi_2 \cdot \int_0^{(1-d_k) T_s} e^{-\mathbf{A_2} \cdot \tau} \cdot d\tau + \mathbf{I_3} \right) \cdot \mathbf{B_2} \qquad (40)$$

In this algorithm the singularity of the matrix $\mathbf{A_1}$ does not matter.

B. Tustin Transformation

For deriving the continuous time model, the system was sampled obtaining discrete time equations. Then the reverse transformation from discrete to continuous time should be performed. The first approach uses Tustin transform [5]:

$$z = \frac{1 + \dfrac{T_s}{2} s}{1 - \dfrac{T_s}{2} s} \qquad (41)$$

The perturbed state vector in Z-domain is:

$$\hat{x}(z) = (z \cdot \mathbf{I_3} - \mathbf{\Phi})^{-1} \cdot (\mathbf{\Psi} \cdot u_r(z) + \gamma \cdot \hat{d}(z)) \qquad (42)$$

From (42) the audiosusceptibility in z is identified as:

$$H_g(z) = (z \cdot \mathbf{I_3} - \mathbf{\Phi})^{-1} \cdot = \frac{\gamma}{z \cdot \mathbf{I_3} - \mathbf{\Phi}} \qquad (43)$$

Similarly, the control to output discrete transfer function:

$$H_c(z) = (z \cdot \mathbf{I}_3 - \boldsymbol{\Phi})^{-1} \cdot \quad = \frac{\boldsymbol{\Psi}}{z \cdot \mathbf{I}_3 - \boldsymbol{\Phi}} \qquad (44)$$

Now z form (41) is substituted in (43) and (44) and the continuous time transfer functions are obtained. Because of the length of the expression, it is not reproduced here, but it was implemented in MATLAB™ and the results will be provided below.

C. Padé Approximations

The parameter z from the discrete transfer function can be substituted by elevated degree approximations using higher degree polynomials, like Padé approximations [6]. The first degree Padé polynomial is the same as Tustin polynomial. The higher order ones were used:

$$z^{(2)} = \frac{12 + 6 \cdot s \cdot T + (s \cdot T_s)^2}{12 - 6 \cdot s \cdot T + (s \cdot T_s)^2} \qquad (45)$$

$$z^{(3)} = \frac{120 + 60 \cdot s \cdot T_s + 12 \cdot (s \cdot T_s)^2 + (s \cdot T_s)}{120 - 60 \cdot s \cdot T_s + 12 \cdot (s \cdot T_s)^2 - (s \cdot T_s)} \qquad (46)$$

$$z^{(4)} = \frac{1680 + 840 \cdot s \cdot T_s + 180 \cdot (s \cdot T_s)^2 + 20 \cdot (s \cdot T_s)^3 + (s \cdot T_s)^4}{1680 - 840 \cdot s \cdot T_s + 180 \cdot (s \cdot T_s)^2 - 20 \cdot (s \cdot T_s)^3 + (s \cdot T_s)^4} \quad (47)$$

Obviously, the use of Padé approximations increases the transfer functions order and involves more computations.

V. THE CONTINOUS TIME MODELING

The final approach tries to cut the cause of the singularity that appears in the system matrix $\mathbf{A}_1$ in the first state. Thus, the capacitive loop that is formed when C_1 and C_o are connected in parallel must be interrupted. This can be done by intercalating a small resistance r inside the loop. It may be considered the equivalent series resistance (ESR) for one of the capacitors or the loss resistance of the diode D_1.

Then it should be proceeding with computation of additional matrices [4]:

$$\xi_D = (\mathbf{A}_1 - \mathbf{A}_2) \cdot \mathbf{x} + (\mathbf{B}_1 - \mathbf{B}_2) \cdot \mathbf{u} \qquad (48)$$

$$\zeta_D = (\mathbf{E}_1 - \mathbf{E}_2) \cdot \mathbf{x} + (\mathbf{F}_1 - \mathbf{F}_2) \cdot \mathbf{u} \qquad (49)$$

The input to output transfer function will be calculated as:

$$G_g(s) = \mathbf{E}_D \cdot (s \cdot \mathbf{I}_3 - \mathbf{A}_D)^{-1} \cdot \mathbf{B}_D + \mathbf{F}_D \qquad (50)$$

and the control to output transfer function would be:

$$G_c(s) = \mathbf{E}_D \cdot (s \cdot \mathbf{I}_3 - \mathbf{A}_D)^{-1} \cdot \xi_D + \zeta_D \qquad (51)$$

After this calculation, the loss resistance r is nullified by performing the limit to 0. The result is as follows:

$$G_g(s = \frac{(1-D) \, R}{(1-D)^2 \cdot R + L \cdot s + R \cdot L \cdot (C_1 + C_o) \cdot s^2} \qquad (52)$$

$$G_c(s = \frac{\lfloor (1-D)^2 \cdot R - L \cdot s \rfloor \cdot V_g}{(1-D) \cdot \lceil (-D)^2 \cdot R + L \cdot s + R \cdot L \cdot (C_1 + C_o) \cdot s^2} \qquad (53)$$

In canonical form [3] the functions look like:

$$G_g(s = \frac{G_{g0}}{1 + \dfrac{s}{Q \cdot \omega_0} \left(\dfrac{}{} \right)^2} \qquad (54)$$

$$G_c(s = \frac{G_{c0} \cdot \left(1 - \dfrac{s}{\omega_z} \right)}{1 + \dfrac{s}{Q \cdot \omega_0} \left(\dfrac{}{} \right)^2} \qquad (55)$$

By identifying, results:

$$G_{g0} = \frac{1}{1-D} \qquad (56)$$

$$G_{c0} = \frac{1}{(1-D)^2} \qquad (57)$$

$$C_e = C_1 + C_o \qquad (58)$$

$$L_e = \frac{L}{(1-D)^2} \qquad (59)$$

$$\omega_0 = \frac{1}{\sqrt{L_e \cdot C_e}} \qquad (60)$$

$$R_0 = \sqrt{\frac{L_e}{C_e}} \qquad (61)$$

$$Q = \frac{R}{R_0} \qquad (62)$$

$$\omega_z = \frac{R}{L_e} \qquad (63)$$

At a first inspection it is remarked that the small signal model resembles to the one of the classical boost converter, if we consider the single capacitor from the classical one as the equivalent parallel capacitance C_e from the 1L2C converter. It is proved that the denominator of the canonical transfer functions is of 2nd order, equal to the number of reactive elements minus one. This leads to the conclusion that the capacitors in parallel are considered one component.

VI. COMPUTATIONS AND SIMULATIONS

Some MATLAB™ scripts were written to compare the results and to prove that the derived models are equivalent.

The converter that was studied has the following values of the components: $V_g = 10V$, $L = 447\mu H$, $C_1 = 100\mu F$, $C_o = 110\mu F$, $R = 24\Omega$, and the driving signal has $f_s = 100kHz$, $D = 0.6$.

2020 IEEE 26th International Symposium for Design and Technology in Electronic Packaging (SIITME)

These values were used to calculate the transfer functions previously presented. In Fig. 4 the Bodé plots of the calculated audiosusceptibility are depicted. Also, with a vertical line the 50kHz Nyquist frequency is marked. The discrete time transfer function has values only below this frequency. The Tustin transformation leads to a continuous time transfer function, but the amplitude is precise only below the Nyquist frequency, compared to the continuous time audiosusceptibility obtained directly. In practice it is not needed a higher precision. The phase information is accurate, and close to each other, in all cases.

In Fig. 5 the control to output transfer function is shown, using also the three methods of computation. In Z domain, the discrete time transfer function is depicted until the Nyquist frequency, which is marked on the plot. The transfer function converted with Tustin method is precise for frequencies lower that the Nyquist frequency for amplitude and for phase until about 3kHz, compared to the continuous time transfer function obtained directly from the state-space model by computing the

limit to 0 for the loss resistance. The models are useful in practical applications, since only low frequency behaviour is needed.

All the transfer functions look like a low pass filter and are in concordance with the theory.

In Fig. 6 trials to approximate with Padé polynomial ratios are shown in comparison to the continuous time input-to output transfer function. The first Padé approximation is the same as Tustin transform, as stated previously. The higher order approximation does not conduct to a higher accuracy and some artefacts appear at higher frequency. All the approximations lead to a correct transfer function up to the Nyquist frequency, thus being useful in practical design operations.

The same comments are also valid for the control to output transfer function using higher degree Padé approximations and depicted in Fig. 7. Although some artefacts appear at about

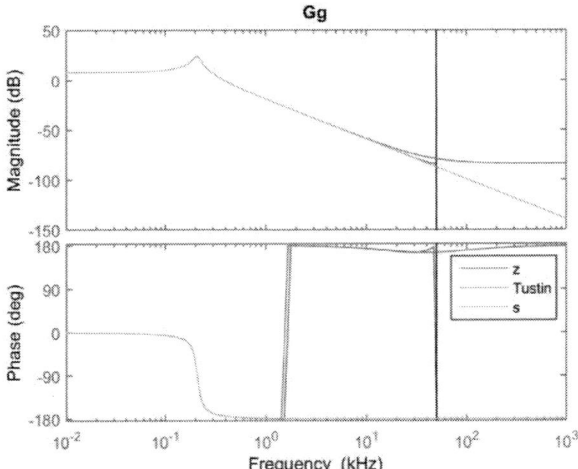

Fig. 4. The audiosusceptibility in Z domanin, converted with Tustin method and in Laplace domain

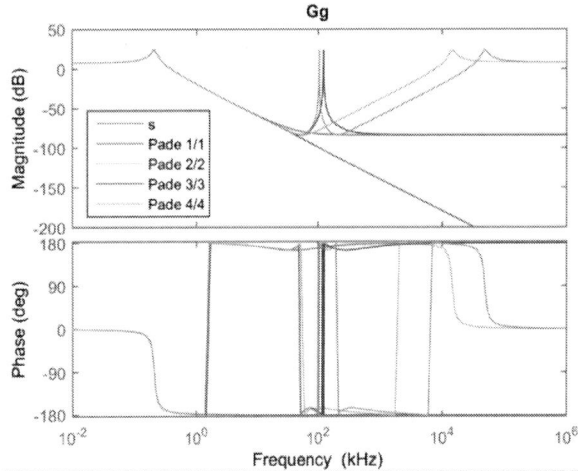

Fig. 6. The audiosusceptibility in Laplace domain and converter from discrete time with Padé approximations

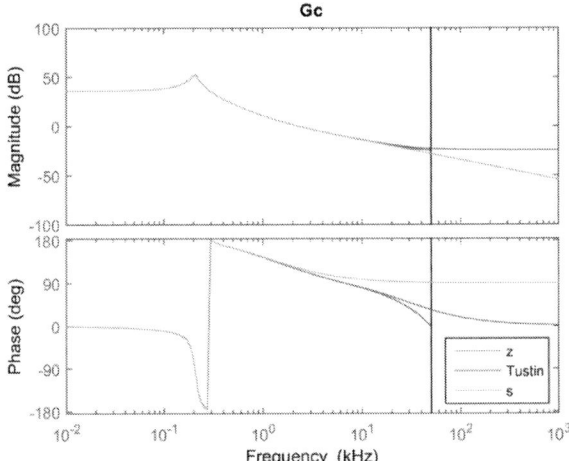

Fig. 5. The control to output transfer function in Z domanin, converted with Tustin method and in Laplace domain

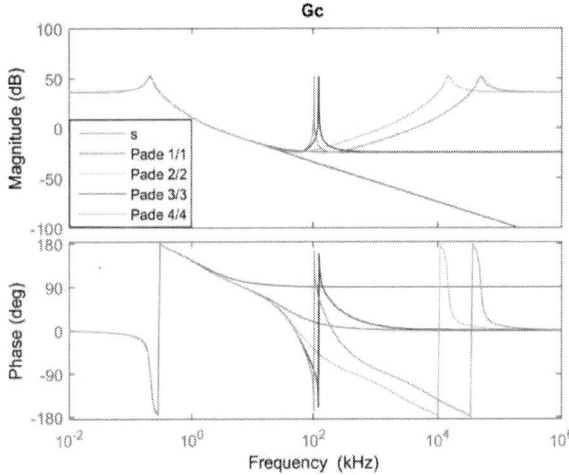

Fig. 7 The control to output transfer function in Laplace domain and converter from discrete time with Padé approximations

100kHz, around the sampling frequency, the results are accurate until 50kHz, which is the Nyquist frequency.

All the presented functions can be used for practical reason since the design, especially of the controller, is made at low frequencies.

VII. EXPERIMENTAL RESULTS

A practical prototype of the converter was designed and constructed using components with the values stated in previous section. The semiconductors were, for S Infineon™ CoolMOS™ Power MOSFET IPD80R450P7 and for D_1 and D_2 Philips Rectifier diodes with Schottky barrier PBYR20100CT.

The prototype of the converter was attached a PWM generator with a sawtooth signal peak-to-peak amplitude of V_{tri} = 5Vpp.

The control to output transfer function was measured with a Ridley Engineering AP300 frequency response analyser (FRA). The input source of the FRA was capacitive coupled to the DC voltage that is compared to the sawtooth triangular signal. The output measurement probe of the FRA is connected to the output of the converter, parallel to the load resistance R. The result of the measurement is depicted in Fig. 8. The measured transfer function contains also the modulator transfer function:

$$H_{PWM}(s) = \frac{1}{V_{tri}} \qquad (64)$$

This is the cause that the measured amplitude is lower than in the theoretical model from previous section, which is the transfer function strictly limited to the converter circuit. Also, the maximum frequency at which the FRA has scanned was 50kHz, the Nyquist frequency in this case, since above it is not necessary to have information and appear some noises also.

VIII. CONCLUSIONS

The analysed converter has two capacitors connected in parallel in one topological state, thus theoretically producing a Dirac-like pulse in the capacitive current and a step signal for capacitive voltages. Also, the number of state variables changes from one topological state to another, since the capacitors are considered to be in parallel in the first state. This behaviour lead to a singular system matrix **A** and the small-signal model could not be derived directly.

The paper proposes two approaches for exact calculation of the small-signal model. The first one uses the discrete time model to obtain the Z-domain transfer functions and then the conversion to continuous time is performed. For the discrete to continuous transformation Tustin method is used. Higher degree

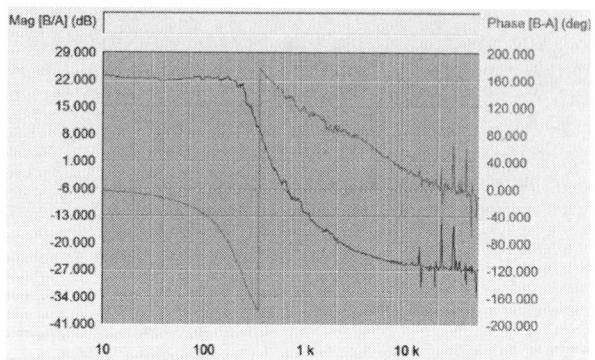

Fig. 8. The measured control to output transfer function

polynomial Padé approximations would not improve the accuracy, as demonstrated. The second approach introduces a small series resistance to break the capacitive loop that produces the high impulse and the singularity. Based on non-singular matrices the transfer functions are calculated. Then, the series resistance is nullified. Both approaches led to same results.

The 1L2C boost converter is useful for high voltage applications, it may be easily coupled to the output a Cockcroft–Walton voltage multiplier cell. It is a high efficiency converter. For a regulated voltage at the output a controller must be used. For a proper design of the controller, the transfer functions are necessary, and they are now available to the user.

Future work will be performed to derive the state-space model of other converters containing a capacitive loop, such as stacked converters. Also, for the second generation Ćuk converters [2], that prolongs the dissipation of the current pulse by resonance, a similar analysis will be carried out.

REFERENCES

[1] D. Zhou, "Synthesis of PWM Dc-to-Dc Power Converters," Ph.D. Thesis, California Institute of Technology, October 1995.

[2] S. Cuk, "True Bridgeless PFC Converter Achieves Over 98% Efficiency, 0.999 Power Factor," Power Electron. Technology, Vol. 36, No. 8, pp. 34-40, August 2010.

[3] R. W. Erickson and D. Maksimovic, Fundamentals of Power Electronics, 2nd Ed., Kluwer Academic, 2001.

[4] J. G. Kassakian, M. F. Schlecht, and G. C. Verghese, Principles of Power Electronics, Addison-Wesley, 1991.

[5] R. Tymerski, "Application of the Time-Varying Transfer Function for Exact Small-Signal Analysis," IEEE Trans. Power Electron., Vol. 9, No. 3, pp. 196-205, March 1994.

[6] M. Vajta, "Some Remarks on Padé-Approximations," 3rd TEMPUS-INTCOM Symp. on Intelligent Syst. in Control and Measurements, Veszprém, Hungary, pp. 1-6, 2000.

Simulating the Evolution of Infectious Agents Through Human Interaction

P.V. Vezeteu and D.I. Năstac

Faculty of Electronics, Telecommunications and Information Technology
POLITEHNICA University of Bucharest, Romania
nastac@ieee.org

Abstract—**Human interaction and behavior were proven to be decisive factors for disease spread among the members of a community. Recent events, determined by the COVID-19 outbreak, forced us to develop methods of control, in order to lower the reproductive number and overall, reduce the number of cases. We propose a software simulation method centered on in what manner a virus spreads inside a closed space. This approach required the implementation of a simplified model of human behavior, while considering the factors that influence the rate of transmission between individuals. Agent-based simulation allowed us to identify different control and prevention means, which are efficient in reducing the natural evolution of a pandemic. The results of the analysis offered insight into the rate of spread and human behavior, or habits that led to a significant increase in the total number of infected agents. Recent literature discussed the simulation of disease spread among the inhabitants of larger areas or studied the way that the pathogen can be spread from person to person through droplets containing the virus. Though we strongly consider that previous studies offer important insight, preventing the spread among smaller groups can let us maintain our activity in a safe and supervised manner. The model has been validated by a series of simulations with respect to COVID-19 real data.**

Keywords—*infectious agent; agent-based simulation; human interaction; COVID-19*

I. INTRODUCTION

Infectious diseases present themselves as a continuous threat, as the density of population is increasing from year to year. In this regard, software tools are used by epidemiologists to control and prevent emergencies while minimizing the impact of such events [1]. In the context of COVID-19, produced by the severe acute respiratory syndrome coronavirus 2 (SARS-CoV-2) [2][3][4], the need of numerical analysis and simulation had become a priority, as it helps authorities to make calculated decisions that overall should decrease the number of cases and deaths.

The literature presents itself with a high number of research topics, with respect to simulation and mathematical modeling, that study infectious diseases from different perspectives. Understanding how a virus is transmitted among high populated areas, as for example cities or countries, allows for prediction and prevention through scenario testing based on a high number of data [5][6]. Another form of numerical simulation has in focus the transmission of SARS-CoV-2 through respiratory droplets. Those studies performed both for indoor spaces [7] and open

areas offer insight into what means of protection can be beneficial or not, or how can we reduce the transmission level in poorly ventilated areas [8][9]. Understanding the efficiency of personal protective equipment (PPE), or other means of protection, is another key component that allows for a safer working environment for clinicians and medical workers, but also for the general public that uses masks to minimize the risk of infection [10][11].

While all this research is highly relevant, we consider that there is a knowledge gap with respect to indoor disease spread that needs more attention, as we find ourselves each day interacting with other people in closed spaces. Human interaction is an important aspect of our lives because we are, mostly in permanent contact, at the workplace, home or public places. Throughout the COVID-19 pandemic measures like social distancing were imposed so people can continue their daily activities in a somehow normal manner.

In this paper we implemented a software simulation that can numerically show the spread of an infectious agent in a closed space. The main objectives were adaptability, flexibility and accuracy. Flexibility refers to the possibility of the software to be modified, such that it can fit different closed spaces (living areas, workplaces, educational facilities). This aspect is important as there are a wide variety of closed areas which differ in structure and dimension. Adaptability takes into consideration the fact that there are many factors that influence the spread or transmission of COVID-19. For this reason, we want to offer the possibility of introducing new parameters that influence the rate of spread. While this study was conducted around COVID-19, the program can be modified such that it can simulate other infectious agents. This can be done by changing the parameters in accordance to a specific virus one wants to simulate. Accuracy refers to the degree of precision the simulation has when running a specific scenario. We want the program to mimic as accurate as possible the reality of infectious agents spreading through human interaction.

For this stage we designed the program in a way that it will fit a real residential building with five floors. We decided that this analysis is of higher importance (compared to a workplace or public spaces) as it can be later readapted for nursing homes, which are considered to be overlooked during COVID-19 pandemic. Studies show that in America nursing home and assisted living facilities accounted for approximatively 42% of COVID-19 deaths [12][13][14]. This phenomenon comes from the vulnerability of people in care, but also from the lack of planning and organization of the staff due to various reasons.

In terms of results and validation, the program managed to simulate based on real parameters computed from the data provided during the COVID-19 pandemic. Using the obtained data, it produced graphs that allow the user to easily draw conclusions and visualize the trend of the infectious agent in a specific predefined building.

II. METHODS

A. Agent based vs equation based simulation

Designing a simulation tool for infectious diseases implies that, one has to decide whether to use agent-based or equation-based modeling. Each comes with advantages and disadvantages [15][16], and to offer the reader a better perspective we are going to discuss both of them below.

Agent based models (ABMs) are comprised of individual entities that interact with the environment based on a set of coded rules. This type of simulation was previously used to study different types of infectious diseases like Ebola, measles and HPV. Compared to equation-based modeling this type needs more computing power and thus, normally, takes longer to run.

Equation-based models (EBMs) are built on a set of predefined equations that describe certain groups of people. Compartmental model is commonly used for virus simulation as it splits the total population and assigns each one of them a specific equation. The simplest model is SIR, susceptible (S), infected (I) and recovered (R) and has associated the following set of equations:

$$\frac{dS}{dt} = \frac{\beta SI}{N} \tag{1}$$

$$\frac{dI}{dt} = \frac{\beta SI}{N} - \gamma I \tag{2}$$

$$\frac{dR}{dt} = \gamma I \tag{3}$$

where N is the total population, β is the rate of transmission, γ is the rate of recovery and $S(t) + I(t) + R(t) = N$.

As previously stated, we aimed at designing a software simulation that is flexible and easily adaptable, while also offering a fair amount of information and complexity. We decided to use an agent-based model while taking advantage of the SIR model compartmentation method. One important advantage of an agent simulation model is that each person can be studied individually. Agents are thus being described by a set of general characteristics while following a simplified model of human behavior. We consider that for smaller groups of people this method is more advantageous.

The three main groups that agents may be distributed into, over the course of simulation, are susceptible, asymptomatic and infected. The recovered set was removed as the purpose of the simulator is to study closed spaces on relatively small populations and thus recovery is unlikely to take place in the same area without introducing other rules. While for this stage ignoring recovery and death, they may be introduced later in combination with a variation of the total number of agents,

mimicking better the procedures of the medical institutions (isolation, quarantine and hospitalization).

B. The simulated environment

The environment refers to the space where interactions agent to agent or agent to environment take place. For this stage of the work we defined a residential building with 5 floors each having 4 apartments. A total of 2 to 3 people per apartment, resulting in a total of 54 independent agents. In terms of spaces we had to consider the interior of the building, floors, apartments and stairs, while also the exterior, as people can commute to work or to other public spaces to satisfy their necessities. This is relevant because agents can contact the virus outside and thus infecting other neighbors, or most likely family and roommates.

When designing a living space or residence an important aspect is to take into consideration its real dimensions. We used AutoCad before the software implementation to understand how the space should be distributed, as it can be seen in Fig 1.

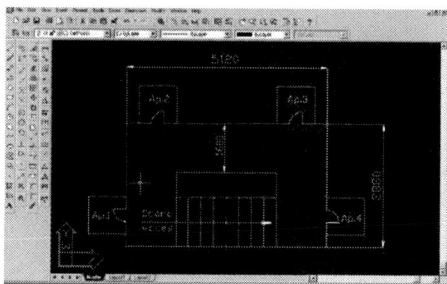

Fig. 1. AutoCad Design of the floors

Overall, the floor had 5.12 m in length and was 2.88 m wide, with a corridor of 160 cm. This accounted for a total surface of 14.74 m^2 without taking into consideration the dimension of the apartments, which for this stage were not of interest.

In terms of code, the area was defined as a discrete set of coordinates where the agents could walk freely depending on their needs. This allowed for more efficiency while not losing much precision and accuracy. Spaces like the interior of an apartment and building exterior were not precisely defined but took into consideration when computing the probabilities of contacting the infectious agent.

C. Susceptible, asymptomatic and infected

Agents were divided into three main categories: susceptible asymptomatic and infected. Each individual entity can pass through those groups based on a set of rules, discussed in the following sections of the paper.

Susceptible refers to the group comprised of healthy individuals who are at risk of becoming infected by a simulated disease. They can contact the virus by being exposed to agents that carry it or going outside the building.

Asymptomatic represents agents that carry the virus, but don't present symptoms yet. Normally, a person will become infected, thus presenting symptoms, after approximative 5 days. This period of time is mentioned in the scientific literature and was computed with respect to data collected during COVID-19 pandemic [17].

The last category, infected or symptomatic, is comprised of people that had contacted the virus while also went through the period of incubation. This group is the most probable to transit the infectious agent to other persons.

D. Rules and parameters of the simulation

A key problem when designing such simulations is that human behavior is unpredictable. There are many decisions that a person could make over the course of a day, so the time had to be reduced at certain important moments. Overall, the time was discretized to capture the following possible schedules: going outside or remaining in their apartments, meeting with other neighbors and coming back home at night. During the day, the following aspects were taken into consideration for each group of people.

Susceptible individuals were at risk of getting infected when in contact with other infected agents, asymptomatic or symptomatic, or when going outside to preform different activities. Each time when a healthy person was in a specified situation the probabilities of contacting the virus were computed. Asymptomatic agents progress through the period of incubation until they start to present symptoms. While in this state, individuals had a lower chance of transmitting the virus. It is important to state that new research regarding COVID-19 uncovered that asymptomatic people only spread the virus two days before showing symptoms [18]. The final group, infected, has a higher chance of spreading the infectious agent. After an agent starts to present symptoms, it will be isolated one day after in his apartment with the possibility of infecting the other people in the studio.

Before the simulation runs the user has the option to set up the parameters that will influence the behavior of the agents and the virus. Rate of spread inside the apartments, staircase, and exterior of the building were firstly considered. Other parameters were the period of incubation and probability of each agent to go outside, computed each day.

As previously stated, while the program was built on a specific set of rules it is up to the user to decide what parameters can reflect different scenarios or introduce other behaviors to develop the model into a more reliable and complex way of analysis. The parameters can be computed or deduced from real data, regarding the infectious agent one wants to simulate.

III. RESULTS

In order to validate our model, we tested the software with the same series of parameters discussed in the above section. The exact chosen values for each one of them were as presented in the Table I.:

TABLE I.   INITIAL PARAMETERS FOR RUNNING THE VALIDATION

No.	Parameter	Value
1	Transmission rate inside the apartments	0.28
2	Transmission rate on the staircase	0.4
3	Transmission rate on the exterior of the closed spaces	0.03
4	Chance of leaving the building	0.28

No.	Parameter	Value
5	Incubation period of the infectious agent	5

After running the simulation, the results showed that the people who went outside to work or buy necessities, represented the principal source of spread. Normally, after one person has caught the virus, the inhabitants of the same apartment were in immediate danger. In most of the cases the spread was due to the interactions with family, and very little because of interactions with the neighbors, as there was a low probability that people will meet when going outside. When the program is running, it will keep track of all tree groups in order to produce graphs. This allows for an easier analysis of virus spread due to human interaction. Fig. 2 and Fig. 3 show the evolution of the three categories during the simulations. This can offer us visual information about how they evolve compared to each other. The first graph is given by unrealistic high probabilities, to better observe the functioning of the simulator, while the results shown in Fig. 2 suggest that the period of incubation is larger than the event of virus transmission from one person to another. This tells us that normally a person will present symptoms before another is infected, allowing us to act before the outbreak takes place in the residence.

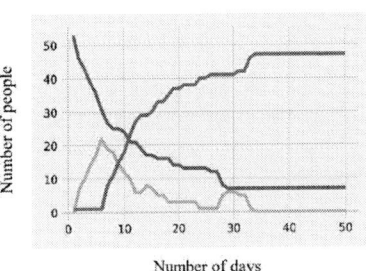

Fig. 2. Rate of change during a period of 50 days for testing purposes (green – susceptible; yellow – asymptomatic; red - symptomatic).

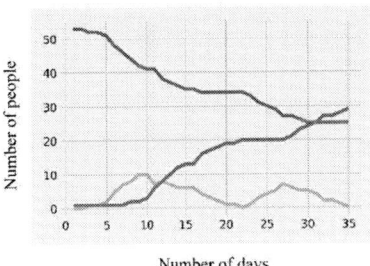

Fig. 3. Rate of change during a period of 35 days based on real data (green – susceptible; yellow – asymptomatic; red - symptomatic).

Depending on the parameters or the data fed into the simulator, we can observe different scenarios. If we want to simulate how a lockdown will affect the residents, we can lower the rate at which people go outside. This of course will result in a smaller chance of catching the infectious agent. Increasing the transmission rate in the hallway area can represent a higher percentage of people meeting with their neighbors. This can allow for a scenario where even if agents are not exiting the building, they can still transmit the virus through extended human interaction. As mentioned before this model can be used to simulate other viruses. This can be done by changing the transmission rates and, or the incubation period.

IV. FUTURE DEVELOPMENT

In this initial stage, while it is possible to modify the program such that it will fit other spaces, it is still not as efficient and intuitive as it could be for an unspecialized user. We consider that a drawing board, through a graphical interface, can facilitate the structuring of an interior. As mentioned before other improvements could refer to the precision of the model and the accuracy with which it reflects a real scenario. This can be accomplished by introducing more factors as for example age, comorbidities, time in contact with other infected individuals. Moreover, while the program is able to produce graphs regarding the spread of the virus, more data like how an agent got infected or where the transmission took place can offer much more insight. If the model run by the program is complex enough it could be used in the future when designing buildings or spaces. Architects or engineers could use similar applications to determine what is the optimum number of people in a certain area such that it will minimize the impact of an infectious agent.

V. CONCLUSION

While the research community has made efforts to understand how an infectious agent is transmitted inside closed spaces, there is still a knowledge gap that needs more attention. The purpose of this paper was to establish the base for a more complex and precise modelling tool. As mentioned before, the software was built such that it offers flexibility and adaptability allowing for higher analysis potential. The results of such modelling tools can allow for a better understanding on how infectious agents spread in closed spaces, offering the possibility of implementing better safety measures and policies.

REFERENCES

[1] L.N. Carroll, A.P. Au, L.T. Detwiler, T. Fu, I.S. Painter, N.F.Abernethy, "Visualisation and analytics tools for infectious disease epidemiology: A systematic review," in Journal of Biomedical Informatics, vol. 51, 2014, pp. 287-298.

[2] C.C. Lai, T.P. Shih, W.C. Ko, H.J. Tang, P.R. Hsueh, "Severe acute respiratory syndrome coronavirus 2 (SARS-CoV-2) and coronavirus

disease-2019 (COVID-19): The epidemic and the challenges," in International Journal of Antimicrobial Agents, vol. 55, 2020, 105924.

[3] T. Singhal, "A Review of Coronavirus Disease-2019 (COVID-19)," in Indian Journal of Pediatrics, vol. 87, 2020, pp. 281-286.

[4] F. Jiang, L. Deng, L. Zhang, Y. Cai, C.W. Cheung, Z. Xia, "Review of the Clinical Characteristics of Coronavirus Disease 2019 (COVID-19) ," in Journal of General Internal Medicine, vol. 35, 2020, pp. 1545–1549.

[5] S. Mei, B. Chen, Y. Zhu, M.H. Lees, A.V. Boukhanovsky, P.M.A. Sloot, "Simulating city-level airborne infectious diseases," in Computers, Environment and Urban Systems, vol. 51, 2015, pp. 97-105.

[6] C.S.M. Currie, J.W. Fowler, K. Kotiadis, T. Monks, B.S. Onggo, D.A. Robertson, A.A. Tako, "How simulation modelling can help reduce the impact of COVID-19," in Journal of Simulation, vol. 14, 2020, pp. 83-97.

[7] V. Vuorinena, M. Aarniob, M. Alavah, V. Alopaeusc, N. Atanasovab, M. Auvinen, et al. , "Modelling aerosol transport and virus exposure with numerical simulations in relation to SARS-CoV-2 transmission by inhalation indoors," in Safety Science, vol. 130, 2020, pp. 1048-1066.

[8] G.A. Somsen, C. van Rijn, S. Kooij, R.A. Bem, D. Bonn, "Small droplet aerosols in poorly ventilated spaces and SARS-CoV-2 transmissions," in The Lancet. Respiratory medicine, vol. 7, 2020, pp. 658–659.

[9] C.J. Noakes, C.B. Beggs, P.A. Sleigh, K.G. Kerr, "Modelling the transmission of airborne infections in enclosed spaces," in Epidemiology and infection, vol. 134, 2006, pp. 1082-1091.

[10] T.M. Cook, "Personal protective equipment during the coronavirus disease (COVID) 2019 pandemic – a narrative review," in Anaesthesia, vol. 75, 2020, pp. 920-927.

[11] M. van der Sande, P. Teunis, R. Sabel, "Professional and Home-Made Face Masks Reduce Exposure to Respiratory Infections among the General Population," in PLoS ONE, 2008.

[12] A. Fallon, T. Dukelow, S.P. Kennelly, D. O'Neill, "Covid-19: the precarious position of Spain's nursing homes," in QJM: An International Journal of Medicine, vol. 113, 2020, pp. 391-392.

[13] The New York Times, "About 40% of U.S. Coronavirus Deaths Are Linked to Nursing Homes," in The New York Times, 16 sept. 2020.

[14] R.A. García, "Covid-19: the precarious position of Spain's nursing homes," in BMJ, 2020.

[15] E. Hunter, B.M. Namee, J.D. Kelleher, "A Comparison of Agent-Based Models and Equation Based Models for Infectious Disease Epidemiology.", AICS, 2018.

[16] M.S. Smolinski, M.A. Hamburg, J. Lederberg, "Microbial Threats to Health: Emergence, Detection, and Response", The National Academic Press, 2003.

[17] S.A. Lauer, K.H. Grantz, Q. Bi, F.K. Jones, Q. Zheng, H.R. Meredith, A.S. Azman, N.G. Reich, J. Lessler, "The Incubation Period of Coronavirus Disease 2019 (COVID-19)," in Annals of Internal Medicine, vol. 172, 2020, pp. 577-582.

[18] E. Gurley, "Covid-19 Contact Tracing", Johns Hopkins University, 2020.

Study of Ceramic Capacitor technology link to Electro Chemical Migration in Automotive Electronics

Szasz Francisc

Human Machine Interface Vehicle Networking and Information (HMI VNI)
Continental Automotive Romania
Timisoara, Romania

Abstract— **Modern Automotive products require the use of high reliability, fail safe MLCC (multilayer ceramic capacitors) on sensitive lines where short circuits and the risk of fire prevent the use of standard MLCC's. It was observed during environmental qualification tests involving high temperature and humidity that the exact technology in which the MLCC is built influences the severity and gravity of ECM (electro-chemical migration). In this paper the performance of 3 types of MLCC's to ECM is studied and the basic characteristic that a high reliability MLCC, form an environmental test point of view, are discussed.**

Keywords— Automotive; Electro Chemical Migration; Ceramic Capacitor; MLCC;

I. INTRODUCTION

During automotive product environmental test validations one of the common issues observed was the formation of ECM (electro-chemical migration). This phenomenon occurs in most tests that involve either high levels of ambient humidity or extreme temperature variations [1][2]. It was observed that one of the most affected component classes are MLCC's (multilayer ceramic capacitors). The ECM acts as a resistor in parallel with the initial capacitor which increases the current leakage of the component [3]. The effect on DUT (device under test) performance is commonly an increase in quiescent current consumption. Depending on the MLCC location in the circuit the effect can of course be very different ranging from no observed effect to extreme cases of DUT destruction. This study focuses on the constructive characteristics of the MLCC that not just enhances the occurrence of ECM but also exacerbate the effect.

It was observed that although ECM occurs on multiple MLCC's all over a DUT, usually there is no influence on performance. In digital circuits the main use case for capacitors is power supply decoupling, therefore adding an extra resistive load in parallel does not influence performance in a noticeable way during normal product operation. The added resistive load caused by ECM was observed to be in the order of Kohm's; even though it is possible to obtain lower resistances, usually the voltage and current available on the capacitor is large enough to destroy the conductive bridge.

One particular issue was observed regarding capacitors placed on permanently supplied power lines. During normal operation the added resistance does not affect performance but in modern electronics one particular requirement is having a low power mode. This is characterized by having the product supplied but not operational and therefor consuming very little power. Common power requirements for automotive units in sleep mode can be as low as 100uA consumption from a 12V supply. In this situation adding an 100Kohm resistor caused by ECM on the 12V supply line results in an increase of 120uA of extra current consumption that causes the unit to exceed the current consumption requirement.

The focus of this paper is to analyze how the constructive proprieties of the MLCC's affect the formation and severity of the ECM. The analysis focuses on capacitors with safety features since these are mandatory to prevent short circuits caused by cracking of MLCC's under mechanical stresses [4].

II. EXPERIMENTAL SETUP AND RESULTS

To test the various types of MLCC's, components of three case sizes where used – 0805, 1206 and 1210 (case size based on the imperial code). These where assembled on a project DUT that was known to exhibit ECM related issues. Three types of "safety" capacitors where tested – Open Mode [5] that has a terminal built with only tin(Sn), nickel(Ni) and copper(Cu) (from here on referred to as Open-Mode); Flexiterm [5] Ag-epoxy type flexible terminal – that has a Ag epoxy layer between the Ni and Cu layers (from here on referred to as Ag-epoxy) and a Flexiterm [5] with Cu epoxy type flexible terminal – that has a Cu epoxy layer between the Ni and Cu layers (from here on referred to as Cu-epoxy).

Regardless of component manufacturer, the structure of the MLCC terminal is the same [7],[8],[9] as seen in Fig. 1

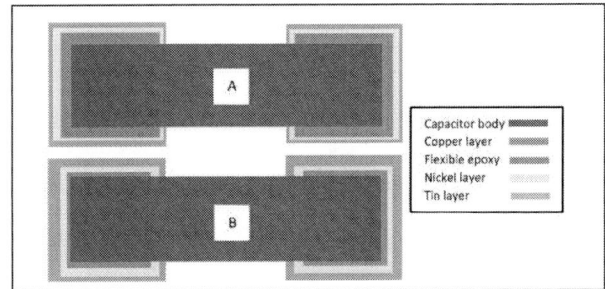

Fig. 1. Terminal structure in Flexible termination (a) and Standard (b) type MLCC's

Each DUT contained a number of 15 case size 0805, 4 case size 1206 and 4 case size 1210 MLCC's. The study was performed on 6 DUT's for each type of capacitor. For each set of 6 DUT's, 7 temperature and humidity cycles where applied according to [1].

A. ECM occurrence and distribution

The first part of the study refers to the occurrence of ECM on the different case sizes and different terminal types. After the test the samples where visually analyzed to see the distribution of ECM based on severity – (1) No trace of ECM; (2) – ECM traces but no clear bridge between the terminals (3) – ECM spanning between the 2 MLCC terminals.

Firstly, the occurrence of ECM on the various units was plotted based on case size and split into types of capacitors as seen in Fig. 2.

All types of studied capacitors showed ECM – none of the analyzed technologies or case sizes prevent the formation of ECM on the tested DUT.

Secondly the distribution based on case size was analyzed. This was done by averaging out all results regardless of MLCC technology. The results can be observed in Fig. 3.

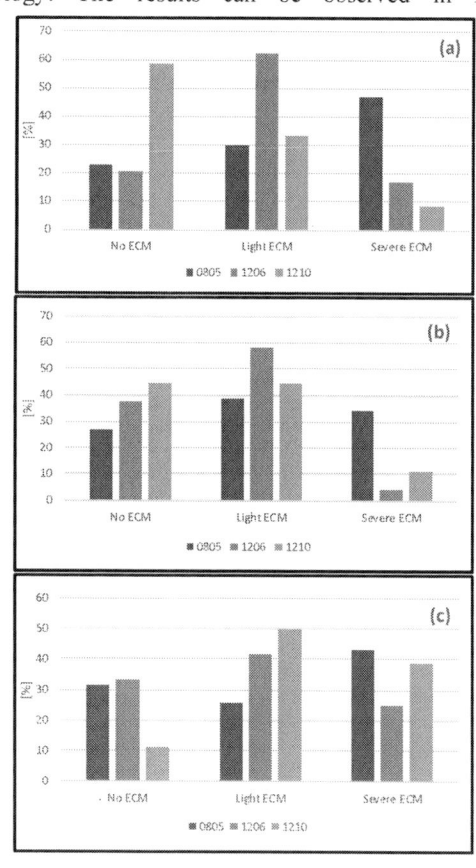

Fig. 2. Occurrence of ECM on (a) Open Mode capacitors (b) Cu-Epoxy capacitors and (c) Ag-Epoxy capacitors

It can be observed that the highest amount of severe ECM is found on the smallest case size – 0805; There is no considerable difference between 1206 and 1210 since both

these case sizes have the same distance between the terminals (the case size is expressed as xxyy where xx is length and yy is width). This confirms that ECM is more likely to occur the smaller the case size.

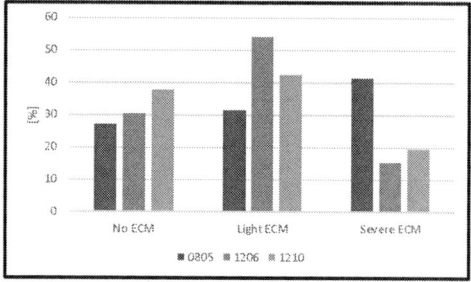

Fig. 3. ECM frquency based on case size

Thirdly the distribution of ECM based on MLCC technology was analyzed. This was done by averaging all the results for each type of capacitor regardless of case size. The results are summarized in Fig. 4.

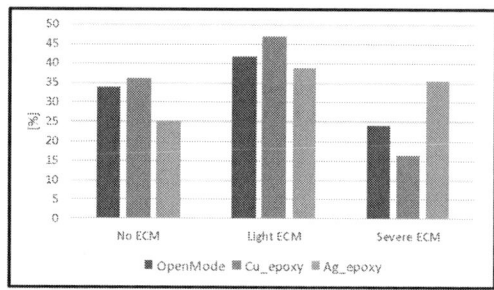

Fig. 4. ECM frequency based on MLCC technology

It can be observed that Ag-epoxy shows more cases of sever ECM compared to the other analyzed technologies.

B. ECM effect on current consumption

To evaluate the impact of the ECM on DUT operation the current consumption was measured on each DUT throughout the test every 10 seconds. The DUT's where supplied from 12V.

The currents where averaged per cycle on all 6 units from each type of MLCC. The results can be seen in Fig. 5.

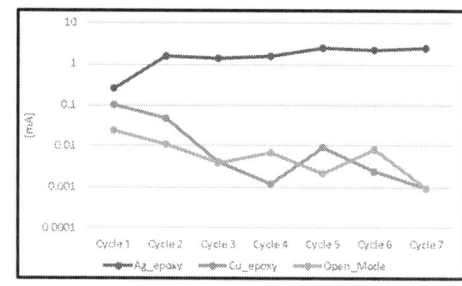

Fig. 5. Average current consumption based on MLCC technology

The samples where measured again after being left at room temperature and humidity for 48h by supplying at 12V and the average DUT current consumption can be found in TABLE I.

TABLE I.   AVERAGE CURRENT CONSUMPTION AFTER 48H

MLCC type	Average current consumption / DUT
Ag-epoxy	2.090mA
Cu-epoxy	0.017mA
Open-Mode	0.007mA

Based on the results it is clearly visible that in DUT's with Ag-epoxy MLCC's there is a significantly higher current consumption. The current consumption increases over the test period as more conductive material is cumulated between terminals because of ECM. On the other hand, with the Cu-epoxy and Open-mode type MLCC's the tendency is for current consumption to decrease.

Another observation is that the current consumption of the Ag-epoxy type MLCC's persists for extended periods of time.

C. EDX analysis

An EDX (Energy Dispersive X-ray) array analysis was performed to identify the atomic species involved in the ECM process and how they disperse over the capacitor body. One capacitor from each test batch was analyzed. The results are summarized in Fig. 6.

The Ag-epoxy capacitor shows a clear silver (Ag) based bridge formed between the terminals. Also tin has migrated but the nickel (Ni) under layer is not visible.

The Open-Mode capacitor only shows that tin(Sn) has electro-migrated. In this case no copper (Cu) was observed but the nickel under layer was exposed.

The Cu-epoxy capacitor also shows only tin migration. In this case the ECM phenomenon was also severe enough to expose the underlying nickel(Ni) layer.

D. Cross section and SEM on Ag-epoxy capacitor

To understand why the Ag-epoxy capacitor shows silver migration even though not all the terminal was previously dissolved – the nickel layer was not exposed, a capacitor was subjected to a cross section analysis to see the layer stack. This can be observed in Fig. 7.

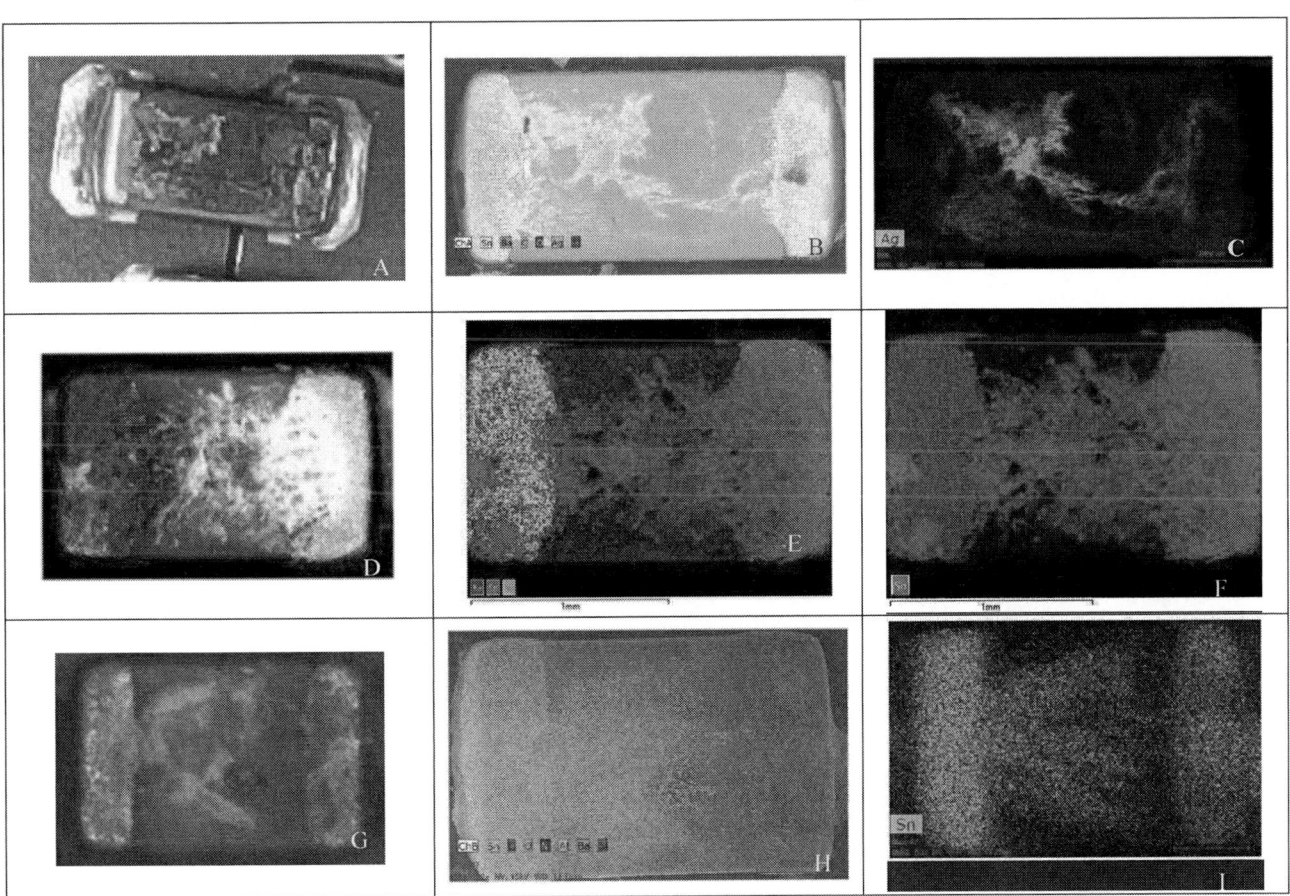

Fig. 6.   SEM-EDX analysis on Ag-epoxy MLCC (A-C), Open Mode MLCC (D-F) and Cu-epoxy (G-I)

$$M \rightarrow M^{n+} + ne^- \qquad (1)$$

To form conductive ECM, the reaction needed at the cathode is the reduction of the metal ion Eq. 2:

$$M^{n+} + ne^- \rightarrow M \qquad (2)$$

Based on how soluble the metal ion is in water the following equilibrium reactions will be shifted to either left – when the hydroxide/oxide is soluble or to the right when the hydroxide/oxide is not soluble[6]:

$$M^{n+} + 2H_2O \rightarrow M_n(OH) + H_3O^+ \qquad (3)$$
$$2M^{n+} + 3nH_2O \rightarrow M_2O_n + 2nH_3O^+ \qquad (4)$$

The solubility of an oxide can be assessed based on its Gibbs free energy of formation. The more negative the energy of formation the less soluble the substance. The characteristic proprieties of the common metallic species found in MLCC terminals are summarized in TABLE II.

In the above experiments the ECM of Ag and Sn was observed but it is known that also Cu migrates during environmental test. Based on the solubility of compounds it can be observed that the compound most likely to form metallic dendrites rather than oxides or hydroxides is Ag (reaction (2) rather than (3) and (4)). Another issue regarding Ag is that unlike Sn and Cu, it's considered a noble metal because of its increased resistance to corrosion and oxidation. Therefor in case of dendrite formation, the Ag dendrites will have a longer lifespan before corroding.

It was observed that the conductivity of the ECM on Ag-Epoxy capacitors increased during the test while in the case of Open-Mode and Cu-Epoxy it decreased or stayed stable. This can be linked to the formation of nonconductive oxides more predominantly than the cathode reduction to conductive metals in the case of Cu and Sn. Any conductive dendrites that did form got oxidized by air since Cu and Sn are not that resistant to corrosion compared to Ag.

TABLE II. MATERIAL CONSTANTS [10]

Compound name	Chemical formula	Gibbs free energy of formation G_f^o [kJ/mol]	Solubility mol/100g H_2O
Silver Oxide	Ag_2O	-11.2	2e-5
Copper (II) oxide	CuO	-129.7	3e-6
Copper (I) oxide	Cu_2O	-146.0	insoluble
Nickel(IV) oxide	NiO_2	-199.0	insoluble
Nickel(II) oxide	NiO	-211.7	insoluble
Tin(II) oxide	SnO	-256.9	5e-7
Tin(IV) oxide	SnO_2	-519.7	1.4e-11

IV. CONCLUSIONS

Based on the experiments performed it was observed that all 3 analyzed MLCC technologies – Open-Mode, Flexible

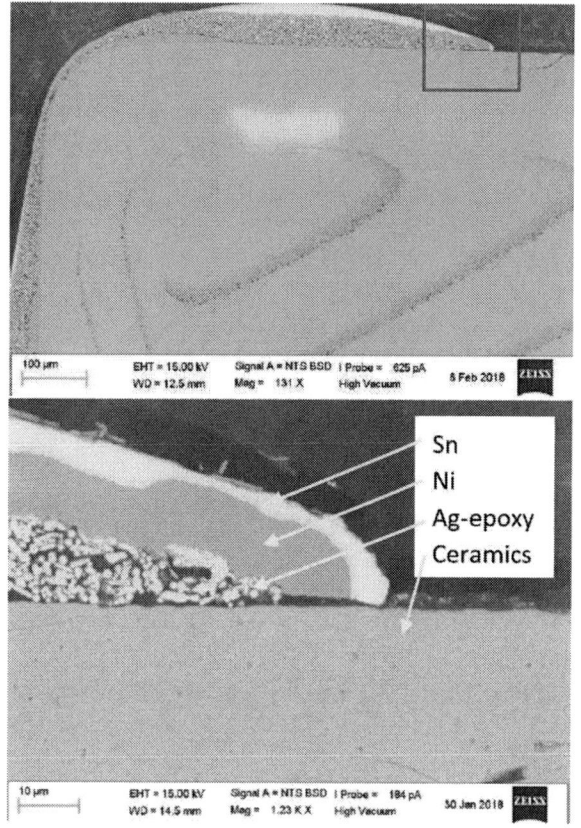

Fig. 7. EDX on cross-section of Ag-epoxy capacitor (a); detail on terminal – ceramic body interface (b)

It can be observed that there is no chemical bond between the Ni/Sn termination and the capacitor body; there is a clear line between the two elements. It is expected that during soldering, thermal and mechanical stresses on the Ni/Sn termination causes it to get lifted in various places and exposes the silver epoxy layer. This would explain why Ag related ECM is observed but the Ni layer under the tin is not visible.

It is expected that a similar behavior can occur on the Cu-epoxy capacitors although Cu migration was not observed.

Unlike the Open mode terminal where the structure is compact and solid – all 3 layers Cu-Ni-Sn are compactly interconnected, with the flexible layer present in the Cu or Ag - epoxy capacitors the high bending resistance feature in the Flexiterm [5] is obtained with the downside of having the possibility of the terminal edge lifting from the capacitor body.

III. RESULTS DISCUSSION

For the ECM phenomenon to happen you need humidity either from condensation during steep thermal transients or it can appear when extremely high humidity causes water to adsorb to the surface.

When two electrodes are placed in a conducting electrolyte and a potential difference is applied the anode will dissolve into the solution based on the oxidation Eq. 1:

termination with Cu-epoxy and Flexible termination with Ag-epoxy show ECM during Automotive environmental tests. The results showed that the smaller the case size i.e. the smaller the distance between the terminals the more likely the formation of severe ECM is. Regarding the formation of electrically conductive ECM the main element responsible was found to be Ag.

It was observed that MLCC's that have Ag in their terminal composition are more predisposed to form not just conductive ECM but the conductive proprieties last for extended periods of time. Therefore to increase robustness of electronic circuits to ECM, based on the above test results it is recommended to use a capacitor where the terminal does not contain Ag like Open-Mode or Cu-epoxy flexible termination type capacitors. It is expected that positive results can be obtained from capacitors with Ag content but which feature some other mechanism to prevent the extraction of silver.

REFERENCES

[1] IEC60068 Environmental testing – Part 2-38: Test Z/AD: Composite temperature / humidity cyclic test; 2009

[2] IEC60068 Environmental testing – Part 2-30: Test Db: Damp heat, cyclic (12h+12h cycle); 2005

[3] K.K. Ng and M. Rajaratnam "Failure Analysis on Multilayer Ceramic Capacitor (MLCC) with Leakage Failure Caused by Silver (Ag) Migration in Molded Plastic Package; 2012

[4] John Maxwell "Cracks: The Hidden Defect", AVX Corporation; 1988

[5] Rutronik "Passive Components – Ceramic Capacitors" p.7

[6] P.E. Tegehall and B.D. Dunn "Evaluation of Cleanliness Test Methods for Spacecraft PCB Assemblies"; ESA STM-275; 2006

[7] Murata, GCJ series, Soft Termination Chip Multilayer Ceramic Capacitors for Automotive, https://www.murata.com/en-eu/products/capacitor/mlcc/smd/gcj

[8] AVX, MLCC with Flexiterm, http://datasheets.avx.com/FlexitermMLCC.pdf

[9] Samsung, Multilayer Ceramic Capacitors, https://www.samsungsem.com/global/product/passive-component/mlcc.do

[10] Inorganic Compounds: Physical and Thermochemical Data http://www2.ucdsb.on.ca/tiss/stretton/database/inorganic_thermo.htm

Electromigration in lead-free solder joints on ceramic PCB substrates

Dániel Straubinger, Dániel Rigler, Attila Géczy
Department of Electronics Technology
Faculty of Electrical Engineering and Informatics
Budapest University of Technology and Economics
Budapest, Hungary

Beata Synkiewicz-Musialska
Łukasiewicz Research Network
Institute of Electron Technology
Warsaw, Poland

Abstract—**The effect of electromigration in lead-free solder joints of SMD components on ceramic substrates were examined in our study. A ceramic composite printed circuit board was created with 0402 and 0603 chip sized components and soldered into daisy-chain configurations. Reference samples, thermally loaded samples and current stressed samples were created to differentiate the influence the high-temperature caused by the Joule heating in case of the current stressing from the effect of electromigration. Destructive and non-destructive methods were used for analysis: X-ray, cross-section, optical microscopy, scanning electron microscopy (SEM), and SEM-EDX, and shear strength measurements were performed. It was found that after 1100 hours of loading process, the shear strength of the investigated samples showed no significant difference between the sample groups. However, cross-sectional sample preparation revealed excessive Cu_6Sn_5 IMC formulation in the solder joints in case of the current stressed samples.**

Keywords—lead-free soldering, electromigration, SMD, components

I. Introduction

Investigating high current density in electronics is a demand for modern applications. With the continuous development of technology, the electronics circuits reach higher integration levels, resulting in a higher current density at interconnects, printed circuit board (PCB) tracks and solder joints. Chip-sized electronics components became small enough to reach current density levels - due to the reduced conductive area – as high so that electromigration can cause reliability problems [1]. Thus long-term reliability concerns can be raised (especially in critical industries such as automotive and space technologies). Electromigration can have an effect on the structure and composition of the solder joints; therefore, it can change its mechanical properties. Formation of different intermetallic compounds can be attributed to electromigration.

Zhang et al. showed in an experimental solder ball configuration that due to the EM-induced atomic transport can cause rapid growth of Cu6Sn5 intermetallic layer formation at the anode side [2]. B. Chao et al. in their experiment showed electromigration induce Cu_6Sn_5 growth from the cathode to anode side in a Pb-free microbump configuration with copper under bump metallisation (UBM) [3]. C Wei et al. found that Cu_6Sn_5 IMCs have higher electromigration rate than Cu_3Sn. They attributed this to the higher resistivity and lower solidus temperature of Cu_6Sn_5 [4].

Y. Shen et al. showed on a Cu-Sn-2.3Ag-Cu microbump structure that Cu6Sn5 formulation can occur in case of specific grain orientations due to electromigration (rapid formation in low-α-angle grains, Cu dissolution and migration in case of intermediate α-angles) [5].

In our previous works [6] a PCB with daisy-chained jumper (0 Ω) resistors were designed to test electromigration in SMD component lead-free solder joints. To reduce MTTF (Mean Time to Failure), elevated temperature were used to accelerate the investigation. However, standard FR4 PCB material turned out to be insufficient to carry the cumulated temperature from the power dissipation and the elevated temperatures, thus limiting the applicable current (and current density). Only the phenomena of intermetallic layer (IML) thickening was observed in the previous studies. In this current work, a new Design of Experiment (DoE) approach is realised, which allow extending the parameter window of the investigation. The change from FR4 to the ceramic-epoxy material (widely used in aerospace, automotive and high-frequency applications) enabled the use of elevated temperature due to the higher thermal diffusivity of the PCB.

II. Experimental

Rogers RO4000 PCB substrate was used to create the circuit for the investigations. A total number of 945 components (thus 1890 solder joints) was soldered for the experiments, containing 540 pieces of 0603 and 405 pieces of 0402 high power thick film jumper resistors. The resistors were selected to withstand the elevated temperature resulting from Joule heating during current stressing, with a maximum operating temperature of 155 °C. The PCB due to its high thermal conductivity and heat capacity (compared to standard FR4) can also withstand the experiments.

The PCB wiring design was the same as in our previous works [6], it consists of two different types of wire angles into the component pads, 90° and 135° (the current density reaches the highest level in those cases).

To achieve the most accurate and even soldering quality, the solder paste was deposited with stencil printing (DEK 248), and the placement was performed with an automatic pick and place machine (TWS DVC Quadra Evo). The solder alloy used for the experiment was a conventional lead-free SAC305. The reflow soldering was performed with an Asscon Quicky 450 vapour phase soldering (VPS) machine, using GALDEN LS230 fluid.

The samples were divided into three different categories, according to Table I.

TABLE I.   DIFFERENT SAMPLES FOR INVESTIGATING ELECTROMIGRATION

Investigated cases and samples		
Ambient temperature [°C]	Current load [A]	Sample types
room	not applied	(A)
120	not applied	(B)
40	2 (0402)/ 2,5 (0603)	(C)

Sample (A) is used as reference samples for solder joint structure without any environmental stressing. Sample (B) is used to differentiate the effect of elevated temperature compared to the current stressed ones (sample (C)). For sample (C) an ambient temperature of 40 °C was applied with an oven to maintain the component temperature constant during the experiments. The applied current on the component results in elevated temperatures reaching approximately 110 °C and 130°C for the 0402 and 0603 sizes, respectively. The current stressing was performed with a regular laboratory power supply, with the components soldered into a daisy-chain configuration.

The investigation involved the measurement of resistance, X-Ray analysis, shear strength of solder joints and optical/SEM analysis on cross-sectioned samples.

III. INVESTIGATION METHODS AND RESULTS

A. X-Ray

Every component was investigated with X-Ray after reflow soldering to have a baseline of voiding in the solder structure. Electromigration could possibly induce void growth which is only detectable via X-Ray without destruction of the samples.

Fig. 1. shows X-Ray images of a 0603 component after soldering, and after approximately 1100 hours of current stressing.

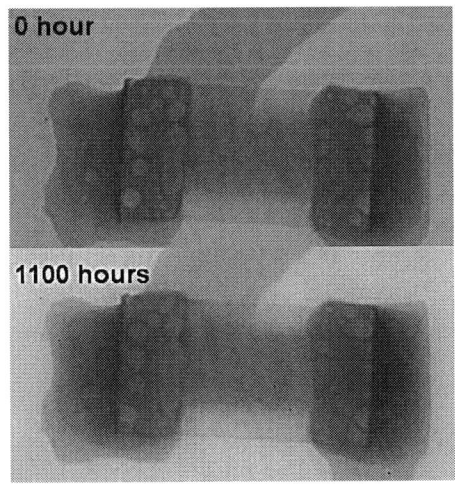

Fig. 1. X-Ray image of the same 0603 component before and after 1100 hour of current stressing

The voids which were present after the soldering remained the same size and number and are distinguishable on the images. Results show that this phase of current stressing did not cause the electromigration to move into the void growth phase.

B. Measurement of resistance

A commonly used indicator of electromigration is a change in resistance. Different studies use a 10-20% elevation in the resistance value as a defect criterion in electromigration. Sample (C) was measured several times during the actual testing period via 4-wire resistance measurement (AGILENT 4338B) made on the daisy chain circuitry, with the samples cooled down to constant laboratory temperature. In the approximately 1100 hours testing period, no change was observed in the resistance, meaning that the electromigration is still in the initial void nucleation and IMC growing phase.

C. Shear strength of solder joints

The shear strength of the solder joints between the different sample groups was tested to see the effect of each thermal loading and current stressing on the joints shear strength. Sheat test was performed with a Dage 2400 shear tester. The position of the shearing knife was checked with an optical microscope through the process and set to the centre of the components. The shearing speed was set to 100 μm/s. For this test method, only 0603 components from all of the different samples were selected due to the larger available number.

Together, 60 shear tests were performed, 2 results from removed from the data because the copper delaminated from the PCB. The fractured surface was investigated via optical microscopy, which can be seen on Fig. 2.

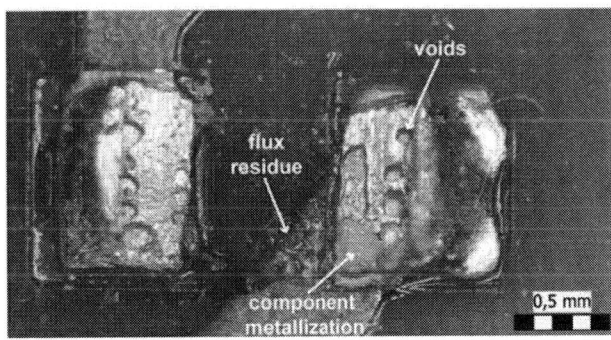

Fig. 2. Fractured surface of a 0603 component after the shear testing

The results showed that in most cases a large number of voids were observable at the surfaces. In several cases, parts of the component metallisation were also observable. In the case of sample B and C, the flux residue's colour change to dark brown, due to the elevated local temperature.

Figure 3. shows the shear strength measurements from 0603 sized components.

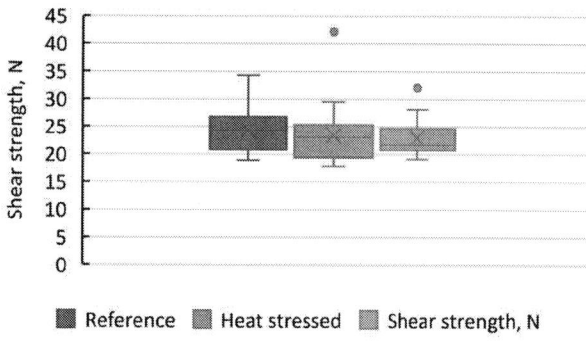

Fig. 3. Shear testing of 0603 components (sample A, B, C)

The results show a very slight decrease in average shear strength, but the standard deviation and spread of the data show no difference between the different samples. The reason can be attributed to the several observed voids (X-Ray, an optical image of the fractured surface) which affects more the shear strength than the elevated temperature and the current state of electromigration.

D. Cross-section analysis

The metal structure of the solder joint was analysed. Samples were prepared with epoxy for cross-sectional analysis. In each case, 5 components were investigated in one sample preparation. A total number of 25 components were selected: both component size and (A, B, C) sample cases were represented. A Struers Knuth Rotor 3 grinding and ATA Saphir 520 polisher were used for preparation. Optical images were taken with an Olympus BX51 metallurgy microscope. The number of samples was defined to maintain a statistically significant number of components for further experiments (due to the destructive method). Fig. 4. shows the selected plane for the cross-section investigation. This plane contains the highest current density region within the solder joints (at the 135° wire connections).

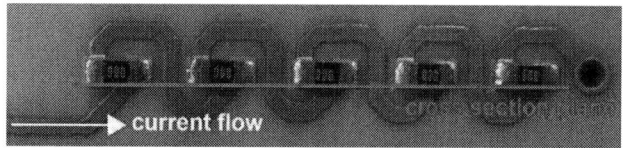

Fig. 4. Grinding plane for the cross-section analysis and the current direction in the solder joints

In Fig. 5. a solder joint of a 0603 resistor can be seen. In case of the selected joint, the electron flow enters from this shown side of the component, from a 135° angle at the pad which was found to be critical in our earlier works [6]. A large region of intermetallic compound was found around the middle of the solder meniscus. The cross-sectioned samples revealed the structure of the jumper resistors, which included two, conductive layers, one at the top and one at the bottom of the component, covered with a passivation layer.

Fig. 5. Solder joint of a cross-sectioned 0603 resistor

On Fig. 6. the other solder joint of the same component (as in Fig. 5.) is presented.

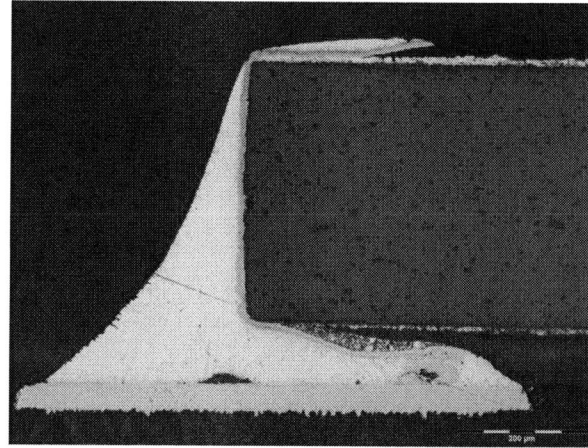

Fig. 6. Other side of the same component from Fig. 5.

The cross-section did not reveal any large IMC formations similar to in the joint shown in Fig. 5.

Based on all of the cross-section samples, a slight increase was found in the intermetallic layer thickness, and in case of the current stressed samples, specifically around the high current density region, under the component. Quantitative analysis requires further processing with image processing, which is planned for future works. The alloy composition of the IMCs was analysed with SEM-EDX.

E. Scanning Electron Microscopy

Based on the optical images of the cross-sections, some samples were selected for further analysis with scanning electron microscopy (SEM). Energy-Dispersive X-ray (EDX) analysis was used for the determination of the alloy composition of the preparations. An SEM-EDX image of the component from Fig. 5 can be seen on Fig. 7.

2020 IEEE 26th International Symposium for Design and Technology in Electronic Packaging (SIITME)

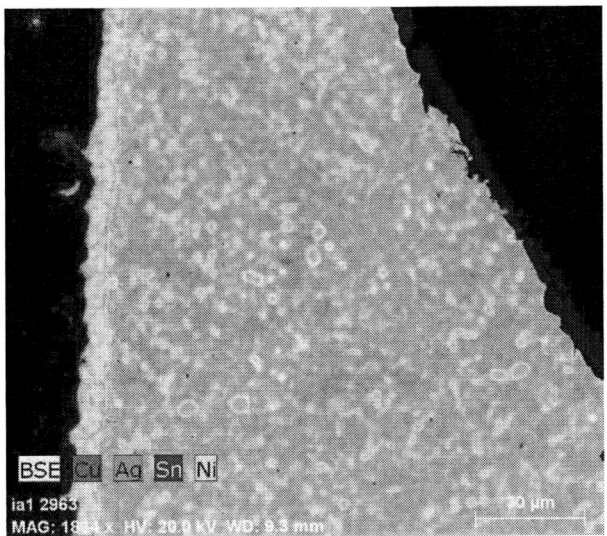

Fig. 7. SEM-EDX image of a solder joint

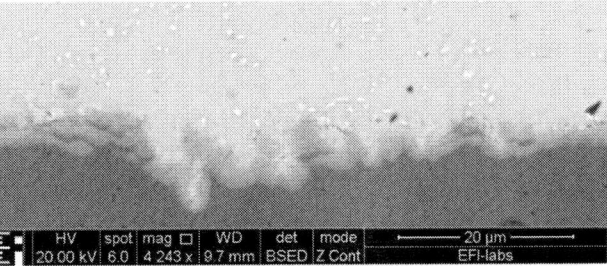

Fig. 9. 0402 component size, copper pad and IML

The EDX image shows that the component metallisation is made of nickel and does not contain a copper layer under it. Therefore the significant copper formations are not dissolved from the metallisation. The conductive metal in the thick film is silver. The extensive IMC found in the joint turned out to be Cu_6Sn_5 (the composition of atomic percentages are 55.17 % and 44.83 % respectively). It is not typical for Cu_6Sn_5 intermetallic compound to be created during reflow soldering in the given area. The same type of intermetallic formation was found on other components too, only in those joints from where the current flow enters the component.

Fig. 8. shows a 0402 sized component with the same type of intermetallic compound formation.

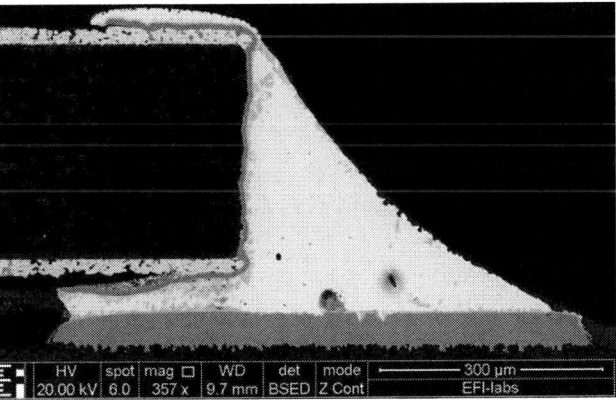

Fig. 8. 0402 component size, current stressed solder joint

It can be seen that the aforementioned IMC formation is found in the high current density areas, under the component and at the top of the solder meniscus. These are the directions, to where the current flows.

Fig. 9 shows the intermetallic layer of the same component.

The intermetallic layer alongside the copper pad is present throughout the boundary, and it can be confirmed. The copper layer was visibly thinned in this region, which might be the source of the previously discussed IMC formations. The copper could migrate from the pad due to electromigration. This phenomenon was not observable on the investigated reference (A) and thermally loaded (B). It is also supposed, that such large zones of Cu_6Sn_5 formation are not expected from the small Cu content of the original alloy composition.

IV. CONCLUSIONS

In our work, a ceramic-composite PCB was created in order to examine the electromigration in lead-free solder joints for an excessive period of time. The prepared samples were divided into three categories: reference, thermally loaded and current stressed. Based on the 1100 hours of heat- and current stressing the samples, the following results can be stated:

- There was no significant difference in the shear strength of the 0603 sized components solder joints in between the reference, thermally loaded and current stressed samples.

- The void growth phase of the electromigration was not reached within our experiments.

- An extensive amount of Cu_6Sn_5 IMC was found in several solder joints where the current flow enters the component. Comparable size of the same IMC composition was not present at the other side of the component where the current leaves the joint, neither at the references.

- The source of the IMC could be originated from the copper pad since the component metallisation is only nickel, the solder paste composition only contains 0.5 % copper. Also, in the reported cases, the boundary was found to be severely distorted in some cases at the copper-solder interface at the pads, suggesting the source of copper in the far side of the given joints.

In the future, the intermetallic layer thickness and formation between the different sample groups are planned to be investigated quantitatively with image processing. The heat annealing and current stressing are continued, a further stage of the electromigration process is going to be analysed. The effect of the Cu_6Sn_5 IMC formulation on the solder joint reliability and strength needs further investigation.

978-1-7281-7507-2/20 $31.00 © 2020 IEEE   21-24 October, Pitesti, Romania

ACKNOWLEDGEMENT

The help of Rogers by supporting the substrate, ASM by the stencil and of Robert Bosch Elektronika Kft. by the X-Ray imaging is highly appreciated.

Supported by the ÚNKP-20-3 New National Excellence Program of the Ministry for Innovation and Technology from the source of the National Research, Development and Innovation Fund."

REFERENCES

[1] Chang, Y.-W., Hu, C., Peng, H.-Y., Liang, Y.-C., Chen, C., Chang, T., ... Juang, J.-Y. (2018). A new failure mechanism of electromigration by surface diffusion of Sn on Ni and Cu metallisation in microbumps. Scientific Reports, 8(1). https://doi.org/10.1038/s41598-018-23809-1

[2] Zhang, Z., Cao, H., Xiao, Y., Cao, Y., Li, M., & Yu, Y. (2017). Electromigration-induced growth mode transition of anodic Cu 6 Sn 5 grains in Cu|SnAg 3.0 Cu 0.5 |Cu lap-type interconnects. Journal of Alloys and Compounds, 703, 1–9. doi:10.1016/j.jallcom.2017.01.292

[3] Chao, B., Chae, S.-H., Zhang, X., Lu, K.-H., Ding, M., Im, J., & Ho, P. S. (2006). Electromigration enhanced intermetallic growth and void formation in Pb-free solder joints. Journal of Applied Physics, 100(8), 084909. doi:10.1063/1.2359135

[4] Wei, C. C., Chen, C. F., Liu, P. C., & Chen, C. (2009). Electromigration in Sn–Cu intermetallic compounds. Journal of Applied Physics, 105(2), 023715. doi:10.1063/1.3072662

[5] Shen, Y.-A., & Chen, C. (2017). Effect of Sn grain orientation on formation of Cu 6 Sn 5 intermetallic compounds during electromigration. Scripta Materialia, 128, 6–9. doi:10.1016/j.scriptamat.2016.09.028

[6] Straubinger, D., Géczy, A., Sipos, A., Kiss, A., Gyarmati, D., Krammer, O., ... Harsányi, G. (2019). Advances on high current load effects on lead-free solder joints of SMD chip-size components and BGAs. Circuit World. doi:10.1108/cw-11-2018-0088

Change Detection in the Complexity of Time Series with Information-based Criteria

Dorel Aiordachioaie
Department of Electronics and Telecommunication
CCETIC Research Center
Dunarea de Jos University of Galati
Galati, Romania

Sorin Marius Pavel
Department of Electronics and Telecommunication
CCETIC Research Center
Dunarea de Jos University of Galati
Galati, Romania

Abstract—**The objective of the work is to emphasize the need for a complexity measures or coefficients, in time series analysis. Such measures must consider the structure of the model of the signal and the computational effort to process him (i.e. analysis, processing, transmitting, storing, etc). The final goal is to build an automated recognition and classification system for complex signals, i.e. mixtures of various types. In this work the complexity is measured with Renyi entropy of high order, i.e. from 5 to 10. Three types of basic signals are considered, in an independent or mixed approaches: determinist, random and chaotic. Mixed approaches means a weighted summation of the basic components. Change detection in the structure of the time series is of equal importance, because the change is an effect of changing in the structure of the model or system, which have generated the time series. Change detection is based on fusion of some criteria based on information extraction and processing. The results of the experiments indicate the feasibility of the method to discriminate the change in the complexity of the time series.**

Keywords—complexity; time series; signals; entropy; information; systems; models; analysis.

I. INTRODUCTION

The analysis of the scientific and engineering literature in the field of signals analysis and processing reveals a relatively new concept, especially when various signals are compared for processing needs and change detection in their structure. We observe also so many fields of science and engineering, which promote and use complexity to describe and classify various physical phenomena, trends or behaviors.

Complexity is a multidisciplinary subject and has many points of views. There are dedicated journal for such subject as Complexity, from Hindawi [1] and Wiley [2], Journal of Complexity, from Elsevier [3], Journals of Physics: Complexity, from IOPscience [4]. Special editions of journals, as e.g. [5], and review papers, e.g. [6–9], are promoted.

The interest in complexity of time series is exemplified in general terms by [10], [11], [12], and [13]. Examples of results based on complexity of the signals are used in:
- medicine (biomedical signals, [14-18]);
- engineering (in rolling bearings [19]);
- time series analysis (prediction of rainfall series [20], detection of change via sliced cross-spectrum in [21]);
- change and anomaly detection, as in [22], [23], and [24].

The strongest connection of complexity is with entropy, of various types as in [25] (Range entropy), [26] (Approximate entropy), [27], [28] (Renyi and Tsallis entropies), and [29] (Multiscale entropy).

In the case of systems it is possible to explain what means complexity and properly defining a measure of complexity by analyzing the size and the connectivity of elements inside of the model. The system engineering point of view comes with many considerations, e.g. with [30] (What is a complex system?), [31] (system performance). In the case of signals, without any a priori information, this is more difficult. In almost all cases, the degree of the complexity is the degree of the randomness.

The objective of the work is to emphasis the importance of the complexity for signals description, in order to choose right analytic and software tools for analysis, and later to detect and to identify the sources of the signal's components. The work is of exploratory type, which try to respond to the question of the importance of the complexity in the description of time series, as a superior/general concept comparing with other common measures based only on infomation (entropies) processing. The next section continuous the introduction with some comments and observations related to complexity and the entropies of time series. Section 3 introduces the equations for the main information-based measures, i.e. entropy and complexity. Section 4 presents the signals involved in this study, from rationale and generation point of views. The last section is for experiments and results.

II. FACTS ABOUT COMPLEXITY AND ENTROPY

For the beginning of the work, some simple pairs of questions-answers are addressed in terms of engineering tasks and objectives. The starting questions and short answers could be:

Q1: What means complexity? A: A quantitative measure / function, which describe a signal or system in terms of difficulty to analyze or to model.

Q2: What is a complex signal? A: A complex signal has a complex model, with deterministic and random components, with non-linear structure (weights) and a number of variables.

Q3: Could we use entropy to describe de complexity of the signals? A: Not always. It is a common choice but with many drawbacks. Commonly, we can use the variation of the entropy to show or to detect the changes in the signals.

Q4: Which properties should have the complexity variables/criteria? A: (1) quantitative/numeric, if possible; (2). monotonicity (Let be two objects: *A*, and *B*, and a complexity measure *C*. If *A* is included in *B* then $C(A) < C(B)$); (3) stability ($C(A \cup B) = max\{C(A), C(B)\}$); (4) sensitivity to the relative variations of the number of the components and of their parameters; (5) insensitivity to some physical parameters, e.g. amplitude of the components, time sliding/translation, etc.. This could be decomposed further in

- translation invariance ($C(k+A) = C(A), k=1,2,\ldots$);
- amplitude invariance ($C(A) = k\,C(A)$;
- expansion invariance ($C(A) = C(kA), k=1,2,\ldots$).

Complexity is discussed in computer science and informational systems as well, e.g. to describe the efficiency of the algorithms and various computations. Complexity can be:
- computational (hardness of computation);
- algorithmic (also called Kolmogorov Complexity) (the complexity of a string of digits or of an expression is equal to the length of the shortest algorithm that may generate it based on an universal computer);
- structural (complexity of the model, functions, size and connections);
- engineering design (number and level of the functions) as in [32].

The next two figures present two cases/examples, in order to understand better the significance of the complexity and the limitations of the entropy (E). Fig. 1 presents an example of two signals ($s_1(t)$ random and $s_2(t)$ determinist, linear with time) with the associated relative entropies, Shannon (SE) and Renyi (RE) of order three, represented in percent, based on equations below with P_i the probabilities of the samples/symbols:

$$\Delta E[\%] = \frac{E(s1) - E(s2)}{E(s1)} 100 \qquad (1)$$

$$SE = -\sum_{i=1}^{k} P_i \cdot log(P_i) \qquad (2)$$

$$RE(\propto) = log\left(\sum_{i=1}^{k} P_i^{\propto}\right) \Big/ (1-\propto) \qquad (3)$$

There is no relevant difference between the Shannon entropies in order to use them as primary descriptors of complexity of the signals. In fact, as the length of the signal is raising, the Shannon entropies become closer. This means that Shannon entropy cannot be used directly to describe or to classify different signals. Something else is happening with Renyi entropy, which maintains a relatively high difference between the Renyi entropies of the considered signals, and thus could be used to asses complexity of the signals and to distinguish them. In Fig. 2 an example is presented of an increasing complexity of a network of five nodes (vertices),

where complexity is described quantitatively by an information index [33].

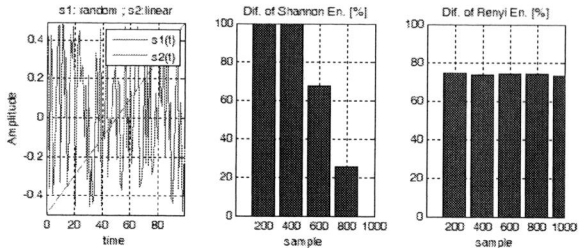

Fig. 1. Shannon and third order Renyi entropies

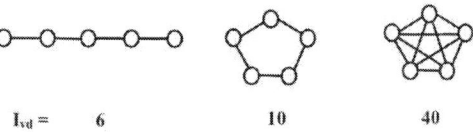

Fig. 2. An example of increasing network (or equivalent graph) complexity (extracted and adapted from [33])

These examples shows that complexity needs special criteria, possible with the contribution of various types of entropies, adapted to the analyzed signals (in general case, objects).

Fig 3. presents the structure of the solution for the problem under study, in the sense that we are looking to find complexity measures adapted to the analyzed signals and their components, in order to optimize the change detection on the complexity of the analyzed signals. This must be done in an automated way and could include adaptation, as presented in the next figures, by including the possibility of correction/change at the level of complexity variables or/and at the change detection criterion.

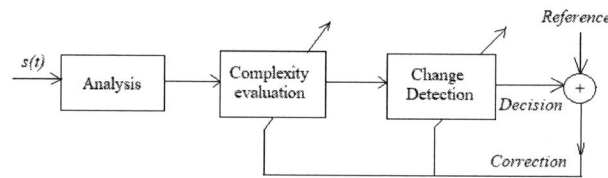

Fig. 3. The structure of the solution for change detection in the complexity

III. COMPLEXITY MEASURES

The most common complexity measures includes various types of entropy and some synthetic variables as variables/markers as Lempel - Ziv Complexity (LZC), inspired from computational complexity or Lyapunov exponent, used in the classification of the chaotic systems and to measure the sensitivity to initial conditions. Some results of these markers are available in [27].

Dependency on the length of time series and the sensitivity on the amplitude of signal are two important properties to consider. The measures of the complexity should be less sensitive at the length of the time series. The minimum length necessary to have in order to detect a change in the complexity of time series is also important to consider.

Based on the relative simplicity of the computation, and some of the previous results, e.g. [24] and [27], in this work we consider only the Renyi entropies of higher order up to 10. In all cases, and from simplicity reasons, the values of the probabilities are estimated in a special preprocessing stage, which is not considered here. The order α of the Renyi entropy controls the degree of sensitivity of entropy towards particular probability density functions (pdfs), [37]. The Shannon entropy is a particular case, when $\alpha = 1$. Other special cases of Rényi entropy include collision entropy ($\alpha = 2$) and minimum entropy (α goes to infinity).

Tsallis entropy (TE) has a similar parameter. Tsallis entropy extends Shannon's entropy for non-additive systems. Both approaches include Shannon's entropy as a special case [28]. Both have some nice properties related to e.g. non-negativity, minimum and maximum values, and monotonicity for $\alpha > 1$, [38], [39]. There is also a relation between Renyi and Tsallis and this is the reason to keep only the Renyi type, with an order varying from 2 to 10. The higher the order, the higher is the variance of the entropy, which could be useful for the sensitivity properties of the change detection criterion especially in noisy working conditions. Other approaches include the Multi Scale Entropy (MSE) and spectral entropy as well, the later in the frequency domain, with some advantages to MSE.

The main computation steps in the method for change detection in the complexity of time series is presented in Fig. 4. Data samples are stored in a data buffer of length N for batch processing. A sliding window of length n<<N is used to extract the model of the signal via an identification block, which provides a statistical model composed of elements /samples/ symbols **S** and their probabilities **P**. The length of the sliding window generates the accuracy of the point change detection. Based on model [**S, P**] the next block compute the complexity measures. There could be various entropies and complexity coefficients as Lempel-Ziv or Lyapunov. The final decision in the change is made by fusing the complexity measures. The output should indicate a value of the decision (True or False) and a time moment of the change if it is the case.

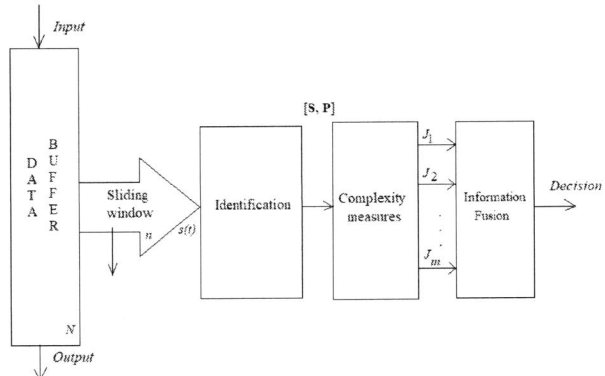

Fig. 4. Singal processing main computation steps for change detection

From complexity point of view, an coefficient CC could be used, with sub unitary positive values, defined as

$$CC = \frac{1}{M} \sum_{i=1}^{M} J_i / max(J_i) \qquad (4)$$

where J_i is an entropy or other entropy-based criterion. The maximum values for such entropies is given form theory or from computer-based experiments for each component and type of the signal.

IV. TEST SIGNALS

The model for the generation of the test signals is presented in Fig. 5, which contains the normalized signal generators (made off-line or as batch processing, by dividing the signal amplitude by its standard deviation), and a variable weighting sum. Thus all signals have unitary variance and the effect of the amplitude variation on various complexity measures can be discarded. Three types of basic signals are considered: determinist $d(t)$, random $r(t)$, and chaotic $c(t)$. In the random type, white and $1/f$ noise are considered. Other interesting types of noise signals which could be used are harmonic noise or fractional Brownian motion [13].

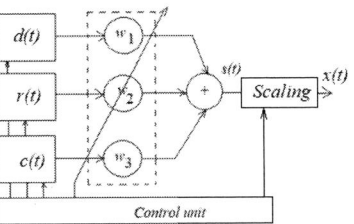

Fig. 5. The model of the test signals generation

The control unit imposes the parameters of the involved signals and the constraints for the weights, including the sum to 1 and positivity. Some time, these constraints supervises convex combination, which is inspired from convex analysis [34]). The basic signals are weighted by $\mathbf{w} = [w_1 \, w_2 \, w_3]$ before summation. The generating equation is

$$s(t) = [d(t) \, r(t) \, c(t)] \cdot [w_1 \, w_2 \, w_3]^T \qquad (5)$$

with

$$\sum_{i=1}^{3} w_i = 1 \qquad (6)$$

We are interested to observe that the changes in the values of the weights are detected by some entropy or complexity measures and if mixture of random and chaotic signals could be also detected.

The test signals are presented in Table I. Some classical very used noise signals are promoted, as Gaussian white and pink noise. Two classical models of chaos are considered. The first one is the logistic map, described by [36] as

$$x(k + 1) = r \cdot x(k) \cdot \big(1 - x(k)\big), k = 0,1,2,\dots \qquad (7)$$

with r as parameter (*growth rate*). Fig. 6 shows an example of the logistic map signal with $r = 4$ and the bifurcation diagram.

TABLE I – THE COMPONENTS OF THE COMPLEX SIGNALS

No	Category	Type	Label	Weights [w1 w2 w3]
1.	Determinist	Sinus	D	[1.0 0.0 0.0]
2.	Random	White	N_1	[0.0 1.0 0.0]
3.	Random	Pink	N_2	[0.0 1.0 0.0]
4.	Complex	-	$D+N_1$	[0.5 0.5 0.0]
5.	Complex	-	$D+N_2$	[0.5 0.5 0.0]
6.	Chaotic	Logistic map	C_1	[0.0 0.0 1.0]
7.	Chaotic	Lorentz	C_2	[0.0 0.0 1.0]
8.	Complex	-	$D+C_1$	[0.5 0.0 0.5]
9.	Complex	-	$D+C_2$	[0.5 0.0 0.5]
10.	Complex	Determinist+ Random + Chaotic	$D+N_1+C_1$	[1/3 1/3 1/3]
11.	Complex	Determinist+ Random + Chaotic	$D+N_1+C_2$	[1/3 1/3 1/3]
12.	Complex	Determinist+ Random + Chaotic	$D+N_2+C_1$	[1/3 1/3 1/3]
13.	Complex	Determinist+ Random + Chaotic	$D+N_2+C_2$	[1/3 1/3 1/3]

The second model is based on the Lorenz equations, [35] given by

$$\dot{x}(t) = \sigma\big(y(t) - x(t)\big)$$
$$\dot{y}(t) = r \cdot x(t) - y(t) - x(t) \cdot z(t) \qquad (8)$$
$$\dot{z}(t) = x(t)y(t) - b \cdot z(t)$$

where $\sigma = 16$, $r = 45.6$, and $b = 4$ are the parameters values of the model. The simulation parameters are: the simulation range $T = 10$ [s], the number of samples $N = 1,000$, and the sampling period $Ts = 0.01$ [s]. Fig. 7 shows an example of the signals generated by the model (8) under the initial conditions $[x_0, y_0, z_0] = [0.0, 1.0, 1.05]$.

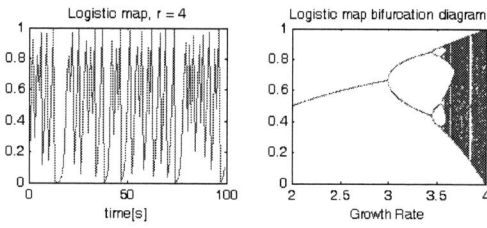

Fig. 6. Chaotic signals of the logistic map

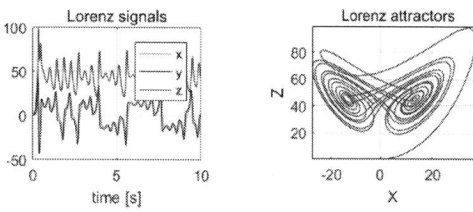

Fig. 7. Chaotic signals of the Lorenz's model

V. RESULTS OF EXPERIMENTS

Some results based on computer simulation are presented. In Fig 8 the evolution of the variance of the Renyi entropy is presented, alpha on horizontal axis, with various signals (#2 and #3 in the left, and #6 and #7 in the right side). The higher the order the higher is the variance. This is the reason to choose alpha equal nine in this work.

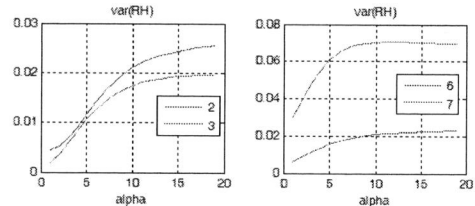

Fig. 8. Variance of the Renyi entropy for random and chaotic signals

Fig. 9 present the evolution of the Renyi entropies for all cases introduced in Table I, and the cases #2 and #6 at the bottom side. There are many cases/combinations which have different Renyi entropies, and thus this allows to recognize and to distinguish among the considered set of signals. Fig 10 presents the average values of the Renyi entropy and the relative differences in the bottom side. The smallest difference is obtained in the case of the white and pink noise signals.

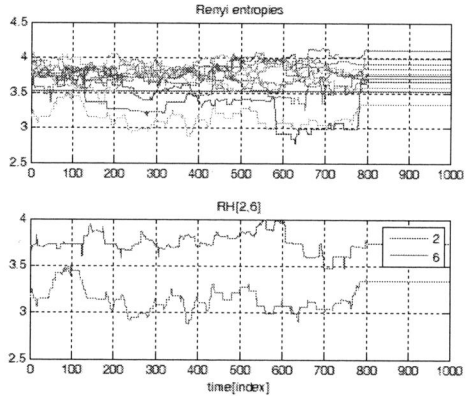

Fig. 9. Renyi entropies for the test signal (see Table I); alpha = 9

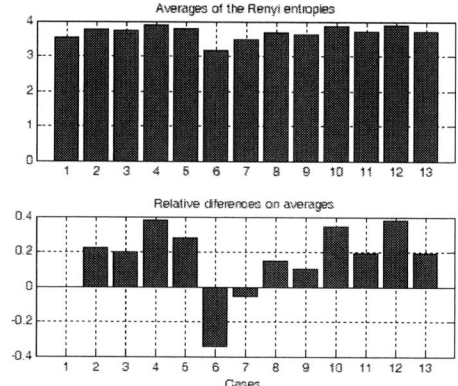

Fig. 10. Renyi entropies for the test signal (see Table I); alpha = 9

Figs. 11, 12 and 13 present - in sets of three sub-figures - the evolution of the analyzed signals, the complexity coefficient (CC) and a cusum variable (cumulative sum; see e.g. [40] and [41]), in order to show the feasibility of the CC to detect a change in the structure of the signal. Fig.11 is for a test signal defined by [X1 X4], which means deterministic, and deterministic plus random noise. We expect to have a change at the middle of record in the evolution of CC, and, secondly, CC greater for the case of the signal X4 (determinist signal + noise). Fig. 12 is for [X4 X8] and Fig. 13 is for the sequence [X2 X3 X6 X7].

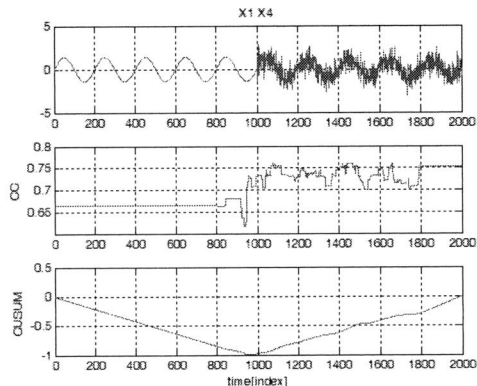

Fig. 11. Renyi entropy for the case X1 and X4

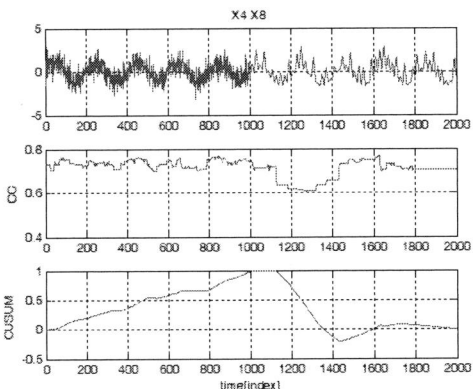

Fig. 12. Complexity CC and cusum variable for the [X4 X8] signal

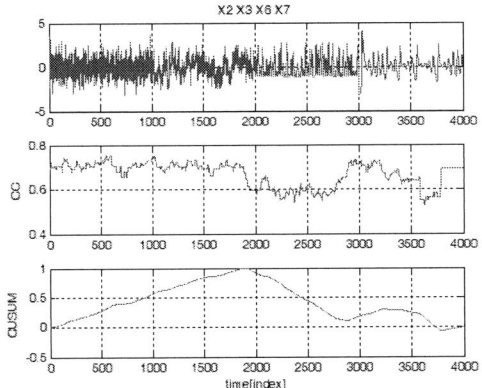

Fig. 13. Complexity coefficient CC and and cusum for [X2 X3 X6 X7]

CONCLUSION

The objective of the paper was to discuss the need for a high-level description measures in the representation of the complexity of signals. Such measures must allow the comparison among signals but also should estimate the computational effort for signal processing and change detection in the structure of the signals.

A change detection method based on Renyi entropies of high order was considered, and a complexity coefficient was defined and used in the description of the test signals. Three basic categories of signals were involved: determinist, random and chaotic. All signals were scaled to a unit variance. Complex signals were used also by using a weighted sum of the basic components/signals. Change detection is exemplified with a variable based on cumulative sum, applied to the complexity coefficient.

In the experiments based on computer simulation, the complexity coefficient was used to detect the change in the structure of the signal. The lowest results were obtained in the detection of the change from white to pink noise and between chaotic signals. For these cases, other complexity measures need developed. For other combination of the signals the results are acceptable.

The present work have used a time domain approach. With Fourier transform, a spectral entropy could be used in an independent or combined with the present entropies. For nonstationary cases, time-frequency spectral entropies could be used also.

Next steps to follow are related to the use of signal-to-noise ratio instead of weights to describe the structure of the signal. A more elaborated complexity criterion must include other variables/measure of the complexity as those coming from both from system engineering (Lyapunov exponent) and computational complexity (Lempel-Ziv coefficient).

REFERENCES

[1] Hindawi, "Complexity", 2020. [Online]. Available: https://www.hindawi.com/journals/complexity/ .

[2] Wiley, "Complexity," Wiley, 2020. [Online]. Available: https://onlinelibrary.wiley.com/journal/10990526.

[3] Elsevier, "Journal of Complexity", Elsevier, 2020. [Online]. Available: https://www.journals.elsevier.com/journal-of-complexity.

[4] IOPscience, "IOP," Journal of Physics: Complexity, 2020. [Online]. Available: https://iopscience.iop.org/journal/2632-072X.

[5] G. Schlotthauer, A.Humeau-Heurtier, J. Escudero, and H.L. Rufiner, "Editorial: Measuring Complexity of Biomedical Signals", Hindawi, Complexity, vol. 2018, ID 5408254, 2018.

[6] Y. Li, X. Wang, Z. Liu, X. Liang and S. Si, "The entropy Algorithm and its variants in the fault diagnosis of rotating machinery: A review," IEEE Access, vol. 6, pp. 66723-66741, 2018.

[7] A.Namdari and Z. Li, "A review of entropy measures for incertainity quantification of stochatic processes," Advances in Mechanical Engineering, vol. 11, no. 6, pp. 1-14, 2019.

[8] J. Sun, B. Wang, Y. Niu, Y. Tan, C. Fan, N. Zhang, J. Xue, J. Wei and J. Xiang, "Complexity analysis of EEG, MEG, and fMRI in mild cognitive impairment and Alzheimer's sisease: A Review", Entropy, vol. 22, pp. 238-260, 2020.

[9] S. Lloyd, "Measures of complexity: a nonexhaustive list," in IEEE Control Systems Magazine, vol. 21, no. 4, pp. 7-8, Aug. 2001.

[10] N. Nagaraj, K. Balasubranabian, and S. Dey, "A new complexity measure for time series analysis and classification", The European Physical Journal - Special Topics, vol. 222, pp. 847-860, 2013.

[11] N. Nagaraj and K. Balasubramanian, "Dynamical complexity of short and noisy time series," The European Physical Journal, vol. 226, pp. 2191-2204, 2017.

[12] F. Serinaldi, L. Zunino and O. A. Rosso, "Complexity-entropy analysis of daily stream flow time series in the continental United States," Stoch Enriron Res Risk Asses, vol. 28, pp. 1685-1708, 2014.

[13] H. V. Ribeiro, M. Jauregui, L. Zunino, and EK. Lenzi, "Characterizing time series via complexity-entropy curves", Physical Review E, No. 6, 2017.

[14] A. H. Husseen Al-Nuaimi, E. Jammeh, L. Sun and E. Ifeachor, "Complexity measures for quantifying changes in electroencephalogram in Alzheimer's disease," Complexity, Hindawi, Article ID 8915079, 12 pages, 2018.

[15] M. Alvez, D.M.Garner, A.M.G.G. Fontes, L.V. de Alcantara Sousa, and V.E.Valenti, "Linear and complex measures of heart rate variability during exposure to traffic noise in healty women", Complexity, Hindawi, Article ID 2158391, 14 pages, 2018.

[16] G.E.A.P.A. Batista, E. J. Keogh, O. M. Tataw and V. M. A. Souza, "CID: an efficient complexity-invariant distance for time series," Data Min Knowl Disc, vol. 28, pp. 634-669, 2017.

[17] J. Bhattacharya and E. Pereda, "An index of signal mode complexity based on orth. transf.", Journal Comp. Neurosci, vol. 29, pp. 13-22, 2010.

[18] D. Labate, F. La Foresta, G. Morabito, I. Palamara, and F. C. Morabito, "Entropic measures of EEG Complexity in Alzheimer's disease through a multivariate multiscale approach," IEEE Sensors Journal, vol. 13, no. 9, pp. 3284-3292, 2013.

[19] S. Wu, P. Wu, C. Wu, J. Ding and C. Wang, "Bearing fault diagnosis based on multiscale permutation entropy and support vector machine", Entropy, vol. 14, pp. 1343-1356, 2012.

[20] C. R. Rivero, J. A. Pucheta, A. D. O. Cañón, L. Franco, Y. J. T. Valdivia, P. S. Otaño, V. H. Sauchelli, "Noisy chaotic time series forecast approximated by combining Reny's entropy with energy associated to series method: application to rainfall series", IEEE Latin America Transactions, Vol. 15, No. 7, July 2017, pp. 1318-1325.

[21] K. Pukenas, "Algorithm for the detection of changes in the dynamics of a multivariate time series via sliced cross-bispectrum", Circuits Systems Signal Processing, vol. 37, pp. 873-882, 2018.

[22] V. Chandola, D. Cheboli, and V. Kumar, "Detecting anomalies in a time series database", Technical Report, University of Minnesota, URL: https://www.cs.umn.edu/research/technical_reports/view/09-004, 2020.

[23] D. Aiordachioaie, Th. D. Popescu and M.S. Pavel, "On change detection in the complexity of time series," in 12th International Conference on Electronics, Computers and Artificial Intelligence (ECAI), Bucharaest, Romania, July, pp. 1-6, 2020.

[24] D. Aiordachioaie, Th. D. Popescu and M.S. Pavel, "On change detection in the complexity of the time series with multiscale Renyi entropy processing", 24th Int. Conf. on System Theory, Control and Computing (ICSTCC-2020), 8 - 10 Oct, Sinaia, Romania, (accepted), 2020.

[25] A. Omidvarnia, M. Mesbah, M. Pedersen and G. Jackson, "Range entropy: A bridge between signal complexity and self-similarity," Entropy, vol. 20, no. 12, p. ID 962, 2018.

[26] S. Pincus, "Approxiate entropy as a measure of system complexity", Proceedings Natl cd Sci, vol. 88, pp. 2297-2301, 1991.

[27] D. Aiordachioaie and Th. D. Popescu, "Aspects of time series analysis with entropies and complexity measures", International Symposium on Electronics and Telecommunications (ISETC-2020), Timisoara, November 05-06, (in press), 2020.

[28] O. Olendski, "Rényi and Tsallis entropies: three analytic examples", European Journal of Physics, Vol. 40, Number 2, vol. 40, no. 2, 2019.

[29] A. Humeau-Heurtier, "The multiscale entropy algorithm and its variants: A review", Entropy, vol. 17, pp. 3110-3123, 2015.

[30] J. Ladyman, J. Lambert and K. Wiesner, "What is a complex system?," Euro Jnl Phil Sci, vol. 3, pp. 33-67, 2013.

[31] L. Riano and T. McGinnity, "Quatifying the role of complexity in a system's performance", Evolving systems, vol. 2, pp. 189-198, 2011.

[32] F. Ameri, J.D. Summers, G.M. Mocko, and M. Poeter, "Engineering design complexity: an ivestigation of methods and measures", Springer, Research in Engineering Design, vol. 19, pp. 161-179, 2008.

[33] D. Bonchev, G.A. Buck, Quantitative Measures of Network Complexity. In: Bonchev D., Rouvray D.H. (eds) Complexity in Chemistry, Biology, and Ecology, Springer, Boston, MA, pp. 191-235, 2005.

[34] R.T. Rockafellar, Convex Analysis, Princeton University Press, 1970.

[35] K.T. Alligood, T.D. Sauer, J.A. Yorke, Chaos, Springer, 1997.

[36] G. Boeing, G. 2016. "Visual Analysis of Nonlinear Dynamical Systems: Chaos, Fractals, Self-Similarity and the Limits of Prediction", Systems, vol. 4 (4), 37. doi:10.3390/systems4040037, 2016.

[37] Z. Huo, M. Martinez-Garcia, Y. Zhang, R. Yan, L. Shu, Entropy Measures in Machine Fault Diagnosis: Insights and Applications, IEEE Transactions on Instrumentation and Measurement, (in press), 2020.

[38] X. Hu and Z. Ye, Generalized quantum entropy, Journal of mathematical physics, 47(2): 023502, 2006.

[39] K.M.R. Audenaert, "Subadditivity of q-entropies for q>1", Journal of Mathematical Physics 48(8): 083507, 2007.

[40] D.C. Montgomery, Introduction to statistical quality control, John Wiley & Sons, 2012.

[41] D. Aiordachioaie and B. Dumitrascu, "On the change detection methods with sensitivity at variance of the processed signal", The IEEE 39th International Conference on Telecommunications and Signal Processing (TSP-2016), in Vienna, Austria, June 27-29, pp. 417-420, 2016.

2020 IEEE 26th International Symposium for Design and Technology in Electronic Packaging (SIITME)

On the Performance of LMS-Based Algorithms for the Identification of Low-Rank Systems

Roxana-Elena Mihăescu, Constantin Paleologu, Silviu Ciochină
University Politehnica of Bucharest
Bucharest, Romania
Email: pale@comm.pub.ro

Jacob Benesty
INRS-EMT, University of Quebec
Montreal, Canada
Email: benesty@emt.inrs.ca

Abstract—**A recent approach has been proposed that exploits the nearest Kronecker product (NKP) decomposition in conjunction with low-rank approximation methods, for efficiently solving low-rank system identification problems. Following this idea, several adaptive filtering solutions were developed. In this paper, we focus on least-mean-square (LMS) algorithms based on NKP, which own the advantage of low computational complexity. The performance features of these LMS-based algorithms are briefly investigated and a variable step-size (VSS) version is derived. The proposed VSS algorithm achieves a better compromise in terms of the main performance criteria.**

Keywords—Adaptive filtering; least-mean-square (LMS) algorithm; nearest Kronecker product decomposition; low-rank approximation; system identification; variable step-size.

I. INTRODUCTION

System identification problems need to be solved in the framework of many applications [1]. A challenging case appears when the impulse response to be identified is of high length, e.g., on the order of hundreds/thousands of coefficients. Such a scenario is associated to the echo cancellation problem, where an adaptive filter is used to model the echo path [2]. In this context, besides the computational amount, the high length of the system significantly influences the convergence rate of the algorithm and the accuracy of the resulting solution.

Recently, an efficient approach was proposed for high-dimension system identification problems [3]. The main idea is to reformulate the original system identification problem based on a combination of low-dimension problems (i.e., shorter filters). This approach exploits the nearest Kronecker product (NKP) decomposition of the impulse response, followed by low-rank approximation techniques. This fits very well for the identification of low-rank systems, like the echo paths. Following this idea, an iterative Wiener filter was developed in [3]. Also, in [4] and [5], several adaptive solutions were proposed, exploiting the recursive least-squares (RLS) algorithm and the Kalman filter, respectively. Lately, in [6], a version based on the least-mean-square (LMS) algorithm was designed.

The main advantage of the LMS-based algorithms consists of a lower computational complexity, as compared to their counterparts from the RLS family or the Kalman filter. On the other hand, LMS algorithms are more sensitive to the character of the input data (e.g., non-stationary or highly correlated inputs). In this paper, we investigate the performance of the

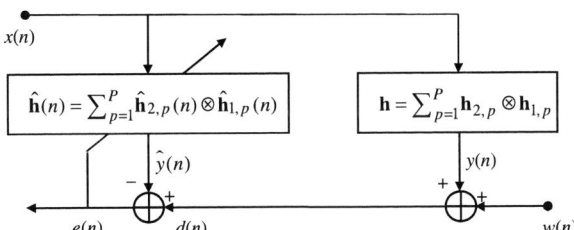

Fig. 1.   Low-rank system identification framework.

LMS algorithm based on the NKP decomposition approach, in the framework of low-rank system identification problems. Moreover, in order to better address the compromise between convergence rate and misadjustment, we design a variable step-size version of this algorithm.

The reminder of the paper is organized as follows. Section II provides the basics of the NKP decomposition approach. The LMS-based algorithm is analyzed in Section III, together with its variable step-size version. Simulation results are provided in Section IV. Finally, the paper is concluded in Section V.

II. NKP DECOMPOSITION-BASED APPROACH

Let us consider the system identification problem depicted in Fig. 1, where the goal is to model an unknown finite impulse response $\mathbf{h}$ (with L coefficients), using an adaptive filter, $\mathbf{h}(n)$, where n is the discrete-time index. In this framework, we assume that all the data is real valued. The output of the unknown system, $y(n)$, is corrupted by a zero-mean additive noise, $w(n)$, which is uncorrelated with the input signal, $x(n)$. Consequently, the desired signal is obtained as

$$d(n) = y(n) + w(n) = \mathbf{h}^T\mathbf{x}(n) + w(n), \qquad (1)$$

where the superscript ^{T} denotes the transpose operator and $\mathbf{x}(n) = \begin{bmatrix} x(n) & x(n-1) & \cdots & x(n-L+1) \end{bmatrix}^T$ is a vector containing the most recent L time samples of the input signal. The adaptive filter is driven by the error signal

$$e(n) = d(n) - \widehat{y}(n) = d(n) - \widehat{\mathbf{h}}^T(n-1)\mathbf{x}(n), \qquad (2)$$

where $\widehat{y}(n)$ represents the estimated output signal.

In the following, we consider that the length of the system is $L = L_1 L_2$, with $L_1 \geq L_2$. Thus, in the context of a low-rank

978-1-7281-7507-2/20 $31.00 © 2020 IEEE

21-24 October, Pitesti, Romania

system identification problem [3]–[5], the unknown system can be expressed as

$$\mathbf{h} = \sum_{p=1}^{P} \mathbf{h}_{2,p} \otimes \mathbf{h}_{1,p}, \tag{3}$$

where $P < L_2$, $\otimes$ denotes the Kronecker product, and $\mathbf{h}_{1,p}$ and $\mathbf{h}_{2,p}$ (with $p = 1, 2, \ldots, P$) are impulse responses of lengths L_1 and L_2, respectively. Consequently, the global adaptive filter can also be decomposed as

$$\widehat{\mathbf{h}}(n) = \sum_{p=1}^{P} \widehat{\mathbf{h}}_{2,p}(n) \otimes \widehat{\mathbf{h}}_{1,p}(n), \tag{4}$$

where $\widehat{\mathbf{h}}_{1,p}(n)$ and $\widehat{\mathbf{h}}_{2,p}(n)$ (with $p = 1, 2, \ldots, P$) are adaptive filters of lengths L_1 and L_2, respectively, as shown in Fig. 1.

This approach in based on the NKP decomposition of the impulse response [7], together with low-rank approximations. As shown in [3]–[5], when $P \ll L_2$, this approach could lead to important advantages in terms of both performance and complexity.

Let us introduce the notation:

$$\underline{\widehat{\mathbf{h}}}_1(n) = \begin{bmatrix} \widehat{\mathbf{h}}_{1,1}^T(n) & \widehat{\mathbf{h}}_{1,2}^T(n) & \cdots & \widehat{\mathbf{h}}_{1,P}^T(n) \end{bmatrix}^T,$$

$$\underline{\widehat{\mathbf{h}}}_2(n) = \begin{bmatrix} \widehat{\mathbf{h}}_{2,1}^T(n) & \widehat{\mathbf{h}}_{2,2}^T(n) & \cdots & \widehat{\mathbf{h}}_{2,P}^T(n) \end{bmatrix}^T,$$

$$\widehat{\mathbf{x}}_{2,p}(n) = \begin{bmatrix} \widehat{\mathbf{h}}_{2,p}(n-1) \otimes \mathbf{I}_{L_1} \end{bmatrix}^T \mathbf{x}(n), \tag{5}$$

$$\underline{\widehat{\mathbf{x}}}_2(n) = \begin{bmatrix} \widehat{\mathbf{x}}_{2,1}^T(n) & \widehat{\mathbf{x}}_{2,2}^T(n) & \cdots & \widehat{\mathbf{x}}_{2,P}^T(n) \end{bmatrix}^T,$$

$$\widehat{\mathbf{x}}_{1,p}(n) = \begin{bmatrix} \mathbf{I}_{L_2} \otimes \widehat{\mathbf{h}}_{1,p}(n-1) \end{bmatrix}^T \mathbf{x}(n), \tag{6}$$

$$\underline{\widehat{\mathbf{x}}}_1(n) = \begin{bmatrix} \widehat{\mathbf{x}}_{1,1}^T(n) & \widehat{\mathbf{x}}_{1,2}^T(n) & \cdots & \widehat{\mathbf{x}}_{1,P}^T(n) \end{bmatrix}^T,$$

where $\mathbf{I}_{L_1}$ and $\mathbf{I}_{L_2}$ denote the identity matrices of sizes $L_1 \times L_1$ and $L_2 \times L_2$, respectively. Thus, (2) can be rewritten as

$$e_1(n) = d(n) - \underline{\widehat{\mathbf{h}}}_1^T(n-1)\underline{\widehat{\mathbf{x}}}_2(n), \tag{7}$$

$$e_2(n) = d(n) - \underline{\widehat{\mathbf{h}}}_2^T(n-1)\underline{\widehat{\mathbf{x}}}_1(n), \tag{8}$$

where we can notice that $e(n) = e_1(n) = e_2(n)$. On the other hand, comparing (7)–(8) to (2), we can reformulate the global system identification problem using a combination of two adaptive filters, $\underline{\widehat{\mathbf{h}}}_1(n)$ and $\underline{\widehat{\mathbf{h}}}_2(n)$, having the "inputs" $\underline{\widehat{\mathbf{x}}}_2(n)$ and $\underline{\widehat{\mathbf{x}}}_1(n)$, respectively. Clearly, as we can notice from (5)–(6), these two filters are "connected," since the input of each filter contains the contribution of the other one (through its coefficients from the previous time index). Nevertheless, these two filters have the lengths PL_1 and PL_2, which can be significantly shorter as compared to L [i.e., the length of the global filter, $\widehat{\mathbf{h}}(n)$], for $P \ll L_2$. In this way, we target to solve a high-dimension system identification problem based on low-dimension "solutions" (e.g., shorter filters) that are combine together to match the original purpose.

III. LMS-NKP Algorithms

The NKP decomposition-based approach was previously exploited in different frameworks [3]–[6]. Most recently, in [6], different versions based on the LMS algorithm were designed, which are computationally appealing as compared to the solutions proposed in [3]–[5].

The updates of the LMS-NKP algorithm [6] are obtained based on (7)–(8) and result in

$$\underline{\widehat{\mathbf{h}}}_1(n) = \underline{\widehat{\mathbf{h}}}_1(n-1) + \frac{\mu_1}{2} \times \frac{\partial e_1^2(n)}{\partial \underline{\widehat{\mathbf{h}}}_1(n-1)}$$
$$= \underline{\widehat{\mathbf{h}}}_1(n-1) + \mu_1 \underline{\widehat{\mathbf{x}}}_2(n)e_1(n), \tag{9}$$

$$\underline{\widehat{\mathbf{h}}}_2(n) = \underline{\widehat{\mathbf{h}}}_2(n-1) + \frac{\mu_2}{2} \times \frac{\partial e_2^2(n)}{\partial \underline{\widehat{\mathbf{h}}}_2(n-1)}$$
$$= \underline{\widehat{\mathbf{h}}}_2(n-1) + \mu_2 \underline{\widehat{\mathbf{x}}}_1(n)e_2(n), \tag{10}$$

where μ_1 and μ_2 are positive constants known as the step-size parameters. The updates (9)–(10) rely on a bilinear optimization strategy [8], i.e., each system is considered fixed while optimizing the other one. Also, we can notice the "connection" between the two systems, i.e., each adaptive filter at time index n depends on the coefficients on both filters at time index $n-1$ [see (5)–(6)].

Next, according to Fig. 1 and based on (3), let us introduce the notation:

$$\underline{\mathbf{h}}_1 = \begin{bmatrix} \mathbf{h}_{1,1}^T & \mathbf{h}_{1,2}^T & \cdots & \mathbf{h}_{1,P}^T \end{bmatrix}^T,$$

$$\underline{\mathbf{h}}_2 = \begin{bmatrix} \mathbf{h}_{2,1}^T & \mathbf{h}_{2,2}^T & \cdots & \mathbf{h}_{2,P}^T \end{bmatrix}^T,$$

$$\mathbf{x}_{2,p}(n) = [\mathbf{h}_{2,p} \otimes \mathbf{I}_{L_1}]^T \mathbf{x}(n),$$

$$\underline{\mathbf{x}}_2(n) = \begin{bmatrix} \mathbf{x}_{2,1}^T(n) & \mathbf{x}_{2,2}^T(n) & \cdots & \mathbf{x}_{2,P}^T(n) \end{bmatrix}^T,$$

$$\mathbf{x}_{1,p}(n) = [\mathbf{I}_{L_2} \otimes \mathbf{h}_{1,p}]^T \mathbf{x}(n),$$

$$\underline{\mathbf{x}}_1(n) = \begin{bmatrix} \mathbf{x}_{1,1}^T(n) & \mathbf{x}_{1,2}^T(n) & \cdots & \mathbf{x}_{1,P}^T(n) \end{bmatrix}^T,$$

which is related to the decomposition of the unknown impulse response. At this point, we should outline that there are inherent scaling factors that appear within the identification process, since $\sum_{p=1}^{P} \mathbf{h}_{2,p} \otimes \mathbf{h}_{1,p} = \sum_{p=1}^{P} \beta_p \mathbf{h}_{2,p} \otimes (1/\beta_p)\mathbf{h}_{1,p}$, where the β_p's are real numbers. However, there is no scaling ambiguity when identifying the global impulse response, $\mathbf{h}$. For the sake of simplicity, in the following analysis, we will omit the scaling factors.

Using the previous notation, the desired signal from (1) can be expressed as

$$d(n) = \underline{\mathbf{h}}_1^T \underline{\mathbf{x}}_2(n) + w(n) \tag{11}$$
$$= \underline{\mathbf{h}}_2^T \underline{\mathbf{x}}_1(n) + w(n). \tag{12}$$

Further, by adding and subtracting $\underline{\mathbf{h}}_1^T \underline{\widehat{\mathbf{x}}}_2(n)$ on the right-hand side of (11), while performing a similar operation with the term $\underline{\mathbf{h}}_2^T \underline{\widehat{\mathbf{x}}}_1(n)$ but on the right-hand side of (12), the desired signal expressions become

$$d(n) = \underline{\mathbf{h}}_1^T \underline{\widehat{\mathbf{x}}}_2(n) + w(n) + w_2(n) \tag{13}$$
$$= \underline{\mathbf{h}}_2^T \underline{\widehat{\mathbf{x}}}_1(n) + w(n) + w_1(n), \tag{14}$$

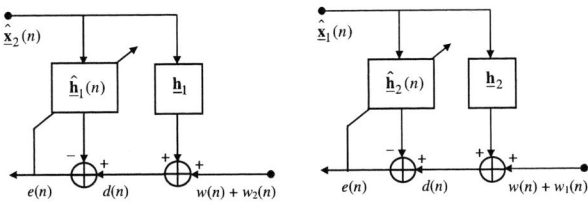

Fig. 2. Equivalent schemes for the identification of the individual impulse responses $\underline{\mathbf{h}}_1$ (left) and $\underline{\mathbf{h}}_2$ (right).

where

$$w_2(n) = \underline{\mathbf{h}}_1^T[\underline{\mathbf{x}}_2(n) - \widehat{\underline{\mathbf{x}}}_2(n)] = \mathbf{c}_2^T(n-1)\underline{\mathbf{x}}_1(n), \quad (15)$$

$$w_1(n) = \underline{\mathbf{h}}_2^T[\underline{\mathbf{x}}_1(n) - \widehat{\underline{\mathbf{x}}}_1(n)] = \mathbf{c}_1^T(n-1)\underline{\mathbf{x}}_2(n), \quad (16)$$

with

$$\mathbf{c}_2(n) = \underline{\mathbf{h}}_2 - \widehat{\underline{\mathbf{h}}}_2(n-1), \quad (17)$$

$$\mathbf{c}_1(n) = \underline{\mathbf{h}}_1 - \widehat{\underline{\mathbf{h}}}_1(n-1). \quad (18)$$

Under these circumstances, the equivalent schemes for the identification of the individual impulse responses $\underline{\mathbf{h}}_1$ and $\underline{\mathbf{h}}_2$ are depicted in Fig. 2. As we can notice in this figure [and also in (13)–(14)], each filter is influenced by an additional "noise," as a result of the contribution of the other filter. According to (15)–(16), these signals depend on the a priori misalignments $\mathbf{c}_2(n)$ and $\mathbf{c}_1(n)$, which are defined in (17) and (18), respectively. Therefore, besides the inherent performance limitation due to the system noise $w(n)$, the LMS-NKP algorithm is also influenced by these "connection" noises, $w_1(n)$ and $w_2(n)$. For LMS-based algorithms, in order to reduce the misadjustment (i.e., to improve the accuracy), a small value of the step-size should be used. In case of the LMS-NKP algorithm, due to the previously discussed limitation, there is a certain lower bound "threshold" in terms of the misadjustment, which cannot be improved despite the value of the step-sizes. We should note that a lower value of the step-size improves the misadjutment, but reduces the convergence rate. On the other hand, a slow convergence rate is not a benefit in terms of the contributions of the additional noises $w_1(n)$ and $w_2(n)$. We should also note that the RLS-based algorithms are less influenced by this issue, especially when using a value of the forgetting factor close to one [4].

The convergence analysis of the LMS-NKP algorithm is not a trivial task, but it is beyond the scope of this paper. For such an analysis, we could exploit some analogies with the tensor LMS [9] or the LMS algorithm for bilinear forms [10].

The normalized versions of the LMS-based algorithms are more robust in terms of selecting the step-size parameters [1]. Consequently, in [6], the algorithm of choice was the normalized version of the LMS-NKP, namely NLMS-NKP. In this case, the time-dependent step-sizes are evaluated as

$$\mu_1(n) = \frac{\alpha_1}{\delta_1 + \widehat{\underline{\mathbf{x}}}_2^T(n)\widehat{\underline{\mathbf{x}}}_2(n)}, \quad (19)$$

$$\mu_2(n) = \frac{\alpha_2}{\delta_2 + \widehat{\underline{\mathbf{x}}}_1^T(n)\widehat{\underline{\mathbf{x}}}_1(n)}, \quad (20)$$

where α_1 and α_2 are the normalized step-size parameters (positive constants smaller than one) and δ_1 and δ_2 are the regularization terms (positive constants for robustness in presence of noise). Nevertheless, in terms of the main performance criteria, there is also a compromise when selecting the normalized step-size parameters (i.e., fast convergence rate versus low misadjustment). In order to address this issue, we propose in the following a variable step-size version of the NLMS-NKP algorithm, namely VSS-NLMS-NKP, which is inspired by the idea from [11].

In our system identification framework, we target to recover the system noise from the error of the adaptive filter. Consequently, we impose this condition in terms of power estimates:

$$E\left[\varepsilon_1^2(n)\right] = \sigma_w^2, \quad (21)$$

$$E\left[\varepsilon_2^2(n)\right] = \sigma_w^2, \quad (22)$$

where $E[\cdot]$ denotes mathematical expectation, σ_w^2 is the estimated power of the noise (e.g., during silences), and

$$\varepsilon_1(n) = d(n) - \widehat{\underline{\mathbf{h}}}_1^T(n)\widehat{\underline{\mathbf{x}}}_2(n), \quad (23)$$

$$\varepsilon_2(n) = d(n) - \widehat{\underline{\mathbf{h}}}_2^T(n)\widehat{\underline{\mathbf{x}}}_1(n) \quad (24)$$

are the a posteriori error signals. Next, we develop (23) and (24) under the conditions (21) and (22), respectively. After some straightforward computations, these result in two quadratic equations that have the solutions:

$$\mu_1(n) = \frac{1}{\widehat{\underline{\mathbf{x}}}_2^T(n)\widehat{\underline{\mathbf{x}}}_2(n)}\left[1 - \frac{\sigma_w}{\sigma_e(n)}\right], \quad (25)$$

$$\mu_2(n) = \frac{1}{\widehat{\underline{\mathbf{x}}}_1^T(n)\widehat{\underline{\mathbf{x}}}_1(n)}\left[1 - \frac{\sigma_w}{\sigma_e(n)}\right], \quad (26)$$

where the power of the error signal can be recursively estimated as $\sigma_e^2(n) = \lambda\sigma_e^2(n-1)+(1-\lambda)e^2(n)$, with $0 \ll \lambda < 1$. Similar to (19)–(20), for practical reasons, the regularization terms δ_1 and δ_2 should also be introduced in (25)–(26). Summarizing, the VSS-NLMS-NKP algorithm is based on the updates of the LMS-NKP counterpart, but using the variable step-sizes from (25)–(26).

IV. SIMULATION RESULTS

Simulations are performed in the context of network echo cancellation [2]. The unknown system is the first impulse response from G168 Recommendation [12], which is padded with zeros up to the length $L = 500$. In this case, the NKP decomposition is performed using $L_1 = 25$, $L_2 = 20$, and $P = 3$ [3]. In order to evaluate the tracking capabilities of the algorithms, the impulse response is abruptly changed in the middle of each experiment, by changing the sign of its coefficients. The far-end signal $x(n)$ is either an AR(1) process [i.e., a white Gaussian noise filtered through the first-order system $1/\left(1 - 0.9z^{-1}\right)$] or a speech sequence, where the sampling rate is 8 kHz. The echo signal $y(n)$ is corrupted by a white Gaussian noise, $w(n)$, such that the signal-to-noise ratio is 10 dB. The performance measure is the normalized misalignment (in dB), defined as $20\log_{10}\left(\left\|\mathbf{h} - \widehat{\mathbf{h}}(n)\right\|_2 / \|\mathbf{h}\|_2\right)$, where $\|\cdot\|_2$ denotes the Euclidian norm.

*2020 IEEE 26ᵗʰ International Symposium for Design and Technology in Electronic Packaging (**SIITME**)*

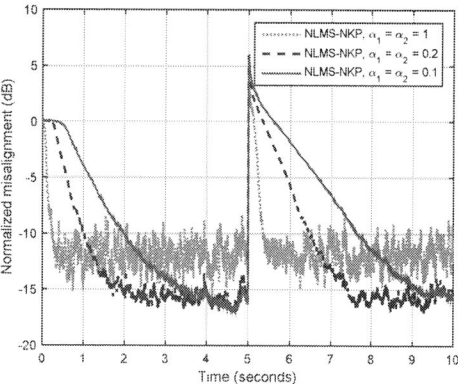

Fig. 3. Performance of the NLMS-NKP algorithm using different values of the normalized step-sizes. The input signal is an AR(1) process.

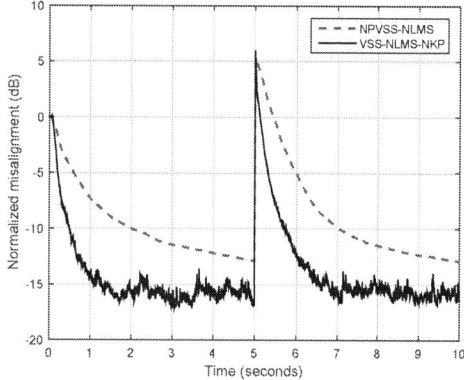

Fig. 4. Performance of the NPVSS-NLMS algorithm [11] and the proposed VSS-NLMS-NKP algorithm. The input signal is an AR(1) process.

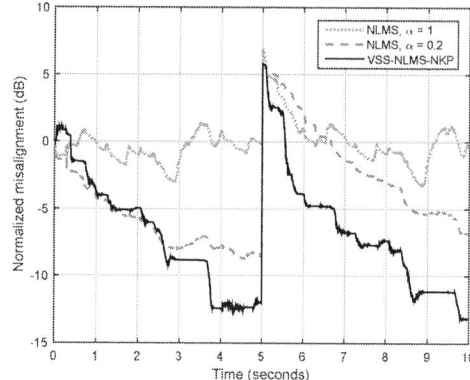

Fig. 5. Performance of the regular NLMS algorithm and the proposed VSS-NLMS-NKP algorithm. The input signal is a speech sequence.

this algorithm have been outlined, together with some of its limitations, which mainly result due to the connection between the component filters. In addition, in order to address the compromise between the main performance criteria, we have developed a variable step-size version, namely the VSS-NLMS-NKP algorithm. Simulation results have indicated the good performance of the proposed solution.

ACKNOWLEDGEMENTS

This work was supported by a grant of the Romanian Ministry of Research and Innovation, CNCS – UEFISCDI, project number: PN-III-P1-1.1-TE-2019-0420, within PNCDI III.

REFERENCES

[1] S. Haykin, *Adaptive Filter Theory*, 4th ed. Upper Saddle River, NJ: Prentice-Hall, 2002.

[2] J. Benesty, T. Gänsler, D. R. Morgan, M. M. Sondhi, and S. L. Gay, *Advances in Network and Acoustic Echo Cancellation*. Berlin, Germany: Springer-Verlag, 2001.

[3] C. Paleologu, J. Benesty, and S. Ciochină, "Linear system identification based on a Kronecker product decomposition," *IEEE/ACM Trans. Audio, Speech, Language Processing*, vol. 26, pp. 1793–1808, Oct. 2018.

[4] C. Elisei-Iliescu, C. Paleologu, J. Benesty, C. Stanciu, C. Anghel, and S. Ciochină, "Recursive least-squares algorithms for the identification of low-rank systems," *IEEE/ACM Trans. Audio, Speech, Language Processing*, vol. 27, pp. 903–918, May 2019.

[5] L.-M. Dogariu, C. Paleologu, J. Benesty, and S. Ciochină, "An efficient Kalman filter for the identification of low-rank systems," *Signal Processing*, vol. 166, id. 107239, 9 pages, Jan. 2020.

[6] S. S. Bhattacharjee and N. V. George, "Nearest Kronecker product decomposition based normalized least mean square algorithm," in *Proc. IEEE ICASSP*, 2020, pp. 476–480.

[7] C. F. Van Loan, "The ubiquitous Kronecker product," *J. Computational Applied Mathematics*, vol. 123, pp. 85–100, 2000.

[8] D. P. Bertsekas, *Nonlinear Programming*, 2nd ed. Belmont, Massachusetts: Athena Scientific, 1999.

[9] M. Rupp and S. Schwarz, "A tensor LMS algorithm," in *Proc. IEEE ICASSP*, 2015, pp. 3347–3351.

[10] C. Paleologu, J. Benesty, and S. Ciochină, "Adaptive filtering for the identification of bilinear forms," *Digital Signal Processing*, vol. 75, pp. 153–167, Apr. 2018.

[11] J. Benesty, H. Rey, L. Rey Vega, and S. Tressens, "A nonparametric VSS NLMS algorithm," *IEEE Signal Processing Lett.*, vol. 13, pp. 581–584, Oct. 2006.

[12] *Digital Network Echo Cancellers*, ITU-T Recommendations G.168, 2002.

In the first experiment, we evaluate the performance of the NLMS-NKP algorithm for different values of the normalized step-size parameters. The results from Fig. 3 confirm the discussion from Section III, regarding the "threshold" on misalignment, despite of lowering the normalized step-sizes.

Next, we compare the performance of the proposed VSS-NLMS-NKP algorithm with its counterpart from [11], i.e., the nonparametric VSS-NLMS (NPVSS-NLMS) algorithm. As we can notice in Fig. 4, the proposed solution outperforms the NPVSS-NLMS algorithm in terms of convergence rate, tracking, and misalignment.

Finally, in Fig. 5, the input signal is speech and the VSS-NLMS-NKP algorithm is compared to the regular NLMS algorithm using different values of the normalized step-size parameter, α. The results show that the proposed algorithm significantly outperforms its conventional counterpart.

V. CONCLUSIONS

In this paper, we have investigated the recently proposed LMS-based NKP algorithm, in the context of low-rank system identification problems. The main performance features of

Investigations at the Interface of a Multilayer Structure Made of Non-conductive and Conductive Resins

Mihai Branzei
Research and Expertise Centre for Special Materials, CEMS
Politehnica University
Bucharest, Romania
mihai.branzei@upb.ro

Ciprian Ionescu, Bogdan Mihailescu, Paul Svasta
Center for Electronic Technology and Interconnection Techniques
Politehnica University
Bucharest, Romania

Gaudentiu Varzaru, Razvan Ungurelu
Syswin Solutions
Bucharest, Romania
gaudentiu.varzaru@syswin.ro

Marin Gheorghe
NANOM-MEMS
Rasnov, Romania

Abstract—**The paper presents the investigation of composite structures made by overlapping two or more layers of resins, including conductive resins, at different time intervals, as well as different temperatures. This situation occurs in the case of a model of implementing an innovative concept of manufacturing electronic modules without using solder alloy, the Occam process. According to this approach, the interconnection structure is arranged on several layers with conductive paths separated from each other by dielectric layers. It is interesting to characterize the interface between layers to determine the possible effect on the reliability of the module. Samples of at least two layers of three types of resins were investigated using Scanning Electron Microscopy/Energy Dispersive X-Ray Spectroscopy analyses. These analyses emphasized there is no chemical bonding at the interface between the layers of resins, but only mechanical. This is a major disadvantage that can lead to the dismantling of the assembly under the action of mechanical forces and environment conditions.**

Keywords: conductive resin, Occam process, SEM-EDS, solderless assembly, additive process.

I. INTRODUCTION

For some time, several techniques for solderless assembly for electronics (SAFE) have been appeared on the market, some of which have already matured. But, because technologies like wire bonding and press-fit have proven to be limited to certain applications, additive processing became the new paradigm [1]. Three directions have been identified in this field: printed electronics [2], 3D printing [3] and Occam process [4]. In a model of carrying out an electronic module based on the latter process several layers of resin must be applied at different time intervals. Some of them may require heat treatments up to 160°C - 180 °C for curing process. A longer time interval between successive layer applications would lead to longer and unprofitable process, while a shorter time would be prone to

incomplete layer adhesion. In multilayer coatings, SEM investigations showed that the interfaces are sources for many problems [5]. For our additive process, the consequences of this imperfection could be: i) affecting the mechanical properties of the electronic module; ii) the penetration of moisture inside the structure; iii) the penetration of chemicals from the environment in which the module operates. The investigations required the production of resin structures in the conditions as close as possible to those of the manufacture of the electronic module. This means the resin deposition and the curing process. Normally, to simplify the manufacturing process, the structures should be from the same resin. However, it is possible to have a base layer made of a certain material over which to build the module by successive deposits of adhesives from another material. The samples corresponding to these situations were created by pouring the resins into copper moulds (Ø18x14 mm). The structure of the samples contains two layers of non-conductive resins deposited at different times, thus creating a single interface; three samples were built with several traces of electrically conductive resin deposited on the first layer, which required a thermal process. Finally, the second layer of resin was deposited over them. After the last curing process, the samples were prepared for SEM investigations.

II. DESIGN OF EXPERIMENTS

A. Materials

There were used the following types of materials: three non-conductive resins, one electrically conductive paste. The materials and the parameters related to the curing are presented in Table I (they are to be referred as Rx in the present work for simplicity purposes). Since all of them are 2-part materials, the ratio between the components are important for the stability of the curing parameters. The dielectric resins are curing at room temperature (RT), but requires different curing times. Their pot life is different, ranging from 25 – 35 minutes (R1), to 4 -6 minutes (R2). As known, the pot life is the time from mixing the two parts together to the point at which the mixed compound is no longer useable. A digital display weighing was used in order to prepare the proper mixing ratios for the R1 compound (100:13).

TABLE I. MATERIALS AND THEIR CURING PARAMETERS

| # | Material | | Curing parameters | | |
	Name, manufacturer	Type of material	Temp [°C]	Time [h]	Mixing ratio[a]
R1	MC62/W363, Elantas	2-part epoxy resin	RT	24	100:13
R2	ADU PU8505, Elantas	2-part polyurethane	RT	8	1:1
R3	SW180, Tatsuta	Silver coated Cu paste	160	0.5	---

[a]By volume

B. Equipment

Although most of the materials are curing at RT, the conductive paste requires a thermal process. This was obtained by programming an oven from the surface-mount technology (SMT) line, MRO250 (Fritsch), unusually used for reflow soldering. The oven started from RT and was set to hold 160° C for 1800 seconds. The target temperatures have been obtain after almost 2 minutes following a slope of 1.2°C/s.

SEM/EDS

The cross section micrographs were investigated / analyzed on a PhenomWorld ProX SEM-EDS at acceleration voltages of 15 kV, backscattered electron detector (BSD) and energy dispersive X-ray spectrometer (EDS) detectors, with nominal resolution of 10 nm, at room temperature. The magnification has been adjusted so that it could see the morphological aspects of resins at their interface(s). Full BSD mode was also selected in order to achieve optimal contrast and clarity of the image. Acquisition time for elemental analysis was 4 min. Field of View (FOV) was about 296 µm. The EDS spectrum shown for each curing time belongs to one of the stars in each figure.

The samples were sectioned, grinded, and polished, on a semi-automatic sample preparation line from BUEHLER, with consumables from the same company. To avoid the identification of residual elements in the EDS analysis, such as Al, Si, for example, these being possible to come from the final emulsion, only polycrystalline diamond suspension have been used.

C. Methods

The 2-component resins were prepared as follows: R1 was manually mixed until the two orange and blue colour components changed to the uniform green colour; R2 was mixed using a twin syringe that pushed the two components of the resin into a mixing tube. The stages of the additive sampling process were: 1 - pouring the first layer of resin; 2 - wait for resin curing (24h); 3 - deposition of conductive paste traces followed by thermal process (160 °C/30 min), only for the samples 3, 6; and 10; 4 - pouring the second layer of the resin; 5 - wait for the second resin layer curing. The samples have been cut in cross section, followed by grinding / polishing, then submitted to SEM-EDS investigations.

Fig. 1. The samples after second layer of resin curring

III. RESULTS

Because it is a first characterization of these "couplings" of 3 resins, one of which is conductive (SW180 - Tatsuta), the characterization have been done by comparing the three interfaces presented in Figures 2, 3 and 4.

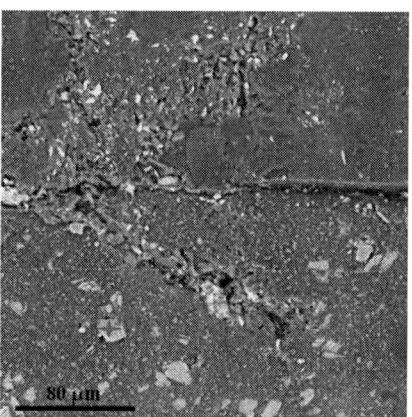

a)

| Spot no. | 1 | | 2 | | 3 | | 4 | |
Element	At%	Wt%	At%	Wt%	At%	Wt%	At%	Wt%
C	71.73	66.21	69.43	63.00	35.03	26.41	75.04	66.48
O	17.14	21.07	19.45	23.51	52.91	53.13	20.17	23.80
N	10.47	11.27	10.06	10.64	0.00	0.00	0.00	0.00
Al	0.51	1.05	0.85	1.74	11.93	20.20	4.31	8.58
P	0.00	0.00	0.00	0.00	0.11	0.21	0.00	0.00
Cl	0.00	0.00	0.00	0.00	0.02	0.05	0.00	0.00
Si	0.00	0.00	0.00	0.00	0.00	0.00	0.21	0.44

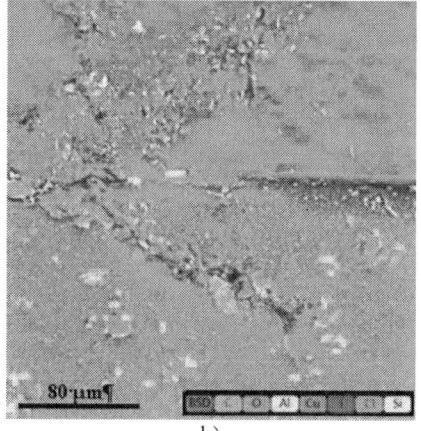

b)

2020 IEEE 26th International Symposium for Design and Technology in Electronic Packaging (SIITME)

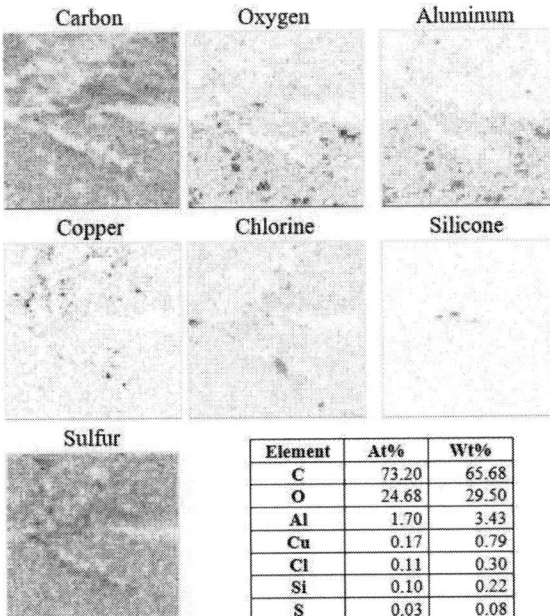

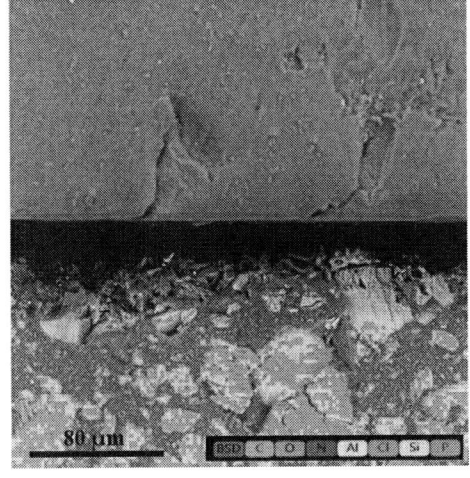

Element	At%	Wt%
C	73.20	65.68
O	24.68	29.50
Al	1.70	3.43
Cu	0.17	0.79
Cl	0.11	0.30
Si	0.10	0.22
S	0.03	0.08

Fig. 2. SEM-EDS images / analysis of the TS2 (R1/R2) cross section bond: a) positioning the EDS analysis spots on the image and the elementary quantitative analysis table; b) BSD (combined map) of the elementary quantitative distribusion and the elementary quantitative analysis table.

The morphology of resins 1 and 2 does not change mostly from one "coupling" to another. Elemental analysis confirms that the resins retain their compositional analysis after curing, at the temperature-time parameters recommended by the manufacturer, the variations have been attributed to local unhomogeneity, where the spot was fixed.

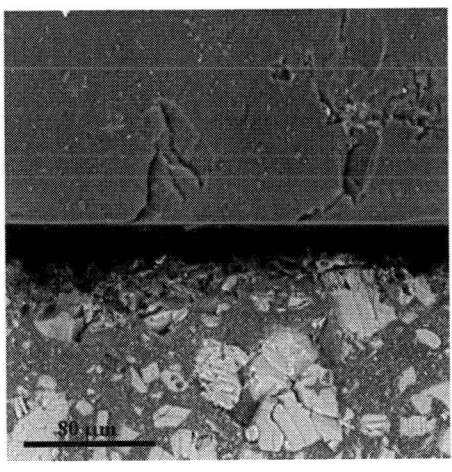

a)

Spot no.	1		2		3		4	
Element	At%	Wt%	At%	Wt%	At%	Wt%	At%	Wt%
C	70.07	64.72	68.92	63.75	55.52	47.29	55.68	54.54
O	16.79	20.66	16.52	20.35	33.91	38.48	30.31	22.29
N	12.70	13.68	14.36	15.49	6.57	6.53	13.89	22.94
Al	0.41	0.84	0.20	0.41	3.80	7.27	0.13	0.24
Si	0.03	0.09	0.00	0.00	0.10	0.21	0.00	0.00
Cl	0.01	0.02	0.00	0.00	0.09	0.22	0.00	0.00

b)

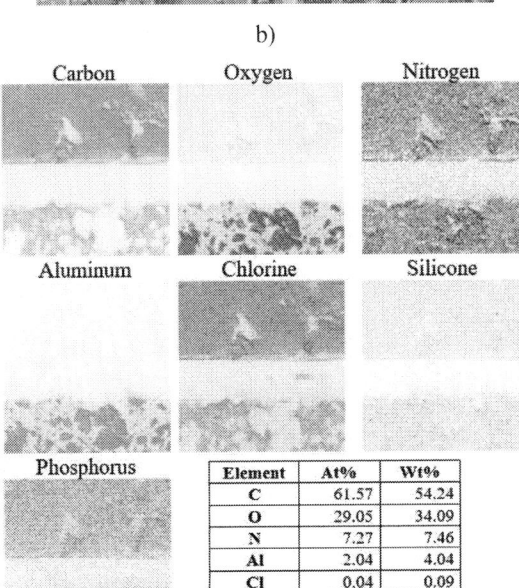

Element	At%	Wt%
C	61.57	54.24
O	29.05	34.09
N	7.27	7.46
Al	2.04	4.04
Cl	0.04	0.09
Si	0.02	0.05
P	0.01	0.03

Fig. 3. SEM-EDS images / analysis of the TS5 (R2/R1) cross section bond: a) positioning the EDS analysis spots on the image and the elementary quantitative analysis table; b) BSD (combined map) of the elementary quantitative distribusion and the elementary quantitative analysis table.

This local unhomogeneity was notified in the case of R1 and R2 resins, at sub-millimeter level, there are spheroidal morphologies with high concentrations of carbon and oxygen. The morphologies of the two resins, belonging to the two categories, epoxy for R1 and polyurethane for R2, respectively, after the baking process, differ essentially, in the sense that the specific reaction products have totally different spatial geometric shapes: spheroidal shapes, unlike polyhedral ones, with flat faces, respectively edges. This will probably result in differences in terms of mechanical strength (shear test for examples).

In the case of SW180 resin, Ag have been also identified, this being in small quantities due to the very thin layer of conductive resins applied at the interface between the two resins, R1 and R2, respectively.

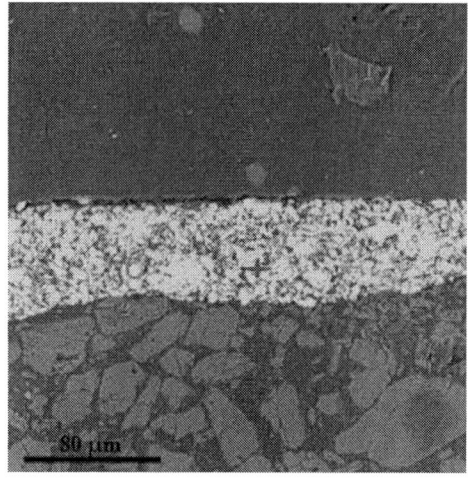

a)

Spot no.	1		2		3	
Element	At%	Wt%	At%	Wt%	At%	Wt%
C	66.78	58.73	24.23	17.18	56.07	23.93
O	17.00	19.92	57.06	53.91	10.68	6.07
N	13.55	13.90	3.96	3.27	0.00	0.00
K	0.99	2.84	0.00	0.00	0.00	0.00
Al	0.65	1.29	13.82	22.01	2.34	2.25
Cu	0.53	2.46	0.70	2.62	0.00	0.00
Na	0.46	0.78	0.01	0.02	0.00	0.00
Cl	0.03	0.07	0.00	0.00	0.00	0.00
Ag	0.00	0.00	0.12	0.78	0.00	0.00
P	0.00	0.00	0.12	0.22	0.70	2.67
F	0.00	0.00	0.00	0.00	1.98	1.34

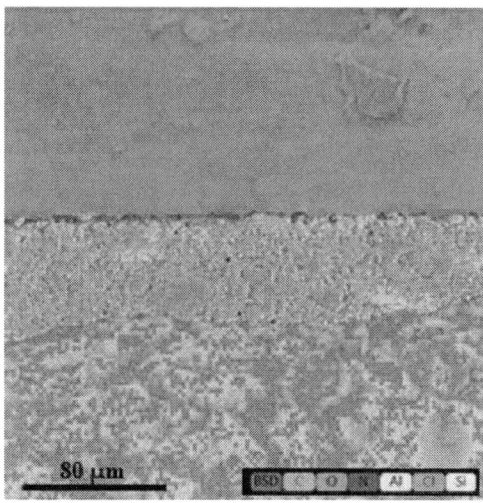

b)

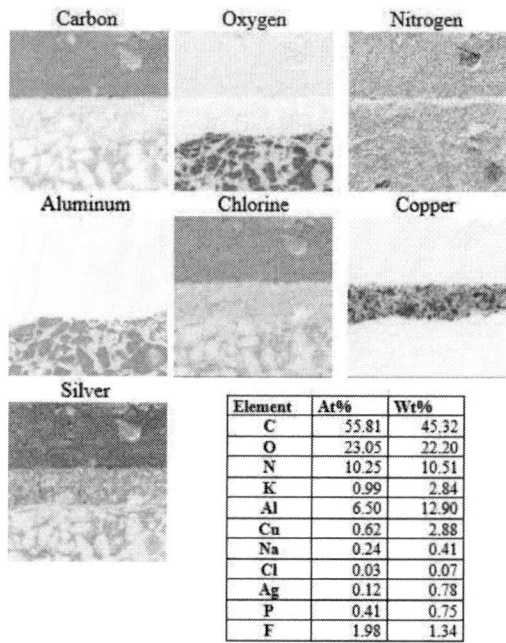

Element	At%	Wt%
C	55.81	45.32
O	23.05	22.20
N	10.25	10.51
K	0.99	2.84
Al	6.50	12.90
Cu	0.62	2.88
Na	0.24	0.41
Cl	0.03	0.07
Ag	0.12	0.78
P	0.41	0.75
F	1.98	1.34

Fig. 4. SEM-EDS images / analysis of the TS6 (R2/R3/R1) cross section bond: a) positioning the EDS analysis spots on the image and the elementary quantitative analysis table; b) BSD (combined map) of the elementary quantitative distribusion and the elementary quantitative analysis table.

IV. CONCLUSIONS

The analysis emphasized there is no chemical bonding at the interface between the layers of resins. Elemental analyses revealed the composition of the structure. In addition to C and O, there is also Al, colored in yellow, predominantly in R2 resin. Aluminum is included in the resin formula to obtain a better thermal conductivity. Another example is presented in figure 5 where it can be observed clearly all of the three materials.

ACKNOWLEDGMENT

This work was supported by the grant of the Romanian National Authority for Scientific Research and Innovation, UEFISCDI, project number 153/2020 "Components' assembly and interconnection through combined solderless technologies", COMPACT within P3 Programme - European and International Cooperation, Subprogramme 3.2 - Horizon 2020 ERANET.

REFERENCES

[1] J. Fjelstad, "Solderless assembly for Electronics. The SAFE approach", I-Connect007, BR Publishing, 2017.

[2] J. Izdebska, "Applications on printed materials", in Printing on Polymers. Fundamentals an Applications, pp. 371-388, 2016.

[3] R. Su, S. H. Park, Z. Li, M. C. McAlpine, "3D printed electronic materials and devices", in Robotic Systems and Autonomous Platforms. Advances in Materials and Manufacturing, pp. 309-334, 2019.

[4] J. Fjelstad, "Electronic assemblies without solder and methods for their manufacture", Patent, US20090008140 A1,Application number US 12/163,870, 8 January 2009.

[5] A. K. Krella, "Degradation of protective PVD coatings", in Handbook of Materials Failure Analysis with Case Studies from the Chemicals, Concrete and Power Industries, pp. 411-440, 2016.

Realization and Testing of a Supercapacitor, Pouch Type Cell

R. Negroiu, P. Svasta, Al. Vasile, C. Ionescu, M. R. Buga[1]

University "Politehnica" of Bucharest, Romania, Centre of Technological Electronics and Interconnection Techniques
[1] The National Research and Development Institute for Cryogenic and Isotopic Technologies – ICSI, Râmnicu Vâlcea
E-mail: rodica.negroiu@cetti.ro

Abstract— The supercapacitor, the relatively new passive component in the electronic field, has begun drawing attention to more and more users in terms of energy storage and delivery applications in a very short time. Researchers are trying to constantly improve and reduce a number of disadvantages for this component, such as: decrease in equivalent series resistance, increase in maximum working voltage or capacity. All these aspects can be improved only by making different variants of components, from different materials and choosing the variant with the best results.

For this, the purpose of this work is to present the realization a supercapacitor, Pouch type cell, by means of the equipment already used in the manufacture of batteries. As materials, will be used for electrodes, aluminum foil on which a porous material will be deposited, and as an electrolyte we will use an ionic liquid, known for its performance regarding much higher working voltage (up to 5V) compared to aqueous electrolytes (maximum 1.2V) or organic electrolytes (maximum 2.7V). Thus, we try to obtain a supercapacitor for which the working voltage exceeds the maximum value encountered so far for a single component and also based on the fact that we have all the information about the realization process we can draw important conclusions about the component made. In this way we can bring improvements to future components, by using other materials instead of those we consider problematic.

Keywords- supercapacitor, Pouch cell, ionic liquid, maximum voltage.

I. Motivation and Description of Work

We already know that the maximum working voltage has always been one of the drawbacks of using supercapacitors in different applications, because its low value of 2.5 V, 2.7 V (organic electrolyte) usually made it impossible to use a single capsule in systems operating at 5 V or 12 V. This is the reason why has always necessary to use a configuration, often mixed (series-parallel) in order to not lose the value of the capacity, but also to increase the value of the desired working voltage in that application. This involves the use of an energy storage system made of supercapacitors, which was robust, with significant weight and size. "Guilty" of this disadvantage of the maximum working voltage is the electrolyte, more precisely the penetration voltage of the various substances that enter its component [1]. For this, the realization of a supercapacitor that uses as electrolyte a material that allows a higher working voltage is a real challenge. Thus, from previous research, we found out that ionic liquid can solve this issue and we decided that together with the specialized team of The National Research and Development Institute for Cryogenic and Isotopic Technologies – ICSI, Râmnicu Vâlcea to make a supercapacitor, Pouch type cell, using an ionic liquid as electrolyte BMI.BF4 - 1-Butyl-3-methylimidazolium tetrafluoroborate ≥97.0% (HPLC) (Sigma-Aldrich). To make the electrodes we used the most common material, namely activated carbon, Carbon Black in a concentration of 95%, the remaining 5% being the binder PVdF - Polyvinylidene fluoride. The separator used was Celgard 2325 (three layers, PP-PE-PP, thickness 25 μm) [2].

As can be seen in the figure 1, the realization process of the supercapacitor, Pouch type cell involves a series of well-defined steps and a long time of realization. Following the realization process, we obtained the supercapacitor presented in figure 1, whose performances were identified following its rigorous testing.

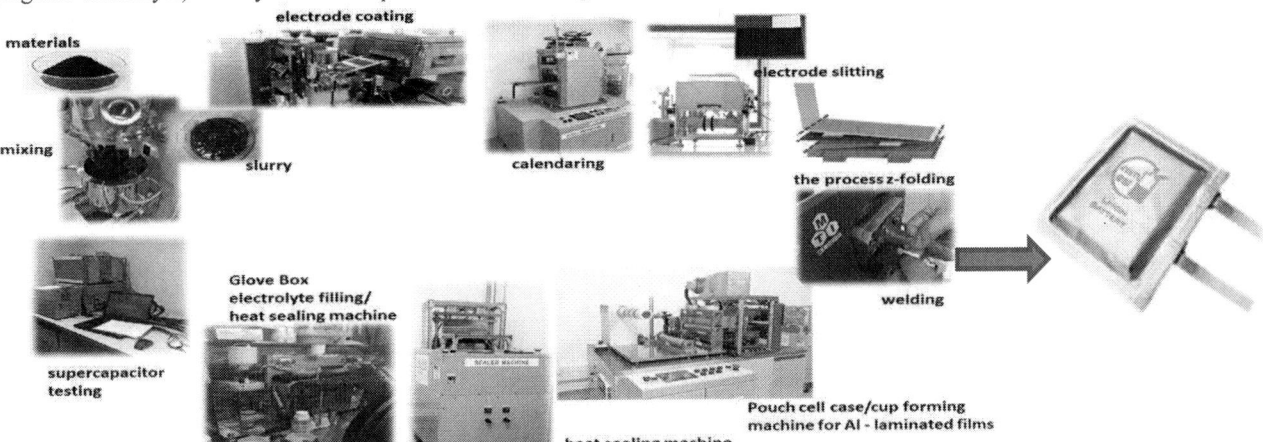

Fig. 1. The realization process of the supercapacitor, Pouch type cell and the cell obtained [3].

II. TEST METHOD

The testing of the Pouch type cell (supercapacitor) performed consisted in charging it with different values of the charging currents, gradually increasing the value of the maximum working voltage, starting with 10 μA, the value of the charging current and 20 mV the value of the voltage. We used this procedure in order to not exceed the maximum voltage threshold and to see what happens during the tests with the freshly made cell [5].

The table below specifies the values of the currents that have been applied to the supercapacitor terminals and the values of the operating voltages for which the cell has been tested.

TABLE I. THE VALUES OF CURRENTS AND VOLTAGES AT WHICH THE SUPERCAPACITOR HAS BEEN TESTED.

I [mA]	0,01	0,1	0,1	0,1	0,2	1	10	1
U [V]	0,02	0,2	0,5	0,75	1	2	2	3

All tests were performed and monitored with the help of high-performance equipment, Solartron Analytical, from the National Research-Development Institute for Cryogenic and Isotopic Technologies - ICSI, Râmnicu Vâlcea (figure 2). As capabilities, this equipment works at values of nanoamps currents, which is useful for my research, because the use of a charging current as low as possible gave me the security of protecting the supercapacitor made. 8 components could be tested simultaneously, given the fact that this equipment had 8 channels, of which on 4 of them we had the opportunity to perform Electrochemical Impedance Spectroscopy. As maximum limits, it could operate up to a maximum voltage of 10 V and a maximum current of 4 A. Experimental data were stored and tabulated using the software in the equipment, Cell Test v. 5. 4. 4. [3]

Fig. 2. Solartron Analytical system used to test the Pouch cell.

III. EXPERIMENTAL RESULTS

The results of this research are based on various experiments that led to the determination of the main parameters as follows:

III.A. Potential monitoring 0 V

The newly made Pouch type cell was subjected to its first test, namely a first monitoring at 0 V in time. In the two images it can be seen that most likely the existing voltage (of very low value) is caused by the thermal agitation that occurs in the cell [6]. Tests were performed in both directions because at that time the polarity of the cell was not established, we did not define an anode and a cathode, given the fact that the two electrodes were identical from a constructive point of view. After performing of these two tests I decided conventionally and marked its polarity on the cell.

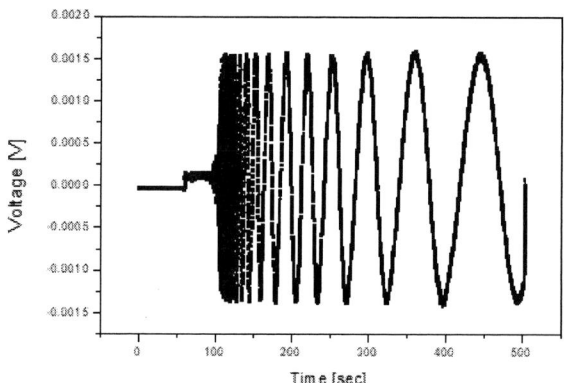

Fig. 3. The first monitoring of the Pouch cell at 0V in one direction.

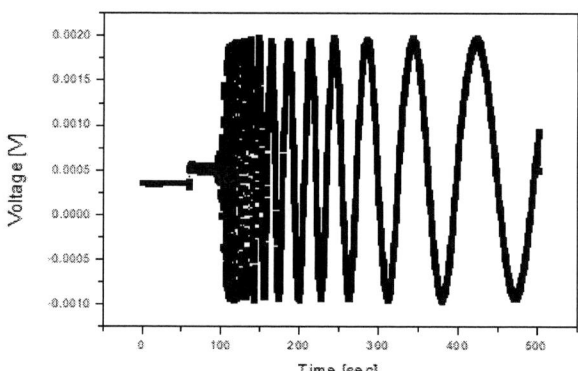

Fig. 4. First monitoring of the Pouch cell at 0V in reverse.

III.B. Self-discharge monitoring in different cases

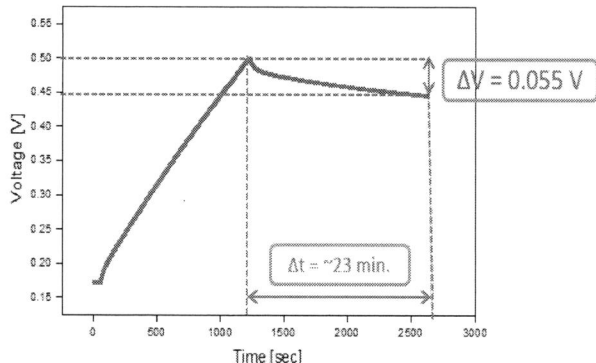

Fig. 5. Monitoring the charge and self-discharge of the Pouch type cell at 500 mV, 100 μA.

Figure 5 exemplifies charging with a current of 100uA up to a maximum voltage of 500mV, charging that it is achieved in about 20 minutes. After a 23-minute open-circuit monitoring of the cell, a self-discharge of 55mV is observed at its terminals. This value represents approximately 10% of the total value of the maximum voltage of 500mV applied to the terminals of our supercapacitor. From the calculations made, most of the self-discharge occurs in the first 5 minutes, respectively 30 mV.

One of the tests performed on the Pouch type cell was charging with a current of 10 mA up to a maximum voltage of 2 V. After performing this test we found that the supercapacitor had a very fast charge and self-discharge, which indicates that from the point of view of the conduction mechanism, the value of the charging current is too high and the double layer of load did not start to form and the ions in the electrolyte solution were not adsorbed by the charges of opposite sign inside the electrode (figure 6) [7].

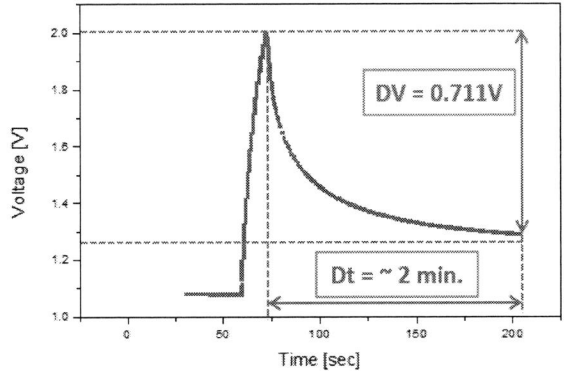

Fig. 6. Monitoring the charge and self-discharge of the Pouch type cell at 2V, 10mA.

As a result, for the next test we decreased the value of the charging current to 1 mA keeping the maximum voltage value of 2 V. As can be seen in figure 7, the charging of the cell was performed in an optimal time. In the case of this test we obtained a self-discharge of 0.613 V after about 12 hours and 30 minutes. So, according to the graph, it can be seen that in the first 30 minutes our supercapacitor "suffered" a self-discharge of about 50% of the total value recorded in the 12 hours and 30 minutes, and in the first 5 minutes of self-discharge monitoring the cell recorded a loss in voltage of approximately 0.258 V.

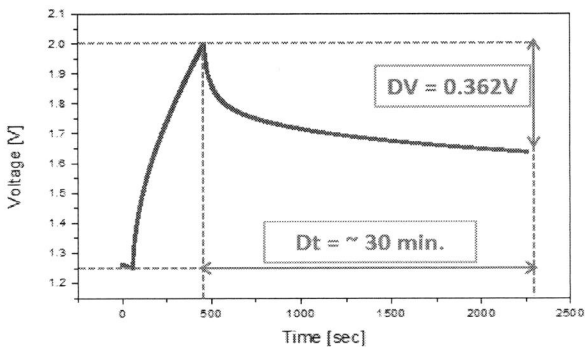

Fig. 7. Monitoring the charge and self-discharge of the Pouch type cell at 2 V, 1 mA.

The graph below represents the charge of the supercapacitor up to the maximum voltage of 3 V keeping the current value at

1 mA in order not to encounter unpleasant situations as in the previous case (2 V, 10 mA) when we recorded a fast charge and self-discharge of the cell .

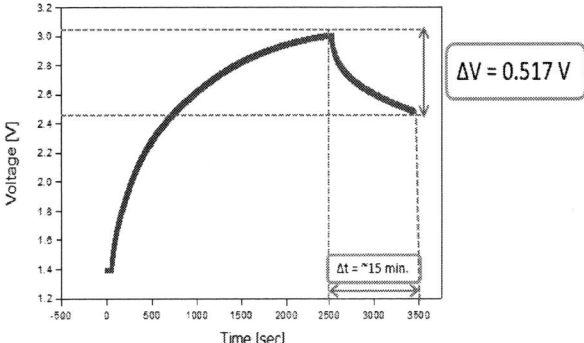

Fig. 8. Monitoring the charge and self-discharge of the Pouch type cell at 3V, 1mA

This is the last test performed on increasing the value of the maximum charging voltage of the supercapacitor because we did not want to be put in the situation of destroying the cell made.

III.C. *Charging the supercapacitor to different working voltage values*

The figure below exemplifies the charging of the supercapacitor at several voltage values, respectively 500mV, 750mV, 2V, 3V. As can be seen, the shape of the graph V (t) takes the form of charging an ideal supercapacitor barely at the voltage value of 3 V. This aspect explains the ability of the electrolyte used, being an ionic liquid, to allow charging with a voltage greater than or equal to 3 V. In order not to destroy the cell, through the process of electrolytic decomposition, we did not increase the voltage value, given that the financial value of the electrolyte used is significant.

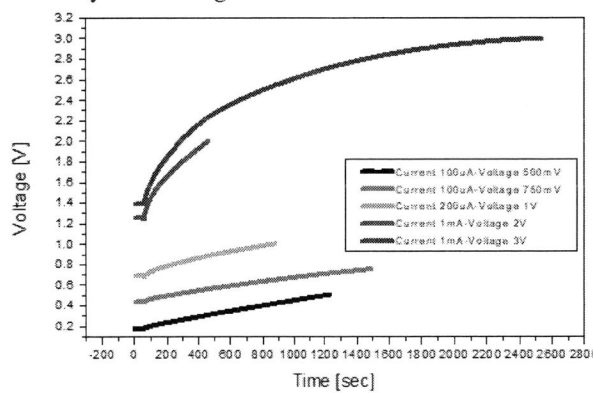

Fig. 9. Charging the supercapacitor to different voltage values.

III.D. *Determination of equivalent series resistance (ESR)*

The table below summarizes the equivalent series resistance (ESR) values calculated using equation 1. It can be seen that, in principle, the value of the equivalent series resistance increases with increasing voltage value. This has a very high value, of the order of ohms, which leads to a significant self-discharge in voltage, so you can seen in the table where specified the self-discharge value after 5 minutes for each case tested.

TABLE II.   VALUES OF EQUIVALENT SERIES RESISTANCE AND SELF-DISCHARGE VOLTAGE AT DIFFERENT CURRENTS AND CHARGING VOLTAGES

Current [mA]	0,01	0,1	0,1	0,1	0,2	1	1
Voltage [V]	0,02	0,2	0,5	0,75	1	2	3
ESR [Ω]	1,81	5,34	5,95	7,63	11,59	10,26	9,76
Self-discharge (after approx. 5 min.) [V]	0,0018	0,019	0,0301	0,042	0,079	0,258	0,354

$$ESR = \frac{U_1 - U_2}{I} \qquad (1)$$

Where U_1 represents the maximum charging voltage and U_2 represents the first value of the voltage from the moment the self-discharge process begins (when the supercapacitor is left in open circuit). Basically, at that moment there is a sudden drop in voltage. I represents the charging current, value used for each case tested separately.

III.E. *Determining the capacity value of the Pouch type cell*

In order to see another gain in the realization of my cell, I should, in addition to the value of the supported working voltage, also find out the value of its capacity. For this process we graphically represented the negative of the imaginary part of the impedance as a function of the inverse frequency (figure 10), data obtained from the electrochemical impedance spectroscopy process when the supercapacitor was charged up to a maximum voltage of 3 V with a charging current of 1 mA. Next, using equation 2 we calculated the value of the cell capacity. For this, I noticed that the graph -Z "(1 / f) is a straight line with a slope of 1 / C. From the calculations we obtained a capacity value of 95.76mF. I mention that for the calculation of the capacity we took into account the values of the negative part imaginary corresponding to times of 200 ms (5 Hz frequency) and 0.02 ms (50 kHz frequency) [4].

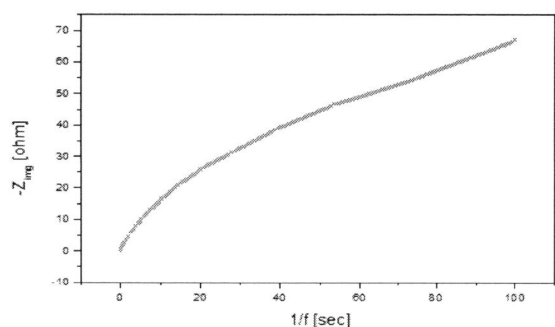

Fig. 10. Graph of the impedance negative imaginary part as a function of the inverse of the frequency.

$$-Z" = \frac{1}{\omega C} \qquad (2)$$

Where ω is angular frequency and $\omega = 2\pi f$.

IV. CONCLUSIONS

1. Obtaining a low capacity value, of the order of mF, can have several causes:
   - ➢ The porosity of the obtained electrode should be too high or too small and thus create gaps or micropores, preventing the access of ions from the electrolyte solution and thus preventing the adsorption and creation of the double charge layer;
   - ➢ The electrolyte solution, being an ionic liquid, has the consistency of a gel with a low ionic conductivity at room temperature impending the mobility of ions and reducing their access to the opposite sign charges at the electrode, thus leading to a low value of capacity.

2. The equivalent series resistance obtained has a very high value, of the order of Ohms due to the low ionic conductivity.

3. In all cases tested on the supercapacitor, most of the self-discharge of the voltage accumulated at its terminals occurs in the first 5 minutes, after these 5 minutes the value of self-discharge registered is lower.

4. Obtaining a supercapacitor that operates at a maximum working voltage of 3 V proved to be an important gain, given the fact that most supercapacitors on the market operate at a working voltage of maximum 2.7 V.

ACKNOWLEDGMENT

The authors of this paper are thankful to Dr. Phys. Mihai VARLAM, general manager of ICSI, Râmnicu Vâlcea for offering high performance measurements conditions.

The work was supported by the Pre-seed project of the Ministry of Education, "Dezvoltare integrată IDev 4.0", MySMIS code 122386.

REFERENCES

[1]   Conway, B. E., "Electrochemical Supercapacitors: Scientific Fundamentals and Technological Applications", Kluwer Academic / Plenum Publishers, New York, 1999.

[2]   Aiping Yu, Victor Chabot, and Jiujun Zhang, Electrochemical Electrochemical Supercapacitors for Energy Storage and Delivery Fundamentals and Applications, CRC Press, 2013.

[3]   https://www.icsi.ro/

[4]   RA Dobre, AE Marcu, M Stanciu, M Vladescu., „Spectroscopic investigation of transparent polylactic acid", Journal IOP Conference Series: Materials Science and Engineering, Volume 572, pp. 012015, Publisher IOP Publishing.

[5]   Ciprian Ionescu, Andrei Drumea, „Extended Current Range of Active Balancing and Monitoring Circuits for Supercapacitor Modules", 2019 IEEE 25th International Symposium for Design and Technology in Electronic Packaging (SIITME 2019), Cluj-Napoca, România, 23-26 octombrie 2016, pp. 272-277.

[6]   Vasile, A,Vasile, I, Ionescu, C,Drumea, A "Trends in fiber optics applications for automotive industry" 2007 30TH INTERNATIONAL SPRING SEMINAR ON ELECTRONICS TECHNOLOGY, International Spring Seminar on Electronics Technology, Pages: 482-486, DOI: 10.1109/ISSE.2007.4432904, Cluj Napoca, ROMANIA.

[7]   Mihaela Pantazica, Andrei Drumea, Cristina Marghescu, "Analysis of self discharge characteristics of electric double layer capacitors" 2017 IEEE 23rd International Symposium for Design and Technology in Electronic Packaging (SIITME), Pages: 90-93, DOI: 10.1109/SIITME.2017.8259864, Constanta, Romania.

Data Mining System Architecture for Industrial Internet of Things in Electronics Production

Reinhardt Seidel, Hassan Amada, Jonathan Fuchs, Nils Thielen, Konstantin Schmidt, Christian Voigt, Jörg Franke

Institute for Factory Automation and Production Systems
Friedrich-Alexander-University Erlangen-Nürnberg
Nürnberg, Germany
reinhardt.seidel@faps.fau.de

Abstract— Data collection and Machine Learning (ML) have already become reality in industrial applications. Also in electronics manufacturing, some successful approaches for the application of ML have been reported. However, for industrial applications, the infrastructure and the respective analysis-models for such approaches need to cope with the occurring data flow and structured storage in production. This contribution presents a cross-vendor data mining infrastructure setup that allows real-time tracking and structured storage of process data during operation in order to complement the material tracking of manufacturing execution systems (MES). Hence, the developed data mining system is the essential basis for real-time prediction use cases of ML in manufacturing.

Keywords—Machine learning; Smart electronics manufacturing; Smart electronic module assembly; Data analytics; Electronics manufacturing

I. Introduction

Modern electronic module assembly faces several challenges including mass customization, an increasing need for cost efficiency to stay competitive in international markets and holistic traceability of products and their components throughout the entire life cycle. In the case of early failure, the question for liability has to be answered which is an important task of the manufacturing execution system (MES) [1]. MES gathers data for many purposes on process control level including material traceability and in some cases quality relevant process data [1]. However, neither are process data archived utterly and structured nor analyzed in real-time and used for process optimization to fully exploit their potential. Although vendor-specific solutions exist for closed-loop control and process optimization cross-vendor machine protocols are not widespread yet. For both, traceability and cost efficiency, gathering and analyzing all available process data promises significant potential for optimization [1–3].

In literature, many potential use cases have been identified and applied on experimental basis. For an operative application of such use cases, a reliable data mining system (DMS) is necessary [3]. In the context of this contribution, the term data mining describes the process of gathering, parsing and storing data in a structured way. For the subsequent utilization of such data, Machine Learning (ML) algorithms can be developed to predict future events based on the mined data and derived patterns. Following the ML use case identification methodology presented by [4], use cases such as predictive maintenance in

solder paste printing (SPP), reduction of false calls with the inspection result from automated optical inspection (AOI) systems and prediction of quality on the basis of solder joint inspection, require process data [5–7].

As a solution to this problem and as an addendum to MES data, this contribution presents an initial brownfield approach to provide an infrastructure setup that allows real-time tracking and structured storage of all process data during production without a need for replacing existing MES tracing in industry. This tackles occurring high data volumes and inhomogeneous data supply caused by non-standardized machine vendor protocols in brownfield as one main challenge to the industrial application of ML [1, 3].

II. State of the Art in Data Acquisition

A. Data requirements for Machine Learning use cases in manufacturing

General requirements of data homogeneity, completeness and volume for ML also apply in surface-mount technology (SMT) manufacturing. Explicit application-specific requirements vary with the corresponding use cases [8]. Further challenges arise from the necessary hierarchical separation of data in pad, component, subpanel and board levels. This leads to challenges for data parsing. Within the inspection routines, care must be taken to ensure coherent identification as can be seen in Fig. 1, especially on component and pad level.

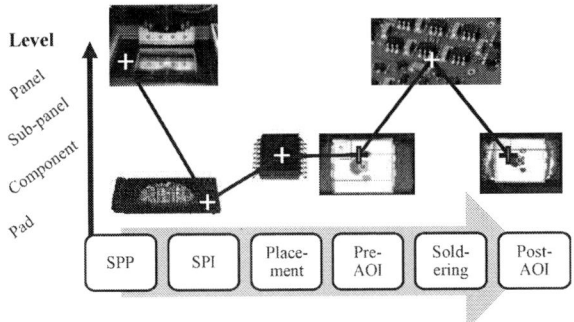

Fig. 1. Hierarchical levels of process and inspection data for the machines in SMT manufacturing; solder paste printing (SPP), solder paste inspection (SPI), placement, soldering and pre-/post reflow soldering automated optical inspection (AOI)

A special focus lies on the conformity of both machine and process data. To prevent risks for incorrect data preparation information, e. g. timestamps, have to be compatible with each other. Hence, system times have to be synchronized. Image data has to be available without artifacts if used for model input in order to avoid the Clever Hans behavior [9]. Otherwise, numerical data can be taken from inspection machines [7]. Furthermore, saving or streaming image data must under no circumstances lead to an increase in the cycle times in industrial application where infrastructure is usually not prepared for high data volume potentially caused by images and processes with a high number of steps, such as component placement for the whole board [1, 3].

In addition to the described technical data quality requirements, the data has to describe the manufacturing process statistically significant. These relevant process influencing factors must be fully identified and recorded with high sensor accuracy [10].

B. MES and cloud-edge computing

MES represent manufacturing from the perspective of separated process steps and are used to trace material and products. They also fulfill control purposes in manufacturing even by process interlocking. As a result, production processes in general and information on key production measures like overall equipment efficiency (OEE) can be displayed in real-time [1].

The MES is a computer-controlled system that optimizes the monitoring and control of industrial processes. MES software integrated into an information system completes the Enterprise Resource Planning (ERP) System, Product Lifecycle Management (PLM) and Supply-Chain Management (SCM) by giving flexibility and agility to an industrial company that wants to increase its responsiveness to market developments. The VDI 5600 defines MES using 10 key functions: Order management, Data acquisition and processing, Quality Management, Performance Analysis, Energy Management, Information Management, Detailed planning and control, Personnel Management, Resource Management, Materials Management [1, 11]. Often the data structure of MES system is timestamp ordered which is suitable for time series-based analysis but challenging in prediction use cases where tracking of process data on product feature level is required.

To handle these huge amounts of data as well as training and optimizing ML models, cloud computing is feasible. Cloud computing is a general term for the provision of on-demand resources and services over the internet. [3] The National Institute of Standards and Technology (NIST) defined cloud computing as "a model for enabling ubiquitous, convenient, on-demand network access to a common pool of configurable computing resources (e.g., networks, servers, storage, applications, and services) that can be rapidly provisioned and shared with minimal management or service provider interaction". [1, 12] This technology provides new opportunities for business models and IT software in manufacturing [12].

Edge computing is a new computing infrastructure that enables integration of basic functions such as networking, data processing and storage close to the equipment on the shop floor. The use of edge computing meets real-time requirements for easy intelligent production and increases network agility and

security [13]. Edge computing emerged with the proliferation of Content Distribution Networks (CDNs), which aim to accelerate web performance by caching web content on edge nodes near the user [14].

C. Infrastructure retrofit in brownfield applications

In context of Industrial Internet of Things (IIoT) and the advancing digitalization, subsequent adaptation of industrial plants to implement the IIoT paradigms plays a special role and is at the same time a major challenge. Although the greenfield scenario represents the ideal for IIoT projects, the brownfield approach is often the reality. This means that old machines have to be enabled to cope with new IT infrastructures and modern industrial communication patterns and protocols. This includes the integration of heterogeneous machinery in a way that homogeneous exhaustive sets of data can be gathered for the application of Machine Learning. However, the effort to integrate older existing assets, also in the context of electronics production, into ubiquitously communicating asset networks is worthwhile. [15, 16]

For the subsequent digital transformation, also called retrofitting, two essential steps must be taken: First, the process running on a production line must be digitally mapped entirely and with great care. Errors or gaps are also transferred from the analog to the digital world and can often only be corrected afterwards with a great deal of effort. In addition, brownfield plants must be networked holistically to enable the required horizontal and vertical integration. Vertical integration describes the transport and integration of machine and process data from the control level into edge- and cloud-based environments.

The integration of communication and computing power allows:

- comprehensive information exchange with IT systems,

- simplified data acquisition without intervention in the control hardware,

- and the provision of plant and process data for downstream services based on standardized interfaces that enable seamless data processing.

The key question of retrofitting is with which sensor or IT component equipment must be retrofitted to be able to exploit the potential of digitalization. The integration of comprehensive and not exclusively function-bound components for extended data acquisition is essential to optimize production processes, develop new data-driven business models and create innovative and smart products and services for customers.

The conversion is also worthwhile for older systems. An example in the context of electronics production is optical inspection. This inspection method generates high-resolution images of soldered circuit boards. With the help of these images, e.g. missing, incorrectly assembled, twisted, or insufficiently soldered components can be identified. Manual evaluation is time-consuming and error-prone, but retrofitting a modern, automated system is associated with high costs. In the course of retrofitting, the resulting images can be automatically forwarded to downstream Smart Services, which use Machine Learning methods for evaluation and assessment.

Fig. 2 shows an example of a system architecture for an automated optical inspection system using software-based services without human involvement. In this example, images are captured by the Imaging Service and communicated to subsequent services via a message-oriented middleware (MOM). The training environment uses the high computing power of the cloud, whereas the ML-based evaluation of the images takes place in the edge, close to the equipment.

The example shown also makes a major hurdle of service-based automation obvious: Often large amounts of data are generated that need to be evaluated. In the AOI, the high-resolution images of a single process comprise several gigabytes for a single panel. For secure and loss-free transmission, efficient data structures and well-planned communication concepts are required. [17]

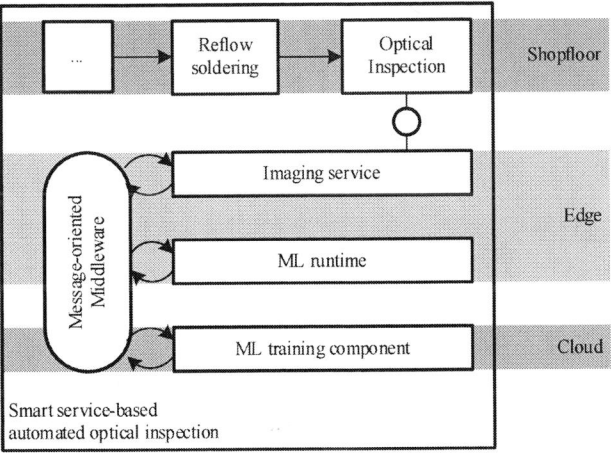

Fig. 2. System architecture for an automated optical inspection system based on software services without human involvement

III. DEVELOPMENT OF A DATA MINING SYSTEM ARCHITECTURE

The DMS infrastructure consists of a cross-machine vendor connectivity solution in brownfield and a SQL-based data storage and parsing schema.

A. Setup of a client-server infrastructure with sockets

The basic requirement for the operation of automatic machine data acquisition is seamless networking. For this purpose, IP networking is established in industry. Devices and sensors must therefore be equipped with systems that enable access to the network. [18] The Transmission Control Protocol/Internet Protocol (TCP/IP) is generally used to establish a network connection between the main computer and the systems. [19]

The transmission of the data is enabled based on the client-server application. This is initially done by process machines and the line computer. The networking communication is carried out using LAN connections based on the TCP/IP protocol. This communication is broken down into data packets with electronic address labels that identify the IP address. Fig. 3 shows the UML sequence diagram of the DMS and its components between one client process machine and the SMT PC which incorporates two services running in parallel and which connect the machine to the SQL Server. The client-server model enables data to be

exchanged between one or more clients and a server. Here, the server can be defined as a service provider offering a client with specific functionality and only waits passively for a request from a client. The client can be referred to as a service user who is actively requesting a service from the server. After request, it waits for the server to provide the requested service. The communication between the two units must work asynchronously as they run independently of each other.

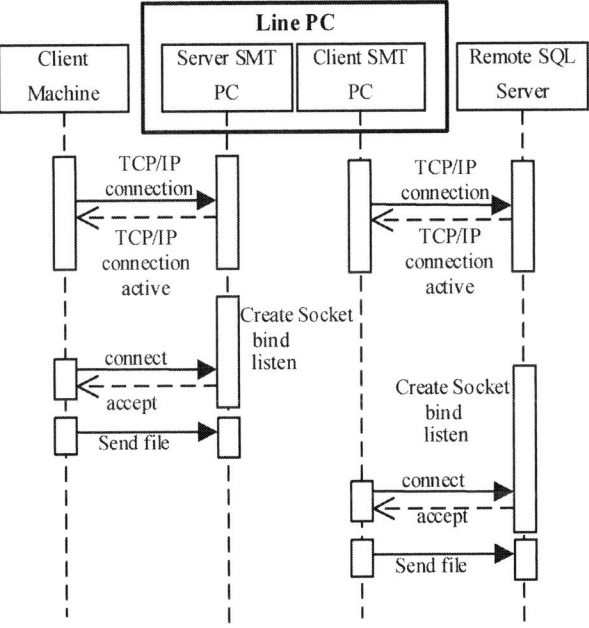

Fig. 3. UML sequence diagram of the full data mining system applied in brownfield.

An application programming interface (API) is required to implement the client/server model. The API most commonly used for creating client/server software is the Socket API, which is intended to allow easy access to the functions of the protocol stack for sending and receiving data. Two sockets are required for data exchange between two software entities. [20]

The protocol guarantees the complete transmission of the data. For this purpose, the gateway provides a socket as a server that accepts the communication partner as the client and starts sending in cycles. Communication in ethernet networks is therefore synchronous [21]. The API corresponding to the TCP should fulfill the following tasks [22]:

- Establishing and closing data transmission connections
- Simultaneous sending and receiving of data
- Connection health checks to detect traffic congestion on the network.
- Preparation and intermediate storage of the data blocks to be transferred
- Data backup

Fig. 4 shows the entire DMS of the SMT process line. Every process module is connected to the central SMT Line PC. By this means in brownfield scenarios even older modules that shall

2020 IEEE 26th International Symposium for Design and Technology in Electronic Packaging (SIITME)

not be connected directly to the internet for security reasons can be integrated. The SMT line PC watches defined file folders on the process modules and copies the files into its directory. Afterwards, data can be pre-processed and ML use cases can be carried out on the SMT PC allowing short latency. From there the data packages are sent to the remote SQL database. On top, a performance dashboard is implemented. From there further less time-critical use cases and long-term analysis can be performed.

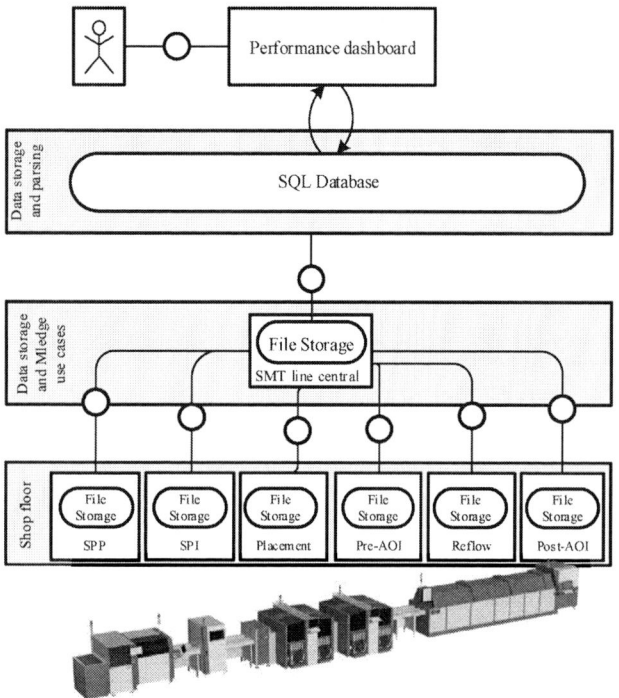

Fig. 4. FMC diagram of the implemented DMS infrastructure

B. Data parsing and storage

Structured data storage and parsing are essential for applications dealing with large amounts of data in smart manufacturing [23]. The data generated in circuit board production are mostly structured or semi-structured. For the automatic storage and quick retrieval of this data, databases provide decisive advantages. The data are stored in a SQL database through a database provider such as MySQL. The MySQL server makes it possible to create several databases on the database management system. Several tables can be created within a database [24]. ML use cases can then be integrated and allow analysis by pattern recognition and the application of predictive models based on the data recorded which e.g. can be used to improve productivity and reduce costs. [2, 25]

C. Task distribution in edge-cloud systems

Under the assumption that complete and coherent data is available, ML use cases can be defined to fully exploit their potential in the production environment. The systematic approach for use case identification in [1] bases on a categorization of use cases regarding the task in terms of description, diagnostics, prediction and prescription. Along the

life cycle and especially the process chain of electronic module assembly, process step internal and cross-process step use cases can be exhaustively identified. This method in cooperation with the domain experts allows systematic identification of the full scope of relevant process data.

However, the crucial aspect of where to perform the data preparation and analysis, model training and maintenance, and the prediction have to be solved. The operation of analytics use cases in the industrial environment, requires the consideration of the three main sub-processes, regarding data handling:

- Collection, storing and processing of data
- Training and evaluation of models
- Deployment, operation and maintenance of models

Distribution of according sub-processes and individual assessment is required. The distribution of these subcategories between machine-oriented edge processing and high-performance on-demand services depends on several factors, categorized into network-, data- and task-related issues, as depicted in Fig. 5.

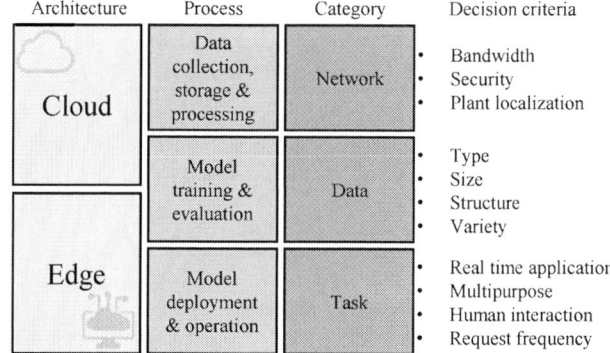

Fig. 5. Processes, categories and critical decision criteria to be considered during the distribution of ML use cases in a cloud-edge environment.

The allocation of the data collection process depends not only on the structure of the data but also on the scope of the application areas (multipurpose). As the training and evaluation of the model in a cloud environment are feasible with limited resources, regarding bandwidth and widely scattered operating locations, the model operation would be unfavorable under the named circumstances. When it comes to ML use cases demanding real-time processing and feedback, a machine-oriented setup is commonly superior. As the quantity of the use cases expands and scaling effects occur, a gradual shift of the sub-processes towards the cloud is conceivable. This is why large companies favor in-house solutions while small- and mid-sized companies rely on standard applications as scaling effects do not come to bear [26].

D. Influence of data processing with DMS

For the analysis of data properties, a sample data set containing about 7000 panels of different product types was prepared. Since MES data in electronics production is mainly timestamp-based for single events and saved in SQL databases, the data volume can be massively reduced by data restructuring

978-1-7281-7507-2/20 $31.00 © 2020 IEEE   78   21-24 October, Pitesti, Romania

either in SQL or other formats. The information density of machine results such as SPI or component placement surpasses the MES as can be seen in TABLE 1.

TABLE 1. Share of the MES and process data volume of the individual hierarchy levels before and after structured storage

Hierarchical Level	Subpanel/ Panel	Component	Pad
Before cleansing	30%	60%	15%
After cleansing	5%	45%	50%

By data cleansing, the average amount of numerical data per panel including components and pads can be reduced from about 2 MB to less than 1 MB. Both the data distribution and the data volume are only to be understood as an orientation. Due to the high dependency on the design of the electronic assembly as well as the IT-infrastructure of individual production lines, generally valid statements on image data can hardly be done and are not taken into account for this analysis.

E. Challenges in flexible matrix production

With the rising demand for individualized solutions for customer-specific issues and the subsequent rise of engineering to order (OTE) processes many manufacturing sectors have been facing a novel form of demand. Traditional linear production systems can face various difficulties in fulfilling this demand. One of these particular difficulties is the corresponding increase in product variances. In electronics manufacturing mixed SMT-THT assembly is still state of the art especially in high mix product ranges. But also partially equivalent SMT products have to go through two different lines or through one line which provides enough feeders for every product.

Using the example of industrial computer systems, customers may request various processors, expansions, connectors or storage options. Consequently, the single product of Motherboard Type A requires several subtypes with varying parts lists. Each product variation requires different mounting programs, a different stencil and various adjusted thresholds for SPI, AOI, automated x-ray inspection (AXI) and flying probe testing. [27, 28]

To combat these forms of waste as a key aspect of lean manufacturing, other industries, especially automotive manufacturing, have already adopted matrix production, in electronics manufacturing, there are several hurdles to overcome to cope with the increased degrees of freedom by complementing MES with the DMS for process data tracking in brownfield. [29]

The necessary data infrastructure in such a perceivable matrix production ought to be separated into manufacturing data, with timestamps, process parameters and mounting positions, testing information such as solder joint form, variable thresholds and testing depth and material infrastructure information that relates to the movement of assemblies, lots, component reels and printing stencils.

When considering this additional dimension it is necessary to distinguish between machines and processes that do require external setup, such as stencil printing, and those that only need the correct production program file for the assembly that is currently being processed. To facilitate the desired improvements on productivity and especially reduction in bottleneck processes accurate correlation of product specifications, location and program data has to be provided at every production and testing step in real-time. Machinery that requires a more extensive setup must be considered a part of a more complex system rather than a standalone machine. Products of similar component mixture must be mounted within a single or a predetermined set of mounting machinery in order to prevent frequent device reel changes.

Therefore, an optimized, reliable and efficient matrix production system in electronics manufacturing requires deep integration of DMS and MES as well as real-time capabilities to allow for conditional routing adjustments eliminating the viability of isolated data management systems segregated by traditional conceptions of development and production. [3, 28, 30, 31]

IV. CONCLUSION AND OUTLOOK

One of the main challenges in the economic use of ML in production is the lack of data availability and homogeneity. Gathering data requires adequate data sources, infrastructure, know-how and methods to exploit the full potential of data.

In this contribution, a data mining system (DMS) is the basis for the brownfield integration of a cross-vendor machine shop floor with SMT line process machines of different ages and type into one connected IT infrastructure. This DMS allows the homogeneous gathering, structured storage and in-time exploitation of data with ML models.

This plug and play system provides the infrastructural precondition to transfer ML use cases from the experimental stage into a real industrial application. Especially in small and midsized companies quick and low-effort applications are required to grain productivity in brownfield applications.

However, future work can be done to streamline the multiple communication channels towards a message-oriented middleware (MOM). [32] Alternatively, the IPC-CFX is an emerging standardization initiative for a world-wide industrial IoT communication standard for assembly manufacturing which is increasingly focused by electronics manufacturing machine vendors.

ACKNOWLEDGMENT

The authors want to thank the Bavarian Ministry for Economy, Media, Energy and Technology (StMWi) and VDI/VDE-IT for funding this research.

REFERENCES

[1] J. Kletti, *MES - Manufacturing Execution System*. Berlin, Heidelberg: Springer Berlin Heidelberg, 2015.

[2] M. Liukkonen, E. Havia, and Y. Hiltunen, "Computational intelligence in mass soldering of electronics – A survey," *Expert Systems with*

Applications, vol. 39, no. 10, pp. 9928–9937, 2012, doi: 10.1016/j.eswa.2012.02.100.

[3] O. Givehchi and J. Jasperneite, "Industrial Automation Services as part of the Cloud: First Experiences," in *Jahreskolloquium Kommunikation in der Automation - KommA, Magdeburg*. [Online]. Available: https://www.researchgate.net/profile/Juergen_Jasperneite/publication/257402460_Industrial_Automation_Services_as_part_of_the_Cloud_First_experiences/links/0deec529371a029616000000/Industrial-Automation-Services-as-part-of-the-Cloud-First-experiences.pdf

[4] R. Seidel, A. Mayr, F. Schafer, D. Kisskalt, and J. Franke, "Towards a Smart Electronics Production Using Machine Learning Techniques," in *International Spring Seminar on Electronics Technology, ISSE 2019 – 2019 42th International Spring Seminar*, 2019.

[5] K. Schmidt, J. Bönig, G. Beitinger, N. Thielen, and J. Franke, "An approach to quality prediction using intelligent SMT solder joint inspection through the use of Machine Learning," in *EBL 2020*.

[6] O. Krammer, T. Al-Ma'aiteh, P. Martinek, K. Anda, and N. Balogh, "Predicting the Transfer Efficiency of Stencil Printing by Machine Learning Technique," in *International Spring Seminar on Electronics 2020*.

[7] N. Thielen, D. Werner, K. Schmidt, R. Seidel, A. Reinhardt, and J. Franke, "A Machine Learning Based Approach to Detect False Calls in SMT Manufacturing," in *International Spring Seminar on Electronics 2020*.

[8] T. Wuest, D. Weimer, C. Irgens, and K.-D. Thoben, "Machine learning in manufacturing: advantages, challenges, and applications," *Production & Manufacturing Research*, vol. 4, no. 1, pp. 23–45, 2016, doi: 10.1080/21693277.2016.1192517.

[9] P. Schramowski *et al.*, "Making deep neural networks right for the right scientific reasons by interacting with their explanations," *Nat Mach Intell*, vol. 2, no. 8, pp. 476–486, 2020, doi: 10.1038/s42256-020-0212-3.

[10] T. Wuest, C. Irgens, and K.-D. Thoben, "An approach to monitoring quality in manufacturing using supervised machine learning on product state data," *J Intell Manuf*, vol. 25, no. 5, pp. 1167–1180, 2014, doi: 10.1007/s10845-013-0761-y.

[11] S. Lee, S. J. Nam, and J.-K. Lee, "Real-time data acquisition system and HMI for MES," *J Mech Sci Technol*, vol. 26, no. 8, pp. 2381–2388, 2012, doi: 10.1007/s12206-012-0615-0.

[12] X. Xu, "From cloud computing to cloud manufacturing," *Robotics and Computer-Integrated Manufacturing*, vol. 28, no. 1, pp. 75–86, 2012, doi: 10.1016/j.rcim.2011.07.002.

[13] Jakob Zietsch, Nils Weinert, Christoph Herrmann, Sebastian Thiede, "Edge Computing for the Production Industry - A Systematic Approach to Enable Decision Support and Planning of Edge,"

[14] L. F. Bittencourt, M. M. Lopes, I. Petri, and O. F. Rana, "Towards Virtual Machine Migration in Fog Computing," in *2015 10th International Conference on P2P, Parallel, Grid, Cloud and Internet Computing (3PGCIC)*, Krakow, Nov. 2015 - Nov. 2015, pp. 1–8.

[15] J. Krüger and A. Verl, Eds., *RetroNet - Retrofitting von Maschinen und Anlagen für die Vernetzung mit Industrie 4.0 Technologie*. Düsseldorf: VDI Verlag, 2019.

[16] T. Lins and R. A. R. Oliveira, "Cyber-physical production systems retrofitting in context of industry 4.0," *Computers & Industrial Engineering*, vol. 139, p. 106193, 2020, doi: 10.1016/j.cie.2019.106193.

[17] J. Fuchs, J. Schmidt, K. Rehman, M. Sauer, S. Karnouskos, and J. Franke, "I4.0-compliant integration of assets utilizing the Asset Administration Shell," in *2019 24th IEEE International Conference on Emerging Technologies and Factory Automation (ETFA)*, 2019, pp.

1243–1247. [Online]. Available: https://ieeexplore.ieee.org/document/8869255

[18] "Maschinendatenerfassung: Grundlagen, Konzepte und Lösungen,"

[19] ASMAS GROUP, "DEK TECHNICAL REFERENCE MANUAL VOLUME 4," 2016.

[20] S. Heinzl and M. Mathes, "Design von Client/Server-Software," in *Middleware in Java: Leitfaden zum Entwurf verteilter Anwendungen — Implementierung von verteilten Systemen über JMS — Verteilte Objekte über RMI und CORBA*, S. Heinzl and M. Mathes, Eds., Wiesbaden: Vieweg+Teubner Verlag, 2005, pp. 71–111. [Online]. Available: https://doi.org/10.1007/978-3-322-80262-0_4

[21] P. Reboredo, "Laufzeitvalidierung einer Plattform zur semantischen Integration von Feldgerätedaten," in *Industrie 4.0 und Echtzeit*, Berlin, Heidelberg, 2014, pp. 81–90.

[22] R. Lerch, "Vernetzung von Messdatenrechnern (Industrie-LAN, WAN)," in *Elektrische Messtechnik*, R. Lerch, Ed., Berlin, Heidelberg: Springer Berlin Heidelberg, 2016, pp. 645–679.

[23] Y. Cui, S. Kara, and K. C. Chan, "Manufacturing big data ecosystem: A systematic literature review," *Robotics and Computer-Integrated Manufacturing*, vol. 62, p. 101861, 2020, doi: 10.1016/j.rcim.2019.101861.

[24] M. Seibold and A. Kemper, "Database as a Service," *Datenbank Spektrum*, vol. 12, no. 1, pp. 59–62, 2012, doi: 10.1007/s13222-011-0075-1.

[25] A. Viloria, G. C. Acuña, D. J. Alcázar Franco, H. Hernández-Palma, J. P. Fuentes, and E. P. Rambal, "Integration of Data Mining Techniques to PostgreSQL Database Manager System," *Procedia Computer Science*, vol. 155, pp. 575–580, 2019, doi: 10.1016/j.procs.2019.08.080.

[26] D. Petrik and G. Herzwurm, "iIoT ecosystem development through boundary resources: a Siemens MindSphere case study," in *Proceedings of the 2nd ACM SIGSOFT International Workshop on Software-Intensive Business: Start-ups, Platforms, and Ecosystems - IWSiB 2019*, Tallinn, Estonia, 2019, pp. 1–6.

[27] F. T. Piller, *Mass Customization*. Wiesbaden: Springer Fachmedien, 2007. [Online]. Available: http://gbv.eblib.com/patron/FullRecord.aspx?p=750687

[28] M. Schönemann, C. Herrmann, P. Greschke, and S. Thiede, "Simulation of matrix-structured manufacturing systems," *Journal of Manufacturing Systems*, vol. 37, pp. 104–112, 2015, doi: 10.1016/j.jmsy.2015.09.002.

[29] P. Greschke, M. Schönemann, S. Thiede, and C. Herrmann, "Matrix Structures for High Volumes and Flexibility in Production Systems," *Procedia CIRP*, vol. 17, pp. 160–165, 2014, doi: 10.1016/j.procir.2014.02.040.

[30] C. Hofmann, N. Brakemeier, C. Krahe, N. Stricker, and G. Lanza, "The Impact of Routing and Operation Flexibility on the Performance of Matrix Production Compared to a Production Line," in *Advances in Production Research*, R. Schmitt and G. Schuh, Eds., Cham: Springer International Publishing, 2019, pp. 155–165.

[31] E. Bogner, C. Kästle, J. Franke, and G. Beitinger, "Intelligent vernetzte Elektronikproduktion," in *Handbuch Industrie 4.0*, G. Reinhart, Ed., München: Carl Hanser Verlag GmbH & Co. KG, 2017, pp. 653–690.

[32] J. Fuchs, H. Herrmann, S. J. Oks, M. Sjarov, and J. Franke, "Increasing Efficiency in Maintenance Processes Through Modular Service Bundles," in *Lecture Notes in Production Engineering, Production at the leading edge of technology*, B.-A. Behrens, A. Brosius, W. Hintze, S. Ihlenfeldt, and J. P. Wulfsberg, Eds., Berlin, Heidelberg: Springer Berlin Heidelberg, 2021, pp. 439–447.

Theoretical and Practical Aspects in the Design and Construction of Active Electrodes for EEG

Daniela Andreea Coman, Silviu Ionita, Ioan Lita

Departament of Department of Electronics Computers and Electrical Engineering
University of Pitesti
Pitesti, Romania
deea.comnan@yahoo.co

Abstract—**This paper gives details about an approach for the development of an active electrode designed to capture electrical signals of the brain. A comparative analysis of the electrode-skin interface made for wet electrodes and also for a dry electrode based on conductive felt is considered. A laboratory model of a dry active electrode was built and evaluated as well in this project.**

Keywords—active electrodes; EEG capturing; low noise amplifier; conductive textile

I. INTRODUCTION

It is known that the brain waves are more difficult to be captured than other biosignals due to their low signal level, their specific low frequency band as well as due to the difficulties in obtaining signals. The electrode-amplifier assembly raises the following problems [1], [2]:

- Contact problems characterized by the following factors: contact resistance, parasitic capacity, contact (impedance) stability, skin preparation and electrode potential,

- Contact problems characterized by the following factors: contact resistance, parasitic capacity, contact (impedance) stability, skin preparation and electrode potential,

- Noise and disturbances: amplifier's own noise, interference (which can be diminished by the constructive solution called active electrode),

- Electrode reliability: corrosion wear, modification of electrochemical characteristics due to the ageing materials,

- Amplifier characteristics: internal noise level, sensitivity (signal-to-noise ratio), frequency band, input impedance, input signal dynamics and power consumption,

- Miniaturization: electrode size, electrode-amplifier assembly size and last but not least, technological and commercial aspects must be taken into account: the difficulty of realization, costs and the need for certain deficient materials.

Under these circumstances, the design and the implementation of an active-electrode system for capturing EEG (electroencephalography) signals requires a preliminary analysis of the electrode-skin interface. Actually, we have to choose between dry electrodes and wet electrodes, considering their both advantages and disadvantages. So, an analysis of the electrical parameters of the electrode is necessary in the design process. The impedance of the electrode-skin contact interface is susceptible of large variations related to different factors depending on the actual measurement such as: (i) the electrode pressure on the skin, which determines the thickness of the conductive film, (ii) the presence of hair in the area, (iii) the physiological moisture level of the skin and (iv) temperature. However, it can be affected by the absorption into the skin and the certain skin sweating level so the contact significantly changes its impedance. Conductive gels are more stable, but having also disadvantages in practice. So, using dry electrodes is worth to be considered [3] from the perspective of conductive textiles. Our project presents an active dry electrode solution with conductive felt. The dry electrode based on conductive felt is compared to wet interfaces to build a more suitable active electrode in a laboratory model. In this paper a circuit analysis is made for a suitable design of the first amplifier stage for an active electrode with the best possible immunity to the variation of the contact impedance.

Wet active electrodes (Ag/AgCl) are preferred in the clinical and research activities because they offer simplicity, stability and low noise characteristics [4]. The presence of the contact gel hydrates the skin, reducing its electrical impedance. So, unwanted disturbance appears, due to the aging phenomenon of the gel and the adhesive. The use of conductive gels is also inconvenient, leaking liquid from one electrode to another in hdEEG (high-density EEG with over 100 electrodes) applications can produce electrical bridges between the electrodes. Therefore the skin needs additional preparations given the long time required for preparation which implies new solutions.

The numerous technological approaches for the development of dry electrodes, as well as the variety of materials, make these dry electrode systems difficult to compare due to a lack of a homogeneous methodology [4].

Dry electrodes were developed in the following directions:

- Microelectromechanical systems MEMS system [5] [6], ensures the signal intensity, the coupling is directly with the skin through the range of microneedles that penetrate the outer skin layer of the scalp. It is an invasive method, not recommended in BCI (brain computer interface) applications considering the presence of the hair and the snapping needles that may result into inaccurate measurements.

- The capacitive electrodes are made without direct contact with the skin. Their use is limited due to the impedance damage which requires pressure. Specific findings in the field suggest alternatives that improve the above mentioned conditions with better results of impedance and of the signal. [7]

- 3D printed shaped electrode [8], made through 3D printing technology, with similar results in terms of time and frequency compared to electrolyte-based electrodes, have disadvantages on the contact – impedance (depending of the mechanical pressure).

- Wearable systems -laminated on the skin [9], and conductive textiles. Smart textiles / electronic textiles (e-textiles) / wearable textiles have enjoyed a continuous growth of 15.5% annually from 2016 to 2022 [10]. The sensors are successfully used in the exploitation of bio-signals and in the BCI interface. In [11] two types of conductive textiles were tested for applications even in neonatology, using an external amplifier of medical instrumentation, in order to demonstrate the feasibility of using textile electrodes in EEG.

II. DESIGN AND IMPLEMENTATION

A. Design

In this section the development of an active electrode with a conductive felt is presented.

The TLC272IDR operational dual amplifier from Texas Instruments, was chosen for its ability to provide the high input impedance and low noise and to combine wide range of input offset voltage grades with low offset voltage drift. Power is at 5V single supply operation. The proposed circuit is depicted in Fig.1. This schematic was used in simulation further.

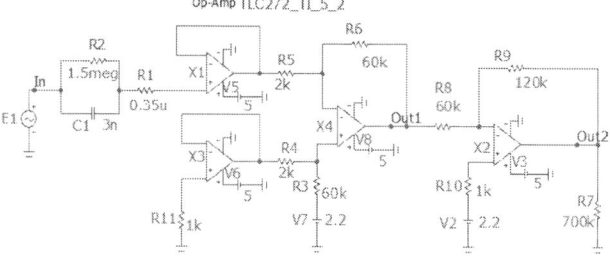

Fig. 1. The simultive active-electrode amplifier circuit

The contact impedance of the electrode with the skin is the critical point that directly affects the signal received at the output of the amplifier. The skin-electrode interface was modeled with a parallel R2C1circuit equivalent to the skin impedance, and the resistance of the conductive material on the electrode surface being of the order of $\mu\Omega$. The conductive material from Holland Shielding Systems BV is made of non-woven polyester fabric with an electrically conductive nickel coating 1.5 mm thick. The value of the skin impedance can vary with frequency between 200Ω at 1Hz to 1MΩ at 1MHz for one square centimeter. The contact between skin and electrode has also variable impedance. The proposed design is immune to the large variation of impedance using inputs through the buffer circuit. The system was mounted on Printed Circuit Boards as is depicted in Fig 2. This laboratory model has the components one-side mounted as is depicted in Fig. 3. The back side contains the proper electrode made of conductive felt as is shown in Fig. 4.

Fig. 2. PBC design   Fig. 3. Front side   Fig. 4. Back side

B. Circuit simulation

The simulations were made for sinusoidal test input signals with values of 83μVpp, 0.84mVpp and 15mVpp at frequencies of 10 and 490Hz. The simulations were obtained with Micro-Cap 12 by Spectrum Software.

The simulation results showed voltages at the output of the circuit (Out2) of 2mVpp, 24mVpp, 440mV at 490Hz, values that can be seen in Fig. 5, Fig. 6 and Fig. 7. The result in Fig.8 was obtained for a sinusoidal signal with the magnitude and frequency in the range of a real EEG signal, for instance 83μVpp and frequency 10Hz (similar to alpha-waves).

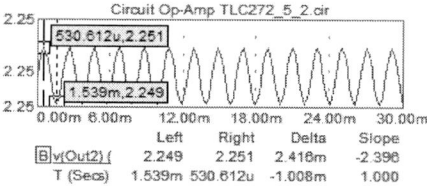

Fig. 5. The response for input of 83uVpp at 490Hz

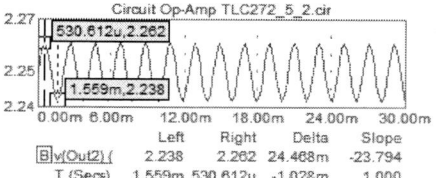

Fig. 6. The response of 84mVpp at 490Hz

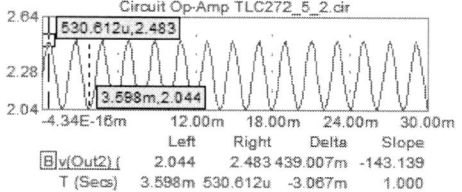

Fig. 7. The response of Vpp 15mV at 490Hz

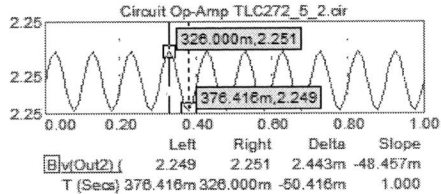

Fig. 8. The response of 83μVpp at 10Hz

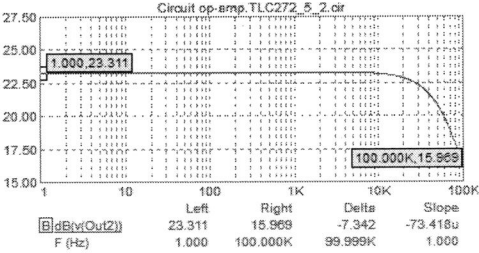

Fig. 9. The gain characteristic in [dB]

C. Testing the laboratory model

The circuit was tested experimentally in the laboratory with the sinusoidal input at 490Hz with same magnitudes 83μVpp, 0.84mVpp and 15mVpp.

Output voltages of 5mVpp was measured when input signal was 83μVpp (see Fig.10a). The 30mVpp were for input of 0.84mVpp (see Fig.10b) and 900mVpp were for input of 15mVpp (see Fig.10c). The average resulted gain of 35dB in practice is in accord with theoretical calculated of the circuit.

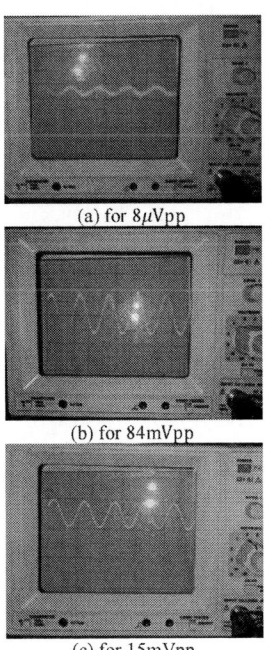

(a) for 8μVpp

(b) for 84mVpp

(c) for 15mVpp

Fig. 10. The measured output

D. Discussions

The proposed model is under development and practical improvements can be made. One of the practical solution for improving the immunity of noise and disturbances is the connection of capacitors around 100nF on the supply lines. In this laboratory model the tests for capturing biocurrents on the human body (head) are being performed and it is expected that some adjustments of the final solution will occur.

III. CONCLUSIONS

In this paper we presented an experimental approach dealing with capturing of weak signals specific of human brain. Some considerations on electrodes were presented and a laboratory amplifier for a dry electrode was experimented. The first evaluations shows the feasibility of the proposed solution, but extended experiments and particular adjustments are necessary to made a prototype. The work is under development.

REFERENCES

[1] D.M.D. Ribeiro, L.S. Fu, L.A. Dias Carlos, and J.P.S. Cunha, "A Novel Dry Active Biosignal Electrode Based on a Hybrid Organic-Inorganic Interface Material", IEEE Sensors Journal, Vol. 11, pp. 2241-2245, October, 2011.

[2] S. Lee, Y. Shin, A. Kumar, K. Kim and H.N. Lee, "Two-Wired Active Spring-Loaded Dry Electrodes for EEG Measurements", Sensors, Vol. 19, pp.1-15, October 2019.

[3] G. Acar, O. Ozturk , A.J. Golparvar, T.A. Elboshra, K. Böhringer and M.K. Yapic, "Wearable and Flexible Textile Electrodes for Biopotential Signal Monitoring: A review", Electronics, Vol.8, pp. 1-25, April 2019.

[4] M. A. Lopez-Gordo, D. Sanchez-Morillo and F. P. Valle, "Dry EEG electrodes", Sensors, Vol.14, pp. 12847-12870, July 2014

[5] N.S Dias, J.P Carmo, A.F. da Silva, P.M. Mendes, J.H. Correia, "New dry electrodes based on iridium oxide (IrO) for non-invasive biopotential recordings and stimulation", Sens Actuators A Phys., pp. 28–34, November 2010.

[6] J. C Chiou, L.-W. Ko, C.-T. Lin, C.-T. Hong, T.-P. Jung, S.-F. Liang, J.-L. Jeng, "Using novel MEMS EEG sensors in detecting drowsiness application", IEEE Biomedical Circuits and Systems Conference, BioCAS 2006, London, UK, pp. 33–36, 29 November–1December 2006.

[7] T.J. Sullivan, S.R. Deiss, G.A. Cauwenberghs," Low-Noise, Non-Contact EEG/ECG Sensor", IEEE Biomedical Circuits and Systems Conference, Montreal, QC, Canada, pp. 154–157, 27–30 November 2007.

[8] A. Velcescu, A. Lindley, C. Cursio, S. Krachunov, C. Beach, C.A. Brown, A.K. P. Jones and A.J. Casson, "Flexible 3D-Printed EEG Electrodes", Sensors,MDPI, Vol. 17, pp. 1650, April 2019.

[9] L. M. Ferrari, U. Ismailov, J-M. Badier, F. Greco and E. Ismailova, "Conducting polymer tattoo electrodes in clinical electro - and magneto-encephalography", Flexible Electronics, Vol. 4, pp.1-9, March 2020.

[10] Electronics and Technology Market by Applications, available online: http://www.marketsandmarkets.com/Market-Reports/wearable-electronics-market-983.html , accessed on April 2020.

[11] J. Lofhede, F. Seoane and M. Thordstein, "Textile Electrodes for EEG Recording- A Pilot Study", Sensors, Vol. 12, pp. 16907-1691, December 2012.

Wearable Smart Prototype for Personal Air Quality Monitoring

Attila Géczy, Lajos Kuglics, László Jakab, Gábor Harsányi

Department of Electronics Technology, Budapest University of Technology and Economics,
Budapest, Hungary
gattila@ett.bme.hu

Abstract — **This work aims to present an advanced prototype of cost-effective wearable air quality monitoring equipment suitable for different use cases due to its small size, easy handling, and smartphone compatibility. In the paper, we present the chosen components and the system design. The prototype development is also presented, with the initial measurements and the aspects encountered during development. We used C and Arduino programming, Android software development. The proposed system comprises of an Arduino based microcontroller, a CCS811 sensor for volatile organic compound (VOC) measurement, a ZPH01 particulate matter (PM) detector, and an HC-05 device for Bluetooth connection. We investigated the prototype both in the laboratory and in a smart urban environment with different scenarios in Budapest, Hungary (street, campus, store, mall, underground). Also, the use of a smartphone connection for data loggings is presented. The paper shows examples of measurement results and the relation to similar results from around the world. The possible application in COVID-19 pandemic related questions is also discussed.**

Keywords— Air quality, sensor cluster, sensor platform, wearable, personal, prototype, cost-effective

I. INTRODUCTION

Air pollution is a serious question of the modern urban environment. The aspect of airborne particulate matter and volatile organic compound content both pose serious threats to health from different aspects - from cities [1] to the insides of living spaces [2]. Also, the industrial workplaces [3] should be investigated from air quality in the ambience around different manufacturing apparatuses.

Recently, questions were raised regarding the airborne diffusion of COVID-19 and various epidemiological threats [4,5] with contradictory results. Summing up, different solutions are required to tackle the problem on multiple scales. Some researchers are more moderate regarding the connection between air pollution and COVID-19 related morbidity and mortality. Comunian et al. [6] highlighted that PM alone could damage pulmonary cells. This can result in inflammation and oxidative stress. However, they point to the importance of further systematic studies, focusing on unfolding the mechanisms behind the connections, and placing PM measurement unit devices around the world in every critical location. They highlighted that the process needs further research concerning Angiotensin-converting enzyme 2 (ACE2) expression after PM exposure.

According to most recent research results (as of 2020 October), the daily PM2.5 and PM10 concentrations and the daily case fatality rate (CFR) numbers showed similarities in Wuhan during the peak of the pandemic. Similarities are with respect to the temporal variation curves, with a noticeable time lag between the results. [7] These results were also extended to the case fatality rate of 49 Chinese cities [8]. Later Yongjian [9] increased the city count under investigation to 120, and with a generalized additive model of pollutants, they linked significant positive associations of PM2.5, PM10, NO2, O3 with the confirmed cases. With the increase of PM2.5, PM10, NO2, and O3 with only 10µg/m3 concentration, the case increase was reported with 2.24%, 1.76%, 6.94%, and 4.76% on a daily basis. Wu reported [10] on conditions in the USA, with 3000 county data, where it was concluded that the 1 µg/m3 PM2.5 is combined with 8% increase in CFR. The extension to Europe focused on French and the gasoline/diesel-based transport sector. The machine learning-based research of Magazzino showed [11] that a pre-determined particulate concentration can foster COVID-19, in the meanwhile increasing the susceptibility of the respiratory system to the infection. As another article highlights [12], imbalanced reductions in primary pollutant emissions facilitated secondary emissions pollutants, creating haze pollution. VOCs were particularly highlighted in the results.

It is also important to note that a pandemic caused lockdown may positively affect the air quality. It was noted that COVID-19 lockdowns caused local and global air pollution to decline - due to transportation reductions and emission decrease. Particulate matter (PM2.5) levels were dropped with 31% on average [13].

The problem of air contamination can be investigated with fixed sensor stations planted by officials, or personal equipment measuring the environment on the scale of the person. For the latter aspect, novel, cheap, commercial sensors and easy-to-develop microcontroller systems might help achieve personal, wearable environment monitoring in personal, urban, healthcare, or industrial environments. In latter areas, the availability of clear air and appropriate ventilation is often not available due to the outdated industrial ambiance. In this paper, we aim to present the prototype of a cost-effective solution with results obtained from different areas, such as a city, public transportation zones, or a laboratory, where electronics manufacturing equipment (PCB fabrication, PCB assembly) is in order.

II. EXPERIMENTAL

A. Designing the hardware

For the central processing module, we chose an Arduino Nano, which is cheap and capable of controlling the given applied sensor modules and wireless communication. For software development, we used Arduino IDE 1.8.10. For the sensors, we chose a CCS811 module, which is used for VOC and eCO2 measurement, focusing on the former data. The sensitivity for the device is eCO2:400-8192 ppm and TVOC: 0-1187 ppm, respectively. The communication is performed via I2C communication. After initial testing, we applied Winsen ZPH01 particulate matter sensor, which has a PWM/UART communication, and it is sensitive for PM above 1 μm. For the communication, we used a cheap HC-05 UART-based Bluetooth module. The power is based on a 2200 mAh power bank with 5V/1A output. The small form factor and functionality rendered this component optimal for power use. The data is gathered via Bluetooth on a Redmi Note 7 Android-based phone. The MIT App Inventor was used for rapid software development and efficient smart application. Figure 1 presents the block diagram of the system.

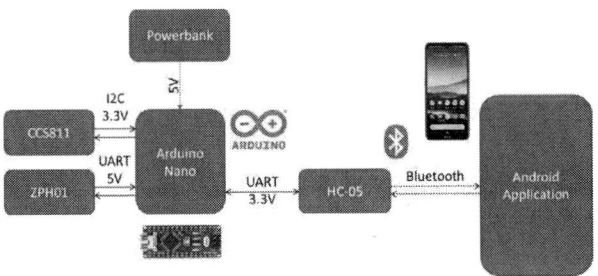

Fig. 1. Block diagram of the system.

In Figure 2, the initial working prototype is presented which was assembled in advance to investigate the capabilities of the given modules. Figure 3 illustrates the assembled prototype shield and Figure 4 shows the completed prototype wore by a person.

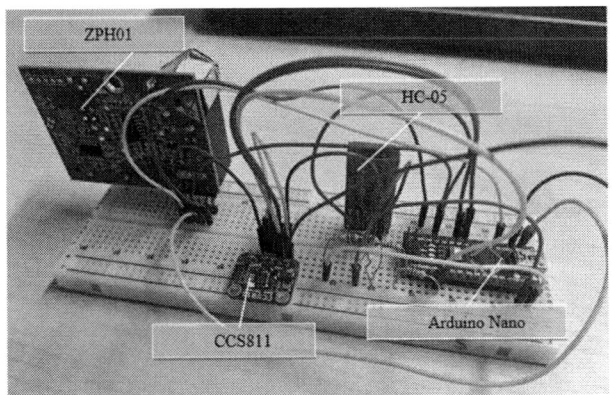

Fig. 2. Early prototype on a breadboard with presented components.

The final prototype adds a developed shield board, containing all essential elements on one board, while maintaining the form factor of a wearable device, which can be worn on the arm, similar to cell phone holders during sports. The construction shows the power bank too, which fits the arm pouch as seen in Figure 3.

Fig. 3. Assembled prototype shield.

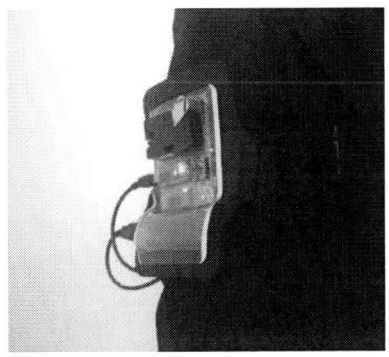

Fig. 4. Assembled prototype positioned on the arm of a person.

B. Experimental scenarios

Two scenarios were drawn at the beginning of our work. We wanted to investigate, how a laboratory engineer or an operator is working around different apparatus in electronics manufacturing environment at our labs. We chose a vapour phase [14] reflow oven (Asscon Quicky 450), where manual loading exposes the operator to the work zone of the oven. The other machine is a semi-closed selective mini-wave [15] soldering oven (Jade S-200), where the operator is also in the same space of the machine. The laboratory has an intricate air ventilation system; both machines have their specific extraction collectors hanging from the ceiling.

The second scenario was in an urban environment in our capital city, Budapest, to reveal the changes of VOC and PM concentration around a pedestrian, who travels with public transport and visits urban locations, such as shops.

III. RESULTS

A. Laboratory Environment

Figure 5 and Figure 6 present the the results of laboratory environment measurements in the vicinity of electronics assembly machines. The VOC levels increase during the working of different machinery. The Vapour Phase Soldering machine is analyzed in Figure 5. From the plot, it is clear that peaks around 300 ppb can be observed, which are in line with the opening of the PCB loading doors. The larger peaks above 1200 s simulate when the operator is leaning above the opened work zone, and the remaining heat transfer medium mist and the vapors of the flux reach the sensor.

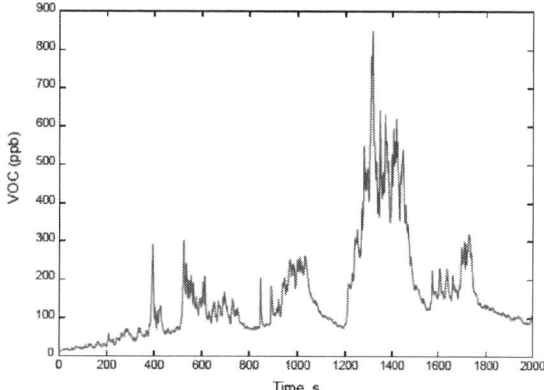

Fig. 5.   VOC results around vapour phase soldering device.

The next figure presents two fluxing cycles in the close vicinity of a selective mini-wave soldering oven. As it is apparent, the values jump up significantly. According to the operator, he did not notice significant changes in the general senses of smelling – so it can be concluded that latent dangers (VOC concentration increase) can be revealed with such sensors in the workspace environment.

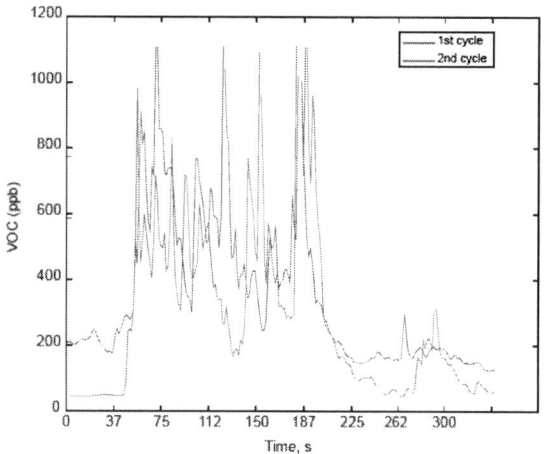

Fig. 6.   VOC results at fluxing stages of a selective mini-wave oven.

B. *Urban Environment*

Figure 3 presents an urban travel scenario (A) which took place on the street, on the tram, and in a store. It is interesting to see that the street has elevated particles and low levels of detected VOCs. The active transportation in the city boosts particles in the air; also, the open-air enables better distribution of VOCs than in closed spaces. The tram has characteristically increased VOC and particle levels, since the people on the tram boost exhausted, used air, and the tram. Also, the particles can

not ventilate correctly. Arriving at a store, further changes are recorded. The air of the store has low particle count, while the modern ventilation and air conditioning systems filtrate the particles. The store also has characteristically increased VOC levels, meaning that the human presence in closed spaces, the furniture, plastic inner decorations, and various cleaning and odor-reducing air fresheners increase the volatile organic compound in the air.

In Figure 8, a similar situation is presented (scenario B), where the urban travel was extended to a longer trip. The street scenario gives similar, almost repeatable results according to the given parts of Figure 7 and 8. Also, it has to be noted that the weather and ambient environment relations were similar in the two winter days after each other. The two visited stores were loaded with VOCs and had a minimal particulate matter count in the air. In the larger spaces in the mall, the situation was somewhat balanced, where both VOCs and PMs were detectable, varying at different locations around the mall. Then the Metro Line 4 gave an interesting result of minimal VOC presence and large PM data. The peaks of the PMs were exactly at the arrival of the underground metro car, meaning that the dynamics of the transportation can be investigated via the rise and fall of PM concentration in the air. Also, the arriving train stirs up the particulate matter in the tunnels and stations. It has to be noted that both PM and VOC levels stayed lower on the street around the campus than in the city. The tram travels had similar aspects as before, with increasing VOC levels and occasional falls in VOC levels when the doors ventilated the air. It is also interesting to see that during the second tram travel, the PM levels rose, which can be attributed to the increasing traffic around the main tram line in the city center, as the day neared the end of the work shifts. The campus was again a quiet location from both aspects; finally, the store and the street gave similar results as before. The possibility of map integration is already realized in the application. The next step can be the building of a common database according to the results obtained from several users. The gathered data can (and in the future must) be evaluated with a comparison of official air quality stations positioned around the city.

It is also interesting to note that while the PM levels peaked in underground stations, the levels were different at different stations - this is due to the different status (state of disrepair, cleaning history of the lines, etc.) of the given stations. It must be noted that the recorded levels are in line with very recent research results, where London underground was investigated [16]. Some results are around the same levels, but the cited research gives significantly larger values too, most probably due to heavy rail and train component wear in the UK capital.

It must be noted that the PM concentration outside can depend highly on weather effects (wind, temperature, and consequential heating of households), too, so further analysis is needed during a whole year time to investigate the changes according to weather situations.

The results also point out that the proposed device would be able to record pandemic-related changes in urban situations as well, such as reducing of traffic and pedestrian density on streets, or inside malls, shops, closed spaces, or professional locations such as hospitals, laboratories, or workplaces.

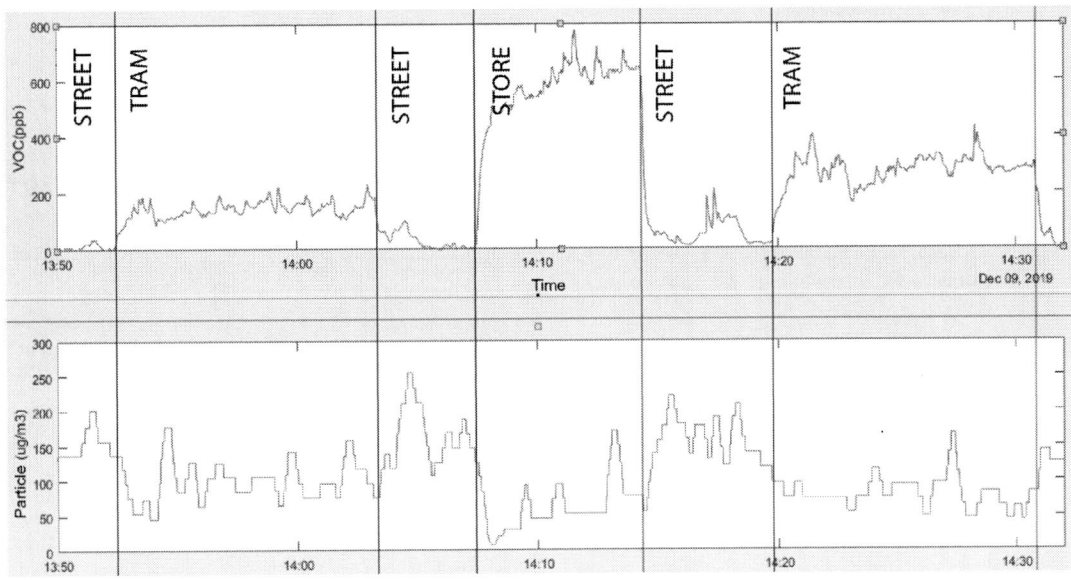

Fig. 7. Volatile Organic Compound (top) and Particulate Matter (bottom) measurements in an urban scenario A.

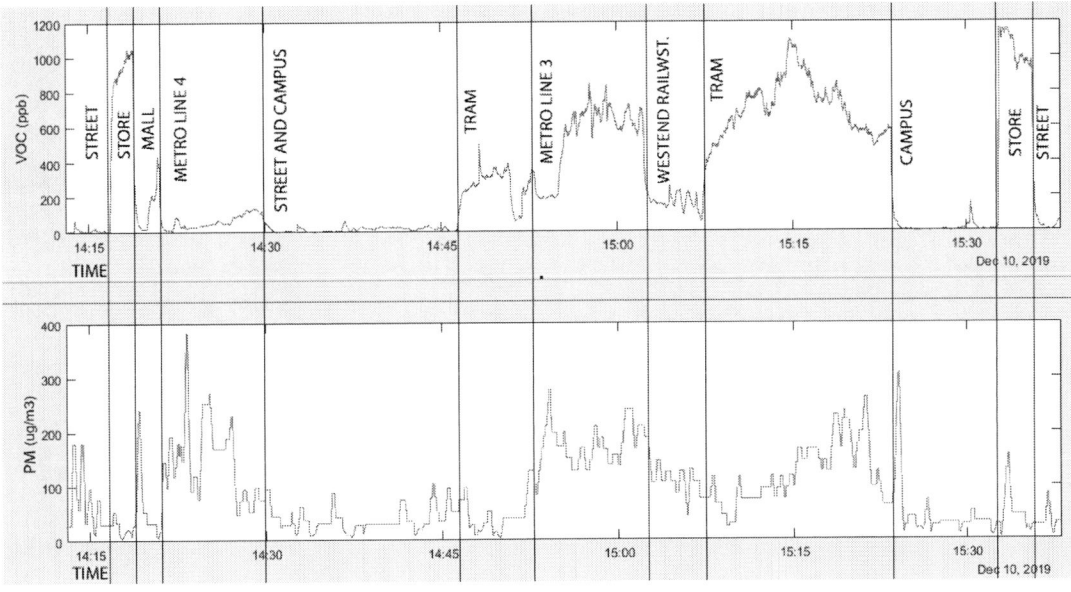

Fig. 8. Volatile Organic Compound (top) and Particulate Matter (bottom) measurements in urban scenario B.

IV. . CONCLUSIONS

The paper focuses on a novel wearable device capable of measuring VOC and PM concentrations around the users. We present a cost-effective solution for the prototype and the possibility of smart connectivity. The setup is tested in different environments, such as electronic manufacturing laboratory (workspace) and urban scenarios in Budapest. It is interesting to see that different results are in line with literature data and also that the characteristic changes are visible across different stages of e.g., an urban scenario. While it is difficult to classify or grade results according to quantitative results for a non-professional user, future investigations may need to point to standardized Air Quality Index referencing (AQI [17]), or a given ergonomic scaling (practical levels of air quality [18]) for easier commercial application.

The device also points to applicability in pandemic-related scenarios (e.g., COVID-19), where the caused changes in the transportation and resulting air quality, and more importantly, the knowledge of air quality itself is essential to avoid highly polluted areas. With connected devices, the aid of governmental control over pollution can also be visualized for future use.

ACKNOWLEDGMENT

Hereby we acknowledge the support of EFOP-3.6.1-16-2016-00014, „Diszruptív technológiák kutatás-fejlesztése az e-mobility területén és integrálásuk a mérnökképzésbe" project.

The research reported in this paper and carried out at the Budapest University of Technology and Economics has been supported by the National Research Development and Innovation Fund based on the charter of bolster issued by the National Research Development and Innovation Office under the auspices of the Ministry for Innovation and Technology.

V. REFERENCES

[1] J.D. Smith, B.M. Barratt, G.W. Fuller, F.J. Kelly, M. Loxham, E. Nicolosi, M. Priestman, A.H. Tremper, D.C. Green (2019). PM2.5 on the London Underground; Environment International, Volume 134, 2020

[2] Miles Brignall, Our MDF furniture brought toxic fumes into our home; The Guardian, Feb 9. 2019.

[3] Oliverio Cruz-Mejía, Alberto Márquez, Mario M. Monsreal-Barrera, Product Delivery and Simulation for Industry 4.0, in M. M. Gunal, Simulation for Industry 4.0, Springer Series in Advanced Manufacturing. 2020.

[4] Yongjian Zhu, Jingui Xie, Fengming Huang, Liqing Cao, Association between short-term exposure to air pollution and COVID-19 infection: Evidence from China, Science of The Total Environment, Volume 727, 20 July 2020, 138704, 10.1016/j.scitotenv.2020.138704

[5] E. Bontempi, First data analysis about possible COVID-19 virus airborne diffusion due to air Particulate matter (PM): The case of Lombardy (Italy), Environmental Research Volume 186, July 2020, 109639, 10.1016/j.envres.2020.109639

[6] Silvia Comunian, Dario Dongo, Chiara Milani, Paola Palestini, Air Pollution and COVID-19: The Role of Particulate Matter in the Spread and Increase of COVID-19's Morbidity and Mortality Int. J. Environ. Res. Public Health 2020, 17(12), 4487; https://doi.org/10.3390/ijerph17124487

[7] Ye Yao, Jinhua Pan, Zhixi Liu, Xia Meng, Weidong Wang, Haidong Kan, Weibing Wang, Temporal association between particulate matter pollution and case fatality rate of COVID-19 in Wuhan, Environmental Research, Volume 189, October 2020, 109941

[8] Ye Yao, Jinhua Pan, Weidong Wang, Zhixi Liu, Haidong Kan, Yang Quiu, Xia Meng, Weibing Wang, Association of particulate matter pollution and case fatality rate of COVID-19 in 49 Chinese cities, Science of The Total Environment, Volume 741, 1 November 2020, 140396

[9] Z. Yongjian, X. Jingu, H. Fengming, C. Liqing, Association between short-term exposure to air pollution and COVID-19 infection: Evidence from China, Science of the Total Environment (2020) Volume 727, 20 July 2020, 138704

[10] Wu, X., Nethery, R.C., Sabath, B.M., Braun, D., Dominici, F., 2020. Exposure to air pollution and COVID-19 mortality in the United States. medRxiv.

[11] Cosimo Magazzino, Marco Mele, Nicolas Schneider, The relationship between air pollution and COVID-19-related deaths: An application to three French cities, Applied Energy Volume 279, 1 December 2020, 115835

[12] C. Huang, Y. Wang, X. Li, L. Ren, J. Zhao, Y. Hu, Clinical features of patients infected with 2019 novel coronavirus in Wuhan, China, The Lancet, 395 (2020), pp. 497-506

[13] Zander S. Venter, Kristin Aunan, Sourangsu Chowdhury, Jos Lelieveld, COVID-19 lockdowns cause global air pollution declines, PNAS August 11, 2020 117 (32) 18984-18990; first published July 28, 2020; https://doi.org/10.1073/pnas.2006853117

[14] Lubomir Livovsky, Alena Pietrikova, Measurement and regulation of saturated vapour height level in VPS chamber, Soldering & Surface Mount Technology, Vol. 31 No. 3, pp. 157-162.

[15] Zoltán Oláh, Miklós Ruszinkó, Réka Bátorfi, Zsolt Illyefalvi-Vitéz, Process parameter optimization of selective soldering, 2012 IEEE 18th International Symposium for Design and Technology in Electronic Packaging (SIITME), 25-28 Oct. 2012, 10.1109/SIITME.2012.6384359

[16] J.D.Smith, B.M.Barratt, G.W.Fuller, F.J.Kelly, M.Loxham, E.Nicolosi, M.Priestman, A.H.Trempera, D.C.Green, PM2.5 on the London Underground, Environment International, Volume 134, January 2020, 105188

[17] Air Quality Index (AQI) - https://www.airnow.gov/aqi/ (accessed at 2020. 10. 06.)

[18] Air Quality Guide for Particle Pollution - https://www.airnow.gov/sites/default/files/2018-04/air-quality-guide_pm_2015_0.pdf (accessed at 2020. 10. 06.)

Decision support platform for intelligent and sustainable farming

Mihaela Bălănescu, Andreea Bădicu, George Suciu,
Carmen Poenaru, Adrian Pasat
Research & Development Department
Beia Consult International
Bucharest, Romania
george@beia.ro

Alexandru Vulpe, Marius Vochin
Research & Development Department
BEAM Innovation, Bucharest, Romania
Bucharest, Romania
{alex.vulpe, marius}@beaminnovation.ro

Abstract — The growing demand for animal products led to an increase of the number and size of animal farms across the world. In this context, the livestock farming industry faces the challenge of integrating high complexity processes, while ensuring to meet sustainability standards. Therefore, decision support systems in the livestock sector must be established to help farmers assess the impact of different strategies before putting them into practice. In this paper, we present a monitoring and management platform that allows farmers to optimize their activities, to manage resources, to reduce costs and to minimize the environmental footprint. The FarmSustainaBL platform follows a farm-level approach for modelling and simulation of specific activities to assist decision makers and farmers in the decisional process, by providing insights into their managerial practices.

Keywords — sustainable farming, GHG emissions, livestock monitoring platform.

I. INTRODUCTION

Most individual farms have low economic power and limited opening to the market, both in terms of inputs and outputs required. These farms are characterized by a very diverse production structure, as well as by insufficient and inadequate technical equipment, which hinders productivity. Given the context of the current economic downturn and limited access to capital, small market-oriented farms need to be supported by innovation actions, in order to achieve goals such as increasing the number and size of farms, more efficient livestock management and reduction of greenhouse gas (GHG) emissions [1]. For this purpose, the development of livestock monitoring, and management platforms can enable farmers to optimize their activities, to minimize costs and to reduce the environmental footprint. These decision support systems based on modelling and simulation help stakeholders in the livestock sector to take appropriate actions towards more productive and sustainable farming [2].

This paper presents a monitoring and management platform that assists farmers in the decisional process, by facilitating the optimization of specific activities and better resource management, to reduce costs and to minimize the environmental footprint. The following sections of the article will approach the subsequent matters: Section II examines the current context and investigates related work, Section III details the use cases, Section IV showcases how the system requirements are established and section V describes the system

design. Finally, Section VI presents the conclusions and future work.

II. AGRICULTURE AND GHG EMISSIONS

The livestock sector is the main source of CH4 emissions in agriculture, as a cause of both enteric fermentation and manure management.

A. Sources of greenhouse gas emissions on livestock farms

The Agriculture sector contributed by adding 16.85% to total GHG emissions in 2017, reaching the equivalent of 19231.76 kt CO2. In the context of GHG emissions in the agricultural sector, CH4 emissions make the largest contribution, followed by N2O emissions. Between 1989 and 2017, GHG emissions from the agriculture sector in Romania decreased by 50.86%. The number of animals decreased during this period, regardless of species and type of operation. Cattle raised for beef, milk or traction power are the animal species responsible for the most emissions, accounting for about 65% of the emissions of the livestock sector. With reference to farming activities, the production (including processing) of feed and enteric fermentation from ruminants are the two main sources of emissions, accounting for 45% and 39% of total emissions. The storage and processing of manure accounts for 10% of all emissions. Approximately 44% of animal emissions are represented by methane (CH4). The remaining part is almost evenly divided between nitrous oxide (N2O, 29%) and carbon dioxide (CO2, 27%). This means that animal supply chains emit: 5% of anthropogenic CO2 emissions, 44% of anthropogenic CH4 emissions and 53% of anthropogenic N2O emissions [4].

B. GHG estimation methodologies

The internationally recognized methodology for estimating GHG emissions from livestock farming is the one proposed by the Intergovernmental Panel on Climate Change (IPCC) and is the method used by the countries included in Annex I to the United Nations Framework Convention on Climate Change (UNFCCC) for international reporting of these emissions. The methodology is structured on three levels, depending on complexity and accuracy [5]. The basic characterization for level 1 is probably sufficient for most animal species in most countries. For this approach, it is good practice collecting the following data: Species and categories of animals, annual population, feed intake, etc.

III. PRECISION LIVESTOCK FARMING

In recent years, the growing demand for animal products led to an increase of the number and size of animal farms across the world. Farmers are faced with a multitude of challenges in order to properly manage their business, as they have to consider increasingly complex factors within their decisions, such as: more diverse and larger numbers of animals to manage, more demanding production requirements, complex business models to choose from, environmental restrictions, etc. Thus, the transition from conventional agricultural practices to more technology-based approaches is mandatory. In this regard, the newest advancements in technology will enable the development of Precision Livestock Farming (PLF) [6]. This concept refers to tools such as real-time and continuous livestock monitoring, farm-related data collection, data analysis, predictions and incident reporting. Using sensors that measure biosignals, PLF can also be used to monitor animal welfare, their overall condition (physical and mental), changes in behavior. Other common sensors employed in PLF collect data parameters for different purposes:

- stable environment conditions: (temperature, humidity, light, gas concentrations)
- animal motions: (gait, speed, position, weight, temperature, sounds)
- animal feed: (composition, flow, weight)

Aside from monitoring, Precision Livestock Farming is useful for more complex tasks such as: modeling, simulation and decision support using machine learning models. Commonly, the main elements of PLF include: continuous or real-time processing of sensor data, data integration and storage, data analysis and modeling, event detection and signaling [7].

A. The use of AI and Blockchain for precision livestock farming

In recent years, since the complexity of the systems involving animals has grown, their monitoring became more difficult and the gathered data increasingly heterogeneous. Therefore, the need for decision support is essential for decision makers, who are constantly faced with making crucial management decisions to meet their objectives: improving animal welfare and reducing GHG emissions.

The incorporation of artificial intelligence algorithms (machine learning and / or deep learning) in decision support for PLF helps stakeholders to achieve several objectives, due to the characteristic usually attached to intelligence - the ability to learn from the environment. Artificial Intelligence in livestock farms helps in accumulation and analyzing data for accurate prediction of consumer behavior (i.e. buying patterns, leading trends). In this way, with the necessary investments, farmers will be able to automate processes, reduce high costs and get a better quality of livestock products, like milk. Taking into consideration that AI and ML are becoming more available and are expected to advance the automation of most of the farm processes and, at the same time, produce information based on the farm's operational history. AI allows producers to interpret data collected by sensors and hardware technologies to provide solutions by mimicking human decision-making - potentially transforming how a smart farm operates [8].

Blockchain is used in many applications of precision agriculture. The technology plays a significant role, the one of replacing the basic methods of storing, sorting and also sharing the agricultural data into a more immutable, reliable, transparent and decentralized way. Blockchain technology is a digital register in which the details of transactions made between users are stored. A blockchain can be considered a dynamic list of records that can only grow, because changes such as deleting or changing previous records are not possible [9]. In the context of precision farming, blockchain technology mixed with the Internet of Things will make the transitions from only smart farms to the internet of intelligent farms and also will increase the control in supply-chains networks. This combination will result in the autonomy and intelligence in managing precision farming in an efficient and optimized way [10]. Blockchain represents the source of truth about the state of farms and contracts in agriculture; here, the collection of this information is very costly. It facilitates the use of data-driven technologies to make farming smarter - reliable data of farming processes are essential for developing data-driving facilities and insurance solutions, facts that make farming smarter and less vulnerable [11].

B. Precision livestock farming Platforms. Related work

The widespread adoption of the Precision Farming concept within the livestock farming industry has the potential to bring more efficiency and real-time capabilities for the supply chain management, to significantly increase productivity, to strengthen the marketing strategies and to reduce the environmental impact of the farming sector. This concept has been investigated within several scientific studies and has been successfully applied in various use cases, presented in the following paragraphs.

Cisco proposes a framework for the implementation of a precision farming platform based on artificial intelligence software which lowers the data volume that needs to be sent. The platform relies on measuring animal movements by processing collar information. With the approach mentioned above, operating lifetime is more than five years without the need for a battery change. Longer battery life is also due to the LoRa protocol instead of LPWAN technologies. By using a simple measurement of the acceleration on the neck, the platform can identify the animal's behavior. An experiment built on a GPS-based collar for cattle management was performed in sub-Saharan Africa countries. The demonstration shows good tracking skills of the used software and registered low-cost solutions compared to other existing PLF platforms [12].

The e-Pasto platform [13] can deal with the most common challenges of the PLF platforms, such as adequacy of the communication protocols and energy usage. A study presents an accelerometer-based solution that improves the energy efficiency of embedded geolocation devices, allowing at the same time an essential animal activity identification. The Machine Learning algorithm used for the suggested solution is based on the moving average method and due to its accuracy, it can successfully identify different animal movements.

A study conducted in 2017 in France [14] explored the viability of the architecture of a farming platform and its performances. The innovative result of the experiment was the validation of a virtual fencing solution that helped farmers to decide and draw the size and shape of the virtual fences with the point-in-polygon geometric computation principle. The scope of the virtual fence is to warn the farmer that the animal went out of the limits and to offer decision support to solve the problem.

One study published in 2020 proposes a blockchain-based framework for managing the agricultural supply chain [15]. The authors propose a framework that integrates IoT-based solutions to enable food traceability and transparency. Another study [16] presented in 2018 in China proposes a public blockchain of the agricultural supply chain system to demonstrate that the advantages of blockchain such as confidentiality, openness and security of information can greatly strengthen the credibility of upcoming service platforms in the field and the efficiency of the system as a whole. In 2019, the article [17] examined whether blockchain is a suitable solution for supply chain management and concluded that it can lead to lower error rates at various stages of the supply chain and improve customer support. For the study, weight sensors were used to measure product weights to enable tracking the transaction between the trader and the customer, which is visible to both parties involved. Another study carried out in 2018 analyzed the implementation of ICT solutions, systems and functions (such as Cloud/Fog and IoT) to make the food supply chain possible [18] and demonstrated that intelligent technology solutions will increase transparency, information flow and management capacity to enable farmers to better interact with other stakeholders within the supply chain. The study proposes a new business model regarding on-demand foods based on the new values of Quality of Experience (QoE) Food Metrics, eliminating the gap between subjective experience and objective indicators based on quality standards.

IV. USE CASE DESCRIPTION

To demonstrate the functionalities of the decision support capabilities of the PLF platform, certain aspects of the use cases have be taken into account: the main objectives of the platform (seen through the eyes of users), the business model (actors involved), the main functional characteristics, the functional components and the requirements associated with the use case (from the main users of the system).

The FarmSustainaBl platform will be tested in two case studies that were selected considering the amount of greenhouse gas (GHG) emissions produced and the processes from which they result. As such, one of the case studies aims to manage GHG emissions from manure, while the other case study addresses the management of CH4 emissions from enteric fermentation. The approach to GHG emissions at farm level (Figure 1) includes direct emissions (from enteric fermentation and manure), indirect emissions (associated with electricity consumption and fuel used in internal processes, e.g. for agricultural machinery) and emissions of indirect greenhouse pollutants (PMx, NH3, etc.).

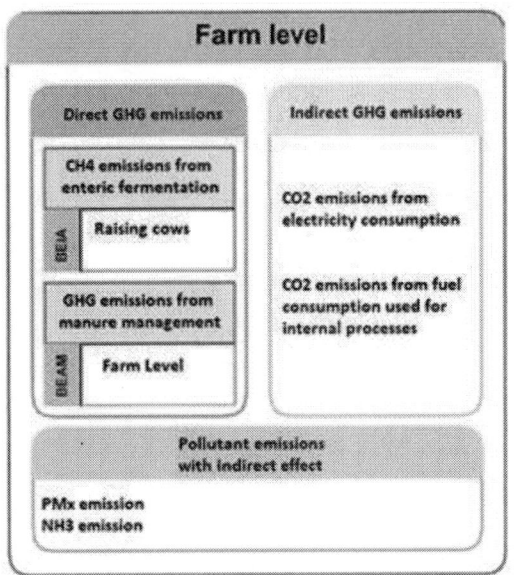

Figure 1. Types of emissions included in the FarmSustainaBL platform.

The developed platform will collect and analyze all these data to provide recommendations to the livestock farming stakeholders (such as farmers, consultants etc.) in order to take management and operational decisions for reducing GHG emissions.

V. DEFINITION OF SYSTEM REQUIREMENTS

The process of defining the system requirements for the proposed solution must consider different sets of information:

- Requirements of stakeholder that can refer to: primary users (requirements from use cases), operator/beneficiaries (efficiency requirements, management and control possibilities, operation, commercial requirements, etc.) or third parties (secondary users) - (requirements arising from use cases).

- Constraints: pre-defined initial design data (eg use of only certain technologies, maximum system size, cost, etc.).

- Conditions: end-user features, availability of resources, including business / commercial. Changing these conditions can influence the behavior and results of the system.

- Dependencies: functional or performance elements of other systems, on which the correct functioning of the system to be performed depends (if the latter has external interactions).

After closely considering the information presented above, the system requirements will be established using the general sequence of steps presented in Figure 2.

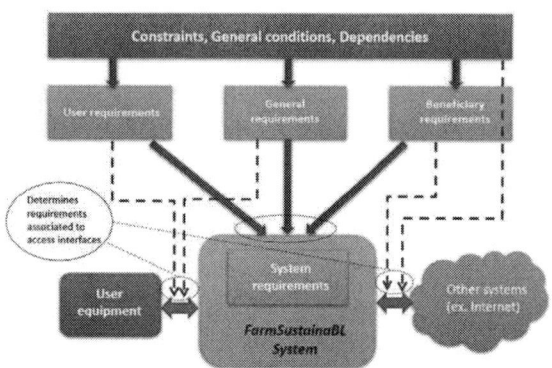

Figure 2. Illustration of the methodology of analysis and collection of system requirements.

VI. SYSTEM DESIGN

The concept of the FarmSustainaBl platform consists of three main components that will be developed on existing farm infrastructure, as detailed below.

A wireless sensor network that will measure specific parameters such as environmental, animal and feed data. The sensors will transmit information via a wireless technology (LoRaWAN) to a sensor measurement aggregator which will communicate with the FarmsustainaBl Cloud.

The FarmSustainaBl Cloud which will enable the realization of the GHG reduction as well as the traceability of the feed and livestock management. It consists of the following components:

- the gateway - the main entry point to the FarmSustainaBl Cloud, responsible for receiving processed and indexed measured sensor data and for sending control and configuration data to the sensor farm environment.

- the data analytics module is responsible for providing meaningful insight on livestock and farm data by performing analytics and business intelligence.

- the blockchain component that will enable smart contracts for automatically offering the best prices for low-GHG emission farm produce.

- the decision support system that will use collected data and data analytics to build models and simulate the farm performance according to different metrics with the purpose of testing various scenarios to determine optimal configurations of farm systems.

The web-based FarmSustainaBl platform that will be a one-stop shop for the platform services such as: business decisions related to low-GHG emission livestock farming, smart contract management, GHG-emission management and tracing, etc.

VII. CONCLUSIONS

Through this paper we highlighted the current technological possibilities of advancing an innovative decision support platform that can optimize operation for livestock farmers by providing valuable information for decision making. The research work performed within the FarmSustainaBl project shows an indisputable need for solutions that can help the livestock farming industry face the challenges of integrating increasingly higher complexity processes, while ensuring to meet sustainability standards. Aside from the decision support system, within this paper, we have presented the process of defining the system requirements of the solution and the main components that will be developed during the project.

The future work will involve the integration of the presented system on existing farm infrastructure and the test field validation of the decision support platform, through the use case demonstrators planned to take place in a cattle farm in Romania.

ACKNOWLEDGMENT

The work presented in this paper has been funded by MCI and European Union, under ERANET-ERAGAS-ICT-AGRI3 program, through FarmSustainaBL project (contract no. 119/2019). We would like to express our gratitude to Tel-MONAER project (subsidiary contract no. 1223/22.01.2018, from NETIO Project ID: P 40270, MySMIS CODE: 105976) and WINS@HI project (PN-III-P3-3.5-EUK-2017-02-0038) for the work presented in this paper.

REFERENCES

[1] S.H. Conrad. "The dynamics of agricultural commodities and their responses to disruptions of considerable magnitude." In Proceedings of the International Conference of the System Dynamics Society, 13, 2004.

[2] N. Hostiou, et al. 'Impact of precision livestock farming on work and human-animal interactions on dairy farms. A review', 2017.

[3] H. Dong , J. Mangino et. al., "2006 IPCC Guidelines for National Greenhouse Gas Inventories", Volume 4: Agriculture, Forestry and Other Land Use , UN, IPCC.

[4] Romania's Third Biennial Report under the UNFCCC. https://unfccc.int/files/adaptation/application/pdf/3178045_romania-br3-1-br3-romania.pdf.

[5] IPCC. Source: https://www.ipcc-nggip.iges.or.jp/. Accessed Oct 2020.

[6] E. Tullo, A. Finzi, and M. Guarino. "Environmental impact of livestock farming and Precision Livestock Farming as a mitigation strategy." Science of the total environment 650 (2019): 2751-2760.

[7] C. Rojo-Gimeno, et al. "Assessment of the value of information of precision livestock farming: A conceptual framework." NJAS-Wageningen Journal of Life Sciences 90 (2019): 100311.

[8] Laloë, Denis. "Artificial Intelligence and Livestock New data, new approaches." (2019).

[9] G. Suciu, C. Nădrag, C. Istrate, A. Vulpe, M. Ditu and O. Subea, "Comparative Analysis of Distributed Ledger Technologies," 2018 Global Wireless Summit (GWS), Chiang Rai, Thailand, 2018, pp. 370-373.

[10] M. Torky and A. Hassanien, 2020. "Integrating Blockchain And The Internet Of Things In Precision Agriculture: Analysis, Opportunities, And Challenges".

[11] H. Xiong, T. Dalhaus, P. Wang and J. Huang. "Blockchain Technology For Agriculture: Applications And Rationale. frontiersin", 2020.

[12] I. Andonovic, C. Michie, P. Cousin, A. Janati, C. Pham and M. Diop, "Precision Livestock Farming Technologies," 2018 Global Internet of Things Summit (GIoTS), Bilbao, 2018, pp. 1-6, doi: 10.1109/GIOTS.2018.8534572.

[13] G Terrasson, A Llaria, A Marra and S Voaden, "Accelerometer based solution for precision livestock farming: geolocation enhancement and animal activity identification", IOP Conference Series: Materials Science and Engineering, Volume 138, II International Congress of Mechanical Engineering and Agricultural Science (CIIMCA 2015) 7–9 October 2015, Floridablanca, Colombia, doi: 10.1088/1757-899X/138/1/012004.

2020 IEEE 26th International Symposium for Design and Technology in Electronic Packaging (SIITME)

[14] G. Terrasson, E. Villeneuve, V. Pilniere, A. Llaria. "Precision Livestock Farming: A Multidisciplinary Paradigm". SMART INTERFACES 2017, The Symposium for Empowering and Smart Interfaces in Engineering, IARIA, Jun 2017, Venise, Italy. pp.55-59.

[15] R. Iqbal and T.A. Butt (2020). "Safe farming as a service of blockchain-based supply chain management for improved transparency". Cluster Computing. doi:10.1007/s10586-020-03092-4.

[16] K. Leng,et al. (2018). "Research on agricultural supply chain system with double chain architecture based on blockchain technology". Future Generation Computer Systems, 86, 641–649. doi:10.1016/j.future.2018.04.061.

[17] S. Kale, A. Apte, S. Raut, S. Dorage, S. M. Bhadkumbhe. "Blockchain based Smart Agri-Food Supply Chain Management". International Journal of Research in Engineering, Science and Management Volume-2, Issue-6, June-2019.

[18] D. Davcev, L. Kocarev, A. Carbone, V. Stankovski, K. Mitreski , 2018. "Blockchain-based Distributed Cloud/Fog Platform for IoT Supply Chain Management".

Study on Unmanned Surface Vehicles used for Environmental Monitoring in Fragile Ecosystems

Mihaela Bălănescu, George Suciu, Andreea Bădicu, Andrei Bîrdici, Adrian Pasat, Carmen Poenaru, Ionel Zătreanu

Research & Development Department
Beia Consult International
Bucharest, Romania
andreea.badicu@beia.ro

Abstract— **The advancement of Unmanned Surface Vehicles (USV) leads to new uses of these tools, such as environmental monitoring in areas where mobile sensors couldn't be deployed, or in-situ sampling took too long. Thus, high quantities of geographical and temporal information can be obtained. The need for such monitoring devices is accentuated in fragile ecosystems that suffer from regular pollution events, eutrophication, or frequent microbiological pollution episodes. The main goal of this paper is to provide a comprehensive analysis on how USVs had helped before in these kinds of environments in order to help us propose a better new solution, based on Artificial Intelligence (AI), that will be able to identify pollution sources, map the environment impact and be an analysis tool for the further research of these ecosystems. The degree of innovation must increase in a domain with so much potential. We propose to use the blockchain technology to add trust and traceability to the flow of sensor data, ensuring at the same time privacy conforming with the GDPR. Our solution aims to use an embedded AI system to provide automated piloting with the assistance Machine Learning (ML) algorithms.**

Keywords—USV; environment' monitoring; IoT; blockchain

I. INTRODUCTION

Nowadays there is a growing public and scientific concern towards the disappearing rate of habitats and species from fragile ecosystems. Coastal waters, deltas, rivers and lakes are a core component for biodiversity conservation, yet considerable amounts of pollutants are still released in areas where continuous monitoring is impossible due to the geographical characteristics.

USV's main purpose is to analyze water quality from areas where static sensors would not perform well or in places that are difficult to access. With these devices the monitoring time would be greatly reduced, and the quality of the tests would be higher. Their reliability shows that they can be used in many places that have a water source.

Most of the USVs that have been designed for research purposes had as primarily role the gathering of oceanographic data such as pollution tracking or bathymetry. One famous USV is the Springer, developed in the United Kingdom by the University of Plymouth and the Marine and Industrial Dynamic Analysis Research Group with the goal to operate hydrographic and environmental surveys in coastal waters [1]. In Germany, University of Rostock developed Measuring Dolphin [2], while in Genova, Italy, The Institute of Intelligent Systems for Automation proposed an autonomous catamaran prototype

vessel called Charlie that was used in Antarctica to gather sea surface samples [3].

The paper includes the following sections: Section II express the motivation for the research and investigates related work, Section III details the use cases and scenarios, and Section IV present the dataflow and PIMEOA AI proposed architecture. Finally, Section V presents the conclusions and future work.

II. MOTIVATION AND STATE OF THE ART

As the water quality is a massive environmental problem, and one of humanity's serious matters, the role of remote IoT sensors and blockchain-based technologies is very important [4-6]. These solutions mainly focus on measuring water levels and simple quality metrics, but there are no currently available solutions on the market that target more sensitive pollutant detection. Paper [4] presents a pilot for real-time groundwater monitoring using satellites for data transmission. A blockchain-based technology is proposed in [5] within which the collected environmental data is unified and data integrity is ensured. The system depicted in [6] uses IoT to measure wastewater volume and quality metrics (pH, hardness and oil content) and blockchain technology for storing data and developing a model.

The Surface Water Quality Monitoring Networks (WQMN) design was used to study the pollution level of the surface water in countries and regions such as the USA, Iran, Taiwan, and China. Monitoring activities of the surface water (rivers, lakes) are frequently made for protecting and improving aquatic habitats. WQMNs can be used for monitoring the water environment status and managing any pollution event emergency. The highly recommended methods for WQMN design are multivariate statistical techniques. They are valuable because of the advantage of analyzing water quality data without other data inputs. The methods which do not require data over a certain period are called geostatistical methods, and they can also execute analyses with only one data sample [7]. PIMEO AI meets new administration requirements and expands the already mentioned ones: network for non-point source management and a pollution source identification system.

Since USVs have many advantages over manned ships, like maneuverability, weight, operation costs, not being prone to accidents due to human fatigue, demand has been very high [8].

Many researchers have used several models to improve the accuracy of water quality predictions. All these models could be organized under two categories: the conventional models and artificial intelligence (AI)-based models. The most frequently used methods for modelling river water quality are based on regression techniques. MLR, MNLR (Multiple Linear and Non-Linear Regression), and ARIMA (Auto-Regressive Integrated Moving Average) models are utilized for this purpose. In the SVR approach (Support Vector Regression), the original data points are mapped into a high dimensional feature space using an appropriate kernel function, in which linear regression is implemented. On the other hand, one of the ARIMA's models purposes is to describe the auto-correlations in the analyzed data. Different types of single and hybrid AI models were considered. Results have shown that the ANN (Artificial Neural Network) model has the best modelling performances, while for the SSL (suspended sediment load), the ARIMA model performed reasonably compared to the ANN model [9].

Currently, deep analysis of suitability, precision, and methods within different surface water quality models is necessary as due to technological advancements, a multitude of software models exist, with huge differences between the results. Also, it is stated that standardization of water quality methods would guarantee consistency of applications, and that most developing countries need to understand better the water quality models and their methods of calculation and how to progress in the model standardization in order to effectively apply these models. Currently, standardization of surface water quality models based on each country's actual conditions is considered a challenge.

Considering that PIMEO-AI is an European project combining innovative technologies such as AI, blockchain, augmented piloting using ML, etc. we could approach the matter of water quality models standardization at European level in order to bring an impact on a market which is currently limited. The steps for this would include a deep research on existing such initiatives which combine AI in water quality, mobility, communication, etc. [10].

An open-source USV (Unmanned Surface Vehicle) platform for measuring water quality in natural water resources was made to improve the already existing USVs made for commercial or personal use. By having less than 4mm/year of rain, the southwestern region of Peru lives with insufficient water resources which are also contaminated by the mining industry. The platform was tested at the Tuti intake station in Peru, and the results show an adequate level of mobility and stability of the USV when it was controlled via the smartphone application. In this demonstration, the mean turbidity, temperature, and pH measured with the onboard sensors were 2180.87 NPU, 6.190 C, and 8.557, respectively [11].

Starting with the 2012 bathing season, the monitoring of the analyzed parameters in the bathing waters was reduced to two microbiological parameters: Intestinal enterococci, Escherichia coli. The evaluation of the bathing water quality allowed the classification of the bathing waters from the natural areas in 4 quality categories: excellent (30.61%), good (44.90%), satisfactory (22.45%) and unsatisfactory (2.04%) [12].

The most used procedure for analyzing the values of the measured parameters is to compare the normal values of each mentioned category with the measured value.

Thus, if the analyzed bathing water, based on the water quality data set, has a percentile value close to the standard value of a main category, it will be included in that category [13].

In the case of inland waters, the official values of intestinal Enterococci / 100ml vary from 200 (excellent quality) to 400 (good quality). Regarding coastal waters and transitional waters, the values of intestinal Enterococci / 100 ml have amounts between 100 and 200. In Romania, in 2019, studies showed excellent figures, so bathing was allowed in all the studied areas. The smallest amount of E-coli 00 mlwas measured in the following areas: Mamaia, Vama Veche, Eforie Nord, Olimp and others. The highest value in June 2019 is 330 and was recorded at Navodari. In 2020, the minimum recorded value of E-coli / 100ml remained the same, and the maximum amount was reached in July (159). Compared to the previous year, the current year has brought significant variations in the number of intestinal enterococci in bathing waters [14].

III. USE CASES AND SCENARIOS

Three fragile ecosystems were selected for the testing of the USV that will be developed in the PIMEO AI project: a lake in a dense urban environment in France (Use Case #1), Danube Delta (Use Case #2) and coastal water in Romania (Use Case #3).

Créteil lake is situated in a dense urban environment, it is an anthropogenic lake that started as a project in 1968 and currently is used for recreational activities but is also exposed to regular pollution events.

The Danube Delta is one of the largest and the best-preserved European river deltas, but also subject to multiple types of pollution and eutrophication, while the coastal waters in the area of the city of Constanta are affected by frequent microbiological pollution episodes.

The scenarios addressed, their scope and relation with the Use Cases are presented in the next table.

Table 1 – The relationship between scenarios and Use Cases

No.	Scenarios	Scope	Use Case
1	Environmental monitoring	Conservation of the flora and fauna	All use cases
2	Identification and prediction of water pollution (e.g. in ports)	Water pollution reduction	Use case #2 Use case #3
3	Monitoring of the bathing waters	Human Health protection	Use case #1 Use case #3

In all the scenarios, the USV offers its services through a web application allowing any user to demand a service. The mission generates parameters such as position of the measured water quality parameters, nutrient loading, microbial pollution and chemical or hydrocarbon leaks. The Private Blockchain Infrastructure is used for communication with the USV. The USV will offer its services to users (citizens or public employees) through the website. The Private Blockchain

Infrastructure will provide a record of the sensor measurements and protects the historical data from possible falsification.

The data flows concerning the designed scenarios are presented in the Figures 1-3.

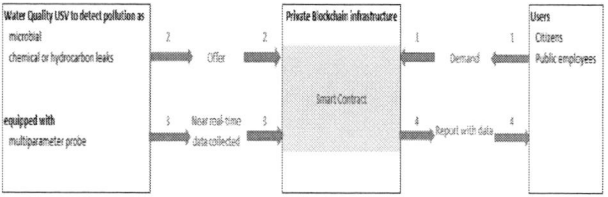

Fig. 1. Scenario "Environmental monitoring" data flow

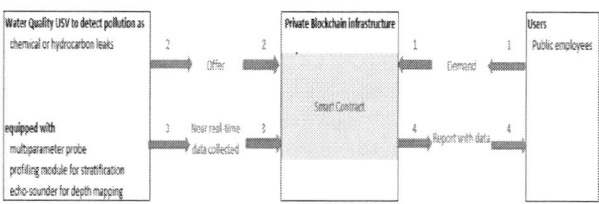

Fig. 2. Scenario "Water pollution prediction" data flow

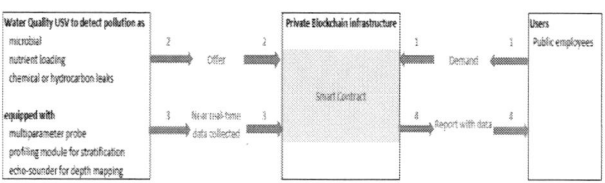

Fig. 3. Scenario "Bathing water" data flow

For the identification of PIMEO AI value, the infrastructure needed, the potential customers and finances a Business canvas model was developed (see Figure 4).

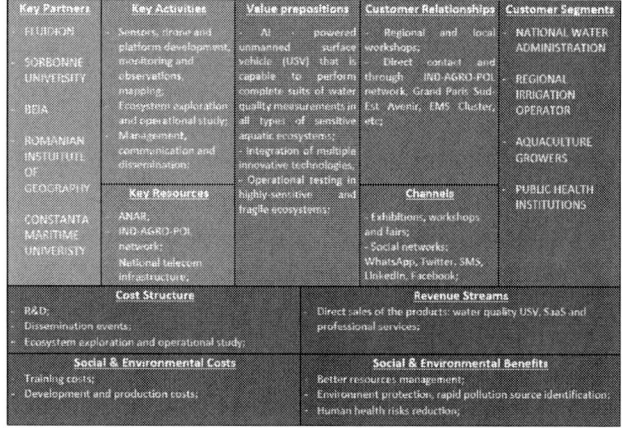

Fig. 4. PIMEO-AI Business Canvas

This model will be further extended to provide a more detailed information considering the important stakeholders as society and natural environment.

IV. ARCHITECTURE

To provide a complete operational integration of the cloud platform with the PIMEO AI drone a data infrastructure needs to be developed (Figure 5).

Based on the identified requirements, an internal research to identify the most suitable technologies for PIMEO AI application was conducted. Also was developed a modular data flow architecture based on compact technologies such as MQTT and Grafana was developed. Grafana [15] is an open source platform for viewing, monitoring, and analyzing data. For each data source, Grafana has a custom query editor and specific syntax.

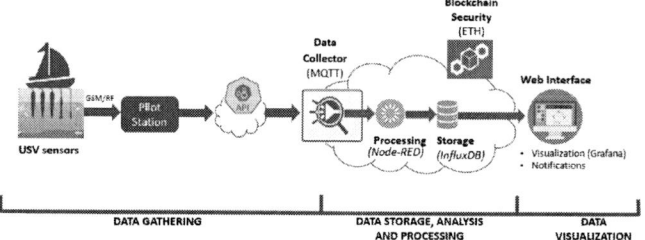

Fig. 5. The schematic data infrastructure of the PIMEO AI application.

In addition to this, the access to the cloud will be implemented using blockchain technology. Blockchain technology is a digital register known as the foundation of Bitcoin transactions, in which the details of transactions made between users are stored. A blockchain can be considered a dynamic list of records that can only grow, because changes such as deleting or changing previous records are not possible.

Blockchain merges transactions into blocks with a unique identity (hash), which are linked in a chain. Each block from the chain contains information about the previous block, apart from the initial block which is called the genesis block. Because the information is not stored in a single location, the records are easily checked. To evaluate the initial transaction, the network components execute mathematical algorithms, to check if the transaction is valid (i.e. at least 51% of the computers involved agree that the transaction is accurate) [16].

The data gathered from the USV sensors is sent to the piloting station and becomes accessible through an API. The data collector is an MQTT broker, through which the data can be sent to the private cloud, to be stored permanently and / or to be further analyzed. The output can be visualized in Grafana, within a private organization with editable permissions. Within the data visualization module, multiple data sources can be queried into multiple tables and graphs, and notifications and alarms can be configured.

V. CONCLUSIONS

PIMEO AI data flow and architecture are designed to ensure the water monitoring to provide a more granular view of the environmental impact in the fragile ecosystems. PIMEO AI will address also the missing critical components (e.g. security and privacy of the monitored data) of the existing

USV used on environmental monitoring. Thus, the project aims to increase the trust and traceability, as well as raising the security of collecting, transferring and management of the data by implementing a safe system based on blockchain technology.

Future work will focus on another novel characteristic represented by the implementation of a machine learning algorithm to develop a probability map of the pollution source position that will be available in real-time for the USV.

ACKNOWLEDGMENT

The work presented in this paper has been funded by the MarTERA partner UEFISCDI and co-funded by the European Union under PIMEO AI project (contract no. 154/2020) and by UEFISCDI under SWAM project (contract no.128/2019).

REFERENCES

[1] W. Naeem, R Sutton, J. Chudley, "Modelling and control of an unmanned surface vehicle for environmental monitoring", UKACC International Control Conference, United Kingdom, Glasgow, 1 September 2006.

[2] J. Majohr, T. Buch, C. Korte, "Navigation and automatic control of the Measuring Dolphin (MESSIN)", in Proceedings of the 5th IFAC Conference on Manoeuvring and Control of Marina Craft, pp 405-410, Aalborg, Denmark, 23-25 August 2000.

[3] M. Caccia, M. Bibuli, Ga. Bruzzone, Gi. Bruzzone, R. Bono, E. Spirandelli, "Charlie, a testbed for USV research" in 8th IFAC Conference on Manoeuvring and Control of Marine Craft, Guaruja, Brazil, 16-18 September 2009.

[4] Chohan, Usman W. "Blockchain and Environmental Sustainability: Case of IBM's Blockchain Water Management." Notes on the 21st Century (CBRI) (2019).

[5] Yan, Jinghai, Fan Zhang, Junwen Ma, Xinxin An, Yuan Li, and Yadong Huang. "Environmental Monitoring System Based on Blockchain." In Proceedings of the 4th International Conference on Crowd Science and Engineering, pp. 40-43. 2019.

[6] Iyer, Sreerag, Snehal Thakur, Mihirraj Dixit, Rajneesh Katkam, Ashish Agrawal, and Faruk Kazi. "Blockchain and Anomaly Detection based Monitoring System for Enforcing Wastewater Reuse." In 2019 10th International Conference on Computing, Communication and Networking Technologies (ICCCNT), pp. 1-7. IEEE, 2019.

[7] Jiping Jiang, Sijie Tang, Dawei Han, Guangtao Fu, Dimitri Solomatine, Yi Zheng, "A comprehensive review on the design and optimization of surface water quality monitoring networks", Environmental Modelling & Software, Volume 132, 2020, 104792, ISSN 1364-8152, https://doi.org/10.1016/j.envsoft.2020.104792.

[8] Zhixiang Liu, Youmin Zhang, Xiang Yu, Chi Yuan,"Unmanned surface vehicles: An overview of developments and challenges", Annual Reviews in Control, Volume 41, 2016, Pages 71-93, ISSN 1367-5788, https://doi.org/10.1016/j.arcontrol.2016.04.018.

[9] Taher Rajaee, Salar Khani, Masoud Ravansalar, "Artificial intelligence-based single and hybrid models for prediction of water quality in rivers: A review", Chemometrics and Intelligent Laboratory Systems, Volume 200, 2020, 103978, ISSN 0169-7439, https://doi.org/10.1016/j.chemolab.2020.103978.

[10] Wang, Qinggai & Lu, Xixi & Jia, Peng & Qi, Changjun & Ding, Feng. (2013). "A Review of Surface Water Quality Models", TheScientificWorldJournal. 2013. 231768. 10.1155/2013/231768.

[11] Wonse Jo, Yuta Hoashi, Lizbeth Leonor Paredes Aguilar, Mauricio Postigo-Malaga, José M. Garcia-Bravo, Byung-Cheol Min, "A low-cost and small USV platform for water quality monitoring", HardwareX, Volume 6, 2019, e00076, ISSN 2468-0672, https://doi.org/10.1016/j.ohx.2019.e00076

[12] https://cnmrmc.insp.gov.ro/images/rapoarte/Raport-Apa-Imbaiere-2015.pdf, Accessed 01.10.2020

[13] http://www.dreptonline.ro/legislatie/hotarare_gestionare_calitate_apa_imbaiere_546_2008.php, Accessed 01.10.2020

[14] https://dspct.ro/monitorizarea-calitatii-apei-de-imbaiere-in-sezonul-estival/, Accessed 01.10.2020

[15] https://grafana.com/, Accessed 01.10.2020

[16] G. Suciu, C. Nădrag, C. Istrate, A. Vulpe, M. Ditu and O. Subea, "Comparative Analysis of Distributed Ledger Technologies," 2018 Global Wireless Summit (GWS), Chiang Rai, Thailand, 2018, pp. 370-373.

A Pupil Detection Algorithm Based on Contour Fourier Descriptors Analysis

Petronela Bonteanu, Radu Gabriel Bozomitu
Telecommunications and IT Department
"Gheorghe Asachi" Technical University of Iasi
Iasi, Romania
petronela.bonteanu@etti.tuiasi.ro, bozomitu@etti.tuiasi.ro

Arcadie Cracan, Gabriel Bonteanu
Fundamentals of Electronics Department
"Gheorghe Asachi" Technical University of Iasi
Iasi, Romania
acracan@etti.tuiasi.ro, gbonteanu@etti.tuiasi.ro

Abstract— A high detection rate pupil detection algorithm is presented. This algorithm includes a preprocessing phase of the dark pupil image consisting of blurring, binarization and morphological operations. Next, during the processing phase, for each pupil candidate contour the Fourier descriptors of the convex-hull enclosure are determined. Each Fourier descriptor is used to obtain a metric using the Total Harmonic Distortion (THD) measure. This metric describes the similarity degree of the considered contour to an ideal ellipse. The algorithm selects the contours that fulfill the THD based metric condition and chooses the largest one as the pupil. Finally, to accurately estimate the pupil center, the Least Square Fit to Ellipse (LSFE) procedure is used. The proposed algorithm achieves over 85% average five pixels detection rate for three representative databases. The main advantage of the method is that the proposed THD based metric can be used for both circular and elliptical shaped pupil detection.

Keywords— Fourier descriptor; total harmonic distortion; image processing; pupil detection algorithm;

I. INTRODUCTION

The human-machine interface is that part of the application that allows users to express their intentions to operate on the computer and interpret the results of actions performed by the machine.

The interface is the meeting place of man and computer and in this interaction, the user can decide if the solution used is really efficient, and the hardware and software components are perfectly harmonized for a pleasant and useful experience. In order to design a good human-machine interface, there must be a transdisciplinary collaboration in which computer science, psychology, sociology, engineering and linguistics specialists are present. Unfortunately, since the introduction of the touchscreen, no obvious steps have been made to improve the comfort of using these interfaces in everyday life.

In the context of the 2020 pandemic, the importance of a contactless human-machine interface has been seen better than ever. Among the different implementation variants of this type of interfaces, like voice recognition, facial mimetic recognition or body gesture recognition, eye tracking is distinguished as a robust and natural solution. Up to this moment these interfaces were used for specific applications like communication with

neuromotor disabled patients, military or gaming [1-3]. In the current background in which a pandemic event affect everyone's life, such an interface that proves to be reliable is becoming extremely important for ordinary people and for ordinary activities and not just for specific applications.

All algorithms for estimating the gaze direction are based on the acquisition of an eye image. Two main classes of eye imaging techniques come from the use or absence of additional light sources: passive imaging and active imaging. In the case of passive images, the eye image is acquired by using the lighting available in the surroundings. Active imaging is the class of eye imaging in which artificial lighting is used to acquire the image. Based on the position of the infrared source in relation to the optical axis of the camera, there are two types of pupil images that can be acquired: light pupil and dark pupil. If the light source is close to the optical axis of the camera, the retina reflects the light coming from the lighting device in the room, making the pupil appear brighter than the iris. If the light source is far from the optical axis of the camera, the reflection from the retina moves away from the camera and the pupil appears darker than the iris.

Infrared imaging is a modern method of determining eye movement and involves the use of infrared light sources that cause reflections on the surface of the cornea. As the eye is illuminated with an infrared light source, it can be considered that this method of estimating gaze is non-invasive, because infrared radiation is invisible to the human and does not disturb his interaction with the computer. The great advantage of using IR light is that the strongest feature of the image obtained by a properly placed camera is the pupil outline rather than the other elements of the eye. This is due to the fact that both the sclera and the iris strongly reflect infrared light, while in the case of light in the visible spectrum, only the sclera has the same characteristic. We can also talk about another advantage of the method compared to the detection of the iris given by the fact that, thanks to the significantly smaller dimensions of the pupil, the difficulties caused by the obstruction made by the eyelid are much diminished.

In the literature there are proposed many algorithms for pupil detection: Starburst algorithm [4], algorithm based on the cumulative distribution function [5], integral projection

2020 IEEE 26th International Symposium for Design and Technology in Electronic Packaging (SIITME)

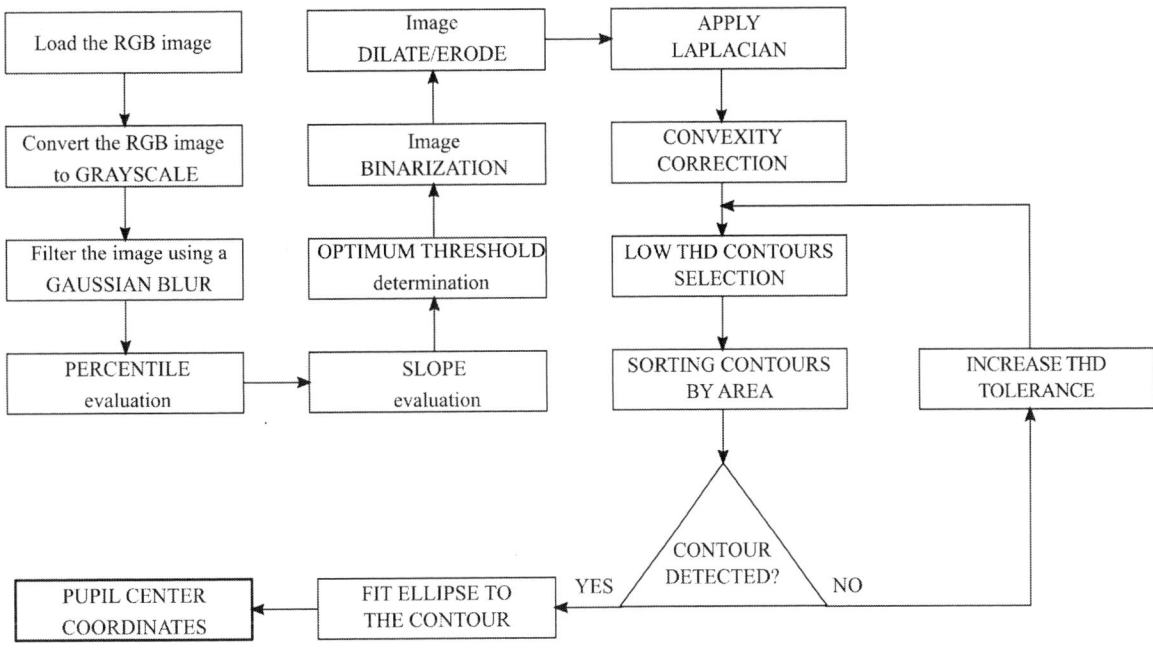

Fig. 1. Flowchart of the proposed algorithm

function algorithm [6], contour analysis [7], circular Hough transform [8], algorithms based on fitting an ellipse (LSFE) [9], the RANSAC paradigm [10], the Harris corner detector [11] and other robust and complex state of the art algorithms, such as ExCuSe [12] or PURE [13], capable of detecting the pupil in real scenarios.

The validation of a pupil detection algorithm (PDA) must be done by checking the operation on a very large number of manually labelled eye images, captured in real environmental conditions.

Although many implementations of eye tracking algorithms reported in the literature are based on the assessment of pupil circularity [8-9, 14], for real-time applications the performance of the algorithm should be independent of features such as pupil shape or orientation. The algorithm proposed in this paper aims to offer a robust solution in this regard.

II. THE PROPOSED ALGORITHM

The processing steps of the proposed algorithm are illustrated in the Fig. 1 flowchart.

Due to the fact that color information does not help the algorithm to identify important features of the image, such as edges, but also because the complexity of the code would increase substantially if the processing was based on luminance and chrominance, the first stage of the algorithm is a conversion from the RGB to the gray scale.

In order to average out rapid changes in pixel intensity, in the second phase the algorithm removes outlier pixels that may be noise in the image by filtering it using a Gaussian blur.

Next, for decomposing the image into foreground and background, the binarization technique is used. Using a technique widely presented in [14], the optimal binarization threshold is determined taking into account the first decrease in the slope of the percentile. The basic idea of this method is that the pupil is the darkest object in the eye image. Consequently, in the representation of the percentile function, immediately after the value corresponding to the pupil intensity, a decrease in the rate of slope variation occurs.

Further, due to the fact that binary images almost always contain numerous imperfections, two morphological image processing operations, dilatation and erosion, are applied to improve the shape and structure of the image resulting from binarization.

Another processing step useful in cases with pupil-attached artifacts (eyelashes, eyebrows as well as corneal reflection due to IR illumination of the eye) consists of applying the two-dimensional Laplace operator [9]. Then, in order to reconstruct the pupil, a convexity correction of the detected contours is made using the convex hull function.

For all these contours the THD metric is computed according to a technique illustrated in the section III of the paper. As pupil candidates the contours that have a THD value higher than an empirically imposed value are selected. From this set the contour with maximum area is selected as best candidate for pupil contour.

The optimum THD threshold used by the proposed metric is refined by a feedback loop as it is shown in the flowchart represented in Fig 1.

III. THE THD METRIC FOR PUPIL CONTOUR SELLECTION

Descriptors are mathematical functions that transform higher order data sets to lower order data sets that retain meaningful properties of the original data. Shape descriptors map image shape information to numerical values.

By the location of the points used to compute the descriptors, shape descriptors can be categorized into contour based descriptors and region based descriptors. Descriptors that use the points on the boundary of the shape, ignoring the shape interior content are called contour based descriptors. On the other hand, descriptors that consider the interior points of the object are called region based descriptors. There are also classes of descriptors that take a mixed approach and use a combination of boundary and interior points to produce the measures.

Generally, the contour based techniques can be more efficient for the shapes which are describable by their object boundary [15].

Fourier descriptors are a class of contour based descriptors. To obtain the Fourier descriptors the contour of the shape is represented as a parameterized curve and the resulting parametric functions are transformed into the Fourier domain using either the real Fourier transforms on the pair coordinate functions, $x(k)$ and $y(k)$ (where k is typically the path length along the contour), or the complex Fourier transform on the complex coordinate function $z(k) = x(k) + j \cdot y(k)$.

For closed contours the coordinate functions are inherently periodic (the start and end points of the contour are the same), and one can use Fourier series to compute contour descriptors.

In order to obtain a metric that describes the ellipticity, we propose computing the Total Harmonic Distortion metrics for the two Fourier series obtained from the periodic coordinate functions.

Assuming that the boundary of a particular shape has N equal length segments (this condition is assured by uniformly re-sampling the contour) numbered from 0 to $N - 1$, the k-th pixel along the contour has position (x_k, y_k). So, we can describe the contour as two parametric equations:

$$x(k) = x_k \qquad (1)$$

$$y(k) = y_k. \qquad (2)$$

By taking the Discrete Fourier Transform of each function, two frequency spectra are obtained:

$$S_x(\omega) = \mathcal{F}\{x(k)\} \qquad (3)$$

$$S_y(\omega) = \mathcal{F}\{y(k)\} \qquad (4)$$

known as Fourier descriptors [15]. In order to obtain an overall metric for the contour similarity degree to an ideal ellipse, the root mean square of the THD of the two spectra is determined:

$$THD_c = \sqrt{\frac{THD_x^2 + THD_y^2}{2}} \qquad (5)$$

where THD_x and THD_y are the total harmonic distortions for the $S_x(\omega)$ and $S_y(\omega)$ respectively.

IV. EXPERIMENTAL RESULTS

The proposed PDA was developed in the Python 3 programming language by using the OpenCV functions and has been tested on three representative databases of eye images. The following databases have been used in our analysis: DB_1 consists in 400 eye images having a 480x640 pixels resolution captured with a head-mounted eye tracking device developed by our team; DB_2 includes 400 eye images with a 480x640 pixels resolution from publicly available Iris Casia Lamp database [16]; DB_3 resents 524 eye images from the set XII provided in [12] with a resolution of 288x384 pixels captured in bad illumination conditions.

The experimental obtained results have been compared to those obtained by the state-of the-art ExCuSe algorithm [12] and also to those obtained by Circular Hough Transform (CHT) based algorithm proposed by the authors in [8]. Fig. 2 shows the detection rates depending on the Euclidian distance for the proposed algorithm compared to those obtained by the opensource ExCuSe algorithm. In Fig 3 these results are compared to those obtained by the CHT algorithm. In Fig 4 the Euclidian distance obtained by the proposed PDA for the three databases are illustrated.

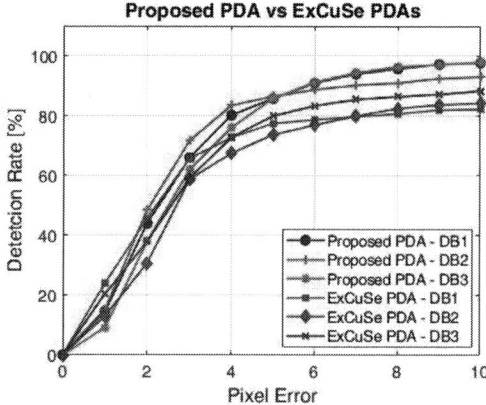

Fig. 2. Detection rate depending on pixel errors achieved by ExCuSe and the proposed algorithm on each database of eye images.

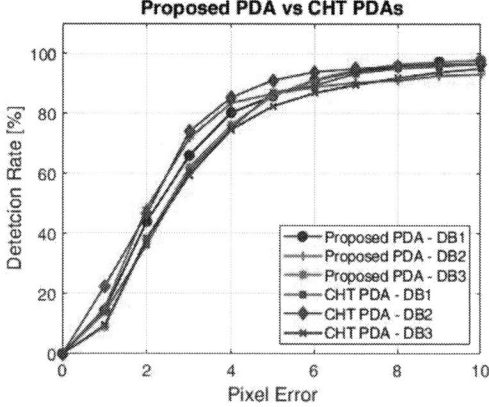

Fig. 3. Detection rate depending on pixel errors achieved by CHT and the proposed algorithm on each database of eye images.

TABLE I. EXPERIMENTAL RESULS

Statistical Measures	ExCuSe [12]			CHT PDA [8]			Proposed PDA		
	DB_1	DB_2	DB_3	DB_1	DB_2	DB_3	DB_1	DB_2	DB_3
DR_5 [%]	77.50	73.75	79.96	87.00	91.00	82.44	85.75	86.5	85.87
$\bar{d}$ [px]	11.56	14.68	12.38	3.37	2.91	3.53	2.87	3.40	3.10
σ_d [px]	68.87	50.69	38.10	3.47	3.89	4.57	2.39	4.62	2.51

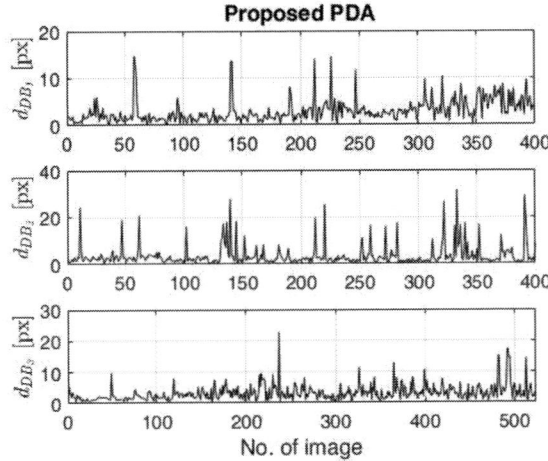

Fig. 4. Euclidian distance obtained by the proposed PDA for the three databases.

In Table 1 several statistical indicators like the detection rate at 5 pixels, the mean value and the standard deviation of the Euclidean distance for all the algorithms analyzed are presented. The better results obtained by the CHT algorithm for DB_2 database are due to the fact that in all eye images the pupil has a circular shape. According to the experimental results, the proposed algorithm provides better results for DB_3 database which is most representative for real time applications, including images acquired from a video recording taken in bad illumination conditions.

The proposed algorithm exhibits similar detection rates for images from three representative eye image databases, proving to be accurate, robust, fast and suitable for real time applications.

V. CONCLUSIONS

In this paper a robust pupil detection algorithm based on Fourier descriptors has been proposed. The algorithm includes well known processing steps (blurring, binarization and morphological operations.) for pupil contour detection.

The contour that best approximates the pupil edge is selected using a metric based on Total Harmonic Distortion measure. Due to the fact that the metric describes the similarity degree of the detected contour to an ideal ellipse, the performance of the algorithm is less dependent by the shape of the pupil compared to the benchmark implementations.

The algorithm has been tested on three representative databases and the resulted accuracy and running time confirm that it is suitable for real time applications.

REFERENCES

[1] R. G. Bozomitu, L. Niță, V. Cehan, I. D. Alexa, A. C. Ilie, A. Păsărică and C. Rotariu, "A New Integrated System for Assistance in Communicating with and Telemonitoring Severely Disabled Patients", Sensors, ISSN 1424-8220, Vol. 19, Issue 9, Article Number 2026, 2019,

[2] D.B. Mohan,G. Prabhakar, J. Shree, L R D. Murthy and P. Biswas, "Eye Gaze Tracking in Military Aviation", 2019.

[3] Wojciech Wojcikiewicz, "Hough Transform, Line Detection in Robot Soccer", Coursework for Image Processing, 14th March 2008;

[4] D. Li and J. D. Parkhurst, "Starburst: A robust algorithm for video-based eye tracking," in Proceedings of the IEEE Vision for Human-Computer Interaction Workshop, 2005.

[5] Mansour Asadifard, Jamshid Shanbezadeh, "Automatic Adaptive Center of Pupil Detection Using Face Detection and CDF Analysis", in Proc. of the International MultiConference of Engineers and Computer Scientists 2010, Vol. I, IMECS 2010, March 17 – 19, 2010, Hong Kong;

[6] Z.H. Zhou and X. Geng, "Projection Functions for Eye Detection", Journal of Pattern Recognition (ELSEVIER), Vol. 37, No. 5, 2004, pp. 1049-1056;

[7] D. Zhu, , S. T. Moore, and T. Raphan. "Robust pupil center detection using a curvature algorithm." Computer methods and programs in biomedicine 59.3 (1999): 145-157;

[8] P. Bonteanu, R. G. Bozomitu, A. Cracan, G. Bonteanu, "A New Pupil Detection Algorithm Based on Circular Hough Transform Approaches", SIITME 2019 proc., pag. 171-172, Cluj-Napoca, October 23 – 26, 2019;

[9] P. Bonteanu, A. Cracan, R. G. Bozomitu and G. Bonteanu, "A New Robust Pupil Detection Algorithm for Eye Tracking Based Human-Computer Interface", ISSCS 2019 proc., Iasi, Romania, 2019, pp. 1-4;

[10] M. Fischler, R. Bolles, "Random sample consensus: a paradigm for model fitting with applications to image analysis and automated cartography", Communications of the ACM 24 (6), 1981, pp. 381-395;

[11] Das, Alak, and Dibyendu Ghoshal. "Human eye detection of color images based on morphological segmentation using modified Harris corner detector.", ICETEEEM, 2012;

[12] W. Fuhl, T. Kübler, K. Sippel, W. Rosenstiel, and E. Kasneci, "Excuse: Robust pupil detection in real-world scenarios," in International Conference on Computer Analysis of Images and Patterns, 2015, pp. 39–51;

[13] T. Santini, W. Fuhl, E. Kasneci. "PuRe: Robust pupil detection for real-time pervasive eye tracking", Comput. Vis. Image Underst. 2018, 170, 40–50;

[14] P. Bonteanu, A. Cracan, R. G. Bozomitu and G. Bonteanu, "A Robust Pupil Detection Algorithm Based on a New Adaptive Thresholding Procedure", EHB 2019 proc., Iași, Romania, November 21-23, 2019;

[15] F. Mokhtarian and M. Bober, "Curvature Scale Space Representation: Theory, Applications, and MPEG-7 Standardization", Norwell, MA, USA: Kluwer Academic Publishers, 26, 2003;

[16] "Casia-Iris-Lamp", http://biometrics.idealtest.org/dbDetailForUser.do?id=4.

Intelligent Warning System for Drivers

Loredana-Maria BURCIU
ETTI, University Politehnica of Bucharest
Bucharest, Romania
loredanamaria0809@gmail.com

Radu-Petru FOTESCU
ETTI, University Politehnica of Bucharest
Bucharest, Romania
radufotescu@yahoo.com

Rodica Constantinescu
ETTI, University Politehnica of Bucharest
Bucharest, Romania
constantinescu.rodica@gmail.com

Paul SVASTA
ETTI, University Politehnica of Bucharest
Bucharest, Romania
paul.svasta@cetti.ro

Abstract— **Nowadays the vehicles contains a lot of electronic tehnology and already exists many smart systems mounted on them. The direction of drive the vehicle is determined with the help of the accelerometer. The proposed system aims to increase sensory sensitivity in identifying where the car is in a perimeter when a student learns to drive. This system help the drivers with less experience, such as beginners, so that they acquire driving skills as soon as possible and this system help the drivers to drive safely. It ensures distance monitoring four directions: front, back, left and right of the car. The system alerts the driver both optical and acoustic when the car is very close to an obstacle when the car moves with a speed lower than the proposed threshold.**

Keywords— Microcontroller, sensors, parking.

I. INTRODUCTION (*HEADING 1*)

This system activates a certain set of ultrasonic sensors depending on the direction of drive the vehicle and display the distances on an LCD display. A red LED will light and the buzzer will sound a warning when one of the four distances is lower than a threshold value. For development this system the following were used: Raspberry Pi, four ultrasonic sensors for distance, an accelerometer, LCD display, two LEDs for warning and a buzzer like in figure 1.

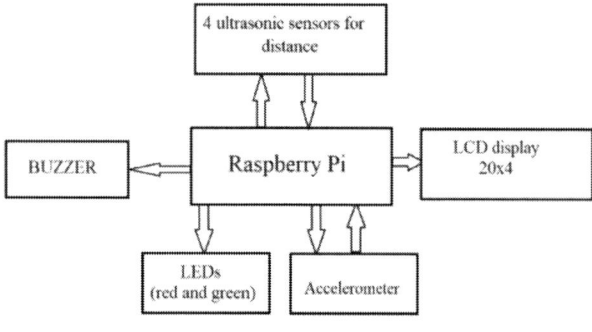

Fig. 1: Block diagram

II. HARDWARE IMPLEMENTATION

A. Raspberry Pi

Raspberry Pi 4 Model B is the latest product in the popular Raspberry Pi range of computers. It offers ground-breaking increases in processor speed, multimedia performance, memory, and connectivity compared to the prior-generation Raspberry Pi 3 Model B+, while retaining backwards compatibility and similar power consumption. For the end user, Raspberry Pi 4 Model B provides desktop performance comparable to entry-level x86 PC systems. [1]

B. Ultrasonic sensor HC-SR04

Four HC-SR04 ultrasonic distance sensors were used to develop this system. This type of sensor is well known in the electronics market due to its stable performance and high accuracy.

In figure 2 it can be seen that this type of sensor has 4 pins, namely: VCC, TRIG, ECHO and GND.

The dimensions of this sensor are as follows: 45 mm x 20 mm x 15 mm. It is good to know these values for realization the project because their position is very important for proper operation and the dimensions must be taken into account.

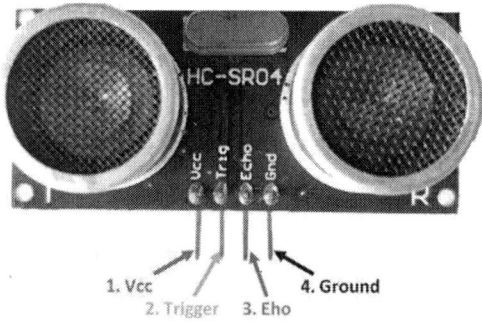

Fig. 2: Ultrasonic Sensor HC SR04 Pin Diagram [2]

. Fig. 3: Ultrasonic Sensor HC SR04 operation [2]

The ultrasonic sensor for distance can measure distances between 2cm and 4m [3]. The measurement error is between 0mm and 1.5cm. This sensor measures with an angle of 15° and the main advantage of this sensor is the low energy consumption (consumed current is about 15mA). The sensor emits a signal with frequency equal with 40 KHz like in figure 3 and determines the time until this reflected signal is received back by the sensor. Based on the previously determined time, the distance is calculated.

C. Digital Accelerometer ADXL345

The ADXL345 is a small, thin, low power, 3-axis accelerometer with high resolution (13-bit) measurement at up to ±16 g. Digital output data is formatted as 16-bit twos complement and is accessible through either a SPI (3 or 4 wire) or I2C digital interface. The ADXL345 is well suited for mobile device applications. It measures the static acceleration of gravity in tilt-sensing applications, as well as dynamic acceleration resulting from motion or shock. It is high

resolution (4 mg/LSB) enables measurement of inclination changes less than 1.0°. Several special sensing functions are provided. Activity and inactivity sensing detect the presence or lack of motion and if the acceleration on any axis exceeds a user-set level. Tap sensing detects single and double taps. Free-fall sensing detects if the device is falling. [4]

D. Buzzer and LCD

A buzzer and an LCD were used for visual and audible warning. It is important to have both types of warnings so as not to distract the driver from driving. The driver is acoustically alerted when he approaches an obstacle and then he check the distances to the obstacle that are displayed on the LCD.

E. Electrical scheme

The electrical scheme of this system was developed in the EasyEDA platform. This platform is a suite of EDA tools that allow engineers to design, simulate and share publicly or privately their works. Also within this online platform, BOM (Bill of materials) can be generated Gerber files and others in various formats, such as: PDF or PNG.

This electrical scheme of the warning system can be seen in figure 4.

After the virtual realization of the connections, the physical connections were also made on a breadboard support. Before the assembly of the system, the functionality of each component was checked.

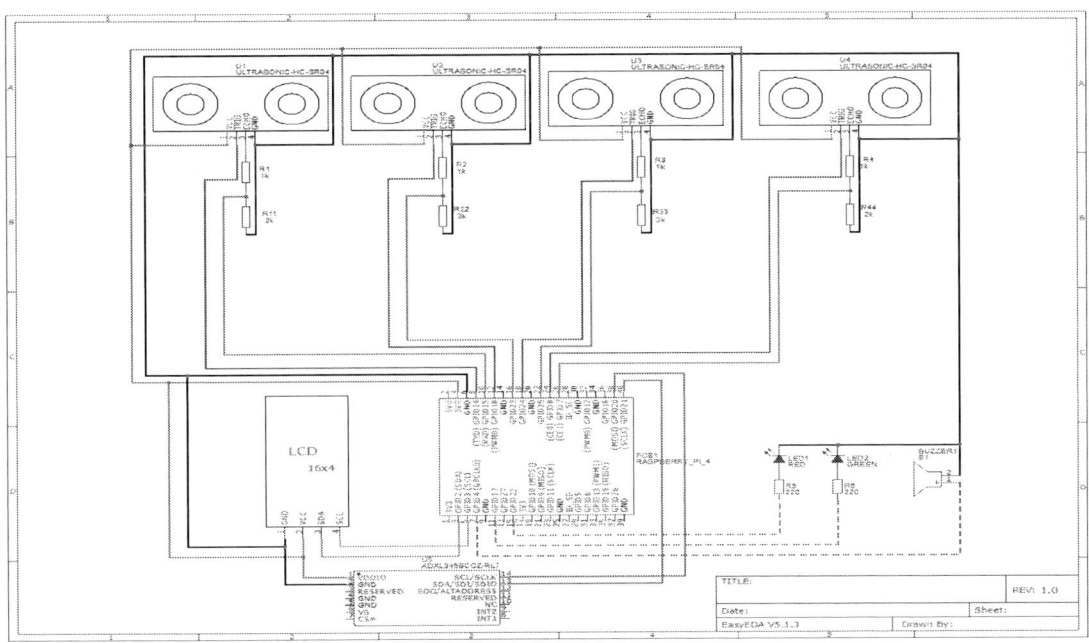

Fig. 4: Electrical scheme of system

III. SOFTWARE IMPLEMENTATION

The source code for this system was developed in the Python programming language (version 3). The source code contains several functions that control the system components and which are called in the main program.

Both the LCD and the motion detection sensor ADXL345 need the SDA (Serial Data Line) and SCL (Serial Clock Line) pins of the Raspberry Pi board. Because the Raspberry Pi minicomputer has defined only one set of SDA and SCL pins, it was necessary to define two more SDA and SCL pins so that the LCD and the motion detection sensor ADXL345 can be connected simultaneously.

Thanks to the Raspberry Pi microcomputer, all these distances from possible objects can be retained and different statistics can be made in order to be able to characterize the performance of a driver.

IV. CONCLUSIONS AND RESULTS

In present exists similar systems that consist of either an ultrasonic distance sensor and two LEDs, either an ultrasonic distance sensor and an LCD display or an ultrasonic distance sensor and a buzzer.

From a hardware point of view, the novelty of this system consists of the simultaneous use of a number of four distance ultrasonic sensors, an accelerometer, an LCD screen, a buzzer and two LEDs.

From the software point of view, the functions "get_distance1()", "get_distance2()", "get_distance3()" and "get_distance4()" for the ultrasonic sensors have been adapted, the functions "green_light() "and" red_light() " for LEDs and has been developed the main program, written in the Python programming language has been developed such that running time will be reduced to a minimum.

Various tests were performed to verify that the distances measured using the ultrasonic sensor are correct. The results of these tests are shown in Table 1 and Table 2.

TABLE I.   MEASUREMENT RESULTS FOR DISTANCE

Real distance [cm]	Distance measured [cm]		Error	[cm]
7.5	7.345	0.155
25	25.003	0.003
10	10	0
100	98.543	1.457
4	4.278	0.278

TABLE II.   MEASUREMENT RESULTS FOR ANGLE

Obstacle placement angle to the sensor	Correct measurement?
5°	YES
15°	YES
17°	YES
20°	NO
30°	NO

In present, this system is at test vehicle level (figure 5) and it can be mounted on a toy car with medium size (figure 6). This system will be developed so that it can be mounted on a real car.

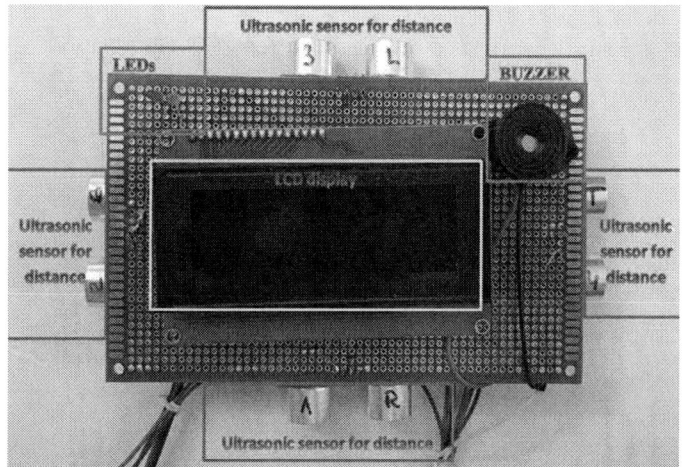

Fig. 5: Test vehicle

Fig. 6: Toy car with medium size

Even if there are currently similar systems on the market, the system presented in this paper has the advantage that it uses a microcomputer and with its help the following can be retained:

- all distances from possible objects determined by ultrasonic sensors;

- the speeds at which the vehicle was driven determined by means of the motion and speed detection sensor ADXL345.

With the help of the previously mentioned and saved data, different statistics or characteristics of the driver can be made.

For example, if this system were connected to another system that determines how nervous the driver is (by measuring the pulse or by the music he listens to), speed statistics can be made according to the driver's mood.

REFERENCES

[1] https://static.raspberrypi.org/files/product-briefs/200521+Raspberry+Pi+4+Product+Brief.pdf accessed on 7 October 2020;

[2] https://components101.com/ultrasonic-sensor-working-pinout-datasheet accessed on 7 October 2020;

[3] https://datasheetspdf.com/pdf/1380136/ETC/HC-SR04/1 accessed on 29 June 2020;

[4] https://www.sparkfun.com/datasheets/Sensors/Accelerometer/ADXL345.pdfr accessed on 7 October 2020;

[5] Derek Molloy, "Exploring Raspberry Pi", John Wiley And Sons Ltd, 2016;

[6] Muftah Fraifer, Mikael Fernstrom,"Investigation of Smart Parking Systems and their technologies", Thirty Seventh International Conference on Information Systems, Dublin 2016.M. Young, The Technical Writer's Handbook. Mill Valley, CA: University Science, 1989;

[7] Atanas Dimitrov ; Dimitar Minchev, "Ultrasonic sensor explorer", 2016 19th International Symposium on Electrical Apparatus and Technologies (SIELA), 2016;

[8] Azmyin Md. Kamal, Salman Habib Hemel, Mohsan Uddin Ahmad, "Comparison of Linear Displacement Measurements Between A Mems Accelerometer and Hc-Sr04 Low-Cost Ultrasonic Sensor", Advances in Science Engineering and Robotics Technology (ICASERT) 2019 1st International Conference on, pp. 1-6, 2019.

[9] Yuan Xiuping ; Li Jia-Nan ; Fang Zuhua, "Hardware Design of Fall Detection System Based on ADXL345 Sensor", 2015 8th International Conference on Intelligent Computation Technology and Automation (ICICTA), 2015;

[10] Jin S. Jang, Moon G. Joo, Won Chang Lee, Dong Won Jung and Zhong Soo Lim, "Identification and distance detection for ultrasonic sensor by correlation method", The International Federation of Automatic Control Seoul, Korea, July 6-11, 2008;

[11] David Pérez-Morales, Salvador Domínguez-Quijada, Olivier Kermorgant, Philippe Martinet, "Autonomous parking using a sensor based approach", 2016 IEEE 19th International Conference on Intelligent Transportation Systems (ITSC), 2016;

[12] Karzan A. Raza ; Wrya Monnet, "Autonomous parking using a sensor based approach", 2016 IEEE 19th International Conference on Intelligent Transportation Systems (ITSC), 2016.

Algorithm to Design Conductive Mesh for Tamperproof Envelope

Sorin Chiţu, Daniel Ciprian Vasile, Tudor Ioan Honceriu, Paul Svasta

UPB-CETTI
University POLITEHNICA of Bucharest
Bucharest, Romania
sorin.chitu@cetti.ro

Abstract—*Protection of the Critical Security Parameters is a permanent concern for the designers, but also for the users of cryptographic equipment. The usage of a conductive mesh is a sensitive and efficient solution in order to protect the firmware, keys or any other sensitive data that could be contained in a cryptographic module. In order to improve the security provided by this principle, based on the flexibility of common technology that can be used to produce PCBs, an algorithm to produce particular designs of conductive mesh on PCBs starting from random bit strings is present in this article. Random design of conductive mesh is useful in order to increase the unpredictability of its electrical characteristics so, in addition to the sensitivity of this conductive mesh which will detect and react even to any attempt of measuring it by probes, an attacker will not have any information which can be exploited. The proposed innovative algorithm provides filling of the full area of envelope which cover the cryptographic module, even if its perimeter is irregular, according to necessary dimensions and profile, keeping traces on a dense grid, without any uncovered areas. The main advantage of the proposed solution consists of the possibility to implement a fully automated production flux, without human participation, and with an increased level of security due to unpredictable electrical characteristics of conductive mesh generated from a true random bit string.*

Keywords—tamperproof; random; mesh; algorithm; security;

I. INTRODUCTION

In order to protect Critical Security Parameters [1][2] or any sensitive data which can be stored inside of a cryptographic module, many solutions were been developed during time but, most of them, involve a long term energy source which provides the necessary energy for the electronic circuits that supervise and protect the data during power-off mode. A solution that doesn't involve an energy source is based on a conductive mesh (CM) and it was presented in work [3]. The conductive mesh, that act like a sensor, is connected to a dedicated electronic circuit that inject different signal pulses on CM and calculate associated spectral power at its output, being capable to detect any intrusion attempt which changes the electrical characteristic of CM in frequency domain. One of the advantages of this principle is that the electrical characteristic of conductive mesh has, besides its sensitivity even to touch of an oscilloscope probe, a nondeterministic form, due to the technological dispersion of

manufacturing, which makes impossible for an attacker to compensate the effect of any physical intrusion or to simulate an adequate signal at the output of the conductive mesh.

According to the presented aspects, a better overall security performance could be obtained by extending the designs of conductive mesh, with tacking into consideration the possibility of ease fabrication of particular models of Printed Circuit Board (PCB) and, also, the low cost of the involved technologies. This extension, that significantly reduce the possibility for an attacker to know or to predict the electrical characteristic of a conductive mesh, can be achieved using a True Random Number Generator (TRNG), whose output bit string will be used in order to create an associated route form on a PCB.

The objective of the work was to develop an algorithm to provide the filling with a continuous route of an area defined by a given perimeter, according to necessary dimensions of the tamperproof envelope, which pass through all the dots of a dense grid. The distance between the grid dots shall be close enough to avoid any intrusion attempt by any means, even cut or bypass any part of the route. Also, a true random number generator was used, as an input to the algorithm, in order to obtain various route designs, which will have unpredictable electrical characteristics. The main effort in algorithm designing was focused on identifying all criteria that has to be taken into consideration in route design in order to obtain the desired objective.

II. DESCRIPTION OF THE SOLUTION

A. Abstracted Problem to be Solved

Summarizing the statement of the problem, a solid line must be drawn in an area defined by a predetermined perimeter, without intersecting with itself. As a prerequisite, a grid is defined on the mentioned area, the distance between the neighboring points being established as a security parameter against any physical intrusion attempts which can occur due to an attacker's actions. In this context are included cuts or bypasses of the routes, or cuts between adjacent routes which could permit the access inside the box which protect the cryptographic module. The route has to pass through all the grid points, between entry and exit. In Fig. 1 is presented a template design of a tamper proof box.

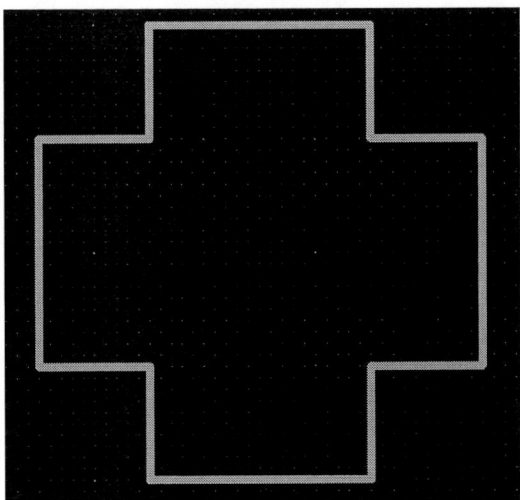

Fig.1: Perimeter of a tamper proof box design template

The entry and exit points of the conductive mesh will be established in accordance with connectivity necessities to main board.

Also, a random bit string is used to conduct the conductive mesh design. The result will be an unpredictable form of the conductive mesh. In these conditions, the electrical characteristics of the conductive mesh will be unpredictable so, according to [3], it will increase the level of security provided by this tamper proof solution.

B. Conductive Mesh Design Principle

Considering a long enough random bit string together with a set of rules, a conductive mesh route will be drawn inside of a predefined perimeter.

In this scope, successive pairs of bits are interpreted according to associations of the four possible directions:

00 – up

01 – right

10 – down

11 – left

The route of the conductive mesh will start from the entry point and it will evolve guided by random bit pairs. Obviously, depending on the neighborhoods of each point, not every pair of bits could be implemented so supplementary rules has to be adopted.

First of all, the availability of the indicated direction has to be checked and, if the movement it's not possible, the corresponding pair of bits will be dropped.

But also, different situations could appear. The main issue is the prevention of creation of potentially closed areas, which once entered there is no exit. Such a situation is presented in

Fig.2 where the evolution from yellow arrow marked point could create a closed area.

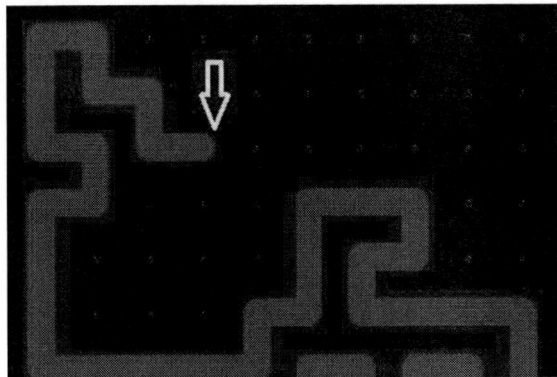

Fig.2: A potentially closed area

To avoid such a situation, the route compulsory must go to the down direction being the unique solution, so all the other pairs of bits will be dropped until a "10" combination will be found.

Based on the presented situation, another helpful rule appears: the pre-reservation route. As shown in Fig.3, traced with green line, some grid points can be pre-reserved for future evolution of the based route (red line). This rule prevents the usage of these points in the configuration of other routes which certainly will produce blocking situations. Integration of the pre-reserved route will be done by consuming the corresponding bits from the string.

Fig.3: Preserved route

Another possible situation consists in the formation of two areas, separated by a two-point wide passage, in which case, the decision of continuing the conductive mesh route has to be taken, mandatory, to the other area than the one containing the exit point. At the same time, the other point of the passage, and other neighborhood points if appropriate, will be pre-reserved for returning route.

In order to implement an algorithm capable of designing conductive mesh, a set of metadata will be associated to each point of the grid. So, besides the information regarding the availability, pre-reservation or occupied status of each point, by indexing the route points, a trace-back option is available in a blocking situation, until the last decision point.

In such situations, if the trace-back doesn't succeed, a permutation between entry and exit points could be a solution. If the problem persists, other entry and exit points must be chosen.

The last rule, in fact, a derogation from the movement rule, consists in the possibility, if it is necessary, of connecting the last two points of the route by a diagonal line. Even in this situation, the security level provided by the tamper proof solution will not decrees.

A practical implementation which exemplifies the mentioned rules is presented in Fig. 4.

Fig.4: Example of a conductive mesh designed based on a random bit string

For this design it was used the random bit string presented in Fig.5.

```
01111000 01101100 00110100 01111000 01001100 11110111 10001010 01001010
11101110 11010011 00100100 10101100 10111111 11101101 11011001 11100011
00011110 01000110 10000110 10000011 11101101 00001110 10110110 01100100
10001110 01010101 11111110 10110110 11110100 11101100 00100011 10111010
00100000 00001001 10100011 00111011 00111101 10110010 01011011 10000111
00100110 10100000 00100010 11011110 01000000 10111101 11011111 11111100
10010000 11010101 10101100 00110110 01101100 00000111 11001110 11101010
01001011 11010101 01001010 00110001 00111110 11101101 10011000 01000010
11111011 11010111 10000010 11000100 00101010 11100011 11111111 01110111
10101001 00110111 00101010 11011010 11011101 00010011 10111101 11001000
11011011 00011001 10010110 00010101 10110111 01010110 01100111 10001000
10010000 00110100 11100111 00010111 00011001 01100001 10111100 00111011
11010111 01110000 00000011 11010100 00110000 10111110 10010110 00010111
00100100 11101000 01101111 01101010 11010011 00001000 00011010 00100101
00011101 00010000 11010100 00100010 10100111 11111001 00000101 11000010
01000100 01111000 00100110 01110000 00111001 01110100 11100110 00001100
```

Fig.5: Random bit string used for conductive mesh design

III. RESULTS

Comparing the results of signal analysis on different conductive mesh designs it can be observed significant differences between electrical parameters. For relevant results to demonstrate these unpredictable differences several designs of conductive mesh were studied. There were analyzed different constructive forms of conductive mesh drawn inside of the same perimeter, a square with 30mm side. The used grid is 0,4mm and the trace with is 0,2mm. A typical conductive mesh design is presented in Fig. 6.

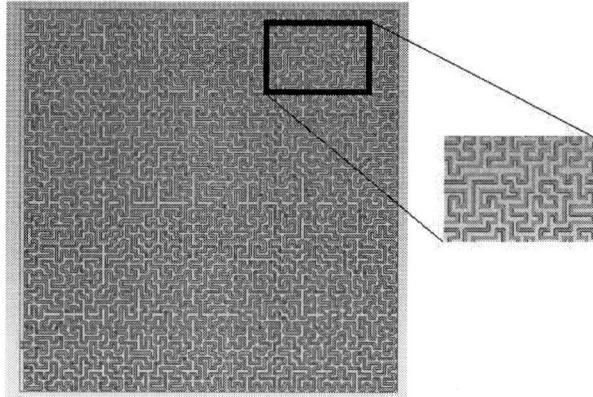

Fig.6: 30x30mm Conductive mesh based on random line segments

IV. CONCLUSIONS

This approach will have the effect of a variation of the electrical characteristic of the conductive mesh in a series production and, implicitly, an unpredictable response to probing signals. Also, using random line segments, instead of a pattern, increase the difficulty of any physical intrusion.

The proposed algorithm provides a solution to complete digitization and automation of tamper proof design and production processes. By implementing a design and production flux without human intervention increase the security level, by a higher trust level in maintaining the confidence of the tamper proof electrical parameters.

ACKNOWLEDGMENT

This paper and research activities were supported from the funds of the project "Integrated Development 4.0" (iDEV40) funded from the ECSEL Joint Undertaking (JU) under grant agreement No 783163 , respectively the "Dezvoltare integrată 4.0" project having the SMIS code 122386 within the Competitiveness Operational Program (POC) - Romania. The JU receives support from the European Union's Horizon 2020 research and innovation programme. It is co-funded by the consortium members, grants from Austria, Germany, Belgium, Italy, Spain and Romania. The information and results set out in this publication are those of the authors and do not necessarily reflect the opinion of the ECSEL Joint Undertaking.

REFERENCES

[1] "FIPS PUB 140-2: Security Requirements for Cryptographic modules". NIST. 2002-12-03.

[2] PROPOSAL FOR A REGULATION OF THE EUROPEAN PARLIAMENT AND OF THE COUNCIL, https://ec.europa.eu/transparency/regdoc/rep/10102/2017/EN/SWD-2017-500-F1-EN-MAIN-PART-6.PDF.

[3] D.C. Vasile, P. Svasta, "Active Tamper Detection Circuit Based on Statistical Analysis", 2017 IEEE 23rd International Symposium for Design and Technology in Electronic Packaging (SIITME) DOI:10.1109/SIITME.2017.8259884.

Machine Learning algorithms for air pollutants forecasting

Marius Dobrea, Andreea Bădicu, Marina Barbu, Oana Șubea, Mihaela Bălănescu, Geroge Suciu, Andrei Bîrdici, Oana Orza
Research & Development Department
BEIA Consult International
Bucharest, Romania

Ciprian Dobre
Department of Computers
University POLITEHNICA, Bucharest, Romania
Bucharest, Romania

Abstract— **Air pollution represents an issue that raises many concerns nowadays, as it has various negative effects on the environment and the economy worldwide. Because of the rapid urbanization, cities are suffering from polluted air, so it is important to predict future air quality. For this purpose, new applications of artificial intelligence should be employed. In this paper, we will present several Machine Learning algorithms, the possible software that can be used for them and the applications used in the field of air quality. Based on the research in the field, we propose SVR, ARIMA and LSTM, 3 Machine Learning models, which can be used to predict air pollution. These algorithms have been tested using time-series for PM$_{10}$ and PM$_{2.5}$ particles. The results showed that SVR and ARIMA algorithms are the most suitable in forecasting air pollutant concentrations.**

Keywords—Machine Learning, Air Pollution, Forecasting, Time Series

I. INTRODUCTION

Air pollution is an issue that concerns a lot of people nowadays and has a significant influence on human health worldwide. It has a great impact on human well-being, the environment and the economic advancement around the world. Recent studies [1] have shown that in 2015, 6.5 million premature deaths worldwide were caused by air pollution. Its impact on health depends on the concentration of the pollutant and the exposure levels. Particulate Matter (PM) represents one of the most important pollutants in regard to the effects on health. Among the best-known PMs are the PM$_{10}$ (PM with a diameter lower than 10 m) and PM$_{2.5}$ (PM with a diameter lower than 2.5 m). Even in small concentrations, these PMs can have many adverse effects on human health [2]. Besides the type of pollutant and its concentration, the duration and the frequency of the exposure are also important factors in the negative impact on human's well-being [3].

Considering that traditional air quality prediction methods require more computational power for the estimation of pollutant concentration, many people are trying to apply Artificial Intelligence (AI) algorithms (machine learning, deep learning), which can lead to better results. There is an increased interest in the machine learning methods for forecasting non-

linear time series information, such as meteorological and pollution data. Machine Learning (ML) is a data technique which teaches a computer to create a model using training data. It is a subfield of artificial intelligence and enables software applications to be increasingly precise in predicting results. ML can review a wide range of data and discover patterns and specific trends.

The literature on this topic proposes a wide range of tools for ML classifiers to deal with time-series data. This paper will consider three models of ML algorithms to forecast air pollution and will assess their performance on forecasting air pollutant concentrations. The models used are SVR, ARIMA and LSTM and will be described in the next paragraphs. Support Vector Regression (SVR) is a version for Support Vector Machine (SVM) algorithm characterized by the use of kernels, hyper planes, boundary lines and support vectors [4]. It is shown to be an effective tool in real-value function estimation and its application presented in chapter 4.1. Chapter 4.2. focuses on the application of the Autoregressive Integrated Moving Average (ARIMA) model, which is a class of models that describe a time series based on its past values [5]. It is characterized by the parameters p (lag order), d (degree of differencing) and q (order of moving average). Long Short-Term Memory (LSTM) is a neural network model used for classification, processing and predicting based on time series data, in which there are lags of unknown duration between important events [6]. The algorithm is characterized by a cell, an input gate, an output gate and a forget gate. The application of this model in the present paper is presented in chapter 4.3.

II. RELATED WORK

Authors in article [7] use 50 time series for monthly temperature and precipitation in Greece, and they want to analyse the quality of the forecast. The proposed methods are based on two ML algorithms: one of them uses a Neural Network (NN), and the other uses SVM. From the experiment, it results that by using the SVM algorithm, it is possible to obtain a better performance. In the end, they concluded that, based on their score, there is no relationship between the parameters that result from the time series and the forecast quality. In their paper [8], the authors propose an online learning algorithm for estimating Auto-Regressive Integrated Moving Average (ARIMA) models. ARIMA models are one of the most general classes of models for forecasting time series. The presented method involves the use of recursive formulation in an online learning environment. According to the experiments, the authors conclude that the proposed online ARIMA model is probably as good as the best ARIMA model. Consequently, the estimation of the parameters can be done online in a scalable and efficient way. In article [9], the authors develop a hybrid method based on ML algorithms, which is supposed to detect jumps in financial time series. The model inputs slow frequency market data and uses a convolutional neural network and LSTM (Long Short-Term Memory). The limitations of the model are also mentioned (very high computational complexity, hence high computation time). Paper [10] presents different deep learning models for predicting future PM_{10} value. The results suggested that the performance of Gated Recurrent Unit (GRU) network is better than the Recurrent Neural Networks (RNN) and LSTM networks. They also discovered a pattern for PM_{10} concentration. In the article [11], the main goal is to obtain a method based on machine learning for $PM_{2.5}$ concentration, using regression modelling with Scikit-learn. The results showed that traffic density is directly correlated with air pollution, especially in the earlier hours of the day: the highest rate of pollution was reached at eight o'clock in the morning. Secondly, the meteorological factors influence the accuracy of the prediction. It was concluded that the best prediction model should be a hybrid data source that includes gas monitoring. The study demonstrated that the performance of the prediction depends on different factors such as traffic, the meteorological factors and the number of trace gases that the hybrid source is based on.

III. DATASET

In order to find the best Machine Learning algorithm to use for the forecasting of air pollution, several analyses were preliminary performed on a dataset that includes the concentration values of PM_{10} and $PM_{2.5}$ air pollutants (Fig. 1).

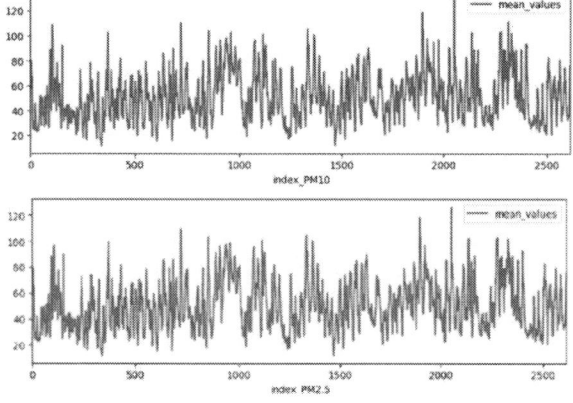

Fig. 1. Concentration values of PM_{10} and $PM_{2.5}$

These parameters have been measured using Libelium sensors every 15 minutes, in Bucharest (Beia office) between 1st of November 2018 – 28th of January 2019 [12]. As it can be observed from the Figure above, the values for the concentration of PM_{10} and $PM_{2.5}$ are very similar.

The first stage of this research consisted in a data analysis that was based on the comparison with requirements of standard data (for example, the values of the concentration must be a null or a positive value). Secondly, a preliminary analysis was made, where more statistical descriptive methods were used. Some of these methods are mean value, variation, or standard deviation. Humidity threshold was selected using the Pearson correlation coefficients, the dimension of the sub-datasets and the ratio between PM_{10} and $PM_{2.5}$.

The next step implied the computation of the average values for the registered data after their validation and for each parameter was obtained a dataset containing 2133 values. Moreover, in **Error! Reference source not found.** is presented the first analysis of the statistical parameters performed on this dataset. As the results show, there is a high variability of the humidity and of the PMs concentrations [12].

TABLE I.   STATISTICAL PARAMETERS VALUES

Statistical parameter	Measured parameter				
	$PM_{10} (\frac{\mu g}{m^3})$	$PM_{2.5} (\frac{\mu g}{m^3})$	Temperature(°C)	Pressure (Pa)	Humidity(%)
Mean	50.81	50.35	23.66	100984	34
Median	48.45	47.96	23.44	101124.4	33.32
Standard error	0.37	0.37	0.01	15.61	0.08
Sample Variance	359.39	351.85	0.41	637127.7	15.21
Minimum value	11.79	11.74	22.85	98605.14	24.47
Maximum value	130.7	126.7	26.18	102865.9	49.27
Confidence level for mean	±0.73	±0.72	±0.02	±30.61	±0.15

IV. EXPERIMENTS

Depending on some external factors like traffic, weather and others, the pollution level can vary significantly so that by the end of October 2018 till the beginning of February 2019 the average values of PM_{10} and $PM_{2.5}$ pollutants for each hour oscillate very much.

A. SVR MODEL

1) Application of the algorithm

The data is stored in the X and y variables. Variable y contains the average values of PM_{10} concentration (pollutant measured in a specific hour), and X contains the indexes to that value, as it is displayed in the example below. Using the Numpy library, the data is shaped in the appropriate way to be used in methods from the Scikit-learn library.

Example of the code used to store the data in X and y variables:

```
X, y = [], []
for i in range(len(PM10)):
    X.append(i)
    y.append(PM10[i][3])

y_data = y
y = ['%.2f' % elem for elem in y_data]
X = np.array([X]).T
y = np.array(y).ravel()
y = np.array(y)
```

To split the data into train and test dataset, it was used a method from Scikit-learn library, which allows choosing the test of train size as a percent. The resulted data is stored in test and train variables.

Example of the code where the data is split into train and test dataset:

```
X_train, X_test, y_train, y_test = train_test_split
(X, y, test_size=0.2)
```

2) Selection of gamma and C parameters

The method used to choose the most suitable values for gamma and C parameters was represented by the value of the correlation coefficient method (which must be as high as possible) and the mean squared error method (which has to be as low as possible). In order to obtain those values, it was necessary to test several values for each parameter using the .fit() method. This method can take two arguments: X_train, which represents the indexes for the training dataset, and y_train, which represents the target values for the training dataset. These are real numbers in regression and the function returns an object stored in the "svr_rbf" variable. Then, after the model is trained, it can be used to make predictions. In order to predict the data, the object created before is used by applying the .predict() method resulting the y_pred variable based on the test dataset. When the prediction is done, the test dataset is compared with the predicted dataset by using the correlation coefficient and calculating the mean squared error.

Example of code used to predict data and to compare the resulted dataset with the test dataset:

```
gamma_list = ['auto', 0.5, 0.3, 0.1, 0.05, 0.04,
0.03,0.02,0.01, 0.005, 0.004, 0.003,0.002,0.001]
c_list = [1, 10, 1e2, 1e3, 1e4, 1e5]

for c in c_list:
    for g in gamma_list:
        svr_rbf = SVR(kernel='rbf',
C=c,gamma=g).fit(X_train, y_train)
        y_pred = svr_rbf.predict(X_test)
        y_pred = ['%.2f' % elem for elem in
y_pred]
        y_pred = [float(e) for e in y_pred]
        corr_coef = np.corrcoef(y_test,
y_pred)[0,1]
        mse = mean_squared_error(y_test, y_pred)
        print("for C = ", c, " and gamma = ", g,
" correlation coefficient is ", corr_coef,
                " and mean squared error = ", mse)
```

3) Results of the algorithm

After applying the algorithm using different values for gamma and C it was chosen the most appropriate values for the coefficients of the model, more exactly C = 100 (or 1e2), gamma = 0.1, kernel='rbf', a correlation coefficient of 0.966 and a mean squared error which equal to 25.3. In Fig. 2 is presented a comparison between the estimated data and the measured data.

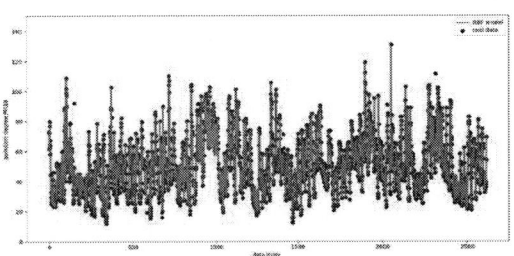

Fig. 2. Comparison between estimated data and measured data

4) Observations

After running the algorithm, for the polynomial kernel, the model wasn't built even after 30 minutes. For C = 10000 (or 1e4) and C = 100000 (or 1e5), the time to create the model took more than 20 minutes. For the same values for gamma and C parameters, the results could be different depending on the training dataset because of the train_test_split method which split the data randomly each time.

B. ARIMA Model

1) Application of the algorithm

The ARIMA algorithm has been used in the process of estimating the parameters PM_{10}, respectively $PM_{2.5}$. 80% of the data set was used for the purpose of training the Machine Learning method. The rest of 20% of data was employed for testing the algorithm in order to estimate the next value, according to the algorithm presented below:

```
series        =        read_csv('file.csv',        header=0,
parse_dates=[0],        index_col=0,        squeeze=True,
date_parser=parser)

X = series.values
size = int(len(X) * 0.80)
train, test = X[0:size], X[size:len(X)]
history = [x for x in train]
predictions = list()
for t in range(len(test)):
    model = ARIMA(history, order=(5,1,0))
    model_fit = model.fit(disp=0)
    output = model_fit.forecast()
    yhat = output[0]
    predictions.append(yhat)
    obs = test[t]
    history.append(obs)
    print('predicted=%f, expected=%f' % (yhat, obs))
error = mean_squared_error(test, predictions)
print('Test MSE: %.3f' % error)
```

For each estimation, the absolute deviation will be calculated and the real (measured) value will be added to the active data set. By employing this method, the average of the set will be modified with each step. The return of the ARIMA algorithm is presented below:

```
predicted=31.517350, expected=29.401999
predicted=36.911160, expected=34.691999
predicted=36.469997, expected=37.681998
predicted=30.688989, expected=32.451999
predicted=31.873749, expected=31.535999
predicted=31.305169, expected=28.096000
predicted=27.492168, expected=27.797500
predicted=27.930571, expected=27.735999
```

```
predicted=34.024295, expected=32.325000
predicted=32.334449, expected=33.146000
predicted=33.602038, expected=34.793999
predicted=35.424062, expected=35.745000
predicted=36.107573, expected=34.813999
predicted=34.594491, expected=35.084000
```

2) Results of the algorithm

ARIMA model was applied using the same dataset as for the SVR algorithm. The results obtained using ARIMA algorithm are presented in Fig. 3.

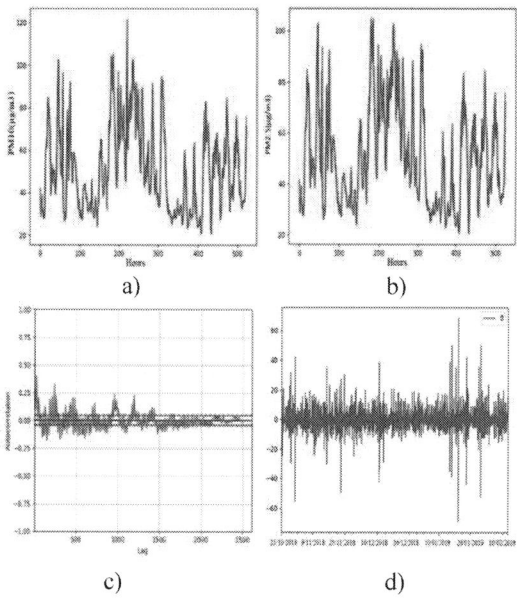

a)   b)

c)   d)

Fig. 3. The results obtained using the ARIMA model for estimating the concentrations of PM_{10} (a), $PM_{2.5}$), the degree of autocorrelation of the algorithm (c), the absolute deviation (d)

The correlation coefficient between the measured values of the concentration of PM_{10} and the estimated ones is 0.921, and for $PM_{2.5}$ parameters is 0.935.

C. LSTM Model

1) Vanilla LSTM Model

Long Short-Term Memory Model (LSTM) Model is a kind of RNN. This model is suited for processing the correlation within time series for both short and long term. The leading innovation of LSTM is the memory cell c_t that acts as an accumulator of the state information. Several self-parameterized controlling gates are accessing, writing and clearing the cell. If the input gate is activated, a new input is provided and the cell is loaded with the information from the new input. If the forget gate f_t is activated, the past cell status c_{t-1} will be no more taken into consideration. Even if the latest cell output c_t will be transmitted to the final state h_t, it is controlled later by the output gate o_t [13]. A LSTM model with a single layer of LSTM units and an output layer used for prediction is called Vanilla LSTM. The architecture of a Vanilla LSTM algorithm can be seen Fig. 4 [14].

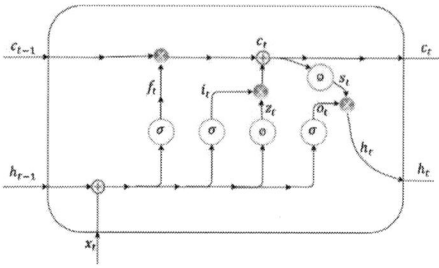

Fig.3. Representation of a vanilla LSTM block structure and its internal information forward flow

2) Application of Vanilla LSTM algorithm

As it was mentioned before, a Vanilla LSTM model is characterized by the fact that the LSTM model has a single hidden layer of LSTM units and another output layer which is used for the prediction process. In this application was used the LSTM model from the layer's library of Keras module. This module can be imported from TensorFlow framework. A Vanilla LSTM model can be defined as it is illustrated in the code below:

```
# define model
model = Sequential()
model.add(LSTM(50,activation='relu',
input_shape=(n_steps, n_features)))
model.add(Dense(1))
model.compile(optimizer='adam',loss='mse')
```

For this experiment, it was used a univariate series, which means that there is only one feature (n_features=1), for one variable, in this case one variable for PM_{10} and one variable for $PM_{2.5}$. In this way it was defined a model which has an output layer used to predict a one numerical value and 50 LSTM units in the hidden layer. The mean column of the csv file for each PM parameter was used as input sequence in format of a list of values. It is presented in the sequence below:

```
# define input sequence
raw_seq = list(dataset_pm2_5['media_valori'])
```

After the data was processed and prepared the fit() method was applied. Vanilla LSTM algorithm uses the Adam version of stochastic gradient descent based on the mean squared error. Fitting stage is necessary to be able to make a prediction. The way in which fit() function is used is illustrated in the sequence of code below:

```
# fit model
model.fit(X,y,epochs=200, verbose=0)
```

3) Results of the algorithm

The last step to be performed is the prediction one. In order to predict the next value of the sequence it is necessary to provide an input. In this case it was provided a dataset of 23 values and it was expected the prediction of the 24th value. For PM_{10} pollutant was provided the first 23 values from the dataset. These values are presented in the sequence of code below.

```
# demonstrate prediction
x_input = array ([72.78399734, 72.07199707,
61.25749779, 61.4219986,
76.73399811, 79.6059967, 64.15999985, 58.09199905,
63.07399902, 67.22666677, 44.46199875, 32.70399933,
29.05599899, 25.27499962, 25.60399933, 26.6279995,
25.3324995, 26.4859993, 27.71199951,
31.55999947,28.41999912, 27.93199921,
24.259999sssss47])
    x_input = x_input.reshape((1, n_steps,
n_features))
    yhat = model.predict(x_input, verbose=0)
    print(yhat)
```

Based on these values, after the first prediction, the model gave as output the next value as equal to *25.366713*, while the expected value was *25.26*. As it can be observed the two values are very close. After performing a 2nd prediction, using the same 23 values, the predicted value was *33.58304*, and after the 3rd prediction the obtained value was *24.5032*.

For $PM_{2.5}$ the first 23 values used as input for the prediction stage are presented in the sequence of code below:

```
# demonstrate prediction
x_input = array ([72.46999,71.90800, 61.13999,
60.39599915, 76.55199738, 79.315998, 63.994999,
57.931999, 62.66399, 64.82333, 40.195998,31.84199,
28.861999, 24.48499, 25.32399,26.17599,  25.0625,
25.97399979, 27.27199936, 29.9239994, 27.99999905,
27.6659996, 23.88599968 ])
x_input = x_input.reshape((1, n_steps, n_features))
yhat = model.predict(x_input, verbose=0)
print(yhat)
```

Using the values presented above the 24th predicted value was: *24.167881*, while the expected output (the real 24th value of the dataset) was *24.3725*. One more prediction test was realised, and the obtained result was: *25.13438*.

CONCLUSIONS

After performing a comparative study of the methods used for data analysis and of the Machine Learning algorithms used for estimating the atmospheric pollutants (PM_{10} and $PM_{2.5}$), it was demonstrated that SVR and ARIMA algorithms are the most suitable in forecasting the air pollutants concentrations, because the correlation coefficient was 0.966 and 0.921 respectively for PM_{10} concentration.

For these methods, it was developed and tested specific algorithms using a dataset that was gathered during the period between the 1st of November 2018 and 28th of January 2019. The results have shown that both algorithms can be used in such examples of air pollution forecasting.

In the future, we want to forecast the air pollutants concentrations for 24 hours with a correlation coefficient greater than 85% using the two proposed algorithms.

ACKNOWLEDGMENT

We would like to express our gratitude to Tel-MONAER project (subsidiary contract no. 1223/22.01.2018, from NETIO Project ID: P 40270, MySMIS CODE: 105976), FarmSustainaBL project (contract no. 119/2019, from ERANET-ERAGAS-ICT-AGRI3 program) and WINS@HI project (PN-III-P3-3.5-EUK-2017-02-0038) for the work presented in this paper.

REFERENCES

[1] P.J. Landrigan, et al. "The Lancet Commission on pollution and health". The Lancet, 391(10119), pp.462-512, 2018.

[2] K.H. Kim, E. Kabir, and S. Kabir. "A review on the human health impact of airborne particulate matter". Environment international, 74, pp.136-143, 2015.

[3] A.Y. Watson, R.R. Bates, and D. Kennedy. "Assessment of human exposure to air pollution: methods, measurements, and models". In Air pollution, the automobile, and public health. National Academies Press (US), 1988.

[4] E. Fradinata, Z.M. Kesuma, S. Rusdiana, and N. Zaman. "Forecast Analysis of Instant Noodle Demand using Support Vector Regression (SVR)". In IOP Conference Series: Materials Science and Engineering (Vol. 506, No. 1, p. 012021). IOP Publishing, april 2019.

[5] J. Contreras, R. Espinola, F.J. Nogales, and A.J. Conejo. "ARIMA models to predict next-dayelectricity prices". 2003.

[6] F.A. Gers, J. Schmidhuber and F.Cummins. "Learning to forget: Continual prediction with LSTM". 1999.

[7] G. Papacharalampous, H. Tyralis, and D. Koutsoyiannis. ""Univariate time series forecasting of temperature and precipitation with a focus on machine learning algorithms: A multiple-case study from Greece. Water resources management, 32(15), pp.5207-5239, 2018.

[8] C. Liu, et al. ""Online arima algorithms for time series prediction. In Thirtieth AAAI conference on artificial intelligence, February 2017.

[9] J.F.A. Yeung, et al. ""Jump detection in financial time series using machine learning algorithms. Soft Computing, pp.1-13, 2019.

[10] V. Athira, P. Geetha, R .Vinayakumar, K.P. Soman. ""DeepAirNet: Applying recurrent networks for air quality prediction. Procedia computer science, 132, pp.1394-1403, 2018.

[11] Y. Rybarczyk and R. Zalakeviciute. ""Regression models to predict air pollution from affordable data collections. Machine Learning—Advanced Techniques and Emerging Applications, 2018.

[12] M. Balanescu, I. Oprea, G. Suciu, M.A. Dobrea, C. Balaceanu, R.I. Ciobanu, and C. Dobre. A Study on Data Accuracy for IoT Measurements of PMs Concentration. In 2019 22nd International Conference on Control Systems and Computer Science (CSCS) (pp. 182-187). IEEE, May 2019.

[13] S.H.I. Xingjian, et al. ""Convolutional LSTM network: A machine learning approach for precipitation nowcasting. In Advances in neural information processing systems (pp. 802-810), 2015.

[14] A. Nigri, S. Levantesi, M. Marino, S. Scognamiglio, and F. Perla. ""A Deep Learning Integrated Lee–Carter Model. Risks, 7(1), p.33, 2019.

Intelligent System for Vehicle Recognition

Radu-Petru FOTESCU
ETTI, University Politehnica of Bucharest, Bucharest,
Romania
radufotescu@yahoo.com

Paul SVASTA
ETTI, University Politehnica of Bucharest, Bucharest,
Romania
paul.svasta@cetti.ro

Loredana-Maria BURCIU
ETTI, University Politehnica of Bucharest, Bucharest,
Romania
loredanamaria0809@gmail.com

Rodica CONSTANTINESCU
ETTI, University Politehnica of Bucharest, Bucharest,
Romania
constantinescu.rodica@gmail.com

Abstract— **The system is able to identify a vehicle according to the license plate and the model of the car and it decides if the vehicle has or not has access in a limited access area. The access of the vehicle in this area is made by a barrier actuated by a servomotor. If the vehicle information exists in the implemented database, a green LED will light for indicate that the vehicle has access in the area and the barrier will be raised. If the vehicle is not in database, a red LED will light and the position of the barrier will not change.**

Keywords—Vehicle,   microcontroller,   actuator,   LCD.
Introduction (Heading 1)

I. INTRODUCTION

This project is useful in identifying and storing vehicle types and registration numbers in an area with limited access. A database will be accessed and it contains the registration numbers and the types of several vehicles. This will be used for access in the area, so the security level of the restricted access area will be increased because vehicles that do not have access can't enter.

The system contains the following main elements: Raspberry Pi, a passive infrared motion sensor, a camera compatible with Raspberry Pi and an LCD screen with a size of 16 x 2. The motion sensor is based on infrared technology, high sensitivity, widely used in various auto-sensing electrical equipment [6].

The operating principle of the system (fig. 1) is as follows: the motion sensor, when it detects the presence of the vehicle, start the camera which takes a JPG photo. From the photograph taken, the necessary information can be extracted which is compared with the information stored in the database, and with the help of the realized source code that accesses the database, it is checked if the registration number and type of vehicle can have access in the area with limited access.

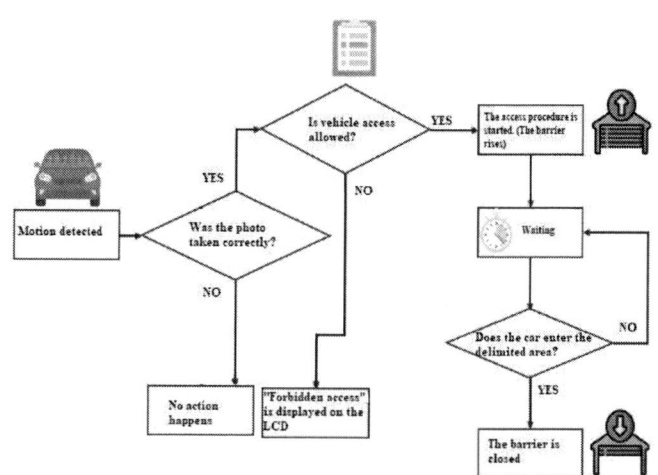

Fig. 1: System operating diagram

II. HARDWARE IMPLEMENTATION

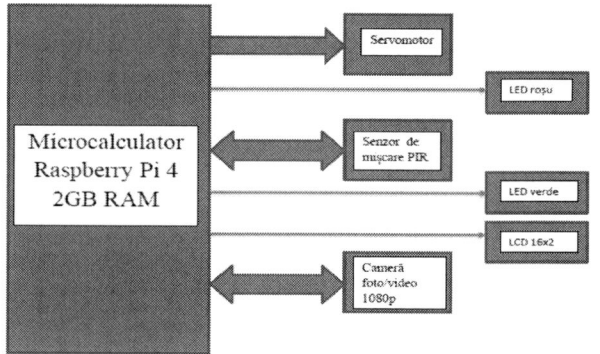

Fig. 2: Block diagram of the system

A. Raspberry Pi 4 Model B/2GB

Fig. 3: Raspberry Pi 4

Raspberry Pi is a series of SBC (Single-board computer) the size of a credit card. It is inspired by BBC Micro and produced in the UK by the Raspberry Pi Foundation. The goal was to create a low-cost, device that would improve programming skills and hardware understanding at the undergraduate level.

The new microcomputer, the Raspberry Pi 4 Model B, comes with a 1.5 GHz quad-core processor, 2 GB LPDDR4 RAM, which will result in a much higher working speed. On the connectivity side we have the new Bluetooth 5.0 standard, wireless LAN b / g / n / ac and a gigabit port. The board also has 2 micro-HDMI ports, so you can enjoy an extensive workspace and thanks to the fact that the board also has 4k support.

B. PI camera module v2

The Raspberry Pi camera has 5 megapixels, a Sony IMX219 sensor and is designed specifically for Raspberry Pi cards. It is capable of still images of 3280 x 2464 pixels and also supports video with a number of frames up to 1080p at 30 fps, but also lower values such as 720p and 640p. It attaches to the Raspberry Pi through the port on the top surface of the board and uses the dedicated CSi interface, designed specifically for interacting with video cameras.

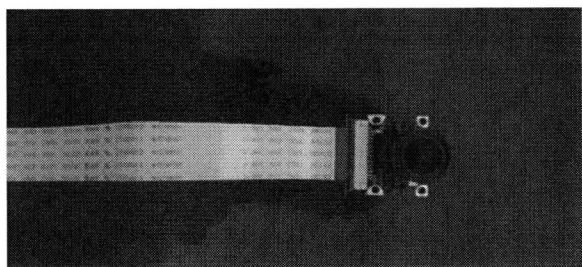

Fig. 4: PI camera module v2

C. Motion sensor PIR HC-SR501

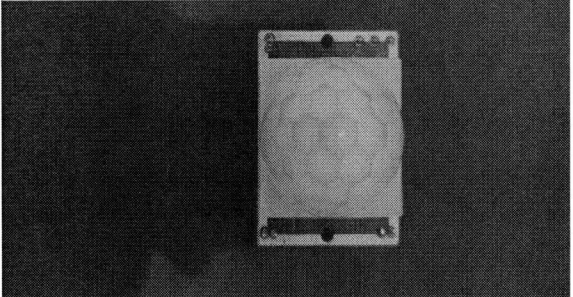

Fig. 5: PIR HC-SR501

A passive infrared motion detector (PIR) first works by adjusting to the "normal" IR level in its detection area. We then look for sudden changes to this IR emission, changes that occur due to the fact that an object or a living being has entered or moved in the monitored area. To detect infrared energy, the detector uses a pyroelectric sensor. This is a device that generates an electric current in response to IR power reception. Because the detector does not emit a signal it is called "passive". He just sits there listening to a change in the energy level in the IR environment. When a change is detected, the PIR motion sensor will trigger an alert by changing its output signal.

The following electronic components were used to create the electrical diagram:

- Raspberry Pi 4 Model B 2 GB

- Infrared motion sensor

- A video camera with night mode dedicated to the Raspberry Pi

- 16 x 2 LCD screen

- green and red LEDs

- two 220 ohm rezistor

- actuator

After adding the hardware components used in the project, we made the physical connections of the components to the Raspberry Pi pins.

The first component that connects to the Raspberry Pi is the dedicated video camera for it. The camera connects with a ribbon cable to the CSI (camera serial interface) port. In the wiring diagram, the CSI port of the Raspberry Pi is on the left and is represented by the name "CSI PORT". The connection of the dedicated camera module is made only when the Raspberry Pi board is turned off.

The infrared motion sensor connects to the pins of the Raspberry Pi board using 3 wires. The VCC port of the motion sensor is connected to the power

2020 IEEE 26th International Symposium for Design and Technology in Electronic Packaging (SIITME)

supply terminal 4 of the board. The GND port is taken to the peace ground pin, and the data port connects to pin 7 of the GPIO4 board.

The display part consists of a 16 x 2 LCD screen. For connection are used 4 wires. The VCC pin connects to pin 2 on the Raspberry Pi board and the GND pin connects to the ground pin to the board. For data transmission, connect the SDA pin to the I2C1 board pin, SDA and SCL to the I2C1 SCL pin.

The light warning part of the system consists of two LEDs, one red and the other green. On the LED circuit are connected two 220 ohm resistors that have the role of limiting the current and ensuring the protection of the LEDs. The two LEDs are connected to the pins of the Raspberry Pi 16 and 18 board respectively GPIO23 red and GPIO 24 green. The LEDs were glued on a perfoboard together with the 2 resistors. Also on the perfoboard board were glued the 2 wires that ensure the connection to the Raspberry PI pins and another wire that connects the two cathodes of the LEDs to the ground pin of the board.

The last component of this system is the actuator that connects to the Raspberry Pi board via 3 wires Insert 3 wires into the base of the actuator up to the pins on the plate. In addition to the VCC and GND power pins, the control pin of the PWM actuator connects to pin 11 and GPIO17, respectively. By means of this wire the actuator control from the realized source code is performed.

To describe the connection of the modules to the Raspberry Pi we made a wiring diagram in EasyEDA program.

EasyEDA is a well-known online platform that allows the realization of electrical diagrams.

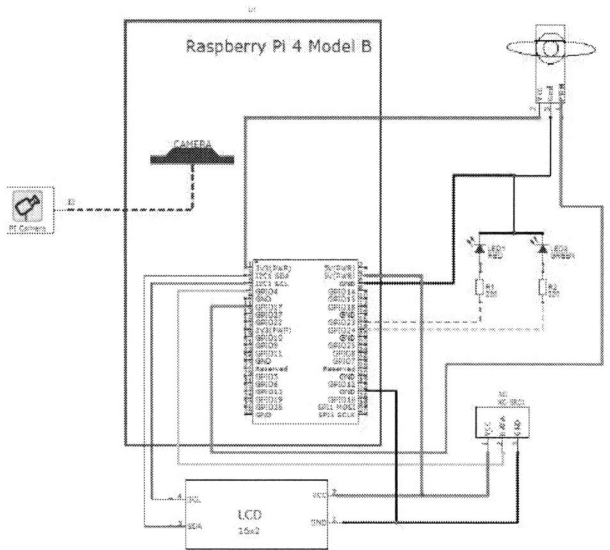

Fig. 6: Wiring diagram of the system

III. SOFTWARE IMPLEMENTATION

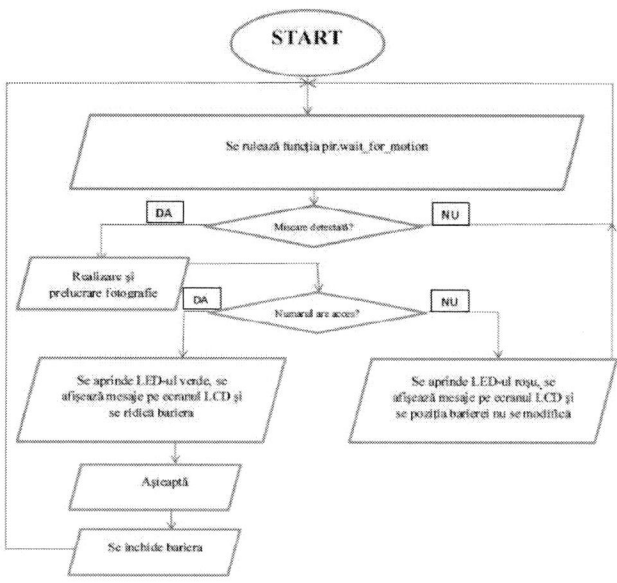

Fig. 7: Logic diagram of source code

In order to create the intelligent vehicle recognition system, several functions have been integrated in a single source code made in the Python programming languageFrom a software point of view, my contribution to this project was to make the main program in the Python programming language, a program that works like this: first it checks if the motion sensor detects the presence of the vehicle, this is done using the function "Pir_motion_detected", and if the vehicle has been detected, the camera will take a photo using the "camera.capture (IMAGE)" function. This photo is used as an argument for the "ocher (IMAGE)" function that will return the license plate number and display the car's brand in the terminal. An if function has been created to check if the registration number returned by the "ocher (IMAGE)" function is the desired one, and after analyzing the result on the LCD screen, different messages will be displayed, different LEDs will light up using the functions create "red ()" and "green ()" and control the movement of the barrier with the help of the servomotor operated by the function "p.ChangeDutyCycle (number)".

IV. OPENALPR PLATFORM. CHARACTER RECOGNITION ALGORITHM

For the automatic recognition of license plates on vehicles we used OpenALPR library written in C ++, with links in C #, Java, Node.js and Python. This library has an API (Open ALPR Cloud) which is a web service that does the analysis images with vehicles transmitted to it and after obtaining the information in the photograph (number of registration, mark) returns them to the user.

Optical Character Recognition (OCR) algorithms allow computers to analyze photos sent by the camera and process the image by extracting the information of interest from this one.

This algorithm works as follows: Any scanned image / document is a file graphic, ie a pixel pattern. The algorithm locates, detects, and recognizes the characters of a images, turning the image into a text file. From here it becomes easy to extract information, browsing the file and found the desired information.

CONCLUSION

The system has been tested using several license plates and different types of vehicles, and some of the results are shown in table 1. A total of 30 tests were performed and the success rate in reading and identifying the registration number and the type of the vehicle is 93.33%. These tests were performed using printed photographs brought within the range of the system and the small errors that occur when recognizing the registration number and the type of vehicle are due to the following factors: incorrect focus of the photo, insufficient light in the room or the movement of the printed photo. Following the tests performed it was found that the time for reading and allowing access of an authorized vehicle is on average 13.322 seconds and for an unauthorized or unidentified vehicle the reading time is on average 5.21 seconds.

Tab 1: Test results

Table results with car's photo		
Nr.	Registration number and type of vehicle	Registration number and type of vehicle detected
1	GJ 64 MRC PEUGEOT 407	GJ 61 MRC PEUGEOT PARTNER
2	GJ 93 RFO VOLKSWAGEN PASSAT	GJ 93 RFO VOLKSWAGEN PASSAT

It is difficult to fool the system with a false license plate because simultaneous checks are made both by the make and model of the vehicle and by the color and registration number.

Due to the fact that this system is implemented on a Raspberry Pi microcomputer, it can be connected with a facial recognition system that makes a connection between the vehicle with access inside and the person behind the wheel.

Existing systems on the market exceed the price of 2000 euros. The presented system has a production cost of 200 euros in the prototype phase. The production cost for the system to be put on the market is 800 euros. The system can be improved so that it can be used to monitor road traffic in various cities. It will be able to send images and information in real time to a database and in addition it could record the speed of movement of the vehicles.

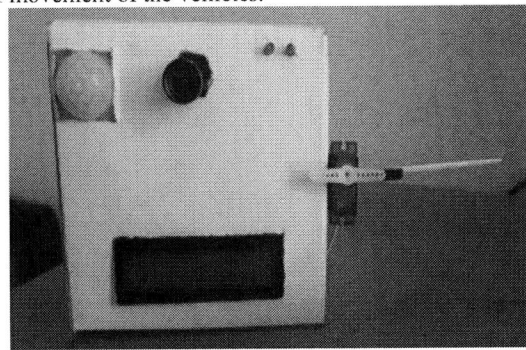

Fig. 8: The prototype of the vehicle recognition system

REFERENCES

[1] Simon Monk, "Programming the Raspberry Pi", McGraw-Hill Education TAB; (October 5,2015)

[2] D. F. Llorca, D. Colas,I. G. Daza, I. Parra, M. A. Sotelo, "Vehicle model recognition using geometry and appearance of car emblems from rear view images", 2014 IEEE 17th International Conference;

[3] R Parisi, E.D.Di Claudio, G Licarelli, G Orlandi, "Car Plate Recognition By Neural Networks and Image Processing."

[4] S.L. Chang, L.S.Chen, Y.C.Chung, S.W.Chen., 2004. Automatic license plate recognition., In: IEEE Transactions on Intelligent Transportation System., vol. 5., pp. 43-53

[5] Plate Extraction and Character Segmentation, 2010, IEEE International Conference on Computational Intelligence and Computing Research

[6] https://datasheetspdf.com/pdf/775434/ETC/HC-SR501/1 accessed on 20 June 2020.

[7] https://www.raspberrypi.org/documentation/hardware/raspberrypi/bcm2711/rpi_DATA_2711_1p0_preliminary.pdf

[8] https://dronebotworkshop.com/using-pir-sensors-with-arduino-raspberry-pi/

2020 IEEE 26th International Symposium for Design and Technology in Electronic Packaging (SIITME)

Investigation on modified SRR for accurate dielectric measurements

R. Gavrilă
Telecommunication Department,
University Politehnica of Bucharest
raluca.gavrila97@yahoo.com

I.A. Mocanu
Telecommunication Department,
University Politehnica of Bucharest
mihai.iulia83@yahoo.com

Abstract— **Based on the revolutionary metamaterials, this paper investigates the frequency behavior of modified split ring resonators (SRRs) to create a sensor for measuring electrical parameters of dielectrics. Focused on simulations, the best cases of SRRs configurations (including the number of double SRRs and strips that were added) will be identified in order to obtain a good result in frequency which is translated into a good selectivity (narrow band) and a peak sharp enough to accurately read the frequency. Moreover, changes will be made to the substrate, both for dielectric characteristics, varying its loss tangent, as well as for its thickness. Using the shift of the resonance frequency method, the best configuration will be selected and analyzed for a sensor application.**

Keywords— Sensors, metamaterials, Split Ring Resonator (SRR)

I. INTRODUCTION

Metamaterials are new artificial materials that have special properties which cannot be found in naturally occurring materials. Metamaterials are 3-dimensional, periodically arranged metallic structures which are smaller than the wavelength of the incident electromagnetic (EM) wave [1], designed to produce an optimal combination of two or more responses to a given excitation. The main properties of these structures are: negative electrical permittivity and negative magnetic permeability *(ε < 0 and μ < 0)*, negative refractive index *(n < 0)*, triplet electric field intensity, magnetic field intensity and wave vector oriented in the opposite direction of the Poynting vector, antiparallelism between phase and group velocities [2].

Metamaterials have attracted a great deal of attention and research interest from microwave engineers in recent years because of their wide applications exploring unknown physical phenomena such as cloaking, reversal of Vavilov-Cherenkov effect, reversal of Goos-Hänchen effect, reversal of Doppler effect, negative refraction, perfect lens, concentrator and negative compressibility [2]. So far, research regarding fabrication, design and application of metamaterials have been extended to a wide range of the EM spectrum including far, medium and near infrared regimes and even optical frequencies [1]. The structures can exhibit a strong localization and enhancement of fields, thus they can be used to improve the sensor property of detecting nonlinear substances and to detect extremely small amounts of analytes [3].

II. MODIFIED SRR STRUCTURES FOR SENSOR APPLICATIONS

A. Split Ring Resonator (SRR)

Split Ring Resonators (SRRs) are resonance structures that are used widely in electromagnetics. For example, these structures are used in periodic configurations to design metamaterial structures. Also, because of their resonance model, SRRs and CSRRs can be used to design slow wave transmission lines, phase shifters, various kinds of microstrip filters, sensors, absorbers, antennas, etc. The circuit model of SRR elements in the waveguide is parallel capacitance and inductance, placed in series in the transmission line [3]. These structures when excited by suitable electromagnetic fields have resonance model and show unusual properties such as negative permeability and negative permittivity near the resonance frequency region.

Double SRR consists of two enclosed metallic loops (rings) separated by a gap and both having splits at opposite sides (Fig. 1. (a). Magnetic resonance is induced by splits of the rings and by the gap between the inner and outer ring [4]. If the excitation magnetic field is perpendicular to the plane of the magnetic field in order to induce resonating currents in the ring and generate equivalent magnetic dipole moment, the structure will exhibit a negative permeability, $\mu < 0$.

Due to the small dimensions of the SRR structures when used for planar microwave applications, they exhibit low radiation losses and very high-quality factors. Another advantage is that SRR has strong magnetic coupling compared with other materials that can be found in nature. Also, the pronounced magnetic response in these materials evidences an advantage over naturally occurring materials [4].

There are several configuration variants, depending on the number of loops and their shapes. The double square SRR, depicted in Fig. 1a) will be further analyzed.

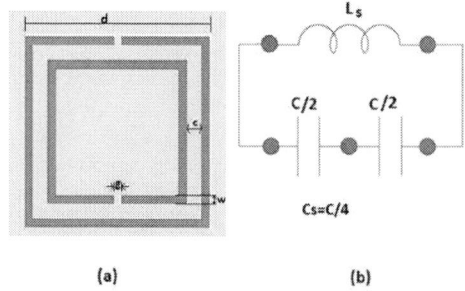

Fig. 1. (a) The double square SRR. (b) Equivalent circuit of SRR.

978-1-7281-7507-2/20 $31.00 © 2020 IEEE   118   21-24 October, Pitesti, Romania

The two rings of SRR are coupled by a strong capacitance caused by the gap between the rings. If a time harmonic field is applied such that the magnetic field is along x axis, there will appear an electromotive force experienced by the SRR. Following the quasi-static model, the induced current lines will pass from one ring to another through the capacitive gaps. Thus, the total current intensity flowing on both the rings remains the same for any cross-section of the structure independent of the angular coordinates. The structure behaves as a LC circuit like the one in Fig. 1. (b). The resonance frequency is given by:

$$f_0 = 1/2\pi\sqrt{LsCs}, \qquad (1)$$

where Cs is the series capacitance of the two halves, upper and lower, and Ls is the total inductance of the SRR [4]. Therefore, the resonance frequency can be adjusted by varying the dimensions of the structure, which will affect both the capacitance and inductance.

B. Modified Split Ring Resonator (SRR)

Starting from the model presented in Fig. 1. (a), the double square SRR, several configurations will be implemented, including the ones with lateral strips placed at a distance s (Fig. 3) from the resonator's rings. These configurations will be simulated and analyzed, using the HFSS ANSYS program.

The data used for the dimensions of the structures that will be used are presented in Table 1:

TABLE I. DIMENSIONS OF THE PARAMETERS USED TO CREATE THE STRUCTURE

Parameters	d	w	c	g	s
Dimensions [mm]	18.4	1.52	1.52	1.22	0.15

In the designing part, some other parameters must be included. They are the following: the thickness of the rings and of the strips: 35 μm, the thickness of the substrate: 1.6 mm, the material used for the rings and the strips: metallization copper, the material used for the substrate: FR4-epoxy (ε_r= 4.4), the dimensions of the substrate [mm]: 27.6 x 75.52 x 1.6, the dimensions of the air box [mm]: 39.6 x 100 x 21.6.

In Fig.2. the structure including the access transmission lines and the frequency response are presented. The parameters S_{11} and S_{21} are highlighted at a resonance frequency of almost 2.07 GHz. It can be observed that the frequency cannot be exactly read, not even with a marker because the shape of the peak is not sharp enough. This is related to selectivity because the more selective it is, the narrower the bandwidth should be and the sharper the peak would be. That is why the quality factor would increase.

In order to increase the quality factor of this structure, lateral strips with the same width as the rings of the SRR from Fig. 2., are placed at a distance of s, from the rings. The new structure is shown in Fig. 3. The small distance between the strips and the rings assures a tight coupling., which can be seen in the simulation results from Fig.3. Also, it can be observed that the resonance frequency decreases because of the extra capacitance given by the lateral strips and a second resonance frequency well-emphasized appears.

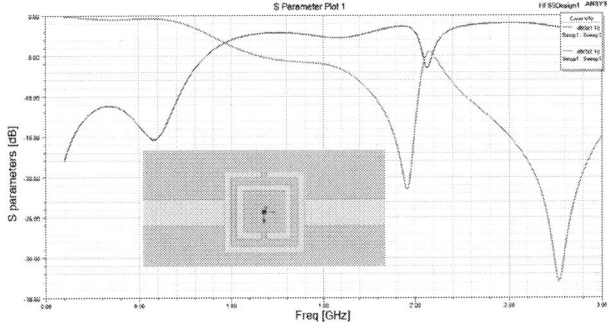

Fig. 2. Simulation results of the scattering parameters S_{11} (red curve) and S_{21} (green curve) for the SRR with the dimensions in Table I.

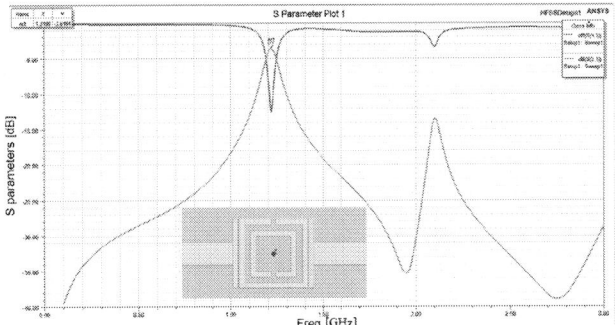

Fig. 3. Simulation results of the scattering parameters S_{11} (red curve) and S_{21} (green curve) for the modified SRR, with the dimensions in Table I.

To improve even more the performances, a structure made of two modified SRRs placed at a distance s from each other and a strip in the middle is investigated. Analyzing the results in Fig.4, we can see that increasing the number of double rings resulted in increasing the number of resonance frequencies and improving the selectivity for both resonance frequencies than in the previous case. Next, we want to add one more modified SRR to see if the selectivity increases even more. As Fig. 5 shows, no such increase happens in comparison to the results in Fig.4., so adding extra SRRs does not improve considerably selectivity. Therefore, for our further investigations, we will consider the structure in Fig.4, with two modified SRRs.

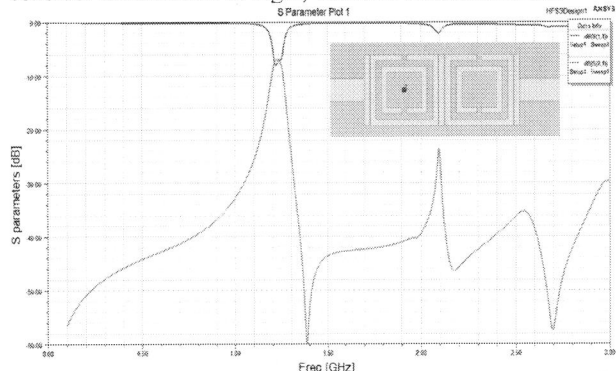

Fig. 4. Simulation results of the scattering parameters S_{11} (red curve) and S_{21} (green curve) for the two modified SRRs and a strip between the rings and lateral strips, with the dimensions in Table I and a 1.6 mm thickness of FR-4.

2020 IEEE 26th International Symposium for Design and Technology in Electronic Packaging (SIITME)

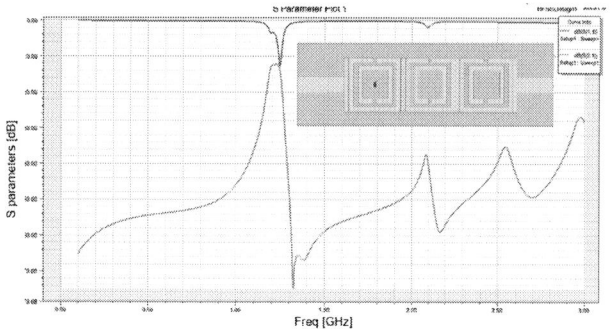

Fig. 5. Simulation results of the scattering parameters S_{11} (red curve) and S_{21} (green curve) for 3 SRRs with a strip in both lateral sides and one strip between the rings, with the dimensions in Table I and a 1.6 mm thickness of FR-4.

III. INVESTIGATION OF DIELECTRIC PARAMETERS OVER PERFORMANCES

A. The influence of dielectric's thickness over performances

In our study we want to investigate how the thickness of the dielectric influences the selectivity. So, we consider the structure from Fig.4, made of two modified SRRs. The thickness is now the lowest one available at the producer: 0.5 mm. The material is the same, FR4-epoxy, with $\varepsilon_r = 4.4$ and loss tangent $\tan\delta = 0.017$. The results in Fig.6 show that the resonance frequency remains the same, as expected, but the resonance is not so well emphasized.

Next, we will consider two other cases with extreme thicknesses, but for a different material: Rogers RO3010, which has $\varepsilon_r = 10.2$ and $\tan\delta = 0.0022$ (Fig. 7 and Fig. 8).

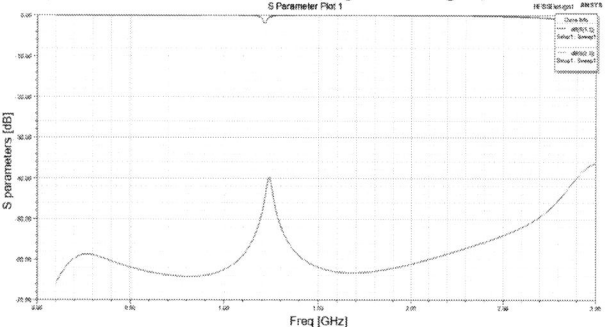

Fig. 6. Simulation results of the scattering parameters S_{11} (red curve) and S_{21} (green curve) for the two modified SRRs and a strip between the rings and lateral strips, with the dimensions in Table I and a 0.5 mm thickness of FR-4.

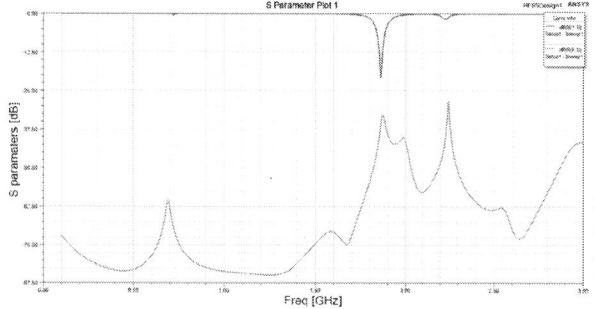

Fig. 7. Simulation results of the scattering parameters S_{11} (red curve) and S_{21} (green curve) for the two modified SRRs and a strip between the rings and lateral strips, with the dimensions in Table I and a 0.13 mm thickness of Rogers RO3010.

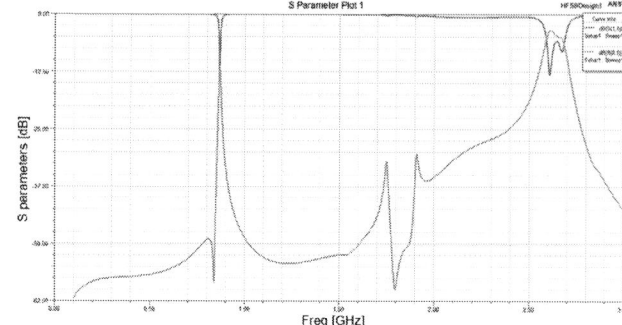

Fig. 8. Simulation results of the scattering parameters S_{11} (red curve) and S_{21} (green curve) for the two modified SRRs and a strip between the rings and lateral strips, with the dimensions in Table I and a 1.26 mm thickness of Rogers RO3010.

The results for the other analyzed dielectric, Rogers RO3010, show the same conclusion: the thicker the dielectric, the better results in terms of selectivity.

So, for further investigations, we will consider the greatest thickness available of the producer for that specific type of dielectric.

B. The influence of the dielectric's loss tangent over performances

Further, we want to make a comparison with a material that has a relative permittivity close to the one of FR4-epoxy, but a smaller loss tangent. We chose Rogers RO3003 with $\varepsilon_r = 3.03$ and $\tan\delta = 0.001$.

As proven in the previous analysis, we will consider the largest thickness available, 1.52 mm, which is very close to the one chosen for FR-4 in Section II.

If we compare the results in Fig. 9 with the results in Fig. 4, we can notice a higher amplitude for S_{21} parameter near the resonance frequency. Still, the response is not as sharp as expected.

In conclusion, the best structure from the ones analyzed so far, which takes into account all advantages and drawbacks is the two modified SRRs with lateral strips, implemented on FR-4 substrate, with a thickness of 1.6 mm. This will be used for our further investigations regarding sensor applications.

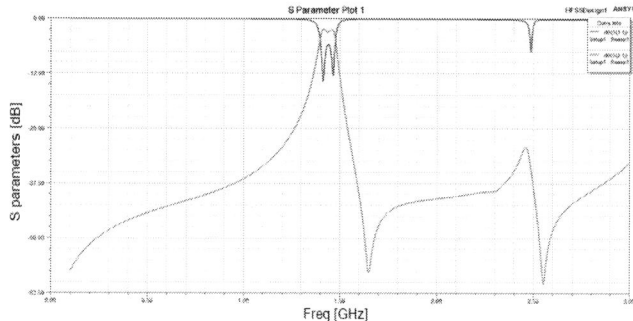

Fig. 9. Simulation results of the scattering parameters S_{11} (red curve) and S_{21} (green curve) for the two modified SRRs and a strip between the rings and lateral strips, with the dimensions in Table I and a 1.52 mm thickness of Rogers RO3003.

978-1-7281-7507-2/20 $31.00 © 2020 IEEE
21-24 October, Pitesti, Romania

2020 IEEE 26th International Symposium for Design and Technology in Electronic Packaging (SIITME)

IV. SENSOR APPLICATION

The structures investigated in this paper are suitable for a sensor application which measures the dielectric constant of a material under test (MUT).The best structure, both from the perspective of performances and low-cost manufacturing is the one analyzed in Fig.4. We will use the "resonant cavity method", which implies having a cavity resonator loaded with MUT (Fig.10) and measuring the shift in frequency and the variation in the quality factor. This method works better for dielectrics, so for MUT, we will consider two types of dielectrics with different relative electrical permittivity and quite similar loss tangent: Rogers RO3003 ($\varepsilon_r = 3$, tan$\delta = 0.001$) and Alumina ($\varepsilon_r = 9.2$, tan$\delta = 0.008$). The results are given in Fig. 11, respectively in Fig. 12.

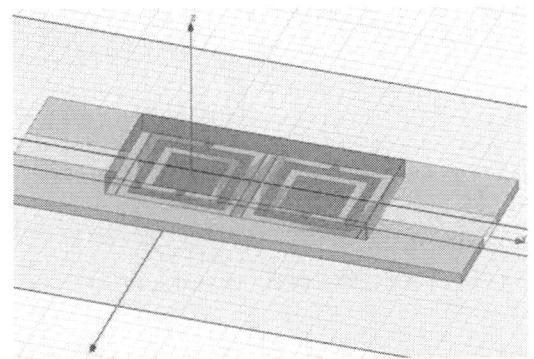

Fig. 10. The target structure with the MUT on top.

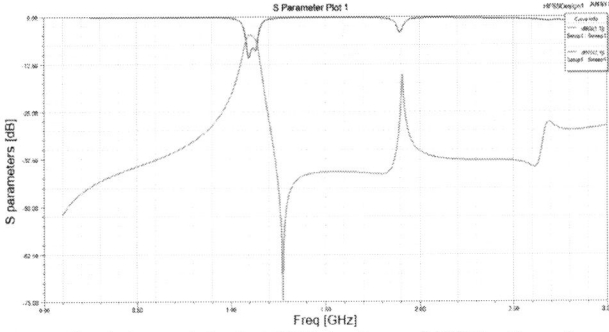

Fig. 11. Simulation result for the MUT being Rogers RO3003 with $\varepsilon_r = 3$.

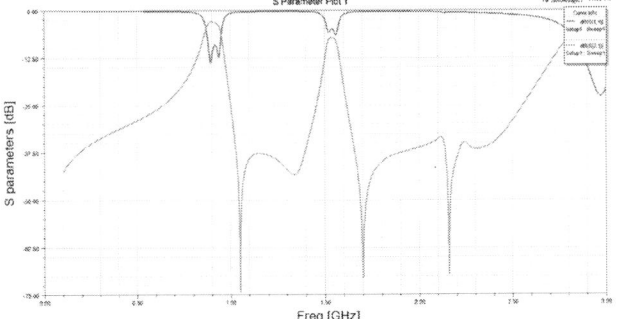

Fig. 12. Simulation result for the MUT being Alumina with $\varepsilon_r = 9.2$.

As expected, the shape is the same as in the simulation for the target structure, Fig.4. For the simulation without the MUT, the resonance frequencies were 1.23 GHz and 2.1 GHz. When placing a MUT of Rogers RO3003, the resonance frequencies are 1.1 GHz, respectively 1.9 GHz, so the shift in frequency has occurred.

In the case of Alumina being the MUT, the behavior is the same, a frequency change also occurs, but to lower frequencies. The first resonance frequency occurs at around 0.9 GHz and the second resonance frequency appears at around 1.55 GHz. Also, the bandwidths around the resonance frequencies become broader and a peak is no longer emphasized. Still, one can observe that in the Alumina case, the second resonance has the amplitude of S_{21} almost the same as for the first resonance frequency.

So, it can be demonstrated that the proposed structure made of two SRRs with lateral strips added can be used for sensor applications, when it comes to measuring dielectric constants of different types of dielectrics.

V. CONCLUSIONS

The optimal structure was identified in terms of number of SRRs and strips. Also, it was made an investigation regarding the best parameters for the substrate, which are large ε_r and small loss tangent, tanδ, for thick substrate. Because the selectivity, respectively the sharpening of the peaks, were followed and analyzed in most cases, it is desired to develop a sensor where, using the resonant cavity method, one can follow these two aspects to compute the relative permittivity of a material under test, respectively to identify it. It was observed that the frequency shifted in the same direction for two materials under test, Rogers RO3003 and Alumina, without modifying the shape of the parameters. Thus, a relationship can be deduced for a training set, following several simulations on different test materials, in order to find out what permittivity the material under test has, and what material it is.

ACKNOWLEDGMENT

This article is based upon work from COST Action CA18223, supported by COST (European Cooperation in Science and Technology).

REFERENCES

[1] T. Chen, S.Li, H.Hui, "Metamaterial Application in Sensing" , Sensors, vol. 1 , issue3, pp.2742-2765, mar. 2012

[2] C. Caloz and T. Itoh, "Left-Handed Transmission Lines and Equivalent Metamaterials for Microwave and Millimeter-Wave Applications," 2002 32nd European Microwave Conference, Milan, Italy, 2002, pp. 1-4

[3] M. S. Boybay and O. M. Ramahi, "Material characterization using complementary split-ring resonators", IEEE Trans. Instrum. Meas., vol. 61, no. 11, pp. 3039–3046, Nov. 2012.

[4] A. N. Reddy and S. Raghavan, "Split ring resonator and its evolved structures over the past decade," 2013 IEEE International Conference ON Emerging Trends in Computing, Communication and Nanotechnology (ICECCN), Tirunelveli, 2013, pp. 625-629

IoT Based Automatic Electronic System for Monitoring and Control of Street Lighting

Seher Kadirova, Teodor Nenov, Daniel Kajtsanov

Department of Electronics, Faculty of Electrical Engineering, Electronics and Automation,
Ruse University "Angel Kanchev", 8 Studentska str., 7000 Ruse, Bulgaria
skadirova@uni-ruse.bg, t.nenov@mail.bg, danielkaytsanov@gmail.com

Abstract—The aim of the article is to design a system for yearly automatic control of street lighting. The idea is to increase the energy efficiency by switching ON the lights only in the dark part of the day. Using GPS coordinates and the trajectory of the sun, the time for switching the lights ON and OFF varies depending on the day of the year. The electronic system for monitoring and control of street lighting system is designed and developed on the basis of the IoT concept. The fundamental principle of the operation is based on the transmission and reception of information via GPRS, GPS and GSM network to a built physical server, connected in one way or another to the Internet network and receiving position data and accurate time from the GPS navigation system.

Keywords— IoT, Automatic street lighting, GPS, GSM, energy efficiency

I. INTRODUCTION

IoT is one of the fast growing fields in technology which handles high date in real time applications. IoT leads to developments nowadays. The Internet of Things (IoT), also known as the Internet of Things or the Internet of Objects, is a concept for a computer network of physical objects, buildings and other objects and possessions having built - in electronic devices for interaction with each other or with the external environment. This concept considers the organization of such networks as a phenomenon capable of reshaping economic and social processes so as to exclude the need for human participation in some actions and operations [1].

In [2] is developed a manual system where the street lights will be switched ON in the evening before the sunsets and they are switched OFF in the next day morning after there is sufficient light on the outside. But the actual timing for these lights to be switched ON is when there is absolute darkness. With this, the power is wasted up to some extent. This project gives solution for electrical power wastage. Also the manual operation of the lighting system is completely eliminated. The proposed system provide a solution for energy saving [2].

Automation systems have the advantage over the manual systems because it increases the productivity, efficiency and reliability, and minimizes the usage of resources to save energy, and reduce the operating cost etc. These automation systems play an essential role in the term "smart home" to make our daily life more comfortable, and to facilitate users

from ceiling fans to ovens, and in other applications. Among all exciting applications, streetlights play a vital role in our environment and also play a critical role in providing light for safety during night-time travel. When the streetlights are in a in ON mode over the whole night, which consumes much energy and reduce the lifetime of the electrical appliances such as a light-emitting diode lamps, incandescent light bulb, gas discharge lamp, and high-intensity discharge lamps. Especially in cities' streetlights, it is a severe power consuming factor and also the most significant energy expenses for a city. In this regard, an automation system is required to control the lights according to needs [3].

In another research of the authors the developed electronic system eliminates the disadvantages of the existing systems by taking date and time from the GPS, as it also gives information about the position of the system. Based on the results the microcontroller calculates and automatically detects geographical area and retrieve relevant data for sunrise and sunset in the area, respectively ensures very precise ON/OFF mode of the lighting system. It doesn't need operator maintenance and initial installation setup. The developed electronic device increases bulb life in result of the dimming effect. On the other hand this decrease of the illumination leads to reduce in the energy consumption [4].

A methodology for iterative optimization of optical systems for street LED luminaires intended for roads designed according to luminance requirement is proposed in another research [5].

A special role in the Internet of Things is played by the measuring instruments, ensuring the transformation of the data of the external environment into machine-readable data, at the same time filling the computing environment with significant information. A wide range of measuring instruments is used, from elementary sensors, consumption meters to complex integrated measuring systems. Within the concept of the "Internet of Things", measuring instruments are generally integrated into networks (e.g. wireless sensor networks, measuring systems), as a result of which it is possible to build systems for inter-machine interaction.

The modular electronic system for monitoring and control of street lighting system is designed and developed on the basis of the IoT concept.

The fundamental principle of its operation is based on the transmission and reception of information via GPRS, GPS and GSM network to a built physical server (for example located in the control room), connected in one way or another to the Internet network and receiving position data and accurate hour from the GPS navigation system.

The aim of the paper is to develop an IoT Based Automatic Electronic System for Monitoring and Control of Street Lighting.

II. DESIGN OF IOT BASED AUTOMATIC ELECTRONIC SYSTEM FOR MONITORING AND CONTROL OF STREET LIGHTING

A. Design of the system

Nowadays, automation systems are preferred because they reduce the use of energy. In modern society it is necessary to use the energy as efficiently as possible. These automation systems play an essential role in making our daily life more comfortable. Among all exciting applications, street light play a vital role in our environment and also a critical role in providing light for safety during night-time travel. Inefficient lighting wastes significant financial resources every year, and poor lighting creates unsafe conditions. Energy efficient technologies and design mechanism can reduce cost of the street lighting drastically. In this regard, an intelligent lighting control system can decrease street lighting costs up to 70% and increase the durability of the equipment. In this paper another approach will be used to control street lighting. The proposed electronic system is built on a modular principle in order to adapt to functionality in different work environments. The main modules are Power supply, Central microcontroller module, and Microcontroller module measurements. The block diagram (Fig. 1.) of the purposed system is based on microcontroller, GPS, and GSM modules.

The power supply module is designed to provide the necessary current and voltage to operate the system. It is realized as a separate module of the system, built on a separate printed circuit board and mounted in a separate housing. This can allow easy replacement if necessary during system operation.

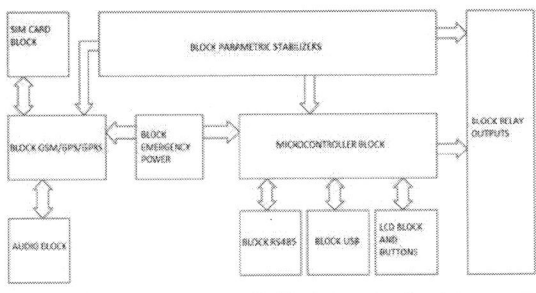

Fig. 1. The block diagram of the purposed system

Microcontroller Block:

BLOCK RS485- The RS485 interface channel is used for communication between the central microcontroller module and the measuring microcontroller modules in the system.

BLOCK USB- PIC18F47J53 has a fully integrated hardware unit supporting USB2 interface. This interface is used by the system setup to set various operating parameters via portable flash memory.

BLOCK LCD - From the input and output ports of the microcontroller of connectors J1 and J4 are output the necessary signals for indication and reading of the buttons for monitoring, setting and control of the system.

BLOCK RELAY OUTPUTS - A relay output block is made up of three identical channels built with a bipolar transistor controlled by the output port of the microcontroller. Relays K1, K2 and K3 are included as a load in the collector of these transistors. Through the contact feathers of these relays the switching of the powerful contactors for switching on and off of up to 3 terminals from the street lighting system is carried out.

BLOCK PARAMMETRIC STABILIZERS - To ensure stable operation of the central microcontroller module, the incoming supply voltage from the power supply unit is supplied to parametric stabilizers that produce the necessary voltages for operation of the central microcontroller unit. Given the high frequency communication signals in the system, the use of as few sources of frequency interference as possible in the topology of the printed circuit board is taken into account.

BLOCK GSM, GPS, GPRS - Built on the basis of a hybrid integrated circuit SIM808, which includes:

- GSM - unit for GSM telephony via mobile operator

- GPS - receiver for reading data from the satellite navigation system.

- GPRS - a unit fully hardware and software provided for Internet connection.

The SIM808 communicates with the microcontroller via a UART data bus. The reading and writing of SIM808 is performed by means of AT commands.

AUDIO BLOCK - An audio block is provided in the system for telephony with a control room.

SIM CARD BLOCK - A SIM card of the selected mobile operator for access to a mobile network is connected to SIM808 via slot J5. All lines to the slot are protected by TVS diodes

BLOCK EMERGENCY POWER - It is planned that in case of power failure the system will receive power from the built-in battery and record the event in the EPROM memory of the microcontroller, after which it will go into emergency mode at low power consumption. When the power is restored, the recorded data is sent to the control room.

B. Main Algorithm and Interaprions of operation of the central microcontroller module

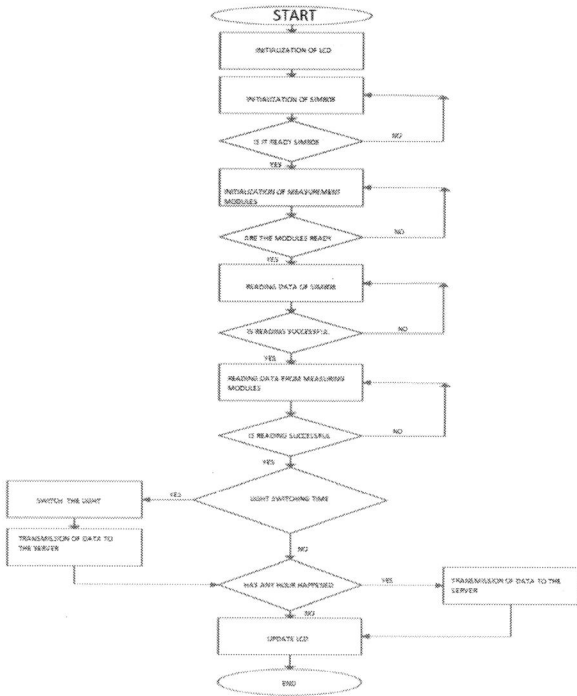

Fig. 2. Main logic diagram of the system

The interruptions of the program are presented in Fig. 3.

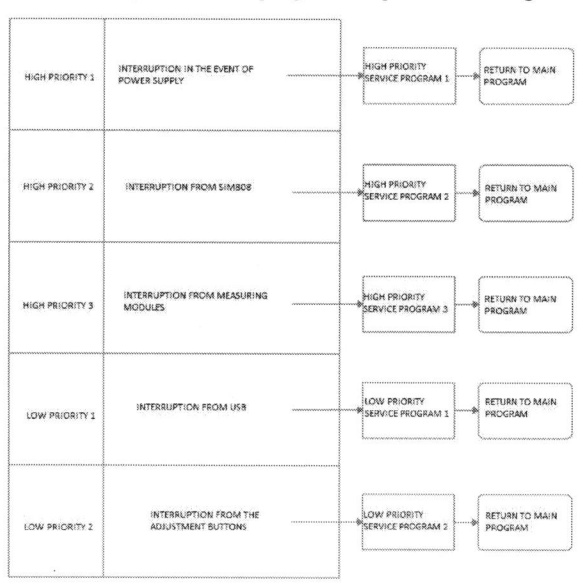

Fig. 3. Interraptions of the program

Three methods for controlling street lighting are provided:

• According to the set mode from the control room: The system has the ability to program ON and OFF hours for each day of the year from the control room.

• On command from the control room: At any time, different branches or the entire lighting system can be switched ON / OFF from the control room

• Autonomous mode at the zenith angle of the sun: In this mode, the system automatically, based on data read by the satellite navigation system, determines the geographical position of the system, calculates the zenith angle of the sun in real time and determines the time of switching lighting.

The printed circuit board of the central microcontroller unit, as well as the circuitry are designed with the help of Altium Desinger 20.0.11. In Fig. 4 is presented the general view of the printed circuit board.

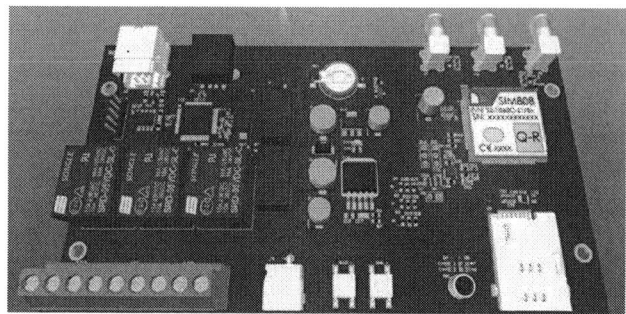

Fig. 4. A general view of the developed printed circuit board.

C. Design of Measuring Module

The measuring module is developed on the basis of IC MCP39F501 of the company Microchip. MCP39F501 is a single-phase IC for monitoring the parameters of power lines with a function for detecting events. In Fig. 5 is presented a block diagram of a measuring module.

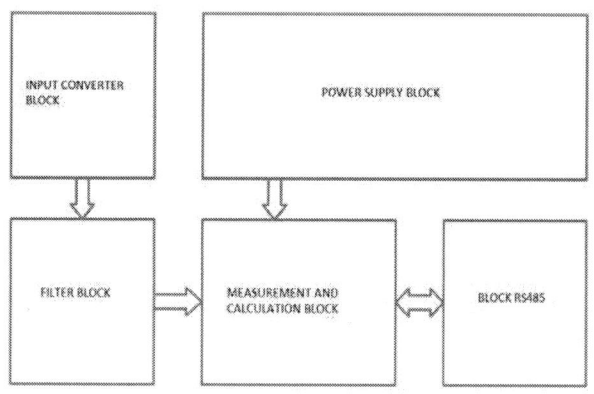

Fig. 5. A block diagram of a measuring module.

The main functional advantages of MCP39F501 are: Measurement of active, reactive and real power; Measurement of the actual effective value of the current; Measurement of the actual effective voltage value; Network frequency measurement, Measurement of the power factor, Measurement error up to 0.1%.

D. Measurement methodology and equations for calculating the measured parameters

– Measurement of effective values of voltage and current

Two 24-bit analog-to-digital converters are built into the integrated circuit for simultaneous measurement of the effective values of current and voltage. The root mean square calculation is performed on the basis of 2^n measurements done, where n - value set in the register of IC.

The equations for measuring the effective values of current and voltage are as follows:

– Equation for the current:

$$I_{rms} = \sqrt{\frac{\sum_{n=0}^{2^n-1}(i_n)^2}{2^n}} \qquad (1)$$

– Equation for the voltage:

$$V_{rms} = \sqrt{\frac{\sum_{n=0}^{2^n-1}(v_n)^2}{2^n}} \qquad (2)$$

- Calculation of the power factor

The power factor is calculated as the ratio of the active power value relative to the effective power value with the following equation:

$$PF = \frac{P}{S} \qquad (3)$$

- Calculation of reactive power

The calculation of the reactive power is performed in a similar way as the calculation of the active power, but when measuring the voltage out of phase at 90° to the current.

III. RESULTS

A. SIMULATIONS AND SOFTWARE DEVELOPMENT

The software is implemented using the development environment of MICROCHIP, MPLAB X IDE v5.05. The software code is written in C ++ and compiled in machine code with an XC8 compiler developed by MICROCHIP. The simulations of the program execution are made with PROTEUS 8 Professional. Waveforms were downloaded from the development boards during tests of different blocks of the circuits.

In PROTEUS 8 Professional program, which has built-in mathematical models of elements, types of communications, simulations of execution of parts of the program have been accomplished.

In Fig. 6 are presented the simulation results and tuning of the communication with the USB module. In this case is presented the process of set the data buffer.

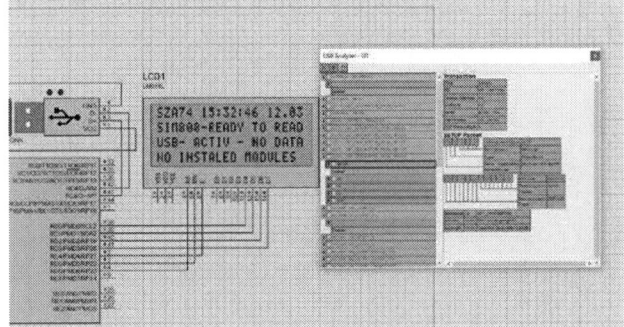

Fig. 6. Simulation results.

In Fig. 7 is presented the setting up the communication of the EUSART2 module between the microcontroller and SIM808, using the Virtual Terminal module built into PROTEUS 8 Professional.

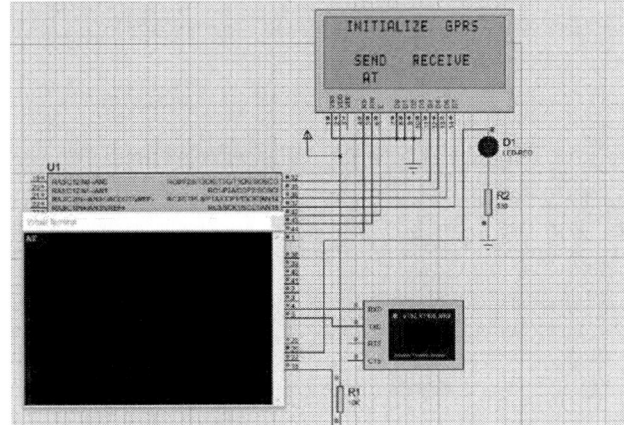

Fig. 7. Setting up step from the communication of the EUSART2 module between the microcontroller and SIM808.

B. EXPERIMENTAL RESULTS

In Fig. 8 are presented the waveforms of RS485 bus in communication between the measuring module and the central microcontroller module.

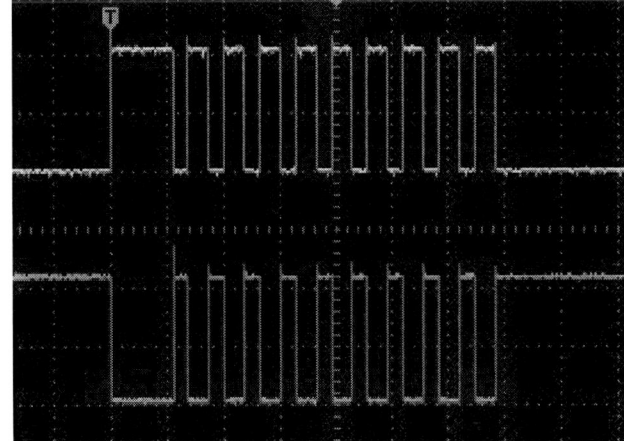

Fig. 8. Waveforms of RS485 bus.

In Fig. 9 waveform from pin 1PPS (one pulse per second) of IC SIM808 is presented. The NMEA (navigation message) from the satellite navigation system flows every second.

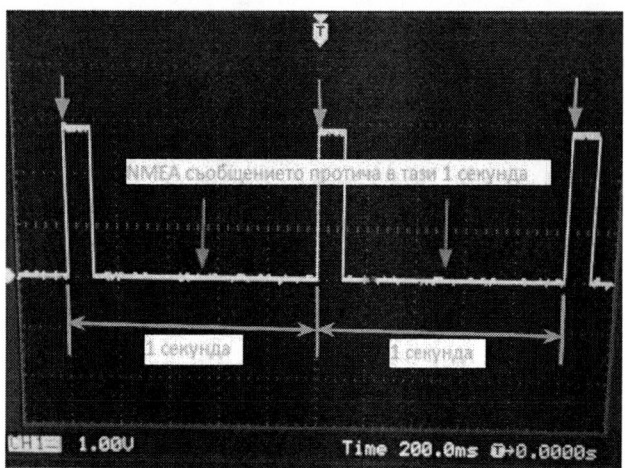

Fig. 9. Waveform from pin 1PPS.

IV. CONCLUSIONS

The system is designed to optimize the management of street lighting in populated areas. It creates statistics for the operation of street lighting systems in real time, on the basis of which to make improvements to the model of operation of the same.

The system provides data on the parameters of the power supply network of street lighting.

The advantage of the system is the possibility to control all branches of the lighting system in the settlement from a dispatcher to a control room or from any place, as long as there is access to internet connection.

Last but not least advantage of the system is the possibility for remote reprogramming or real-time adjustment.

ACKNOWLEDGMENT

The article presents results from Project No 2020 - EEA - 05, with financial support of the National Science Fund of University of Ruse.

REFERENCES

[1] J. Mathew, R. Rajan, and R. Varghese, IOT BASED STREET LIGHT MONITORING & CONTROL WITH LoRa/LoRaWAN NETWORK, International Research Journal of Engineering and Technology (IRJET), Volume: 06 Issue: 11 | Nov 2019

[2] P. Tambare, P. Venkatachalam, and D. Rajendra, Internet Of Things Based Intelligent Street Lighting System for Smart City, Internet Of Things Based Intelligent Street Lighting System for Smart City, Vol. 5, Issue 5, May 2016

[3] Z. Mumtaz, S. Ullah, Z. Ilyas, et al. An Automation System for Controlling Streetlights and Monitoring Objects Using Arduino. Sensors (Basel). 2018;18(10):3178. Published 2018 Sep 20. doi:10.3390/s18103178

[4] S. Kadirova, and D. Kajtsanov, (2017). A real time street lighting control system. Paper presented at the 2017 15th International Conference on Electrical Machines, Drives and Power Systems, ELMA 2017 - Proceedings, 174-178. doi:10.1109/ELMA.2017.7955426

[5] P. Tsankov, M. Yovchev, H.and Ibrishimov, (2019). LED luminaire optical lens optimization for road lighting designed according to luminance requirement. Paper presented at the 2019 2nd Balkan Junior Conference on Lighting, Balkan Light Junior 2019 - Proceedings,

Dynamic adaptation of power emissivity for mobile microstrip antennas in a variable impedance environment with microcontroller

L. Baicu[1], B. Dumitrascu[1], M.Culea[2] and N. Nistor[1]
[1]Department of Electronics and Telecommunications
[2] Department of Computer Science and Information Technology
"Dunarea de Jos" University of Galati
Galati, Romania
nicusor.nistor@ugal.ro

Abstract—In this paper a qualitative and quantitative study of emissivity control for micro-strip type microwave antennas using adaptive correction of the transmission power when the wave impedance of the propagation medium is not constant and it is influenced by nearby moving conductive objects or by objects with different reflection or absorption ispresented. We propose a dynamic adaptation module controlled by a microcontroller for the microwave transmission circuit. The microcontroller reads the current variation for a fixed current level of the emitter which contains the information about the propagation efficiency of the electromagnetic wave and generates a RF carrier amplitude control signal. The mathematical model of a micro-strip antenna working in a certain emission band and the dynamic hardware adaptation possibility for the power of the emission so that the medium influence will not affect the minimum useful power needed by the receiver are considered. The main idea is to create an algorithm to reduce the supplemental RF emitter energy consumption working in open waveguide (air or vacuum) and whose antennae are not perfectly adapted or are moving, such as wireless devices or GSM antennae. For these particular cases, the surrounding objects having the physical dimension of the emitting antenna behave like receivers and will absorb totally or partially the energy emitted. The phenomenon considered is the one of reappearing of the reflected waves in the antenna, which will change the antenna circuit similar to a supplemental reactive load in the emitting circuit. The effect in this case is the increase of the absorbed electrical power in the final RF stage. Measuring the current is a challenge because two steps are needed; a pre-scaling and a final scaling using a voltage zoom by temporally modifying the voltage references of the ADC converter. Finally, the solution proposed in this paper consists in the very precise periodic measurement in the final RF circuit by using a shunt resistor connected to the ground; when a supplemental charge appears, the microcontroller will change (increase or reduce) the amplitude of the RF carrier so that the reflected wave's amplitude returning to the antenna is subsequently smaller. The system uses an experimental model with a process microcontroller with a 10 bits ADC.

Keywords— *Process monitoring, micro-strip antenna, microcontroller process, wave reflexions, impedance adapter.*

I. INTRODUCTION

Starting from the concept of a RF transmission chain in which the emitter must be adapted to the transmission conditions in order to send a maximum amount of energy to the useful load, we can notice the following.

Transmission power is a function dependent on the square voltage of the emission line:

$$P_{em} = \frac{|U_{em}|^2}{Z_0}$$

(1)

Where Pem is the transmission power; Uem is the RF line voltage; Z0is the impedance of the propagation medium. The power reflected at the load is dependent on the same square law:

$$P_r = \frac{|U_r|^2}{Z_0}$$

(2)

Where Pr is the reflected transmission power from the propagation medium; Ur is the reflected voltage of the RF line. We get a RF power that propagates to the receiver that is dependent on the dynamic level of the reflexion represented by the reflexion coefficient Γ :

$$P_{load} = P_{em} - P_r = P_{em}(1 - |\Gamma|^2)$$

(3)

Conductive discontinuities of the propagation medium that generate dynamic variations of the impedance, Z_0, that appear while moving the transmission module influence the reflexion coefficient Γ that will modify the transmission efficiency and lower the RF voltage reaching the receiver for a fixed transmission power.

If we assume that the attenuation coefficient in the propagation line α is different from zero and we consider the losses on the waveguide, which in this case is spatially open, the power at the waveguide entry is:

$$P_{in} = P_{em}e^{2\alpha Z} - P_r e^{-2\alpha Z} = P_{em}(e^{2\alpha Z} - |\Gamma|^2 e^{-2\alpha Z}) \quad (4)$$

The transmission efficiency on the line can be expressed as:

$$\eta = \frac{P_{load}}{P_{em}} = \frac{(1 - |\Gamma|^2)}{(e^{2\alpha Z} - |\Gamma|^2 e^{-2\alpha Z})} \quad (5)$$

Where Z is the propagation distance of the electromagnetic wave between the emitter and the receiver

In the case of a perfect adaptation the efficiency's formula has a maximum which can be approximated as:

$$\Gamma = 0 \rightarrow \eta_{max} = e^{-2\alpha Z} \cong 1 - 2\alpha Z \quad (6)$$

An expression which shows the linear dependence between the efficiency and the constant line attenuation, alpha. When materials with a conductivity constant close to zero are used then the transmission efficiency is very closed to 1.

The article proposes to monitor the variation of the fixed bearing current absorbed from the power supply of the transmitter. If in the transmission chain there are factors producing zonal or dynamic loss of adaptation by changing the impedance of the environment, there will be reflected waves which will increase the consumption of the transmitter above a standard emission consumption level.

In this context the present article proposes a temporary increase or decrease of the emission power through the variation the amplitude of the RF carrier so that the receiver receives a level constant of useful voltage. Some interesting ideas in the realization of the present article were read in the works: [1], [2], [3], [4].

II. PRELIMINARY EXPERIMENT

A preliminary experiment of implementing the idea of the article was to use an RF broadcast module with a micro-strip antenna of a flat circular shape working on the 433 MHz frequency. This antenna, in constant RF carrier emission, was placed in the vicinity of a conductive body, having dimensions comparable to the size of the antenna (adaptation in relation to the wavelength). The antenna was rotated with an angle sweep between -90 and +90 degrees, with permanent monitoring of the current variation in relation to a fixed bearing.

This involves using an electronic circuit or a pre-calibration of acquisition of analogue data in the microcontroller so that we observe small variations in current or voltage that overlap with the value of the current consumed in the emission.

The schematics of the experimental assembly and the device which can achieve the adaptation of the emission power in relation to the emission in dynamic conditions of the antenna are presented below:

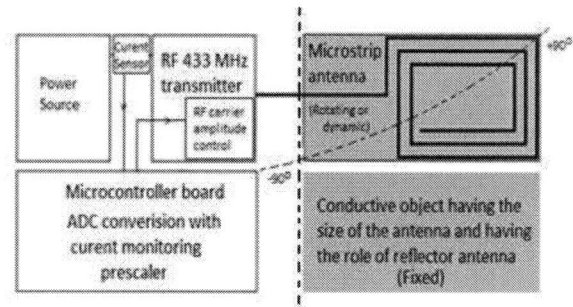

Fig. 1. Schematic of a rotation experiment

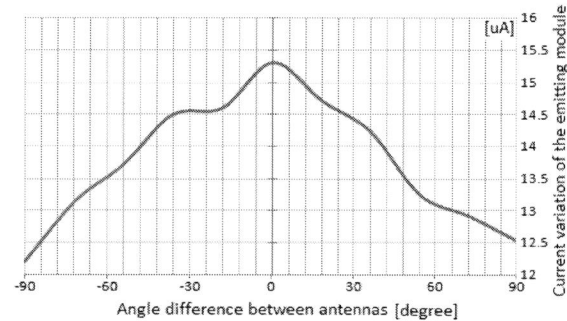

Fig. 2. Current varriation monitorring during the rotation experiment

The experimental results consider a maximal reflected wave for parallelism between the antennae's planes. The experiment was performed in the laboratory by separating the continuous current component, only the gradient of the alternative current being measured. The experiment was performed with a constant rotating speed between the two microstrip antennae.

III. SYSTEM DESIGN AND RESULTS

The easiest way of reducing the reflected wave is to use a attenuation device connected on the propagation line right near the load so that a small part of the emission power is lost but double the power of the reflected wave is lost. This is represented in fig. 3.

In particular for the open systems the reflected wave cannot be stopped by using attenuators close to the load by requires that the emitting antennae shouldn't be placed close to conductive objects with similar or comparable sizes. In the case of moving wireless antennae this is not possible and this paper addresses the problem of dynamical adaptation of the emitting power.

The experiments confirm the initial hypothesis and in the following we present two experimental results, one in which the angular position between the two antennae is taken into account and the second one where the distance between them is changed. In both of the cases we have obtained high resolution data proving the initial assumption.

2020 IEEE 26th International Symposium for Design and Technology in Electronic Packaging (SIITME)

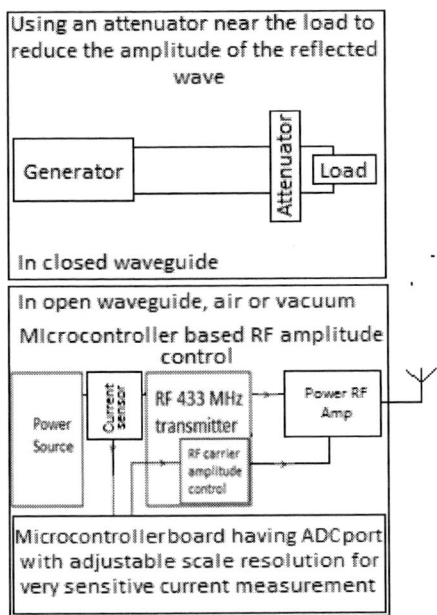

Fig. 3. The reflected wave atenuation problem for closed and open waveguide

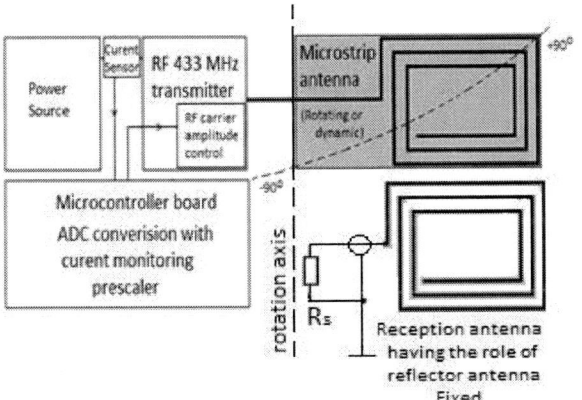

Fig. 4. Experimental circuit with microcontroler nr 1. Antennae Rotation

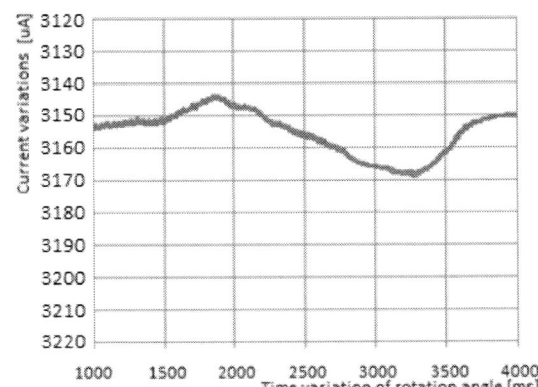

Fig. 5. Experimental results for circuit no. 1

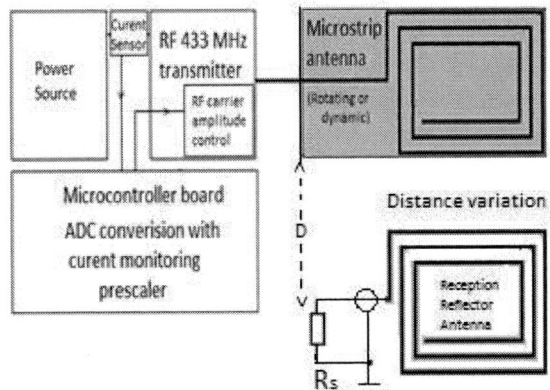

Fig. 6. Experimental circuit nr 2. Distantance between antennae planes

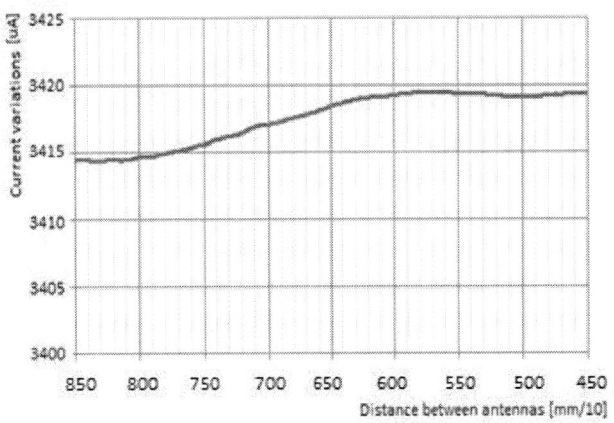

Fig. 7. Experimental results for circuit no. 2

In the figure it can be observed that the emmiting circuit with a CE bipolar transistor generate RF power on a complex load and in the same there is a suplemental stress created by the reflected wavewhich is percieved by the circuit as a supplemental load, much smaller than the one for which the circuit was designed.

The supplementary energy consumption can be found in the global current consumption because there are time intervals in a period of the oscillation of the final RF stage when the oscillation generated by the direct wave will consist in a positive voltage while the reflected wave will have a negative voltage. The oscillator will impose the highest amplitude but it will have to use a supplemental power consumption from the power supply.

The difference in power consumption is not big, but to reduce this phenomen the solution is to temporary reduce the amplitude, thus reducing the emitting power of the final RF stage.

978-1-7281-7507-2/20 $31.00 © 2020 IEEE 129 21-24 October, Pitesti, Romania

2020 IEEE 26th International Symposium for Design and Technology in Electronic Packaging (SIITME)

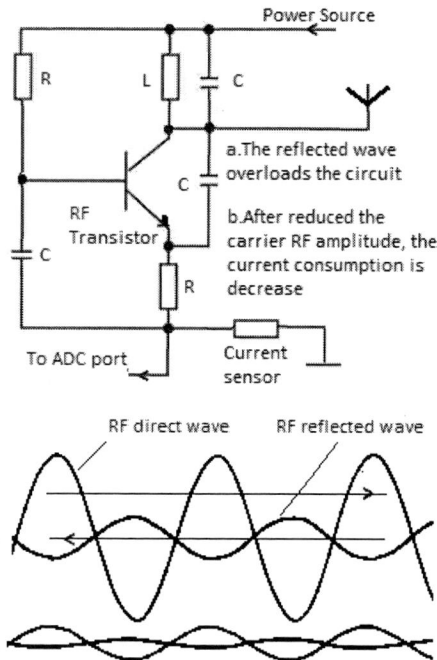

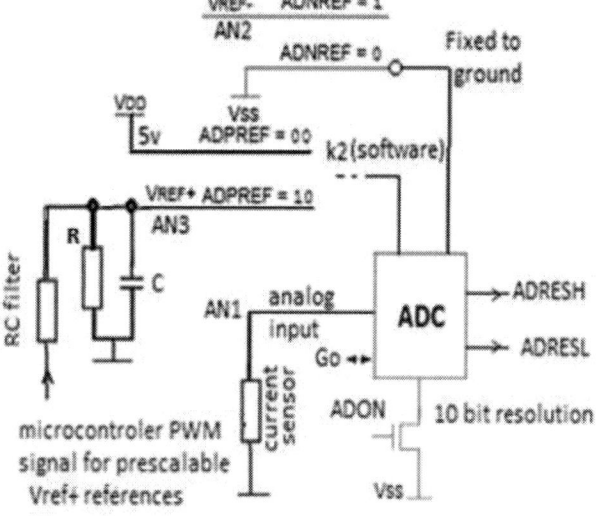

Fig. 8. Two situation with and without reducing the RF amplitude

The typical structure of the ADC port of the microcontroller allows the setting of the reference values Vref+ and Vref- via user or a special hardware device. The typical structure of this device are from the document [5] and is presented in the next figure.

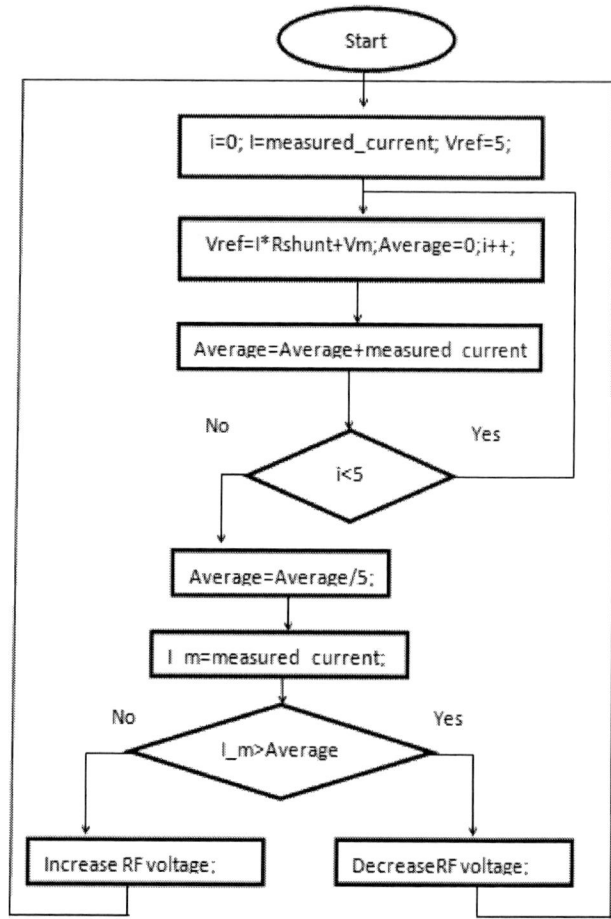

Fig. 9. Schematic for implementing variable refrence of ADC.

The data acquisition and current optimisation algorithm is presented in fig. 10.

Fig. 10. Current optimisation algorithm

The result of running this algorithm is that in the case of the appearance of a conductive object and dimensions comparisons with the length of the antenna that produces reflections in the emission circuit, the amplitude of electromagnetic waves will be decreased. If the object that produces the reflections moves away or changes its position relative to the antenna, the amplitude of the wave returns to the previous step.

VI. CONCLUSIONS

The authors are proposing a model which can be used in a dynamic regime to solve the problem of wave reflexions. Theoretical notions are presented as well as some possibilities of solving the problem are presented.

The problem of supplemental energy consumption in non-adapted RF circuits doesn't concern the actual power consumption but rather the increased electromagnetic pollution and protecting the emission components against fatigue generated by overload while increasing the reliability and endurance.

An interesting issue that remains to be studied in the future by theauthors is that of optimizing the level of emission power in contextobtaining the useful effect, in the sense that the amplitude level theRF transmitter cannot decrease as much, even in reflective conditionsproducts of neighboring objects, since in this case the receiver havinga limited sensitivity will no longer be able to receive the signal innominal conditions.

A cumulative study comprising a functional compromise between the two phenomena is useful, provided that the transmitter is able to have a feedback at the signal level from thereceiver. The present study is used experimentally and implementable in the case mobile devices running on batteries for which energy resources are very closely monitored and managed. An additional emission consumption due to the neighboring reflections and which is more power than the receiver needs leads to a decrease incomponent and device reliability.

References

[1] T. Ikeda, S. Saito and Y. Kimura, "A Frequency-Tunable Dual-Band Single-Layer Shorted Multi-Ring Microstrip Antenna Fed by an L-probe with Varactor Diodes," 2019 IEEE International Symposium on Antennas and Propagation and USNC-URSI Radio Science Meeting, Atlanta, GA, USA, 2019, pp. 905-906, doi: 10.1109/APUSNCURSINRSM.2019.8888596;

[2] R. Sarvendranath and N. B. Mehta, "Exploiting Power Adaptation With Transmit Antenna Selection for Interference-Outage Constrained Underlay Spectrum Sharing," in IEEE Transactions on Communications, vol. 68, no. 1, pp. 480-492, Jan. 2020, doi: 10.1109/TCOMM.2019.2950680.

[3] Félix Vega-Stavro ; Fernando Albarracín-Vargas, "Variable Impedance Feed Structure for Impulse Radiating Antenna", 2019 International Conference on Electromagnetics in Advanced Applications (ICEAA)

[4] Jie Hu ; Fang Liu ; Yuanan Liu. "Joint energy-efficient relay antenna selection and power adaption for two-way full duplex MIMO relay"

2017 3rd IEEE International Conference on Computer and Communications (ICCC)

[5] http://ww1.microchip.com/downloads/en/devicedoc/41291d.pdf

Software Controlled Radio Receiver for Versatile Wireless Communications

Daniel Alexandru Visan, Mariana Jurian, Ioan Lita, Laurentiu Mihai Ionescu, Alin Gheorghita Mazare
Electronics, Computers and Electrical Engineering Department
University of Pitesti
Pitesti, Romania
ioan.lita@upit.ro

Abstract—**In this paper is presented the design and the implementation of a software controlled reconfigurable receiver that is intended to be used for wireless communications using analogue or digital modulation techniques. The hardware part of the system is realized around the specialized circuits RTL2832U and R820T2 which together forms a wideband radio receiver with an extended frequency domain of operation, ranging from 42MHz to 1000MHz. The versatility of the proposed equipment is achieved by using a novel architecture based on the software defined radio technology. The digital signal processing of the received signals is realized with a sampling frequency of 2,56Ms/s at a resolution of 8 bits. Through the USB interface, the receiver module is controlled by a computer that runs a proprietary Labview application. By using a very flexible configuration based on high performance circuits and software defined radio architecture, the proposed receiver represents a reliable solution for many applications including data transmission in wireless sensor networks or personal communications.**

Keywords—receiver, software control, wireless communications.

I. INTRODUCTION

The diversity of the existing communication standards and the wide range of modulation techniques that are used nowadays generated an increased need for developing a new technology capable to offer an extended flexibility in implementation and exploitation of wireless communication systems. The realization of the transition from the existing radio systems to digital architectures implies also the use of a new approach and modern solutions for implementing suitable equipment for long distances communications. In this context, the software-defined radio become a key technology that has the potential to play an important role in the process of redesigning the classical radio communication systems. Although the main characteristics of this technology were defined theoretically with many years ago, the practical implementation of the concept was not possible until nowadays, when the digital circuits and electronic components integration known a tremendous development. Integrating smart processing methods into a software defined radio provide a way to realize a further developing step, creation of a cognitive radio. Taking into account the abovementioned elements, in this paper is approached this important research topic by proposing a versatile receiver based on the innovative concept of software defined radio [1].

II. THE SOFTWARE-DEFINED RADIO TECHNOLOGY

Software-defined radio refers to a wireless system that relies mainly on software to implement processing functions which in classical communication systems are usually implemented in hardware. The main advantage of this approach is given by the fact that the radio system can be very easily adapted to operate with various types of communication protocols or modulations, simply by updating the internal software. Multi-standard terminals in mobile radio systems can be efficiently implemented by using software-defined radio technology. As can be seen in Fig. 1, in a typical structure of a software-defined radio, the analog-to-digital and digital-to-analog converters are placed just after the radiofrequency input stage, meaning that all subsequent processing are realized in software. This configuration is most desirable but in practice can be difficult to be implemented [2], [3].

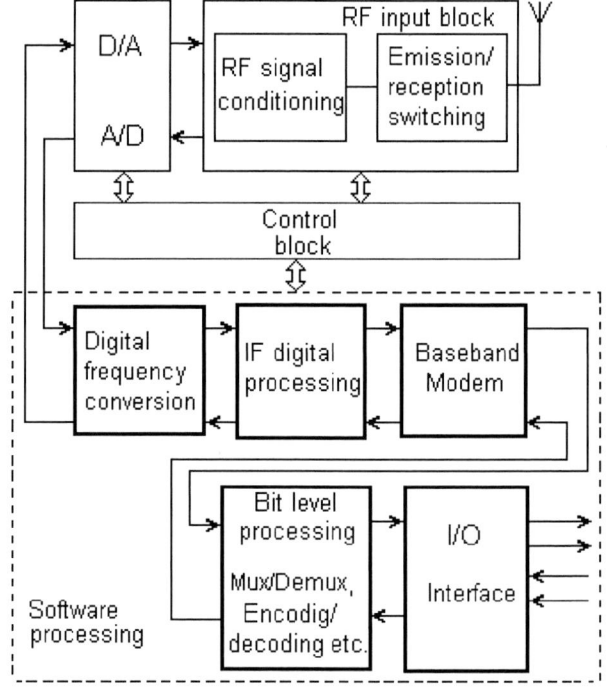

Fig. 1. The main elements of an emission/reception system based on software-defined radio technology.

III. The Structure of the Receiver

In Fig.2 is presented the block diagram of the system. The main elements used for implementing the receiver are the RTL2832U and R820T2. These specialized circuits represent a digital signal processor and a tuner. The operation of the entire receiver is controlled by a LabVIEW application running on a local computer. Although the R820T2 was created initially for implementation of TV receivers and for streaming applications, subsequently it was proved that this circuit can also be successfully used as an excellent radiofrequency digital tuner for software defined radios. The main signal processing blocks that compose the R820T2 tuner includes, among others, a performant fractional phase-locked-loop (PLL) based frequency synthesizer, an efficient variable gain amplifier, a low noise radiofrequency amplifier, a high-speed mixer and a tracking filter [2]. As can be observed in Fig. 2, the R820T2 tuner operate as a radiofrequency input block for the digital signal processor RTL2832U.The input signal for the R820T2 is coming directly from the antenna. The first low noise amplifier ensures a noise figure to the input of around 3,5dB and an image frequency rejection of 64dBc.The overall sensitivity of the tuner is of minimum -79,5 dBm, which can be consider an acceptable value for the proposed design. Also, the tuner is capable to operate in a wide frequency domain ranging from 42 to 1002MHz. From this point of view the chosen tuner is capable to fulfill one of the main objectives imposed to software-defined radio systems, namely to operate with an extended range of frequencies that are specific to multiple wireless communication standards. The internal mixer integrated in the tuner realizes the frequency shift of the spectrum of the input signal toward the intermediate frequency of approximatively 3,57 Mhz. But the intermediate frequency doesn't have a fixed value been dependent of the received channel bandwidth [4].

The frequencies that are necessary for the down-conversion process are generated with an internal frequency synthesizer that operates together with a reference local oscillator controlled by a quart crystal (28,8 MHz). With this arrangement the tuner achieve a frequency resolution of 1 Hz. This parameter reflects in the performances of the entire receiver, allowing a very good selectivity in all input bandwidth. The overall gain of the tuner is established by an amplification gain control loop (AGC). The gain control is realized both on the final variable gain amplifier and on the low noise amplifier situated at the RF input of the circuit. The second most important component of the receiver is the digital signal processor R2832U. The R2832U operates according to the principles of a coded orthogonal frequency division multiplexing (COFDM) demodulator.

Overall, the main task of the R2832U is to capture the quadrature (I/Q) sample values of the received signal and to pass them to the PC for further processing and for user presentation in audio format. Although the circuit can process data from tuners operating at intermediate frequency (IF) or at low-IF, in the proposed application is more convenient, from the digital processing point of view, to use direct conversion to zero intermediate frequency. On this way are eased the constraints related to the processing speed of the receiver's circuits [5].

The block diagram of the R2832U circuit can be observed in Fig. 2. The built-in, high-speed, analog-to-digital converter (ADC) operates with a resolution of 8 bits and is preceded by an anti-aliasing filter. The sample values from the ADC are separated into two different data streams, obtaining two quadrature output signals denoted I (in phase) and Q (quadrature). The quadrature strings are down-converted with a set of two digital mixers and the resulting signals are digitally low-pass filtered and decimated. The result of these processing steps is sent to the PC through the USB interface. Among the above-mentioned operations, the circuit realizes additional tasks for improving the reception: data synchronization, co-channel interference cancelation, channel estimation and correction and accurate frequency adjustment using an internal numerically controlled oscillator.

Considering the operation of the entire receiver, consisting of the R820T2 tuner connected to the RTL2832Udigital signal processor, the input signal coming from the antenna can be expressed as a sum between the information signal s(t) and noise n(t).Because the receiver is realized for a wideband operation, at the input of the tuner will be present a multitude of informational signals. This creates important issues regarding the separation of the desired channel using software processing algorithms. This problem is solved in the proposed receiver by using dedicated algorithms implemented in a LabVIEW application [6], [7].

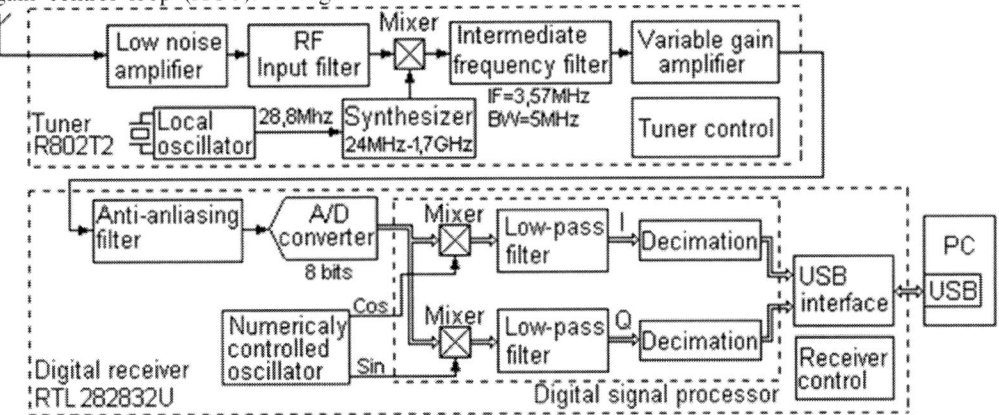

Fig. 2. The simplified structure of the proposed receiver.

2020 IEEE 26th International Symposium for Design and Technology in Electronic Packaging (SIITME)

IV. IMPLEMENTATION AND RESULTS

For verifying the correct operation of the proposed receiver, a prototype board was implemented and tested. The hardware structure of the software-controlled receiver can be observed in Fig. 3. In this picture can be remarked the input ports for the antenna, the USB communication interface and the specialized circuits RTL2832U and R820T2 which are provided with small heat sinks. The 28,8 MHz quart crystal is also visible near the R820T2 tuner. The software routines which have already been implemented allow a wideband reception of radio signals that use various communication techniques. In the actual configuration, the proposed receiver is capable to operate with multiple communication standards such as the amplitude modulation with suppressed or conserved carrier, wideband or narrowband frequency modulation, amplitude shift keying, frequency shift keying, phase shift keying, double-sideband and single-sideband modulations etc. These techniques were tested in this phase of development of the project, but the proposed system can be easily adapted to operate with a much broad range of communications standards. As already been mentioned in the previous sections, the receiver has an important part of its functionalities implemented in software. Regarding the programming language, in this case was chosen LabVIEW, because of its inherent advantages. For example, the time required for realizing the control interface of the receiver was considerable shorter in comparison with an implementation realized in a text programming language such C or Matlab. Nevertheless, the functionality and flexibility offered by the LabVIEW graphical programming allowed a simple and straightforward realization. In Fig. 4 is presented the block diagram for a part of the application used to manage the operation of the receiver. As can be observed, the whole program runs in a while loop which is continuously executed if the receiver is detected on the USB port and the internal event detection subroutine is not activated. If the external hardware module of the receiver is connected to the PC, subsequently is opened a communication session. The internal buffers of the RTL2832U and R820T2 are reset and the registers are initialized. In the next step is read the buffer containing the user settings. These settings are introduced through the graphical interface of the application. Also, through a short negotiation between the PC application and the external hardware module, are established other important parameters such is the sampling rate, the number of the samples that will be acquired etc. Subsequently, three important steps follow, namely the acquisition of radiofrequency signal, the demodulation in accordance with the selected standard, and finally the conversion of the data from internal format to the audio stream that can be played using the audio system of the PC. In Fig. 5 is presented the front panel of the interface for the proposed receiver. As can be remarked, the application is structured in two sections. The section in the left side is dedicated to the basic controls of the receiver: the tuned frequency, the type of the modulation, the AGC control, the filter type and the bandwidth of the selected filter. The section in the right side of the front panel contains four graphical displays which allow a real time analysis of the most important signals. In the captured image presented in Fig. 5 are displayed the waveforms for the radiofrequency signal received from the tuner and the audio signal at the output of the demodulator. Both waveforms are represented in time and frequency domain. The case presented in Fig. 5 is for a frequency modulated signal with a central frequency of 102,2MHz. The reception quality is very good and the tuning capabilities are much better in comparison with a classical system [8].

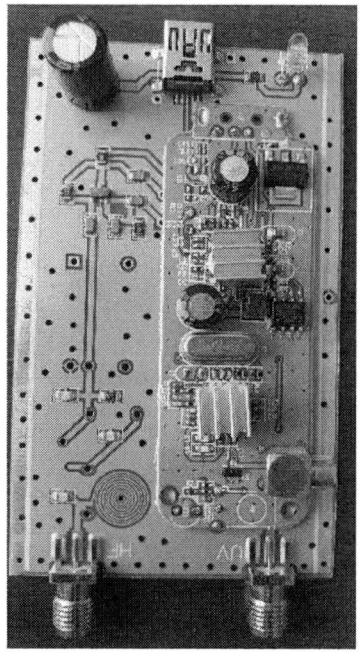

Fig. 3. The hardware structure of the receiver.

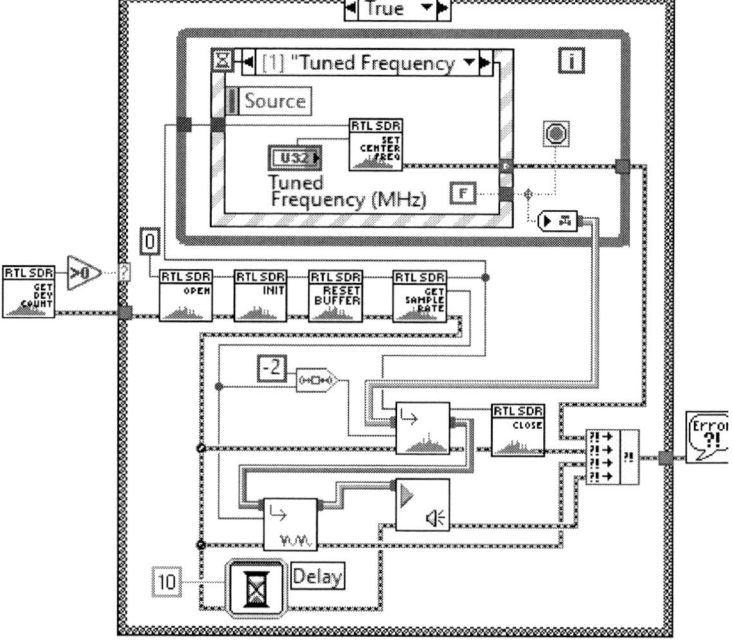

Fig. 4. The LabVIEW block diagram of the subroutine used for controlling the receiver.

978-1-7281-7507-2/20 $31.00 © 2020 IEEE   134   21-24 October, Pitesti, Romania

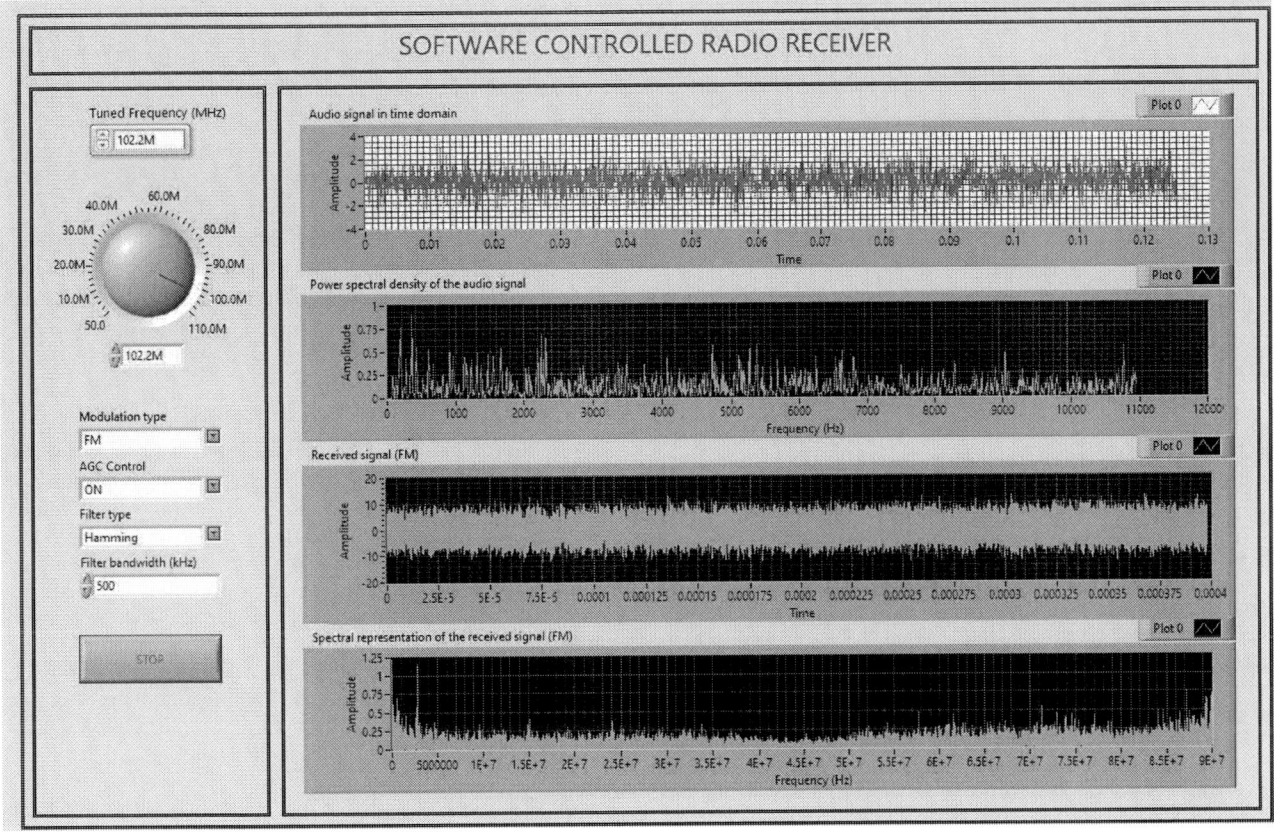

Fig. 5. The interface of the receiver, containing the graphical displays and the tuning controls.

V. CONCLUSIONS

The implementation of essential functionalities of wireless systems in software allows the operation according with multiple communication standards, thus reducing the need to use specialized equipment for specific applications.

In this paper was presented an enhanced receiver based on software-defined radio technology. Using a very efficient hardware architecture based on highly integrated circuits along with a control application implemented in LabVIEW led to obtaining a versatile radio receiver capable to operate with many communication standards.

Also, the proposed receiver is suitable for testing new wireless communication methods because the development of the new projects is more rapid and don't require to design and fabricate dedicated hardware.

ACKNOWLEDGMENT

This work was supported by a grant of the Romanian National Authority for Scientific Research, CNCS/CCCDI - UEFISCDI, project number PN-III-P1-1.2-PCCDI-2017-0332, "Increasing the institutional capacity for research in bioeconomics for innovative using of the autochthonous vegetal resources for obtaining high value added horticultural products (BIOHORTINOV)".

REFERENCES

[1] R. Chen, H. Hashemi, "A 0.5-to-3 GHz software-defined radio receiver using sample domain signal processing", IEEE Radio Frequency Integrated Circuits Symp. (RFIC), 2013.

[2] A . H. Ramadan, J. Costantine, Y. Tawk, K. Y. Kabalan, C. G. Christodoulou, "A reconfigurable RF front-end receiver for autonomous spectrum sensing cognitive radios", 10th European Conf. on Antennas and Propagation (EuCAP), 2016.

[3] Adrian Tulbure, Calin Petrascu, Marian Vladescu, "Calibration Methodology for a 3D Measurement System of Electromagnetic Radiation", International Semiconductor Conference (CAS), 2019.

[4] M. Dholu, K.A. Ghodinde, "Internet of Things (IoT) for Precision Agriculture Application", 2nd Int. Conf. on Trends in Electronics and Informatics (ICOEI), 2018.

[5] A. A. N. Azlin, H. Mansor, A. Z. Hashim, T. S. Gunawan, "Development of modular smart farm system", IEEE 4th Int. Conf. on Smart Instrumentation, Measurement and Application (ICSIMA), 2017.

[6] K. Vachhani, R. A. Mallari, "Experimental study on wide band FM receiver using GNU Radio and RTL-SDR", Int. Conf. on Advances in Computing, Communications and Informatics (ICACCI), 2015.

[7] C. Murtaza, S. Cicioğlu, A. Çalhan, "Performance analysis of software-defined network approach for wireless cognitive radio networks", 26th Signal Processing and Communications Applications Conference (SIU), 2018.

[8] J. Liu, X. Wang, K. Gong, G. Zhou, X. Ma, Y. Yuan. "A Software-Defined Radio Oriented Approach in Constructing RF Front-End Circuit Based on Fully Differential Amplifier", IEEE 9th Int. Conf. on Electronics Inf. and Emergency Communication (ICEIEC), 2019.

2020 IEEE 26th International Symposium for Design and Technology in Electronic Packaging (SIITME)

Automation Module for Precision Irrigation Systems

Ioan Lita, Daniel Alexandru Visan, Alin Gheorghita
Mazare, Laurentiu Mihai Ionescu
Electronics, Computers, Electrical Engineering Department
University of Pitesti,
Pitesti, Romania
ioan.lita@upit.ro

Adrian Ioan Lita
Axiplus Engineering SRL
Engineering Department
Bucharest, Romania
adrian.lita@ieee.org

Abstract—**In this paper is presented an improved design for an automation module which is dedicated for applications in irrigation systems used in precision agriculture. The architecture of the automation module relies on Arduino Uno development board containing the microcontroller ATmega328P which performs the majority of the control tasks required in the operation of the proposed system. The long-distance data transmission and remote monitoring of the important parameters of the system is realized with the connectivity ensured by SIM800L quad-band GSM/GPRS module. Also, the local connection between the automation module and a PC running a software control application is realized with the CH340G USB to USART converter which is integrated on the Arduino board. The automation module controls a precision irrigation system containing a set of sprinklers, a humidity sensor, a flow meter and a variable speed pump. By using a modern implementation approach, based on versatile ATmega328P microcontroller, together with the communication capabilities offered by the GSM/GPRS module SIM800L, the proposed design represents an efficient and cost-effective solution for implementing smart irrigation systems for modern agriculture industry.**

Keywords— automation, precision irrigation, microcontrollers.

I. INTRODUCTION

In the context of climatic changes and population growth, the use of modern electronic technologies in the agricultural domain for improving the efficiency of this industry has become almost mandatory. Precision agriculture based on electronic automation modules and distributed sensor networks represents a very efficient solution for solving many problems regarding the farming resources optimization. In the broad spectrum of applications that are specific to the precision agriculture, the irrigation systems occupy a very special place because the water is a key resource that can significantly influence the efficiency of the farming systems. Although the irrigation systems were constantly improved over the years, only in the late period was realized a real progress in the implementation of the complex control systems for agriculture, by using powerful informatic systems, microcontrollers and new transmission technologies for the sensors. Precision irrigation technology allows the implementation of a new management methods for the water resources to meet the need of the soil and plants. In this context, this paper presents the structure of an enhanced automation module for testing the concept of precision irrigation. The design is based on an Arduino Uno development board and GSM/GPRS module for long distance data transmission [1], [2].

II. PRECISION IRRIGATION SYSTEMS

Precision irrigation systems represents a technology applied in agricultural industry with the purpose to improve the utilization of the water and soil resources, without compromising the productions. A precision irrigation system must acquire and analyze data from a distributed network of sensors and, based on a predefined program and generate commands in a feedback loop for activating different types of control devices. The predefined program can contain elaborated algorithms that can operate in accordance with soil characteristics and with the local climatic conditions. Because usually the control algorithm of the precision irrigation system is implemented in software, it is obtained an enhanced flexibility and configurability. Also, a precision irrigation system can operate with multiple software models and algorithms that can be easily interchanged depending on the needs. This create conditions for a great reusability of the automation part of the system, including also the hardware elements [3].

In Fig. 1 is presented a typical structure of a precision irrigation system. As can be observed, the system is structured in multiple sections. The monitoring area is the first section of the system which comprise sensors for specific parameters (moisture, temperature etc.) and actuators for performing various commands like dripping activation, electro-valves control etc. The second section is represented by the local data acquisition and processing module that usually contains also wireless communication capability. Finally, the last section is composed by a gateway that connects all system to the Internet for additional data processing, adaptive control, decision support and cloud computing [4].

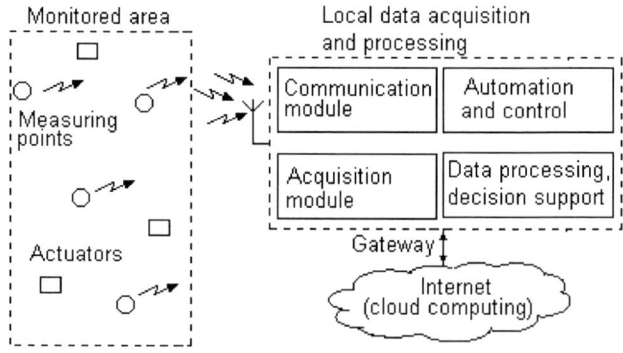

Fig. 1. Typical structure of a precision irrigation system

978-1-7281-7507-2/20 $31.00 © 2020 IEEE

21-24 October, Pitesti, Romania

2020 IEEE 26th International Symposium for Design and Technology in Electronic Packaging (SIITME)

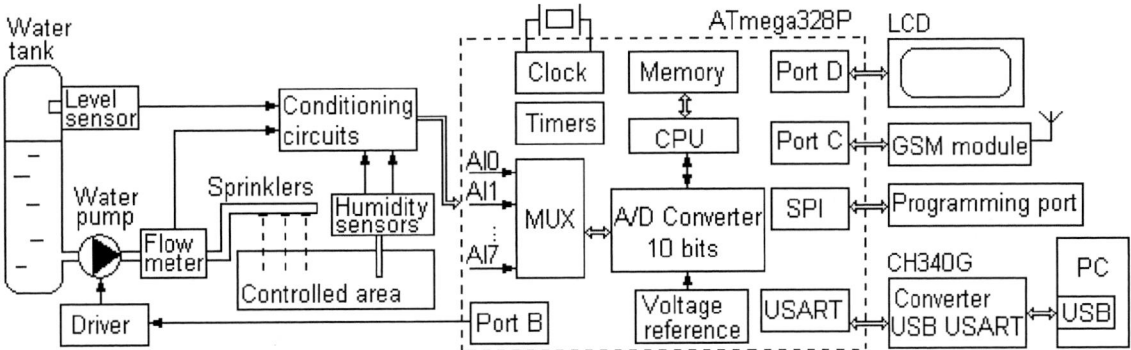

Fig. 2. The block diagram of the automation module used for the implementation of the precision irrigation systems.

III. THE STRUCTURE OF THE AUTOMATION MODULE

The simplified block diagram of the proposed system is presented in Fig. 2. The proposed design contains a section of automation and control, realized with the microcontroller ATmega328P, a section containing the sensors and the execution elements, and a section responsible with the transmission of the acquired data. Dedicated sensors are used for measure the humidity in the soil and the water quantity that is provided to the sprinklers. As can be observed in Fig.2, the system uses a water tank as buffer for alleviating the variable character of the supply sources that are existent in the site where the system is installed [5]. The level of water in the tank is permanently monitored and the information is sent to the automation module through a conditioning circuity. A key parameter that is used by the automation module in the process of decision-making process is soil and air humidity. In this case, both parameters are constantly monitored and the acquired information is sent to the microcontroller through a set of conditioning circuits that are connected to its analog inputs (AI0-AI7). The water quantity that is sent to the irrigation system is accurately measured with a flowmeter based on turbine principle. The output signal of the flowmeter is a train of impulses with the frequency proportional with the volume of the water that passthrough the flowmeter. In order to accommodate the impulses from the flowmeter with the logical levels of the microcontroller, a dedicated circuit is used. The same circuit realizes also the galvanic insulation between the internal circuit of the flowmeter and the board containing microcontroller. Observing the Fig.2 can be seen that the GSM/GPRS SIM800L communication module and the local display are connected with the ATmega328P microcontroller through the digital ports denoted C and D respectively. For ensuring the local connection between the automation module and a PC is used the internal USART block of the ATmega328P microcontroller and the circuit CH340G which operates as a bidirectional USART to USB to converter. The operation of the software application that was programmed into the memory of the ATmega328P microcontroller can be analyzed by looking to the simplified state diagram presented in Fig. 3. As can be observed this figure, the program starts with a general initialization of microcontroller's internal variables and registers. In the next step are read the outputs of the sensors and a user menu displayed on the local LCD. Subsequently, this data will also be

sent to a remote location via the GSM module. If the user wants to establish new threshold values for parameters of the automation module then the application enters in a specific loop that allows to realize this procedure [6], [7].

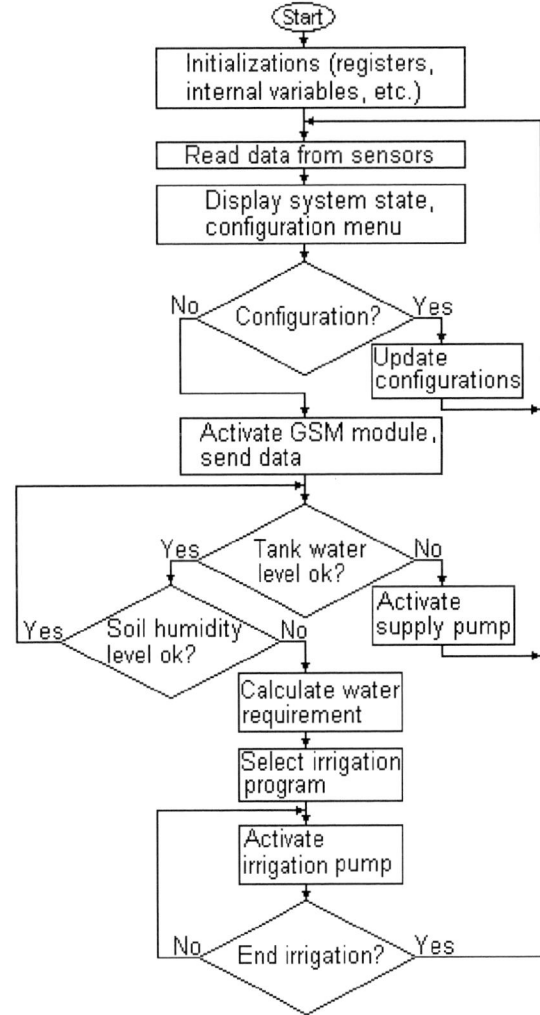

Fig. 3. The simplified state diagram of the software application for the automation module.

978-1-7281-7507-2/20 $31.00 © 2020 IEEE   137   21-24 October, Pitesti, Romania

IV. IMPLEMENTATION AND RESULTS

For verifying the correct operation of the proposed automation module, a small-scale precision irrigation system was implemented. The electrical diagram of the board containing the Atmega328P microcontroller, the local LCD and the interface circuits for sensors is detailed in Fig. 4. The air humidity was measured with H25k5A resistive sensor that is capable to operate in a range of 20-90% relative humidity, has a relatively small hysteresis of only 3% at full-scale and can realize accurate determination in the temperature domain 0-60°C. The moisture in the soil was measured with SEN-13637 sensor [8]. This device consists of two pads that operate as a variable resistor, depending on the conductivity of the soil, which in turn is proportional with the water content in the soil. The water volume transferred from the local tank to the irrigation system is monitored by the PRZ-1800/L flowmeter which has a precision of 2% and can operate with liquids having a maximum temperature of 65°C. The measurement resolution of the flowmeter is of 1880 impulses/liter [9].

The data transmission from the automation system to a remote PC is realized through the SIM800L quad-band GSM/GPRS module. The device is integrated in a dedicated extension board (shield) for the Arduino Uno development system contains the ATmega328P microcontroller. The wireless transmission is made at a relatively low speed of 85,6kbps, according to General Packet Radio Service (GPRS), but taking into account the extended availability of the mobile network, the automation module can be monitored virtually from everywhere. In Fig. 5 is presented the practical realization of the experimental prototype of the automation module and the reduced scale irrigation system. After tests realized with this reduced scale model it was found that the implemented system operates in accordance with the designated parameters. The flexibility in operation, scalability and the rapid reconfiguration offered by the proposed design is also due to the software application used to control the automation module. These characteristics represents major advantages of the proposed design. [10], [11].

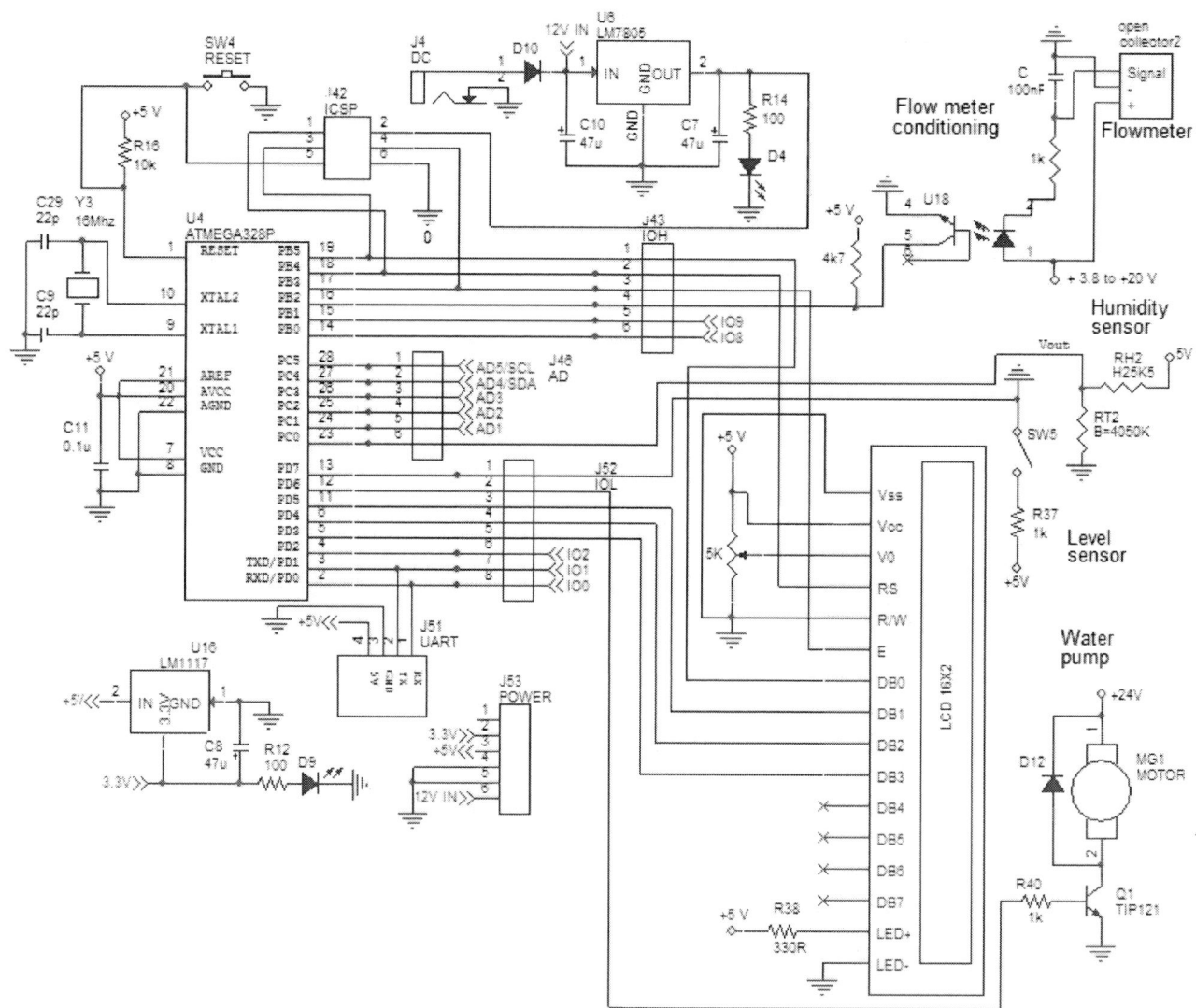

Fig. 4. The electronic schematic of the main board of the automation module for precision irrigation systems.

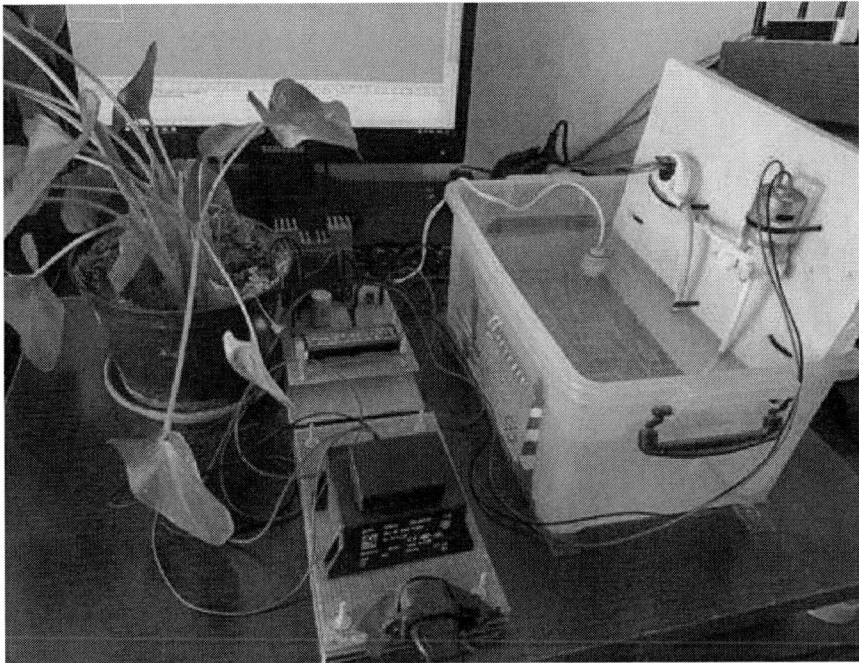

Fig. 5. Small-scale experimental setup, containing the automation module for irrigation systems used in precision agriculture.

V. CONCLUSIONS

Although it has always been an important issue, the management of water resources has now become, in the context of climate change, of vital importance for the development of a modern, much more efficient agriculture. Precision irrigation systems are a modern solution for optimizing the use of water in agriculture. Although the technology has many advantages, it is not widely adopted due to the initial development costs that are required for implementation and at the same time due to the requirements for the use of qualified personnel to operate the systems. This paper presented an enhanced design for implementing a reliable and efficient automation module for precision irrigation systems. Because is based on microcontroller and software programming, the proposed module is suitable for adaptive control and can be integrated in a larger system due to its communication capability using GSM network. Also, due to the implementation of the control algorithm in software, the proposed module has intrinsic flexibility in operation, scalability and allows rapid reconfiguration to suit any specific application. Tests realized with the reduced scale model of the automation module revealed that the implemented system operates in accordance with the designated parameters.

ACKNOWLEDGMENT

This work was supported by a grant of the Romanian National Authority for Scientific Research, CNCS/CCCDI - UEFISCDI, project number PN-III-P1-1.2-PCCDI-2017-0332, "Increasing the institutional capacity for research in bioeconomics for innovative using of the autochthonous vegetal resources for obtaining high value added horticultural products (BIOHORTINOV)".

REFERENCES

[1] I. Mat, M. R. M. Kassim, A. N. Harun, "Precision irrigation performance measurement using wireless sensor network", Sixth Int. Conf. on Ubiquitous and Future Networks (ICUFN), 2014.

[2] R. Nageswara Rao, B. Sridhar, "IoT based smart crop-field monitoring and automation irrigation system", 2nd Int. Conf. on Inventive Systems and Control (ICISC), 2018.

[3] M. Jhuria, A. Kumar, R. Borse, " Image processing for smart farming: Detection of disease and fruit grading ", Int. Conf. on Image Information Processing (ICIIP-2013), pp. 521 - 526, 2013.

[4] A. Cabarcas, C. Arrieta, D. Cermeño, H. Leal, R. Mendoza, C. Rosales, "Irrigation System for Precision Agriculture Supported in the Measurement of Environmental Variables", 7th Int. Engineering, Sciences and Technology Conf. (IESTEC), 2019.

[5] Adrian Tulbure, Emilian Ceuca, Dirk Turschner, "Arduino based power semiconductors tester for urban traction systems", Proceedings of the 37th International Spring Seminar on Electronics Technology, 2014.

[6] R. Teymourzadeh, S. Addin Ahmed, K. Wai Chan, M. Vee Hoong, "Smart GSM based Home Automation System", IEEE Conference on Systems, Process & Control (ICSPC), 2013.

[7] Prashant B. Wakhare, S. Neduncheliyan, Gaurav S. Sonawane, "Automatic Irrigation System Based on Internet of Things for Crop Yield Prediction", Int. Conf. on Emerging Smart Computing and Informatics (ESCI), 2020.

[8] Z. Feng, "Research on water-saving irrigation automatic control system based on internet of things", Int. Conf. on Electric Information and Control Engineering, 2011.

[9] I. N. R. Hendrawan, L. P. Yulyantari, G. A. Pradiptha, P. B. Starriawan, "Fuzzy Based Internet of Things Irrigation System", 1st Int. Conf. on Cybernetics and Intelligent System (ICORIS), Vol. 1, 2019.

[10] N. Agrawal, S. Singhal, "Smart drip irrigation system using raspberry pi and arduino", Int. Conf. on Computing, Comm. & Automation, 2015.

[11] G. Elizabeth Rani, S. Deetshana, K. Yaswanth Naidu, M. Sakthimohan, T. Sarmili, "Automated Interactive Irrigation System - IoT Based Approach",IEEE Int. Conf. on Intelligent Techniques in Control, Optimization and Signal Processing (INCOS), 2019.

LoRa and Bluetooth-based IoT alarm clock device for hearing-impaired people

Cătălina Mărculescu, Alina Machedon
Biotechnology and Bioengineering Department
University Politehnica of Bucharest
Bucharest, Romania
marculescu.catalina@yahoo.com

Ana-Maria Claudia Drăgulinescu, Ioana Marcu,
Ciprian Zamfirescu
Telecommunications Department
University Politehnica of Bucharest
Bucharest, Romania
ana.dragulinescu@upb.ro

Abstract— **Waking up in the morning represents a daily routine activity for the most part of the people who use common sound alarm clocks or who use their mobile phone as an alarm clock. For hearing-impaired people these solutions are not suitable. Other systems are equipped also with vibration motors and, thus, let the user choose if vibrations or sound signal are triggered when the alarm starts. These systems, though, require the time adjustment whenever the device is unplugged or when power outage takes place. Moreover, there is no solution available that gives a feedback to the caregivers, in order to be noticed that the user has been indeed waken by the alarm system. Through this paper, we propose an alarm clock system based on vibrations motors, ultrasonic sensors, Bluetooth and LoRa technology and server to wake the user, to ascertain the awakening and to send notifications to the caregivers. In addition, to mitigate the need to adjust the time in the case of a power failure, through the connection to LoRa server and to the battery operation, the time can be automatically updated.**

Keywords— *deafness, LoRa, Bluetooth Low Energy, alarm clock, ultrasonic sensor, vibration*

I. INTRODUCTION

People contending with hearing loss find very difficult to wake up for performing the daily activities. There are approximatively 51 million hearing-impaired people in Europe, that is, almost one out of ten Europeans suffers because of the hearing loss [1]. All age categories are afflicted as well. In Romania, according to the Government reports [2], 3 out of 1000 viable new-borns are diagnosed with hypoacusis. In US, 20% of teenagers exhibit forms of hearing loss [3]. The problem of absence of the classes, work and other important meetings is mitigated through the proposal and trading of various alarm clock systems [4].

Kumari et al [5] proposed a system for hard of hearing people based on Raspberry Pi, vibration motor and camera in order to notify people when a visitor presses a doorbell. The image of the visitor is captured and sent to a so-called wearable device worn by the user comprising of another Raspberry Pi board, an LCD screen and a vibration motor. When the image is received, the vibration motor is triggered and the image is displayed on the LCD. The wearable device, though, is still too sizable and depends on a wired power supply.

Another proposed device by Conley at al [6] was made to detect falls for visually and hearing-impaired people. It is worn around the neck and have two basic functions: to respond with a vibration to alert the user and to send an emergency signal when a button is pushed. It has advanced locating technologies and two-way voice communication. The major components include a bracelet and a mobile application installed on the central manager's smartphone or tablet.

Another step forward resides in the use of a hearing-dog robot [7], [8] that follows a user in order to transmit the alarms by touching the user as a real dog would do. Despite its advantages as autonomy and capability of following the user, the alarms are hardly understood from the point of view of the bumping behaviour.

Zohari [9] proposed a system with two separate operational modes: work (monitors heart rate) and sleep (with vibrating alarm clock).

The commercially available devices, though, are dependent on a power supply or run on battery. Moreover, they are based on sound amplification principle, being a source of disturbance for the others. In addition, when they are not supplied anymore or when they run out of battery, the user must set again the correct time in order to be notified about the awakening moment. Through this paper, we propose an alarm clock system and a mobile application for the hearing-impaired people that can alert people to wake up using vibrations and can notify the family about the event. The user interacts with the alarm clock system through an ultrasonic sensor in order to stop the alarm in a contactless manner, even from a distance of about 2 m. The most important aspect related to our proposal resides in the use of low power devices, especially on Low Power Wide Area Network communication technology such as LoRa to assure the time-synchronization of the alarm system when the system is restored from the absence of a power supply. Using another low power technology as Bluetooth Low Energy (BLE), the user can connect to a smartphone to set the alarm time and the family can receive notifications.

The paper is organized as follows: Section II reviews the two popular IoT communication technologies BLE and LoRa. Section III describes the system workflow and hardware emphasizing the role of LoRa communication, whereas Section

IV reveals the testing scenarios we proposed and the experimental results we obtained. Finally, Section V concludes the paper and highlights the future directions.

II. LoRa and Bluetooth Low Energy Communication Technologies

LoRa is low power wide-area networks wireless technology, essential for Internet of Things (IoT) and machine-to-machine (M2M) applications. LoRa uses the entire channel bandwidth to transmit a signal which makes it resistant to noise, long term relative frequency, doppler effects and fading [10]. As a spread spectrum modulation technique derived from chirp spread spectrum technology, its performances depend on parameters as spreading factor (SF), transmission power (P_{tx}), difference between the height of the transmitter and receiver antennae (Δh). For example, a lower spreading factor results in a faster transmission, but on a lower range.

BLE is an emerging wireless technology developed for short-range control and monitoring applications that is expected to be incorporated into billions of devices in the next few years [11]. The advantages of BLE are longer connectivity between devices due to its low power consumption (more efficient than classic Bluetooth, ZigBee and WiFi), bit rate, latency, scalability, confidentiality, authentication mechanisms and error correction [12]. The disadvantages consist in security vulnerability due to its wireless connections, small transmission distances (approx. 50 m indoor and 150 m outdoor) and lower transfer rates. BLE is successfully used in applications including indoor localization [13], human presence detection and e-health solutions (such as portable medical sensor kit platform with cloud connectivity [14]). Considering the latter, BLE may represent an optimum solution for connectivity among devices for a hearing-impaired people application.

III. Proposed System

A. Workflow

In Fig. 1, the workflow of the proposed system is depicted. The alarm clock can function autonomously or dependent on the user's smartphone. In Fig. 1, the case when the alarm time is set using the mobile application installed on the smartphone is emphasized. Thus, after selecting the date, hour and minute for the alarm, the application sends the data to the alarm clock system using BLE communication. If the user previously sent an alarm, then the data concerning the previous alarm time is stored and the new alarm time is set. This action prevents the alarm clock system to continue alerting when the user stops the alarm system. Further, the new alarm time is compared with the real time provided by RTC (Real-Time Clock) module.

If the alarm time is reached, the voltage on the vibration motors input pins is increased until it reaches a pre-defined value (maximum voltage). The system is equipped with an ultrasonic sensor that measures the distance when the alarm is triggered. When the user is still asleep, the measured distance is constant, but when the user wakes up and waves the hand, the distance between the sensor and the hand will decrease under a specified threshold and the alarm is deactivated and the voltage on the vibration motors pins becomes 0 (LOW).

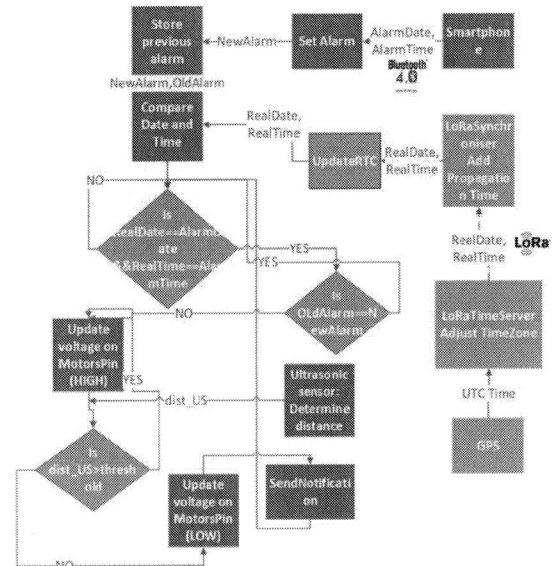

Fig. 1 Workflow of the proposed system

Because the time can be corrupted by interruptions in power supply (either battery or wired) and because during the interval between the alarm time and the alarm set time the link between the smartphone and the alarm clock system can be unavailable, the RTC update will be independent on the smartphone.

B. RTC Update using LoRaTimeServer

The RTC will be updated as follows: the system component called LoRaTimeServer receives the UTC (Coordinated Universal Time) time from the GPS module connected to the LoRa server and adjusts the TimeZone in accordance with the location of the server. The time-related data is sent to the LoRaSynchroniser block using LoRa communication. Due to the limitations of the transmission technology (very low data rate and latency), the block LoRaSynchroniser estimates the propagation delay and properly adjusts the time of the RTC. UpdateRTC block enables the automatic updating of the RTC module based on the output of the block LoRaSynchroniser.

C. LoRa hardware components

The LoRa communication hardware modules we propose are Semtech SX1276, with a sensitivity ranging between -111 dBm and -148 dBm. They are integrated in Lopy v4 development boards (Pycom) and operate at a frequency of 868 MHz. To track GPS position of the mobile node, we use Pytrack board equipped also with an SD card slot and interface.

IV. Tests performed on LoRa technology

On the strength of the specific of LoRa networks and communication, the LoRaTimeServer can be used by many devices at a time, with a coverage domain of approximatively 5.24 km in urban area and 45 km in rural area [15].

To determine the performances of LoRaTimeServer and LoRaSynchroniser and the proper time offset between them, four scenarios were proposed and several tests were conducted.

LoRa TimeServer equipment was located and the experiments took place in a hilly town area, in Prahova County, Romania, at an altitude of 504 m. The antennae height was varied according to Table I scenarios. Antennae Δh represents the difference between the LoRa network antenna and the wearable device's antenna.

TABLE I.   DESCRIPTION OF LoRa SCENARIOS

Scenario	Mobility	Distance T-R	Antennae Δh	SF	P_{tx}
I	No	1.5 m	0 m	7	
II	Yes	1.5 m-100 m	0-7.14 m	7	14 dBm
III	Yes	1.5 m-350 m	<1 m	12	
IV	Yes	1.5 m-650 m	3 m	12	

In the first scenario, called *no mobility scenario*, one assumes that the LoRaTimeServer and the LoRaSyncroniser are not changing their positions one with respect to the other. The distance between them is 1.5 m. In the second scenario, the alarm system is mobile and, thus, the LoRa Synchroniser position with respect to server will vary from 1.5 m to 100 m in line of sight conditions (LOS). The height difference between the antennae varies from 0 to 7.14 m. In the third scenario, the alarm system is also mobile, thus the LoRaSynchroniser position with respect to server will vary again from 1.5 to 350 m, this time, in non-line of sight (NLOS) conditions, in the unfavorable case when the antennae height difference is less than 1 m. For the first three scenarios, we chose SF=7.

In all four scenarios, LoRaTimeServer will send the time coordinates based on UTC time and on the time difference according to the time zone. LoRaSynchroniser will store its GPS coordinates on the SD card available on Pytrack board. For all tests, for each packet comprising the time received, parameters relevant for LoRa communication as RSSI (Received Signal Strength Indicator), SNR (Signal to Noise Ratio), transmission power and number of retransmissions are captured and stored on server side. RSSI is the parameter of interest for all scenarios as it depicts the relative quality of the LoRa signal.

A. No mobility scenario (I) and LOS scenario (II)

The non-mobility scenario appears in the case when the person using the proposed solution, sets the alarm right before falling asleep, being located at bed level. However, this favorable scenario is not always present. Many people are setting the alarm in advance or with a certain recurrence, avoiding inconveniences by omission. In the meantime, people are performing dynamic activities, most of them at a certain distance from home proximity. This is the reason why data synchronization and already set alarm must remain available.

The distributions of RSSI values with respect to the packets received for no mobility and first mobility scenario (maximum distance lower than 100 m) are ploted in Fig. 2. When the mobile node is closer to LoRaTimeServer, the values of RSSI increase, whereas when the mobile node move away from the LoRaTimeServer, the values of RSSI start decreasing.

In addition, analysing the histograms, one can distinguish between the no mobility scenario and mobility scenario because

in the first scenario, the range of RSSI values is narrower than for a mobility scenario.

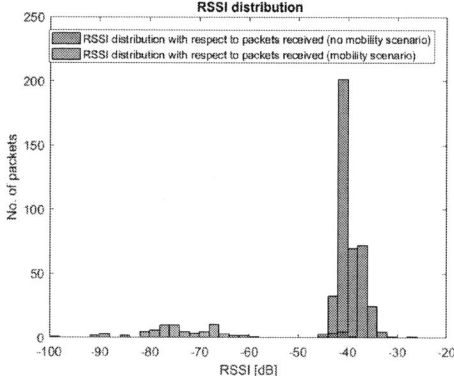

Fig. 2 RSSI distribution with respect to the packets received

B. NLOS Scenarios (III, IV)

In *Scenario III*, we simulated the worst situation when Δh is lower than 0.5 m. The mobile node performed its movement on the following trajectory: P0-P8-P0-P17-P0-P22 (Fig. 3).

Fig. 3 Mobile node route, packets received and maximum distance achieved (Δh<0.5, SF=12)

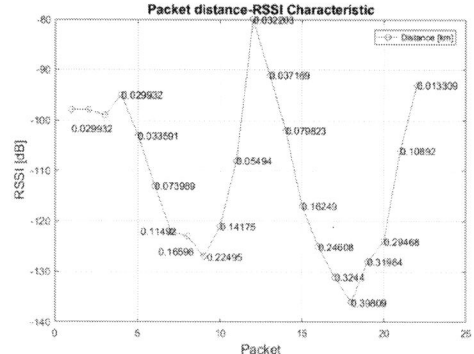

Fig. 4 RSSI Characteristic in Scenario III

In Fig. 4 one can observe the maximum distance achieved is 0.398 km (computed based on GPS coordinates and Heaversine Equation), with the lowest RSSI value of -136 dB (P0-P17). For P0-P8 trajectory, the maximum distance achieved is 0.224 km, with RSSI=-127 dB. Fig. 4 reveals the geographical distribution of the packets received and the maximum range determined.

In Scenario IV, the height of the transmitter antenna was increased such that Δh becomes approximatively 3 m. The two routes started in the same point P0. First route goes towards P25 where it reaches the maximum distance of 598.22 m (Google Earth), returns from P25 and finishes at P42. The second route reached a maximum distance of 615.98 m (Google Earth) at P65, returns from P65 and finishes at P83.

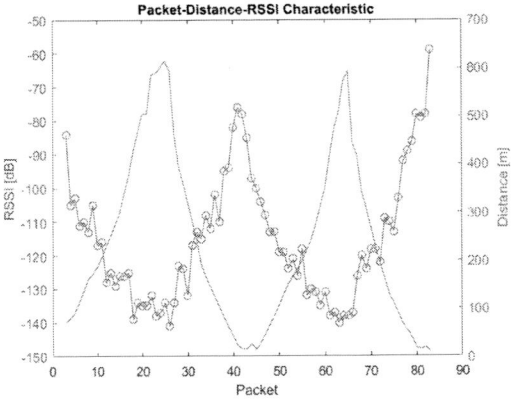

Fig. 5 RSSI Characteristic in Scenario IV

Table II compares Scenarios III and IV. The maximum range for Scenario III was obtained for RSSI=-136 dB. For a similar RSSI the range increases with 54.76% when Δh increases from 1 to 3 m (Scenario IV).

TABLE II.   COMPARISON BETWEEN SCENARIOS III AND IV

Scenario	Δh	Max. range	Min. RSSI
III	< 1 m	398 m	-136 dB
IV	3 m	615.98 m	-137 dB

Also, by plotting the histograms of the time differences between packets, in Scenario III the packets are received usually with a delay of 12 s, whereas for Scenario IV the delay increases to 18 s, values that are allowed for non-critical devices, such as alarm clocks.

V. Conclusions and Future Directions

The proposed framework enabled by low power devices and Low Power Wide Area Network technologies such as LoRa is the main differentiator from other existing solutions because it has the capability to assure the time-synchronization of the alarm system when the system is restored from the absence of a power supply. Moreover, the very low energy consumption feature of the wearable device does not require repetitive and frustrating recharging cycles. Another important outcome is the radio benefits in terms of signal coverage (almost 55% increase) emphasized by changing the height of the antennae with only 2 m. Next steps will be addressed by installing more antennae at increased heights for further studies and also an in-depth wearable device architecture analysis, as the current paper focused on LoRa time-synchronisation.

Acknowledgment

This research was funded by the Operational Programme Human Capital of the Ministry of European Funds through Financial Agreement 51675/09.07.2019, SMIS Code 125125.

References

[1] European Federation of Hard of Hearing People, „Global Hearing Loss Statistic," [Interactiv]. Available: https://efhoh.org/wp-content/uploads/2017/04/Hearing-Loss-Statistics-AGM-2015.pdf.

[2] Guvernul României, „Actualitate,". Available: https://www.gov.ro/ro/stiri/ministerul-sanatatii-a-lansat-programul-national-de-screening-auditiv-pentru-nou-nascuti-i-un-sistem-informatic-national-de-evidenta-a-rezultatelor-testelor-auditive-dezvoltat-cu-ajutorul-govithub&page=1.

[3] Center for Audiology, „Teenage hearing loss on the rise," Available: https://centerforaud.com/blog/teenage-hearing-loss-on-the-rise.

[4] S. Kasim, H. Hafit, . T. H. Leong, R. Hashim, H. Ruslai, K. Jahidin și . M. S. Arshad, „SRC: Smart Reminder Clock," în IOP Conference Series: Materials Science and Engineering, Melaka, Malaysia, 2016.

[5] P. Kumari, P. Goel și D. S. R. N. Reddy, „PiCam: IoT based Wireless Alert System for," în 2015 International Conference on Advanced Computing and Communications, 2015.

[6] K. Conley, A. Foyer, P. Hara, T. Janik, J. Reichard, J. D'Souza, C. Tamma și C. Ababei, „Vibration Alert Bracelet for Notification of the Visually and Hearing Impaired," Journal of Open Hardware, 2019.

[7] M. Furuhashi, T. Nakamura, M. Kanoh și K. Yamada, „Haptic communication robot for urgent notification of hearing-impaired people," în 2016 11th ACM/IEEE International Conference on Human-Robot Interaction (HRI)., 2016.

[8] S. Furuta, T. Nakamura, Y. Iwahori, S. Fukui, M. Kanoh și K. Yamada, „Efficient User-Searching of A Hearing-Dog Robot in Consideration of user's Life Rhythm," în 2018 Joint 10th International Conference on Soft Computing and Intelligent Systems (SCIS) and 19th International Symposium on Advanced Intelligent Systems (ISIS), 2018.

[9] T. Zohar, „Vibrating, pulse-monitored, alarm bracelet". USA Brevet US9318013B2, 2013.

[10] U. Noreen, A. Bounceur and L. Clavier, "A study of LoRa low power and wide area network technology," 2017 International Conference on Advanced Technologies for Signal and Image Processing (ATSIP), Fez, 2017, pp. 1-6, doi: 10.1109/ATSIP.2017.8075570.

[11] C. Gomez, J. Oller, Josep Paradells, "Overview and Evaluation of Bluetooth Low Energy: An Emerging Low-Power Wireless Technology", in Sensors 2012, 12, 11734-11753

[12] E. Georgakakis, S. A. Nikolidakis, D. D. Vergados, C. Douligeris, "An analysis of Bluetooth, ZigBee and Bluetooth Low Energy and their use in WBANs," in Wireless Mobile Communication and Healthcare, Springer, 2011, pp. 168-175

[13] K. NonAlinsavath, L. Edi Nugroho, W. Widyawan, K. Hamamoto, "Integration of Indoor Localization System using Wi-Fi Fingerprint, Bluetooth Low Energy Beacon and Pedometer Based on Android Application Platform", in International Journal of Intelligent Engineering and Systems, Vol.13, No.4, 2020

[14] T. Md. Siham Sayeed, Md. T. Rayhan, S. Chowdhury, "Bluetooth Low Energy (BLE) based portable medical sensor kit platform with cloud connectivity", in International Conference on Computer, Communication, Chemical, Material and Electronic Engineering (IC4ME2), ISBN: 978-1-5386-4775-2, Bangladesh, 2018

[15] A. M. C. Drăgulinescu, A. F. Manea, O. Fratu și A. Drăgulinescu, „LoRa Based Medical IoT System Architecture and Testbed," Wireless Personal Communications, vol. March, pp. 1-23, 2020

Android Application for Data Processing from a Gas Detection Sensor in Atmosphere

A. Alexandrescu and D.I. Năstac

Faculty of Electronics, Telecommunications and Information Technology
POLITEHNICA University of Bucharest, Romania
nastac@ieee.org

Abstract—**We live in an environment where it is debated whether there is a causality between pollution and climate change, which has a negative impact on both human health and nature. Because of this, determining and improving air quality has become a priority in today's society. Combustion of fossil fuels, industrialization or meteorological factors play a significant role in air pollution. Consequently, the society becomes interested in the existence of an application that provides a real-time air quality index of a certain geographical area, as well as a forecast on this air quality index for a few hours or days. Therefore, this paper proposes a practical solution for a faster adaptation of the human user by creating a database with information about polluted areas in the city. As a result, a person will be able to know the air quality both in the area where he finds himself, as well as in another location where he wants to move.**

Keywords—AQI; pollution; sensor; prediction; correlation; meteorological values

I. Introduction

By air pollution is meant the presence in the atmosphere of substances harmful to the human body, which depending on the concentration and the time to which the human is subjected cause disturbances in the natural balance, affecting health but also the environment. The main substances that affect us are sulfur and nitrogen oxides, chlorofluorocarbons, carbon dioxide and monoxide [1].

According to the World Health Organization [2], air pollution is the biggest risk to the health of citizens in the European Union. In the European Union, pollution causes around 400,000 premature deaths each year, and costs related to hospitalization and treatment amount to hundreds of billions of euros. Citizens in urban areas are mostly affected by this problem. In 2015, about a quarter of Europe's population living in urban areas was exposed to low air quality [2], with levels of air pollutants harmful to health. This air pollution is increased in urban areas compared to rural areas because the population density is higher, the amount of air pollutants released is higher, and the dispersion is more difficult to produce in cities [2][3].

In addition, from a report published by National Health from Italy in March 2020 [4], the rapid increase of contagion rates that has affected some areas of Northern Italy could be tied to atmospheric particles pollution acting as a carrier and booster there, the people from these zones are more sensitive to this disease.

Because of this, several applications have been developed to inform the population about the level of air quality. A first application is offered by the Ministry of Environment. Within

this application the population has the possibility to access a platform in which the data contains information regarding the air quality from several areas of the country. They are displayed on a map, which depending on the value is also displayed in a different color. Although the idea of these stations is very good, the reality is completely different [5]. A simple search on Google can display a lot of surveys that show the real situation regarding these measurements. There are many situations in which due to lack of data the information was not displayed on the map. For example, a station in the Ploiesti area that had been defective since March 2012 was repaired in September 2016 [6]. So, for 4 years the data were missing from the system and because of this the population did not know the level of pollution in this area.

Another application available to the population is the one made by Philips. It consists of 15 sensors located in Bucharest and Ploiesti, permanently recording, from October 1, 2018, the level of air pollutants, and the data are available to the public on the website [7] or in the Airly application for smartphones, through the first independent air quality monitoring network in Romania. This system is consisting of 15 sensors, 14 of them located in the Capital, and one in the city of Ploiesti, which permanently monitors the level of suspended particulate emissions, PM 10, PM2.5 [6].

The application has been built so that anyone who has a device for measuring the air quality index to find out its value and publish the values in its location. In addition, the data is displayed on a map with different colors depending on the value of air quality. The connection between the device and the mobile phone is made through Bluetooth technology. In addition to displaying real-time data on the map, there is a need to find a way to predict the air quality index based on weather data, so as to observe how pressure, temperature, wind direction or other weather parameters influence the quality index of the air.

In addition, a device capable of measuring the air quality index has been built. It has been designed to meet two characteristics: easy to use, and low price.

II. Air Quality Index

The air quality index abbreviated AQI is a standard used internationally and allows the assessment of the level of air pollution based on a hierarchy of pollutants according to the ratio with the maximum permitted concentration and the degree of danger. Depending on the degree of danger, the noxious substances are divided into 4 categories [8]:

- class I: very dangerous noxious substances, ozone,

chlorine, mercury (n =1,7);

- class II: hazardous pollutants, hydrogen sulphide, nitrogen oxides, formaldehyde (n=1,3);

- class III: moderately dangerous noxious substances, sulfur dioxide, soot, suspended dust (n=1);

- Class IV: least hazardous pollutants, carbon monoxide, aliphatic hydrocarbons (n=0,9).

The air quality index is calculated according to formula (1) for each pollutant, at the end the AQI value being determined by the arithmetic mean of all the results obtained previously.

$$AQI = 100 * \left(\frac{C}{CMA}\right)^n \qquad (1)$$

Where:

- C represents the concentration of the noxious substance obtained from the sensor;

- CMA is the maximum permissible concentration for noxious substances;

- n is a hazard index.

The values obtained will be compared with the data provided by the United States Environmental Protection Agency, data that also refer to the impact of air quality on the health of the population. Data show in Figure 1.

Air Quality Index Levels of Health Concern	Numerical Value	Meaning
Good	0 to 50	Air quality is considered satisfactory, and air pollution poses little or no risk.
Moderate	51 to 100	Air quality is acceptable; however, for some pollutants there may be a moderate health concern for a very small number of people who are unusually sensitive to air pollution.
Unhealthy for Sensitive Groups	101 to 150	Members of sensitive groups may experience health effects. The general public is not likely to be affected.
Unhealthy	151 to 200	Everyone may begin to experience health effects; members of sensitive groups may experience more serious health effects.
Very Unhealthy	201 to 300	Health alert: everyone may experience more serious health effects.
Hazardous	301 to 500	Health warnings of emergency conditions. The entire population is more likely to be affected.

Fig. 1.   Interpretation of AQI values [8]

Specific air quality index is the method used in Romania for the air quality index, being a system for monitoring the concentrations recorded for several pollutants, such as SO2, NO2, PM10, O3, CO. The general index is established for each station as the highest of the specific indices corresponding to the pollutants monitored by the authority of the Ministry of Environment [8].

There is a possibility that the information provided may be erroneous or even missing, as obtaining a result is constrained by the measurement at the station level of at least 3 specific indices, an incident that occurs quite often. For example, in the case of Bucharest on the evening of 10-06-2019 out of a total of 7 stations measuring the level of pollution than 2 stations provide a final result of the specific air quality index [6].

In addition, the value of a single specific index is displayed, which can influence the final result by neglecting the values of

other pollutants that have much more dangerous effects on human health.

III.   THE INFLUENCE OF AQI ON THE POPULATION

The level of air quality is one of the elements related to the environment that affects human health. According to studies conducted by the European Commission together with the World Health Organization, approximately 400,000 deaths are due to poor respiratory air quality and because of this, in addition to increased mortality, there are problems with health costs amounting to hundreds billion annually [2]. The industrialization of large cities and with this the construction of factories inside the town, have led to an increase in the concentration of certain substances dangerous to the human body, such as benzene, toluene, and thus a decrease in air quality and an increase in the number of people suffering from respiratory diseases.

The increase in the number of cars and the reduction in green space have led to an increase in the air quality index in terms of dust particles in suspension or PM as they are abbreviated. They are 2.5 microns or 10 microns in size. Due to their very small size, they manage to pass through the natural filter in the nose and enter the alveoli of the lungs, causing lung cancer over time. In the case of young children, breathing is achieved in a higher percentage than adults through the mouth. So, the nasal filter is avoided, and even larger particles, PM 10, reach the lungs which are much more fragile, having a sensitive lung tissue in development. This results in an aggravation of diseases that can even lead to death if the person is exposed in the long term to a high concentration of dust in the suspension.

The existence of an application that provides real-time information on the level of pollution is very useful for society, so that people avoid areas with low air quality and move to areas with much higher air quality.

IV.   METEOROLOGICAL ELEMENTS THAT INFLUENCE AQI

Local weather and climate affect air quality, and therefore great care must be taken in drafting this project. The level of pollution can change from one hour to another due to the increase in the concentration of the respective pollutant but also due to some changes in the meteorological data. For example, certain regions in Romania are subject to different weather conditions. Constanta is much more exposed to air currents than another locality, and because of this the concentration decreases due to this. The weather can therefore directly influence the level of air quality in a certain location. For example, sun, rain, air temperature and wind can influence AQI for the following reasons [9]:

- The sun causes certain pollutants to undergo chemical reactions, thus producing smog ;

- Rain makes it much easier for pollutants to pass into drinking water ;

- Temperatures also cause an acceleration of reactions in the atmosphere ;

- Wind speed causes dispersion and dilution of pollutants. In addition, strong winds can cause an increase in dust particles ;

- Usually, as the altitude increases, the air cools. But there are also situations in which an upper layer of air is warmer than a lower one. This process is called inversion. This process influences the air quality quite a lot because the top layer behaves like a lid and thus the pollutants are not dispersed in the atmosphere, and their concentration increases. To better understand this process, we have exemplified in Figure 1 how the presence of a warmer air layer is influenced [9].

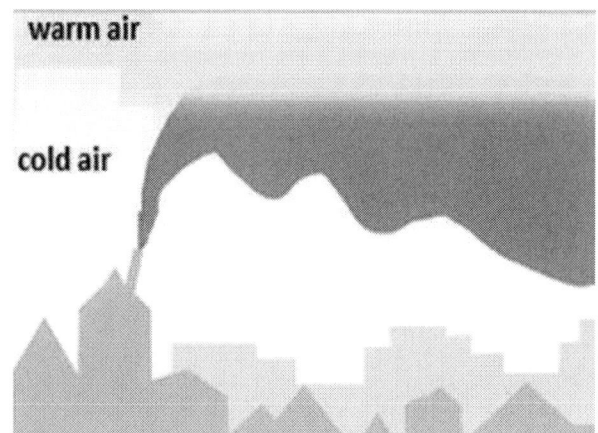

Fig. 2.   How smoke propagation is influenced by air layers [9]

In addition to these meteorological elements, we continued to study them during January - May and we discovered another element that has an important role in increasing the level of pollution. Pressure is an element that determines the increase or decrease of this air quality index. During this period, we noticed that with the increase of the pressure it determines an increase of the pollution level. To demonstrate this, we performed an analysis in the form of graphs, from which it can be seen that these variations of AQI are influenced by pressure.

We tried to find the variable on which the air quality index depends the most such that the forecast obtained contains as few errors as possible, because a false value of an AQI can alarm the population for no reason and so the application will lose users in the future.

The data come from the website of the Ministry of Environment and are values obtained at various measuring stations in Ploiesti [5].

Figure 3 shows a graph where is the value of NO2 in μg / m3, cherry color, pressure value, green color and blue temperature value. In this graph it can be seen that in the period when the pressure has a high value and the air quality index has an increased value. This has been observed in other situations. Therefore, pressure plays an important role in the variation of AQI.

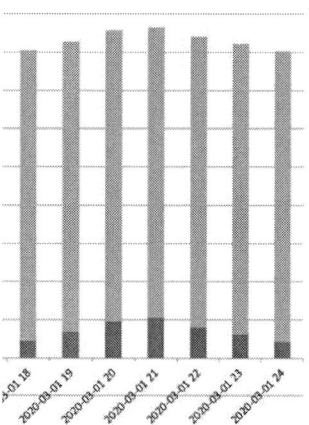

Fig. 3.   Graph NO2 value, pressure, temperature.

Another element that has an important role in the variation of AQI is represented by the wind direction. It was observed both in the field, for example in the case of the city of Ploiesti when the wind blows from the direction of the Lukoil refinery a presence of the smell of oil in the breathed air, as well as in the measurements by increasing the level of benzene in the atmosphere. Therefore, wind direction also plays an important role in forecasting the air quality index.

V.   AIR QUALITY MEASURING DEVICE

The device has been designed to meet two characteristics: to be easy to use, and its price is not high. An Arduino Uno board, a Bluetooth Master Slave HC-05 module and an MQ-135 gas sensor were used.

The MQ-135 sensor is a sensor used in applications to determine the level of pollution based on certain substances such as carbon dioxide, carbon monoxide, ethanol, toluene, acetone and ammonium. Its operation is based on a compound of tin, tin oxide, which in clean air has a low conductivity, and in polluted air depending on the increase in the concentration of the substance an increase in conductivity is obtained. The MQ135 sensor was used to design the device for measuring and transmitting data to the phone because it provides fairly accurate results at operating temperatures between 15-30 degrees Celsius, but also because of its price.

The Bluetooth Master Slave HC-05 module is used in projects where it is desired to transmit data, without using the cable, over distances not exceeding 10 meters. The communication channel is based on the UART protocol. The transfer rate is 1 Mbps, using the 2.45 GHz frequency band, with low power consumption. Its advantage is determined by the fact that it can be set according to the utility in the project as "master" or "slave" [10].

This device is shown in Fig. 4 and also contains an external battery so that it can be used in any location.

A library provided by Arduino, MQ135.h, was used to determine the air quality level. This provides the value of a resistance obtained from the sensor depending on the concentration of the airborne nuisance. Based on this value, the concentrations of each pollutant will be calculated later, to be obtained based on the results of the air quality index.

Fig. 4. Atmospheric gas measuring device

VI. ANDROID APPLICATION

The application has been designed to allow the transfer and addition of data from the device to a database. In addition, they are displayed in a way that is easy for the user to read via a map. This map contains markers at which we find information about longitude, latitude and AQI. Depending on the air quality around this marker will create a circle of a certain color.

Figure 5 shows the interface for adding data to the system by the user. The user connects to the device by pressing button 1. This is done through 2 steps. The first step is to check the phone to see if it supports Bluetooth technology or if it is turned on, and if not, the user is informed of this issue. Once this stage is completed, a single communication channel will be created between the two devices. Once these two steps are completed, the data will be transmitted character by character to the mobile phone and will be displayed in box 6 in Figure 5. In addition to this information, the user's coordinates will also be displayed in box 7 in Figure 5. In order to add the data in system users will click the Share button, and this data will be published to the system.

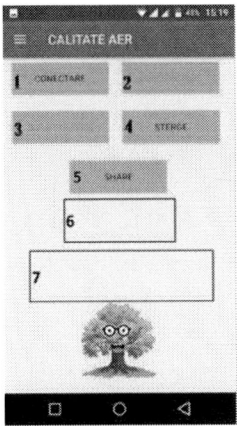

Fig. 5. Data addition interface.

At the interface level, the delete, disconnect and pollution degree buttons are inactive as long as the connect button is active. This has been developed to avoid errors that may occur by accidentally pressing a button involving the connection between the two devices.

With the existence of data in the system there must be a mechanism by which they can be displayed in a much easier way for the user. The solution consists of an interactive map containing markers that provide information about the air quality index and a method implemented by Google Maps called heatmap. Depending on the level of pollution, a circle of a certain color will be created on the map, which provides information related to the air quality index. If the AQI is less than 100 a green circle and a message will be displayed, Air quality is acceptable. If the AQI is between 100 and 150 a yellow circle and a message will be displayed, sensitive people may be affected. If the AQI is between 150 and 200, an orange circle and a message will be displayed, it affects the population. If the AQI exceeds 200, a red circle and a message will appear on the map, alert level exceeded.

When placing the marker, the value of the air quality index and an information message for the user are displayed, depending on the severity of the situation. In addition, it is possible to start navigation to a selected marker at which the pollution level is low.

This interface is presented in figure 6, where you can see several markers for the Ploiesti and Bucharest area, where the air quality is within the legal norms without endangering human health.

Fig. 6. Map interface

VII. AIR QUALITY PREDICTION BASED ON METEOROLOGICAL DATA

Regarding the data on the air quality level, they were exported from the platform provided by the Ministry of Environment [5] and are values obtained from a station located in Ploiesti. Meteorological data were exported from a platform that provides information in this area [11].

A crucial step in developing a forecast model is choosing the input parameters. In principle, any set can be introduced in the Neural Networks for training and assessment. However, the number of possible parameters and the number of ways in which they can be presented is too large to test all possible combinations. We must limit ourselves to a limited number of tests, based on available experience and knowledge of the phenomenon [12].

For this reason, a presentation of the elements related to the practical implementation in terms of predicting the level of air quality was made.

As we presented in the previous chapter, the choice of parameters on which to base the AQI prediction has a decisive role. For this reason, in addition to the actual prediction, an analysis was performed to observe how the meteorological parameters influence the level of air quality.

In this paper we used the following parameters measured between 01.01.2020 and 05.05.2020. At the level of data provided by those from the Ministry of Environment, it is not possible to export data from previous years.

Parameters used:

- Date (year / month / day) and Time(hours);

- The value of NO2 in micrograms / cubic meter measured at a station of the Ministry of Environment in Ploiesti marked PH-2;

- Temperature is measured at 2 m above the ground;

- Pressure represent the low sea pressure;

- Wind direction in degrees 0 degrees corresponds to N, 180 corresponds to S denoted wind direction;

- Pa precipitation liter / square meter;

- Humidity in the atmosphere;

- Po ground pressure at the station level;

- VV is the range of ground visibility, measured in km;

- Td dew point temperature measured at 2 m above the ground;

- Ff represents the wind speed in m/s;

- Clouds represent the percentage of cloudiness;

- The baric tendency denoted Pa represents the dynamics of atmospheric pressure in the last 3 hours.

The database contains a total of 2994 rows and 13 columns with various parameters. We studied how the variables depend on each other by calculating a correlation index for each parameter.

These values were represented both by a heatmap for a quick interpretation and by presenting a table containing correlation coefficient for each parameter. From an experimental point of view, we obtained the way in which the variables are closer to PH-2, because if the correlation coefficient will be higher than the prediction will be much better.

From Fig. 7 it can be seen that the value of air quality is influenced by a correlation coefficient of over 0.2 for temperature, wind speed, the time at which measurements were made and humidity.

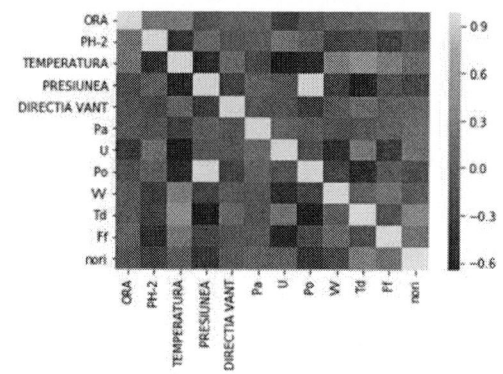

Fig. 7. Heatmap correlation parameters

These things can also be demonstrated physically or chemically. A high temperature causes an acceleration of reactions in the atmosphere. Wind speed determines the dispersion and dilution of pollutants. In addition, strong winds can cause an increase in dust particles. The time is an important element in predicting the air quality index. In the morning 7-9 and in the evening 16-18 the number of cars in traffic increases, and because of this the air quality decreases. Humidity is another important element because it affects the speed of reactions in the atmosphere and the increase in the smell of pollutants in the air.

So, at this point we know how the parameters influence the air quality level, so in the future we will be able to eliminate the elements that do not influence the element that is to be predicted.

For this we must separate the element to be predicted from the rest of the parameters. For this reason we named target (PH-2), the element that is to be predicted and features(time, temperature, pressure, wind direction, Pa, U, Po, VV, Td, Ff, clouds), the parameters on the basis of which the prediction is made.

The division of this data batch into training batch and test batch was done through the *train_test_split* function which performs this distribution randomly, having the ratio 70-30, 70% of the data will be from the training batch, and the rest up to 30 from the test batch.

At the beginning we applied the linear regression and displayed what is the score in terms of predicting the level of air quality. The result obtained is 0.35. It was observed that in the case of sudden increases in data, the prediction in turn increases the air quality index. I also tried to compare in terms of returned data. In the Figure 8 the first 10 data that were recorded by the Ministry of Environment and the data that were predicted based on linear regression. It can be seen that in some situations the data are similar.

AQI measured	AQI predicted
0,061451	0,09550299
0,058491	0,0996766
0,012538	0,05216827
0,16412	0,15607864
0,083997	0,06345524
0,23122	0,26239965
0,260116	0,12882307
0,181877	0,13901339
0,171707	0,20806007
0,069307	0,14661624

Fig. 8. Comparison of real data with data obtained from the prediction

Achieving a high prediction score is difficult because the air quality index is largely influenced by human intervention. For example, in the city of Ploiesti there are sudden increases in the level of air quality even if the meteorological parameters and the time at which it was measured are identical to another case in the past. This is because either there is a problem at a refinery in the city area and thus certain substances are released into the atmosphere, or fires can occur which can also affect the air quality index. Therefore, the prediction of air quality can be made in certain areas where human intervention is low and thus any increase or decrease will be influenced by meteorological parameters.

In order to obtain this score, several variants were tried for which the scores were not the best. The number of elements that train the network must be taken into account. In addition, with the increase in the number of meteorological parameters, an increase in the prediction score was observed.

Another important element in obtaining a high prediction score is determined by the correlation of the data and their normalization. There have been situations where this method increases the score by up to 5%.

VIII. CONCLUSION

The idea of his application started from the knowledge of a real value of the air quality index, without the need to depend on the systems provided by the competent institutions in this field. Thus, this project aims to solve a fairly common problem in society and to determine an improvement in air respiration by developing solutions.

The application aims to inform the user about the level of pollution, both in the area where it is located, through the measuring device, and in other areas of the country through an interactive map containing data sent by other users.

Therefore, it was possible to design a device capable of measuring the level of air quality in any area and send this data to a phone, for which the user can later publish this information in the system. In addition to this, an Android application has been created that takes data from the device, adds it to a database and displays it at the level of an interactive map at the user's request.

In this paper we also observed how meteorological parameters can influence the level of air quality. A prediction was also made starting from these hypotheses obtained following the determination of the correlation matrix.

The prediction model will work as long as humans do not influence these parameters. If refineries are present in the area, the prediction will be very difficult to make based on meteorological parameters.

In the future we will try to extend the number of data that will be used to train the prediction model and thus increase the prediction rate.

REFERENCES

[1] D. Dinu, V. Sandu, "The air pollution", "Poluarea aerului", Editura Tehnică, Brașov, 2005.

[2] J. Wojciechowski, "Air pollution: our health is not yet sufficiently protected", Report number 23/2018 from European Comision, Luxembourg.

[3] D.I. Nastac, "Intelligent processing of multidisciplinary information for adaptive forecasts in the context of globalization", "Prelucrarea inteligentă a informațiilor multidisciplinare pentru prognoze adaptive în contextul globalizării", Editura Muzeul Național al Literaturii Române, București, 2013.

[4] L. Setti, F. Passarini, "Report on the effect of air pollution and the spread of viruses in the population", "Relazione circa l'effetto dell'inquinamento da particolato atmosferico e la diffusione di virus nella popolazione", Societa Italiana di Medicina Ambientale, 2020.

[5] ***,"Air quality". Accessed on: 08.10.2020. Available: http://www.calitate aer.ro/public/home-page/?__locale=ro.

[6] Andra Alexandru, "Air monitoring", Company Philips, Nov 09, 2018.

[7] ***,"Map Airly". Accessed on: 08.10.2020. Available: https://airly.eu/map/en/.

[8] I.C. Iojă, "Methods of research and assessment of the state of the environment", "Metode de cercetare și evaluare a stării mediului", Editura Etnologică, București, 2013, pp. 59-68.

[9] D. Smith, "Temperature Inversions: How Weather Can Trigger Air Pollution", Kaiterra, Jul 14, 2020.

[10] ***,"Datasheet HC-05 Serial Bluetooth Module", iteadstudio, Jun 18, 2010.

[11] ***,"Weather history download", Accessed on: 08.10.2020. Available: https://www.meteoblue.com/en/weather/ archive.

[12] D.I. Nastac, N. Tanase and P.D. Cristea, "Smart predictive model for air pollutants", in Proceedings of GSP 2011 - 2nd International Workshop on Genomic Signal Processing, Bucharest, Romania, 27-28 June 2011, Paul Dan Cristea, Ed., pp. 131-134.

Application of Ultrasonic Sensors in Mapping Vineyard Parameters

E. Szilagyi, S. Meza[1], D. Petreus, T. Patarau and R. Etz

Department of Applied Electronics, Technical University of Cluj-Napoca, Romania
[1] Communication Department, Technical University of Cluj-Napoca, Romania
eniko.lazar@ael.utcluj.ro

Abstract—**In our increasingly environmentally conscious age, agriculture faces major new challenges in order to reduce its energy inputs, optimize the application of phytosanitary treatments, minimize its environmental impact and increase its production. Smart agriculture represents an inevitable trend in which ultrasonic sensors also play an important role. This paper presents parts of a smart mapping system for vineyards. As of now, the system contains 14 ultrasonic sensors, two accelerometer and gyroscope modules, one stepper motor and an Arduino MEGA 2560 development board. Its aim is to map several vineyard parameters: the system will allow the determination of the distribution and density of the vineyard foliage horizontally and vertically. This work concentrates on the detection mode of the system for the vineyard, future work will present data processing techniques for the optimization of the application of phytosanitary substances in the vine culture.**

Keywords—vineyard, precision agriculture, ultrasound sensors, microcontroller, mapping

I. INTRODUCTION

In agricultural production, the proper application of pesticides and nutrients is a vital process, but it is also one of the most dangerous agricultural operations [1]. On one hand, phytosanitary treatments can have very serious environmental consequences, on the other, spraying plays an important role in improving productivity and reducing harvest losses. Therefore, optimizing the use of pesticides represents an essential contribution to agricultural production.

One of the biggest and most expensive challenges for farmers is figuring out the optimal way of using sprayers systems [2]. Trends in the modern world focus on the significance of precision agriculture (PA) for the efficient use of pesticides. Although there are many complex definitions of precision agriculture (PA), Gebbers and Adamchuk described it in a simple way: PA means "to apply the right treatment in the right place at the right time" [3]. PA represents an entire farming management approach that combines improved machinery, sensors, satellite positioning and information technology to optimize production by taking into consideration the uncertainties and variability within agricultural systems [4].

Over the last few decades, the development of different types of sensors (such as ultrasonic and light sensors and rangefinders [5,6,7]) has greatly contributed to facilitating the collection of information in orchards or groves (leaf height, width, foliage density and other structural parameters) [8]. This information

can help growers develop a more efficient and sustainable way of spraying, helping to optimize agricultural tasks such as irrigation and fertilization, as well as cutting and cultivation techniques. Tractor mounted sprayers are illustrated in Fig. 1. In Section A a standard air-blast sprayer with constant spray output and manual operation can be observed. Section B illustrates an optimized way of using the sprayer: the nozzles of a canopy actuated (on/off) sensor sprayer with constant outputs are automatically turned on and off as a plant's presence is sensed. In this way, the amount of the phytosanitary substances used can be reduced [9].

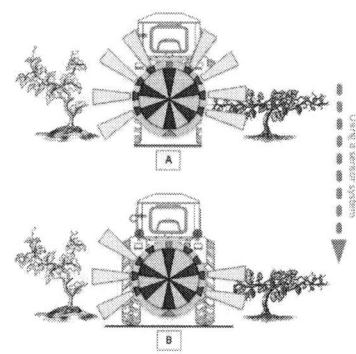

Fig. 1. Tractor mounted sprayers

There exist different methods to detect and map tree crops and vines, but not all of them can be used effectively to improve phytosanitary treatments. Reference [10] compares three methods in order to determine the fundamental parameters of the canopy structure of plants: 1. the traditional manual measurement method, 2. the use of LIDAR and 3. ultrasonic sensors. Results show that ultrasonic sensors are better at defining the average canopy characteristics, while LIDAR sensors offer detailed information about the canopy. Reference [11] presents a method for mapping vineyard leaf area with multispectral satellite imagery. Reference [12] presents a brief outline of technologies used in precision viticulture such as monitoring technologies (geolocation, remote sensing, proximal sensing), VRTs (variable-rate technologies) and "agbots" system.

The authors' proposed method for mapping a vineyard uses 14 ultrasonic sensors which are mounted on a rigid plastic bar and rotated using a stepper motor. The accelerometer and

978-1-7281-7507-2/20 $31.00 © 2020 IEEE

21-24 October, Pitesti, Romania

gyroscope modules determine the tilt angle and the translation of the sensor system. Their measurements can be integrated to obtain position and orientation information in order to locate the grapevines.

The remainder of this paper is structured as follows: Section 2 briefly describes the components of the sensor system and the hardware implementation. Section 3 outlines the software implementation. Section 4 presents the data processing method and some preliminary results of the plant detection. Section 5 contains the conclusions.

II. THE SENSOR SYSTEM

A. Ultrasonic sensors

Ultrasonic sensors are widely used in precision viticulture to gather information about the canopy characteristics of a vineyard. In this project, 14 ultrasonic sensors (HC-SR04) were used to measure the distance between the plastic bar and the grapevines. The plastic bar has two parts: the lower part where the sensors are mounted, and the upper part that covers the wire connections, Fig. 2.

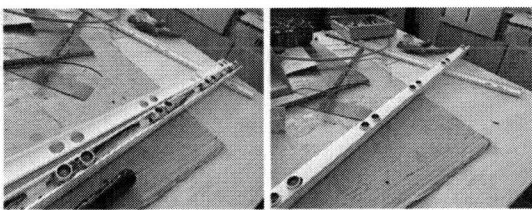

Fig. 2. Work in progress: plastic bar with ultrasonic sensors

This plastic bar is fixed to a wooden bar and rotated by a stepper motor.

A HC-SR04 ultrasonic ranging module can measure distances between 2cm and 400 cm. It includes an ultrasonic transmitter, a receiver and the control circuit [13]. The module has four pins, VCC, GND, TRIG and ECHO.

B. Accelerometer and Gyroscop modules

The MPU-6050 combines a MEMS 3-axis gyroscope and a 3-axis accelerometer with an onboard DMP (Digital Motion Processor). This module is mounted on the end of the wooden bar, Fig. 3.

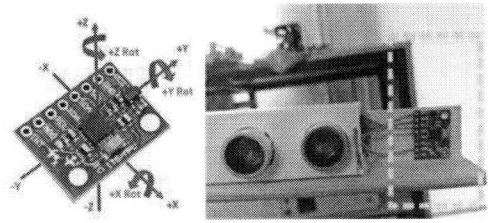

Fig. 3. MPU-6050 module

By extracting information from both the accelerometers and the gyroscope (raw data), the tilt angles can be determined: roll, pitch and theta [14]. Combining the raw gyro and accelerometer data by using filters, a better estimate of angles can be realized.

In Fig. 4., the roll angle (φ) was calculated using different methods. The first method uses a mathematical formula and the raw data from the accelerometer:

$$\phi = \arctan(\frac{accY}{\sqrt{accX^2 + accZ^2}}) \qquad (1)$$

Where: accY is the Y-axis acceleration, accX is the X-axis acceleration, accZ is the Z-axis acceleration.

To include the angle measured by the gyroscope the gyro data has to be integrated. Because the gyro gives the angular rate, to calculate the angle the angular rate has to be multiplied by the time:

$$\phi = \phi + gyroX \cdot dt \qquad (2)$$

Where: gyroX is the X-axes gyroscope data

The third method is a Kalman filter based on information fusion between (1) and (2).

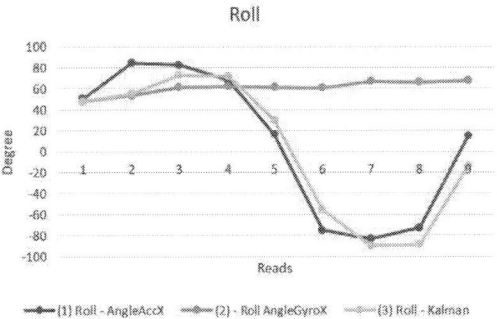

Fig. 4. Roll Angle Representation

The pitch and theta angles can be calculated the same way.

C. Stepper Module

The choice of stepper motor depends on the application's torque and speed requirements. By analyzing the weight of the wooden bar and knowing that very high speed is not required, a Robotron 52/60 motor was chosen to rotate the bar. It is a 4-wire stepper motor with a 1.8° step angle (200 steps/revolution) and the front end is equipped with a plastic disc (pulley). A belt is looped around this pulley and goes to the second wheel which will rotate the wooden bar, Fig. 5.

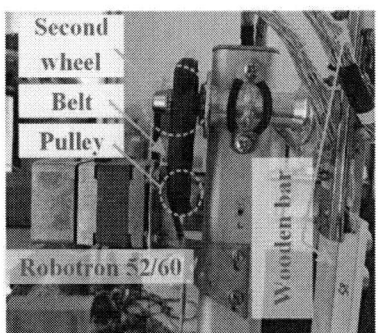

Fig. 5. Work in progress: Robotron 52/60

The stepper motor is controlled by a LN298N motor driver, Fig. 6. It is a dual H-Bridge motor driver which allows speed and direction control.

Fig. 6. LN298N Motor Driver

D. Microcontroller

The chosen microcontroller in this project is an Arduino Mega 2560. One of the main considerations for choosing this development board was the number of pins required:

- Ultrasonic Sensors - 16 pins: 14 x Echo, 2 x Trigger. Two different trigger signals are used for the left and right sides of the plastic bar. The Vcc and GND signals are connected together.

- MPU-6050 Module: 2 x 3 pins: SCL, SCK, AD0. The SCL and SCL pins are used for I2C communication. Pin AD0 determines the address of the slave device. The Vcc and GND signals are connected together.

- Stepper motor and driver module: 4 pins (IN1, IN2, IN3, IN4).

- Total min requirement: 26 pins.

The Arduino Mega 2560 is based on the ATmega2560 microcontroller, has 54 digital input/output pins and 16 analog pins. If further expansion of the project is required (addition of new sensors, modules), it is not necessary to replace the entire board, the remaining free pins can be used at any time.

E. The proposed system

The pin diagram of the proposed system is presented in Fig. 7.

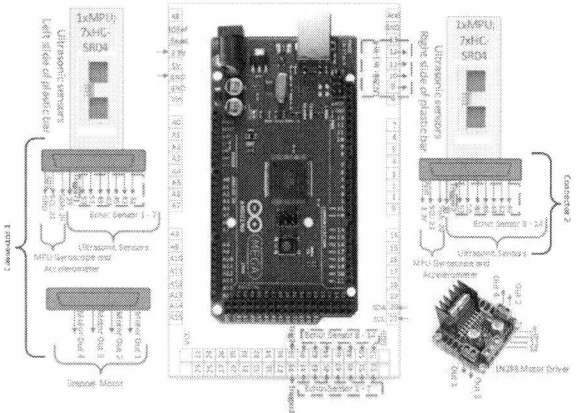

Fig. 7. Pin diagramm of the proposed system

Two connectors were used to connect the wires of the components together, Fig. 8:

1. The first connector (Connector 1) contains the trigger signal and the echo signals of the first 7 ultrasonic sensors. The first MPU-6050 module is also connected in this connector. The remaining pins are connected to the output pins of the stepper motor.

2. The second connector (Connector 2) includes the pins for the other 7 sensors and the second MPU module.

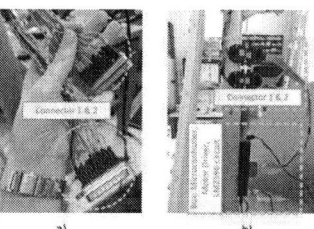

Fig. 8. Work in progress: Connectors

A box containing the development board, the motor driver circuit and the LM2596 DC to DC step down regulator is fixed at the bottom of the support bar, Fig.8 b) and Fig. 9.

Fig. 9. Arduino Mega 2560 and its box which is mounted on the rigid bar

The whole proposed system is presented in Fig. 10:

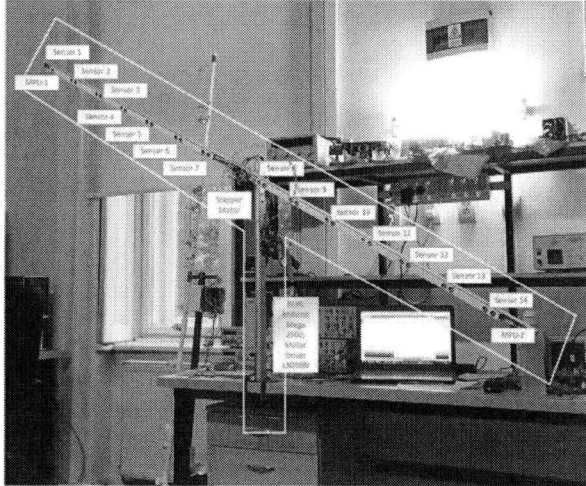

Fig. 10. The proposed sensor system

14 ultrasonic sensors are mounted on two plastic bars on the right and left sides. These plastic bars are fixed on a wooden bar for support and are rotated using a stepper motor. Using the MPU modules the pitch, roll and theta angles can be determined. Also, the displacement of the sensor bar can be accounted for.

III. SOFTWARE IMPLEMENTATION

The Arduino Software (IDE) was used to develop the code. The proposed algorithm for the sensor system is presented in Fig. 11.

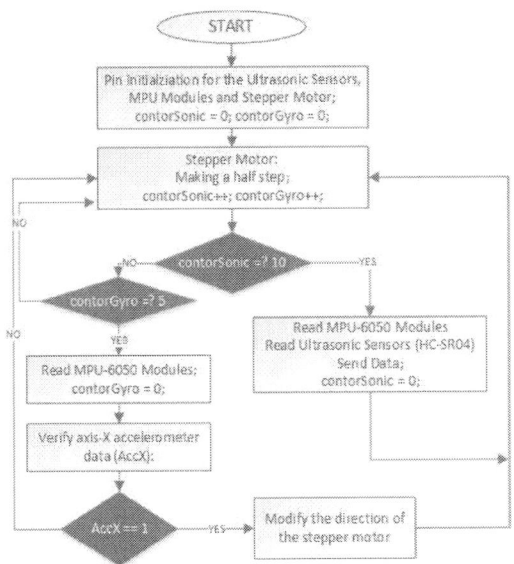

Fig. 11. Proposed algorithm for the sensor system

The process starts with the initialization phase: pins, sensors and stepper motor initialization. After that, the program enters an infinite loop:

1. The stepper motor takes a half step. Half steps were used over the full step procedure to reduce the motor vibration and to get more consistent rotating moves. This smaller step angle provides smoother operation due the increased resolution of the angle. Each time the motor makes a half step, two counters are incremented: counterSonic and counterGyro. These counters influence the timing of when the sensors, the accelerometer and the gyroscope have to be read. Reading the data at every step would make the rotation lose continuity.

2. For this reason, the ultrasonic sensors and the raw data of MPU models are read when the counter for the sensors reaches the value 10. The data is then transmitted over the serial port and is processed by Matlab.

3. When the counterGyro (counter for the accelerometer and gyroscope) is 5, the raw accelerometer and gyroscope data are read. If the axis-X acceleration is 1, the stepper motor's direction will be modified, so a full rotation of the wooden bar can be achieved. Fig. 11 presents the values of axis-X and Y.

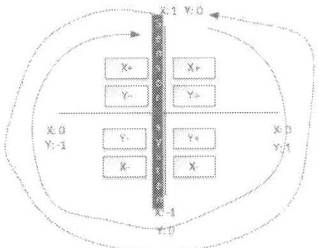

Fig. 12. Acceleromter: axis-X and axis-Y

As can be seen on the figure, the gyroscopes are placed in a way so that the axis-X has the value 1 at the top of the motion. The green continuous line is the clockwise direction, the blue dotted line represents the anticlockwise direction of the stepper motor.

IV. DATA PROCESSING AND PRELIMINARY RESULTS

The data transmitted over the serial connection is in the following format, Fig. 13. The raw data from the accelerometer is transmitted first: axis-X, axis-Y and axis-Z. This data is followed by the distances between the ultrasonic sensors and the bar. All the measurements are in cm. The HC-SR04 sensors can measure distances between 2cm and 400cm, therefore the values above can be interpreted as if nothing was detected. For example, if sensor 1 measures a distance of 517.12 cm, it means that it has detected no obstacle.

```
(X,Y,X): (-0.28, -0.73, 0.12)
S1:517.12  / S2:518.56
S3:517.18  / S4:391.50
S5:518.94  / S6:515.41
S7:1030.27 / S8:107.77
S9:105.54  / S10:104.03
S11:103.55 / S12:102.61
S13:104.31 / S14:107.82
```

Fig. 13. Data over the serial

The first step to process this data and to map the area is to convert polar coordinates to cartesian coordinates. The distances between the ultrasonic sensors and the origin are known, the α angle can be calculated. Using simple trigonometric equations, the cartesian x and y coordinates can be determined. Using these coordinates and Matlab, different maps can be drawn.

Fig. 14 depicts a map in which the sensor system makes an entire turn. Every sensor reading represents a point in a circle.

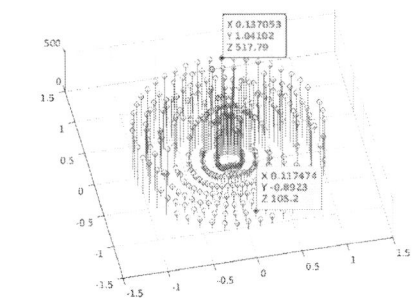

Fig. 14. Mapping object – points - Matlab results

In this figure, two points are marked:

- Point 1: The x and y coordinates are 0.13 and 1.04, respectively. Sensor 1 does not detect any object.

- Point 2: The x and y coordinates have the values 0.11 and -0.89. Sensor 2 detects an object (in this case, a grapevine).

In Fig. 15., color coding is also used to better represent the distances between the objects and sensors. Blue color means that an object has been detected within 1m, Green color means that the plants/objects detected are farther. Yellow color means that nothing has been detected.

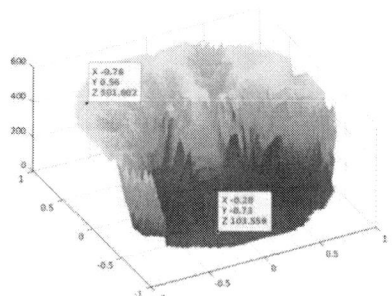

Fig. 15. Mapping object – Meshgrid - Matlab results

These preliminary results are based on the fact that the sensor system is fixed in space, is not mounted on a tractor or other agricultural machine, therefore the generated map has a simple circular form. Further results will be published.

V. CONCLUSIONS AND FUTURE DEVELOPMENT

This paper presents the hardware implementation of an ultrasonic sensor system which can be used in mapping vineyards. Further, one possible solution for processing the extracted data has also been presented.

Moreover, to improve the proposed system, future developments are needed: (1) The control of the stepper motor was realized using the half step movement mode. Microstepping could offer greater control and smoother operation for the stepper motor. (2) Gear systems require more precise alignment than pulley systems but with microstepping, this is not a problem. A gear system would also offer an increase in power and in speed. (3) Using a gear system, the steps taken can be counted and so the change of direction algorithm of the motor can be simplified. (4) Centering the wooden and plastic bars also presents a problem. This must be improved. (4) Reducing the reading time of the ultrasonic sensors is needed. (5) The use of LIDAR sensors instead of ultrasonic sensors is recommended because they offer a more precise image of the detected objects. (6) To further study the methods of phytosanitary treatments and possible applications of this system to improve these treatments.

There are issues to overcome before widespread use of the proposed sensor system could be implemented. These are not only related to the further development of the system, but above all to the fact that the farms will have to want to use these systems and improve their phytosanitary treatments.

ACKNOWLEDGMENT

This work was supported by a grant from the Romanian Ministry of Research and Innovation, CCCDI-UEFISCDI, project number PN-III-P1-1.2-PCCDI-2017-0251/4PCCDI/ 2018, within PNCD III.

REFERENCES

[1] I. Abbas, J. Liu, M. Faheem, R. S. Noor, S.A. Shaikh, et al, "Different Real-Time Sensor Technologies for the Application of Variable-Rate Spraying in Agriculture", Sensors and Actuators A: Physical, pp. 1-35, Available online 9 August 2020, https://doi.org/10.1016/j.sna.2020.112265

[2] A. J. Landers, "The answer is blowing in the wind. In Aspects of Applied Biology", Vol. 66, International advances in pesticide application, pp. 177–184, 2002

[3] R. Gebbers, V. I. Adamchuk, "Precision Agriculture and Food Security", Science, Vol. 327 (5967), pp. 828-83, February 2010

[4] Joint Research Centre (JRC) of the European Commission, P. J. Zarco-Tejada, N. Hubbard, P. Loudjani, "Precision agriculture: an opportunity for EU Farmers-potential support with the CAP 2014-2020", Study, Catherine Morvan, PE 529.049, European Union, 2014

[5] D. Petrovic, M. Jurisic, V. Tadic, I. Plascak, Z. Barac, "Different Sensor Systems for the Application of Variable Rate Technology in Permanent Crops". TECHNICAL JOURNAL, Vol. 12 (3), pp. 188-195, 2018

[6] S. Azfar, A. Nadeem, A.B. Alkhodre, et al., "Monitoring, detection and control techniques of agriculture pests and diseases using wireless sensor network: A Review", International Journal of Advanced Computer Science and Applications, Vol. 9 (12), 2018

[7] Z. Zhang, X. Wang, and Q. Lai, and Z. Zhang, "Review of Variable-Rate Sprayer Applications Based on Real-Time Sensor Technologies", Automation in Agriculture-Securing Food Supplies for Future Generations. InTech; 2018. doi:10.5772/intechopen.73622

[8] A.F. Colaço, J. P. Molin, J.R. Rosell-Polo, A. Escolà, "Application of light detection and ranging and ultrasonic sensors to high-throughput phenotyping and precision horticulture: current status and challenges", Horticulture Research, Vol.5/35, pp. 1-11, July 2018

[9] B. Warneke, J. W. Pscheidt, R. Rosetta, L. Nackley, "Sensor Sprayers for Specialty Crop Production", A Pacific Northwest Extension Publication - OSU Extension Catalog, PNW 2019, https://catalog.extension.oregonstate.edu/pnw727/html [Accesed on 30.09.2020]

[10] J. Llorens, E. Gil, J. Llop, A. Escolà, "Ultrasonic and LIDAR Sensors for Electronic Canopy Characterization in Vineyards: Advances to Improve Pesticide Application Methods", Sensors (Basel, Switzerland), Epub, Volt 11(2), pp. 2177-2194, 2011

[11] L.F Johnson, D.E Roczen, S.K Youkhana, R.R Nemani, D.F Bosch, "Mapping vineyard leaf area with multispectral satellite imagery", Computers and Electronics in Agriculture, Volume 38, Issue 1, Pages 33-44, 2003

[12] A. Matese, S. F. Gennaro, "Technology in precision viticulture: A state of the art review", International Journal of Wine Research, Vol. 7(1), pp. 69-81,2015

[13] Ultrasonic Ranging Module HC - SR04 datasheet, https://cdn.sparkfun.com/datasheets/Sensors/Proximity/HCSR04.pdf [Accesed on 30.09.2020]

[14] M. Pedley, "Tilt Sensing Using a Three-Axis Accelerometer", Freescale Semiconductor Application Node, Document Number: AN3461, https://www.nxp.com/files-static/sensors/doc/app_note/AN3461.pdf [Accesed on 0.10.2020]

Electronic system for measuring frequency in GHz range

F. Vasile, A. Craciun, M. Vladescu, P. Schiopu,
V. Feies
Optoelectronics Research Center (UPB-CCO)
University Politehnica of Bucharest
Bucharest, Romania
vasile_florentin@yahoo.com

N. D. Codreanu
Center for Technological Electronics and Interconnection
Techniques (UPB-CETTI)
University Politehnica of Bucharest
Bucharest, Romania
norocel.codreanu@cetti.ro

Abstract—**Ultra-sensitive TFBAR (Thin Film Bulk Acoustic Wave Resonator) mass sensors can be used for the detection of dangerous substances. The chemically-sensitive part of such a sensor consists in a thin film of gold (Au) which covers the sensing area of the TFBAR and on top of which a layer of specific antibodies is deposited. TFBAR is a volume acoustic wave resonator that works in longitudinal vibration mode (thickness extension) and hence its resonance frequency is determined by the inverse of the piezoelectric layer thickness and directly proportional to the material dielectric constant. This paper presents the design and implementation of an electronic system based on a differential approach for measuring the chemical contamination-induced frequency deviation of a TFBAR sensor based on homodyne mixing of the output signals of two Pierce oscillators: a sensing one and a reference one.**

Keywords—homodyne detection, Thin-Film Bulk Acoustic Wave Resonator (TFBAR), sensor, integrated electronic and photonic systems

I. INTRODUCTION

Ultra-sensitive TFBAR sensors has found important applications as chemical sensors, for detecting slight mass changes due to contamination with small amounts of a target substance, such as an explosive or dangerous gas [1], [2].

The detection of an explosive or dangerous substance in the vapour state is performed with the help of antibodies – prepared specifically for each target substance – which are immobilized on the sensing surface of a TFBAR. When the vapour of the targeted explosive substance comes in contact with the TFBAR sensor, it binds to the antibodies, producing a slight mass change which, in turn, induces a deviation of the resonance frequency of the TFBAR.

This paper presents an electronic system based on a differential approach for measuring the frequency deviation of a TFBAR sensor as a result of its chemical contamination. The measurement method is based on homodyne mixing of the output signals of two Pierce oscillators [3]: a sensing one (which contains the sensor exposed to the target substance) and a reference one.

The Pierce oscillators employ a new type of sensor which is based on a micro-electro-mechanical system (MEMS) approach [5] associated with the manufacture of a thin

piezoelectric layer [6] that allows the operation of the resonator at high frequencies (1 ÷ 3 GHz) and with a high quality factor Q.

In addition to the problem of integrating the TFBAR sensor with the Pierce oscillator, another condition is necessary to be achieved: the spectral width of the two oscillators needs to be less than the frequency deviation to be measured. This guarantees that the signal obtained from the mixing - although it will have a wider spectral width due to the convolution of the spectra of the two oscillators - will not contain in the spectrum negative frequencies that "mirror" in the positive frequency range, which could disturb the correct measurement of the frequency deviation.

II. DEVELOPMENT OF THE PIERCE OSCILLATOR

The sensing system for detecting explosive substances is based on two identical TFBAR sensors manufactured in MEMS technology, one being active (i.e. exposed to the dangerous substances) and the other one being passive, used as a reference.

These two sensors are included as resonators in two identical Pierce oscillators. The oscillator containing the passive TFBAR sensor will oscillate will oscillate on the reference frequency f_0 (which is about 2 GHz due to the manufacturing process of the TFBAR structure), while the oscillation frequency of the other oscillator will be $f_0 + \Delta f$, where Δf depends on the vapor concentration of the substance to be detected.

The frequency deviation, Δf, between the two oscillators is obtained by using a mixer to combine the outputs of the two oscillators [7] and a low-pass filter (with the 3-dB cutoff frequency of approx. $100 \div 200$ kHz) to select only the low frequency component Δf corresponding to the difference $(f_0 + \Delta f) - f_0$.

Furthermore, using a properly designed frequency discriminator, the frequency deviation is transformed into a voltage which can be then converted to a digital signal and processed by a microcontroller-based electronic circuit.

An example of a Pierce oscillator based on CMOS transistors and using a TFBAR (or FBAR), shown in Figure 1, can be found in [4].

2020 IEEE 26th International Symposium for Design and Technology in Electronic Packaging (SIITME)

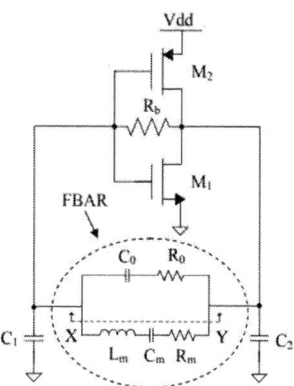

Fig. 1. An example of a Pierce oscillator based on CMOS transistors and using a TFBAR structure, as it is presented in [4].

The general block diagram of the Pierce oscillator which was implemented for the electronic system of measurement presented in this paper is shown in Figure 2.

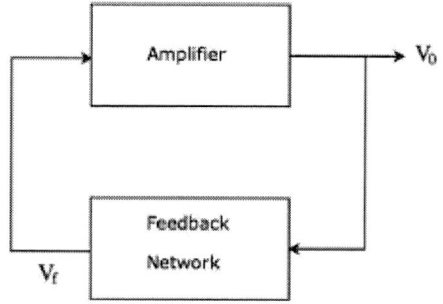

Fig. 2. General block diagram of the Pierce oscillator.

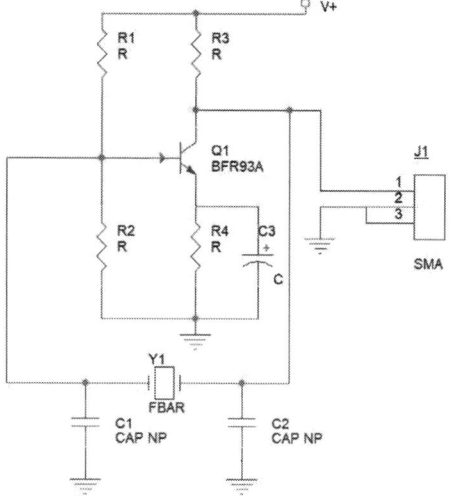

Fig. 3. Schematic diagram of the Pierce oscillator.

The amplifier is based on the BFR93A NPN bipolar junction transistor (BJT), which has a maximum working frequency of 6 GHz, large enough with respect to the oscillating value of around 2 GHz. The feedback network has a π-type structure, with the FBAR resonator being connected

between the output and the input and two high-quality NPO ceramic capacitors placed between each end of the resonator and the ground, as depicted in Figure 3, where the amplifier and the feedback network can be easily distinguished.

The power supply voltage was selected to be 5 V [8]. For designing the oscillator, the following parameters were considered: the collector voltage to be approx. 3.5 V, the gain of the amplifier approx. 20, with possibility to be further increased if the capacitor C3 is placed in parallel with the resistor R4. In the current design, this capacitor is not assembled, but a corresponding printed circuit board (PCB) footprint was used and placed at the correct location. Figure 4 presents the final PCB layout design before manufacturing.

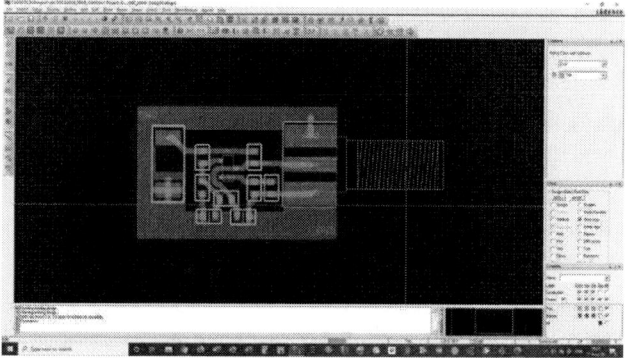

Fig. 4. The PCB layout of the Pierce oscillator.

In order to maximize the performances of the oscillator and minimize the parasitic effects, the passive components were chosen to be in 0603 chip packages. Additionally, the BFR93A transistor was selected to be packaged in a classical small size SOT23 plastic case.

The technological development and manufacturing of the Pierce oscillators was made in the laboratories of "Politehnica" University of Bucharest, Department of Electronics Technology and Reliability, Center for Technological Electronics and Interconnection Techniques (UPB-CETTI) and Optoelectronics Research Center (UPB-CCO), using Press-n-Peel Blue transfer foil from Techniks, the etching process being done in turbulent and warm (approx. 50°C) ferric chloride ($FeCl_3$) with a concentration of 38%. The final aspect of the manufactured oscillator is shown in Figure 5.

The interface of the oscillator with the mixer is done based on a SMA (SubMiniature version A) connector, which is a classical semi-precision coaxial RF connector used as interface for coaxial cables with screw-type coupling mechanism.

The SMA has a 50Ω characteristic impedance and is designed to work in the range $0 \div 18$ GHz, fully matched with the necessities of the current Pierce oscillator.

According to the theoretical calculus, the DC values of the oscillator are as follows: $V_C = 3.55$ V, $I_C = 1.5$ mA, the current absorbed from the power supply being of approx. 3.2 mA.

Because the real values of the passive components acquired were a bit different than the theoretical ones, the measured values after assembling and testing are $V_C = 3.95$ V and $I_C = 1.1$ mA. The FBAR resonator which will be used for generating the oscillation frequency of approx. 2 GHz (shown in Figure 6) is under manufacturing, the location of it being in the positive feedback circuit, from the collector to the base of the BFR93A transistor (Y1 in Figure 3).

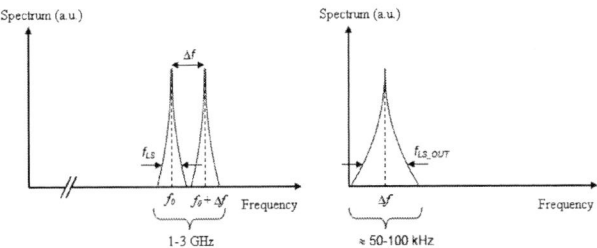

Fig. 7. The illustration of the signal spectra at inputs (left) and output (right) of the mixer.

In Figure 8 is illustrated the measured spectrum of an FBAR Pierce oscillator whose resonance frequency is 1.88165 GHz and the spectral bandwidth is of the order of 2 kHz [5].

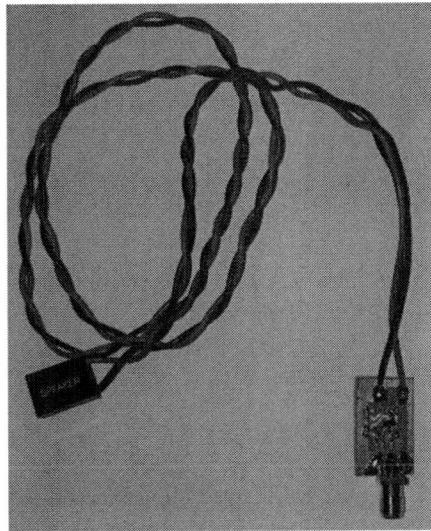

Fig. 5. The oscillator after assembling.

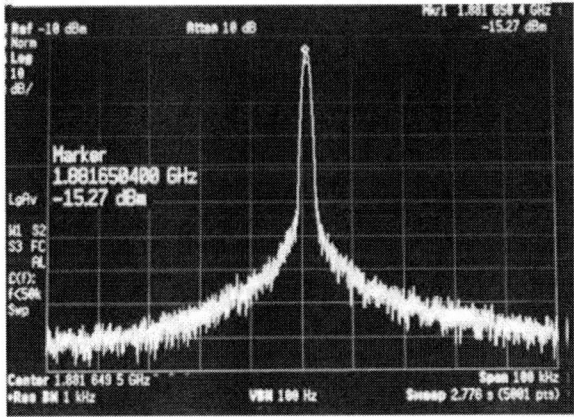

Fig. 8. The measured spectrum of an FBAR oscillator (Pierce oscillator) [5].

The chosen block diagram based on measuring the resonance frequency deviation of the TFBAR sensor by the differential method by mixing the signals of two Pierce oscillators (homodyne version) is a simple design scheme and requires low investment costs. The results can also be useful for different studies related to optoelectronics, in general [10], [11].

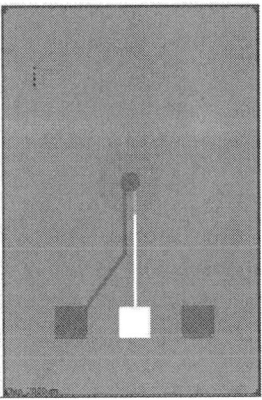

Fig. 6. One of the masks of the FBAR resonator.

Acknowledgment

This article is supported by the Complex Project "Sensors and Integrated Electronic and Photonic Systems for People and Infrastructures Security" (SENSIS), Component Project No. 1 "Portable Micro-Systems based of Sensors with TF-BAR areas for multiple detection of Explosive Gases", Project Code PN-III-P1-1.2-PCCDI-2017-0419, Financing Contract No. 71 PCCDI/2018.

A. Abbreviations and Acronyms

TFBAR (Thin Film Bulk Acoustic Wave Resonator), MEMS (Micro-Electro-Mechanical System), BJT (Bipolar Junction Transistor), (Printed Circuit Board), SMA (SubMiniature version A), SENSIS (Sensors and Integrated Electronic and Photonic Systems for People and Infrastructures Security), UPB-CETTI (Center for Technological Electronics and Interconnection Techniques), UPB-CCO (Optoelectronics Research Center).

References

[1] Z. Bielecki, J. Janucki, A. Kawalec, et al., "Sensors and Systems for the Detection of Explosive Devices-An Overview", Metrology and Measurement Systems, vol. 19 (1), pp. 3-28, 2012.

[2] S. Singh, "Sensors-An effective approach for the detection of explosives", Journal of Hazardous Materials, 144 (1-2), pp. 15-28, 2007.

III. Results

In Figure 7 are illustrated the signal spectra at input (left) and output (right) of the mixer. For the homodyne variant [9], the oscillation frequency is determined and not the voltage level of the Pierce oscillator.

[3] A. Liapine, "Resonant Cavities as Beam Position Monitors: Part 3. Analog signal processing. Laboratory measurements", 2005, available at: www.hep.ucl.ac.uk/~liapine/.

[4] Y.H. Chee, A.M. Niknejad, and J. Rabaey, "A Sub-100μW 1.9-GHz CMOS Oscillator Using FBAR Resonator", 2005 IEEE RFIC Symposium [Digest of Papers, INSPEC Accession Number: 8479113, 2005].

[5] A.M. Niknejad, "MEMS Reference Oscillators, EECS 242: Lecture 25", University of California, Berkeley 2009, available at: http://rfic.eecs.berkeley.edu/~niknejad/ee242/pdf/eecs242_lect25_xtal.

[6] M. Akiyama, K. Kano, and A. Teshigahara, "Influence of growth temperature and scandium concentration on piezoelectric response of scandium aluminum nitride alloy thin films", Appl. Phys. Lett. 95, 162107, 2009.

[7] A. Tulbure, C. Hutanu, and Gh. Brezeanu, "Small Signal Impedance Analysis of High Efficient Power Devices", Proceedings of the 38th International Semiconductor Conference (CAS 2015), Sinaia, pp. 229-232, October 2015.

[8] Gh. Serban, L. Ionescu, and A. Mazare, "The possibility of optimisation for power supply consumption using evolvable power regulator", Revue Roumaine des Sciences Techniques, Serie Electrotechnique et Energetique, Tome 57, No. 2, pp. 222-231, Avril-Juin 2012.

[9] F. Vasile, A. Craciun, M. Vladescu, P. Schiopu, V. Feies, and N. Codreanu, "The use of homodyne detection for measuring small frequency differences between two RF oscillators", under publication at Advanced Topics in Optoelectronics, Microelectronics and Nanotechnologies ATOM-2020 conference.

[10] M. Vladescu, and P. Schiopu, "Advanced Educational Program in Optoelectronics for Undergraduates and Graduates in Electronics", Proc. SPIE 9258, 92580B, 2015.

[11] P. Schiopu, O. Iancu, and M. Vladescu, "Optoelectronics-Theory and applications", Nautica Publishing House, Constanta, 2013.

An Approach for Calculating the Temperature at a Point in the Cross Section Formed by Temperature Sensors

Snezhinka Lubomirova Zaharieva
Department of Electronics
University of Ruse "Angel Kanchev"
Ruse, Bulgaria
szaharieva@uni-ruse.bg

Iordan Ivanov Stoev
Department of Electronics
University of Ruse "Angel Kanchev"
Ruse, Bulgaria
istoev@uni-ruse.bg

Adriana Naydenova Borodzhieva
Department of Telecommunications
University of Ruse "Angel Kanchev"
Ruse, Bulgaria
aborodzhieva@uni-ruse.bg

Svilen Ivanov Stoyanov
Dobrudza Technological College, Dobrich, Bulgaria
Technical University of Varna
Varna, Bulgaria
svilen.stoyanov@tu-varna.bg

Abstract—**This paper presents a mathematical model developed in MATLAB that calculates the temperature in a cross section formed by six temperature sensors. The calculated temperature can be obtained for each point on the basis of geometric coordinates and time at an interval of ten minutes. The cross section is a section of a room where temperature data are measured. A given geometric point of the section will serve as input data for temperature when calculating the energy model of the premises. Through the presented mathematical model, premises with larger geometry and larger temperature differences in the measured temperatures between the floor and the ceiling can be evaluated by the method of similarity.**

Keywords—digital temperature sensors, Fourier series, mathematical model, MATLAB

I. INTRODUCTION

A. Systems for monitoring, management and control processes based on WSN

Wireless technologies today are the preferred solution for building systems for monitoring, management and control of various processes. Wireless sensor network (WSN) is at the heart of these systems. It consists of sensors scattered in space, through which the users can monitor and record the physical conditions of the environment, as well as organize the collection of measured information.

These systems have the following advantages:

- Provide continuous online monitoring of the points where the sensors are installed;

- Monitor several points simultaneously;

- Quickly detect conditions inconsistent with the normal, in which case an early warning could be carried out,

through means of notification at several levels and actions can be taken to prevent future errors;

- Extend maintenance intervals;

- Easy installation and low price;

- Open communication network.

The principle of operation consists of wireless transmission of data from the sensor to the receiver connected to a computer via a serial interface or Internet connection. The computer collects and analyses the measurement data using appropriate software. The processed information is disseminated in the described ways to the end users.

Continuous monitoring provides means to assess the current condition of the equipment and detects anomalies at an earlier stage. The use of wireless technology eliminates the need for special cables and provides lower installation costs compared to other types of online condition monitoring equipment.

B. Electronic system for energy flow management in residential premises

An electronic system for energy flow management has been developed and installed in a laboratory at the University of Ruse "Angel Kanchev" [1, 2, 3]. Fig. 1 and Fig. 2 show the East and West walls, where the temperature sensors are positioned – points 2, 3, 4, East wall and 6, 7, 8, West wall, which form a cross section [3]. For the needs of the energy model for control of the energy flow in the laboratory, it is necessary to obtain information about the temperature at any point of the cross section based on the geometrical arrangement of these points.

The number of temperature sensors is six and as it can be seen from the figures, they are positioned against each other. In this way, symmetry is achieved and the temperature is measured

---

The paper presents results from Project 2020 - FNI - FEEA - 02, „Research and development of specialized electronic industrial process control systems based on statistical analysis", with financial support of the National Science Fund of University of Ruse.

in close proximity to the floor (temperature sensors 4 and 8), in the middle (temperature sensors 3 and 7) and close to the ceiling of the room (temperature sensors 2 and 6).

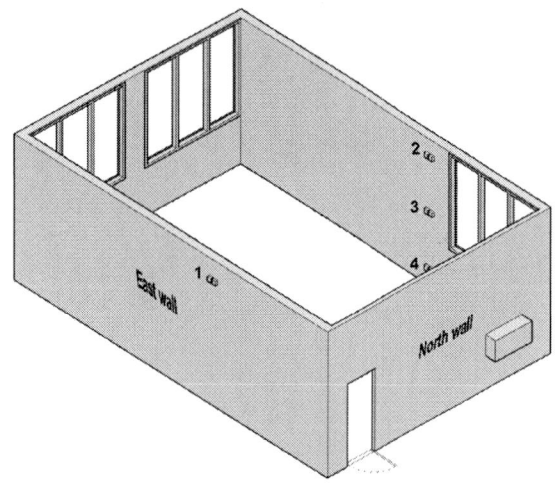

Fig. 1. Digital temperature sensors, West wall

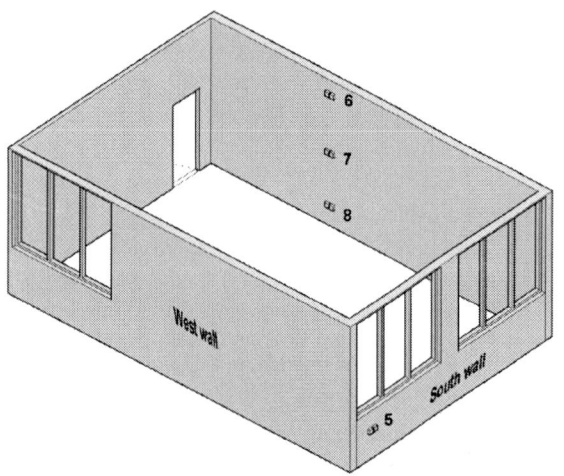

Fig. 2. Digital temperature sensors, East wall

The purpose of this paper is to present a mathematical model in MATLAB, used to calculate the temperature in a cross section formed by six temperature sensors. The calculated temperature can be obtained for each point based on geometric coordinates and time in an interval of ten minutes.

A given geometric point of the section will serve as input data for temperature when calculating the energy model of the premises. Through the presented mathematical model, premises with larger geometry and larger temperature differences in the measured temperatures between the floor and the ceiling can be evaluated by the method of similarity.

II. THEORETICAL JUSTIFICATION OF THE PROPOSED APPROACH FOR TEMPERATURE EVALUATION AT A POINT OF THE CROSS SECTION

In a cross section formed by the geometric location of the sensors, a linear regression model is proposed, giving the temperature at each point of the section, depending on its x and y coordinates.

$$T = \alpha_0 + \alpha_1 x + \alpha_2 y \tag{1}$$

where: $\alpha_0, \alpha_1, \alpha_2$ are constants to be determined depending on the measurements.

The proposed regression model describes the temperature only at specific time of day. Temperature measurements are for a period of one day, due to which large differences in temperatures occur. If the time of day in which the measurements were made is not taken into account, the model would not be accurate. For this reason, a model is proposed in which the time of day in which the measurement is made is included as a variable and the model takes the form:

$$T = \alpha_0(t) + \alpha_1(t)x + \alpha_2(t)y \tag{2}$$

where: $\alpha_0(t), \alpha_1(t), \alpha_2(t)$ are functions that depend on the moment of measurement (time).

If two consecutive days with absolutely identical parameters are considered, the principle of continuity concludes that the temperature at the end of the first day at each point of the section must coincide with the temperature at the beginning of the second day at the corresponding point of the section. This shows that in the considered model the functions $\alpha_0(t)$, $\alpha_1(t)$ and $\alpha_2(t)$ are necessary to be periodic with a period of twenty-four hours. For such functions it is recommended to use Fourier type functions [4, 5, 6]:

$$\alpha_i = \alpha_0^i + \sum_{k=1}^{n_1} \alpha_k^i \cos \frac{2k\pi}{24} t + \sum_{k=1}^{n_2} b_k^i \sin \frac{2k\pi}{24} t \tag{3}$$

where: $i = 0, 1, 2$, or:

$$\alpha_i = \alpha_0^i + \alpha_1^i \cos \frac{2\pi t}{24} + \alpha_2^i \cos \frac{4\pi t}{24} + \dots + \alpha_{m_i}^i \cos \frac{4mt}{24} +$$
$$+ b_1^i \sin \frac{2\pi t}{24} + b_2^i \sin \frac{4\pi t}{24} + \dots + b_{n_i}^i \sin \frac{2nt}{24} \tag{4}$$

where: $i = \overline{0,1,2}$

The coefficients $\alpha_0^i, \alpha_{k_1}^i, b_{k_2}^i$, $i = \overline{0,3}$, $k_1 = \overline{1, m_i}$, $k_2 = \overline{1, n_i}$ are subject to determination, and their number $m + n + 1$ is determined by considerations of accuracy, simplicity of the model and the number of measurements. In this case, a criterion for minimizing the error is required. This criterion can be the sum of the errors squared to be minimal, i.e. the least squares method, the maximum error can be minimal, i.e. the minimax method or others.

The following model is considered for $m_i = 2$, $n_i = 2$, $i = \overline{0,3}$:

$$T = \alpha_0^0 + \alpha_1^0 \cos\frac{2\pi t}{24} + \alpha_2^0 \cos\frac{4\pi t}{24} + b_1^0 \sin\frac{2\pi t}{24} + b_2^0 \sin\frac{4\pi t}{24} +$$
$$+ \left(\alpha_0^1 + \alpha_1^1 \cos\frac{2\pi t}{24} + \alpha_2^1 \cos\frac{4\pi t}{24} + b_1^1 \sin\frac{2\pi t}{24} + b_1^1 \sin\frac{4\pi t}{24}\right)x + \quad (5)$$
$$+ \left(\alpha_0^2 + \alpha_1^2 \cos\frac{2\pi t}{24} + \alpha_2^2 \cos\frac{4\pi t}{24} + b_1^2 \sin\frac{2\pi t}{24} + b_2^2 \sin\frac{4\pi t}{24}\right)y$$

The least squares method is chosen as the criterion minimizing the error. The coefficients should be calculated $\alpha_0^i, \alpha_k^i, b_k^i$, as well as their confidence intervals guaranteed with probability $\gamma = 0.95$.

In order to verify the statistical significance of these factors, it is necessary to use T-Student's test. Another equivalent approach is through their confidence intervals. If a confidence interval of a given coefficient, constructed with guaranteed probability $\gamma = 0.95$, contains the number zero, then the coefficient is statistically insignificant at the level of significance probability [7].

III. RESULTS OF THE EVALUATION OF THE TEMPERATURE MODEL AT ANY CROSS SECTION POINT IN THE RESIDENTIAL PREMISES

Based on the theoretical justification for estimating the temperature at any point of the cross section in the premises, a criterion minimizing the error was chosen and this is the method of the least squares.

The coefficients $\alpha_0^i, \alpha_k^i, b_k^i$ are calculated, as well as their confidence intervals, guaranteed with probability $\gamma = 0.95$ (Table I). In order to check the statistical significance of these coefficients, it is necessary to use their confidence intervals. If a confidence interval of a given coefficient, constructed with guaranteed probability $\gamma = 0.95$, contains the number zero, then the coefficient is statistically insignificant at a significance level of probability $\alpha = 1 - \gamma = 0.05$ [7].

After removing the statistically insignificant coefficients (Table I, in red), the new coefficients in the simplified model are calculated. After re-removal of the statistically insignificant coefficients, the simplified model is as follow:

$$T = \alpha_0^0 + b_2^0 \sin\frac{4\pi t}{24} + \alpha_0^1 x + b_1^1 \sin\frac{2\pi t}{24}x + b_2^2 \sin\frac{4\pi t}{24}y \quad (6)$$

The calculated coefficients from Eq. 6 are given in Table II, as it is seen that all coefficients are statistically significant.

TABLE I. RESULTS FROM REMOVING STATISTICAL INSIGNIFICANT COEFFICIENTS – FIRST TIME

Value of the coefficients in the model	Trust confidence interval guaranteed by probability, $\gamma = 0.95$
$a_0^0 = 24.4189$	(24.3843;24.4535)
$a_1^0 = -0.0143$	(-0.0632;0.0347)
$a_2^0 = -0.0373$	(-0.0862;0.0117)
$b_2^0 = -0.0479$	(-0.0969;0.0010)
$b_2^0 = 0.1346$	(0.0856;0.1835)
$a_0^1 = -0.0105$	(-0.0165;-0.0045)
$a_1^1 = -0.0133$	(-0.0217;-0.0048)
$a_2^1 = 0.0025$	(-0.0060;0.0109)
$b_1^1 = -0.0091$	(-0.0175;-0.0006)
$b_2^1 = 0.0086$	(0.0002;0.0171)
$a_0^2 = 0.0110$	(-0.0028;0.0247)
$a_1^2 = -0.0123$	(-0.0318;0.0071)
$a_2^2 = 0.0041$	(-0.0153;0.0236)
$b_2^2 = -0.0206$	(-0.0401;-0.0012)
$b_2^2 = 0.0209$	(0.0015;0.0404)

TABLE II. RESULTS FROM REMOVING STATISTICAL INSIGNIFICANT COEFFICIENTS

Value of the coefficients in the model	Trust confidence interval guaranteed by probability $\gamma = 0.95$.
$a_0^0 = 24.4382$	(24.4124;24.4639)
$b_2^0 = 0.1598$	(0.1158;0.2038)
$a_0^1 = -0.0105$	(-0.0167;-0.0043)
$b_1^1 = -0.0235$	(-0.0297;-0.0127)
$b_2^2 = 0.0209$	(0.0007;0.0412)

The coefficient of determination $R^2 = 0.83$ is also calculated. The determination is checked for statistical significance. The coefficient of determination is statistically significant, which means that model presented in Eq. 6 is adequate. The largest positive error is 0.76 and the largest negative error is 0.64.

Fig. 3, Fig. 4 and Fig. 5 show the calculated temperature in the cross section at different times of the day. The graphics are built in the MATLAB software environment and aim to get a visual idea of the developed model. In the three-dimensional reference system, the units of X and Y axes are in meters, and the Z axis is the measured room temperature.

2020 IEEE 26th International Symposium for Design and Technology in Electronic Packaging (SIITME)

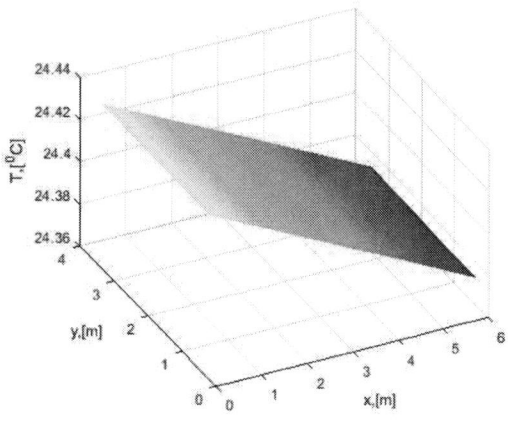

Fig. 3. Representation of temperature, calculated in the mathematical model for 0:00 h

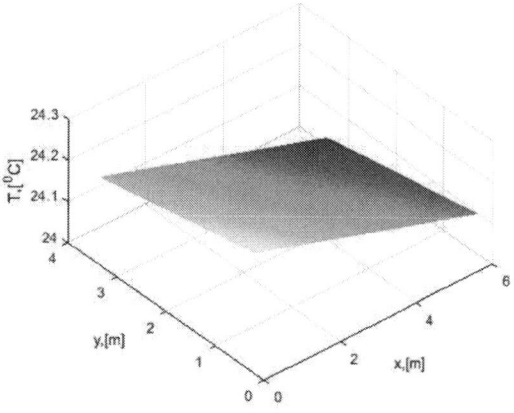

Fig. 4. Representation of temperature, calculated in the mathematical model for 8:00 h

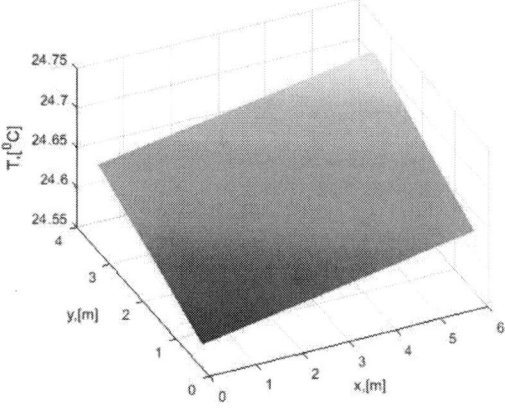

Fig. 5. Representation of temperature, calculated in the mathematical model for 16:00 h

Warmer colours (yellow and orange) indicate increased values of the room ceiling temperature. The turquoise colour shows the temperature in the middle of the room, and in dark blue it is evident from the three graphs that the value of the floor temperature is the lowest.

IV. CONCLUSION

A mathematical model has been developed in MATLAB, used to calculate the temperature in a cross section formed by six temperature sensors. The calculated temperature can be obtained for each point based on geometric coordinates and time in an interval of ten minutes. The cross section is the part of the room where the temperature data is measured. A given geometric point of the section will serve as input data for the temperature when calculating the energy model of the premises.

The coefficient of determination was also calculated, and after verification was proved to be statistically significant. This shows that the model is adequate, with the largest positive error calculated at 0.76 and the largest negative error calculated at 0.64.

Graphical interpretations of the approximated temperature in the cross section at different times of the day are presented.

Through the presented mathematical model, the premises with larger geometry and larger temperature differences in the measured temperatures between the floor and the ceiling can be estimated by the method of similarity.

REFERENCES

[1] I. Stoev and V. Mutkov, "Microclimatic data collection multisensor system for design of energy model in residential buildings", 2018 20th International Symposium on Electrical Apparatus and Technologies (SIELA), Bourgas, 2018, pp. 1-3, doi: 10.1109/SIELA.2018.8447124.

[2] I. I. Stoev, S. L. Zaharieva and V. A. Mutkov, "Evaluation of Gross Errors in Measured Temperature with an Electronic System for Management of Residential Energy Systems", 2019 27th Telecommunications Forum (TELFOR), Belgrade, Serbia, 2019, pp. 1-4, doi: 10.1109/TELFOR48224.2019.8971309

[3] I. I. Stoev, S. L. Zaharieva and A. N. Borodzhieva, "An Approach for Assessment of the Synchronization Between Digital Temperature Sensors", 2019 27th Telecommunications Forum (TELFOR), Belgrade, Serbia, 2019, pp. 1-4, doi: 10.1109/TELFOR48224.2019.8971271

[4] A. Asenov, V. Pencheva and I. Georgiev. "Modelling passenger service rate at a transport hub serviced by a single urban bus route as a queueing system", IOP Conference Series: Materials Science and Engineering, 2019, No Volume 664.

[5] D. Grozev, V. Pencheva, I. Georgiev, I. Beloev, "Investigation of the operation mode at Ruse-Danube Bridge border checkpoint considered to be a mass service system with incoming flow of automobiles at a non-stationary mode of operation",/ MATEC Web of Conferences, 2018, No 234, 06003.

[6] D. Grozev, M. Milchev and I. Georgiev, "Analysis of the load on the taxi system in a medium-sized city", IOP Conference Series: Materials Science and Engineering, 2019, No Volume 664.

[7] V. Pencheva, A. Tsekov, I. Georgiev, S. Kostadinov, "Analysis and assessment of the regularity of mass urban passenger transport in the conditions of the city of Ruse", Transport Problems, 2018, No Volume 13.

Analyzing the RFID Failure Impact on Availability of IoT Services

C. Corches, I. C. Donca, O. Stan and L. Miclea
Automation Department
UTCN / Faculty of Automation and Computer Science
Cluj-Napoca, Romania
Cosmina.corches@aut.utcluj.ro

M. Daraban
Applied Electronics Department
UTCN / ETTI
Cluj-Napoca, Romania
mihai.daraban@ael.utcluj.ro

Abstract—**Internet of Things (IoT) spread would not have been possible without Cloud Computing. To have small gadgets it is needed to move the processing and storage components out of the device and into Cloud. However, having all devices connected to data centers can create traffic jams and delays in accessing process data, which can pose real problems for real-time applications. A solution was to bring the resources closer to the IoT devices, by creating intermediary layers as Fog and Edge Computing. Through this approach, delay issues for critical applications are solved, but other problems are rising instead. Adding more nodes in the communication chain can have a negative impact over reliability of an IoT device in performing its purpose. The paper presents stochastic models to analyze the availability and reliability of RFID IoT sensors when sending data to the Cloud. Through the proposed models, the components and communication topology that are having the most significant impact on the availability and reliability will be identified.**

Keywords—*dependability, reliability, availability, IoT, RFID, Edge Computing, Cloud*

I. INTRODUCTION

It is no surprise that the evolution of IoT and their spread among the users has got the interest of both industries and university researchers. The purpose is to either monitor (e.g. monitoring a person's health, an industrial work in progress process), or building context-aware based on user actions and gathered data. Even if at a first glance, having multiple devices interconnected working and acting together with the help of Cloud computing, sounds great, this can also cause tremendous problems.

The heterogeneous environment (e.g. IoT sensors, devices, Edge and Fog Computing systems from different manufacturers) represents a challenge in guaranteeing interconnection and interoperability between devices [1]. Another issue is caused by the reliability and availability of the IoT devices or IoT sensors in providing the data needed for Cloud computing. The paper presents a scenario analyzing the communication chain at the Edge layer: IoT sensor → Microcontroller → Router (Edge layer) ↔ Pepper robot.

The heterogeneous environment applies not only to the IoT devices but also to the protocols and wireless technologies used by the devices to send data. One technology that is adopted at a large scale is Radio-Frequency Identification (RFID). Passive RFID tags are relying on the electromagnetic pulse from the readers as a power source to transmit the identifier. As there is no internal battery, passive RFID tags are the most cost effective, theoretically having an unlimited life span [2]. The paper proposes an analysis on availability of IoT services because of RFID failure in the communication chain.

II. DEPENDABILITY MODELING

The dependability of a system is its ability to avoid failures, characterized by a frequency rate and a severity higher than an accepted threshold, in the provision of services [3]. In a practical situation the dependability of a system should be sufficient to characterize the "dependency" associated with the system in question. For example, the dependence of system A on system B can be interpreted as the extent to which the dependence of system A can be affected by that of system B.

According to the IEC 192-01-22 standard, the dependency property integrates the following attributes [3] [4]: availability, reliability, safety, integrity, maintainability. Regarding communication networks, the most important attributes are reliability and availability [5].

A. Reliability Block Diagram (RBD)

The dependability modeling through Reliability Block Diagram (RBD) is based on the graphical representation of systems success paths through block or modular structures [6]. The advantages of this technique are the ease with which one can implement and visually verify the success of a system. The RBD division is done in such a way as to emphasize the operating logic of the analyzed system.

In a scenario in which all the blocks structures need to be available, for the system to be functional, it is necessary to have the blocks in the RBD diagram placed in series, Fig. 1a. When the failure of a block in the diagram does not affect the operation of the system (there is at least one path from input to output), then there are redundant components, and their representation is made using parallel structures, Fig. 1b.

When it is desired to determine the availability of a system including the relationships between conditions and events, it is necessary to use Petri Nets. Initially, the analysis model based on Petri Nets did not contain the time component, but this was added later in Stochastic Petri Net (SPN).

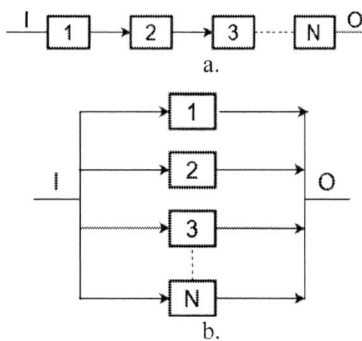

Fig. 1. RBD Representations: a. Series, b. Parallel

In SPN it is possible to model the state of the components that make up the system, so that the system's state derives from the components' state and no longer being explicitly expressed. [5]. In RBD modeling, the blocks structures represent certain components of the analyzed system. However, in SPN the events, the transitions and the states of the system are described using symbols [5] [6], TABLE I.

TABLE I.   SYMBOLS USED IN SPN

Symbol	Symbol description
⭕	Event (location) within the analyzed system
—	Instantaneous transition, propagation of the event without delay
▭	Time-conditioned transition, the event propagates after a certain time
↷	The arc interconnects events to transitions
●	The token is found in the locations, represents the data, but at the same time acts as an indicator for the state of the system.
⟋°	Inhibitor arc, realizes the interconnection of events to transitions.

The transitions are triggered if the input events meet the condition imposed during the transition [7][9]. When a transition is triggered, one token will be absorbed from each event connected to the transition input, and then tokens will be generated to the output events.

Compared to RBD diagrams, SPN can analyze the reliability and respectively monitor the failure rate, but also allow the observation of the system's dynamic behavior. This helps in tracking the fault propagation and analyzing the system in case of failure. In the paper, the analysis of the SPN proposed model was performed within the Mercury tool [9][10].

III. RFID MONITORING ARCHITECTURE

The evolution of technologies as well as of the processing realized at Cloud layer, but also within IoT devices represents the ideal context in the development of smart devices [11]. One area where there are opportunities for the development of such devices is in medical care, where intelligent sensors can monitor the condition of patients (e.g. blood pressure, blood oxygenation, body temperature, blood sugar) [8]. Although the information received from the sensors is useful in alerting

medical staff in emergency situations (e.g. high blood pressure, low oxygen levels), sometimes it is still desirable to monitor over a longer period of time and even create an overview of the patient's condition by incorporating information received from several sensors.

To achieve the level of integration mentioned above, the storage and processing exceed the resources available at the sensor level [8]. For this reason, it is necessary to transmit the information to Edge, Fog or Cloud layers.

Through advanced image processing (Computer Vision) it is possible to monitor the elderly. Computer Vision and advanced image processing are very useful in identifying the gestures made by the person concerned, but by combining them with Radio-Frequency Identification (RFID) tags, it is possible to monitor, for example, the type of drug administered [12]. Fig.2.

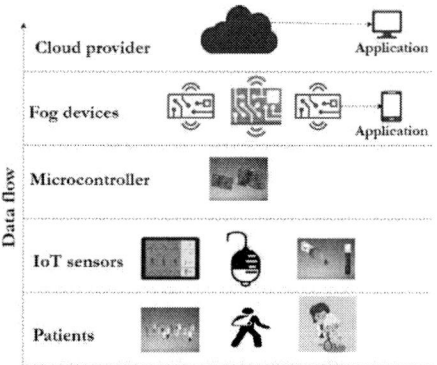

Fig. 2. Example of architecture for patient monitoring

A solution for implementing the monitoring system for the elderly is using a social humanoid robot Pepper. The robot cannot read RFID tags, so it must be connected to a development board to provide this feature. To model the dependability of the proposed system, its description was used using RBD diagrams and SPN.

A. RBD Analysis of the Proposed RFID System

Fig.3 shows the RBD diagram of the proposed system following the implementation in the Mercury tool.

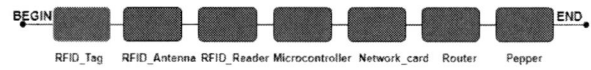

Fig. 3. RBD diagram representation for monitoring the parameters transmitted by an RFID tag via the Pepper robot

To determine the reliability of the system analyzed in Fig.3, it is necessary to know Mean Time to Failure (MTTF) parameters, respectively Mean Time to Repair (MTTR). Because passive RFID tags do not contain permanently powered electronic components, and the processes performed are mostly read only, the lifespan of these components is about years. When an RFID tag fails, it will be replaced with a new one due to the low costs. Precisely for this reason, in the literature we found information only about Mean Time Between Failures (MTBF), when it comes to the reliability of these components used according to design specifications [6] [13] [14].

The values used in RBD block modeling of the communication chain using RFID technology are those from TABLE II [8] [13] [14].

TABLE II.   NECESSARY PARAMETERS FOR RBD ANALYSIS OF THE PROPOSED SYSTEM

	MTBF	MTTF	MTTR
RFID_tag	13,148,730	13,148,725.49	5
RFID_antenna	8,765,820	8,765,815	5
RFID_reader	262,974.6	262,969.6	5
Microcontroller		44957	5
Network_card		8,765,820	5
Router		35064	4

Regarding the values used for MTTF in the case of microcontroller and router, respectively, they have been identified in the literature [8] [15] [16]. The RBD represented in Fig. 4 was used in the dependability analysis of the Pepper humanoid robot.

Fig. 4.   RBD diagram representation for Pepper robot dependability analysis

To the best of our knowledge, we could not identify values regarding MTTF and MTTR for analyzing the Pepper robot's reliability. Hence, values found in the literature regarding MTTF and MTTR were chosen for Pepper_OS, respectively Pepper_App [8][15]. Regarding Pepper_HW, the robot production started with 2014, and the model purchased by the research center in 2019 is refurbished. This led us to consider a value of 5 years for MTTF, TABLE III. In TABLE IV are presented the values obtained for MTTF and MTTR associated with the Pepper robot, following the analysis of the diagram from Fig. 4.

TABLE III.   PARAMETERS FOR PEPPER ROBOT RBD ANALYSIS

Block	MTTF	MTTR
Pepper_HW	43829	5
Pepper_OS	1440	1
Pepper_APP	6865.3	0.167

TABLE IV.   RBD ANALYSIS RESULTS FOR PEPPER ROBOT

Parameter	Value	Unit
MTTF	1158.85	hours
MTTR	0.96	hours
Availability	0.99917	%
Uptime	8758.5	hours/year
Downtime	7.29	hours/year

Following the analysis of the RBD diagram in Fig. 3 the values in TABLE V were obtained, representing the availability of the monitoring system using RFID tags.

TABLE V.   DEPENDABILITY PARAMETERS OBTAINED FROM THE ANALYSIS OF THE PROPOSED MODEL

Parameter	Value	Unit
MTTF	1089.57	hours
MTTR	1.18	hours
Availability	0.99892	%
Uptime	8756.33	hours/year
Downtime	9.48	hours/year

The sensitivity analysis aimed to identify the parameters that have the greatest impact on the availability of the modeled system, TABLE VI.

TABLE VI.   IMPACT OF COMPONENT RELIABILITY PARAMETERS ON SYSTEM AVAILABILITY

Impact Factor	Parameter	Sensitivity
Major	MTTR Pepper	-8.613×10^{-4}
	MTTR Router	-2.848×10^{-5}
	MTTR Microcontroller	-2.222×10^{-5}
Minor	MTTF RFID_antenna	6.5×10^{-14}
	MTTF Network card	6.5×10^{-14}
	MTTF RFDI Tag	2.889×10^{-14}

The variation of the parameters MTTR Pepper, MTTR Router and MTTR Microcontroller on the availability of the monitoring system with RFID tags is represented in Fig. 5, Fig. 6 and Fig. 7.

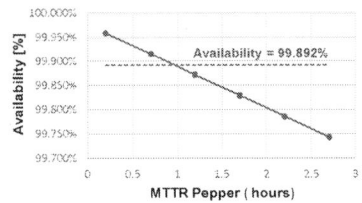

Fig. 5.   MTTR Pepper vs. Availability

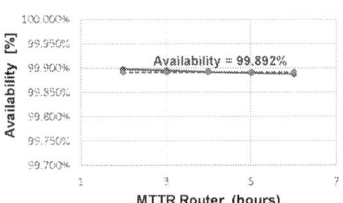

Fig. 6.   MTTR Router vs. Availability

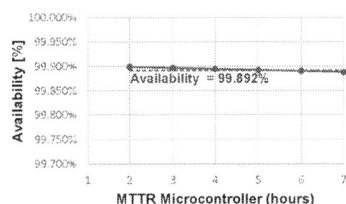

Fig. 7.   MTTR Microcontroller vs. Availability

The analysis performed through RBD diagrams considers aspects such as the behavior over time of hardware components, operating systems and applications running on them, but the proposed model neglects the communication process between RFID reader and tag/tags.

B. SPN Analysis of the Proposed RFID System

When transmitting data from RFID tag to RFID reader, their reception may be affected by electromagnetic interference, requiring data retransmission. The probability that a bit will be received correctly, when using On-Off Keying (OOK) modulation, can be described by the following equation [17]:

$$P_{bit} = 1 - \frac{1}{2} \cdot e^{-\frac{E_b}{2N_\sigma}} \qquad (1)$$

, where $-\frac{E_b}{2N_\sigma}$ represents Signal-to-Noise Ratio (SNR).

Regarding the probability of correctly receiving the unique identifier associated with a tag (UID), this can be described by the equation:

$$P_{tag\_UID} = P_{bit}{}^N \qquad (2)$$

, where N represents the number of bits used in encoding the UID. For example, in the case of Texas Instruments tags, the UID is encoded via 64 bits [18] [19].

In addition to the effect of electromagnetic interference, another situation that affects the correct reception of data from an RFID tag is the occurrence of a collision [20]. When a collision is detected by the RFID reader, the anti-collision algorithm it is launched. One such algorithm is the one proposed by the ISO15693 standard, which involves identifying RFID tags and reading their contents one by one based on the Unique Identifier (UID).

From the point of view of the algorithm for identifying the UIDs of RFID tags, the probability that a slot will be selected by a tag is 1/L [21] [22], where L represents the number of slots used within the frame to identify the tags in the vicinity of the RFID reader. Hence, the probability of a slot being selected by a certain number of tags is described by a binomial distribution [22] [23]:

$$P_{N,\frac{1}{L}}(n) = C_N^n \left(\frac{1}{L}\right)^n \left(1 - \frac{1}{L}\right)^{N-n} \qquad (3)$$

, where N is the number of tags in the vicinity of the RFID reader, and n is the number of tags that select the slot in question for transmitting the UID.

Therefore, the probability that within a UID identification frame of N tags, that a tag will occupy a certain slot is given by the equation:

$$P_{tag\_slot} = P_{N,\frac{1}{L}}(1) = N \cdot \frac{1}{L}\left(1 - \frac{1}{L}\right)^{N-1} \qquad (4)$$

Applying the principles of equation (3) and (4) it can be deduced that the probability that a slot in the frame is not selected by any tag is given by equation:

$$P_{slot\_idle} = P_{N,\frac{1}{L}}(0) = \left(1 - \frac{1}{L}\right)^N \qquad (5)$$

In conclusion, the probability of a collision occurring within a frame is given by the equation:

$$P_{collision} = 1 - P_{tag\_slot} - P_{slot\_idle} \qquad (6)$$

To model the processes described above, which occur during the identification of tag UIDs in the vicinity of the RFID reader, a model was used using SPNs within the Mercury tool.

Within the proposed model, the *Start* location represents the initial number of slots that will be analyzed. As one slot is analyzed, the *TI_0* instantaneous transition removes one token from the *Start* and produces one in the *Slot* location. The introduction of a guard expression conditions the triggering of the *TI_0* instantaneous transition:

$$(\#Slot + \#P1 + \#P2 + \#P3) = 0 \qquad (7)$$

In the proposed model the probability that a frame slot is in one of the available states (i.e. single tag (*P1*), idle (*P2*), collision (*P3*)) is modeled through instantaneous transitions *T_Tag*, *T_Idle*, *T_Coll*. The values obtained by means of equations (4), (5) and (6) were applied to the mentioned instantaneous transitions as weight parameters. For example, if we assume that there are 3 tags near the RFID reader, then the probability that a certain slot in a frame of 16 slots will be chosen by a single tag is 16.48%, a value that has been associated in the form of weight to *T_Tag* transition. As for the values given for the other two transitions (i.e. *T_Idle*, *T_Coll*) those are: 82.40% and 1.12%.

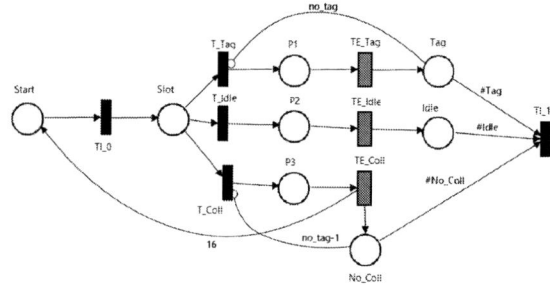

Fig. 8. Modeling the process of identifying UID tags via SPN

An analysis of the proposed model shows that the triggering of the *T_Tag* or *T_Coll* transitions is conditioned by the presence of a slot for analysis (token in the *Slot* location), but also by an inhibitor arc that has the following meanings depending on the transition:

- *T_Tag* transition: because the weight associated with the arc is identical to the number of tags analyzed in the system (*no_tag*), thus modeling the maximum number of slots occupied by tags during the query;

- *T_Coll* transition: by means of the inhibitor arc the number of collisions solved is limited to *no_tag* − 1. This limitation assumes that at least one tag is identified after each anti-collision process.

In addition to the use of the weight and trigger conditions mentioned above, the monitoring expressions described in TABLE VII were also used.

By means of time-conditioned deterministic transitions (*TE_Tag*, *TE_Idle*, *TE_Coll*) the time required for slot processing is modeled. Once a tag is associated with a slot, it should be possible to receive the bits associated with its UID. According to equations (1) and (2) depending on the SNR value, the probability of erroneous reception of the UID code associated with the tag can be determined. In the proposed model it was considered a probability of 0.001% [24] that a bit would be erroneous during reception, Fig. 9.

TABLE VII. INSTANTANEOUS TRANSITIONS MONITORING EXPRESSIONS: *T_IDLE, T_COLL*

Transition	Purpose	Monitoring expression
T_Idle	The transition can only be activated if the number of slots left is greater than the number of tags to be identified	$(\#Start + \#Slot)$ $> (no\_tag - \#Tag)$
T_Coll	Prevents triggering if only one more tag needs to be identified	$(\#Tag)$ $< (no\_tag - 1)$

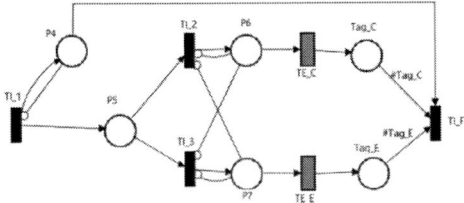

Fig. 9. Modeling the erroneous reception of the tag UID's bits

The conflict between the instantaneous transitions *TI_2* (correct reception) and *TI_3* (erroneous reception) was solved by allocating the weight according to equations (1) and (2): 0.937975 and 0.062025. In location *Tag_C* will be found after processing the tokens corresponding to the correctly identified tags and in location *Tag_E* the tokens associated with the tags whose UID was received incorrectly. To analyze the dependability of the entire system by SPNs, the proposed model from Fig. 10 was analyzed in the Mercury tool.

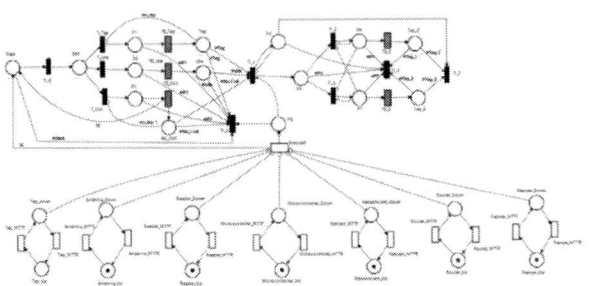

Fig. 10. Modeling the system for reading UID tags via SPNs

To model the operation of the components of the analyzed system, the values from TABLE II and TABLE IV were used in the time-conditioned MTTF and MTTR (associated to each component) time transitions. These parameters were modeled as in the case of RBD analysis by means of negative exponential distributions. The identification process of tags is done through the *Request* transition, but only when all components in the circuit are functional. The average trigger duration of the *Request* transition has been set to 10 minutes. The transition will not be triggered if the tag identification process is in progress, which is signaled by the following guard expression:

$$(\#Start + \#P5 + \#P6 + \#P7) = 0 \qquad (8)$$

Within the physical system, the failure of one of the components of the model during the process of identifying and reading the UIDs associated with the tags causes the loss of information and of the communication process. One solution for modeling the situation described above is to introduce the instantaneous transitions *TI_4* and *TI_5*. These two transitions will absorb all the tokens from the locations of the structures that model the identification (Fig. 8) and modeling of the bit reception (Fig. 9) associated with the UIDs. To condition the triggering of the aforementioned instantaneous transitions only in case of failure of the system components, the following monitoring expression has been introduced:

$$((\#Tag\_down + \#Antenna\_Down + \#Reader\_Down +$$
$$\#Microcontroller\_Down + \#Networkcard\_Down +$$
$$\#Router\_Down + \#Pepper\_Down) >= 1) \qquad (9)$$

IV. RESULTS

The Fig.11 represents the values of the availability of the analyzed system through SPNS.

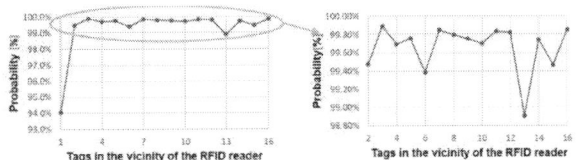

Fig. 11. Availability of the analyzed system depending on the number of RFID tags

Unless there is a single tag in the vicinity of the RFID reader, in all other cases the availability of the system is on average over 99.8%, comparable to the result obtained from the analysis of the RBD diagram. The following metric was used to determine availability:

$$P\{(Tag\_C > 0)\} \qquad (9)$$

The major difference between SPN and RBD is the modeling of the erroneous reception of the bits associated with a tag, which determines that the availability of the system decreases to 94% when we have a single tag in the system. In Fig.12 the availability of the system according to the following metrics is represented:

$$P\{(Tag\_C > (no\_tag - 1))\} \qquad (10)$$

$$P\{(Tag\_C > (no\_tag - 2))\} \qquad (11)$$

Analyzing the results of Fig. 12 we can observe that it is indicated to have a number of tags which represents half of the number of slots in a frame in the close vicinity of an RFID reader. This approach would obtain only half of the available data rate. However, this allows a better control in terms of the duration of identification of UIDs through the anti-collision process, and the probability of correct reception of all tags in a single query is 59.33%.

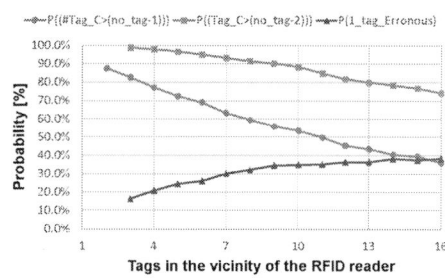

Fig. 12. Availability of the system analyzed according to metrics (10) and (11)

In the proposed configuration there is a 32.06% probability that one tag will be erroneous out of the eight. If the number of tags is equal to the number of slots, then the probability that one tag is erroneous (38.20%) is higher than the probability that all 16 tags will be read correctly in a single query (36 %).

V. CONCLUSIONS

The paper presents a scenario analyzing the communication chain at the Edge layer: IoT sensor → Microcontroller → Router (Edge layer) ↔ Pepper robot. The communication chain was described and analyzed through RBD and SPNs using Mercury tool. The SPNs model describes the communication processes that take place during anti-collision algorithm and the tag UID error probability. Through the performed analysis, the optimal tag number versus available slots in a RFID reader was determined. Even if the maximum data throughput is obtained when the number of tags is equal to the number of slots, the probability of encounter errors during tag reading and also the delays needed to resolve the anti-collision between tags UID can affect the system. Therefore, it is needed to find a compromise between tags number and available slots to have a better control over system timings. In future work we plan to consider the probability of not having a proper alignment between RFID reader antenna and the RFID tags and to analyze the Edge-→Fog→Cloud layer communication chain.

ACKNOWLEDGMENT

This research was funded by the Ministry of Education and Research of Romania, section CCCDI - UEFISCDI, grant number PN-III-P1-1.2-PCCDI-2017-0734 / ROBIN - "Roboții și Societatea: Sisteme Cognitive pentru Roboți Personali și Vehicule Autonome" (Robots and Society: Cognitive Systems for Personal Robots and Autonomous Vehicles) within PNCDI III.

REFERENCES

[1] G. Mujica, R. Rodriguez-Zurrunero, M. R. Wilby, J. Portilla, A. B. Rodríguez González, A. Araujo, T. Riesgo and J. . J. Vinagre Díaz, "Edge and Fog Computing Platform for Data Fusion of Complex Heterogeneous Sensors," *Sensors*, vol. 18, no. 11:3630, pp. 1-26, 2018.

[2] K. Ding and P. Jiang, "RFID-based production data analysis in an IoT-enabled smart job-shop," *IEEE/CAA Journal of Automatica Sinica*, vol. 5, no. 1, pp. 128-138, 2018.

[3] A. Avizienis, J.-C. Laprie, B. Randell and C. Landwehr, "Basic concepts and taxonomy of dependable and secure computing," *IEEE Transactions on Dependable and Secure Computing*, vol. 1, no. 1, pp. 11-33, January-March 2004.

[4] I. E. C. (IEC), "http://www.electropedia.org," International Electrotechnical Commission (IEC), February 2015. [Online]. Available:

http://www.electropedia.org/iev/iev.nsf/display?openform&ievref=192-01-22. [Accessed 28 Aprile 2020].

[5] W. Ahmad, O. Hasan, U. Pervez and J. Qadir, "Reliability modeling and analysis of communication networks," *Journal of Network and Computer Applications*, vol. 78, pp. 191-215, 2017.

[6] A. K. Verma, S. Ajit and D. R. Karanki, Reliability and Safety Engineering, 2 ed., London: Springer-Verlag London, 2016.

[7] P. D. T. O'Connor and A. Kleyner, Practical Reliability Engineering, vol. Fifth Edition, United Kingdom: John Wiley & Sons, 2012.

[8] G. L. Santos, P. T. Endo, G. Gonçalves, D. Rosendo, D. Gomes, J. Kelner, D. Sadok and M. Mahloo, "Analyzing the IT subsystem failure impact on availability of cloud services," in *2017 IEEE Symposium on Computers and Communications (ISCC)*, Heraklion, Greece, 3-6 July 2017.

[9] B. Silva, R. Matos, G. Callou, J. Figueiredo, D. Oliveira, J. Ferreira, J. Dantas, A. Lobo, V. Alves and P. Maciel, "Mercury: An Integrated Environment for Performance and Dependability Evaluation of General Systems," in *45th Annual IEEE/IFIP International Conference on Dependable Systems and Networks (DSN 2015)*, Rio de Janeiro, Brazil, 2015.

[10] V. Lira, E. Tavares, S. Fernandes, P. Maciel and R. M. Silva, "Virtual Network Resource Allocation Considering Dependability Issues," in *2013 IEEE 16th International Conference on Computational Science and Engineering*, Sydney, NSW, Australia, 2013.

[11] H. Bangui, S. Rakrak, S. Raghay and B. Buhnova, "Moving to the Edge-Cloud-of-Things: Recent Advances and Future Research Directions," *Electronics*, vol. 7, pp. 1-31, 2018.

[12] F. M. Hasanuzzaman, X. Yang, Y. Tian, Q. Liu and E. Capezuti, "Monitoring activity of taking medicine by incorporating RFID and video analysis," *Network Modeling Analysis in Health Informatics and Bioinformatics*, vol. 2, no. 2, pp. 61-70, 07 February 2013.

[13] C.-T. Huang, S.-J. Wang, W.-L. Wang and Y.-S. Wang, "Construction of an Online RFID Enabled Supply Chain System Reliability Monitoring Model," in *2012 International Symposium on Computer, Consumer and Control*, Taichung, Taiwan, 2012.

[14] Siemens, "RFID Systems SIMATIC RF600," Siemens AG, NÜRNBERG, Germany, 2016.

[15] G. L. Santos, P. T. Endo, M. F. Ferreira da Silva Lisboa Tigre, L. G. Ferreira da Silva, D. Sadok, J. Kelner and T. Lynn, "Analyzing the availability and performance of an e-health system integrated with edge, fog and cloud infrastructures," *Journal of Cloud Computing: Advances, Systems and Applications*, vol. 7, no. 16, pp. 1-22, 2018.

[16] M. A. d. Q. Lima, P. R. Maciel, B. Silva and A. P. Guimarães, "Performability evaluation of emergency call center," *Performance Evaluation*, vol. 80, pp. 27-42, 2014.

[17] W. Su, K. M. Beilke and T. T. Ha, "A Reliability Study of RFID Technology in a Fading Channel," in *2007 Conference Record of the Forty-First Asilomar Conference on Signals, Systems and Computers*, Pacific Grove, CA, USA, 2007.

[18] Texas Instruments, "RI-I17-112A-03 Tag-it™ HF-I Plus Transponder Inlays," Texas Instruments Incorporated, 2014.

[19] Texas Instruments, "TRF7960EVM ISO15693 Host Commands," Texas Instruments Incorporated, 2008.

[20] C. Corches, O. Stan, L. Miclea and M. Daraban, "Embedded RTOS for a Smart RFID Reader," in *2019 IEEE 25th International Symposium for Design and Technology in Electronic Packaging (SIITME)*, Cluj-Napoca, Romania, 2019.

[21] J.-R. Cha and J.-H. Kim, "Novel Anti-collision Algorithms for Fast Object Identification in RFID System," in *11th International Conference on Parallel and Distributed Systems (ICPADS'05)*, Fukuoka, Japan, 2005.

[22] Z. Qu, X. Sun, X. Chen and S. Yuan, "A novel RFID multi-tag anti-collision protocol for dynamic vehicle identification," *PLOS ONE*, vol. 14, no. 7, pp. 1-25, 05 July 2019.

[23] J. Park, M. Y. Chung and T.-j. Lee, "Identification of RFID Tags in Framed-Slotted ALOHA with Robust Estimation and Binary Selection," *IEEE Communications Letters*, vol. 11, no. 5, pp. 452-454, May 2007.

[24] D. Jiang, N. Mei, D. Liu and Z. Zhang, "System study on direct sequence spectrum of radio-frequency identification," *Journal of Communications Technology and Electronics*, vol. 59, no. 11, pp. 1200-1205, 201

A Metamodel Residual-based Stopping Criterion for Adaptive Verification of Integrated Circuits

Ingrid Kovacs[1], Marina Țopa[1]
[1]Technical University of Cluj-Napoca, Romania
ingrid.kovacs@staff.utcluj.ro

Monica Ene[2], Andi Buzo[3], Georg Pelz[3]
[2]Infineon Technologies, Bucharest, Romania
[3]Infineon Technologies AG, Neubiberg, Germany

Abstract— **Adaptive verification is gaining increasing use as solution to overcome the coverage problem of complex integrated circuits' verification. As its name suggests, it places more points in regions of interest by learning the information from previous data, offering great potential to the engineer to understand the behavior of the system under study and to identify interesting regions in the design space. Since it is an iterative task, it should be stopped at the point that is optimum or near-optimum. This paper proposes a stopping criterion for planning adaptive experiments (simulations/measurements) for the identification of extreme regions (minimum/maximum) of the response. It is based on the concept of metamodeling and helps to efficiently characterize the targeted extreme region of complex integrated circuits with the advantage of low experimental effort. The approach was evaluated on synthetic test functions and an analog integrated circuit. The stopping criterion shows substantial improvements in the number of simulations/measurements while maintaining high resolution in the region of interest.**

Keywords— *adaptive sampling, integrated circuit verification, stop criterion*

I. INTRODUCTION

Computer simulation models of modern electronic applications are gaining increasing use for engineering problems such as verification space exploration, global optimization, sensitivity analysis or metamodeling (also known as surrogate models or response surface models) [1-2]. The verification of integrated circuits using simulation models is becoming more and more complex, time-consuming and expensive. A proper coverage of the verification space would require thousands or even millions of simulations, which is often not affordable.

To handle the complexity, experiment controllers based methods are used, also known as Design of Experiment (DoE) [2]. It is found that the choice of the observed samples by DoE is crucial and at the same time beneficial for further applications. For example, a high prediction quality of a metamodel can help in convergence speedup in metamodel-based optimization or accurate identification of influential factors at the sensitivity analysis. Thus, DoEs that deal with how to perform experiments for a better understanding of a given phenomenon with as few samples (simulations/measurements) as possible have been intensively studied [3-6].

In general, the DoEs can be classified into two categories [6]: *one-shot sampling* and *sequential sampling* (also known as *adaptive sampling* or *active learning*). *One-shot sampling* approaches determine the sample size and experiment runs at the beginning of the experiment. Examples are Monte Carlo, full- and fractional factorials, D-optimal, Taguchi Orthogonal Arrays [2], [5]. However, their disadvantage is that with no prior knowledge, it is almost impossible to determine an optimal and suitable sample size for a specific engineering problem. Moreover, the *one-shot sampling* approaches are model-independent, meaning that the determination of sample points is independent of the model output. Thus, the sampling method does not gain benefits from any new findings from a subsequent experiment process.

Hence, flexible *sequential sampling* approaches have been introduced, which sequentially add new experiment runs (simulations/measurements) using the information from previous iterations. By doing so, one can obtain better resolution at the region of interest with fewer experiments, because the sampling process is able to adapt to the properties of the target function or the region of interest [3], [6-7]. Depending on the application, the region of interest can be a targeted extreme region [8], high uncertainty region [9], border search using support vector machines (SVMs) [10], failure region characterization using metamodels [11] etc.

Some *sequential sampling* methods show, however, a limitation. The choice of the optimal number of experiment runs depends on the application and is difficult to be known in advance. An arbitrary choice of the number of experiments to run can lead to a too high number of simulations/measurements, with no new information brought by the last few ones, i.e. wasted experiment runs. Thus, stopping criteria have been proposed for some adaptive sampling methods [9], [12].

For verification space characterization, the *metamodel find extreme* adaptive sampling technique was proposed in [8]. Its scope is to find the targeted extreme region (minimum/maximum) of a response function using the concept of metamodeling [2]. However, its limitation is the lack of a stopping criterion.

In this paper, we propose a stopping criterion for the *metamodel find extreme* adaptive sampling technique that overcomes this limitation. It is based on the learning of the metamodel, quantified in an information metric. This measures how much the metamodel and the targeted extreme change from one iteration to the other and stops the sampling process when the targeted extreme and the residuals among the last *window size* iterations do not change anymore. The concept is applied to a set of 200 synthetic test functions of different shapes and types.

Results show improvements in both the coverage and the number of simulations required. In addition, the concept is applied at lab verification of an analog integrated circuit and proves its advantages.

The paper is organized as follows: Section II describes the proposed approach of stopping criterion, followed by the description of the results in Section III. Conclusions are drawn in Section IV.

II. APPROACH DESCRIPTION

To accomplish the proper coverage and to increase the probability of detecting the targeted extreme region with as few experiment runs as possible, the proposed adaptive verification approach combines the sequence of steps illustrated in Fig. 1. For each sample, the model is simulated and it is assessed for coverage. Afterwards, the results are fed to the stop sampling metric. If the targeted extreme region is reached, the adaptive sampling process stops. The responses are handed to the post-processing phase, where visualizations are done or subsequent engineering problems can be treated, e.g. building accurate global metamodels.

A. Metamodel find extreme

The steps of the *metamodel find extreme* adaptive sampling approach are illustrated in Fig. 2 and basically consist of a metamodeling technique and a selection of the next sample.

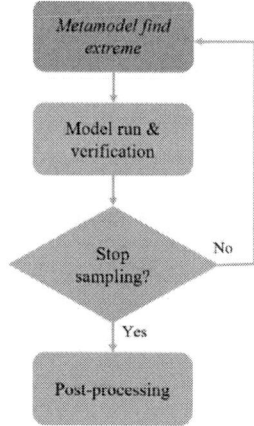

Fig. 1 Proposed adaptive verification plan

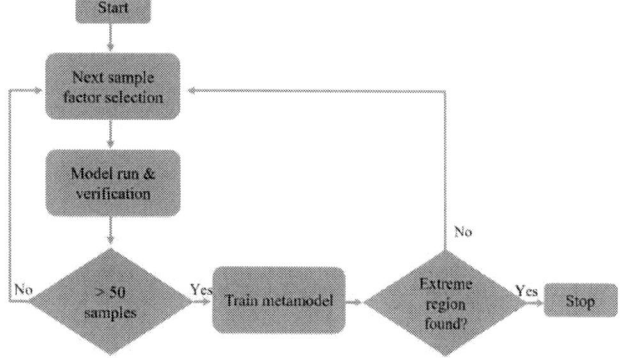

Fig. 2 *Metamodel find extreme* methodology steps

A.1. Train metamodel: To find the targeted extreme region, a metamodeling technique is used after each new added sample. Therefore, the given number of input factor configurations with the simulated response values are used to train a metamodel [13] and to search for the targeted extreme region. The metamodel's complexity (polynomial: linear, quadratic, cubic, Gaussian process, etc.) depends on the test-case and chosen factors [13]. For example, if a polynomial regression metamodel is used, its general form is described as:

$$y(f) = c_0 + \sum_{o=1}^{m} \sum_{i=1}^{n} c_i^{(o)} \cdot f_i^{o} + \sum_{j=1}^{n-1} \sum_{k=j+1}^{n} c_{jk} \cdot f_j \cdot f_k \qquad (1)$$

where f is the n-dimensional vector of input factors, c are the coefficients of the metamodel and represent the effects of factors on the response, m is the order of the polynomial. With each experiment run (simulation/measurement), the prediction of the metamodel gets better. Note that the first training of the metamodel is done after 50 runs.

A.2. Next sample factor selection: experiment runs which are closer to the targeted extreme region are of more interest than those which are further away. To select the next experiment run from the input factor variation space with a higher probability to be close to the targeted extreme region, a validation set of 10000 random samples is taken from the input factor space. The sample size of this validation set was chosen empirically based on the observation that a higher sample size does not increase the model accuracy. Then, the metamodel is used to predict the values of the response on the validation set. All points from the validation set are getting a weight that is correlated to the prediction value. For example, if *targeted extreme=max*, predictions with higher values will get higher weights. Finally, the next sample is chosen accounting for the weight of each sample. Therefore, the probability of the samples from the targeted extreme region is increased.

In order to visually assess the advantages of the adaptive sampling process performed by the *metamodel find extreme* method over the *one-shot sampling* approaches, we used the MATLAB peaks function (two factors X_1 and X_2 and one response) and performed 500 runs with each sampling approach. Fig. 3 a) illustrates the resulting response surface using a uniform Monte Carlo DoE, while Fig. 3 b) presents only the 2D projection on the factors space. Note the random spread of the samples, independent of the model output.

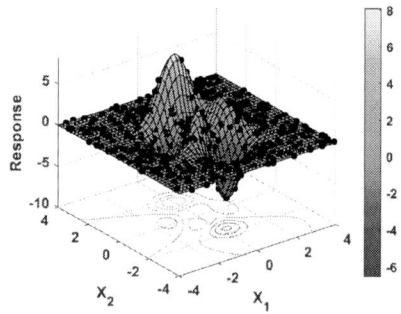

Fig. 3 a) Illustration of the response surface using *one-shot sampling*

2020 IEEE 26th International Symposium for Design and Technology in Electronic Packaging (SIITME)

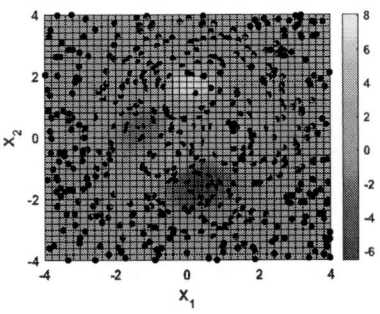

Fig. 4 b) 2D projection on the factor space (*one-shot sampling*)

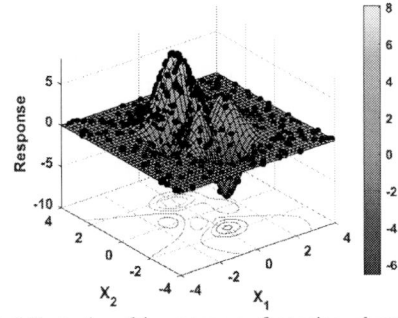

Fig. 5 a) Illustration of the response surface using *adaptive sampling*

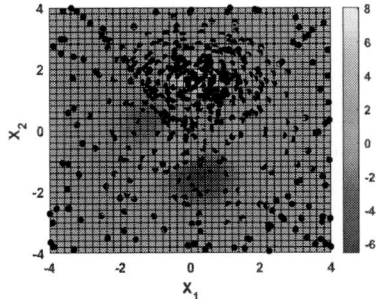

Fig. 6 b) 2D projection on the factor space (*adaptive sampling*)

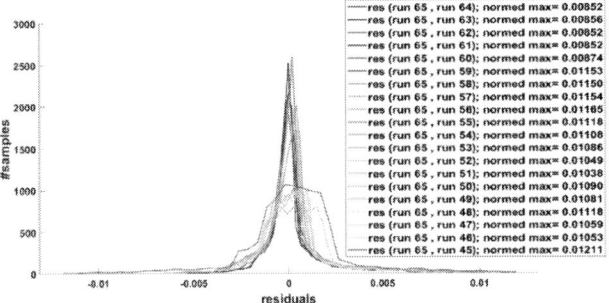

Fig. 7 Illustration of the learning process for the stop criterion

Fig. 4 a) illustrates the resulting response surface using the *adaptive sampling* approach (*targeted extreme=max*). Note how the algorithm places more samples in the targeted extreme region. In the 2D view from Fig. 4 b) one can notice how the samples are concentrated in the maximum region.

The stop criterion of the *adaptive sampling* scheme comes into play now, as it would be of great interest to stop the sampling as soon as no new information is added by new samples, instead of running an a-priori defined number of experiment runs.

B. Stop sampling

After the model run and verification phase, stop sampling checks if more experiment runs are needed. To decide when to stop sampling, we need to define how much is learned about the targeted extreme region. To quantify this, we use the learning of the metamodels built on the last *window size* runs. If the metamodels do not change over the last *window size* runs, new samples will not bring new information, thus the sampling can stop.

We quantify the learning of the metamodel in run N by computing the distances of its prediction with the predictions of the metamodels from runs $N-1$, ..., $N-window size$. These distances are also called residuals [13-14] and their normed value is computed using (2):

$$res = \frac{|\hat{y} - y|}{max(y) - min(y)} \qquad (2)$$

where $\hat{y}$ is the predicted response in runs $N-1$, ... , $N-window$ size and y is the predicted response in run N.

The threshold to use for concluding about residuals depends on the experimental phase. 1% was found a good threshold for the mean of absolute residuals, while 1% to 5% is chosen for the absolute maximum [13]. The condition to stop is to have a decreasing trend in the absolute maximum residuals over the last *window size* runs, with *threshold* = 5%. The learning process of the metamodels is illustrated in Fig. 5, with current run N=65 and *window size*=20.

Thus, the residuals are computed between the metamodel from this run and *window size* previous runs. Note the tightening trend of the residuals' distribution and maximum residuals value from older metamodels (dark red) to recent metamodels (light blue).

III. RESULTS

First, the proposed stopping criterion for adaptive sampling was tested on 200 synthetic functions including two factors (X_1 and X_2). The functions were of different nature: with/without noise, monotonic/ non-monotonic, including local minima/maxima, etc. Fig. 6 illustrates the values of the first synthetic test function with 10^6 uniform Monte Carlo runs. The factor-response function contains an exponential term, several logarithmic and quadratic terms and some random Gaussian noise.

The next step was to apply the adaptive sampling approach with the stopping criterion and to inspect to which extent the targeted extreme region was detected. Fig. 7 shows the results for the *targeted extreme=max* in green dots, while the light gray surface is plotted only for a better understanding of the response surface. Note that the targeted extreme region was already detected after 72 experiment runs when the adaptive sampling stopped. For comparison purposes, Fig. 8 shows the information from the next extra 50 samples after the stop (runs=73:122, blue dots) and a rest of maximum 900 samples (runs=123:900, red dots). For most applications, it is difficult to know in advance the number of experiments, because it is strongly dependent on the application. For this specific test function, the arbitrary choice of experiment runs (N=900) would have led to a too high number of simulations to detect the targeted extreme region, with most of the samples wasted.

Fig. 9 illustrates the values of the second synthetic test function with two factors (X_1 and X_2) with 10^6 uniform Monte Carlo simulation runs. Fig. 10 depicts the result of the adaptive sampling approach with and without the stop criterion. Again, the adaptive sampling algorithm stopped when the optimum number of experiments was reached.

Then, the adaptive sampling methodology with the stop criterion was applied also on experimental data, i.e. at the lab verification of an analog integrated circuit. The considered circuit, its input factors and analyzed response cannot be disclosed in this paper (due to confidentiality reasons of a binding NDA). However, it is worth mentioning that the methodology is applicable to almost any circuit that involves continuous variables. The system under study included three factors (X_1, X_2 and X_3) and one response. The targeted extreme of the response was the maximum. Similar to the synthetic test function case, adaptive sampling was run with and without the stopping criterion. Fig. 11 presents the results. As a 3D visualization of the factor-response surface is not possible, the scatter plots of each factor vs response were inspected and grouped as function of three categories: the samples until the stop criterion was met (group 1 – green, run=181), the next extra 50 samples after the stop (group 2 – blue, runs=182:231), and a rest of maximum 900 samples (group 3 – red, runs=232:1000).

For visualization purposes, Fig. 12 illustrates the factor-response space using only two factors and the response, with the same three categories of experiment runs.

Notice that the algorithm succeeded in detecting the maximum region prior to stopping the adaptive sampling approach. The new samples (runs 182:1000) were considered only for validation purposes; they do not bring additional information. Thus, using the stopping criterion, the adaptive sampling stopped at the point that is optimum, saving approximately 80% of the experimental effort, which translates into lower experimental time and costs.

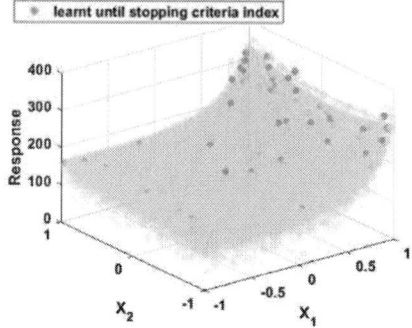

Fig. 9 Illustration of adaptive sampling with stop criterion on synthetic test function 1

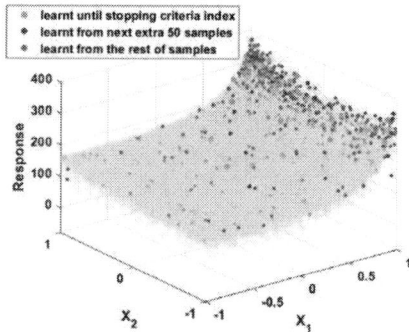

Fig. 10 Illustration of adaptive sampling with stop criterion vs. without stop criterion on synthetic test function 1

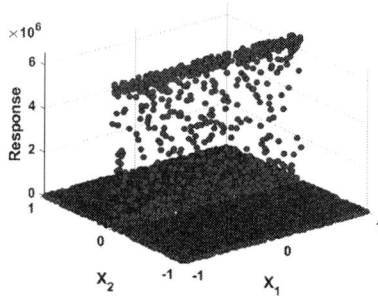

Fig. 11 Illustration of the factor-response space for synthetic test function 2

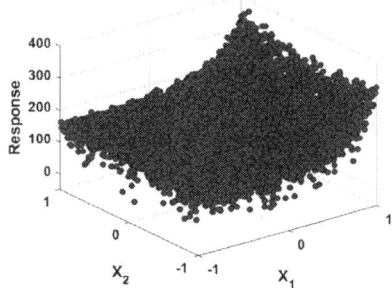

Fig. 8 Illustration of the factor-response space for synthetic test function 1

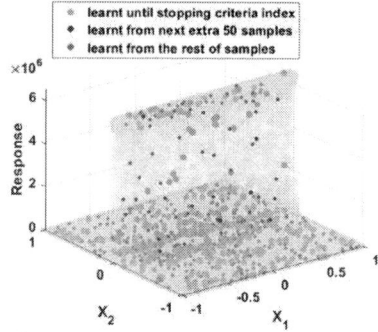

Fig. 12 Illustration of adaptive sampling with stop criterion vs. without stop criterion on synthetic test function 2

2020 IEEE 26th International Symposium for Design and Technology in Electronic Packaging (SIITME)

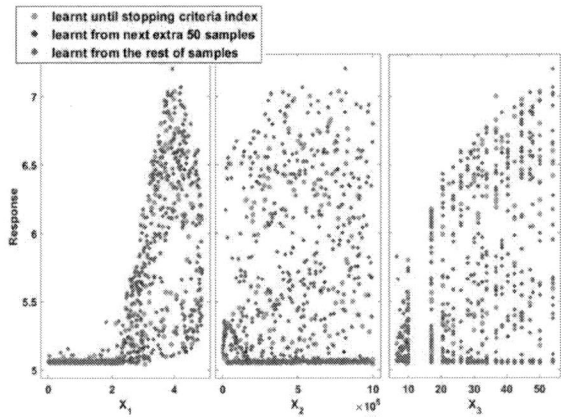

Fig. 13 Illustration of adaptive sampling with stop criterion on IC verification in lab

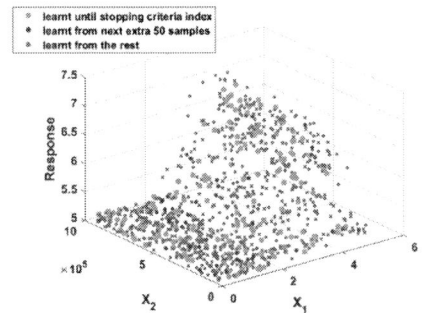

Fig. 14 Illustration of adaptive sampling with stop criterion on IC verification in lab (2 factors vs. response)

IV. CONCLUSIONS

The paper proposes an approach to determine the optimal point when to stop experiment runs (simulations/measurements) of the *metamodel find extreme* adaptive sampling technique used for characterization purposes and coverage assessment at early verification of analog integrated circuits.

The stopping criterion is based on the learning of the metamodel and uses the information of residuals to detect the point when new experiment runs do not bring useful information. Thus, the running of unnecessary simulations/measurements is avoided.

The approach was tested on several synthetic functions and at the lab verification of a real-life analog integrated circuit. In all cases, the stopping criterion proved its efficiency. More precisely, at the use case from the lab, the stopping criterion saved 80% of the experiment runs which would have been run without the stopping criterion setting. This means a significant reduction of the experimental time and cost.

ACKNOWLEDGMENT

This work was co-funded by the European Regional Development Fund through the Operational Program "Competitiveness" POC -A1.2.3-G-2015, project P_40_437, contract 19/01.09.2016.

The author would like to pay special regards to Andrei Sandu from Infineon Technologies for providing his synthetic function generator tool.

REFERENCES

[1] L. Wang, S. Shan, GG. Wang, "Mode-pursuing sampling method for global optimization on expensive black-box functions", Engineering Optimization, vol. 36, no. 4, pp.419-438, 2004.

[2] Douglas C. Montgomery, Design and Analysis of Experiments, John Wiley & Sons, 2017.

[3] H. Liu, Y. S. Ong, J. Cai, "A survey of adaptive sampling for global metamodeling in support of simulation-based complex engineering design", Structural and Multidisciplinary Optimization, vol. 57, no. 1, pp. 393-416, 2018.

[4] K. Crombecq, "Surrogate modelling of computer experiments with sequential experimental design", PhD thesis, University of Antwerp, 201

[5] S. Garuda, I. Karimia, M. Kraft, "Design of computer experiments: A review", Computers & Chemical Engineering, vol. 106, pp. 71-95, 2017.

[6] D. Gorissen, K. Crombecq, I. Couckuyt, T. Dhaene, P. Demeester, "A Surrogate Modeling and Adaptive Sampling Toolbox for Computer Based Design", Journal of Machine Learning Research, vol. 11, pp. 2051-2055, July 2010.

[7] K. Crombecq, D. Gorissen, D. Deschrijver, and T. Dhaene, "A novel hybrid sequential design strategy for global surrogate modelling of computer experiments", SIAM Journal of Scientific Computing, vol. 33, no. 4, 2011.

[8] M. Dobler, et al., "Rapid Design Space Exploration of a State-of-the-art PSI 5 Controller", MBMV, pp. 3-12, 2013.

[9] I. Kovacs, M. Topa, M. Ene, A. Buzo, G. Pelz, " A Jackknife Variance-based Stopping Criterion for Adaptive Verification of Integrated Circuits", 43rd International Spring Seminar on Electronics Technology (ISSE), Demanovska Valley, Slovakia, pp. 1-6, 2020.

[10] M. Dobler, M. Harrant, M. Rafaila, G. Pelz, W. Rosenstiel and M. Bogdan, "Bordersearch: An Adaptive Identification of Failure Regions", Design, Automation & Test in Europe Conference & Exhibition (DATE), Grenoble, pp. 1036-1041, 2015.

[11] I. Kovacs, M. Topa, M. Ene, A. Buzo, G. Pelz, "A metamodel-based adaptive sampling approach for efficient failure region characterization of integrated circuits", accepted to IEEE DCIS 2020.

[12] J. Stricker, B. Koeppl, A. Buzo, J. Kirscher, L. Maurer and G. Pelz, "Efficient Simulative Pass/Fail Characterization Applied to Automotive Power Steering", First International Conference on Artificial Intelligence for Industries (AI4I), Laguna Hills, CA, USA, pp. 69-72, 2018.

[13] M. Rafaila, "Planning experiments for the validation of Electronic Control Units", Doctoral Thesis, Faculty of Electrical Engineering and Information Technology, Vienna, 2010.

[14] NIST/ SEMATECH, The e-Handbook of National Institute of Standards and Technology (NIST), http://www.itl.nist.gov/div898/handbook/NIST

Prediction algorithms using specialized software tools for steel industry equipment

E. Raducan
Department of Automation and
Electrical Engineering
"Dunarea de Jos" University of
Galati, Romania
elena.raducan@ugal.ro

V. Nicolau, M. Andrei, G. Petrea
Department of Electronics and
Telecommunications
"Dunarea de Jos" University of
Galati, Romania
viorel.nicolau@ugal.ro

G.M. Vlej
Department of Automation-
Digitalization
Liberty Steel Group
Galati, Romania
elena.raducan@ugal.ro

Abstract—The paper aims to present a model for predictive maintenance applicable in steel industry for critic equipment. This paper investigates the application of multi-step time prediction to sustain the turbo blowers (TB) equipment prognostics using software analytics algorithms which describe forecasting models, statistical approach or the formulas specifies for industrial equipment developed based by DAX formulas. This application represent a new method for realize a predictive maintenance describe as an industrial revolution characterized by smart systems and Internet-based solutions.

Keywords—Turbo blowers, maintenance, predictive models, machine learning, steel industry

I. INTRODUCTION

The application of prediction for industrial equipment become a vital function to orderly deliver of failure information in advanced to the maintenance equip to have time to adjust production line flow and prepare the maintenance action. Generally, prediction can be divided into three principal approaches: model and experience based, and data driven. For this, one of the solutions to forecast the failure of the industrial equipment is by applying the time series method. This technique has become significantly approved in condition-based of predictive maintenance [1].

Maintenance activities combine technical, administrative, and managerial actions during the lifetime of an equipment to preserve it in the standard-state conditions. Maintenance strategies could be categorized in corrective maintenance (CM), preventive maintenance (PM), and predictive maintenance (PdM) [2]. Generally, in the industry, it always has been an increased competition in cost reduction, which is linked with the production availability and the status of the equipment. The key goal of prognostic is to reduce cost of maintenance and downtime under the premise of zero failure manufacturing, achieving the self-aware, self-predict and self-maintain abilities.

Predictive maintenance and prediction, through the availability of data from industrial processes, could optimize spare parts usage, increase equipment lifetime, reduce plant number of accidents. Yet, with all this benefits, creating a model that considers all the conditions, integrates all measurements from the field, reduces uncertainty in diagnosis and computes agile decision-making is very challenging [3].

The furnaces represent the equipment most important from steel industry and it produce hot metal using coke and sinter as main raw materials. The sinter is made by iron ore and coke from metallurgical coal in the coke plant. [4].

In sintering process, the human operator is constantly monitoring both the machine and the process. In a non-automated system, the determination of combustion points is another important aspect for the operator, but the decision based on operator observation is usually sub-optimal [5]. In steel and ferrous industry, where sintered ore is the main raw material for the blast, the stability and quality of sintered material are important factors in the efficiency of furnace production [6]. The sintering process is multivariate, non-linear and complex, with parameters that vary over time.

In blast furnace, the liquid iron is produced by the reduction of ore with reducing gases. The reduction is obtained by the reaction of oxygen with coke and coal and this oxygen is a component of enriched hot air blast which is distributed at the bottom of the furnace through the blowpipes. The oxygen with air composes the volume necessary for the process in blast furnace which is provided by the air blowers [7]. These turbo blowers (TB) take the air from atmosphere and compress to the required pressure. The turbo blowers are the first equipment in the air blast system and this equipment is named "critical" because the air blower is required to be managed to meet the air flow limits which in turn are determined by the operating limits of the blast furnace.

The goal of the paper focuses on models for predictive maintenance applicable in steel industry for critic equipment, like a it blast system. The approach investigates the application of multi-step time prediction to sustain the turbo blowers equipment prognostics using software analytics algorithms. Forecasting models and statistical approach are based on Data Analysis Expressions (DAX) software application formulas. DAX is a library of functions and operators that can be combined to build formulas and expressions in Power BI, Analysis Services, and Power Pivot in Excel data models.

The paper is organized as follows. Models of air blast system and turbo blowers equipment in furnaces control are studied in section 2. In section 3, predictive analytics tools are presented. Experimental results are pointed out in section 4. Conclusions are presented in section 5.

II. AIR BLAST SYSTEM MODELS

In blast furnace, turbo blowers take the air from atmosphere and compress to the required pressure. The compressed air at about up to 200 °C temperature, after compression is supplemented with oxygen and blown into the HS (Hot Stoves), where the temperature is raised to interval 1200-1250°C. The air blast systems of high capacity BF (blast furnace) operate with temperatures of up to 1350°C [7].

The whole process for air blast systems is composing by turbo blowers (air blower), hot stoves, bustle pipe, tuyer stocks, valves and control instruments. Two variants of air blast system are illustrated in Fig. 1.

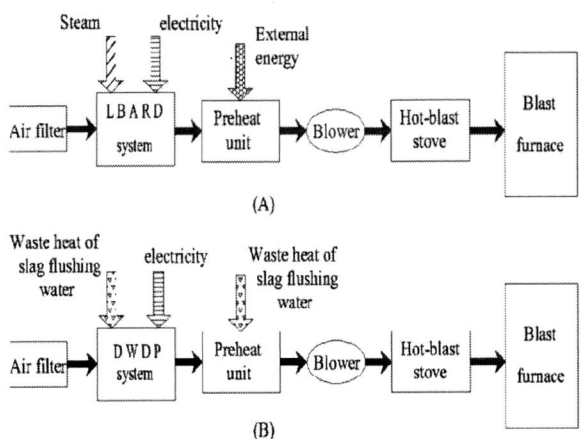

Fig. 1. Variants of air blast system

Turbo blowers equipment represents the main part of air blast system. It needs to have a high level of reliability, because the performance of the high capacity blast furnace depends a lot on the performance of the turbo blowers. The basic diagram of turbo blowers with auxiliary equipment is represented in Fig.2.

For generating the air blast, the furnace is equipped with centrifugal turbo blowers, which it are built in single equipment, with four air gears. Two of them are used for compressing, where the air is controlled to cooler air, and the third and fourth gears are placed on a single workflow.

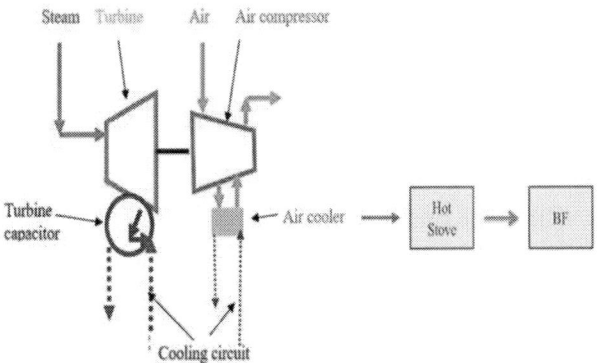

Fig. 2. Basic diagram of turbo blowers

III. PREDICTIVE ANALYTICS TOOLS

Prediction is a technique that uses actions to monitoring tools and techniques to record the performance of TB equipment during normal operation to detect the possible errors and fix them before they result in failure. This forecast regards an industrial problem of condition created with predictive maintenance and that forecast is important because a timely alert for a potential future error could lead the employees to preventive measures on the equipment [8].

Predictive analytics tools offer the possibility to create an algorithm based on forecasting models, statistical approach or the formulas specified for TB equipment and for air blast system. It is necessary to apply advanced analytic techniques on software tools for prediction about technical condition, environment, usage, maintenance history.

During a prediction is necessary to run in sequence the following steps [9], which are shown in Fig. 3:

- Sensors – Data collecting
- Data communication
- Central data store, where the data are stored, processes and analyzed
- Predictive analytics algorithms applied to data
- Determine the corrective action to equipment based on forecast, like dashboards and alerts.

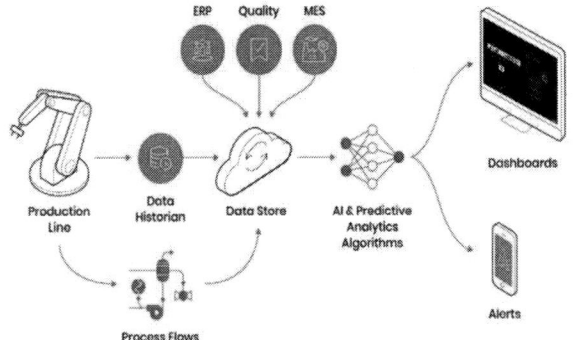

Fig. 3. Steps in predictive maintenance

To create the predictive algorithms, which can generate dashboards and alerts, it is necessary to have sophisticated software, like Power Bi developed by Microsoft, as in Fig. 4.

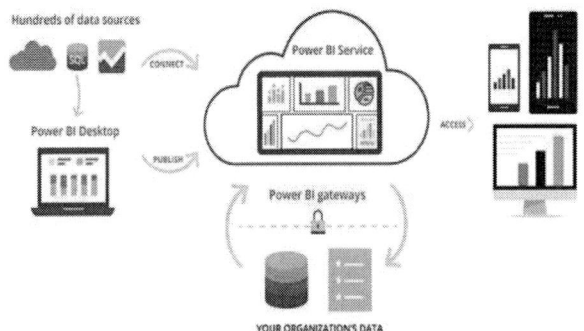

Fig. 4. Steps in predictive maintenance

Power Bi is business analytics software with interactive visualizations. The interface "Power BI Desktop" is an interface for desktop, which offer a warehouse for data cleaning, generate formulas in Data Analysis Expressions and interactive dashboards.

A model has been implemented for turbo blowers equipment, starting with acquiring data from process and store into a database. Data contains condition parameters to be predicted. An example of acquired data is illustrated in Fig. 5.

Data	Debit aer	Presiune	Turatie	Temp aer ref	Temp asp
1/20/2020 12:00:00 AM	114330.9859375	2.85732312635942	2960	124.31713104248	17.6953147888184
1/20/2020 12:15:00 AM	115690.6671875	2.77891088284944	2958	122.93402557373	19.3663162231445
1/20/2020 12:30:00 AM	115251.0265625	2.75075561299044	2943	122.034403298118	19.2933999633789
1/20/2020 12:45:00 AM	113949.6828125	2.73345832824707	2940	121.637734476725	19.5138885498047
1/20/2020 1:00:00 AM	114142.2671875	2.78880205154419	2943	121.99218711853	19.2158579508464
1/20/2020 1:15:00 AM	115381.36875	2.74432892004649	2940	121.609911639874	18.7673625946045
1/20/2020 1:30:00 AM	111408.1375	2.83362944546858	2939	122.557506752014	18.1524843851725
1/20/2020 1:45:00 AM	111627.6859375	2.86363810963101	2943	122.950876393038	17.1983509063721
1/20/2020 2:00:00 AM	111789.70390625	2.86411606348478	2942	123.050347290039	17.6888042449951
1/20/2020 2:15:00 AM	115394.56875	2.75081009229024	2943	121.456162643433	18.3810768127441
1/20/2020 2:30:00 AM	113896.45	2.82938951327477	2945	122.233795674642	18.5850685119629
1/20/2020 2:45:00 AM	112027.8171875	2.86203139781952	2944	122.938985988072	17.8732650756836
1/20/2020 3:00:00 AM	116848.53125	2.92449850506253	2986	125.545141906738	17.0312469482422
1/20/2020 3:15:00 AM	116681.0953125	2.82547506533171	2971	124.161323664739	17.1961807250977
1/20/2020 3:30:00 AM	114860.075	2.84965297381083	2956	123.359803040576	18.0729187011719

Fig. 5. Sequence of data acquisition

DAX include statistical and forecasting aggregation functions as average, standard deviation, variance, and his possibility is to create a calendar or make a forecasting with the line chart. In this paper, the prognosis was realized with formulas implemented with DAX and with the preview version of a forecasting feature for a line chart. Examples of DAX formulas are represented in Fig. 6.

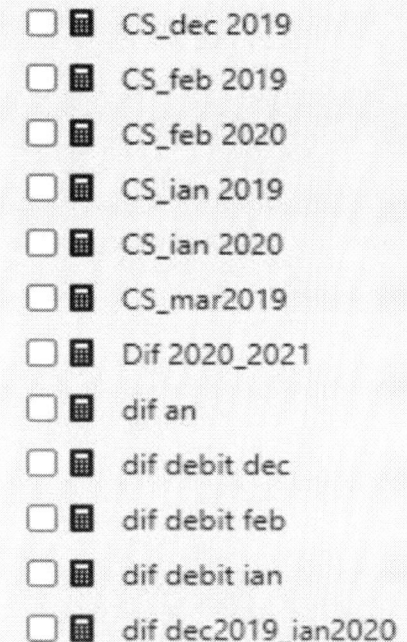

Fig. 6. DAX formulas examples

IV. EXPERIMENTAL RESULTS

To determine the prediction model, the evolution of air flow was observed on a time horizon, depending on engine speed. It revealed the evolution, the variation for speed controller, under conditions of suction and pressure.

Data acquisition is made in 3 timeframes and it is based on monitoring and analysis of the process inputs and outputs for air flow, pressure in accord with speed controller. The goal is to determine if the process respects the imposed limits and to observe the evolution of the equipment. In fig.6 is presented an example for the evolution of turbo blower compressor.

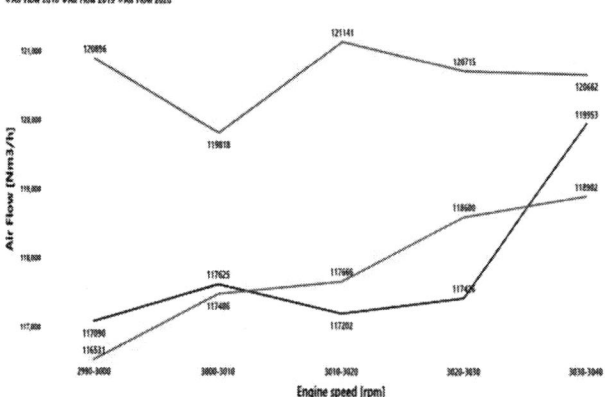

Fig. 7. Evolution of turbo blower compressor in 3 timeframes

The potential degradation of turbo blower equipment can be estimated by comparing the variation of the system functioning state. This can be done by computing state variations between timeframes using DAX formulas, as shown in Fig. 8.

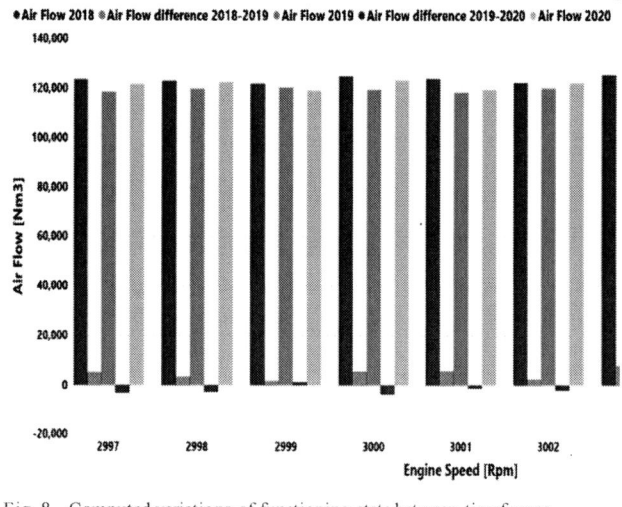

Fig. 8. Computed variations of functioning state between timeframes

In the proposed prediction method, the condition of monitoring data, which is determined by the expert, is transformed into "failure probabilities" (FPS). The role of this FPS method is to calculate the grade of degradation probability of the equipment.

The model allows the use of regression based on the eq. (1):

$$p(x) = \frac{e^{g(x)}}{1 + e^{g(x)}} = \frac{1}{1 + e^{-g(x)}} \qquad (1)$$

where $p(x)$ is the grade of the probability of degradation, x is a vector corresponding with the independent variable and $g(x)$ is logic of the model.

The data model used for generating the line chart with prediction contains 4 tables with multi time series. The tables are linked with a set of relationships, as can be seen in Fig. 9, with the role to accurately calculate results and display the correct information in the report.

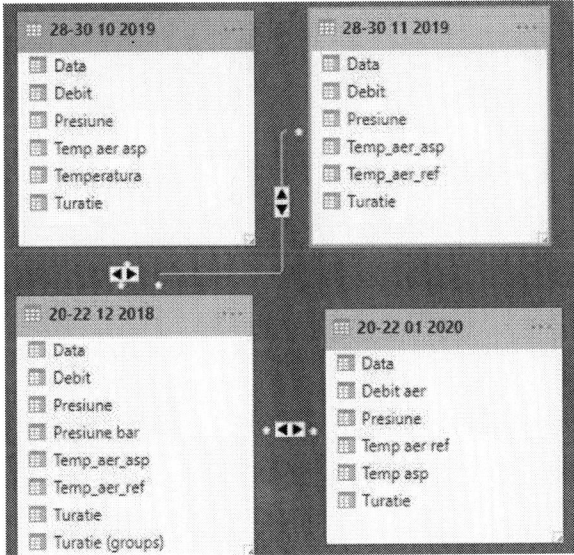

Fig. 9. Tables with multi time series for prediction

The line chart with forecast is illustrated in Fig. 10. By having this line chart which illustrate the prognostics, it gives more time to the engineers to plan and decide if continue the operation of turbo blowers equipment with high risk of failure or to break up the production and perform the maintenance action. This model of prediction can be more attractive when can be realized for long range forecast.

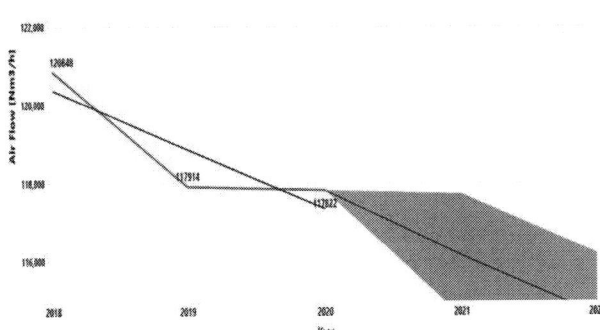

Fig. 10. Line chart with forecast

The model has the role to characterize the degradation of the turbo blowers equipment, which is vital for blast furnace and by default for steel industry. It was observed from the graphs made above that there is a degreasing evolution for turbo equipment and the trend for two next years is downward. With the help of the maintenance team it will be planned the equipment maintenance.

V. CONCLUSIONS

In this paper, forecast models for air blast system are studied, using specialized software and an application of predictive model is create. The determined probabilities have potential to demonstrate the trend of equipment, calculate the life estimation, assisting maintenance resolution making and development of techniques in smart systems and Internet-based solutions.

REFERENCES

[1] Type of Conveyer Belts used for Industrial Purposes, Plant Automation Technology Platform, Global B2B Industrial Automation Platform, https://www.plantautomation-technology.com/articles/types-of-conveyor-belts-used-for-industrial-purposes

[2] Zhe Li and Kesheng Wang, Yafei He, "Industry 4.0 – Potentials for Predictive Maintenance", Conference Paper, January 2016.

[3] Borui Li a, Andrew Lauden a, Jonathan Davis a and Klaus Stohl b, "High Quality Predictive Maintenance Through Advanced Analytics".

[4] Rodrigo Rezende Amaral, "A new SPC tool in the steelshop at ArcelorMittal Gent designed to increase productivity"

[5] X.H Fan, H.D. WANG, "Mathematical model and Artificial Intelligence of sintering process", Central South University Press, Vol. 47, 2002. pp. $064 - 072$.

[6] R. Nicole, "Implementation of an Advanced Process Automation System for Improvement of Sinter Machine # 3 Operation at Bokaro Steel Plant", Journal of Mechanical Engineering and Technology, Vol. 1, 2013, pp. 88-91.

[7] Satyendra, Air Blast System for Blast Furnace, 2016.

[8] Nikolaos Kolokas, Dimosthenis Ioannidis, Thanasis Vafeiadis, "Forecasting faults of industrial equipment using machine learning classifiers", Conference Paper, July 2018.

[9] Mainnovation, "Predictive Maintenance 4.0 - Predict the unpredictable", June 2017.

[10] F. Jin, J. Zhao, C. Sheng and W. Wang, "Causality diagram-based scheduling approach for blast furnace gas system," in IEEE/CAA Journal of Automatica Sinica, vol. 5, no. 2, pp. 587-594, Mar. 2018, doi: 10.1109/JAS.2017.7510715, 2018.

2020 IEEE 26th International Symposium for Design and Technology in Electronic Packaging (SIITME)

Numerical Models of the Electrochemical Migration: a short review

A. Gharaibeh, B. Illés, A. Géczy and B. Medgyes

Department of Electronics Technology,
Budapest University of Technology and Economics
Budapest, Hungary,
medgyes@ett.bme.hu

Abstract— **Electrochemical migration has attracted more attention from researchers due to its high risk of electronics reliability. It is a multistage process that may result in catastrophic failure of electronics. The growing interest of this phenomenon requires precise modeling to predict the time-to-failure, which consists of two main stages, incubation and growth of dendrites. This paper presents a short overview regarding the ECM process, main factors affecting the process, the main historical numerical models for electrochemical deposition in electrochemistry, electrochemical migration on electronics, and the importance of machine learning in building up more precise lifetime models compared to physics of failure models.**

Keywords—Electrochemical Migration; Time-to-Failure; Electronics Reliability; Dendrite Growth

I. INTRODUCTION

The phenomenon of electrochemical migration (ECM) was first reported in the 1950s; silver migration was first observed on the silver-plated terminals of telephone switchboard due to its easiness oxidation and reduction [1]. Due to the continuing miniaturization of electronic systems and the increase in their usage, ECM has attracted more attention from researchers. It is defined as a process by which the metal is dissolved in ionic from its initial location at the anode, migrated via the electrolyte in the presence of bias voltage, and deposited at the cathode side.

It is a kind of humidity-induced failure accompanied by lower Surface Insulation Resistance (SIR) due to dendrite growth [2], [3]. When the dendrites bridge the two electrodes, the short circuit will occur and subsequently may lead to catastrophic failure in electronics. Therefore, Time-to-Failure (TTF) [4]–[6] is an important characterization parameter used to assess the ECM susceptibility of materials quantitatively.

ECM process consists of four consecutive steps: (1) electrolyte layer formation, by absorbing water molecules or water condensation onto the substrate. (2) dissolution of metal, releasing electron(s) by the oxidation and become a positive ion (3) ion transport, by migration, diffusion and/or convection, and (4) deposition of metal ions (dendrite growth), by the sequence of reduction at the cathode side. All of these steps are illustrated in figure 1.

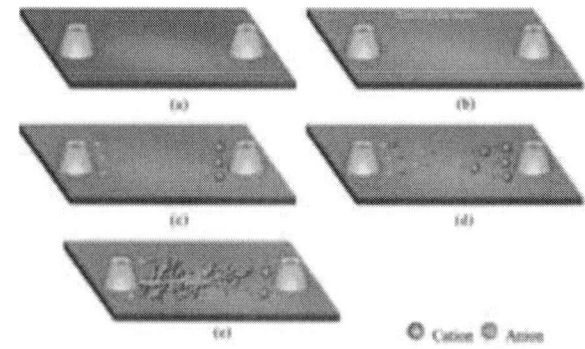

Fig. 1. The schematic diagrams illustrate (a) ECM sample, (b) electrolyte layer formation, (c) dissolution of metal, (d) ion transport and (e) deposition of metal ions (dendrites growth) [7].

Rapid formation of water layer formation significantly affects the overall TTF; this can be observed when comparing Water Drop (WD) test and water condensing tests such as Thermal Humidity Bias (THB) tests. WD test avoids the water condensation process, and the TTF is usually expressed in seconds [8], [9] while the TTF in THB tests are typically expressed in hours [10], [11].

ECM process modelling can be divided into two main stages, incubation and growth of dendrites [12]–[15] the former one covers the application of bias voltage to incubate the metal ions while the latter one covers dendrite initiation until the electrical short formation. Their summation is the overall TTF; they need to be modelled either numerically or by the electrochemical processes of the failure. The numerical modelling of ECM is still not deeply addressed in the literature. Therefore, this work aims to present a short review of the ECM process, the various applied ECM numerical models/simulations, and machine learning TTF models to highlight the open questions, possible research ways.

II. MAIN FACTORS AFFECTING ON ECM PROCESS

A. Temperature and relative humidity

The adsorption of water on the metal surface is a function of temperature and relative humidity, and it is important to provide electrolyte layer formation [16]. The resulted moisture film is expressed in terms of monolayers of water, with the increase in the number of monolayers the conductivity of surface increases as well. Moreover, to act as a liquid, there is a critical number of monolayers of water must be condensed on the surface [17]. At constant temperature, the increase in RH will increase the surface conductivity, but there is a critical RH above which the surface conductivity increases significantly [15].

The change in RH is correlated with the the temperature. If the temperature drops to below the dew point compared to the surrounding air, the relative humidity will be high and exceed 100% at some issues which will result in the formation of a water layer (condensation) [18]

B. Voltage bias and spacing

Voltage bias is essential for ECM to occur since it mainly effects on the metal dissolution. Once it is applied across the two electrodes, an electric field is developed between the two electrodes which are mainly responsible for ionic migration, as described in (1) [19] and (2) [20]:

$$E = \frac{V}{d} \tag{1}$$

$$v = \mu E \tag{2}$$

Where E is the magnitude of the electric field (V/m), V is the applied bias voltage (V), d is the spacing between two electrodes (m), v is the migration velocity of an ion (m/s), μ is the ion mobility (m^2/V. s). ECM to occur needs a critical voltage bias range which is approximately (2-100 VDC) [21], this range contains potentials greater than the required potential for electrochemical oxidation/reduction of the metal and lower than the potentials at which the failure tend to change from ECM to other failures.

Normally, if the spacing is fixed the higher voltage bias, the shorter TTF is expected [22], [23]. However, it has been found that higher voltage bias at the same spacing may result in slower TTF by generating more precipitation which is formed during ion migration step will inhibit the ion migration and dendrite growth [24], while the growth of dendrites might be hindered or even destroyed due to heavy gas evolution at the anode and cathode sides [24], [25]. Longer spacing plays an important role in increasing TTF since it takes longer time for the continuous electrolyte layer formation and ion transport and thus increasing the incubation period [2].

C. Materials of electrodes

For decades lead-bearing solders (Sn-Pb) have been widely used in electronics manufacturing. However, due to the toxicity of lead, the restriction of hazardous substances (RoHS) banned the use of Pb in electrical products [26]. The advent of lead-free solder alloys required further research to investigate the ECM susceptibility of these types compared to Sn-Pb alloys, among lead-free alloys Sn-Ag-Cu (SAC) alloys are the most promising candidate, the very high difference in TTF can be found even for the very similar composition low Ag content SAC alloys [27]. Moreover, further investigations were executed to know the impact of other alloying elements in the ECM process. It has been found that some elements accelerated the ECM process when being doped to lead-free alloys [28]–[30], while other elements hindered the ECM process [31], [32]. With the development in nano-technology, alloying lead-free solder alloys with nano-particles enhanced its ECM resistance [33], [34].

D. Type of contamination

Contaminants have a significant impact on the overall ECM process, especially hygroscopic types. They can accelerate the electrolyte layer formation by lowering the critical relative humidity needed for the condensation, increasing the anodic dissolution rate and the conductivity of the electrolyte [7]. Chloride is the most common contaminants where it can be originated from dust in the air, human sweat and fingerprints and salt spray from the seaside [35]. More chlorides may enhance anodic dissolution. However, it resulted in more precipitates as well which affect the overall ECM process by hindering the ion migration and acting against the dendrite growth depending on their concentration in the electrolyte [25], [36], [37].

Flux residues are an important type of contamination since they are composed of ionic substances [38], [39] and other species. They provide a migration path between the electrodes before the testing [13]. Introducing no-clean flux made a misconception that the cleaning after the soldering process is not required assuming all additives and solid are completely evaporated during the process. However, this is not always true [40], [41]. Furthermore, improper rinses might increase the contamination level of PCB after soldering [42]. However, recommendations for making no-clean flux reliable and safe are published elsewhere [43].

III. MODELS OF ECM

ECM on electronics is different from electrochemical deposition (ECD) in the electrochemistry, the differences can be summarized in three aspects. Firstly, the thickness of ion transport medium on electronics restricts the role of convection since it is mainly as adsorbed moisture film, or bulk water generating from the condensation. Secondly, the presence of reacting ions and possibly supporting electrolyte in ECD solutions make the ionic strength in the medium relatively higher compared to the solutions in the ECM process. Thirdly, in the ECM process, there are no reacting ions in the transport medium before the application of voltage bias while they are easily available for deposition in electrochemistry. All of these differences may restrict the application of ECD theories to ECM process [13].

A. Introduction of numerical models

To achieve a good numerical model of ECM ions transport in the electrolyte, reactions at the electrodes and suitable boundary conditions should be introduced to complete the model. In ECM process the effect of convection can be neglected since there is no supporting electrolyte and electroactive ions before the application of potential difference, the ionic flux of ionic species (k) in the electrolyte can be simulated by Nernst-Planck equation [13], [44]:

$$J_k = -D_k \nabla C_k - \mu_k C_k \nabla \emptyset \tag{3}$$

Where J_k is mass flux, D_k is diffusion coefficient, C_k is concentration, μ_k is ionic mobility of metal ions and ϕ is applied electric potential. The first term in the right-hand side in (3) indicates the diffusion which is driven by the concentration gradient (∇C_k) while the second term indicates the migration which is driven by the potential gradient ($\nabla\emptyset$). In the absence of homogenous reactions, the variation of the concentration is governed by the continuity equation [13], [44]:

$$\frac{\partial C_k}{\partial t} = -\nabla.J_k \tag{4}$$

Combining (3) and (4) to get (5):

$$\frac{\partial C_k}{\partial t} = \nabla.D_k\nabla C_k + \mu_k \nabla C_k.\nabla\emptyset \tag{5}$$

Assuming the electroneutrality in the electrolyte so that:

$$\sum z_k C_k = 0 \tag{6}$$

Where z_k the electric charge number of species k in the electrolyte. By assuming electroneutrality in the electrolyte, the surface concentration of species at the cathode could be set to zero [13], [44]. The assumption of electroneutrality reduces the Poisson equation to the Laplace equation:

$$\nabla.\sigma\nabla\emptyset = 0 \tag{7}$$

Where σ is the electrical conductivity of the electrolyte. The coupled set of (5) and (7) is called Poisson-Nernst-Planck (PNP) equations. In the case of a binary electrolyte system, in the absence of convection and assuming electroneutrality in the electrolyte a simplified diffusion equation can be obtained [45], [46]:

$$\frac{\partial C}{\partial t} = D \nabla^2 C \tag{8}$$

Where, $C = \frac{-c_1}{z_2} = \frac{c_2}{z_1}$ and $D = \frac{z_1\mu_1 D_2 - z_2\mu_2 D_1}{z_1\mu_1 - z_2\mu_2}$

In the ECM process, the main reaction is the anodic dissolution of the metals, the current density for the metal dissolution is expressed by Butler-Volmer equation [47]:

$$i_m = i_{0,m} \, e^{\frac{\emptyset - E_m^0}{\beta_m}} \tag{9}$$

Where

$i_{0,m}$: exchange current density.

E_m^0: equilibrium potential.

β_m: the slope of the Tafel curve for the system with given metals.

In the case of binary alloys (A-B), the expression for the anodic current in terms of mole fractions N_A and N_B of the two metals can be written as (10) [48]:

$$i_{tot} = \frac{i_A \, i_B \, (z_B N_B + z_A N_A)}{z_B N_B i_A + z_A N_A i_B} \tag{10}$$

B. ECM on electronics

There is a lack in the literature regarding the numerical modelling of ECM on electronics. However, He et al. [13] developed a physicochemical model using Nernst-Plank equation (3) and electrochemical impedance spectroscopy (EIS) measurements to simulate ECM of copper electrodes in a deionized (DI) water. With assuming diffusion and migration, Cu^{2+} is the reacting ion. The incubation time can be calculated by integrating the flux density at the cathode with the time and equating it to cumulative deposited amount with a critical radius of r_{crit} while the growth time can be calculated by integrating the flux density of the reacting ions with time and equating it to cumulative deposited amount with the distance between the two electrodes [13].

Numerical simulation was performed for ECM of Sn-9Zn solder alloy in DI water at 3 and 5VDC [44] by applying Nernst-Planck and Poisson equations, concentration profile showed Zn ions are the first species to reach the cathode surface and form deposits. Furthermore, by observing concentration of ions at mid-length, it has been found in case of 3V the concentration of Zn^{+2} was much lower than the concentration of Sn^{2+} while in case of 5V Zn^{2+} already achieved the anodic surface concentration of 1.0×10^{-5} mol/L by the start of ECM process whereas Sn^{2+} just reached the value of nearly 9.0×10^{-7} mol/L [44].

The previous two studies provided examples of how to build up a numerical model for ECM simulation. However, further work is essential to provide a better knowledge of ECM such as inserting flux residues in the modelling, modelling ECM process using different electrolytes (rather than DI water), modelling ECM process from the solder alloys point of view using in the electronics industry, modelling the metal dissolution processes, reduction processes of metal ions, etc to list a few aspects.

C. Machine learning as an alternative to the physics of failure modelling

In addition to numerical modelling of ECM, overall TTF models based on the physics of failure were introduced. Zhou et al. [49] developed a TTF model based on the Faraday law of immersion silver surface finished copper electrodes with various concentration of NaCl to simulate the dust contamination. The model was validated through results of WD and SIR tests [49]

Yang et al. [15] developed a TTF model for silver migration as a function of temperature, relative humidity, bias and material properties as illustrated in (11). The model based on the analytical equation of surface resistivity between the anode and cathode as a function of the adsorbed water and ion concentration in the interelectrode region. The model was validated through the THB and SIR tests, and it has been found that SIR degradation can occur before the initiation of dendrite growth and TTF is determined when the ion concentration reaches a threshold value. For the TTF models based on the physics of failure, there was a deviation between experimental and predicted TTF values due to the regression analysis for the corresponding factors in the TTF formula.

$$TTF = nF \times \frac{m_0}{M} \times \beta \times \frac{1}{V} \times \frac{(1-RH)(1+(c-1)RH)}{cRH}$$
$$\times \exp\frac{E_\sigma}{RT} \qquad (11)$$

Where TTF is the time-to-failure in seconds, n is the metal valence, F is Faraday's constant, m_0 is the mass of precipitated metal at the cathode, M is the atomic weight of metal which migrated in the process, β is a parameter related to monolayers of the adsorbed water on the insulation surface, V is the bias voltage, RH is the relative humidity, E_σ is the activation energy for electrolyte conduction, c is a parameter related to the heat of vaporization and condensation, R is the gas constant, T is the temperature.

Recently, building up TTF models using machine learning showed more efficiency to predict TTF values than the physics of failure models in complicated environment conditions by having much lower normalized mean square error (NMSE), Zhou et al. [50], [51] developed random-forest-regression-based failure models using Python programming language for THB test results and got lower NMSE than the physics of failure model developed in [15]. However, modelling using machine learning requires a large amount of data to achieve accurate prediction. Moreover, more significant errors appeared when TTF is longer (e.g. at a lower temperature, RH and voltage bias) therefore more corrections are needed for the used algorithms in case of machine learning TTF modelling.

Conclusion

With the continuing miniaturization of electronic products and progress in industrial electronic technology have largely increased the risk of ECM. ECM process is a consecutive four-step process: (1) electrolyte layer formation, (2) dissolution of metal, (3) ion transport and (4) deposition of metal ions. TTF is the summation of all previous steps, and as a characterization parameter of the ECM process, it cannot be easily predicted due to the complicated contribution of its influencing factor. However, it needs to be modelled either by regression or numerically, modelling studies usually divided the ECM process into two stages; incubation and growth of dendrites. Most of the studies in the literature are related to ECD in electrochemistry which may restrict their application to ECM on electronics. Numerical modelling of the ECM process is still not deeply addressed in the literature. The numerical modelling consists of equations for ion transport in the electrolyte, reactions at the electrodes and suitable boundary conditions to solve the proposed set of equations. Beside numerical modelling, overall regression models based on the physics of failure were introduced. However, there was a deviation between the experimental and expected values of TTF. Moreover, the complicated contribution of the influencing factors of the ECM process complicates building up physics of failure model. Instead, building up TTF models using machine learning showed better life prediction performance by showing lower NMSE values. Further research is required for numerical modelling and correction of the machine learning algorithms.

Acknowledgment

The research reported in this paper has been supported by the National Research, Development and Innovation Office – NKFIH, FK 132186. This research was also supported by the Higher Education Excellence Program of the Ministry of Human Capacities in the frame of Nanotechnology and Materials Science research area of Budapest University of Technology and Economics (BME FIKP-NAT).

References

[1] G. T. KOHMAN, H. W. HERMANCE, and G. H. DOWNES, "Silver Migration in Electrical Insulation," *BELL Syst. Tech. J.*, vol. 34, no. 6, pp. 1115–1147, 1955.

[2] S. Zhan, M. H. Azarian, and M. G. Pecht, "Surface Insulation Resistance of Conformally Coated Printed Circuit Boards Processed With No-Clean Flux," *IEEE Trans. Electron. Packag. Manuf.*, vol. 29, no. 3, pp. 217–223, 2006.

[3] Y. Zhou, P. Yang, C. Yuan, and Y. Huo, "Electrochemical migration failure of the copper trace on printed circuit board driven by immersion silver finish," *Chem. Eng. Trans.*, vol. 33, pp. 559–564, 2013.

[4] P. Yi, C. Dong, Y. Ji, Y. Yin, J. Yao, and K. Xiao, "Electrochemical migration failure mechanism and dendrite composition characteristics of Sn96.5Ag3.0Cu0.5 alloy in thin electrolyte films," *J. Mater. Sci. Mater. Electron.*, vol. 30, no. 7, pp. 6575–6582, 2019.

[5] L. T. F. Mendes, V. F. Cardoso, and A. N. R. da Silva, "Electrochemical Migration on Lead-Free Soldering of Pcbs," *ECS Trans.*, vol. 23, no. 1, pp. 271–278, 2009.

[6] X. Qi, H. Ma, C. Wang, S. Shang, X. Li, Y. Wang, and H. Ma, "Electrochemical migration behavior of Sn-based lead-free solder," Journal of Materials Science: Materials in Electronics, vol. 30, no. 15, pp. 14695–14702, 2019.

[7] X. Zhong, L. Chen, B. Medgyes, Z. Zhang, S. Gao, and L. Jakab, "Electrochemical migration of Sn and Sn solder alloys: A review," *RSC Adv.*, vol. 7, no. 45, pp. 28186–28206, 2017.

[8] B. Medgyes, E. Román, Á. Bohnert, S. Szurdán, X. Zhong, and G. Harsányi, "Electrochemical Migration Investigations on SAC-Bi-xMn Solder Alloys," in *2018 IEEE 24th International Symposium for Design and Technology in Electronic Packaging, SIITME 2018 - Proceedings*, 2019, pp. 80–83.

[9] S. Ádám, B. Medgyes, D. Rigler, B. Szabó, and L. Gál, "Electrochemical Migration of SAC305 Solders and Tin Surface Finish in NaCl Environment," in *2018 IEEE 24th International Symposium for Design and Technology in Electronic Packaging, SIITME 2018 - Proceedings*, 2019, pp. 71–75.

[10] S. Oh, D. Kim, W. Hong, K. Kim, and C. Oh, "Copper electrochemical migration growth in an air HAST," *Microelectron. Reliab.*, vol. 100–101, p. 113394, 2019.

[11] X. He, M. H. Azarian, and M. G. Pecht, "Effects of solder mask on electrochemical migration of tin-lead and lead-free boards," *IPC APEX EXPO Tech. Conf. 2010*, vol. 2, pp. 1297–1335, 2010.

[12] J. L. Barton and J. O. Bockris, "The electrolytic growth of dendrites from ionic solutions," *Proc. R. Soc. London. Ser. A. Math. Phys.*

Sci., vol. 268, no. 1335, pp. 485–505, 1962.

[13] X. He, M. H. Azarian, and M. G. Pecht, "Analysis of the kinetics of electrochemical migration on printed circuit boards using Nernst-Planck transport equation," *Electrochim. Acta*, vol. 142, pp. 1–10, 2014.

[14] J. C. Bradley, S. Dengra, G. A. Gonzalez, G. Marshall, and F. V. Molina, "Ion transport and deposit growth in spatially coupled bipolar electrochemistry," *J. Electroanal. Chem.*, vol. 478, no. 1–2, pp. 128–139, 1999.

[15] S. Yang and A. Christou, "Failure model for silver electrochemical migration," *IEEE Trans. Device Mater. Reliab.*, vol. 7, no. 1, pp. 188–196, 2007.

[16] S. Lee and R. W. Staehle, "Adsorption study of water on gold using the quartz-crystal microbalance technique: Assessment of BET and FHH models of adsorption," *Zeitschrift fuer Met. Res. Adv. Tech.*, vol. 88, no. 11, pp. 880–886, 1997.

[17] B. Da Yan, S. L. Meilink, G. W. Warren, and P. Wynblatt, "Water Adsorption and Surface Conductivity Measurements on α-Alumina Substrates," *IEEE Trans. Components, Hybrids, Manuf. Technol.*, vol. 10, no. 2, pp. 247–251, 1987.

[18] R. Hienonen and R. Lahtinen, "Corrosion and climatic effects in electronics," *VTT Publ.*, no. 626, pp. 3–242, 2007.

[19] E. Bumiller, M. Pecht, and C. Hillman, "Electrochemical migration on HASL plated FR-4 printed circuit boards," in *Presented at 9th Pan Pacific Microelectronics Symposium Exhibits & Conference*, 2004, pp. 37–41.

[20] J. O. M. Bockris and A. K. N. Reddy, *Modern electrochemistry, Plenum press, New York.* 1970.

[21] J. A. Jachim, G. B. Freeman, and L. J. Turbini, "Use of surface insulation resistance and contact angle measurements to characterize the interactions of three water soluble fluxes with FR-4 substrates," *IEEE Trans. Components Packag. Manuf. Technol. Part B*, vol. 20, no. 4, pp. 443–450, 1997.

[22] M. S. Jung *et al.*, "Improvement of electrochemical migration resistance by Cu/Sn intermetallic compound barrier on Cu in printed circuit board," *IEEE Trans. Device Mater. Reliab.*, vol. 14, no. 1, pp. 382–389, 2014.

[23] B. I. Noh, J. W. Yoon, W. S. Hong, and S. B. Jung, "Evaluation of electrochemical migration on flexible printed circuit boards with different surface finishes," *J. Electron. Mater.*, vol. 38, no. 6, pp. 902–907, 2009.

[24] X. Zhong, G. Zhang, Y. Qiu, Z. Chen, and X. Guo, "Electrochemical migration of tin in thin electrolyte layer containing chloride ions," *Corros. Sci.*, vol. 74, pp. 71–82, 2013.

[25] V. Verdingovas, M. S. Jellesen, and R. Ambat, "Influence of sodium chloride and weak organic acids (flux residues) on electrochemical migration of tin on surface mount chip components," *Corros. Eng. Sci. Technol.*, vol. 48, no. 6, pp. 426–435, 2013.

[26] Directive of the European Commission for the Reduction of Hazardous Substances, Directive 2000/0159 (COD) C5-487/2002, LEX 391, PE-CONS 3662/2/02 Rev 2, ENV581, CODEC 1273, 2003.

[27] B. Medgyes, B. Illés, and G. Harsányi, "Electrochemical Migration of Micro-alloyed Low Ag Solders in NaCl Solution," *Period. Polytech. Electr. Eng.*, vol. 57, no. 2, pp. 49–55, 2013.

[28] W. Dai, L. S. Duan, and C. Y. Zhong, "Electrochemical Migration and Electrochemical Corrosion Behaviors in 3wt. % NaCl Solution of 64Sn-35Bi-1Ag Solder with In doping for Micro-nanoelectronic Packagings," pp. 1372–1376, 2012.

[29] L. Hua, G. K. Yang, and H. Q. Zhang, "Effects of Ge doping on electrochemical migration, corrosion behavior and oxidation characteristics of lead-free Sn-3.0Ag-0.5Cu solder for electronic packaging," *Adv. Mater. Res.*, vol. 146–147, pp. 953–961, 2011.

[30] L. Hua and J. S. Zhang, "Corrosion Behavior of 64Sn-35Bi-1Ag Solder Doped with Zn in NaCl Solution and its Electrochemical Migration Characteristics in High Humid Thermal Condition for Electronic Packaging," in *12th International Conference on Electronic Packaging Technology and High Density Packaging*.

[31] N. K. Othman, E. M. Salleh, C. Sarveswaran, and F. Che Ani, "Effect of phosphorus and nickel on electrochemical migration of Sn-3Ag-0.7Cu solder paste in simulated body fluid," *Solid State Phenom.*, vol. 273 SSP, no. April, pp. 61–65, 2018.

[32] L. Hua, M. W. Sou, W. J. Zhang, and Q. L. Hu, "Electrochemical migration and rapid whisker growth of Zn and Bi dopings in Sn-3.0Ag-0.5Cu solder in 3wt.% NaCl solution," *Adv. Mater. Res.*, vol. 239–242, pp. 1751–1760, 2011.

[33] N. K. Othman, F. R. Omar, and F. C. Ani, "Electrochemical migration and corrosion behaviours of SAC305 reinforced by NiO, Fe2O3, TiO2 nanoparticles in NaCl solution," *IOP Conf. Ser. Mater. Sci. Eng.*, vol. 701, p. 012044, 2019.

[34] Fakhrul Rifdi Omar, Emee Marina Salleh, Norinsan Kamil Othman, Fakhrozi Che Ani, and Zambri Samsudin, "The Effect of Electrochemical Migration of Pb-free Sn-3.0Ag-0.5Cu Solder Reinforced by NiO Nanoparticles," *J. Mater. Sci. Eng. A*, vol. 8, no. 5, pp. 185–189, 2018.

[35] G. Harsányi, "Irregular effect of chloride impurities on migration failure reliability: Contradictions or understandable?," *Microelectron. Reliab.*, vol. 39, no. 9, pp. 1407–1411, 1999.

[36] D. Minzari, M. S. Jellesen, P. Møller, and R. Ambat, "On the electrochemical migration mechanism of tin in electronics," *Corros. Sci.*, vol. 53, no. 10, pp. 3366–3379, 2011.

[37] X. Zhong, G. Zhang, Y. Qiu, Z. Chen, W. Zou, and X. Guo, "In situ study the dependence of electrochemical migration of tin on chloride," *Electrochem. commun.*, vol. 27, pp. 63–68, 2013.

[38] V. Verdingovas and M. S. Jellesen, "Solder Flux Residues and Humidity-Related Failures in Electronics : Relative Effects of Weak Organic Acids Used in No-Clean Flux Systems Solder Flux Residues and Humidity-Related Failures in Electronics : Relative Effects of Weak Organic Acids Used in No-," *J. Electron. Mater.*, vol. 44, no. 4, pp. 1116–1127, 2015.

[39] K. Piotrowska, R. U. Din, F. B. Grumsen, and M. S. Jellesen, "Parametric Study of Solder Flux Hygroscopicity : Impact of Weak Organic Acids on Water Layer Formation and Corrosion of Electronics," *J. Electron. Mater.*, vol. 47, no. 7, pp. 4190–4207, 2018.

[40] K. S. Hansen, M. S. Jellesen, P. Møller, P. J. S. Westermann, and R. Ambat, "Effect of solder flux residues on corrosion of electronics," in *Proceedings - Annual Reliability and Maintainability Symposium*, 2009, pp. 502–508.

[41] J. E. Sohn and U. Ray, "Weak Organic Acids and Surface Insulation Resistance," *Circuit World*, vol. 21, no. 4, pp. 22–26, 1995.

[42] K. Rendl, V. Wirth, and F. Steiner, "Impact of No – Clean Fluxes Cleaning on PCB Ionic Contamination," in *2015 38th International Spring Seminar on Electronics Technology (ISSE)*.

[43] P. Isaacs and T. Munson, "What makes no-clean flux residue benign?," in *2016 Pan Pacific Microelectronics Symposium (Pan Pacific)*.

[44] H. Ma, A. Kunwar, J. Chen, L. Qu, Y. Wang, X. Song, P. Råback, H. Ma, and N. Zhao, "Study of electrochemical migration based transport kinetics of metal ions in Sn-9Zn alloy," Microelectronics Reliability, vol. 83, pp. 198–205, 2018.

[45] Y. K. Kwok and C. C. K. Wu, "Numerical simulation of electrochemical diffusion-migration model with reaction at electrodes," *Comput. Methods Appl. Mech. Eng.*, vol. 132, no. 3–4, pp. 305–317, 1996.

[46] Y. . Choi and K.-Y. Chan, "Exact solutions of transport," *J.*

Electroanal. Chem, vol. 334, pp. 13–23, 1992.

[47] T. M. Amorim, C. Allély, and J. P. Caire, "Modelling coating lifetime: first practical application for coating design," in *Proceedings of the COMSOL Conference1*, 2008, pp. 1–7.

[48] R. F. Steigerwald and N. D. Greene, "The Anodic Dissolution of Binary Alloys," *J. Electrochem. Soc.*, vol. 109, no. 11, p. 1026, 1962.

[49] Y. Zhou, Y. Li, Y. Chen, and M. Zhu, "Life Model of the Electrochemical Migration Failure of Printed Circuit Boards under NaCl Solution," *IEEE Trans. Device Mater. Reliab.*, vol. 19, no. 4, pp. 622–629, 2019.

[50] Y. Zhou, Y. Chen, Q. Xie, and Y. Li, "Modeling Research of Electrochemical Migration Failure on Printed Circuit Board," in *Prognostics and System Health Management Conference, Chongqing, China, Oct. 26-28, 2018*, pp. 209–214.

[51] Y. Zhou, L. Yang, Y. Li, and W. Lu, "Exploring the Data-Driven Modeling Methods for Electrochemical Migration Failure of Printed Circuit Board," in *Proceedings - 2019 Prognostics and System Health Management Conference, PHM-Paris 2019*, 2019, pp. 100–105.

SoC based IoT sensor network hub for activity recognition using ML.net framework

Alexandru Alexan
Department of Electric, Electronic and
Computer Engineering, Technical
University of Cluj-Napoca,
North University Center Baia Mare,
Baia Mare, Romania
alexanalexandru@gmail.com

Anca Alexan
Department of Electric, Electronic and
Computer Engineering
Technical University of Cluj-Napoca,
North University Center Baia Mare,
Baia Mare, Romania

Oniga Ştefan
Department of Electric, Electronic and
Computer Engineering
Technical University of Cluj-Napoca,
North University Center Baia Mare,
Baia Mare, Romania

Abstract—**Nowadays single-board computers - SBC and System on a Chip SoC's get smaller and provide more processing power, yielding perfect alternatives to low-cost local hubs for small sensor networks. Even though sensors can now be connected directly to the cloud, for example, the ESP32 SoC, a local hub can provide a simple interface for multiple sensors to reach the IoT cloud. One of the best options for a local hub is the Raspberry Pi SBC, based on the Broadcom BCM2837 SoC, due to its small form factor and power consumption. We propose a system based on the Raspberry Pi SBC that interconnects connects two SensorTag CC2650 Bluetooth modules to the cloud, a MongoDB database. The obtained data is then processed offline using Microsoft's ML.NET machine learning framework to detect the user's activity. This framework was chosen as it's free, cross-platform, and open-source.**

Keywords— SBC, sensor network hub, IoT, SoC, ML.NET, sensor gateway, cloud, sensor network, Sensortag CC2650, Raspberry PI, python, MongoDB

I. INTRODUCTION

One of the most important resources nowadays is data, and data gathering and storing is being done from any source possible. This way the IoT environment in which we are always online, connected, and synced, is created. An environment that is safe and has the potential to improve many aspects of our lives [1].

The basic flow for data in an IoT environment is started by the collection phase, data is captured from any everyday device [2]. The next phase is the data analysis to obtain results, based on the obtained data, as quickly as possible. The last step is represented by the storage process, as data is saved for reference or additional processing. Some activity domains benefit more from the expansion of this omnipresent IoT cloud, for example, healthcare and home automation, as cloud-enabled devices are becoming more and more used [3].

The IoT cloud is formed by the great mesh of devices (IoT nodes) used every day by us, devices that are meant to help and keep us up-to-date.

Since we have an exponential growth over the number of IoT devices, storing very large amounts of data from distinct many endpoints is required. The only platform that is scalable enough, can provide vast amounts of storage capacity and fault tolerance at a high speed is the distributed cloud database. Using this kind of a data storage system allows a small sensor network to grow

To upload data into the cloud, the IoT node must be connected to the internet. A couple of sways this can be accomplished are direct connection & local hub. For a direct connection to the internet, the IoT node is connected directly to a computer network that has internet access. For a local hub, the IoT node is connected only to a local hub and the local hub is in turn connected to the internet

We have advantages and disadvantages in both options presented, choosing the correct one mainly depends on the system's overall architecture and restrictions, like performance. Future system extensibility must also be taken into consideration.

Our objective is to set-up a local hub as a gateway device and that will gather, upload to the cloud, and process data from sensors. Our chosen local data hub is an SBC device due to size, price, and performance restrictions.

We use the acquired data from the hub with Microsoft's machine learning framework ML.NET, via offline processing, to recognize the activity type. We analyze and detect 5 common activities: walking, sit on a chair, sitting on a chair, get up from a chair, lying on the bed.

II. RELATED WORK

A. Sensors

Activity detection can be handled by lots of different sensors types, and three of the major types are vision-based sensors, ambient sensors, and wearable sensors.

Vision sensors can be used to detect a wide range of objects and activities but they mainly rely on video cameras and in some cases controlled environments. To improve accuracy, other sensor types are used along with video cameras, [4].

Ambient sensors are used to detect activities via PIR or ultrasonic sensors and even pressure sensors.

Wearable sensors [5] have usually more restrictions, as they have to be small and are usually battery powered. Unfortunately, since these sensor types are mainly battery-powered, they have a limited communication range to increase the run time. Very frequently used sensors are accelerometers, gyroscopes, and magnetometers.

B. Sensor network hub and cloud

In most of the cases where a local network hub is used, this is under the form of a mobile device, like a smartphone. This device mainly bridges the sensors and provides internet access to them, sending also the received data into the cloud. In some

978-1-7281-7507-2/20 $31.00 © 2020 IEEE

cases, the processing/pre-processing of the data can be done on the hub itself, provided that the computational power and energy restrictions are met [6].

Custom hardware hubs can be constructed with performance and extensibility in mind but they imply a higher cost since they mostly used custom modules and components compared with a simple device like a smartphone. One of the main advantages of a custom hardware hub is the operating system support, as they tend to support multiple operating systems [7].

C. Machine learning framework

Many frameworks can be used for machine learning that varies in complexity and available features. One of them is TensorFlow, a complex open-source deep-learning library [8] that can support multiple programming languages like python, C, GO, and JAVA. Multiple libraries that provide high-level functionalities are built on top of more complex ones, like TensorFlow. For example, InferPy is a high-level Python API framework build on top of TensorFlow that specializes in probabilistic modeling [9]. For the .net framework, one of the best options is ML.NET, a Microsoft framework that allows usage of complex machine learning pipelines in new and existing .net based applications. Since .net core works also on other non-Microsoft operation systems, this library can thus be used on a wide variety of platforms, proving its cross-platform characteristic.

The ML.NET framework is open source thus increasing the security and extensibility of the final code by allowing developers and security consultants to examine the existing code.

Also one of the most important aspects is that this framework has been tested and used for Microsoft products over the past decade, proving the fact that this is indeed a technology that is used by hundreds of millions of users worldwide [10].

The fact that is free is a bonus, making this a great candidate that can rival other machine learning frameworks.

III. PROPOSED SYSTEM

We propose a small sensor network consisting of two sensors that connect to a local hub that is used for data collection and cloud relay, shown in figure 1. The data is then downloaded and analyzed locally using a machine learning

The chosen sensors are Texas Instruments CC2650 that connect to a Raspberry PI local hub via Bluetooth. All components were chosen due to their low cost and power design and create a simple yet powerful data acquisition and processing system.

A. Sensors

The uses sensor modules are CC2650 SimpleLink SensorTag manufactured by Texas Instruments. The supported connectivity is Bluetooth and they provide a very wide range of built-in sensors. One of the two most important aspects is the size, as they have a very small footprint and low energy Bluetooth connectivity.

The system uses two identical sensors for the data acquisition process, the first two components of the sensor network. Due to an updated firmware, we obtain between 20 and 50 samples per second.

Even though the sensors can identify themselves via Bluetooth by name, we use the Bluetooth MAC address for authentication and the hub assigns a friendlier name for each sensor. From the wide range of sensor data available, we use the accelerometer and gyroscope data.

B. The sensor network hub

The hardware platform used for the local data hub was a Raspberry PI SBC. This device was chosen due to its price, flexibility, and operating system support, as this SBC can handle Windows 10 IoT Core, different Linux flavors including Ubuntu and Raspbian to name a few. The Broadcom BCM2837 chip is the SoC heart of this small unit, based on an ARM Cortex A53 quad-core architecture. This module features rich connectivity, allowing different kinds of sensors to communicate with the hub direct without any additional hardware add-ons components. For low range, we have Bluetooth low energy & Bluetooth 4.1 classic connectivity. The chosen operating system is Raspbian OS as the bluepy Python library, an interface to Bluetooth LE, is used to connect the two sensor modules. For DateTime capabilities, the DateTime python module is added to precisely timestamp each received data. The pymongo packages are used to connect to the used Mongo cloud database to upload the

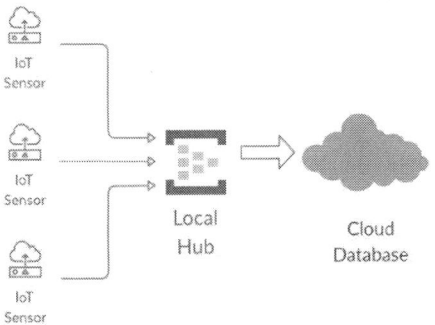

Fig. 1. Sensor network hub

framework, ML.NET.

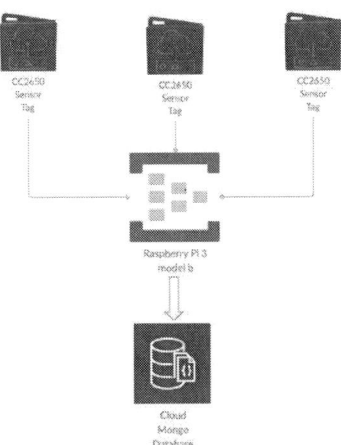

Fig. 2. The implemented sensor network

sensor data. The implemented sensor network is shown in figure 2.

The main program running on the hub is written in Python and waits for each SensorTag to be available, the software flowchart is shown in figure 3. Whitelisted Bluetooth MAC addresses are allowed only for connection.

Flowchart

1 Connect

Connect to the SensorTag modules with MAC address authentication.

2 Activate & read

Enable acceleration/gyroscopic sensors
Read sensor data

3 Upload to DB

Upload sensor data to
Mongo Cloud Database(DB)

Fig. 3. Software flowchart

The acceleration and gyroscope sensors are enabled for each sensor in for the reading process to occur. The data read process occurs continuously and without any delays. Each record receives a timestamp upon arrival and this is used further for processing.

C. Cloud database

The chosen cloud database is MongoDB since it is scalable and NoSQL document-based. We can start with just a few sensors, and grow exponentially if needed. A collection was created in the database for storing the sensor data, collection that is composed of multiple documents with the following fields: unique id, sensor type, timestamp, movement data & activity type.

Another implementation option for improving storage capabilities would be to use a collection of data including

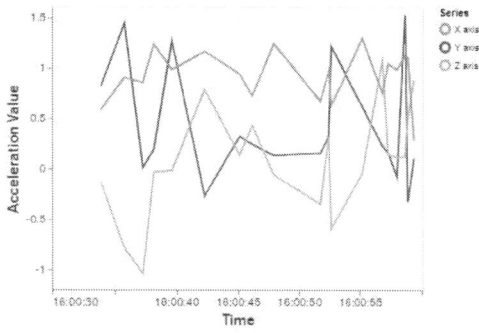

Fig. 4. MongoDb accelerometer data

multiple sensor data types per sensor unit. This option would improve the storage and simplify processing if multiple sensor

data types are read for each sensor module. Figure 4 shows the database saved acceleration data for 25 seconds.

The system stores about 20 samples, or documents, per second for each sensor.

D. Data pre-processing

To obtain the two usable input files for training and testing the neural network, the raw data needs to be preprocessed. The raw data is obtained by exporting data from the MongoDB database.

This preprocessing is done in another .net console application that extracts and reformats the data. As stored and exported from the MongoDB database, each document, or row, has a sensor type, timestamp, and values for each of the three axes. The sensor type can be one of the two accelerometers or gyroscopes used. We need these values grouped to have for each record all the values from all four sensors, each with values for each axes. To achieve this, we match records based on the timestamp and allow matching records with a difference of a maximum of 25 milliseconds. We seek the minimum difference in time to obtain the best possible match within the allowed 25 milliseconds difference. Since we obtain about 20 records per second, one record at every 50 milliseconds, we chose a safe interval of 25 milliseconds for matching data from the four sensors. Data matching is allowed only between data gathered for the same activity. The final resulting data contains an activity type, a timestamp, and movement data for each of the four sensors used, two acceleration, and two gyroscope sensors. The data in this format is then split into the two files used for training and testing, using about 70% of data for training and the rest for testing. The format used for the two files is tab-separated value text.

To achieve a good data consistency for the two files, each activity type data was processed and split separately, to ensure an even distribution for both learning and testing.

E. Machine learning using ML.NET framework

Since we wanted to implement our machine learning processing using the .net core framework we choose Microsoft's ML.NET framework. Currently, we are using a .net core console application as the main app for offline recognizing activities and leveraging the power of the ML.NET framework.

This console application uses the two preprocessed text files as input, training, and testing tab-separated value files. These files were obtained from exported MongoDB sensor data as previously stated and about 70% of data is used to train the neural network and about 30 % for testing. The ML.NET framework tests multiple trainer algorithms and chooses the best-found one for the final implementation. In our case, using accelerometer and gyroscope data, the best-found one is "FastTreeOva" out of 74 tested, as shown in figure 9.

The training process was done over 300 seconds and the chosen trainer algorithm had the best accuracy of 99.68% as shown in figure 5.

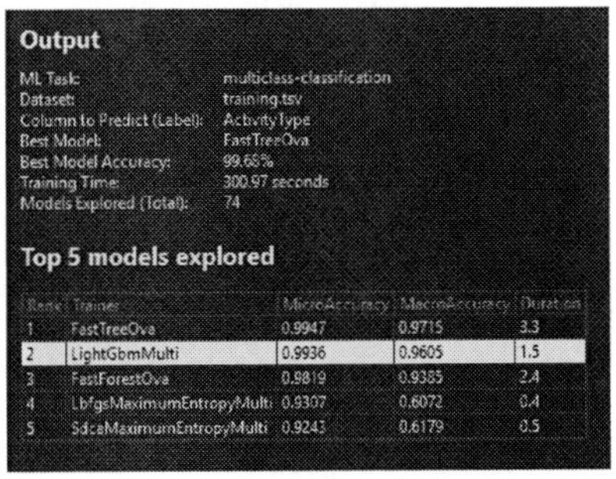

Output

ML Task: multiclass-classification
Dataset: training.tsv
Column to Predict (Label): Activity Type
Best Model: FastTreeOva
Best Model Accuracy: 99.66%
Training Time: 300.97 seconds
Models Explored (Total): 74

Top 5 models explored

Rank	Trainer	MicroAccuracy	MacroAccuracy	Duration
1	FastTreeOva	0.9947	0.9715	3.3
2	LightGbmMulti	0.9936	0.9605	1.5
3	FastForestOva	0.9819	0.9385	2.4
4	LbfgsMaximumEntropyMulti	0.9307	0.6072	0.4
5	SdcaMaximumEntropyMulti	0.9243	0.6179	0.5

Fig. 5. Best training algorithm found

After the network is trained, we use the testing values that are labeled and predict the activity type for them saving the obtained result for each record. After that, we compare the results, and as shown in figure 6 we obtained 3404 correct predictions out of the 3881 total records, about 87.70% accuracy.

IV. THE HUB, CLOUD AND SECURITY

Since today the usage of distributed computing environments for storing and processing data is increasing, the security of these systems must be carefully reviewed. The well-known term for this virtual environment is the cloud, an extendible and powerful tool that can be useful at virtually any task and most importantly at any scale. Since the cloud can be used even with the smallest project to aid in storing and even processing data, this system can grow with the application that it serves, since more resources and services can be allocated if needed almost instantly. But even though using the cloud can yield multiple advantages, special care must be taken for keeping the operation secure, and most importantly, keep the data safe. Even a small vulnerability in any part of the system can spell disaster, especially if security best practices are not met. Possible security problems can target the sensor hub, either at the connection with the sensors or at the connection with the cloud level. The local sensor network is using a MAC white list to allow only trusted sensor nodes to connect. This means that even if a similar fake sensor exposes the same capabilities as a genuine one, it will not be able to connect to the sensor network. The connection with the cloud, in our case the MongoDB is secure by default. Atlas encrypts the data, both in-transit, and at-rest, and makes it easy to control access with role-based user management. Also yet another layer of security, only explicit IP addresses are allowed to connect to the database, based on a whitelist. This ensures a high degree of security and ensures that only the terminals that need to access the data can access the data.

Since the implemented machine learning component uses offline data exported from the MongoDB, there is no security risk associated with the current implementation. Also currently the main processing is done offline, so the security characteristics of the used system are inherited by the processing program.

V. CONCLUSIONS

In this paper, we present the implementation and testing of an SBC based sensor network hub that gathers, preprocess, and upload data into the cloud. Two sensors created a small sensor network used for testing.

The used Microsoft's ML.NET framework proved to be a valuable machine learning tool that leverages one of the most popular frameworks and development environments, the .net framework with Visual Studio. We managed to obtain good activity recognition (87.70% accuracy) with a relatively small code footprint and great processing speed.

To improve the activity recognition process, the next step would be to do additional data pre-processing and use data frames for combining a small snapshot of sensor data for processing. This way we would probably be able to improve the accuracy of more complex activities that have a longer period.

Since currently, we analyze the activities locally, with data downloaded from the MongoDB database, for future development we can move the local application to the cloud and start feeding data directly from the MongoDB database.

The middle database layer plays an important role and allows, in this case, to extend the system with other components that need access to movement/activity data.

REFERENCES

[1] Wang, Juan & Hao, Shirong & Wen, Ru & Zhang, Boxian & Zhang, Liqiang & Hu, Hongxin & Lu, Rongxing. (2020). IoT-Praetor: Undesired Behaviors Detection for IoT Devices. IEEE Internet of Things Journal. PP. 1-1. 10.1109/JIOT.2020.3010023.

[2] Singh K.J., Kapoor D.S. Create Your Own Internet of Things: A survey of IoT platforms. IEEE Consum. Electron. Mag. 2017;pp. 6:57–68

[3] N.Q. Mehmood, R. Culmone, L. Mostarda, "Modeling temporal aspects of sensor data for MongoDB NoSQL database", J. Big Data, 2017, pp. 4:8

[4] Sergiyenko, Oleg & Tyrsa, Vera & Flores-Fuentes, Wendy & Rodríguez-Quiñonez, Julio & Mercorelli, Paolo. (2018). Machine Vision Sensors. Journal of Sensors. 2018. 1-2. 10.1155/2018/3202761.

[5] Kalatturu, U. & Abirami, G.., Offline smart phone based human-fall detection system. International Journal of Pure and Applied Mathematics. 118. 195-199, 2018

[6] Vamos Daniel, Oniga, Stefan , Alexan Anca. (2018). Personal data acquisition IOT gateway. Carpathian Journal of Electronic and Computer Engineering. 11. 44-47. 10.2478/cjece-2018-0008.

[7] Yuejiao Cheng, Yuejiao & Chenglong Jiang, Chenglong & Shi, Jiong. (2016). A Fall Detection System based on SensorTag and Windows 10 IoT Core. 10.2991/mse-15.2016.4

[8] S. W. D. Chien, C. P. Sishtla, S. Markidis, Z. Jun, I. B. Peng, and E. Laure, "An Evaluation of the TensorFlow Programming Model for Solving Traditional HPC Problems," in Proceedings of the 5th International Conference on Exascale Applications and Software, 2018, p. 34.

[9] Rafael Cabañas, Antonio Salmerón, Andrés R. Masegosa, InferPy: Probabilistic modeling with Tensorflow made easy, Knowledge-Based Systems, Volume 168, 2019, Pages 25-27, ISSN 0950-7051

[10] Lee, Yunseong, Alberto Scolari, Byung-Gon Chun, Markus Weimer and Matteo Interlandi. "From the Edge to the Cloud: Model Serving in ML.NET." IEEE Data Eng. Bull. 41 (2018): 46-53.

Machine learning activity detection using ML.Net

Anca Alexan
Department of Electric, Electronic and
Computer Engineering
Technical University of Cluj-Napoca,
North University Center Baia Mare,
Baia Mare, Romania
anca.alexan@cunbm.utcluj.ro

Alexandru Alexan
Department of Electric, Electronic and
Computer Engineering, Technical
University of Cluj-Napoca,
North University Center Baia Mare,
Baia Mare, Romania

Oniga Ștefan
Department of Electric, Electronic and
Computer Engineering
Technical University of Cluj-Napoca,
North University Center Baia Mare,
Baia Mare, Romania

Abstract—our living environment is becoming more and more aware of our presence and starts to react and interact with us. The smart house has long moved from concept to reality, as our homes are now digitalized with all sorts of smart devices. Since we now have more data available the ever, an important part is to be able to analyze the data and provide the user with useful information. Neural networks are a good way of providing systems capable of handling large volumes of data that are able to learn based on users' data. Our proposed neural network system consists of a .NET software application that processes data from the CASAS activity detection system. The software application uses the ML.NET NET machine learning framework for predicting the users' activity. The ML.NET framework was chosen since its open-source and cross-platform.

Keywords— CASAS, machine learning, neural network, activity detection, ML.NET

I. INTRODUCTION

Nowadays smart places are getting popular by the minute and their usage is increasing significantly. The emerging trend of smart areas was generated by the relatively recent IoT revolution, as the Internet of Things devices slowly became part of our everyday life.

With these new emerging technologies, base elements of residential and commercial areas can now be easily connected, simplifying the process of transforming our environment into a smart one. But these new hyper-connected environments generate a huge volume of data, data that needs to be stored, processed, and interpreted[1]. The already huge volume of data increases exponentially if a lot of persons are present in the same area, at the same time. Data manipulation for pre-processing and processing presents its own challenges, as different types of data manipulations need to be used, like extremity analysis, data flow analysis, and database analysis[5]. Choosing either one of these data analysis processes highly depends on the application and desired results. One combination will work for one kind of data but fail miserably with a different kind of data [11].

Machine learning-based data analysis is a very popular and effective data analysis mechanism. If machine learning is desired to be used for the main data analysis, the data acquired for these smart areas need to be preprocessed, processed, and manipulated so that a machine learning algorithm can use them[9].

One very important aspect of using machine learning techniques is that behind them, the analytic models should learn permanently from the continuous data flow[10].

In this paper we present the dataset preprocess flow and the obtained results for base activity recognition. The used dataset is generated by the CASAS system. Using the machine learning development tool from Microsoft, ML, we will obtain a dataset for recognizing basic human activities. These obtained results will be compared with the data generated by the Matlab neural network algorithms. The data is preprocessed and divided into dynamic windows alongside multiple window indexes in order to increase the activity recognition rate.

A. CASAS as the data acquisition system

The CASAS system [2] represents a smart house in a box, a smart house kit with a very minimalistic and simple design that can be easily installed and offers out of the box, smart functionalities, without any tweaks or changes.

Activity recognition, in this case, consists of mapping a sequence of sensor data to a corresponding activity tag. The activity recognition software, called AR, does this labeling process in real-time when new sensor data is recorded by the system [3]. This data can be exported, and the exported file format contains the tagged data with the following fields:

- Date
- Hour
- Location
- State

B. Machine learning tools

We've focused our attention on two machine learning instruments, one created by Matlab and the other one created by Microsoft.

Machine learning (ML) is an IT subdomain type of artificial intelligence(AI) that adds the learning ability to a system without explicit programming[4].

ML has evolved from model recognition and the computation theory. This technology has greatly evolved lately, as numerous tools have been created to easily implement ML-based solutions. Nevertheless, it's very difficult to find a tool that can suit any application and scenarios [12]. Some tools have a higher processing speed, others can be easily customized and extended for the desired application [13].

Scalability is another important aspect, as depending on the application, this can be as important as [4]:

1. the usability
2. speed
3. coverage

4. extensibility

5. programming language support

Each application has some very important factors, that dictate the approach used; for example, the processing speed is important when the application has real-time data and especially if the data flow is substantial and at high speed [14].

When choosing a tool is very important for the tool to be extendable, to be able to integrate it with other existing/new applications or system, and to be well documented [15].

A learning algorithm takes an input dataset as a training set. There are three main types of learning types: supervised, unsupervised, and reinforced [5].

In supervised learning, the training dataset is generated from the input data entries alongside the data tags. Supervised learning consists of learning to predict the correct output data for the presented input data.

In the case of unsupervised learning, the previously mentioned data tags are not required for the training dataset [16].

Consolidation learning handles the problem of learning what is the proper action or action sequence for maximizing detection gain [17].

When the applied machine learning tags are part of a finite number of categories, the applications are called classification tasks. When the tags are composed of one or multiple variables, the machine learning application is called regression tasks.

ML.NET is an open-source library for machine learning that supports the creation of complex pipelines, quick evaluation, and direct prediction usage. The pipelines are composed of multiple steps that transform the input data and then translates that data into multiple ML models.

ML.NET represents Microsoft's approach to machine learning and was developed to bring the power of ML closer to the developers, allowing them to use only one framework for adding, testing, and deploying ML pipelines.

ML.NET framework's goal is to provide a simple and effective tool that is scalable over very big datasets and yet provides good performance. The unification of all data processing transformation into a single API alongside great machine learning models yielded a product that is simple yet very powerful.

The ML.NET framework supports at least 80 features alongside 40 machine learning models [6]. Also, this framework can be used with other machine learning tools like Accord.NET, TensorFlow, and CNTK.

Out of the box, ML.NET includes several standard operators that can also be composed due to the DataView abstraction. The ability to expand the operators by composition generates very efficient machine learning pipelines.

The main operator class is "transform" and applies to a DataView to be able to generate a derived DataView. This operator is used to prepare the data for training, testing, or prediction generation.

"Learners" are the components that are trained on the data and generate predictive models. They are machine learning specific algorithms.

The "evaluator" components use the scored datasets to generate metrics for indicating key result features like precision or recall.

"Loaders" are used to create an abstraction, to represent a data source as a DataView.

The "savers" components are used for data serialization, to be able to serialize DataViews to a form compatible with the loader.

ML.NET offers the ability to quickly add a machine learning component to any .NET applications, even web or desktop applications. Due to this feature, automatic prediction can be obtained with the data available in the application. Machine learning applications use existing patterns in analyzed data to make predictions. This means that using machine learning algorithms doesn't need explicitly programmed logic [7].

The MATLAB programming tool also offers lots of advantages over other languages or tools sets, as the main basic structure data element is a matrix. MATLAB also allows the existing functionality to be expanded by using toolboxes. Toolboxes are collections of functions that offer more specialized functionalities. MATLAB even allows for audio, video, and image data processing, supporting domain-specific operations. Any issues with the neural network can be easily detected by data visualization and data checks even before the network training using the "Deep Network Designer" app. This app allows the building of very intricate network architectures, easily update existing ones, and even modify already trained networks for transfer learning purposes [8].

The "Deep Learning Toolbox" is a tool to create and use deep neural networks with all the associated algorithms, trained models, and applications.

With small data sets, transfer learning can be applied from existing neural networks or even models imported from 3rd party apps like TensorFlow-Keras and Caffe [8].

II. IMPLEMENTATION DESCRIPTION

In figure 1 we present the activity recognition process for base activities. This contains three stages:

1. The raw data extraction phase from the CASAS system
2. Data preprocessing
3. Activity recognition

For processing, a dataset from the Center for Advanced Studies in Adaptive Systems (CASAS) was used. This Washington State University's dataset contains data from multiple users that executed the same activity types. From this available data, we've chosen the Kyoto dataset.

For preprocessing, the window size chosen was dynamic and it takes into consideration the activity begin and end. For a certain activity, the first and last active sensors can be determined from the labeled data. The sensor's activation order can also be identified for a particular activity.

For optimizing the preprocessing phase, for speed and quick swap of datasets, a build-in tool was created. This allows us to quickly test different datasets and avoid running manual data updates. This tool is in the same project as the main machine learning processor, decreasing the number of

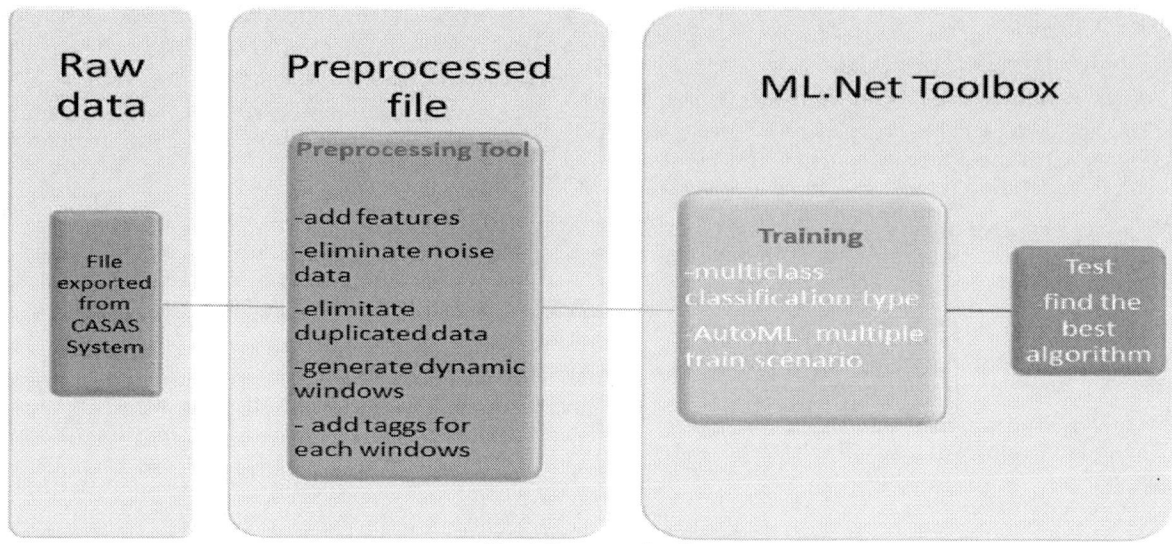

Fig. 1. Data processing diagram

components used by the system, and generates the training and testing files used by the neural network. The preprocessing tool uses the raw data provided by the CASAS system and generates the completely preprocessed output files.

Multiple sub-phases are executed as part of the preprocessing operation. The first pre-process sub-phase is to eliminate the "OFF" status records. This eliminates the noise generated by this kind of record and the neural network will not take them into account for processing. The second pre-process sub-phase is to eliminate duplicate records, the ones that are generated by the sensor's own reactivation process.

To improve the accuracy of activity detection, in the preprocessing phase, each record has the location and activity type added. This was done by normalizing the timestamp length.

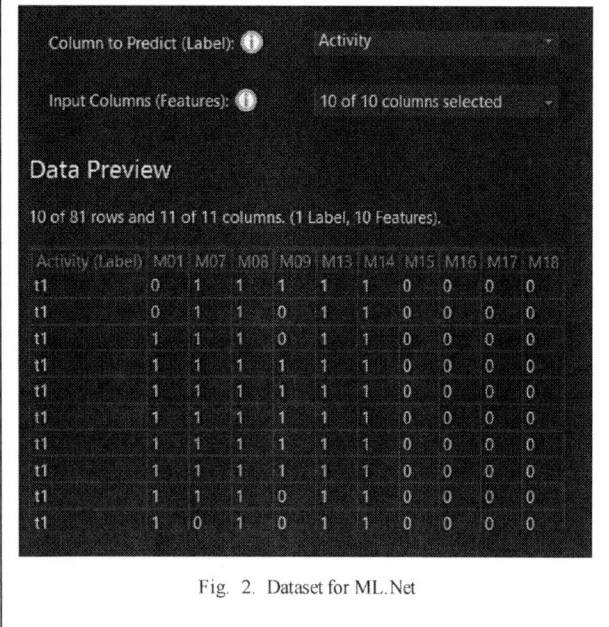

Fig. 2. Dataset for ML.Net

The overall total number of events, per location, was calculated and the value was used to determine the probability of a certain reported event per location. A grouping is generated containing three-time indicator values, in hours, which are attached to each record. These three values represent the top three hours that have the largest number of events per location. Each activity has the duration calculated, and a duration category is assigned. The activity category is also added to each record alongside general sensor data for the previous sensor and the next one. The data is then formatted as data windows and tagged. The output generated file is created for historical purposes and its feed to the machine learning main processor app.

The main processing unit is implemented based on the ML.NET framework. The framework has multiple templates that can jumpstart development. We didn't use a default scenario, we've created a custom one that allows the most flexible implementation. The next step is to provide the training dataset alongside the column to predict; as shown in figure 1, the column to predict is the Activity and it's followed by the 10 input data columns. The activities are t1: Make a phone call in the dining room; t2: Wash hands; t3: Cook; t4: Eat; t5: Clean.

AutoML supports data files with or without a header line supplying a header line that is more convenient than not supplying one. The goal of a multiclass classification problem is to predict a discrete value where there are three or more possible values to consider. For the training phase, the machine learning task chosen was "multiclass-classification". For the training phase, short training time was chosen, 60 seconds as shown in figure 2. We've increased the default training time from 10 seconds to 60 seconds to increase the system's recognizing precision. The entire data was split into training and testing data, about 70% of the initial data was used for training with the rest of 30 % used for verifying the algorithm's precision.

In the training phase, the system tests multiple algorithms to identify the best one for the particular training scenario. AutoML automatically creates and evaluates several different machine learning models using different algorithms.

For the input data pre-processed with dynamic windows width, we've obtained a prediction precision of 71,43 % with the Averaged Perceptron algorithm as shown in figure 2. This precision was calculated by the ML.NET framework based on the training data. Multiple algorithms were tested in this phase and based on the recognition precision, the best one was chosen; in figure 2 some of the different trainers used for determining the best one are shown alongside the micro-precision.

Using the remaining 30% data, representing the testing data, we've checked the final machine learning algorithm implementation and managed to obtain a recognition rate of 71.5%.

III. RESULTS

Our machine learning implementation managed to obtain a prediction precision of 71,5 %, using the AveragedPerceptronOva algorithm.

In [18] the same CASAS dataset was used for activity recognition with a software-based on Matlab. The best recognition rate for the Matlab algorithm was 67, 78 % with the dynamic window with. The same Matlab algorithm yields the worst results with fixed-width windows data, with a recognition rate of only 52,77%.

TABLE I.   RESULTS COMPARISON OF ML TOOLS

	Machine Learning Tool	Activity recognition percent
1	Matlab NN tool[18]	52.77%
2	ML.NET AVERAGED PERCEPTRON OVA	71.5%
3	ML.NET SDCA MAXIMUM ENTROPY MULTI	67%
4	ML.NET LIGHT GBM MULTI	67%
5	ML.NET SYMBOLIC SGD LOGISTIC REGRESSION OVA	30%
6	ML.Net FAST TREE OVA	63%
7	ML.Net LINEAR SVM OVA	67%
8	ML.Net SGD CALIBRATED OVA	52%

IV. CONCLUSIONS

For data preprocessing, a tool was created that generates the input data file for the recognition system. This external tool extracts and computes the raw characteristics and creates dynamic windows that are then exported into a file.

Our obtained results using the ML.NET machine learning framework are better than our previous algorithms implemented in Matlab [18]. The recognition rate was improved by 9.72 %.

In order to increase the recognition rate further, we plan to improve the filtering algorithm and extract new characteristics in the preprocessing phase.

REFERENCES

[1] Tinghui Wang, Diane J. Cook sMRT: Multi-Resident Tracking in Smart Homes with Sensor Vectorization, IEEE Transactions on Pattern Analysis and Machine Intelligence, 2020

[2] Diane J. Cook, Aaron S. Crandall, Brian L. Thomas, Narayanan C. Krishnan, "CASAS: A Smart Home in a Box", Computer, 2013

[3] .J. Cook, M. Schmitter-Edgecombe, Aaron Crandall, Chad Sanders, and Brian Thomas. "Collecting and Disseminating Smart Home Sensor Data in the CASAS Project"

[4] Qolomany, Basheer & Al-Fuqaha, Ala & Gupta, Ajay & Benhaddou, D. & al-wajidi, Safaa & Qadir, Junaid & Fong, "Leveraging Machine Learning and Big Data for Smart Buildings: A Comprehensive Survey", Alvis 2019

[5] Mohammad Saeid Mahdavinejad, Mohammadreza Rezvan, Mohammadamin Barekatain, Peyman Adibi, Payam Barnaghi, Amit P. Sheth, "Machine learning for internet of things data analysis: a survey" Digital Communications and Networks 4 (2018) 161–175

[6] Zeeshan Ahmed, Saeed Amizadeh, Mikhail Bilenko, Rogan Carr, Wei-Sheng Chin, Yael Dekel, Xavier Dupre, Vadim Eksarevskiy, Eric Erhardt, Costin Eseanu, Senja Filip, Tom Finley, Abhishek Goswami, Monte Hoover, Scott Inglis, Matteo Interlandi, Shon Katzenberger, Najeeb Kazmi, Gleb Krivosheev, Pete Luferenko, Ivan Matantsev, Sergiy Matusevych, Shahab Moradi, Gani Nazirov, Justin Ormont, Gal Oshri, Artidoro Pagnoni, Jignesh Parmar, Prabhat Roy, Sarthak Shah, Mohammad Zeeshan Siddiqui, Markus Weimer, Shauheen Zahirazami, Yiwen Zhu "Machine Learning at Microsoft with ML.NET", arXiv:1905.05715, 2019

[7] Matteo Interlandi, Sergiy Matusevych, Misha Bilenko, Saeed Amizadeh, Shauheen Zahirazami, Markus Weimer, "Machine Learning at Microsoft with ML.NET", 32nd Conference on Neural Information Processing Systems (NIPS 2018), Montréal, Canada

[8] Phil Kim "MATLAB Deep Learning With Machine Learning, Neural Networks and Artificial Intelligence ", ISBN 1484228456, 2017

[9] J. Suto, S. Oniga, C. Lung, I. Orha, Comparison of offline and real-time human activity recognition results using machine learning techniques, Neural, Computing and Applications, March 2018., https://doi.org/10.1007/s00521-018-3437-x

[10] F. Monori, S. Oniga, Processing EEG signals acquired from a consumer grade BCI device, Carpathian Journal of Electronic and Computer Engineering, ISSN: 1844-9689, Volume 11, Number 2, 2018, pp. 29-34

[11] T Majoros, B Ujvári, S Oniga, EEG data processing with neural network, Carpathian Journal of Electronic and Computer Engineering 12 (2), 33-36

[12] Iram Fatima, Muhammad Fahim ,Young-Koo Lee, Sungyoung Lee, "Analysis and effects of smart home dataset characteristics for daily life activity recognition", Supercomput DOI 10.1007/s11227-013-0978-8, 2013

[13] Jun Zhang , ZhongCheng Wu, Fang Li, Chengjun Xie , Tingting Ren , Jie Chen , Liu Liu, "A Deep Learning Framework for Driving Behavior Identification on In-Vehicle CAN-BUS Sensor Data", Sensors, 2019

[14] Eirini Anthi, Lowri Williams, Małgorzata Słowi'nska, George Theodorakopoulos, Pete Burnap, "A Supervised Intrusion Detection System for Smart Home IoT Devices", IEEE Internet of Things Journal, 2019, 10.1109/JIOT.2019.2926365

[15] Xiao Guo , Zhenjiang Shen , Yajing Zhang, Teng Wu, "Review on the Application of Artificial Intelligence in Smart Homes", MDPI, 2019

[16] P. Gupta, R. McClatchey, P. Caleb-Solly, "Tracking changes in user activity from unlabelled smart home sensor data using unsupervised learning methods" Neural Computing and Applications 2020,

[17] Suneth Ranasinghe, Fadi Al Machot and Heinrich C Mayr, "A review on applications of activity recognition systems with regard to performance and evaluation" International Journal of Distributed Sensor Networks 2016, Vol. 12(8), DOI: 10.1177/1550147716665520

[18] Anca Alexan, Alexandru Alexan, Oniga Ştefan, Iuliu Alexandru Pap, "Analysis of activity detection data pre-processing", 2019 IEEE 25th International Symposium for Design and Technology in Electronic Packaging (SIITME), 2019

Embedded System for Smart Controlling Electronic Devices

D. G. Bălan

Faculty of Automatic Control and Computers
Politehnica University of Bucharest
Bucharest, Romania

A. Drumea, A. E. Marcu

Electronic Technology and Reliability Department
Politehnica University of Bucharest
Bucharest, Romania

Abstract—**As technology is evolving day by day, the remote control is starting to be a normality in our lives. In this context, it is described an implementation of a system for controlling the functional time of consumers in a living apartment for daily use. It consists in an electronic module and a soon to be mobile application running on Android devices connected through Bluetooth. The functionality of the system consists in allowing the user to program the running time for a consumer in his home by entering the times in the application.**

Keywords—IoT; embedded; Texas Instruments

I. INTRODUCTION

In a complicated world, technology is meant to keep things simple for humanity, but why not think technology in a simpler way? The system presented is designed to control the functionality time of consumers (e.g. the lights in a room). For that is needed a microcontroller and its additional circuits, a relay module, and a Bluetooth module.

The functionality of the entire system together is shown in the Fig. 1. A signal is transmitted to the microcontroller and it goes two ways. One of them is to show that the entire system is functional by turning on a LED [1]. The other implies the functionality of the consumers.

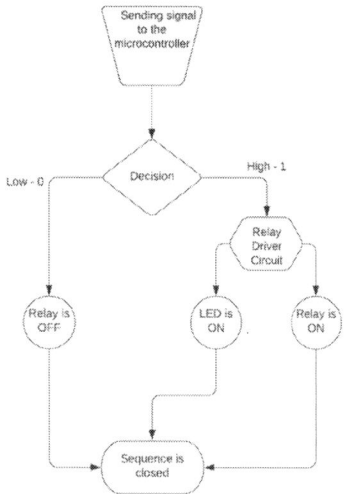

Fig. 1. The system flow diagram.

The signal is transmitted between two times given from the user. The start time, which means the time when the consumer will start to function and the stop time, at this one the consumer will stop functioning. The difference between those times is considered the functionality time of consumers. As soon as the relay driver circuit is receiving the signal, depending on its state (High – 1 or Low – 0) it will open or not the consumer as following: the signal is received in the base of the transistor, if its state is 1 ('on') the relay will switch on the consumer, if it is 0 ('off') the consumer will remain off or will be turned off depends on the previous state [2].

Besides this introduction, the paper is structured as follows: Section II describes the hardware design and development, Section III presents the software development, Section IV presents the system implementation, Section V presents the results and Section VI concludes the paper.

II. HARDWARE DESIGN AND DEVELOPMENT

The design of the electronic module is an important part of the system, especially the printed board circuit (PCB). The main goal for an electronic device is to be efficient. In this context, the designing phase was focused on an optimal thinking approach in order to gain good results. The authors decided to use two voltage planes (for 3.3 V and 5 V) and a ground one (GND) in order to maintain almost the minimum number of routes [3].

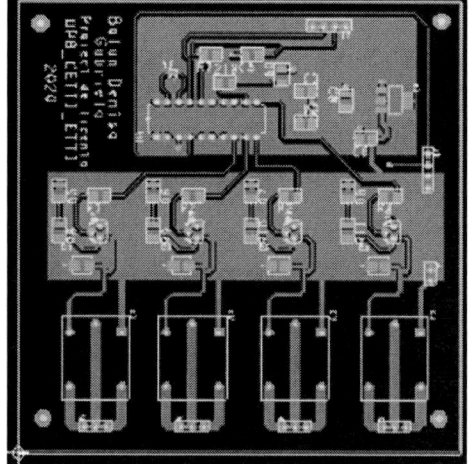

Fig. 2. The TOP view of the system PCB.

2020 IEEE 26th International Symposium for Design and Technology in Electronic Packaging (SIITME)

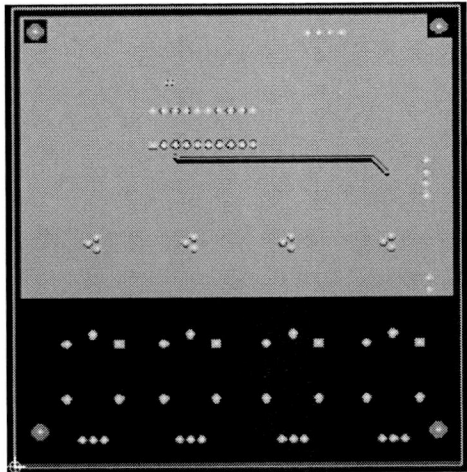

Fig. 3. The BOTTOM view of the system PCB.

Starting from the bottom of the TOP view, shown in Fig. 2, the relay module which consist in four electronic relay devices. In the middle is presented the control circuit for the relays (it consists in diodes, resistors, transistors, and a LED to signal each functionality). As most of the devices are powered by 5 V, the plan was put on the TOP view only over these components.

At the top of it, is presented the 16 bits Texas Instruments Microcontroller and its additional circuits for good functioning. The Bluetooth module is connected externally through the connector with 4 pins positioned between the plane of 5 V and the upper part of PCB.

As the microcontroller is powered by 3.3 V voltage, a voltage regulator was needed and is presented on the upper part as well.

Talking about the BOTTOM view, shown in Fig. 3, the ground plane is in the foreground which is not passing through the relays module as the one behind is controlling high voltages.

As the module is controlling high voltages as well as high currents, the dimensions of the routes were chosen wisely and showed in Table 1. For draining a high current is needed a thicker route than others [4].

TABLE I.   DIMENSIONS OF ROUTES

Voltages Signal Routes	Dimension [mil]
230V	80
5V	32
3.3V	32
Others	20

III. SOFTWARE DEVELOPMENT

The application presented in this project works on a relatively small radius, as a Bluetooth connection is being used between the mobile application and the electronic device [5]. Moreover, the application is expected to work towards four users in a room. For testing, it has been used on a connection of light bulbs.

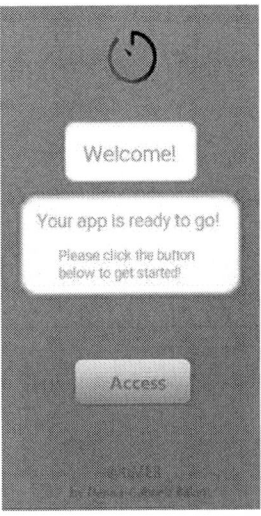

Fig. 4. Welcome Screen.

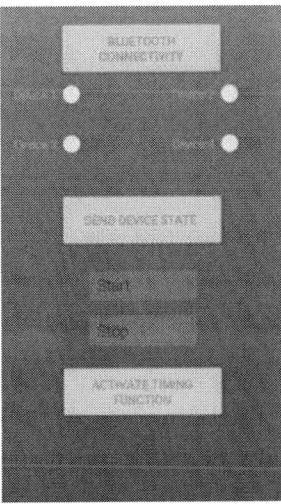

Fig. 5. Principal Menu

Later on, changes can be made for the mobile application to be used upon other consumers in the room fitting the user's needs and wants and also the electronic devices future updates.

The system is controlled remotely by a mobile application which is running on Android Systems for the moment. The application has two screens as following:

- Welcome Screen (shown in Fig. 4).

- Principal Menu (shown in Fig. 5).

The first screen is welcoming and inviting the user to access the command interface which is the next screen by clicking the button 'Access'.

The control is thought to be efficient and intuitive. The first step is to assure the Bluetooth connection between the mobile phone and the hardware system developed through pressing 'Bluetooth Connection' button. As so, the user will choose which of the devices will be programmed and send the data to the microcontroller through the first button entitled 'Start Device

State'. Continuing, the last part is to set the start time and the stop time in 24 h format (*hh:mm*) and send them to microcontroller by pressing 'Activate Timing Function' button. The screen number two is shown in Fig. 4 in progress.

The data is transmitted by Bluetooth connectivity so when the user is pressing the button specified for different functions (as told before), the information is sent by TXD bus and received by the microcontroller through the RXD bus [6].

In this phase, is recommended for the first use of the system to send the data when the minute is changing to have a better synchronization without losses, after that, until the system is unplugged, the starting hour will be kept in the memory of microcontroller and the clock will start ticking from that.

IV. IMPLEMENTATION

In order to minimize the cost of the overall system and as well to be a great candidate for SMD technology of implementation PCBs, the electronic components were chosen accordingly. The result of the electronic devise is shown in Fig. 6.

The system is desired to be user friendly and efficient, so the authors chose not only an easy to read hardware implementation, but also a methodology from where the common user can benefit the accessible results.

Usually, the relay has the NC pin (normally closed) attached to the COMM pin (common). By applying the microcontroller's impulse on the base of the transistor, the latter will open and enable the commute of the COMM pin to the NO pin (normally open), thus ensuring the flow of the current and turning on the consumer linked between the 2pins.

The microcontroller will send a signal which will in turn be captured by the NPN type transistor. Then, this will be amplified and reach the LED on one side, where it will ensure its illuminations during its functioning, and the relay on the other side, where it will ensure the commutation so the timing will start.

The device or devices connected to the relays will remain active until the Android System's clock will hit the stop hour given by the user when the data was sent. The clock made from the crystal quartz from the hardware part made will start ticking as soon as the user is sending the data needed in application and as well the current time of the Android System.

In order to not reboot the entire system every time we want to change the state of the devices connected in the system, the microcontroller will process the data given by pressing 'Activate Timing Function' as following: first characters sent are the current time, in the middle is presented the start time and the last ones are for the stop time. The microcontroller will set only once the current time from the phone to its own so the timing will start at the beginning of setting the environment.

V. RESULTS

For checking the entire system good function, a test board was made with four light bulbs (shown in Fig. 7). The lights are numbered and connected clearly in order with the relays. A test

was made in order to see if the embedded system can provide the main function of timing for one device.

The user chose to activate the first device (as a result the switch button on the graphical interface of the mobile application turned on) and after that the times were set (start time and stop time).

Fig. 6. The PCB mounted.

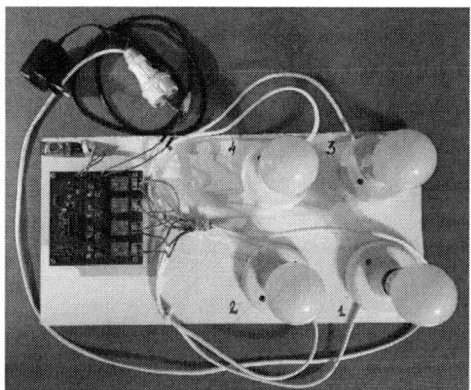

Fig. 7. The test board.

Fig. 8. Setting the times to function of the system.

2020 IEEE 26th International Symposium for Design and Technology in Electronic Packaging (SIITME)

Fig. 9. The functioning of the mobile aplication for the device number 1.

Fig. 10. Result of the device number 1.

As the commands were transmitted, the microcontroller received the start hour and the clock begun ticking. At the specified start time, the system is changing its state to ON, as so the led connected to relay one is lighting (shown in Fig. 9).

The functionality for the device connected to the first relay can be seen in the Fig 9.

The system is thought to be intuitive so the first relay will command the first device connected, the second relay the next one and so on for four devices.

As well the embedded system functionality was tested for four light bulbs in order to see if it can support powering the devices at the same time and the result was positive.

When the time reaches the start hour, the first light bulb is on and it keeps its state until the hour hits the stop time. The clock is ticking at the 32.768 kHz frequency as it is used a crystal quartz.

Android devices are a good choice when talking about controlling simple systems because they can offer enough processing power, standard connectivity options and local storage, as so the embedded system can be efficiently controlled and can give great results [7].

VI. CONCLUSION

The consumers ability to have wireless control over their homes through an external gadget is a well-known application, especially under the name of "SMART HOME", a field seeing constant growth.

These can be categorized under multiple criteria, such as the distance between the consumer and the application, the ability to control various household items and more. Another difference is the type of wireless control, either Wi-Fi or Bluetooth.

When choosing such a system, the main issues are complexity and efficiency, as well as reduced energy consumption. Considering these facts, the designed system is a good prototype accordingly to its small power consumption, simplicity, and efficiency.

Moreover, a decisive factor is also the price, which varies depending on the type of application, with more advance models seeing an increased price. The system designed was thought to be implemented with components which does not imply big expensive in order to be economically proficient.

REFERENCES

[1] C. Ionescu, A. Vasile, N. Codreanu, and R. Negroiu, "Comparative studies on dimming capabilities of retrofit LED lamps," Advanced Topics in Optoelectronics, Microelectronics, and Nanotechnologies VIII, pp. 100102Z, December 2016.

[2] P. Horowitz, W. Hill, The Art of Elecronics' 3rd ed., Cambridge University Press, 2015.

[3] https://www.parallel-systems.co.uk/pcb/, PCB Tutorias.

[4] M. Kraig, Complete PCB Design Using OrCAD Capture and PCB Editor, Newnes, 2009.

[5] R.Piyare, M.Tazil, "Bluetooth Based Home Automation System Using Cell Phone," IEEE 15th International Symposium on Consumer Electronics, 2011.

[6] B. Wirsing, "Sending and Receiving Data via Bluetooth with an Android Device," 2014.

[7] A. Drumea, "Control of Industrial Systems Using Android-Based Devices," 36th Int. Spring Seminar on Electronics Technology, 2013.

Impedance matching for UHF band antennas on ceramic substrate

Călin Mircea
Centre for Technological Electronics and Interconnection
Techniques
University "Politehnica" of Bucharest
Bucharest, Romania
mircea.calin@cetti.ro

Svasta Paul
Centre for Technological Electronics and Interconnection
Techniques
University "Politehnica" of Bucharest
Bucharest, Romania
paul.svasta@cetti.ro

Abstract— This paper explores a practical approach and a simple method for impedance matching that can be used in the case of ceramic substrate UHF antennas for embedded systems design. Through the techniques hereby presented, there was an attempt made to avoid the complex mathematical theory or heuristic methods, because in the real world there are many influential factors that can affect circuit performance. The tuning method would not have much value without a proper measurement system calibration, whose main goal is to cancel out phase shifting due to cable and transmission line length. By these means, we will also clear up a topic that is frequently disregarded in other antenna tuning articles, that is reference plane calibration. Practical examples and experiments were created for the purpose of supporting the techniques implied in this study.

Keywords— Antenna tuning; Impedance matching; VNA calibration; Reference plane; Characteristic impedance; RF measurement setup.

I. INTRODUCTION

The frequent use of antennas in modern electronic application represents a growing need for study and optimization of the matching process, that will ultimately contribute to the performance enhancement of RF systems. Antenna tuning is still an elusive process with many mysteries for most application design engineers, who often report problems which are not accounted in theoretical models and elude perception. In order to assimilate the meaning of the work presented here, users must have only basic knowledge regarding Vector Network Analyzer, RF microstrip circuits, measurement setups and embedded electronics design. The main problem regarding the subject remains hidden for most people that fail to optimize their antenna impedance through measurement and tune up. Even if the tuning method is well acknowledged, improper setup or calibration will totally ruin measurement results.

The concept of impedance matching represents a technical process that allows the designer to optimize RF circuit performance through the minimization of energy losses and increase of power transfer from source to load. This is done by measuring and adapting antenna impedance in order to obtain a near perfect impedance match.

An impedance match can be achieved mainly in two ways: by obtaining a source impedance equal to the complex conjugate of the load impedance or by creating complex impedances that have a null imaginary part, for both source and load. The first method is mostly used when matching power amplifiers with

other RF stages [2], but in the case of antennas and transmission lines it is preferred to match the antenna impedance to the characteristic impedance of the line, that is already close to a purely resistive value [2].

II. DESIGN PRINCIPLES

A. Transmission line losses

When handling frequencies in the UHF band, transmission line length, PCB layout and device enclosure have a great influence over complex impedance value, mostly on the imaginary part of the measured impedance. Signal reflections due to discontinuities, as seen in Fig 1, cause power loss, a frequency dependent impedance and nonlinear signal response.

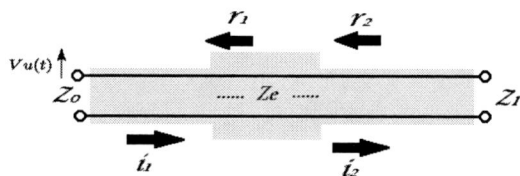

Fig.1 Signal reflections caused by discontinuities in the line impedance

Electronic devices that are manufactured in the form of chips, in SMT or THT technologies, have to be mounted on a PCB or some other kind of support material before they are connected in the measurement setup. This is the case for most ceramic chip antennas, that are produced in a variety of form factors, and require a form of adaptation when connected to the coaxial ports of a vector network analyzer (VNA).

The device under test (DUT), which is the antenna circuit in our case, will be physically connected to the VNA by at least two or three different transmission lines or adapters. In the effort of making these connections, great care must be taken into preserving the same impedance throughout the circuit because by doing so, only some minor reflections will appear. Each additional section that is introduced in the DUT circuit will also cause a certain alteration to the impedance that will translate into load mismatch and standing waves. The example presented in Fig. 1 notes the incident signal as i_1 and i_2, reflected signal as r_1 and r_2, Z_0 being the initial impedance, and Z_1 the resulting impedance, as seen by the VNA connected at Z_0.

2020 IEEE 26th International Symposium for Design and Technology in Electronic Packaging (SIITME)

The analysis conducted in this regard indicates that the calibration procedure for a fixtured measurement as the case presented in Fig. 1 will present additional difficulties if the signal reflections are not taken into account when measuring. By eliminating measurement setup influences, we are able to conduct proper measurements in microstrip or other non-coaxial circuits and achieve a good level of separation between the transmission medium attributes and true DUT characteristics. This is further studied in the form of a VNA port calibration and reference plane calibration, referred to the theory presented in [5] [6].

B. Reference plane calibration

For the accurate measurement calibration of the VNA port, a series of known devices, known as loads, are connected following the procedure recommended by the apparatus manufacturer [4]. System calibration is realized by measuring the real response of the cables and adapters that precede the DUT and comparing it with the known response of the calibration standards. By this means, error coefficients will be automatically calculated and from this moment on, systematic errors will be mathematically removed from the measured response.

Calibration needs to be done for a single port of the VNA, as we will be measuring only the impedance and reflection coefficient of the antenna. Therefore, we will need a set of high quality Short-Open-Load (SOL) standards, that usually equip the VNA set. To this point every procedure is standard and well defined, however the effects of the fixture used for the DUT are not accounted for. This is where the novelty of this procedure comes into place: we will perform a further calibration check and reference plane adjustment using the entire DUT fixture. Otherwise, without a precise reference plane setup, the representation of the complex impedance on the Smith chart will be affected and the user will not be able to conduct a correct matching process.

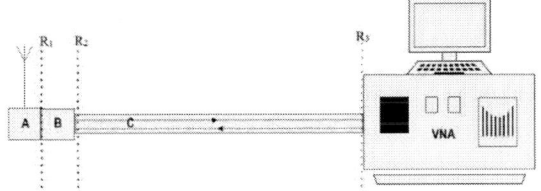

Fig.2 Simplified view of the measurement setup

The usual test setup is represented in Fig. 2 as composed of the VNA, a transmission line or coaxial cable (C), adapter connector (B) and DUT fixture (A). The physical termination planes are represented with a vertical dotted line. Following this representation, the reference plane after one port calibration is believed to be situated at the boundary between the adapter and the fixture (R1). Actually, the reference plane is measured as being inside the calibration loads or adapter connectors [4], and after load removal the VNA reference plane is in fact wrongly calibrated at an inexistent point, as represented in Fig. 3. For this reason, an accurate reference plane calibration process needs to be conducted with the DUT connected. Similar errors will be introduced if modifications are made upon the connection setup after calibration.

The common problem would be created by the need to change a certain line adapter to accommodate the DUT. This kind of alteration will change the length of the line and modify the electrical length of the circuit to measure, so it leads to an error in the value of the complex impedance.

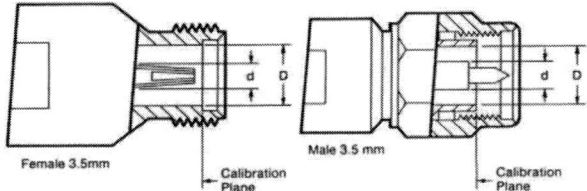

Fig.3 Mechanical construction of 3.5mm type connectors [4]

III. MATCHING PROCESS

In the impedance tuning process, most of the measurements will regard the complex impedance in the cartesian format (R+jX) plotted by the VNA on a Smith chart and the reflection coefficient |S11|. The reflection coefficient is a parameter that quantifies how much of the power injected by the network analyzer is reflected due to an impedance discontinuity in the transmission line and DUT. It is generally accepted that a good value for the reflection coefficient has to be less than -10dB, as it represents a power loss of under 10%. In the experiment section, we will consider the antenna usable bandwidth as being a frequency interval where |S11| is better than -10dB.

The implied impedance matching methods offer a cost efficient and simple implementation by using regular SMD inductors and capacitors for the matching circuit of the antenna following a precise calculation and placement of the passive components on the system PCB [2] [7] [8]. A disadvantage of this framework is that commercially available LC components limit the methods presented here to the UHF band – which is equivalent to a maximum of 3GHz. Beyond this frequency band, stray capacitances and inductances introduced by the component case start to dominate over nominal component value, this situation suggests the need for other types of circuits.

A. Measurement setup and calibration

After the standard VNA one port calibration is completed as described in the equipment manual, we will still need to perform a further calibration of the reference plane. This is in fact the most important part of the process and will be the key to a successful antenna tuning. Although it might be a known matter for some of the most experienced RF design engineers, this exact approach was not discussed in other scientific materials.

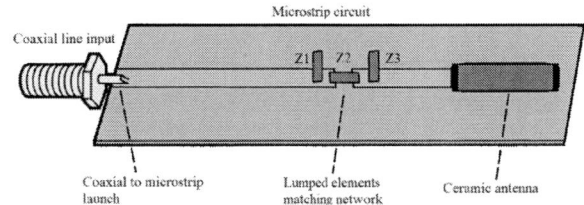

Fig.4 Simplified view of DUT fixture board

Firstly, the LC tuning network represented in Fig. 4 must be populated with a 0ohm resistor in the place of Z2, while Z1 and

978-1-7281-7507-2/20 $31.00 © 2020 IEEE   197   21-24 October, Pitesti, Romania

Z3 will remain unpopulated. Another short circuit, in the form of a 0ohm resistor, must pe temporarily placed in a shunt configuration as closely as possible to the feed point of the antenna. It can be even soldered directly between antenna input terminal and ground. This short circuit will give us the opportunity to measure and adjust the exact electrical length of the circuit before adding the LC network components.

Most vector network analyzers have an automatic feature, known as "auto port extension" [4] which removes the phase effect due to the PCB transmission line length from the measured data. The port extension calibration must be performed after the placement of the short circuit components described above. By using automatic port extension, the VNA adds a time delay to the reference signal path in order to produce a linear phase change [4]. Depending on the equipment or the setup, this automatic feature might not be enough to perform a precise cancellation of phase delay, in this case we will proceed to making a manual adjustment by adapting the electrical delay until the trace line crosses the resistive axes corresponding to a short circuit. After the port extension calibration is completed, the shunt short circuit preceding the antenna must be removed but Z2 will remain a 0ohm resistor.

B. Complex impedance value adjustment

A first measurement of the raw antenna impedance is plotted on the Smith chart and then components are added to the matching network in an effort to get the impedance value close to the ideal characteristic impedance of the system, represented by the centre point of the chart.

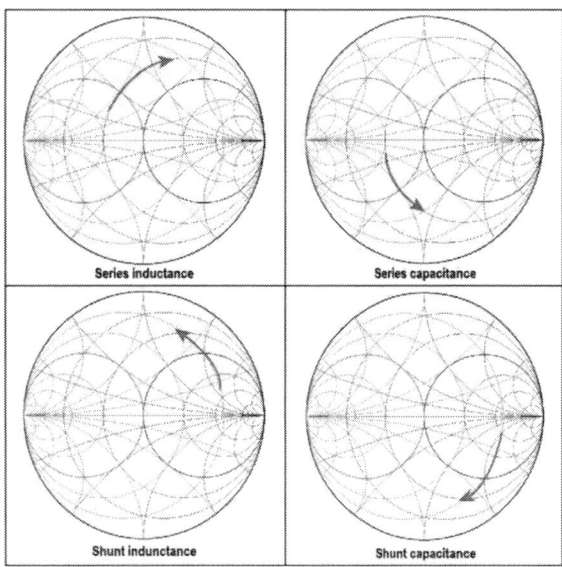

Fig. 5 Matching network component effect over antenna impedance as viewed on a Smith chart

Depending on the location of the initial impedance point, a certain combination of matching elements can convert the impedance to an intermediate value and finally to the characteristic impedance desired.

The matching process and calculus are thoroughly described in well-known scientific work [2] [3] [7] and we will focus only on some particular aspects that could greatly influence the final result. The route that the impedance point has to travel on the Smith chart is usually composed of two or three different sections, depending on the number of components in the network. Usually a bicomponent matching circuit can provide a valid solution to adapt impedance from almost any point on the Smith chart.

When adding components as in Fig. 5, the designer should take into account the parasitic effect the case has on the true value of the component. As a good precaution, values can be chosen a little smaller than what is calculated in order to cancel out parasitic influence. A general rule of thumb for the 0402 case would be to consider 0.2pF stray capacitance and 0.5nH inductance. After the placement of the first component in the matching network, the designer should verify that the impedance modification tracks the expected theoretical model. If there is an important difference between measured and theoretical impedance, then the port extension calibration should be redone.

IV. RESULTS AND DISCUSSION

To validate the antenna matching method proposed, several antenna fixture boards were fabricated, on which a series of measurements and experiments were conducted to prove the validity of the process. Experiments where focused on UHF radio frequency band, with the use of 433MHz and 2.6GHz antennas constructed on ceramic substrate materials. The antenna components selected are not specifically designed for a certain application, thus they can function with good performance in most industrial, telecommunication, scientific and medical domains. In the experiments presented in this paper, the tuning method was applied in the case of a Johansson antenna model no. 0433AT62A0020, which is intended for use in the 433MHz frequency band and a Linx antenna model ANT-2.45, designed for use in the 2.4GHz band, but will be tuned for 2.6GHz. In our experiments, performance was evaluated by measuring signal return loss, SWR, impedance variation versus frequency, bandwidth and temperature stability.

We focused the experiments on antennas fabricated using Low Temperature Cofired Ceramic (LTCC) technology. The fabrication process embeds the antenna conductive element into a ceramic substrate [10].

Modern ceramic materials have increasingly good performance when considering stability over time and temperature. LTCC technology aims at producing miniaturized antenna chips that prove to be reliable for a wide range of applications. The key domains that use such antennas are represented by embedded wireless products for home, medical or military applications. Manufacturers have a wide portfolio of antennas made specifically for Bluetooth, GSM, 802.11, LoRa, ZigBee and other ISM (industrial, scientific and medical) frequency bands [11].

Ceramic antennas are designed to be used on automated assembly lines, but nonetheless, they can also be successfully soldered using manual assembly techniques. Manufacturers recommend a maximum operation temperature of 85°C and a soldering temperature that doesn't exceed 225 °C [11].

2020 IEEE 26th International Symposium for Design and Technology in Electronic Packaging (SIITME)

A. Practical measurements and results

The test circuit seen in Fig. 6 was physically assembled on a double-sided PCB manufactured using regular FR4 TG 130-140 that has a relative permittivity of 4.3, thickness of 1.6mm and platted with 30μm copper. Special care has to be given to the PCB ground plane design, as to clear antenna premises and minimize parasitic capacitance. In order to minimize loss of RF power, the feed line from the SMA connector to the antenna is a transmission line designed for a characteristic impedance of 50 ohm.

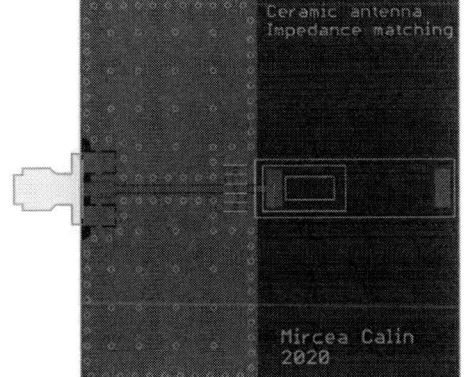

Fig. 6 Antenna universal test circuit PCB design

Care should be taken regarding matching values that are recommended in datasheet, because these only apply to the evaluation board design [10]. Overall board size and components layout may differ, but some important design rules must be kept. Antenna near field (one wavelength) has to be kept clear of traces, components and ground planes [11]. These rules apply for both layers of the board because conductive materials will impact the radiation pattern. In order for the quarter wavelength antenna to function, an opposite ground plane is needed on the circuit side of the board. The dimensions of the ground plane may vary but it has to be taken into account that it will affect antenna response. However, any changes that modify impedance can be canceled out during the tuning process.

The tuning network schematic used in the case of the 433MHz antenna in presented in Fig. 7. It is composed of a SMA connector and two SMD components in a 0402 case: one series inductor and a shunt capacitor that form a second order network.

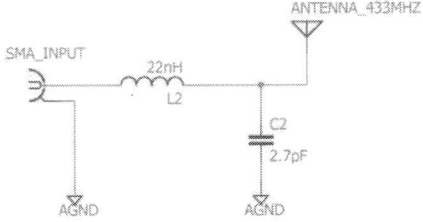

Fig. 7 Matching network configuration and values for the 433MHz antenna

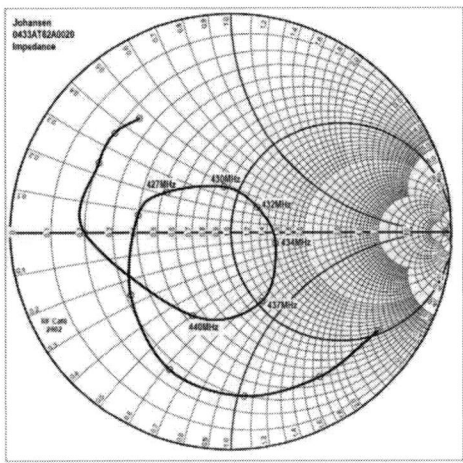

Fig. 8 Antenna impedance measured for the 433MHz band after the matching process

Antenna impedance for the 433MHz Johansson model is plotted on the Smith chart in Fig.8, based on the data points obtained from the VNA in order to evaluate the frequency dependence of the impedance over a wide bandwidth. By this means it can be easily observed that the usable bandwidth is composed of the points that are around the centre of the Smith chart, closest to the ideal impedance of 50Ω.

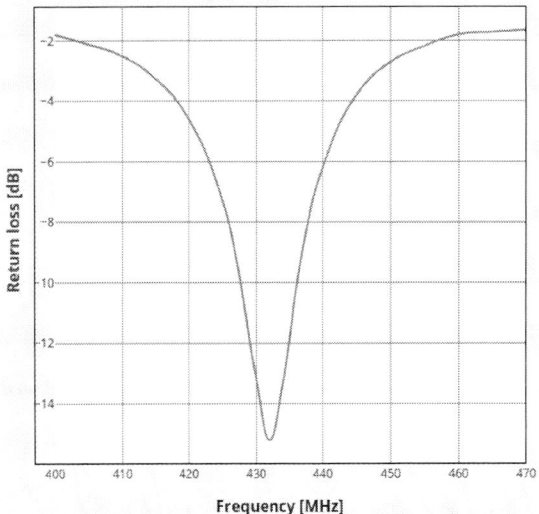

Fig. 9 Antenna impedance reflection coefficient for the 433MHz band

The return loss coefficient after tuning, seen in Fig. 9, confirms very good power transfer to the antenna within an 8MHz bandwidth, this result was determined considering a return loss coefficient better than -10dB and a centre frequency of 433MHz. The best performance is obtained at the resonant frequency of 433MHz, were $|S_{11}|$ reaches a value of -15dB, that corresponds to a transmission efficiency of over 96%.

The tuning network schematic used in the case of the 2.6GHz antenna in presented in Fig. 10. It is composed of a SMA connector and two SMD components in a 0402 case: a series inductor and a shunt capacitor that form a second order network.

978-1-7281-7507-2/20 $31.00 © 2020 IEEE
199
21-24 October, Pitesti, Romania

2020 IEEE 26th International Symposium for Design and Technology in Electronic Packaging (SIITME)

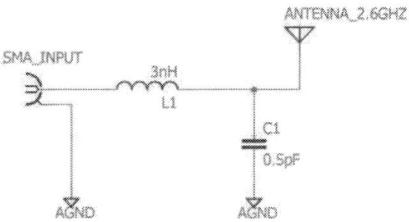

Fig. 10 Matching network configuration and values for the 2.6GHz antenna

Antenna impedance is plotted on a Smith chart in Fig.11 based on the data points obtained from the VNA, in order to evaluate the frequency dependence of the impedance over a wide bandwidth. By this means it can be observed that the majority of frequency points are around the centre of the Smith chart, closest to the ideal impedance of 50Ω. The scatter of measurement points on the chart is reduced, so we are expecting a wide bandwidth response in the case of this antenna.

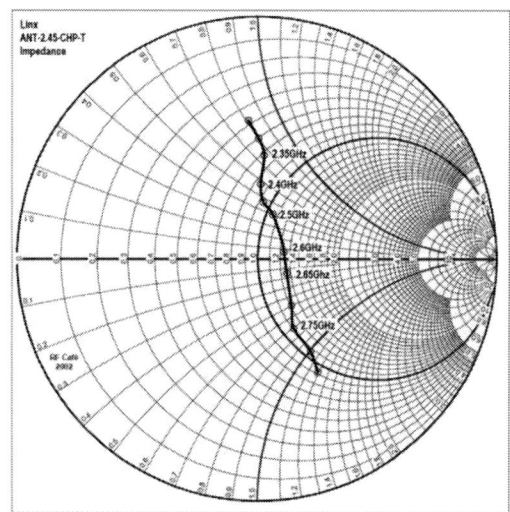

Fig. 11 Antenna impedance measured for the 2.6GHz band after the matching process

The plot in Fig. 12 represents the return loss coefficient after tuning for the Linx ANT-2.45. The results confirm very good power transfer to the antenna within a 300MHz bandwidth that includes both 2.4 and 2.6 GHz bands. Bandwidth was determined considering a return loss coefficient better than -10dB.

The best performance was obtained at the resonant frequency of 2.57GHz, were |S11| reaches a value of -17dB, that corresponds to a transmission efficiency of over 98%. As can it can be observed from the results, very good performance was obtained from an antenna used in a different band that it was originally designed for. This merit is claimed by the tuning process of the input impedance, which can accurately control antenna electrical characteristics and thus modify frequency response as needed by the application. Impedance tuning is the

equivalent of modifying antenna element dimensions, as it would be done in the case of classic wire antennas.

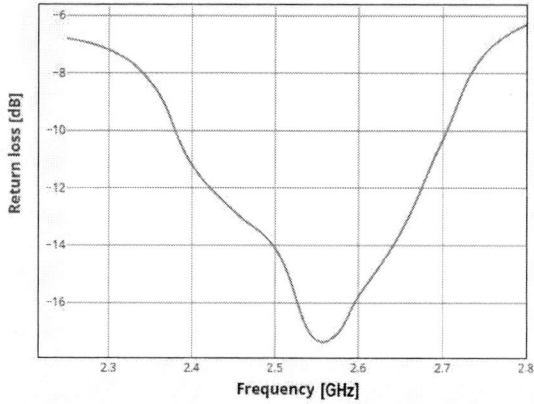

Fig. 12 Antenna impedance reflection coefficient for the 2.6GHz band

B. Further experiments

A further investigation was conducted in order to evaluate the temperature dependence of the antenna impedance. To obtain a clear result, return loss was measured over a 75°C interval, starting from a room temperature of 25°C, and plotted over a wide frequency range. The |S11| parameter measurement was preferred over complex impedance because a Smith chart plot of the data did not offer a sufficiently clear image of the changes that occurred in the antenna impedance.

Temperature variation will change the dielectric constant of the substrate material of PCB and antenna, also affecting the mechanical dimensions of the material which tends to expand or contract, thus modifying the volume of the dielectric which becomes temperature dependent [11].

From the graphic results shown in Fig. 13 for the 433MHz antenna we can conclude that temperature influence could be neglected in most cases but some effects exists and is manifested mostly in a shift of the resonance frequency. In all the experiments conducted, the increase of temperature has an inverse effect on the resonance frequency. By having this experimental data, the designer can counteract temperature influence by moving the usable bandwidth in the higher frequency range as to cancel out temperature shift.

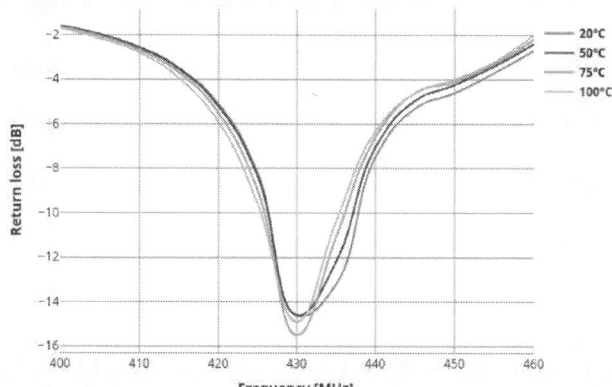

Fig. 13 The effect of temperature over antenna impedance reflection coefficient for the 433MHz band

Reflection coefficient variation for temperatures up to 100°C is plotted in Fig.14 for the 2.6GHz antenna. From this study we can conclude that temperature rise has a relatively small effect on antenna performance and is mostly beneficial. Temperature influence is manifested in the form of a small frequency shift to the lower side of the spectrum, but this does not affect antenna performance in the rated 2.6GHz band. Some minor changes can also be observed in the amplitude of the reflected wave in the form of a 2dB improvement of $|S_{11}|$ over the entire plotted frequency band.

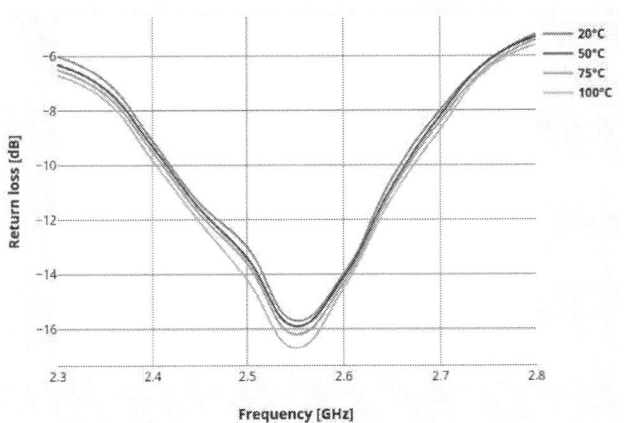

Fig. 14 The effect of temperature over antenna impedance reflection coefficient for the 2.6GHz band

V. CONCLUSION

In this paper, we have introduced a simple wideband impedance matching method for ceramic substrate antennas that are embedded in complex electronic circuits. Test boards and matching circuits for two conventional chip antennas were developed to demonstrate this method and provides a simple guide for achieving the best power transfer of a radio transmission for an embedded system. Design procedures and debates are given to provide clear guidance in the design, measurement and evaluation of functionality. The antenna tuning method described provides sufficiently accurate results to be used in practice. As presented in the work, there are a couple of different ways of validating the correctness of the results and all function well if they are combined with a precise reference plane calibration.

REFERENCES

[1] C. Hoarau, P. -. Bailly, J. -. Arnould, P. Ferrari and P. Xavier, "A RF tuneable impedance matching network with a complete design and measurement methodology," 2007 European Microwave Conference, Munich, 2007, pp. 751-754, doi: 10.1109/EUMC.2007.4405301.

[2] Chris Bowich, "RF Circuit Design", second edition 2009, Newnes publishing.

[3] Ofer Aluf, "Microwave RF Antennas and Circuits", Nonlinearity Applications in Engineering, Springer publishing 2017.

[4] Hewlett Packard "Product Note 8510-5A – Specifying calibration standards for the HP8510 network analyzer".

[5] Agilent Technologies' Product Note 8720-2, "In-Fixture Microstrip Measurements Using TRL Calibration," 1991.

[6] D. Rytting, "Appendix to an analysis of vector measurement accuracy enhancement techniques," RF & Microwave Symposium paper, March 1982.

[7] R. Rhea, "The Ying-Yang of Matching", High Frequency Electronics, Apr. 2006.

[8] Ki R. Shin, Yoon W. Kang, Maurice Piller, Aly E. Fathy, "Broadband antenna matching network design and application for RF plasma ion source", Particle Accelerator Conference, New York, NY, USA, 2011.

[9] K. Carver, and J. Mink, "Microstrip Antenna Technology," IEEE Transaction on Antennas and Propagation, vol. 29, no. 1, January 1981

[10] Johansson antenna datasheet for model no. 0433AT62A0020 https://johansontechnology.com/datasheets/0433AT62A0020/0433AT62A0020.pdf

[11] Proper PCB Design for Embedded Antennas, Application Note AN-00502 by Linx technologies https://linxtechnologies.com/wp/wp-content/uploads/an-00502.pdf

2020 IEEE 26th International Symposium for Design and Technology in Electronic Packaging (SIITME)

Key Expansion in Cryptographic Systems

Sorin Chiţu, Daniel Ciprian Vasile, Ionuţ Daniel Trămândan, Paul Svasta

UPB-CETTI
University POLITEHNICA of Bucharest
Bucharest, Romania
sorin.chitu@cetti.ro

Abstract—The manner in which a secret key is used in symmetric cryptographic systems is a permanent challenge for cryptographers due to its importance in overall security required for information protection. In order to mitigate the success rate of an attacker, a key expansion stage is included in cryptographic algorithms. This article presents an innovative solution that maximizes the key space, without reveal any information to an attacker who might monitor the communication channel. The prerequisites of this approach are the secret keys, a pseudorandom numbers generator and a random number generator, all of these being aggregate by a challenge-response mutual authentication protocol and a permutation function. For an efficient implementation, the already existing conductive mesh and the associated electronic circuits, dedicated to protect the cryptographic module against physical intrusion attempts of an attacker, are used to generate random numbers, without being necessary other electronic circuits. The numbers established as a result of the mutual authentication protocol will be used further to permute the secret key so, after each logical (re)synchronization, different keys will be used in cipher algorithm. Due to this key space extension process, the confusion and diffusion properties belonging to cypher algorithm are improved.

Keywords—key, cryptography, pseudorandom, permutation, algorithm

I. INTRODUCTION

Changing the secret keys in the cryptographic system is always a sensitive operation whether it takes place in symmetrical or asymmetrical ciphers. Also, it has to be taken into account that a too often or too rare key change could be exploited by an attacker in different ways. Applicable in the symmetrical cryptography domain, the proposed solution to extend the lifetime of the secret keys, without increasing the risk in security, consists of identical permutation of the keys in each corresponding terminal according to the result of the challenge-response protocol.

The new main idea introduced in this work is to use the existing conductive mesh that implements a tamperproof solution [1] to generate random numbers. So, inside of each device there will be generated trusted random sequences, which will be used in an authentication protocol. In Fig.1 is presented the block diagram to change the secret keys by permutation.

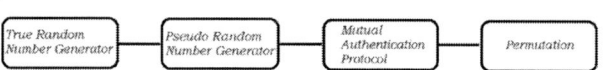

Fig.1 Block diagram of keys space expansion principle

II. DESCRIPTION OF THE SOLUTION

A. Random Number Generator Principle

To implement the mentioned tamperproof solution, 32 different frequency signals are injected into the conductive network and the output signals are analyzed. The resulted values characterize the conductive mesh being dependent on its design. The detection of any tamper intrusion attempt is based on observing any changes of output signals. Performing repetitive measurements on each frequency it was observed a relatively small dispersion of amplitude values, caused by quantization noise and, also, by environmental electromagnetic noise. It is important to notice that an attacker couldn't successfully influence the electromagnetically environment or external temperature in order to control the randomness, due to implicitly affecting of tamperproof sensor.

B. Pseudorandom Number Generator

Even if a True Random Number Generator (TRNG) is implemented based on unpredictable physic phenomena, in order to prevent the effects of possible failures, a test module will analyze the random bit stream output. But implementation of a test module involves hardware and software resources and is time-consuming.

An alternative solution is to implement a Pseudorandom Number Generator (PRNG) after the previous Random Number Generator. So, the random output bit stream consist the seed for the initialization of PRNG. In this situation, the bit stream output of PRNG has guaranteed good statistical characteristics, according to randomness needs in cryptographic applications.

One of the simplest PRNG consists of a Linear Feedback Shift Register (LFSR) which, in order to obtain the maximum period length, it has to implement a prime polynomial as register's feedback.

Moreover, to achieve increased performance, a Gollmann Cascade could be implemented, which is based on a combination of several LFSRs. This structure provides a nonlinear bit stream evolution, due to conditioning of the evolution of each LFSR by the output of previous LFSR.

Furthermore, the random numbers already generated by each communication terminal, RNDa and RNDb, are used in a challenge-response mutual authentication protocol. The protocol could be improved by using time stamp labels against time replay attacks which is targeted to discover secret

Freq [MHz]	50	55	60	65	70	75	80	85	90	95	100	105	110	115	120	125
Min	3295	3229	3185	3135	3062	3051	3107	3181	3206	3201	3152	3079	3026	3050	3086	3121
Max	3303	3237	3194	3144	3068	3058	3113	3187	3213	3208	3159	3085	3033	3056	3093	3127
Average	3299.1	3232.4	3189.9	3140.0	3065.2	3054.8	3110.0	3184.6	3209.9	3203.7	3154.9	3081.6	3029.5	3052.8	3089.6	3123.9
Difference	8	8	9	9	6	7	6	6	7	7	7	6	7	6	7	6
Freq [MHz]	130	135	140	145	150	155	160	165	170	175	180	185	190	195	200	205
Min	3143	3142	3116	3063	2997	2936	2911	2890	2856	2819	2775	2715	2629	2501	2301	2024
Max	3149	3148	3122	3069	3004	2943	2919	2897	2864	2825	2781	2723	2635	2507	2308	2030
Average	3146.4	3144.9	3119.7	3066.4	3000.6	2939.5	2915.0	2893.5	2860.0	2821.8	2778.2	2718.6	2632.1	2503.7	2304.6	2027.0
Difference	6	6	6	6	7	7	8	7	8	6	6	8	6	6	7	6

Fig. 3: The dispersion of amplitude values measured at the output of conductive mesh

authentication key, Ka, supposing the hash function, *h*, is known, according to Kerckhoff's principle.

C. Mutual Authentication Protocol and Key Permutation

The obtained random 64 bits from the output of TRNG and PRNG chain are used in a challenge-response protocol, exemplified in Fig.2, in order to establish, in negotiation with a corresponding terminal, a common known number that will be used, in the next stage, by a permutation function [2].

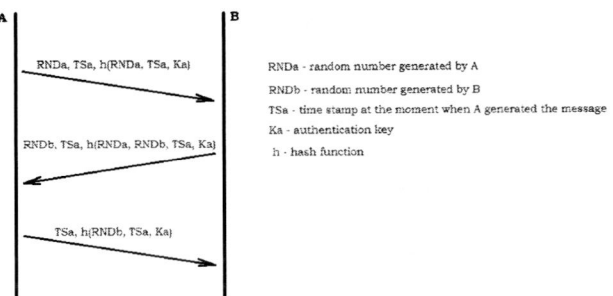

Fig. 2: An efficient implementation of a challenge-response mutual authentication protocol

The permutation function rearranges the elements of the original secret key according to the number that was established between the involved communication terminals, during mutual authentication protocol. It is important to notice that an attacker will not be able to capitalize the established number even he will intercept the communication channel between the two corresponding terminals.

Each time an authentication protocol is used, a new key is generated by rearranging the original bits of the initial key. From a cryptanalyst point of view, this is equivalent with a new key, so his work will be more difficult. Also, the usage of this permutation function provides uniform cover of keys space determined by the all bits combination.

III. RESULTS

To evaluate the signal variation sampled at the output of conductive mesh an experimental test design PCB which covers the whole 30x30mm surface was used. The conductive mesh drawing consists of straight-line segments with a thickness of 0.2mm spaced at 0.2mm from neighboring lines.

In order to analyze the randomness component of the signals measured at the output of the conductive mesh, there were performed numerous samples acquisitions to constitute a database. The calculated average of signal dispersion is about 6.8 quanta of a 12 bit ADC. Coding this dispersion on 2 bits and considering the mentioned 32 frequencies result an unpredictable 64 bit stream that can be used as a random number. The experimental results, presented in Fig. 3, confirm that a conductive mesh could be used in order to generate random numbers as seed for the next module which is a pseudorandom number generator.

The statistical analysis results, presented in Appendix 1, does not validate the random number strings thus obtained for direct use in cryptographic algorithms but, as it was presented in Fig.1, the next Pseudorandom Number Generator module will use these bits as seed so the final result is guaranteed suitable according to its destination.

IV. CONCLUSIONS

Targeting the keys space expansion, by the proposed method, not only the overall security of a cryptographic terminal was achieved, but also confusion and diffusion of a secure cipher were improved [3]. This method could be applied to any cipher type belonging to symmetrical cryptography, either block (like AES) or stream ciphers.

The proposed algorithm provides a solution to complete digitization and automation of tamper proof design and production processes. By implementing a design and production flux without human intervention increase the security level, by a higher trust level in maintaining the confidence of the tamper proof electrical parameters.

The presented solution, besides its contribution to security improvement of the secret keys usage in a cipher algorithm, is costless due to already existing conductive mesh which is integrated in the tamperproof module.

ACKNOWLEDGMENT

This paper and research activities were supported from the funds of the project "Integrated Development 4.0" (iDEV40) funded from the ECSEL Joint Undertaking (JU) under grant agreement No 783163 , respectively the "Dezvoltare integrata 4.0" project having the SMIS code 122386 within the Competitiveness Operational Program (POC) - Romania. The JU receives support from the European Union's Horizon 2020 research and innovation programme. It is co-funded by the consortium members, grants from Austria, Germany, Belgium, Italy, Spain and Romania. The information and results set out in this publication are those of the authors and do not necessarily reflect the opinion of the ECSEL Joint Undertaking.

REFERENCES

[1] D.C. Vasile, P. Svasta, "Protecting the Secrets: Advanced Technique for Active Tamper Detection Systems", SIITME 2019, IEEE 25th International Symposium for Design and Technology in Electronic Packaging, Cluj-Napoca, Romania DOI: 10.1109/SIITME47687.2019.8990877

[2] D.E. Knuth, "The Art of Computer Programming: Seminumerical Algorithms", Vol. 2, pp. 139-140

[3] Claude Shannon, classified report "A Mathematical Theory of Cryptography", 1945, https://www.iacr.org/museum/shannon/shannon45.pdf

Appendix 1

Tests of Distinguishability from Random
TEST: monobit_test
  Ones count   = 1061102
  Zeroes count = 577298
  FAIL
  P=0.0
TEST: frequency_within_block_test
  n = 1638400
  N = 99
  M = 16549
  FAIL
  P=0.0
TEST: runs_test
  prop 0.647645263671875
  tau 0.0015625
  FAIL
  P=0.0
TEST: longest_run_ones_in_a_block_test
  n = 1638400
  K = 6
  M = 10000
  N = 75
  chi_sq = 852.6905259225296
  FAIL
  P=6.326085848635833e-181
TEST: binary_matrix_rank_test
  Number of blocks 1600
  Data bits used: 1638400
  Data bits discarded: 0
  Full Rank Count = 461
  Full Rank -1 Count = 932
  Remainder Count = 207
  Chi-Square = 0.2869478317657266
  PASS
  P=0.8663433976951915
TEST: dft_test
  N0 = 778240.000000
  N1 = 792949.000000
  FAIL
  P=0.0
TEST: non_overlapping_template_matching_test
  PASS
  P=0.9904867968470971
TEST: overlapping_template_matching_test
  B = [1, 1, 1, 1, 1, 1, 1, 1, 1, 1]
  m = 10
  M = 1062
  N = 968
  K = 5
  model = [352, 179, 134, 97, 68, 135]
  v[j] = [8, 15, 33, 30, 35, 847]

chisq = 12555.624363408284
FAIL
P=0.0
TEST: maurers_universal_test
  sum = 1336441.3361360026
  fn = 5.741294613024493
  FAIL
  P=0.0
TEST: linear_complexity_test
  M = 512
  N = 3200
  K = 6
  chisq = 6.42070091130002
  P = 0.3777477696067265
  PASS
  P=0.3777477696067265
TEST: serial_test
  psi_sq_m = 648230.8836718751
  psi_sq_mm1 = 467284.70205078134
  psi_sq_mm2 = 290570.0715917968
  delta1   = 180946.18162109377
  delta2   = 4231.551162109245
  P1   = 0.0
  P2   = 0.0
  FAIL
P=0.0
P=0.0
TEST: approximate_entropy_test
  n   = 1638400
  m   = 3
  Pattern 1 of 8, count = 68658
  Pattern 2 of 8, count = 121313
  Pattern 3 of 8, count = 152965
  Pattern 4 of 8, count = 234362
  Pattern 5 of 8, count = 121313
  Pattern 6 of 8, count = 266014
  Pattern 7 of 8, count = 234362
  Pattern 8 of 8, count = 439413
  phi(3) = -4.246841
  Pattern 1 of 16, count = 22745
  Pattern 2 of 16, count = 45913
  Pattern 3 of 16, count = 44125
  Pattern 4 of 16, count = 77188
  Pattern 5 of 16, count = 44846
  Pattern 6 of 16, count = 108119
  Pattern 7 of 16, count = 74324
  Pattern 8 of 16, count = 160038
  Pattern 9 of 16, count = 45913
  Pattern 10 of 16, count = 75400
  Pattern 11 of 16, count = 108840
  Pattern 12 of 16, count = 157174
  Pattern 13 of 16, count = 76467
  Pattern 14 of 16, count = 157895
  Pattern 15 of 16, count = 160038
  Pattern 16 of 16, count = 279375
  phi(3) = -4.893008
  AppEn(3) = 0.646167

ChiSquare = 153943.93212476827
FAIL
P=0.0
TEST: cumulative_sums_test
FAIL: Data not random
FAIL
P=0.0
P=0.0
TEST: random_excursion_test
J=15
x = -4 chisq = 2.144357   p = 0.828831
x = -3 chisq = 3.000720   p = 0.699875
x = -2 chisq = 5.001500   p = 0.415697
x = -1 chisq = 14.998500   p = 0.010369
x = 1   chisq = 4.205338   p = 0.520247
x = 2   chisq = 3.642156   p = 0.601995
x = 3   chisq = 3.566630   p = 0.613330
x = 4   chisq = 4.801013   p = 0.440644
J too small (J < 500) for result to be reliable
 PASS
P=0.828831423785823
P=0.6998748216021984
P=0.4156971521741977
P=0.010368749105646946
P=0.5202470850491279
P=0.6019950717407222
P=0.6133304972973969
P=0.4406444169768629
TEST: random_excursion_variant_test
J= 15
x = -9 count=0   p = 0.506555
x = -8 count=0   p = 0.479500
x = -7 count=0   p = 0.447521
x = -6 count=0   p = 0.408961
x = -5 count=0   p = 0.361310
x = -4 count=0   p = 0.300623
x = -3 count=0   p = 0.220671
x = -2 count=0   p = 0.113846
x = -1 count=10   p = 0.361310
x = 1   count=13   p = 0.715001
x = 2   count=9   p = 0.527089
x = 3   count=2   p = 0.288487
x = 4   count=2   p = 0.369673
x = 5   count=1   p = 0.394207
x = 6   count=1   p = 0.440900
x = 7   count=1   p = 0.478376
x = 8   count=1   p = 0.509275
x = 9   count=3   p = 0.595163
J too small (J=15 < 500) for result to be reliable
 PASS
P=0.5065551690490404
P=0.4795001221869534
P=0.4475209101304692
P=0.40896134203687456
P=0.36131042852617884
P=0.30062298819690675
P=0.22067136191984682

P=0.11384629800665803
P=0.36131042852617884
P=0.7150006546880892
P=0.5270892568655381
P=0.2884874633234893
P=0.3696734409705522
P=0.3942069504679318
P=0.44089980988590066
P=0.47837563924795834
P=0.5092754371240833
P=0.5951631467095724

SUMMARY
----
monobit_test   0.0   FAIL
frequency_within_block_test   0.0   FAIL
runs_test   0.0   FAIL
longest_run_ones_in_a_block_test
6.326085848635833e-181 FAIL
binary_matrix_rank_test   0.8663433976951915
PASS
dft_test   0.0   FAIL
non_overlapping_template_matching_test
0.9904867968470971 PASS
overlapping_template_matching_test   0.0   FAIL
maurers_universal_test   0.0   FAIL
linear_complexity_test   0.3777477696067265
PASS
serial_test   0.0   FAIL
approximate_entropy_test   0.0   FAIL
cumulative_sums_test   0.0   FAIL
random_excursion_test   0.010368749105646946
PASS
random_excursion_variant_test
0.11384629800665803 PASS

Blockchain-Based Image Copyright Protection System using JPEG Resistant Digital Signature

Robert Alexandru Dobre
Department of Electronic Technology and Reliability
Politehnica University of Bucharest
Bucharest, Romania
robert.dobre@upb.ro

Radu Ovidiu Preda, Radu Alexandru Badea, Mihai Stanciu
Telecommunications Department
Politehnica University of Bucharest
Bucharest, Romania

Alexandru Brumaru
INVITE Systems
Bucharest, Romania

Abstract—A notable increase in the distribution of digital images was fueled by the social media platforms. In the fight for attention, the posts containing images have more chances to make users stop scrolling through the crowded news feed than the posts containing only text. The effect is enhanced if the photos that are used in the posts are professional grade. In the context of the COVID-19 pandemic, more and more businesses have moved to online. Every business makes efforts to catch the attention of potential customers with high quality images. These situations could lead to intended or accidental image copyright infringement. This paper proposes a system that can be used to detect and avoid copyright infringement. Photographers can use it to register their photos and businesses can use it to check if the image they want to use is copyright protected or not. If it is, the system also allows the purchase of the right to use the photo. The solution is based on blockchain because of its immutability property. Businesses can look for images on many sites, thus it is very probable that recompressed versions of the same photo would be stored in different places. Therefore, it is important to develop an algorithm that can extract a signature that is resistant to JPEG compression from the image. The signature should be stored on the blockchain along the identification data of the copyright owner.

Keywords—*blockchain; copyright protection; Hyperledger Fabric*

I. INTRODUCTION

The COVID-19 pandemic caused many countries to declare state of emergency. Businesses were faced with the necessity to move to online commerce or enhance their already existing online presence. Therefore, the competition for the customers' attention is decided by how well the businesses manage their website and social media pages. Images play a decisive role in this situation. Professional shots of products or illustration of services have much higher chances to attract customers and project a favorable image of the company. Doubtlessly, professional photoshoots are costly. Because more companies could sell the same product, the following situation could happen: one company pays for the photoshoot of a product and post the photos on their site. Another company finds these photos and use them for the product they sell. In this case, a

copyright infringement happened, which could be intended or accidental. Because we can assume that the two companies are competitors, they may not want to discuss details with each other. Therefore, the second company has difficulties to find if the image is copyrighted [1-2] or available for free. A distributed, immutable database that stores all the information about the pictures and their rightful owners helps in this situation.

A network with such properties is a blockchain [3-4]. Unfortunately, images are too large to be directly stored on blockchains. Therefore, some features that uniquely identify the photo are extracted from the image and stored. These features represent the digital signature of the image. Usually, when uploaded to websites or social media pages, images are compressed. Thus, it is of utmost importance that the digital signature extraction algorithm [5,6] would extract the same signature from the original image and from recompressed versions of it. The authors already proposed a blockchain-based image authentication system, being ones of the pioneers of image applications involving blockchain [7]. This work extends the blockchain-based image security research to the copyright protection problematic. The paper presents the blockchain network architecture and the digital signature extraction algorithm, representing the image copyright protection system.

The paper is structured as follows: Section II briefly presents the blockchain technology, Section III thoroughly presents the image signature extraction algorithm, Section IV presents the architecture of the proposed system and shows the results, and Section V concludes the paper.

II. BLOCKCHAINS

Blockchains have become popular since the introduction of one of its applications, the first blockchain-based cryptocurrency, Bitcoin [8], in 2009. A blockchain is, as the name suggests, a linked collection of blocks containing information, an emerging technology [9-13]. Every block, besides the general information that it contains depending on the application, it also features a time stamp and some cryptographic data pointing to the previous block that help to demonstrate the integrity of the blockchain at any moment. One of the great advantages of blockchain is that it offers the possibility to record data without needing a central overseeing entity. Besides Bitcoin, there are many famous blockchains up and running, like Ethereum, Ripple, BigchainDB, Hyperledger and many others, each targeting specific use cases.

One important classification of blockchains is based on who can see the information stored in it. There are public and private blockchains. Bitcoin and Ethereum, for example, are public blockchains because anyone can see the data stored at various addresses and the transactions that were performed in the blockchain using a blockchain explorer platform. There are use cases in which a private blockchain is needed. The best example is a blockchain that holds the information about the interaction of various suppliers with other many distributors. It would be an economic disadvantage to have the transactions between suppliers and distributors stored on a public blockchain because the competition could know the terms and take measures to better adapt to the market conditions.

Hyperledger is a large blockchain oriented collaboration of businesses from technology, IoT, finance and other fields, providing blockchain systems for industry. There are many projects that resulted from this collaboration, tuned for different applications like Fabric, Composer, Cello, Explorer, Burrow, and Sawtooth. Hyperledger Fabric is an open source, highly modular, distributed ledger framework for developing general blockchain applications for business use. The current version is Fabric 2.0 [14], launched in January 2020, featuring quicker transactions, refined technology. Given its open source and private nature, Hyperledger Fabric is a good candidate for the application presented in this paper.

III. THE SIGNATURE EXTRACTION ALGORITHM

It has been discussed above the fact that a complete image, at usual resolutions used on the web, is too large to be stored completely on the blockchain. The network would become mostly a storage space and would be very difficult and slow to operate. Therefore, a unique identifier for the images should be extracted from the images that should be small enough to be easy to store on the blockchain, while being also resistant to certain operations that are commonly found when working with photos on the web. Resistance to a certain operation means that the same signature would be extracted from the image before and after the operation has been applied to the image. The most important operation is JPEG (Joint Photographic Experts Group) compression. Almost every site applies its own JPEG compression to the images stored on it to reduce the space occupied by the images. Therefore, the signature extraction algorithm should be resistant to JPEG compression. To achieve this, the algorithm follows closely the steps of JPEG compression and is described further.

The starting point is a color digital image. The first stage of the algorithm is represented by the conversion of the image from the RGB space to the YC_bC_r space. In the next stages, only the Y component is used, that represents the luminance. The Y plane is then split in non-overlapping 8 by 8 pixel blocks, just as the JPEG compression operates, then the two dimensional Discrete Cosine Transform (DCT) is computed for each block. The obtained coefficients are then quantized using the JPEG quantization matrix for the luminance, that is a function of the JPEG quality factor q, denoted with $\Delta(q)$. If we denote the matrix containing the 64 DCT coefficients obtained after applying the two dimensional DCT transform to an 8 by 8 pixel block with $\mathbf{K}$, the quantization operation referred above is computed as:

$$\mathbf{K}_N = \text{round}\left[\quad \oslash \Delta(q) \right], \tag{1}$$

where $\oslash$ is the Hadamard (element-wise) division operator, and

$$\Delta(q) = \begin{cases} \text{round}\left(\dfrac{50 \cdot \Delta(50)}{q} \right), 1 \le q \le 50 \\ \text{round}\left[50 \cdot \Delta(50) \cdot (2 - 0.02 \cdot q) \right], 50 < \quad \le 100 \end{cases}, \tag{2}$$

where $\Delta(50)$ is the standard quantization matrix used in JPEG compression, for a quality factor equal to 50. $\mathbf{K}_N$ is the matrix that stores the newly obtained coefficients, named normalized coefficients.

This quantization operation, described in (1), is the key element that assures the method's resistance to JPEG compression. The quantization function, in general, defined as

$$\delta(x, s) = s \cdot \text{ro} \quad \text{d}\left(\frac{x}{s} \right), \tag{3}$$

where x is the value to be quantized and s is the step size, has the following property, demonstrated in [15]:

$$\delta\left[\delta(x, s_2), s_1 \right] = \delta(x, s_1) + \varepsilon, \text{ when } s_2 < s_1, \tag{4}$$

with $\varepsilon \in \{ -s_1; 0; s_1 \}$. This means that the signature extracted from the image with this method will be the same even if the image suffered a JPEG compression with a quality factor larger than a set value used in the algorithm, denoted with q_{min}. Usually, the quality factors used by websites and other platforms for media transfer or storage that apply JPEG compression are over 70. Therefore, in the algorithm, any value q_{min} of smaller than 70 should assure a sufficient degree of generality to the method.

The steps described to this point assure that the JPEG compression will not change the results in obtaining the normalized DCT coefficients in the matrices $\mathbf{K}_N$ if the quality factor used in the compression operation is greater than the one used in the method, q_{min}, but so far no feature extraction is obtained. The feature extraction algorithm is described further and is obtained after testing many approaches, selecting the one that provided the best results.

$\mathbf{K}_N$ is a matrix with 64 elements, having 8 lines and 8 columns. Therefore, its elements can be addressed with

$$\mathbf{K}_N(i, j), \ i, j \in \{0, 1, ..., 7\}, \tag{5}$$

with $\mathbf{K}_N(0,0)$ denoting the normalized DC coefficient, and the others being named normalized AC coefficient, the names referring to special frequencies. From each 8 by 8 block processed as explained before, 8 bits are extracted, representing the feature for the current block, denoted with $b_0,...,b_7$. They are obtained by threshold comparison in (6) where $\mathbf{K}_N(0,0)_{max}$ is the maximum value that can be obtained for the normalized DC coefficient and, for b_6 and b_7 the threshold value was empirically determined. With this approach, the size of the signature will be only 0.52% of the size of the uncompressed image, making it much better suited to be stored on blockchains as image identifier.

$$b_0 = \begin{cases} 1, & \mathbf{K}_N(0,0) \geq \dfrac{\mathbf{K}_N(0,0)_{max}}{4}, \\ 0, & \text{otherwise} \end{cases}$$

$$b_1 = \begin{cases} 1, & \mathbf{K}_N(0,0) \geq \dfrac{\mathbf{K}_N(0,0)_{max}}{2}, \\ 0, & \text{otherwise} \end{cases}$$

$$b_2 = \begin{cases} 1, & \mathbf{K}_N(0,0) \geq \dfrac{3 \cdot \mathbf{K}_N(0,0)_{max}}{2}, \\ 0, & \text{otherwise} \end{cases}$$

$$b_3 = \begin{cases} 1, & \mathbf{K}_N(0,1) \geq 0 \\ 0, & \text{otherwise} \end{cases},$$

$$b_4 = \begin{cases} 1, & \mathbf{K}_N(1,0) \geq 0 \\ 0, & \text{otherwise} \end{cases},$$

$$b_5 = \begin{cases} 1, & \mathbf{K}_N(2,0) \geq 0 \\ 0, & \text{otherwise} \end{cases},$$

$$b_6 = \begin{cases} 1, & \left|\mathbf{K}_N(0,1)\right| \geq 3 \\ 0, & \text{otherwise} \end{cases},$$

$$b_7 = \begin{cases} 1, & \left|\mathbf{K}_N(1,0)\right| \geq 3 \\ 0, & \text{otherwise} \end{cases},$$

(6)

To ease the understanding of the method, an example of signature extraction from an image block is detailed further. Supposing that the DCT coefficients that participate in the process have the following values: $\mathbf{K}(0,0) = 684$, $\mathbf{K}(0,1) = 44.3$, $\mathbf{K}(1,0) = -18.5$, $\mathbf{K}(2,0) = 27.6$, and we desire that the signature should be resistant to JPEG compression with a quality factor greater than 50, therefore $q_{min} = 50$. With this, the quantization matrix for q_{min} is calculated, $\Delta(q_{min})$ using (2), keeping in mind that $\Delta(50)$ is given in the JPEG documentation. In the chosen case, because $q_{min} = 50$, no calculation is needed, the matrix from the JPEG documentation is used. The values found in the $\Delta(50)$ matrix on the same positions like the coefficients in the $\mathbf{K}$ matrix that participate in the process, namely $(0,0),(1,0),(0,1)$, and $(2,0)$ are equal to 16, 11, 12, and 14. Equation (1) is used to obtain the normalized coefficients:

$$\mathbf{K}_N(0,0) = \text{round}(684/16) = 43,$$
$$\mathbf{K}_N(0,1) = \text{round}(44.3/11) = 4,$$
$$\mathbf{K}_N(1,0) = \text{round}(-18.5/12) = -1,$$
$$\mathbf{K}_N(2,0) = \text{round}(27.6/14) = 2.$$

$\mathbf{K}_N(0,0)_{max}$ is equal to 128, therefore the extracted signature in this example case is: $b_0...b_7 = 10010100$. The process repeats for all the blocks in the image. By concatenating the signatures extracted from each block, it results the signature for the whole image. It is to be noted that, if the dimensions of the image are not equal to a multiple of 8 pixels, there will be two areas of the image that will not participate in the signature extraction. They are placed on the right and the bottom of the image, and their width and, respectively, height are equal to at most 7 pixels. These bands are very small to be considered that they contain an important part of the image, given today's image resolutions.

IV. THE PROPOSED COPYRIGHT PROTECTION SYSTEM AND RESULTS

The block diagram of the system is shown in Fig. 1, and its user interface in Fig. 2. The main two components are

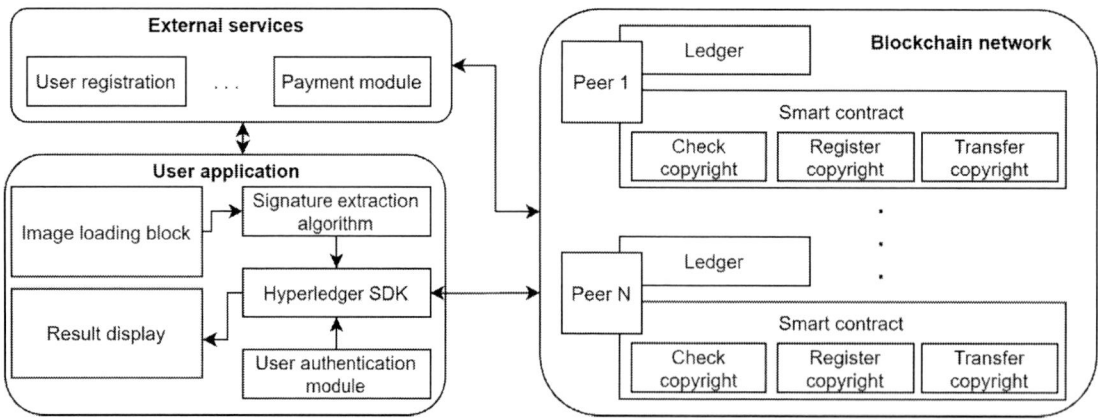

Fig. 1. The block diagram of the proposed blockchain-based image copyright protection system

*2020 IEEE 26th International Symposium for Design and Technology in Electronic Packaging (**SIITME**)*

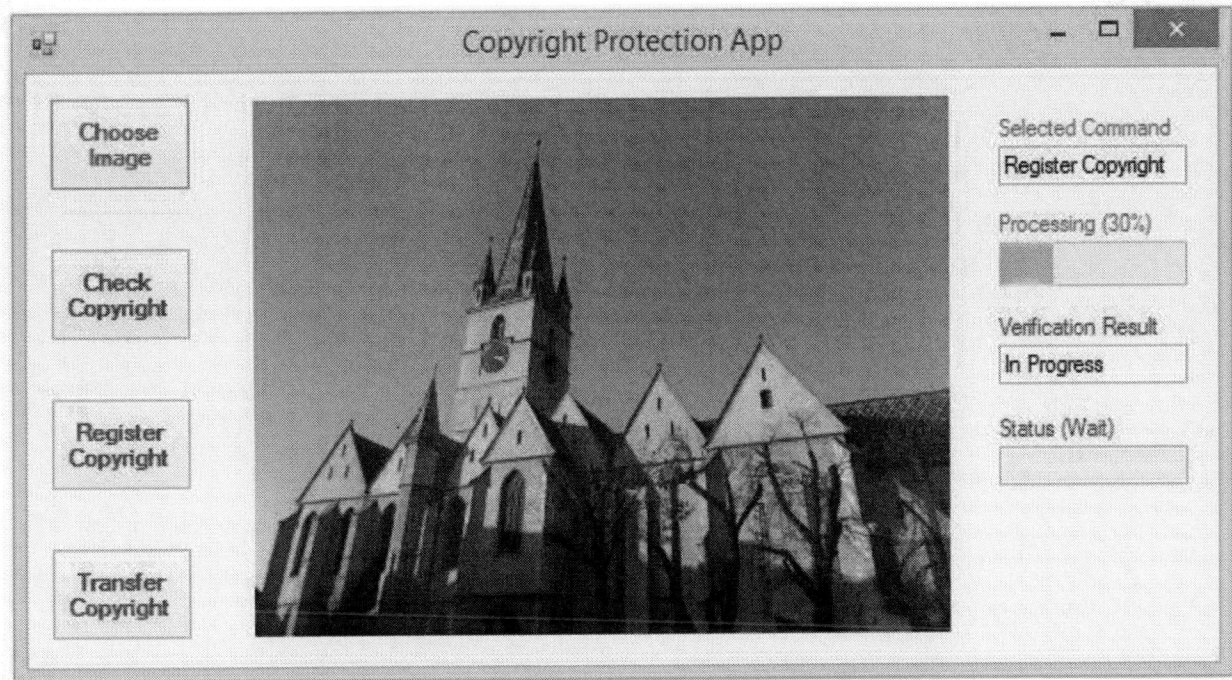

Fig. 2. The user interface of the blockchain-based copyright protection system

represented by the user application and the blockchain network. The user application can load images and apply the signature extraction algorithm described in this paper on them. After this processing, using the extracted signature, user authenticated identity and the Software Development Kit for the blockchain network, the user can invoke certain functions available on the blockchain network. The functions are defined on the blockchain network, so the user can only invoke them and receive results depending on its privileges and resources.

The "Check copyright" function receives the image signature and queries the ledger to determine if it is registered by a photographer, then it informs the user regarding the copyright state of the selected image, as well as if they have the right to use it or they must buy it. The "Register copyright" function allows photographers to write the signatures of their images to the ledger. In this way, when checking for copyright, the pictures having those signatures written on the ledger will be marked as copyrighted. The "Transfer copyright" function allows the copyright transfer from photographers (owners) to the users that bought the right to use the photos. The signature of the image for which the copyright was bought is copied to the data collection of the user that bought it, being marked also with an identifier that shows it is a user, not a photographer that holds the copyright, which must be unique. Photographer role can also be registered to media agencies.

The resistance to JPEG compression was tested with 200 images. The value of q_{min} was set to 60, and the recompression used a quality factor between 70 and 100. The same signature was extracted after recompression in all the cases.

V. CONCLUSIONS

In this paper a blockchain-based image copyright protection was proposed, that uses as the picture identifier a JPEG resistant image signature. The paper briefly presents the context in which such system is needed, it thoroughly describes the image signature extraction algorithm, showing why it can resist JPEG compression and presenting also an example of signature extraction, for ease of understanding, and describes the copyright protection system based on Hyperledger Fabric, discussing the architecture and showing a proposed user interface. The system was tested with 200 images and proved to be functioning as expected. Future work will explore other image signature extraction algorithms, that could resist to other operations too.

ACKNOWLEDGMENT

This work has been supported by the Operational Programme Human Capital of the Ministry of European Funds through the Financial Agreement "Developing the entrepreneurial skills of the PhD students and postdoctoral students - key to career success (A-Succes)" contract number 51675/09.07.2019 POCU/380/6/13, SMIS code 125125, and in part under the research subjects "e-Government" during the sustainability timeframe of the project "Endowing investment at Ad Net Market Media laboratories for R&D in future mobile communications" – FUTURE-NET-LAB, funded by the EU under the Operational Program Competitivity (POC), priority axis 1, action 1.1.1:.Large R&D infrastructures.

REFERENCES

[1] A. Pluszyńska, 'Copyright Management by Contemporary Art Exhibition Institutions in Poland: Case Study of the Zachęta National Gallery of Art',

Sustainability, vol. 12, no. 11, Art. no. 11, Jan. 2020, doi: 10.3390/su12114498.

[2] D. Megías, M. Kuribayashi, and A. Qureshi, 'Survey on Decentralized Fingerprinting Solutions: Copyright Protection through Piracy Tracing', *Computers*, vol. 9, no. 2, Art. no. 2, Jun. 2020, doi: 10.3390/computers9020026.

[3] 'Understanding Blockchain Technology: Abstracting the Blockchain'. https://ieeexplore.ieee.org/courses/details/EDP521 (accessed Sep. 20, 2020).

[4] A. Hafid, A. S. Hafid, and M. Samih, 'Scaling Blockchains: A Comprehensive Survey', *IEEE Access*, vol. 8, pp. 125244–125262, 2020, doi: 10.1109/ACCESS.2020.3007251.

[5] R. A. Dobre, R. O. Preda, and A. E. Marcu, 'Improved Active Method for Image Forgery Detection and Localization on Mobile Devices', in *2018 IEEE 24th International Symposium for Design and Technology in Electronic Packaging (SIITME)*, Oct. 2018, pp. 255–260, doi: 10.1109/SIITME.2018.8599235.

[6] R. O. Preda, 'Semi-fragile watermarking for image authentication with sensitive tamper localization in the wavelet domain', *Measurement*, vol. 46, no. 1, pp. 367–373, Jan. 2013, doi: 10.1016/j.measurement.2012.07.010.

[7] R. A. Dobre, R. O. Preda, C. C. Oprea, and I. Pirnog, 'Authentication of JPEG Images on the Blockchain', in *2018 International Conference on Control, Artificial Intelligence, Robotics Optimization (ICCAIRO)*, May 2018, pp. 211–215, doi: 10.1109/ICCAIRO.2018.00042.

[8] S. Nakamoto, "Bitcoin: A Peer-to-Peer Electronic Cash System", 2008, available online at https://bitcoing.org/bitcoin.pdf.

[9] P. W. Khan, Y.-C. Byun, and N. Park, 'A Data Verification System for CCTV Surveillance Cameras Using Blockchain Technology in Smart Cities', *Electronics*, vol. 9, no. 3, Art. no. 3, Mar. 2020, doi: 10.3390/electronics9030484.

[10] A.-M. C. Drăgulinescu, A. Drăgulinescu, I. Marcu, S. Halunga, and O. Fratu, 'SmartGreeting: A New Smart Home System Which Enables Context-Aware Services', in *Future Access Enablers for Ubiquitous and Intelligent Infrastructures*, Cham, 2018, pp. 158–164, doi: 10.1007/978-3-319-92213-3_23.

[11] A. M. C. Drăgulinescu, A. F. Manea, O. Fratu, and A. Drăgulinescu, 'LoRa-Based Medical IoT System Architecture and Testbed', *Wireless Pers Commun*, Mar. 2020, doi: 10.1007/s11277-020-07235-z.

[12] A.-M. Dragulinescu, A. Dragulinescu, C. Zamfirescu, S. Halunga, and G. Suciu, 'Smart Neighbourhood: LoRa-based environmental monitoring and emergency management collaborative IoT platform', in *2019 22nd International Symposium on Wireless Personal Multimedia Communications (WPMC)*, Nov. 2019, pp. 1–6, doi: 10.1109/WPMC48795.2019.9096192.

[13] I. B. B. Vasile, P. Schiopu, and C. Marghescu, 'Modern techniques and technologies for unbundled access in the local loop', in *Advanced Topics in Optoelectronics, Microelectronics, and Nanotechnologies VII*, Feb. 2015, vol. 9258, p. 92580E, doi: 10.1117/12.2070839.

[14] 'A Blockchain Platform for the Enterprise — hyperledger-fabricdocs master documentation'. https://hyperledger-fabric.readthedocs.io/en/release-2.2/ (accessed Sep. 20, 2020).

[15] C.-Y. Lin and S.-F. Chang, 'A robust image authentication method distinguishing JPEG compression from malicious manipulation', *IEEE Transactions on Circuits and Systems for Video Technology*, vol. 11, no. 2, pp. 153–168, Feb. 2001, doi: 10.1109/76.905982.

High performance interconnecting technique using power line communication

1st Elena Valentina Dumitraşcu
Center for Electronics
Technology and
Interconnection Techniques
University "Politehnica" of
Bucharest
Bucharest, Romania
valentina.dumitrascu@cetti.ro

2nd Paul Mugur Svasta
Center for Electronics
Technology and
Interconnection Techniques
University "Politehnica" of
Bucharest
Bucharest, Romania
paul.svasta@cetti.ro

3rd Madalin Vasile Moise
Center for Electronics
Technology and Interconnection
Techniques
University "Politehnica" of
Bucharest
Bucharest, Romania
madalin.moise@cetti.ro

4th Aurelian Kotlar
Vitesco Technologies Romania
Timişoara, Romania
aurelian.kotlar@continental-corporation.com

Abstract — **The research was based on determining the most efficient techniques to decrease the number of cables in a system. One interconnection technique that allows a significant reduction of the wire length is the use of the power line as a data bus. The paper is focused on the use of power line communication in the automotive field, to control emissions by preheating the catalytic convertor. The catalytic converter is an element with an essential role in reducing the car emissions. Inside the catalytic convertor take place some chemical reactions at a high level temperature (around 600 °C) [1]. The problem occurs when the car is started because the catalytic convertor needs approximately 10 to 15 minutes to reach the optimum operating temperature. By integrating this interconnecting technique in a system of command and control of the heating of a catalytic convertor, it is possible to observe the behavior of the DC power line communication and at the same time to solve the problem of the classical catalytic convertor.**

Keyword s — *automotive, heated catalytic converter, power-line communication (PLC)*

I. INTRODUCTION

Air pollution is a global challenge that affects not only the ecosystems, but also the life. According to the European Environment Agency (EEA), the most significant contribution to air pollution is made by the road transport sector [2].

Emission reductions is an increasingly discussed topic given the evolution of the automotive field. More and more companies have been researching effective ways to improve the performances of the catalytic converter. One study [3] shows that, by introducing a heating resistance before the catalyst, the heating time will be shorter so the catalytic converter will be more efficient.

DC power line communication is a high performance interconnecting technique that allows to replace, in safe and efficient manner, the traditional serial communication protocols.

Using the PLC technology it could be observed a decreasing of the number of cables (that impling a diminution in the cables weight) that involves not only a reduction of the costs but also providing the same performance as using different data buses. [4,5]. Furthermore, by reducing the vehicle weight the greenhouse gases emission will be decreased along with the fuel consumption.

II. DESCRIPTION OF WORK

The system block is illustrated in Fig.1. The developed module is represented in blue and the peripherals that ensure its proper functioning are represented in orange.

The main blocks of the system are:

- *Power supply* is represented by the car battery.

- *Powertrain control module (PCM)* has a role in controlling the activity of resistive load.

- *Communication interface* is a transceiver that facilitates the transmission of information through the power line. It is used both as a transmitter and as a receiver.

- *Electronic control unit (ECU)* it is the interface through the resistive load is controlled.

- *Switching and diagnosis circuit* represents the environment through the switching of the resistive load is performed. The module provides at the same time, the diagnostic signals for the ECU.

- *Load* is the heating resistance of the catalytic converter.

The operation principle of the system is as follows:

The command is transmitted, serial, by the PCM. It is received by the communication interface (SIG100 transceiver) and modulated in phase, through a PSK technique [6]. Thus, the resulted signal is superimposed over the power line, through a coupling capacitor. The coded message is taken over and decoded by a second communication interface, electronically identical to the first one, the receiver has the frequency set identical to the transmitter. Further, the information is transmitted serially to the ECU, which monitors the activity of the switching circuit and transmits back to the PCM, serial, information about the state of the circuit.

The prototype includes:

A. Communication interface – power line

The Power Line Communication (PLC) is the main element of the system. A transceiver (SIG100 [7]) from YAMAR

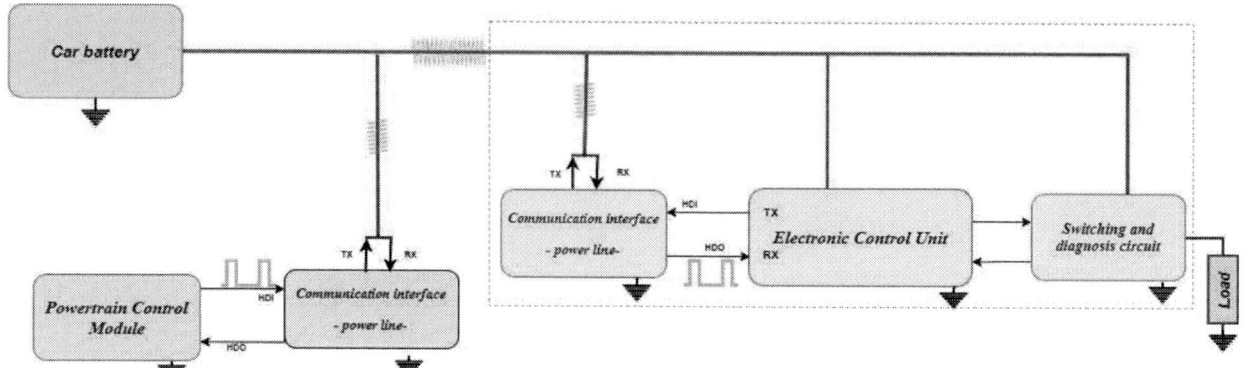

Fig. 1. System block diagram

was chosen to implement this type of information transmission. This is an advanced version of the SIG60 [8] device, respective SIG102 [9], being in full development process. The electronic scheme of the communication interface is the one proposed by the manufacturer [7], to this were added the necessary changes to integrate the circuit into the system.

The communication speed (baud rate) was set at 19200 bits/s, this has an essential role in the secure transmission of information. Within the system, both the microcontroller and the two transceivers have the baud rate set to this value.

B. Electronic Control Unit

The ECU is represented by a module that contain a microcontroller ATmega2560 [10], produced by Microchip.

This is an 8 bits microcontroller with a RISC architecture, it was chosen mainly due to the large number of general purpose inputs/outputs that can be used for the control of the system and for data acquisition.

The ECU consists of several submodules that lead to the proper operation of the microcontroller. The ECU includes: the ATmega2560 microcontroller, a USB-UART transceiver and the power supply unit with both a 5V and a 3.3V outputs.

C. Switching and diagnosis circuit

The block switching is based on an NMOS transistor used in a high side configuration. For a reliable transistor we started with design requirements, namely, developing a module which switches a resistive load of about 120 mΩ, supplied at a voltage between 10V and 16V. In this case, the switch is represented by a TPWR8503NL [11] transistor, produced by Toshiba. This transistor was chosen mainly due to the low RDS (on) resistance, the high drain current, but also the low switching times. The DSOP Advance capsule offers a good heat transfer between the component and the environment.

In order for the transistor to work properly, without being destroyed, it is desired that the specific parameters do not exceed the thresholds presented in the datasheet [11].

The power dissipated by a transistor is calculated using the equation (1). Equation (2) presents the formula for calculating the temperature of the transistor junction, depending on the maximum dissipated power (P$_{Dmax}$), the thermal resistances (R$_{\theta JA}$) between the device and the environment and the ambient temperature (T$_A$). The total thermal resistance of the transistor (from junction to air) is calculated according to relation (3), for which R$_{\theta JC}$ is the junction - case resistance, R$_{\theta PCB}$ is the thermal resistance of the PCB and the resistance R$_{\theta CA}$ is the resistance between the case and the environment.

$$P_D = I_D{}^2 \cdot R_{DS(on)} \qquad (1)$$

$$T_\theta = P_{Dmax} \cdot R_{\theta JA} + T_A \quad (2)$$

$$R_{\theta JA} = R_{\theta JC} + R_{\theta PCB} + R_{\theta CA} \quad (3)$$

Following the calculations performed, we decided to use 4 transistors connected in parallel, and for a better heat dissipation, we chose a heat sink. The heat sink has the following dimensions 31 mm × 31 mm and it has a thermal resistance equal to 4.7°C/W.

After a detailed study on the aspects related to the temperature, we decided to calculate the junction temperature of a transistor, for a value of the ambient temperature of 105°C. The ambient temperature under the hood [12] can reach the maximum value mentioned above. Thus, for the configuration with 4 transistors, the junction temperature can reach the value of 113.71°C.

Transistors are simultaneously controlled through a PWM signal received from the microcontroller.

In order for the transistor to be able to enter the conduction region, its gate-source voltage must be higher than the threshold voltage. In order to sufficiently increase the gate potential, we used an amplifier circuit, a NMOS High-Side Gate Driver FAN7171-F085 [13], from ON Semiconductor.

The diagnostic block consists of a resistive divider and a Hall sensor, ACS770KCB-150B- PFF-T [14], necessary for measuring the current through the circuit.

The voltage divider allows voltage measurement through the microcontroller.The channels of the digital analog converter, included in the microcontroller, allow a maximum voltage of 5V as an input.

As the name suggests, the diagnostic block is used to monitor the operation of the circuit and to determine -the appearance of anomalies in its behavior.

For the blocks presented at points A, B and C, three individual **printed circuit boards (PCB)** were designed, these present the possibility to interconnect through some strategically placed pins.

All PCBs were made on 2 layers, TOP and BOTTOM. For which of them was used an FR4 insulating layer with a thickness of 1.5 mm, the tracks were made of copper with a thickness of 1 oz / ft2 (35 μm).

Both the schematics and the PCBs were designed using EAGLE 9.5.2 [15].

The **programming of the microcontroller** was performed through the SPI serial interface. The code was developed in the C programming language.

In Fig.2 is presented the flow chart diagram of the program, the implementation of the code is based on the use of two functions "pwm_control" and "diagnosis".

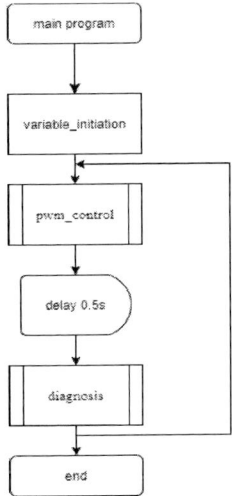

Fig. 2. Flow chart diagram of the main program

The entire operation of the circuit is based on the switching of the transistors Q1, Q2, Q3, Q4. The control over the PWM belongs entirely to the PCM. Thus, the microcontroller receives, through the serial interface, a value between 1 and 100, representing the signal duty cycle. For each change of the PWM value the condition of the circuit is checked, after this process a code is returned to the PCM, the code signals the occurrence of a malfunction or not.

In TABLE I are presented the error codes that may be returned after the diagnosis is performed.

TABLE I.   CORRESPONDING ERROR CODES

Fault	Power supply less than 9 V	Short to GND	Open load	Short to VDD	PWM 100%
Error code	#Err00	#Err01	#Err02	#Err03	#Err04

III. RESULTS

The circuit testing was performed in two phases.

First phase involved the simulation of the functionality of the switching and diagnosis circuit, along with virtual programming of the microcontroller.

The second phase involved the physical testing of the entire system, initially by testing individual modules, then by testing the entire system.

A. Simulation of the switching and diagnosis circuit

To test the functionality of the switch and diagnosis circuit was performed a simulation using Proteus [16] software.

The circuit was supplied with a constant voltage source (12V) and the load used had a fixed value of 120 mΩ. Two operating modes have been simulated:

a) Normal operation:

The circuit was tested at different values of the duty cycle for the control signal, the result was as we expected. It could be observed that, by applying a PWM (5V) with a set duty cycle, a signal with an amplitude of 12V was obtained at the load terminals, having a duty cycle identical to the control signal.

For a duty cycle of 50%, it has been observed a load current of 98.8 A, due to the voltage drop across the transistors, of approximately 160 mV.

b) Malfunction:

In order to test the error detection mechanism, all five errors presented in TABLE I were simulated.

Following the simulations, it was demonstrated the efficiency of the system to detect possible defects that may occur. In all five cases, the user was warned, by displaying, on serial monitor, an error message specific to each case.

B. Physical testing

a) PLC interface test :

The purpose of testing the power line communication interface was to better understand this new interconnection technology, by observing:

- The message frame of the sent and received message.
- The connection mode of the master and the slave within the network.
- The benefits of this type of serial communication.

In Fig.3, it can be seen the waveforms specific to the PLC communication. It can be seen on channel 1 (CH1 - yellow), the signal modulated and superimposed over the power supply signal, and on channel 2 (CH2 - blue), the received signal demodulated.

b) System testing:

Testing the developed prototype highlighted the systems functionality and also helped point out both the benefits that power line communication brings as well as future possible improvements.

In Fig.4 is illustrated the prototype under test, it can be notice the reduction of the number of cables (two wires from power supply) from PCM to ECU. It is also worth to mention that the system was developed to allow the use of a load of 120 mΩ, but it was tested for a higher resistive load. This method

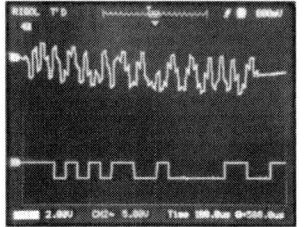 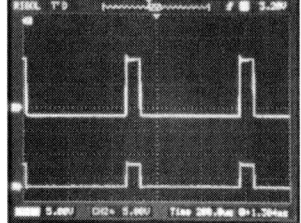
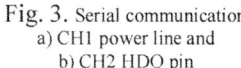

Fig. 3. Serial communication
a) CH1 power line and
b) CH2 HDO pin

Fig. 4. Switching transistors
a) CH2 control signal and
b) CH1 load voltage - PLC
transmission

was used, first of all, because we desired a detailed study about the PLC technology, in the future, the system will be tested by connecting an appropriate load.

In Fig.5, are presented the waveforms resulting from testing the entire system. Thus, on the CH1 channel (yellow color) is illustrated the waveform of the signal obtained at the load terminals, it has an amplitude of 12V, on a positive level (the transistor is in the "ON" state). On the CH2 channel is the control signal of the transistors.

By comparing the result obtained through the PLC method with the result obtained through the classical serial transmission method, it could be observed that the PLC does not influence the transmission of information from PCM to ECU. Moreover, it was possible to observe a decrease in the number of cables in the system and in the same time the performance of the system was maintained.

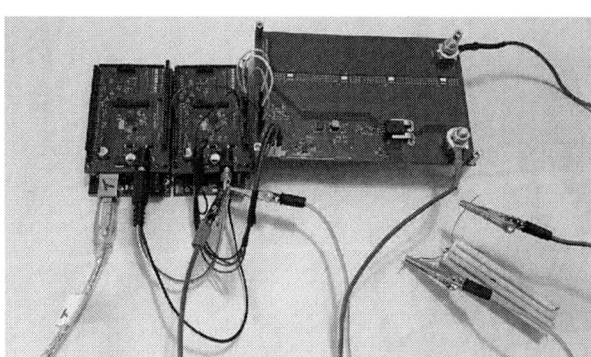

Fig. 5. Experimental setup - developed system

IV. CONCLUSION AND FUTURE WORK

After testing the system, it was demonstrated the possibility of switching a resistive load, through the DC power line.

Through the simulation it was possible to observe the capacity of the system to detect faults, but also to switch the suitable resistive load (120 mΩ).

By using the power line communication, it was highlighted the reduction of the number of cables, without the need of use dedicated wires to transmit data. Because the developed system is a small one, no decrease in costs has been observed, but in the case of a much more complex system, both the price and weight factors will be much easier to be noticed.

The topic of the paper has a great relevance given the context of using the power line communication on the automotive field. Fact that determines us to adapt the system to the LIN communication interface and to integrate the PLC module in a lot more subsystems of the car (by developing a network with several slaves).

ACKNOWLEDGMENT

We are thankful to Yamar Company [7] and Vitesco Technologies Romania [17] for their support in realization of this project

REFERENCES

[1] S.K. Sharma, P. Goyal, R.K. Tyagi, "Conversion efficiency of catalytic converter", International Journal of Ambient Energy, 2015

[2] Emissions of air pollutants from transport, https://www.eea.europa.eu/data-and-maps/indicators/transport-missions-of-air-pollutants-8/transport-emissions-of-air-pollutants-6

[3] Manuel Presti, Lorenzo Pace, "An alternative way to reduce fuel consumption during cold start: the electrically heated catalyst", SAE International, 2011

[4] Yair Maryanka , Ofer Amrani, "MPBUS DC Powerline Communication Eliminates Wiring In Launchers And Spacecrafts", IAC-15,D1,2,2,x27730

[5] Yair Maryanka, "Using Power Line Communication for Harness Reduction in Automotive"

[6] Peter Nisbet, „Study of power line communication modems for automotive communication networks," pp. 1-26, 2014

[7] https://yamar.com/

[8] https://www.yamar.com/datasheet/DS-SIG60.pdf

[9] https://yamar.com/datasheet/DS-SIG102.pdf

[10] http://ww1.microchip.com/downloads/en/DeviceDoc/ATmega640-1280-1281-2560-2561-Datasheet-DS40002211A.pdf

[11] https://toshiba.semiconstorage.com/apen/semiconductor/product/mosfets/12v-300v-mosfets/detail.TPWR8503NL.html

[12] https://www.golledge.com/news/the-aec-q200-standard-what-does-it-really-mean/

[13] https://www.onsemi.com/pub/Collateral/FAN7171_F085-D.PDF

[14] https://www.digikey.com/en/datasheets/allegromicrosystemsllc/allegro-microsystems-llcacs770datasheet

[15] https://www.autodesk.com/products/eagle/overview?plc=F360&term=1-YEAR&support=ADVANCED&quantity=1

[16] https://www.labcenter.com/

[17] https://www.vitesco-technologies.com/en/

Image Compression and Noise Reduction through Algorithms in Wavelet Domain

Cătălin Dumitrescu
Faculty of Transports
University Politehnica of Bucharest
Bucharest, Romania
catalin.dumitrescu@upb.ro

Maria Simona Răboacă, Ioana Manta
National Centre for Hydrogen and Fuel Cells
National R&D Institute for Cryogenic and Isotopic
Technologies ICSI Rm. Vâlcea
Râmnicu Vâlcea, Romania
simona.raboaca@icsi.ro, ioana.manta@icsi.ro

Abstract— **Image compression and noise reduction are two important application in the field of digital image processing. The use of transformed domains plays an important role in both mentioned applications, offering a different representation of the data, in which the main characteristics to be processed become more distinct. Image compression aims to reduce the number of bits needed for the digital representation of the image with or without an acceptable loss of image quality. The underlying idea behind the use of linear transforms in signal processing is a more efficient quantization of the coefficients corresponding to the transformed domain, compared to the case of the coefficients corresponding to the primary space. From this point of view, the wavelet transforms allow, due to the multi-resolution representation that they accomplish and the correlation between the sub-images of this representation, a more efficient quantization and entropic encoding of the coefficients corresponding to the transformed domain. In the field of image compression, the multi-resolution analysis property of the wavelet transform allows progressive quality-scalable and spatially scalable transmissions, very useful in applications that require interactive searching or when bandwidth transmission channels are available. The wavelet transform focuses most of the signal energy in the low pass sub-band with the lowest resolution, the coefficients of this sub-band having high absolute values. The rest of the coefficients obtained after decomposition have small absolute values, being canceled after the quantization process. A good compression rate can be obtained in this case if the absolute values of the coefficients and their locations are encoded separately.**

Keywords— *image compression, wavelet decomposition, image denoise, wavelet noise reduction*

I. INTRODUCTION

The link between lossy image compression and noise reduction algorithms using the wavelet domain has been theoretically demonstrated in the past. Both applications use similar input data processing schemes and are based on the same statistical models of the images. We can therefore speak of a parallelism between the two applications in the field of digital image processing.

However, this connection between the two applications can be regarded differently, considering not the parallelism between them, but a possible complementarity.

Ji and Zhang considered in [1] an adaptive compression algorithm consisting of three phases (training, coding and decoding) which led to a good PSNR value and compression ratio, while removing speckle noise.

In [2] a hybrid method was proposed for reducing the impulse noise in corrupted images. The method is based on fuzzy logic, considering data from neighbor pixels and it was tested on similar gray scale images to the ones addressed in this paper.

Similar pixel neighborhood related algorithms were studied by Tirandaz et al. in [3] for image segmentation, obtaining high performances of over 92% and good noise rates. Image compression based on segmentation was also studied in [4] in the case of brain MRI images.

Gong et al. developed in [5] an encryption algorithm for the compression of optical images, using the Walsh Hadamard transform. The results showed a high compression capability and security to noise attacks.

Noise reduction in JPEG compressed images is important in many fields, for example image forensics, discussed by Kumawat and Pankajakshan in [6]. Using the DCT (Discrete Cosine Transform) coefficients they computed a threshold for detecting JPEG uncompressed images. A similar approach was also considered in [7] by Sun et al for reducing artifacts in JPEG compressed images.

Another field of interest is represented by the ultrasound imaging. In [8], an adaptive bi-lateral filtering was proposed for reducing artifact noise and simultaneously maintain the correct information related to contours. The proposed method managed to increase the PSNR ratio by approximately 130%.

Biomedical signals were also studied by Schanze in [9] and a Singular Value Decomposition method was proposed for compressing and reducing noise. In [10], Ahmed and Sankar developed a method based on IWT (Integer Wavelet Transform) for image compression and a learning algorithm for region extraction.

Compression noise can also be reduced using a parallel convolutional neural network studied in [11] by Amaranageswarao et al. The advantage of using such algorithm is the reduced training time and it is suitable for very degraded decompressed images.

It is a well-known fact that images distributed on social media are subject to noise and poor compression rates. In [12], a solution is proposed based on the wavelet coefficients of the second order Discrete Wavelet Transform.

Gungor and Gencol developed in [13] a procedure for wavelet denoising and JPEG compression, reaching an optimal functioning point between the quality and compression rate.

Similarly, this subject was also addressed in [14], where a learning algorithm was proposed for successful image denoising and compression. Results showed a good noise reduction while preserving original image information lost during the compression.

Image compression is also implemented in surveillance and facial recognition; therefore, it is highly important to a have a denoised image. This aspect was addressed by Chen and He in [15], where they proposed an adaptive learning algorithm with optimizable coefficients, obtaining efficiencies over 70%.

A solution for denoising of spectral near infrared images was proposed in [16], taking into consideration the wavelet transform and other similar methods.

An alternative to JPEG compression was proposed by Kouadria et al. in [17]. The authors developed a low complexity method for compressing images coming from wireless network sensors, based on a ROI (region of interest) and on the Tchebichef transform. The method was also applied in [18] for the particular case of brain medical images used in wireless communication systems.

In [19], a new method is developed for achieving a super resolution image recovered from a noisy JPG compressed image based on a two-stage model (recovery from JPG and generation of super resolution image).

Helbert et al. presented in [20] an algorithm in the wavelet domain, more specifically based on a wavelet regularization, for the particular case of image restoration (inpainting process).

Therefore, in this paper we propose to find an optimal compromise between image compression and noise reduction, two aspects which are highly linked, based on algorithms in the wavelet domain. Our results will further be compared to state-of-the-art results from the literature.

The paper further structured as follows: in Section II, the basics of image compression and noise reduction are presented, as well as a state of the art in the field, followed by Section III were our results are discussed and compared to state-of-the-art results. Section IV presents the main conclusions and gives future research directions.

II. MATERIALS AND METHODS

A. Image compression and noise reduction

The issue of simultaneous image compression and noise reduction will be addressed in the context of compatibility with the JPEG standard. Thus, we consider a degraded image with white Gaussian additive noise to be compressed using the JPEG standard. We will show that by pre-processing the degraded image using noise reduction algorithms, better compression rates can be obtained under the JPEG standard, without losing the final image quality at the decoder.

In this case, we will follow the gain that is achieved during compression and the quality of the restored image.

The problem of improving the quality of the images resulting from a lossy compression – decompression process is also made taking into account the JPEG standard. Thus, on some images compressed using the JPEG standard under the conditions of low bit/pixel rates, we will test several noise reduction algorithms presented in this stud, following the dependence between the compression rate and the gain obtained by post-processing in the quality of the recovered image.

B. Improving the quality of the images compressed using JPEG standard through noise reduction algorithms in the wavelet domain

Efficient static image compression using standardized algorithms, such as the JPEG standard, is important in many applications having bandwidth limitations for image transmission, or memory limitations when it comes to image storage. Using this standard, compression rates of 10:1 or higher can only be achieved with the price of image degradation perceived by the human eye. In such situations, two effects mainly appear: the loss of details and the block effect. The first effect is due to the neglection of the high frequency components in the quantizing process of the DCT coefficients, and the second is generated by performing the DCT transform on 8 by 8-pixel blocks.

In standardized compression of images and video sequences, the discrete cosine transform on blocks is the most used transformation. Each DCT coefficient belonging to transform block (usually 8 by 8 pixels) is quantized before entropic encoding. In order to obtain a certain compression rate, the quantization step is not the same for all coefficients and for all blocks. When using quantization steps having high values, the so-called block effect appears at the limits of the quantization blocks, generated by the fact that the last sample in a block will be processed differently than the first sample in the next block. In order to improve the visual quality of the image obtained after the decoder, it is desired to reduce these artifacts. Both the artifacts generated by the compression process and the success in reducing them depend, first of all, on the spectral content (in details) of the image.

The reduction of the artifacts generated by the block effect was first approached [21], [22] as a low pass filtering process applied on the transition zones from one block to another. In [23], an iterative method is proposed for the reduction of these artifacts, based on the method of projections on convex sets (POCS), which imposes some conditions in both time and frequency domain on the restored image at the decoder. Approaches to reducing the block effect from compressed images based on the theory of projections on convex sets is also done by Y. Yang, A. Katsggelos and N. Galatsanos in a number of papers [24], [25]. In [26], H. Joung, U. Chong and S. P. Kim propose an algorithm for reducing the block effect generated by using the discrete cosine transform on blocks, based on the sub-decomposition of the decoded image and the further processing in two steps: in the first step, by minimizing a cost function representing the discontinuities between blocks, a correction of the DCT coefficients is performed. In the second step, a local filtering is performed, using the whole sub-band image representation. The autors report improvements of about 0.5 dB in PSNR terms in the case of the Lena image, 512 by 512 pixels, at a compression rate of 0.2 bpp.

J. Chou, M. Crouse and K. Ramchandran [27] propose a block effect reduction algorithm based on the estimation of the DCT coefficients applying the principle of maximum likelihood. The results obtained by the authors, considering the Lena image are, in PSNR terms, 1 dB for 0.15 bpp 0.8 dB for 0.24 bpp and 0.45 dB for 0.43 bpp.

Neglecting the details of the implementation and the complexity degree, the algorithms used to reduce the block effect lead both to smoothing the artificial discontinuities between blocks and smoothing the contours from the original image. Given that the wavelet filtering algorithms are characterized by a good image contour conservation, two groups of researchers, one from Rice University in Houston, consisting of D. Wei, R. A. Gopinath, H. Guo, J. E. Odegard, M. Lang [28], and another one from Princetown University, consisting of Z. Xiong, M. T. Orchard, Y. Q. Zhang [29], approach the problem of diminishing the characteristic block effect of compressed images using the JPEG standard by using algorithms to reduce the noise in the wavelet domain.

In [27], the authors process the images resulting from the compression – decompression process based on JPEG/ DCT algorithms by softly truncating the wavelet coefficients resulting from discrete wavelet decomposition, obtaining a significant improvement in the visual quality of the images. The authors considered in the experiments that they carried out two variants of the soft truncation operation: by using the same threshold value for all of the wavelet decomposition detail sub-bands and by using threshold values adapted to each sub-band. The authors report a performance improvement in both cases, showing that the solution of soft truncation with threshold values adapted to each detail sub-band leads to better results.

In [30], the same problem is approached in the context of undecimated wavelet transform, using the hard truncation of the wavelet coefficients. The results obtained in this case are much better. Thus, for the Lena image compressed using JPEG with 0.625 bpp, the authors obtain an increase in PSNR from 35.80 dB to 36.19 dB, for a compression rate of 0.25 bpp they obtain an increase from 30.41 dB to 31.42 dB (as against 31.08 dB) and an increase from 27.33 dB to 28.57 dB (compared to 28.30 dB), in the case of a rate of 0.18 bpp.

J. Lu [31] proposes a method of reducing the block effect using scale-oriented contour representation, according to Mallat's theory [23]. In this way, the author shows that the block effect is removed without further blurring of the contours or the generation of new artifacts. However, the author does not characterize the method compared to other methods and does not indicate PSNR values for the images provided by the proposed method.

In [24], the authors propose an algorithm to reduce the block effect based on the use of the super-complete wavelet representation of the decompressed images. The authors exploit the inter-band correlation in order to accurately identify the contours in the image, to avoid smoothing them by truncating the wavelet coefficients. Considering the same Lena image, under the conditions of its compression using the JPEG standard, the authors report the following results: for a rate of 0.15 bpp and improvement is obtained, in PSNR terms, from 26.44 dB to 27.58 dB, for a rate of 0.24 bpp the improvement obtained is from 29.58 dB to 30.37 dB, and for a rate of 0.43 bpp the improvement is from 32.36 dB to 32.46 dB.

Another approach regarding the improvement in quality of the compressed images by using transformed domains is that of D. Wei and A. Bovik [32]. The authors address the issue of improving the quality of static images subject to the compression – decompression process by determining the best basis, using for the representation of the input data the wavelet transform into invariant packets at translation. Therefore, the algorithm proposed by the two authors acts both in the coding part, where the best basis is searched for the input data, and in the decoding one, where, after the image reconstruction operation, the image is subject to a truncation operation of the wavelet coefficients.

This approach, however, involves designing an entire compression – decompression algorithm, which gives it less practical applicability due to incompatibility with the JPEG standard.

In this paragraph, we will study the effect of several noise reduction algorithms, showing that it is possible to obtain an improvement in the quality of the images that have been subject to JPEG compression – decompression processes at low bit/pixel rates.

In all these cases, it is observed that by post-processing the decompressed images using noise reduction algorithms in the wavelet domain, images characterized by higher PSNR values and better visual quality are obtained.

III. RESULTS AND DISCUSSIONS

We tested two noise reduction algorithms in the wavelet domain, algorithms which, following the simulations performed on the degraded images with white additive Gaussian noise, proved to be the most efficient algorithms among those implemented in this paper, in terms of the amount of noise removed and preservation of the contours. The first algorithm consists in the soft truncation of the wavelet coefficients with local threshold and correction value based on the contour map [21] in translation invariant version, and the second one consists in using the multiple Wiener filter in the wavelet domain [23] with estimation obtained also by soft truncation of the wavelet coefficients with local threshold and correction value based on the contour map.

The results obtained are superior to those reported by Y. Yang, N. Galatsanos and A. Katsaggelos [25] which are considered the best results by both the reasearchers group of Z. Xiang, M. T. Orchard and Y. Zhang [24], as well as that of J. Chou, M. Crouse and K. Ramchandran [27].

Table 1 compares the results obtained, in PSNR terms, by Y. Yang, N. Galatsanos and A. Katsaggelos [25], and those obtained by the algorithms proposed in this paper.

Both tested algorithms use as parameter, when used to reduce the white additive Gaussian noise from the images, its estimated dispersion. In this case, however, we do not deal with such a situation, so it is improper to discuss about a dispersion of noise, let alone proceed to its determination. However, both filters need such a parameter to work. In this case, we obtained the best results by trying different values for this parameter, which we will call equivalent dispersion. A first observation is that the two filters lead to the best results in PSNR terms, for the same images, for different values of the equivalent dispersion. Another observation is that the same filter, in the case of different images but with the same compression rate, obtains the best results for different values of the equivalent dispersion, depending on each image separately. This is similar to the variation of optimal threshold values in the case of soft truncation of wavelet coefficients, depending on the processed image. For a given image, however, a correlation is observed between the optimal value of the equivalent dispersion and bit/pixel rate. Thus, the lower the bit/pixel rate, the higher the optimal value of the equivalent dispersion must be. This is illustrated by the data in Table2.

TABLE I.   COMPARATIVE RESULTS OBTAINED BY Y. YANG, N. GALATSANOS AND A. KATSAGGELOS [25] THROUGH THE TWO METHODS: A – BY SOFT TRUNCATION OF THE WAVELET COEFFICIENTS WITH LOCAL THRESHOLD AND CORRECTION VALUES BASED ON THE CONTOUR MAP IN TRANSLATION INVARIANT VERSION; B – USE OF THE MULTIPLE WIENER FILTER IN THE WAVELET DOMAIN WITH ESTIMATION OBTAINED BY SOFT TRUNCATION OF THE WAVELET COEFFICIENTS WITH LOCAL THRESHOLD AND CORRECTION VALUE BASED ON THE CONTOUR MAP. THE TEST IMAGE IS LENA, 512 BY 512 PIXELS.

Results: Y. Yang, N. Galtsanos and A. Katsaggelos [25]			Own results			
PSNR Compr. image (dB)	Compr. rate (bpp)	PSNR Proces. image (dB)	PSNR Compr. image (dB)	Compr. rate (bpp)	PSNR Proceced image Method A (dB)	PSNR Processed image Method B (dB)
32.36	0.34	32.81	32.39	0.34	33.00	33.06
29.58	0.24	30.43	29.46	0.22	30.56	30.50
26.44	0.15	27.58	26.47	0.16	27.68	27.56

TABLE II.   RESULTS OBTAINED BY THE TWO MENTIONED METHODS, A – BY SOFT TRUNCATION OF THE WAVELET COEFFICIENTS WITH LOCAL THRESHOLD AND CORRECTION VALUES BASED ON THE CONTOUR MAP IN TRANSLATION INVARIANT VERSION; B – USE OF THE MULTIPLE WIENER FILTER IN THE WAVELET DOMAIN WITH ESTIMATION OBTAINED BY SOFT TRUNCATION OF THE WAVELET COEFFICIENTS WITH LOCAL THRESHOLD AND CORRECTION VALUE BASED ON THE CONTOUR MAP. THE TEST IMAGES ARE 512 X 512 PIXELS.

Img	Rate (bpp)	Compressed image		Processed image Method A			Processed image Method B		
		PSNR (dB)	C (%)	σ_e	PSNR (dB)	C (%)	σ_e	PSNR (dB)	C (%)
Peppers	0.37	27.4	52.7	0.035	28.14	51.3	0.050	28.17	48.8
	0.25	24.8	40.6	0.052	25.82	40.9	0.100	25.96	37.9
Goldhill	0.47	27.4	51.1	0.022	27.29	51.5	0.035	27.82	50.2
	0.35	26.4	44.0	0.027	26.82	43.8	0.050	26.88	43.8
Super	0.52	22.7	52.6	0.037	23.06	52.1	0.050	23.10	51.8
	0.68	23.7	59.5	0.025	23.94	58.6	0.050	24.05	58.7

From the data in the two tables, it can be seen that the two algorithms used lead to very close results. However, the multiple Wiener filter in the wavelet domain leads to better results in PSNR terms, but weaker in terms of contour preservation. Figures 1-6 show images obtained by using the two filters on some images that were first subject to a compression – decompression process by JPEG.

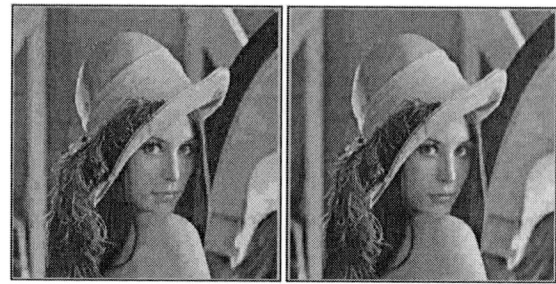

Fig. 1: Lena, 512 by 512 pixels; a. compressed at 0.16 bpp, PSNR = 26.47 dB; b. processed by soft truncation of the wavelet coefficients with local threshold and correction value based on the contour map in translation invariant method, PSNR = 27.68 dB

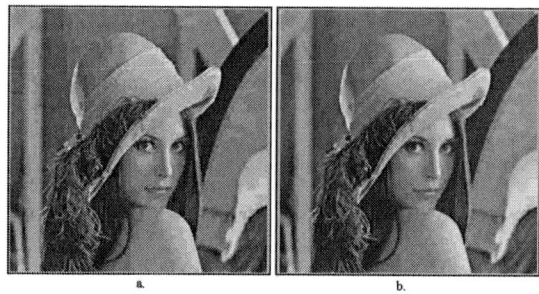

Fig. 2: Lena, 512 by 512 pixels; a. compressed at 0.22 bpp, PSNR = 29.46 dB; b. processed by soft truncation of the wavelet coefficients with local threshold and correction value based on the contour map in translation invariant method, PSNR = 30.56 dB

Fig. 3: Lena, 512 by 512 pixels; a. compressed at 0.34 bpp, PSNR = 32.39 dB; b. processed by soft truncation of the wavelet coefficients with local threshold and correction value based on the contour map in translation invariant method, PSNR = 33.00 dB

Fig. 4: Elaine, 512 by 512 pixels; a. compressed at 0.18 bpp, PSNR = 28.74 dB; b. processed by soft truncation of the wavelet coefficients with local threshold and correction value based on the contour map in translation invariant method, PSNR = 29.74 dB

2020 IEEE 26th International Symposium for Design and Technology in Electronic Packaging (SIITME)

a.   b.

Fig. 5: Peppers, 256 by 256 pixels; a. compressed at 0.378 bpp, PSNR = 27.41 dB; b. processed by soft truncation of the wavelet coefficients with local threshold and correction value based on the contour map in translation invariant method, PSNR = 28.14 dB

a.   b.

Fig. 6: Peppers, 256 by 256 pixels; a. compressed at 0.258 bpp, PSNR = 24.82 dB; b. processed by soft truncation of the wavelet coefficients with local threshold and correction value based on the contour map in translation invariant method, PSNR = 25.82 dB

IV. CONCLUSIONS

In this paper, two aspects are approached in which the noise reduction algorithms in the wavelet domain can improve the image compression performance achieved by the JPEG standard. A first aspect is that in which the image available for compression is degraded by noise. In this case, the compression efficiency decreases dramatically, by pre-processing the image using noise reduction algorithms in the wavelet domain, better compression rates are obtained than in the case of the image not degraded by noise (in the same image quality conditions as those that would be obtained if only the noise reduction were done without compression). The second aspect concerns the case of JPEG compressed images with a low bit/pixel rate, in which case the image degradation occurs at the decoder. Also, in this case, the use of noise reduction algorithms in the wavelet domain shows its efficiency. Several implemented algorithms are tested for this purpose. This time, the gain in PSNR terms is no longer as great as in the case of reducing the white additive Gaussian noise, the degradation of the images never being modeled the same. The most efficient filter in this situation, considering both the performance and the computational effort, is the Wiener filter in the wavelet domain. The results obtained in improving the quality of compressed images using the JPEG standard at low bit/pixel rates together with noise reduction algorithms proposed in this paper, are at the level of the best results reported in the literature [25], sometimes even better than these in PSNR terms.

ACKNOWLEDGMENT

This work was supported by a grant from the Romanian Ministry of Research and Innovation, CCCDI-UEFISCDI, project number PN-III-P1-1.2-PCCDI-2017-0776/No. 36 PCCDI/15.03.2018, within PNCDI III.

REFERENCES

[1] Ji, X., Zhang, G. An adaptive SAR image compression method. *Computers and Electrical Engineering*, **2017**, *62*, 473-484.

[2] Sadeghi, S., Rezvanian, A., Kamrani, E., An efficient method for impulse noise reduction from images using fuzzy cellular automata. *Int. J. Electron. Commun.*, **2012**, *66*, 772-779.

[3] Tirandaz, Z., Akbarizadeh, G., Kaabi, H. PolSAR image segmentation based on feature extraction and data compression using Weighted Neighborhood Filter Bank and Hidden Markov random field-expectation maximization. *Measurement*, **2020**, *153*, 107432.

[4] Sran, P.K., Gupta, S., Singh, S. Segmentation based image compression of brain magnetic resonance images using visual saliency. *Biomedical Signal Processing and Control*, **2020**, *62*, 102089.

[5] Gong, L., Qiu, K., Deng, C., Zhou, N. An optical image compression and encryption scheme based on compressive sensing and RSA algorithm. *Optics and Lasers in Engineering*, **2019**, *121*, 169-180.

[6] Kumawat, C., Pankajakshan, V. A robust JPEG compression detector for image forensics. *Signal Processing: Image Communication*, **2020**, *89*, 116008.

[7] Sun, M., He, X., Xiong, S., Ren, C., Li, X. Reduction of JPEG compression artifacts based on DCT coefficients prediction. *Neurocomputing*, **2020**, *384*, 335-345.

[8] Shao, D., Yuan, Y., Xiang, Y., Yu, Z., Liu, P., Liu, D.C. Artifacts detection-based adaptive filtering to noise reduction of strain imaging. *Ultrasonics*, **2019**, *98*, 99-107.

[9] Schanze, T. Compression and Noise Reduction of Biomedical Signals by Singular Value Decomposition. *IFAC PapersOnLine*, **2018**, *51-2*, 361-366.

[10] Ahmed, A.T., Sankar, S. Investigative Protocol Design of Layer Optimized Image Compression in Telemedicine Environment. *Procedia Computer Science*, **2020**, *167*, 2617-2622.

[11] Amaranageswarao, G., Deivalakshmi, S., Ko, S.B. Blind compression artifact reduction using dense parallel convolutional neural network. *Signal Processing: Image Communication*, **2020**, *89*, 116009.

[12] Latha, P.M., Fathima, A.A. Collective compression of images using averaging and transform coding. *Measurement*, **2019**, *135*, 795-805.

[13] Gungor, M.A., Gencol, K. Developing a compression procedure based on the wavelet denoising and JPEG2000 compression. *Optik – Int. J. for Light and Electron Optics*, **2020**, *218*, 164933.

[14] Lee, J.W., Lee, O.Y, Kim, J.O. Dual Learning based compression noise reduction in the texture domain. *J. Vis. Commun. Image R.*, **2017**, *43*, 98-107.

[15] Chen, Z., He, T. Learning based Facial Image Compression with semantic fidelity metric. *Neurocomputing*, **2019**, *338*, 16-25.

[16] Zhang, C., Zhou, L., Zhao, Y., Zhu, S., Liu, F., He, Y. Noise reduction in the spectral domain of hyperspectral images using denoising autoencoder methods. *Chemometrics and Intelligent Laboratory Systems*, **2020**, *203*, 104063.

[17] Kouadria, N., Mechouek, K., Harize, S., Doghmane, N. Region-of-interest based image compression using the discrete Tchebichef transform in wireless visual sensor networks. *Computers and Electrical Engineering*, **2019**, *73*, 194-208.

[18] Dhouib, D., Nait-Ali, A., Olivier, C., Naceur, M.S. ROI-based Compression Strategy of 3D MRI Brain Datasets for Wireless Communications. *IRBM*, **In Press**.

[19] Li, B., Shi, Y., Wang, B., Qi, Z., Liu, J. RGSR: A two-step lossy JPG image super-resolution based on noise reduction. *Neurocomputing*, **2021**, *419*, 322-334.

[20] Helbert, D., Malek, M., Bourdon, P., Carre, P. Patch graph-based wavelet inpainting for color images. *J. Vis. Commun. Image R.*, **2019**, *64*, 102614.

[21] Saito, N. Simultaneous Noise Suppression and Signal Compression Using a Library of Orthonormal Bases and the Minimum Description Length Criterion. *Wavelets in Geophysics*, **1995**, New York: Academic, 299-324.

[22] Wei, D., Lang, M., Odegard, J.E., Burrus, C.S. Quantization Noise Reduction Using Wavelet Thresholding for Various Coding Scheme. *Proceedings of SPIE Conference on Mathematical Imaging*, **1995**, San Diego, CA, July 12-14 1995.

[23] Xiong, Z., Orchard, M.T., Zhang, Y.Q. A deblocking Algorithm for JPEG Compressed Images using Overcomplete Wavelet Representations. *IEEE Trans. On Circuits and Systems for Video Technology*, **1997**, *7*, 433-437.

[24] Raboaca, M.S., Dumitrescu C., Manta, I. Aircraft Trajectory Tracking Using Radar Equipment with Fuzzy Logic Algorithm. *Mathematics*, **2020**, *8(2)*, 207.

[25] Yang, Y., Galatsanos, N., Katsggelos, A. Projection-based Spatially Adaptive Image Reconstruction of Block-Transform Compressed Images. *IEEE Trans. On Image Processing*, **1995**, *4(7)*, 896-908.

[26] Joung, H., Chong, U., Kim, S.P. Block Artifact Reduction by Optimization Utilizing Subband Decomposition. *7th KSEA Northeast Regional Conference*, New Brunswick, NJ, March 1996.

[27] Chou, J., Crouse, M., Ramchandran, K. A Simple Algorithm for Removing Blocking Artifacts in Block-Transform Coded Images. *IEEE Signal Processing Letters*, **1998**, *5(2)*, 33-35.

[28] Wei, D., Guo, H., Odegard, J.E., Lang, M., Burrus, C.S. Simultaneous Speckle Reduction and Data Compression Using Best Wavelet Packet Bases with Application to SAR Based ATD/R. *Technical Report*, Rice University, CML TR95-02, 1995.

[29] Raboaca, M.S., Dumitrescu, C., Filote, C., Manta, I. A New Adaptive Spatial Filtering Method in the Wavelet Domain for Medical Images. *Appl. Sci.*, **2020**, *10(16)*, 5693.

[30] Jansen, M. Wavelet Thresholding and Noise Reduction. PhD Thesis. *Catholic University Leuven*, April 2000.

[31] Lu, J. Signal Recovery and Noise Reduction with Wavelets. PhD Thesis. *Dartmouth College*, 1993.

[32] Wei, D., Bovik, A.C. Enhancement of Compressed Images by Optimal Shift-Invariant Wavelet Packet Basis. *J. Vis. Commun. Image R*, **1998**, *9(1)*, 15-24.

Usage of ZigBee and LoRa wireless technologies in IoT systems

Vlad-Dacian Gavra , Ovidiu Aurel Pop
Applied Electronics Department
Technical University of Cluj-Napoca
Cluj-Napoca, Romania
vlad.gavra@ael.utcluj.ro

Abstract— **IoT systems are based sensors and actuators to enable ubiquitous sensing to measure environment parameters from delicate ecologies and natural environments to urban environments. By connecting these sensors and actuators to a big network, like internet, an automatization can be performed, and repetitive actions can be done in background by the IoT ecosystem and save a lot of time. IoT can do such things in Home Automation, Smart Cities and even in Air Quality analysis. IoT solution are dependent on the way sensors are transmitting data to cloud or up to the internet. This paper will present the benefits of using Zig Bee instead of using traditional Wi-Fi sensor and present some of the characteristics of LoRa sensors. Cloud computing contributed to the expansion of the IoT systems by offloading local IoT devices of computation intensive tasks. Fog computing brings Cloud closer to the sensors and by doing this minimize communication latencies.**

Keywords—IoT, ZigBEE, LoRa

I. INTRODUCTION

In the last years, Internet of Things technology is booming, estimated connected devices number for 2020 will be 30.73 billion and for 2025 estimations are for 75.44 billion and according to a report made by Gartner, there will be around 8.4 billion connected things worldwide in 2020. This number is expected to grow to 20.4 billion by 2022 of connected devices through IoT technology [6], the estimation the estimations are very different but both of them are talking about billions of devices and tenth of billions of devices.

This concept of Internet of Things is an emerging technology which is getting widley accepted around the world. A definition of this concept is a network of connected embedded objects and devices, with a unique ID, capable to communicate and take actions without human interactions [5]. The main applications for Internet of Things systems are public services, smart commercial buildings, smart homes and transport services but this technology heads also for healthcare services, utilities, smart cities agriculture [5].

For setting up an IoT system the following stages should be fulfilled: should have sensors to collect data from the environment and actuators to take actions, a protocol to transfer data from sensors to gateways and over internet, an application that is used to collect and analyse data to infer knowledge from it and an interface to display, organize and decision making [5].

Data collected from sensorial devices are transmitted in IoT network using different protocols like 3G, Bluetooth, ZigBee, LoRa, etc. depending on the distance range, transmitting timing. For example, environment monitoring for agriculture often uses low data rates wireless protocols such as loRaWAN or 6LoWPAN because information of environments such as temperature and humity does not change rapidly [8][11].

Architecture of the IoT systems has introduced a new concept as Fog-Computing which is a convergence network of interconnected and distributed smart gateways, this network has being developed to reduce the transmission latency of the IoT network and to reduce the load for the embedded devices [6].

For low power application where data rates are not high and timing is relative low ZigBee and LoRa protocol can be used in IoT systems [9][11].

II. IOT ARCHITECTURE

The Internet of Things will be the Internet of future, we have seen a huge increase in different technology areas like smart city, smart homes, smart transportation and smart grids. Due to the reduced computational power, battery life and bandwidth IoT devices are replaced by Cloud application, all the computing and storage tasks are being done in this section. In order to set-up an IoT system this should respect the general IoT architecture presented in Fig 1[10][5].

A. Sensing Layer

This layer is the bottom layer of the IoT technology, the place where data acquisition is made through sensors and actions are made through motors, actuators or any switching devices.

B. Network Layer

A mesh of different communication technologies put together, having the same purpose being to transmit data to the next layer. At this point or layer of the IoT architecture, communication protocols used to transfer information inside of the sensing layer have to converge to the same network this being the internet either through wired transmission or a wireless transmission.

C. Middleware Layer

Used for data conversions or to storage data and also for computing and mathematical calculation. Usually a Cloud server with unlimited computational resources or limited only by user and technology. Cloud servers can be configured to assure the

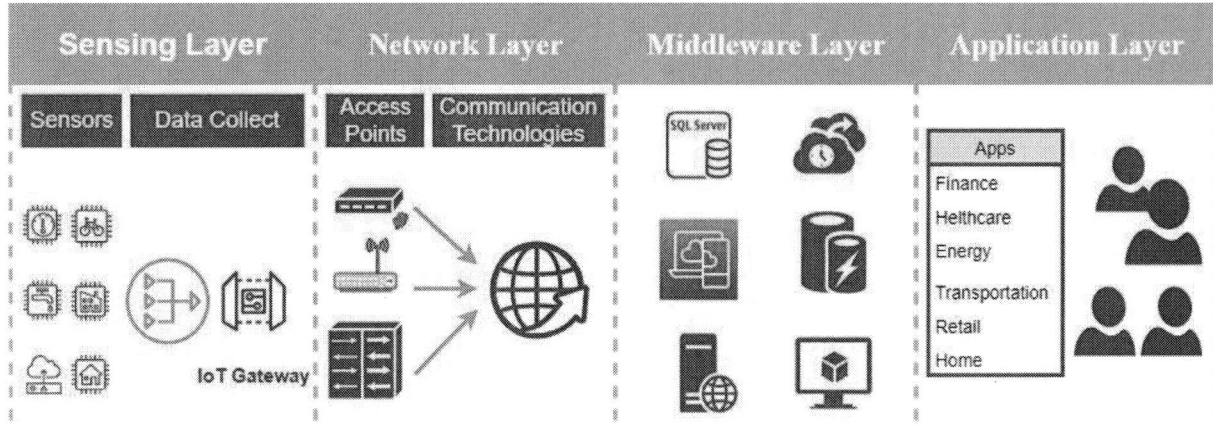

Fig. 1 Basic four-layer IoT Architecture

resources necessary for the IoT functionality application and can be reconfigured if the functionality is growing or is reduced offering scalability for IoT ecosystem.

D. Network Layer

A mesh of different communication technologies put together, having the same purpose being to transmit data to the next layer. At this point or layer of the IoT architecture, communication protocols used to transfer information inside of the sensing layer have to converge to the same network this being the internet either through wired transmission or a wireless transmission.

E. Application Layer

Application Layer is used in order to create a user-friendly platform where users can login and create an easier way to manipulate data.

These applications give the user access to the data from the field in a more orderly way and within its meaning. Graphs can be drawn based on data already stored and transformed, helping in decisions taking or for trend identifiers.

III. FOG AND EDGE COMPUTING

In term of functionality Edge and Fog computing are doing the same thing, both of those technologies are pushing data to the analytic platforms which are located near the sensing devices and actuators [6]. Edge computing technology is implemented usually on the sensing devices or gateways that are close to the sensing devices. Fog technology moves Edge computing to wired devices like gateways but in this case, devices are far away from sensing layer.

With Fog computing, data processing is made by a node or a gateway from the same network while with Edge computing data is processed on the sensing nodes, data not being moved [6].

An advantage of the edge computing is by not transferring data in the network which assure security and discretion for that data, by processing the data on the device that created it, like sensorial devices. Besides security advantages of the Edge

technology, this offer also time and resources in the maintenance of operation by collecting and analyzing data in real time. Of course, Edge computing has also disadvantages. For some applications is very hard to balance the sensing processing and cloud processing. Speaking from costs point of view sometimes is more effective to analyze data in Cloud than on the sensing device [7].

Fog computing present some advantages too, by transferring and manipulating data in the same network, aggregate data form multi-devices into local storage devices. Fog computing architecture allow system scalability despite edge computing from the larger capability of data processing [6].

IV. ZIGBEE PROTOCOL

In order to extend the functionality of an IoT system the big challenge is to embed more wireless communication technologies in one gateway or in the Network Layer. Based on this ZigBee is permissive by allowing multiple routers and end devices, WiFi protocol allows connection to the big network which is the Internet [2][9].

A. ZigBee Alliance

Founded in 2002 and composed by a group of companies that maintain and define the ZigBee protocol. The name ZigBee is a trademark registered for this group. The main activity of this group is not only the definition and publications for this standard but also applications profile for OEM vendors in order to create interoperable products. Over past years members number of the alliance has increased to over 500 between we can see Comcast, IKEA, Legrand, Samsung SmartThing and Amazon [2].

B. IEEE 802.15.4 Standard

This standard has been released in 2003 and was developed by IEEE 802 standard committee. IEEE 802.15.4 defines specifications for PHY and MAC layers of wireless networking, but it does not specify any requirements for higher network layers. It is developed for applications with low transmitting rate and which power capabilities are limited and can't stand a heavy protocol stack [3][4].

C. ZigBee devices

ZigBe devices are providing low-cost solutions, low bit-rate and reduced power consumption capable for running on battery for wireless sensor networks. Zigbee protocol, based on IEEE 802.15.4 wich are operating on 2.4GHz frequency range are very efficient from power consumption point of view.

ZigBee stack is defined by four layers, fisrt two layers being used from IEEE 802.15.4 standard, meaning PHY (Physical) layer and MAC (Medium Access Control) Layer responsible for data transmission and last two layers, Network layer and Application layer which are ZigBee specific. The layers of ZigBee stack are presented in Fig 2[3][4].

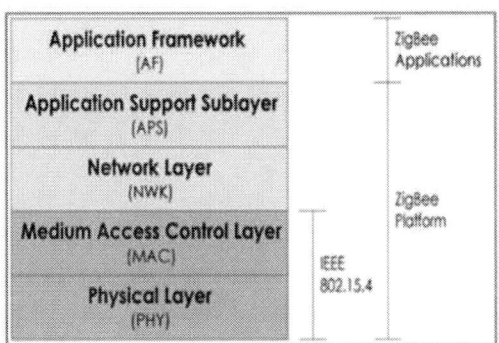

Fig. 2 ZigBee stack layers

Besides low-cost and low energy consumption ZigBee modules bring other advantages like bit-rate up to 250kbps at a 2.4 GHz band, 128-Bit AES security, multiple network topologies modes (Cluster Tree, Mesh and Star) and two mod operation modes in the network as beacon mode or non-beacon mode. Related to the operation modes, in beacon mode the topology of the network is cluster tree or mesh, but data is transmitted cyclical from a network coordinator to the network nodes, in this way consumed energy is controlled. In non-beacon mode, network topology is configured in Peer-to-Peer mode and communication is dual, from coordinator to nodes and vice-versa but in this mode power is not controlled [9].

ZigBee modules can sustain multiple application profiles for different operation like: ZigBee Home Automation, ZigBee Health Care, ZigBee Smart Meters, Weather Stations or Irrigation systems.

ZigBee resembles with Bluetooth but simpler, lower data rate and most of the time the device is snoozing which leads to a very efficient device that can run on an AA battery up to 2 years. In term of data rates ZigBee devices can go up to 250kbps while Bluetooth can go even to 1mbps. When we talk about a master-slave network ZigBee allows up to 254 nodes despite Bluetooth which allows 8 slave nodes [4][9].

V. LoRa Network

LoRa, right now, is a growing technology in the market with the operational field in a non-licensed band below 1 GHz for long-range communication link operation. The technology that it is used by LoRa was derived from chirp spread spectrum modulation (CSS) which was made in 1940 for military application. The CSS implementation trades data rate for sensitivity within a fixed channel bandwidth, with this the data transmission long distances transfer and interference robustness can be achieved. Lora it's the first low-cost implementation for commercial usage [11][12].

Those advantages are achieved by using an adaptive modulation technique with multichannel multi-modem transceiver for receiving multiple number of messages from the channels. This method provides advantages in managing the data rate [12].

Other benefits of LoRa are [11][12][13]:

• Quality of Service – unlicensed spectrum and asynchronous protocol, can handle interference, multipath, and fading.

• Battery life & latency – The sleep functionality is possible with LoRa devices and it can stay in this state for as little or as long as the application desire, in this way making the device very efficient with the energy used. For the application that are insensitive to the latency and do not have large amounts of data to send LoRa is a very good choice.

• Network coverage & range – This is a major advantage of LoRa because the whole city could be covered by one gateway or base station.

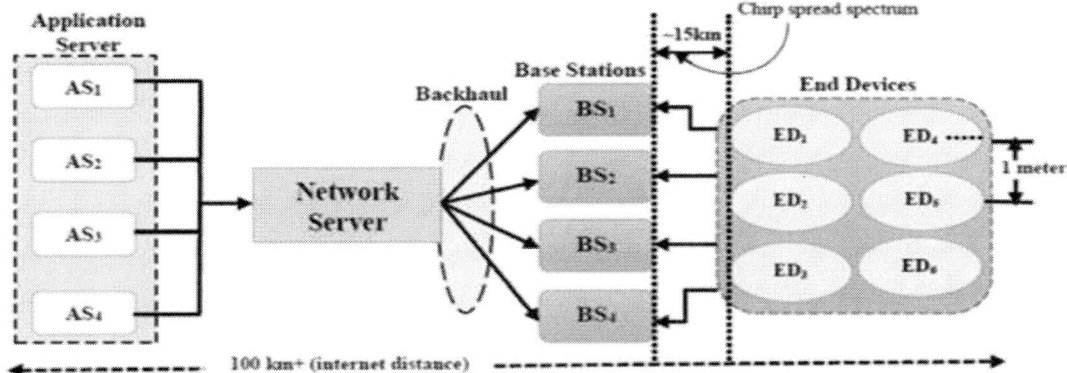

Fig. 3 LoRa WAN Network achitecture with application server and network server, connected with base station and EDs

• Deployment model – LoRa ecosystem is very mature and production-ready.

• Cost – The Spectrum cost is free and the network & Deployment cost varies from 100-1000$/gateway, which is very cost friendly.

VI. RESULTS

In order to extend the functionality of an IoT system the big challenge is to embed more wireless communication technologies in one gateway or in the Network Layer. Based on this ZigBee is permissive by allowing multiple routers and end devices, WiFi protocol allows connection to the big network which is the Internet [2]. Some research from the last few years in sensor networks from IoT systems tried to implement hybrid wireless communication by putting together two or more wireless communication technologies. One of these papers tried to put together ZigBee and WiFi protocols using the same band, 2.4 GHz, by making gateways that can transmit concurrently ZigBee radio signals and Wi-Fi radio signals. [4]. As ZigBee, LoRa is a low power and low-bit rate wireless communication technology but with a larger distance range which can be used to connect the Sensing Layer of an IoT system to connect to Internet.

VII. CONCLUSIONS

As a conclusion, ZigBee technology provide a low power device with a standardized communication protocol, IEEE 802.15.4, capable of embedded application due to ZigBee Alliance Network and Application Layer specification and extending IoT functionality by hybrid communication between ZigBee and other wireless technologies as Wi-Fi or LoRa which brings a lot of advantages to IoT systems.

REFERENCES

[1] Paul Brous and Marijn Janssen. 2015. Advancing e-Government Using the Internet of Things: A Systematic Review of Benefits. In Proceedings of the 14th IFIP WG 8.5 International Conference on Electronic Government - Volume 9248. Springer-Verlag, Berlin, Heidelberg, 156–169.

[2] J Alharbe, N. & Atkins, Anthony & Sheikh Akbari, Akbar. (2013). Application of ZigBee and RFID Technologies in Healthcare in Conjunction with the Internet of Things. 10.1145/2536853.2536904.

[3] Froiz-Míguez I, Fernández-Caramés TM, Fraga-Lamas P, Castedo L. Design, Implementation and Practical Evaluation of an IoT Home Automation System for Fog Computing Applications Based on MQTT and ZigBee-WiFi Sensor Nodes. Sensors (Basel). 2018;18(8):2660. Published 2018 Aug 13.

[4]] Li, Yan & Chi, Zicheng & Liu, Xin & Zhu, Ting. (2018). Passive-ZigBee: Enabling ZigBee Communication in IoT Networks with 1000X+ Less Power Consumption. 159-171. 10.1145/3274783.3274846.

[5] Vijai P, Sivakumar PB (2016) Design of IoT systems and analytics in the context of smart city initiatives in India. In: 2nd international conference on intelligent computing, communication and convergence. Procedia Comput Sci 92:583–588. https://doi.org/10.1016/j.procs.2016.07.386.

[6] T. Nguyen Gia, A. M. Rahmani, T. Westerlund, P. Liljeberg and H. Tenhunen, "Fog Computing Approach for Mobility Support in Internet-of-Things Systems," in IEEE Access, vol. 6, pp. 36064-36082, 2018, doi: 10.1109/ACCESS.2018.2848119.

[7] S. Yi, Z. Hao, Z. Qin and Q. Li, "Fog Computing: Platform and Applications," 2015 Third IEEE Workshop on Hot Topics in Web Systems and Technologies (HotWeb), Washington, DC, 2015, pp. 73-78, doi: 10.1109/HotWeb.2015.22.

[8] Ergen, S. (2004). ZigBee/IEEE 802.15.4 Summary.

[9] Drew Gislason. 2008. Zigbee Wireless Networking (Pap/Onl. ed.). Newnes, USA.

[10] Garcia, L.; Parra, L.; Jimenez, J.M.; Lloret, J.; Lorenz, P. IoT-Based Smart Irrigation Systems: An Overview on the Recent Trends on Sensors and IoT Systems for Irrigation in Precision Agriculture. Sensors 2020, 20, 1042.

[11] Kais Mekki, Eddy Bajic, Frederic Chaxel, Fernand Mayer. A comparative study of LPWAN technologies for large-scale IoT deployment. Volume 5. Issue1, March 2019 Page 1-7

[12] Rashimi Sharan Sinha, Wei Yiqiao and Seung-Hoon Hwang (2017). A survey on LPWA technology: LoRa and NB-IoT. Division of Electronics and Electrical Engineering, Dongguk University-Seoul, Republic of Korea. Received 4 January 2017; accepted 14 March 2017 Available online 21 March 2017

[13] Gaia Codeluppi , Antonio Cilfone , Luca Davoli and Gianluigi Ferrari. LoRaFarM: a LoRaWAN-Based Smart Farming Modular IoT Architecture. Internet of Things (IoT) Lab, Department of Engineering and Architecture, University of Parma, Parco Area delle Scienze, 181/A, 43124 Parma, Italy, 4 April 2020.

Towards real-time and real-life image classification and detection using CNN: a review of practical applications requirements, algorithms, hardware and current trends

Mariana Eugenia Ilas
"Politehnica" University of Bucharest
Dept. of Electronics, Telecom. and IT
Bucharest, Romania
mariana.ilas@upb.ro

Constantin Ilas
"Politehnica" University of Bucharest
Dept. of Computer Science
Bucharest, Romania
constantin.ilas@upb.ro

Abstract—**CNN are already used in production applications in some areas, whereas for many real-life applications this is not possible yet, mainly because of speed and/ or power consumption limitations. In parallel, there are sustained efforts for simplifying CNN structure and for designing new processor architectures. In this paper we analyze the requirements of inference time and power consumption for a large variety of applications, running on all type of platforms, from edge devices to servers. We also review all hardware systems which can be used for CNN implementation, GPUs, TPUs, FPGA, prototype ASICs. Finally, we correlate the performance of these devices with the real-time and power requirements and the complexity of most popular CNN architectures. Thus, we determine what types of practical applications are currently possible, quantify the gaps for the others, and discuss how these gaps can be reduced.**

Keywords—*CNN, real-time, real-life, power consumption, practical applications, detection, classification, inference time, complexity, hardware platforms, GPU, TPU, FPGA, ASIC*

I. INTRODUCTION

Convolutional Neural Networks (CNN) have become very popular in the area of image classification and detection in recent years [1] - [3]. Numerous algorithms have been proposed and used either independently, or in conjunction with handcrafted feature extraction algorithms [1], [4]-[9]. In both cases, the accuracy of the CNN-based methods is superior to that of classical approaches [5], [7]-[9]. However, the complexity of neural networks (NN) and particularly CNN, is generally high and real-time implementation is not easy to achieve, especially for embedded systems. Moreover, in embedded implementations in general, and particularly in battery powered devices, energy consumption is critical. For these reasons, several processor architectures, designed and optimized for efficient implementation of NN (and particularly CNN) have been also introduced [10]-[16]. These complement the existing Graphical Processing Unit (GPU) chips, as well as new GPU architectures, specially designed for NN acceleration [17], [18]. In this paper we analyze what algorithms can already be implemented in real-time, or, if not, what is the gap for such an

implementation, and what practical applications can thus be solved. For this, we review the requirements of most important application areas (in terms of inference time, power consumption, accuracy), the complexity and accuracy of various algorithms, as well as recent progress in new processor architectures, suitable for CNN implementation.

There are several surveys in the area of neural networks and specific on CNN for image processing [1], [3], [10], [19]-[24]. However, they deal either with applications in a particular area (e.g. [20], [21], [23]) or focus on reviewing various network architectures and their performance [10],[19],[24]. In terms of hardware architectures dedicated to running CNN, [22] covers only the reconfigurable ASICs prototypes.

Compared to existing literature, our main original contributions are:

- The analysis of multiple real-life application areas and their main requirements in terms of CNN inference time and power consumption. To the best of our knowledge, this topic has not been covered yet.

- The survey of various hardware platforms, including new dedicated processors, specially designed for efficient implementations of CNN. This topic has been covered only partially in literature.

- The review of complexity (number of operations) of the main types of networks used for classification and respectively detection. This review is based on data available in the literature.

- The correlation between the CNN complexity, processor capability and specific application requirements, in order to determine to what degree, the requirements can be fulfilled, the existing gaps and the possibility of closing them. To the best of our knowledge, this topic has not been covered yet.

II. REQUIREMENTS FOR CNN IN VARIOUS APPLICATIONS

A. General Considerations on Inference Time and Power Consumption Requirements

In this section we will analyze the main requirements for CNN used in several applications areas. The generic CNN requirements, which can be determined for all existing applications in a particular area, are the maximum inference time and the power/ energy consumption. Other requirements, such as accuracy, are specific to each application.

The general relationship, between the time and power/ energy budget for an arbitrary application is presented in Fig. 1. The two main blocks present in all CNN applications are the

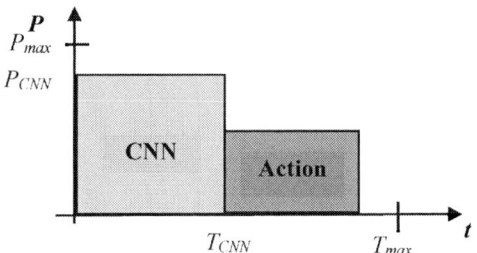

Fig. 1. Power consumption vs execution time for an arbitrary application which includes a CNN block. The maximum power consumption level P_{max} is set for the entire application.

CNN block, performing a specific function (classification, detection, tracking, etc) as well as an Action block, which, based on the CNN output, generates the desired results (e.g. for an autonomous car it can be deciding to steer/ brake and generating the appropriate commands. For a modified reality application it may consist on drawing a hat on a person head and change its position as the person position changes, etc). The total execution time for the two blocks has to be smaller than the maximum application time T_{max}, which depends on the actual application characteristics. Based on T_{max}, one can derive the maximum inference time for the CNN, T_{CNN}. The values for these two times can be as small as *1 ms* for surveillance applications or more than *1 s* for condition detection in medical images exam or satellite image classification.

The maximum power which can be dissipated during the application execution depends on the type of hardware system and its power budget. In some cases, especially when the hardware system is powered on battery, a very small power is critical, but a low power consumption is desirable for all applications, even if they run on desktops or servers, due to the implications on size and cooling [25], [26].

If T_{max} is very large, see Fig. 2, T_{CNN} is limited mostly because the total energy dissipation (the area of the CNN rectangle) cannot be too high. Moreover, a small T_{CNN} (much smaller than T_{max}) is beneficial, because the processor frequency and supply voltage can be both reduced and energy dissipation is further reduced. Indeed, if we compute the energy dissipation E_d from the power dissipation [27], we obtain:

$$E_d = K\,V_{dd}^2 f_{clk}\,T_{CNN} = K\,V_{dd}^2 f_{clk}\,\frac{N_{cycles}}{f_{clk}} \qquad (1)$$

which shows that, for a given processor and algorithm, E_d depends only on V_{dd}. Many processors have a dynamic frequency scaling mechanism, which also decreases the supply voltage V_{dd} when the frequency is decreased, thus reducing E_d [27]. For example, if the frequency is decreased by 50% the voltage can be decreased by around 20%, hence according to (1), the energy is reduced by 36%. Since this mechanism is not identical for all processors, we only discuss the requirements in terms of maximum inference time and maximum power consumption (no matter what frequency is used).

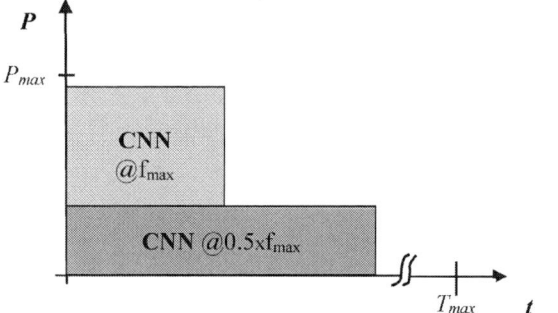

Fig. 2. CNN block energy saving, with dynamic frequency allocation: if processor frequency is reduced at half, inference time doubles, but voltage can also be reduced, resulting in a lower total dissipated energy (rectangle area becomes smaller)

B. Main CNN Application Areas and Their Requirements

The various application areas and their requirements are presented in Tab I. For each area we list examples of typical applications, the main function of the CNN (e.g. classification/ detection), the number of categories (classes) to be classified/ detected, the requirements in terms of total application execution time, CNN inference time, as well as maximum power consumption.

Autonomous vehicles (AV) and Advanced driver assistance systems (ADAS) represent a very challenging area, due to very high vehicle speeds, requesting very small reaction times. Also, the total electrical power demand in an automobile is very high, due to complex electronic systems, hence the power consumption is limited. There are several aspects which determine the requirements for T_{max}, and T_{CNN}. On one hand, the reaction time of the automatic system has to me smaller than that of humans. According to [28], this is in the range of 390-600 ms (depending on age and other factors). On the other hand, if the driver is supposed to act/ take over the control, then the total reaction time consists on the automatic system reaction time plus the driver reaction time [28],[29]. Furthermore, we can correlate the total reaction time with the amount of change in the distance between the car and an object. For instance, considering the relative speed between the car and the object as 80 km/h, the distance changes by 22 m in 0.5 s. Therefore, for this application area T_{max} should be 0.1s or less in order to allow reasonable cruise speeds. For comparison, disengagement T_{max} determined on autonomous vehicle trials in 2014 – 2015 was in the range of 0.6 – 0.8 s [29], while nominal cruising speed for those vehicles was in the range of 20 km/h. More information about AV and ADAS is provided in [30]-[32]. In terms of power consumption, the vehicle alternator provides a relatively high power, but the total power consumption of all the electronics on car is also very high, so a low power consumption for new applications (e.g. CNN) is critical [33]. Considering the typical power consumption of a usual processor used in cars [34], an acceptable level of power consumption between $500 - 2000$ mW is obtained.

CNN have become very popular in medical image processing in recent years [35],[36]. There is a large variety of applications (see also Table 1), including detection and segmentation of brain lesions (tumors, micro-bleeds), as well as many other tissues, detection and classification of chest nodules and breast mass-like lesions, skin conditions, colonoscopy polyps, etc [35]-[39]. The typical operations are classification and detection, as well as image segmentation. In many applications a binary classification/ detection is used, but there is a trend for expanding multi-class operations [35]. In terms of processing time, there are two distinct categories: one which involves separate processing of one/ few images, in which T_{max} can be relatively high (1-2 s or so). This is typical in exam classification, where the algorithm is required to assist the physician who is analyzing a limited number of images at a certain time. In MRI detection, a high number of images (up to 50-100) are generated, in bursts, at intervals of around 5 minutes. In this case, a burst needs to be processed before the next one arrives, hence a detection time of around 2-5 s/ frame is enough. The second class of applications is represented by video processing, where video images are acquired and processed during an examination (e.g. a colonoscopy [37]). In this case

T_{max} has to be smaller than the video frame succession time. Since videos have normally a rate of 25 or 30 frames/ s, T_{max} should be smaller than $30 - 40$ ms. Consequently, T_{CNN} for these applications is in the range of $20 - 30$ ms. The equipment used in medical applications can be grouped in two classes. The first one consists on standard computers or specially designed desktop equipment, which are all supplied from the mains power supply. In this case, reduced power consumption is desirable, but not critical, so desirable P_{max} is comparable to CPU consumption of typical PC applications, which is around 2-10 W [40]. In some cases, usage of GPU cards, with a total dissipated power of 50-250 W may be accepted. The second class of equipment is represented by Edge, portable devices used for Point-of-care (PoC) testing. Examples include ECG and heart health analysis, and ultrasound processing [38]. In this case, a CNN power consumption comparable to that of mobile phones apps (1-2 W) or less is essential.

It is interesting to see that medical applications follow a similar pattern to that of car industry, in the sense that currently CNN-based systems assist humans (in this case doctors/ clinicians), but aim at evolving towards replacing them in the diagnostic process [36]. In fact, especially in binary/ reduced number of categories classification, the accuracy can be better than that of experts [37].

Surveillance applications rely mostly on video processing [41], [42]. Detection is at the core of these applications, but

TABLE I.   CNN MAIN APPLICATION AREAS AND THEIR TIME AND POWER CONSUMPTION REQUIREMENTS

Area	Typical Apps.	Operation	Number of categories (classes)	Max. total app. execution time T_{max} (ms)	Max. CNN inference time T_{CNN} (ms)	Energy consumption importance and power supply type	Max. desired power consumption (W)
Autonomous vehicles, ADAS	Detection of other cars, cyclists, lane detection, traffic sign detection [28] - [32]	Detection	Reduced	< 100	10 – 50	Critical; alternator/ battery	0.5 – 2
Medical	Classification benign/ malign in different images, detection and segmentation of various lesions, skin conditions, colonoscopy polyps, etc [35]-[39]	Classification detection, segmentation	Binary, trend towards (reduced) multi-class	1000-2000 for exam class., 30 – 40 for real-time video	100 – 1000, 20 – 30	Desirable; mains for desktop eq. Critical; battery for edge devices	1 - 10, 0.5-2
Surveillance	Traffic surveillance, person and face detection and recognition, object detection and tracking, crowd behavior, abnormal event detection [23], [41]-[44].	Detection	Reduced	30 – 40	1 - 10	Desirable; mains	1 – 10 desktop or over (server)
Baggage scannung	Detection of arms and dangerous items [45],[46]	Detection	Reduced (Binary)	300 - 600	100 - 400	Desirable; mains	1 – 10 desktop or over (server)
UAV image processing	Traffic monitoring and tracking, search and rescue, fires, natural distasters detection, landsliding crop management [20], [47]-[49]	Detection (tracking)	Reduced	30 – 40 for real-time video	1 – 10	Desirable; mains	1 – 10
Satellite image processing	Crop identification, metereology, disaster and crisis-management, geological and mineral exploration [21], [50], [51].	Classification Detection.	Reduced	100 - 1000	50 - 500	Desirable; mains	1 - 10
Mobile phones (consumer)	User authentication, object detection for augmented reality, activity recognition [52]-[58].	Detection, Classification	Reduced	30 – 40	10 – 20	Critical; battery	0.2 – 2

further processing layers are needed, e.g. fine-grained recognition [43], face recognition, tracking, abnormal activity detection [44], including in crowds [23]. The total application execution time T_{max} is imposed by the video frame succession time, so it has to be 30-40 ms. Nevertheless, the CNN inference time has to be much smaller, due to the delays of data (mostly image) transmission during different layers of the network (especially when fog and cloud architecture is used), as well as due to the processing layers mentioned above, which are also performed. Therefore, the desirable T_{CNN} is 1 – 10 ms. Since normally the processing is done on PCs and servers, power consumption can be relatively high (in line with typical power consumption levels of applications running on these platforms).

In baggage scanning the focus is on finding arms as well as other dangerous goods [45],[46]. The detection is either binary (eg gun detection) but mostly it is a limited-class detection (i.e. different types of forbidden items). Normally the detection is based on individual images of bags moving with a relatively low speed, so T_{max} is around half a second. Since the detection is the main operation, T_{CNN} is close to T_{max}, while power consumption is in the range of typical PC applications.

There is a large variety of Unmanned Aerial Vehicles (UAVs) and of their applications [20], [47] – [49]. In most of these applications, CNN are used for object detection. Some of the applications are real-time (e.g. traffic monitoring, vehicle tracking), whereas for other execution time is less important (land-sliding, crop management). For those which are real-time, T_{CNN} has to be considerably slower than T_{max}, due to other operations that need to be performed, including image transmission. Power consumption is in the range of typical PC applications.

Satellite image processing is somehow related to the previous category (UAV images) [21], [50], [51]. The main difference is that it is not performed real-time, due to large image transmission times from the satellite to the processing center.

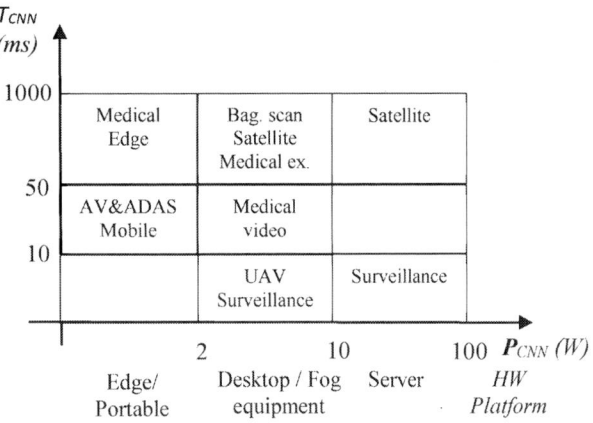

Fig. 3. Summary of inference time and desired power consumption for CNN processing in various application areas and running on different types of hardware platforms

Consequently, both T_{max} and T_{CNN} can be higher than corresponding values in most UAV applications. The processing is done on servers or PCs hence the power requirements are related to those of usual applications running on these platforms.

Achieving reduced complexity CNN that would allow their implementation on mobile phones has been a research trend in recent years, and special architectures have been proposed (e.g. MobileNet, [52]) and improved [53]-[58]. Typical applications range from user authentication [55] to modified reality (MR) and augmented reality (AR) [57]. Since AR is created on real-time video, T_{max} is 30-40 ms. Due to other operations needed in AR, T_{CNN} has to be in the range of 10 – 20 ms. On mobile phones, higher power consumption applications dissipate around $1 - 2W$ [59], so for CNN implementation the power requirements can be maximum 2 W.

There are also papers describing CNN usage in IoT image processing [60]-[62]. The implementations vary from hybrid (IoT + PC) to entire IoT, but using powerful hardware platforms, comparable to mobile phones. Also, the actual applications mentioned (surveillance, navigation, localization) usually do not require a battery with a very long autonomy, but a typical rechargeable one. Consequently, for such applications, both inference time and power consumption requirements can be assimilated to those discussed on mobile phones.

As it can be seen in Table 1, there is a large variety of areas in which CNN are applied, for many practical applications. In terms of CNN inference time requirements, these applications can be classified in very fast (for an inference time of less than 10 ms), fast (10-30 ms) or slow (above 50 ms). In terms of power consumption, the applications can be classified as power critical (for less than 2W) or not critical (for 2-10 W). Fig. 3 presents a summary of the requirements (inference time and desired dissipated power) for CNN in the various application areas. The most challenging areas are autonomous cars/ ADAS and mobile phones, as well as UAV and Surveillance.

III. COMPLEXITY OF DIFFERENT CNN ARCHITECTURES

In this section we will briefly review the complexity, expressed in total number of operations, of several CNN architectures. The operations correspond to the full processing of an input frame (i.e. image) and are usually measured in Giga Operations (GOP). The vast majority of these operations are multiplications and additions [63]. Detailed description of CNN functionality and typical architectures used in classification and detection can be found in [2], [24], and [63]-[77]. An interesting aspect of CNN architectures is the fact that they can be very easily implemented in distinct software of hardware processes, running in parallel. In Table II we represent the complexity (GOP) of most common classification networks, as well as their accuracy, expressed in Top 5% errors. The data is taken from [22], [24], [67]. Note that these CNN architectures are designed for the classification of a large number of categories, usually being trained and tested on ImageNet, which consists on 1000 categories [68]. Hence, the classification error is expressed by Top 5% errors, meaning that the result is consider correct if the target class is among the top 5% predictions of that network.

Detection algorithms are also based on similar CNN architectures, but with a much higher level of complexity. Usually they are evaluated on PASCAL VOC 2007/ 2012 (containing 20 classes) or COCO datasets (containing 80 classes) [68].

TABLE II. COMPLEXITY AND ACCURACY OF MAIN CLASSIFICATION CNN

CNN Classification Network	Complexity (GOP)	Top 5% Error (%)
AlexNet	0.7	16
VGG-16	15	7.4
GoogLeNet	1.4	6.7
ResNet50	3.9	5.3
MobileNet v2	0.3	9
ShuffleNet	0.26	10

Their accuracy is given in Medium Average Precision (mAP), which is the arithmetic mean of the Average Precision computed for each detected class.

TABLE III. COMPLEXITY AND ACCURACY OF MAIN DETECTION CNN

CNN Detection Network	Complexity (GOP)	mAP (%)
Tiny Yolo	6.9	60
Yolo V2	39.4	76.8
Fast RCNN	300	78.2
Faster RCNN	150	76.4
SSD 300 (w/ VGG-16)	34.9	74.3
SSD 512 (w/ VGG-16)	52	76.8
SSD with MobileNet v2	1.2	72.7

The complexity of most popular detection algorithms, as well as their accuracy (for Pascal VOC dataset) is presented in Table III. Data is taken from [63], [70]-[73]. As it can be seen, the complexity of detection algorithms is generally much higher than of classification algorithms. Indeed, considering for example ResNet50 and Yolo V2 (which are both reasonable trade-offs in terms of complexity and accuracy in their categories), we see that the complexity of Yolo V2 is 10x higher than that of ResNet50. The only exception is SSD with MobileNet v2 as backbone, for which a very low complexity is obtained. However, despite the fact that the mAP is still high, it has several areas in which the accuracy is much lower, such as detection of small and even medium objects. In conclusion, this higher complexity implies that detection operations are much more challenging in terms of both inference time as well as power consumption, compared to classification operations.

Finally, we have to note that the values presented in Table II and III above are determined in the respective papers, based on a certain input image dimension, and based on an architecture created to classify, or respectively detect, a high number of classes. In reality, as seen in the previous section, in many practical applications the networks may be used for classifying/ detecting a reduced number of classes. It is therefore possible to partially reduce the complexity of the network to solve the reduced problem. This would affect both the complexity and the accuracy of the final solution designed and used in a specific application. However, the complexity presented in Table II and III provides a relevant range in order to determine approximate inference time and power consumption on particular hardware solutions.

IV. HARDWARE ARCHITECTURES FOR EFFICIENT CNN IMPLEMENTATION

Due to their high complexity, CNN used for object classification and detection are usually tested as prototypes on powerful computing platforms, such as GPU cards. These platforms allow a real-time implementation (for many algorithms), but, with a very high-power consumption. In recent years, specialized processors and hardware accelerators have been designed for CNN efficient implementation, in terms of both inference time and dissipated power. In this section we will review some of the most popular GPU systems as well as specialized new architectures, together with the performance of existing implementations.

A. GPU Systems

GPU chips have a parallel architecture and consist on hundreds of elementary units (referred to as streaming processors or cores, or ALUs) each being able to execute one multiplication and addition per clock [74]. This architecture makes them suitable for CNN implementation, since CNN consists (mostly) on numerous multiplications and additions, which can be (to a large degree) executed in parallel. In a simplistic manner, the total number of operations that a GPU chip can execute per second is obtained multiplying the number of "cores" with the number of operations they can execute in parallel (usually 2, i.e. one multiplication and one addition) and with the chip frequency. This gives the so-called peak performance (measured in tera floating point operations per second, TFLOPS), which is up to 10 TFLOPS [74]. Moreover, some GPUs have also tensor cores included [74]. These are cores are designed to perform very rapidly matrix (tensor) operations. However, the actual inference speed of a particular CNN on a GPU depends not only on the GPU number of cores and clock, but also on other factors. Thus, the memory transfer rate (esp. between GPU and CPU or external memory) is often a limiting factor [75]. Very important is also the number of non-parallel sequences in the code (Amdahl's law) [76]. From a software perspective, most GPU makers, as well as the open source community (OpenCL) offer C as well as Python compilers and support PyTorch, TensorFlow, etc. The programming language, as well as the libraries used also influence the inference time.

The most important drawback of GPU systems is the power consumption, which is, for today's high-performance systems in the range of 200-300 W [63]-[65]. This limits their usage to servers and some desktop-type equipment (if the internal power supply unit can support it). Smaller GPUs for edge usage have also been introduced (e.g. Jetson TX2).

B. VPU and NPU Systems

Vision Processing Units (VPU) are a newer category of chips aiming to "enable demanding computer vision and edge AI workloads with efficiency" [77]. They support floating point formats. There are not many applications reported yet. Available data do not allow a complete evaluation, however it seems that inference time is much larger than on GPUs, but also with a dissipated power in the order of couple of Watts only [78]-[80]. Neural Processing Unit (NPU) is another new category of low-power chips designed for computationally intensive ML applications [81]. It is likely to enter in mass-production for IoT chips [82]. It works on Int8 and Int16 data types and can achieve 1-10 TOPS (tera operations per second), based on 512-4096 8x8 bit MAC units. There are no implementations reported yet.

C. TPU systems

Tensor Processing Units (TPU) are popular NN accelerators, available for use either in standalone chips [11], or within the cloud [83]. The embedded (edge) platforms achieve 4 TOPS in fixed point, whereas cloud-available TPUs are floating point and achieve up to 420 TFLOPS. The cloud-based chips contain multiple cores, whereas the Edge TPU has only one. The cores are based on a systolic array architecture, which for the cloud chips has the size of 128x128 and for the embedded one possibly 64x64. The advantage of this architecture is the fact that it avoids the bottleneck of massive data access, since the output of each MAC operation is the input in the next one. Thus, it is a very specialized architecture, with great results on matrix multiplication type of problems (as are NN). On the other side, it works best as an accelerator, for a main processor, although it is fast also when used stand-alone [84]. In terms of software libraries, The Edge TPUs support only TensorFlow Lite, whereas the cloud TPUs support TensorFlow. The nominal power consumption for Edge TPU is 2 W only.

D. FPGA Systems

FPGA implementations allow achieving good inference speed at a lower power consumption, through efficient designs on powerful FPGA platforms. This continues a trend already existing in classical detection systems, based on handcrafted feature detection and SVM classifiers [85]. A recent study shows very good performance in term of speed (25 fps on Yolo V2 detection) and power consumption 27.2 W [63]. It uses fixed-point data representation, but the decrease in accuracy is negligible. The efficient design is obtained through data dependency analysis and specific operations pipeline. The design is based on OpenCL methodology combined with register-transfer level (RTL) design method. Other papers show slower implementations, e.g. inference times of 4 – 26 ms for classification algorithms [86], using a similar Arria10 FPGA. In terms of computations per second, the implementations range from 61 – 740 GOPS [63]. The drawback of FPGAs is the fact that the CNN has to be implemented in digital design, which requires different skills and may take longer than sw implementations.

E. Prototype ASICs

For even lower power consumption than the values achieved with FPGAs, special ASICs have been proposed [12]-[16]. These can work as stand-alone processors or accelerators, but are at prototype level and very few details are provided on supported programming languages and tools. They are either fixed-structure accelerators, or have a reconfigurable architecture. Thus, [12] presents a reconfigurable processor/accelerator, with processing elements array which can be reconfigured, to facilitate efficient implementation of CNN, full-connected layers, or recursive neural networks. The authors

TABLE IV. IMPLEMENTATION OF REDUCED COMPLEXITY CNN ON VARIOUS HARDWARE PLATFORMS (MAIN VALUES: SPEED, POWER, PWER EFFICIENCY)

HW Platform	CPU E5 2620		GPU Titan X				FPGA Arria 10		CPU+GPU Jetson TX2		EdgeTPU	iPhone8	Reconf. ASIC	Eyeriss ASIC
CNN	Tiny Yolo V2 [63]	Yolo V2 [63]	Tiny Yolo V2 [63]	Yolo V2 [66]	SSD 300 [73]	SSD 512 [73]	Tiny Yolo V2 [63]	Yolo V2 optim. [63]	eSSD MobNet1 [88]	SSD MobNet1 [89]	SSD MobNet2 [87]	SSD MobNet1 [89]	AlexNet [12]	AlexNet [13]
Clock [MHz]	2100		1400				200		2000+1300		700	2390 (in catalog)	200	100-250
GOP	6.97	34.9	6.97	34.9	34.9	52	6.97	29.6	1	1.2	1.2	1.2	3.5 in paper	0.665 in paper
Repr	Float		Float				Fixed		Float		Fixed	Float	Fixed	Fixed
Bits	32		32				8-16		32		8	32	8-16	8-16
mAP % VOC2007	57.1	76	57.1	76	74.3	76.8	56.5	73.6	65.8	72.7	n/a	72.7	n/a	n/a
mAP % COCO	n/a	n/a	n/a	n/a	23.2	26.8	n/a	n/a	18.32 ImageNet	n/a	29.48	n/a	n/a	n/a
Speed (fps)	0.67	0.16	207	67	46	19	105	25	67 (our calc)	73	31.91	22.8	105	34.7
GOPS	4.67	5.6	1442	2338	1605	988	731.9	740	67	87.6	38.2	n/a	368.4	23.16
Power W	80	80	219	219	219	219	27.2	27.2	15 in catalog		2 in catalog	n/a	0.290	0.278
Power efficiency GOPS/W	0.06	0.07	6.58	10.67	7.33	4.51	26.9	27.2	4.44 (our calc)	5.84 (our calc)	19.1	n/a	1270	85.6
Energy J	119.4	500	1.06	3.27	4.76	11.56	0.26	1.09	0.41	0.2 (our calc)	0.06 (our calc)	n/a	0.0028	0.008
mAP/ Energy	0.48	0.15	53.9	23.2	n/a	n/a	218.1	67.65	44.47	363.5 (our calc)	491 (our calc)	n/a	n/a	n/a

claim a peak computational power of 409.6 GOPS (at 200 MHz) and a peak efficiency of 5.09 TOPS/ W. Power consumption is low (290 mW), but achieved speed is also relatively low (105 fps for AlexNet), giving a 1.27 TOPS/ W (based on a very high complexity indicated by authors for AlexNet). Another reconfigurable accelerator, Eyeriss [13] achieves 34.7 fps on AlexNet, with a dissipated power of 278 mW. The peak computational power is 42.0 GOPS, while for AlexNet the obtained efficiency is 85.6 GOPS/ W.

F. Implementation results and correlation with requirments of different applications

In Table IV is presented a summary of representative implementations on these platforms. Data is taken from the indicated sources, except for those which were not provided, but could be computed based on the other available information. For the ASICs chips, only classification algorithms have been implemented (last two columns). For all the others, we selected detection algorithms, because they are more challenging and more frequently needed. As it can be seen, the GPUs, the FPGAs, the Jetson TX2 and the Edge TPUs can all achieve detection (Yolo V2 or SSD) at speeds equal or larger to normal video rates, but at very different power consumption rates. In terms of power efficiency, the highest values are obtained for prototype ASICs, followed by FPGAs and Edge TPU. Highest number of operations per second is achieved by PC/ server powerful GPU systems. They can implement Yolo V2 or SSD300 in 15-20 ms/ frame. This is close to the most aggressive requirements we estimated, of 10 ms/ frame (see Table I and Fig. 3). However, the power consumption is much higher than the desired value, especially for desktop-like (fog) systems, but still possible to accommodate. Consequently, these applications can be successfully implemented in either servers of desktop systems, but further reduction in both inference time and power consumption is still very important. For Edge devices, detection is possible at approximately 13 – 30 ms/ frame. In terms of power consumption, only the Edge TPU provides the desirable value, while Jetson TX2 is still possible to be used in particular systems (eg in some vehicles). Overall, some applications can be successfully implemented today (e.g. medical applications, lower speed object detection for AV/ ADAS), but for very high speed applications, more powerful architectures and/ or improved algorithms are needed. On mobile phone, classification can be easily achieved, while detection is possible only on very powerful phones at speeds which are merely enough for real-time video playing. In such a scenario, the processor bandwidth is probably fully occupied, and the power consumption is, most likely, unsustainable for a long time. Consequently, unless complexity of detection algorithms is further reduced, usage of integrated accelerators is essential. As mentioned before, this is possible to happen in the future [86]. Finally, for medical applications which require relatively high speed and desktop-like platforms, custom equipment can be designed, using existing CPUs and accelerators (e.g. TPU, Jetson) for best power consumption, or, in some applications, using a PC with a lower power consumption GPU card (e.g. 75 W) may be acceptable.

V. CONCLUSIONS

In this paper we analyzed to what extent the execution time and power consumption requirements of most important applications for image classification and detection can be met with the existing hardware platforms and CNN architectures. For this, we started in Section II by discussing the requirements of seven application areas, which we grouped in different categories. We then reviewed the most popular CNN architectures and their complexity, as well as the existing hardware platforms for CNN implementation, from the widely used GPU systems to the prototype ASICs (Sections III and respectively IV). Finally, we collected the performance indicators for several implementations on these platforms, and correlated them with the requirements obtained in Section II. Based on this, we derived several conclusions.

While classification is achievable for most practical applications, the main challenges are for detection applications. Overall, the biggest gaps are on Edge devices requiring very fast inference (e.g. for high speed AV/ ADAS), where substantial progress in hardware performance and/ or algorithm simplification is still needed in order to meet the most aggressive requirements. For mobile phones, rapid detection can be efficiently achieved if accelerators are integrated within the application processor. For desktop/ fog equipment, efficient systems can be designed using existing CPU and accelerator chips, thus fulfilling the requirements for most applications in medicine. Similarly, solutions can also be designed today for medium and lower speed edge systems (e.g. for medical PoC apps), based also on a pair of chips. For all these applications, an integrated CPU with accelerator would be a great enabler. Finally, high speed implementations (e.g for surveillance or UAV apps) can only be achieved on severs (and maybe desktops), using powerful GPU boards but with the cost of very high power consumption. A reduction of the power consumption is desirable for GPU systems in servers and very important for expanding their usage on desktops.

Furthermore, the fastest detection results presented in literature are obtained for low complexity detection algorithms (e.g. Yolo V2 and MobileNet SSD). Depending on the application specific parameters (number of classes to be detected, input image size and quality, etc) their accuracy may not be high enough for some real-life requirements. This is another reason for which further progress in hardware performance and/ or algorithm improvement is essential.

REFERENCES

[1] W. Liu, Z. Wang, X. Liu, N. Zeng, Y. Liu, and F.E. Alsaadi, "A survey of deep neural network architectures and their applications", Neurocomputing, No. 234, pp. 11-26, 2017.

[2] Z. Qin, F. Yu, C. Liu, and X. Chen, "How convolutional neural network see the world - A survey of convolutional neural network visualization methods", Math. Found. of Comp., American Inst. of Math. Sci., vol 1, No 2, May 2018.

[3] O.I. Abiodun, A. Jantan, A.E. Omolara, and K.V. Dada, N. A. Mohamed, "State-of-the-art in artificial neural network applications: A survey", Heliyon, vol. 4, No. 11, November 2018

[4] L. Zheng, Y. Yang, and Q. Tian, "SIFT meets CNN: A decade survey of instance retrieval", IEEE Trans on pattern anal and machine intell, vol 40, No 5, pp. 1224-1244, 2017.

[5] C. Yoo, D. Han, J. Im, and B. Bechtel, "Comparison between convolutional neural networks and random forest for local climate zone

classification in mega urban areas using Landsat images". ISPRS Jnal. of Photogramm. and Remote Sensing, No. 157, pp. 155-170, 2019.

[6] T.J. Alhindi, S. Kalra, K.H. Ng, A. Afrin, and H.R. Tizhoosh, "Comparing LBP, HOG and deep features for classification of histopathology images", IEEE Intl Joint Conf. on Neural Netw. (IJCNN), pp. 1-7, July 2018.

[7] H. Ahamed, I. Alam, and M.M. Islam, "HOG-CNN based real time face recognition". IEEE Intl Conf on Adv in El and Electronic Eng (ICAEEE), pp. 1-4, Nov. 2018.

[8] Y. Wang, Z. Fang, and H. Hong, "Comparison of convolutional neural networks for landslide susceptibility mapping in Yanshan County, China". Sci. of the total env., No 666, pp. 975-993, 2019.

[9] K.A. Nugroho, "A comparison of handcrafted and deep neural network feature extraction for classifying optical coherence tomography (OCT) images". IEEE Intl Conf on Inf and Comp Sci (ICICoS), pp. 1-6, 2018.

[10] W.G. Hatcher, and W. Yu, "A survey of deep learning: Platforms, applications and emerging research trends", IEEE Access, vol. 6, pp. 24411-24432, 2018.

[11] S. Cass, "Taking AI to the edge: Google's TPU now comes in a maker-friendly package", IEEE Spectrum, vol. 56, No 5, pp16-17, 2019.

[12] S. Yin, P. Ouyang, S. Tang, F. Tu, X. Li, S. Zheng and S. Wei, "A high energy efficient reconfigurable hybrid neural network processor for deep learning applications", IEEE Jnal. of Solid-State Circ., vol 53, No 4, pp. 968-982, 2017.

[13] Y.H. Chen, T. Krishna, J.S. Emer, and V. Sze, "Eyeriss: An energy-efficient reconfigurable accelerator for deep convolutional neural networks", IEEE Jrnl. of Solid-State Circ., vol. 52, no. 1, pp. 127-138, 2016.

[14] B. Moons, and M. Verhelst "A 0.3–2.6 TOPS/W precision-scalable processor for real-time large-scale ConvNets", IEEE Symp. on VLSI Circuits, pp. 1-12, June 2016.

[15] B. Moons, R. Uytterhoeven, W. Dehaene, and M. Verhelst, "14.5 envision: A 0.26-to-10tops/w subword-parallel dynamic-voltage-accuracy-frequency-scalable convolutional neural network processor in 28nm fdsoi", IEEE Intl. Solid-State Circ. Conf. (ISSCC), pp. 246-247, February 2017.

[16] R. Mochida, K. Kouno, Y. Hayata, M. Nakayama, T. Ono, H. Suwa, and Y. Gohou, "A 4M synapses integrated analog ReRAM based 66.5 TOPS/W neural-network processor with cell current controlled writing and flexible network architecture", IEEE Symp. on VLSI Tech., pp. 175-176, June 2018.

[17] S. Markidis, S.W. Der Chien, E. Laure, I.B. Peng, and J.S. Vetter, "Nvidia tensor core programmability, performance & precision". IEEE Intl. Parallel and Distrib. Proc. Symp. Workshops (IPDPSW), pp. 522-531. May 2018.

[18] Y.E. Wang, G.Y. Wei, and D. Brooks, "Benchmarking TPU, GPU, and CPU platforms for deep learning", arXiv: 1907.10701, 2019.

[19] W. Wang, Y. Yang, X. Wang, W. Wang, and J. Li, "Development of convolutional neural network and its application in image classification: a survey", Optical Eng., vol. 58, no. 4, 040901, 2019.

[20] C. Kanellakis, and G. Nikolakopoulos, "Survey on computer vision for UAVs: Current developments and trends", Jrnl. of Intell. & Robotic Syst., vol. 87, No 1, pp. 141-168, 2017.

[21] G. Cheng, and J. Han, "A survey on object detection in optical remote sensing images", ISPRS Jrnl. of Photogrammetry and Remote Sens., vol.117, pp. 11-28, 2016.

[22] M.P. Véstias, "A survey of convolutional neural networks on edge with reconfigurable computing", Algorithms, vol. 12, no. 8, pp.154, 2019.

[23] G. Tripathi, K. Singh, and D.K. Vishwakarma, "Convolutional neural networks for crowd behaviour analysis: a survey", The Visual Comp., vol. 35, no. 5, pp. 753-776, 2019.

[24] M.Z. Alom, T.M. Taha, C. Yakopcic, S. Westberg, P. Sidike, M.S. Nasrin, and V.K. Asari, "The history began from alexnet: A comprehensive survey on deep learning approaches". arXiv preprint arXiv:1803.01164, 2018.

[25] S. Lee, D. Pandiyan, J.S. Seo, P.E. Phelan, and C.J. Wu, "Thermoelectric-based sustainable self-cooling for fine-grained processor hot spots", IEEE Intersoc. Conf. on Thermal and Thermomech. Phen. in Electr. Syst., pp. 847-856, May 2016.

[26] G. Da Costa, J.M. Pierson, and C. Fontoura, "Leandro Mastering system and power measures for servers in datacenter", Sust. Comp.: Inf. and Syst., vol.15. pp: 28-38, ISSN 2210-5379, 2017.

[27] D. Suleiman, M. Ibrahim, and I. Hamarash, "Dynamic voltage frequency scaling (DVFS) for microprocessors power and energy reduction", Intl. Conf. on Electr. and Electronics Eng. vol. 12, December 2005.

[28] R. Matheson, "Study measures how fast humans react to road hazards", MIT News, Aug. 7, 2019, https://news.mit.edu/2019/how-fast-humans-react-car-hazards-0807#:~:text=MIT%20researchers%20have%20found%20an,as%20fast%20as%20older%20drivers., last accessed Sept 2020.

[29] V.V. Dixit, S. Chand, and D.J. Nair, "Autonomous vehicles: disengagements, accidents and reaction times", PLoS one, vol. 11, no. 12, e0168054, 2016.

[30] E. Yurtsever, J. Lambert, A. Carballo, and K. Takeda, "A survey of autonomous driving: Common practices and emerging technologies", IEEE Access, vol. 8, pp. 58443-58469, 2020.

[31] P. Li, X. Chen, and S. Shen, "Stereo r-cnn based 3d object detection for autonomous driving", IEEE Conf. on Comp. Vision and Pattern Rec. pp. 7644-7652, 2019.

[32] E. Arnold, O.Y. Al-Jarrah, M. Dianati, S. Fallah, D. Oxtoby, and A. Mouzakitis, "A survey on 3d object detection methods for autonomous driving applications", IEEE Trans. on Intell. Transp. Syst., vol. 20, no. 10, pp. 3782-3795, 2019.

[33] R.K. Mazlan, R.M. Dan, M.Z. Zakaria, and A.H. Hamid, "Experimental study on the effect of alternator speed to the car charging system", MATEC Web of Conf. vol. 90, pp. 01076, EDP Sciences, 2017.

[34] NXP Semiconductors, "i.MX 8M Quad Power Consumption Measurement", App. Note AN12118, 08/2018, last accessed Sep. 2020 https://www.nxp.com/docs/en/nxp/application-notes/AN12118.pdf

[35] G. Litjens, T. Kooi, B.E. Bejnordi, A.A. Setio, F. Ciompi, M. Ghafoorian, et al. "A survey on deep learning in medical image analysis", Medical Img. Anal., vol. 42, pp. 60-88, 2017.

[36] M. Bakator, and D. Radosav, "Deep learning and medical diagnosis: A review of literature", Multimodal Techn. and Interact., vol. 2, no. 3, pp. 47, 2018.

[37] G. Urban, P. Tripathi, T. Alkayali, M. Mittal, F. Jalali, W. Karnes, and P. Baldi, "Deep learning localizes and identifies polyps in real time with 96% accuracy in screening colonoscopy", Gastroenterology, vol. 155, no. 4, pp. 1069-1078, 2018.

[38] M.R. Azghadi, C. Lammie, J.K. Eshraghian, et al "Hardware Implementation of Deep Network Accelerators Towards Healthcare and Biomedical Applications",. arXiv preprint arXiv:2007.05657, 2020..

[39] M. Sajjad, S. Khan, K. Muhammad,et al., "Multi-grade brain tumor classification using deep CNN with extensive data augmentation", Journal of Comp. Sci., vol. 30, pp. 174-182, 2019.

[40] C. Szabó, and E.M.M. Alzeyani, "Measuring energy efficiency of selected working software", Studia Universitatis Babes-Bolyai Informatica, vol. 63, no. 1, pp. 5-16, 2018.

[41] V. Tsakanikas, and T. Dagiuklas, "Video surveillance systems-current status and future trends", Comp. & El. Eng., vol. 70, pp. 736-753, 2018.

[42] G. Sreenu, and M.S. Durai, "Intelligent video surveillance: a review through deep learning techniques for crowd analysis", Jrnl. of Big Data, vol. 6, no. 1, pp. 48, 2019.

[43] Sochor, J., Špaňhel, J., & Herout, A. (2018). Boxcars: Improving fine-grained recognition of vehicles using 3-d bounding boxes in traffic surveillance. IEEE transactions on intelligent transportation systems, 20(1), 97-108.

[44] Coşar, S., Donatiello, G., Bogorny, V., Garate, C., Alvares, L. O., & Brémond, F. (2016). Toward abnormal trajectory and event detection in video surveillance. IEEE Transactions on Circuits and Systems for Video Technology, 27(3), 683-695.

[45] S. Akçay, M.E. Kundegorski, M. Devereux, and T.P. Breckon, "Transfer learning using convolutional neural networks for object classification within x-ray baggage security imagery", IEEE Intl. Conf. on Img. Proc. (ICIP), pp. 1057-1061, September 2016.

[46] S. Akçay, M.E. Kundegorski, C.G. Willcocks, and T.P. Breckon, "Using deep convolutional neural network architectures for object classification

and detection within x-ray baggage security imagery", IEEE Trans. on Inf. Forensics and Security, vol. 13, No. 9, pp. 2203-2215, 2018.

[47] C. Kyrkou, and T. Theocharides, "Deep-Learning-Based Aerial Image Classification for Emergency Response Applications Using Unmanned Aerial Vehicles", In CVPR Workshops, pp. 517-525, June 2019.

[48] D. Giordan, M.S. Adams, I. Aicardi, et al. "The use of unmanned aerial vehicles (UAVs) for engineering geology applications", Bull Eng Geol Environ, 2020. https://doi.org/10.1007/s10064-020-01766-2

[49] L. Yang, and W. Zhang, "Beam tracking and optimization for UAV communications", IEEE Trans. on Wireless Comm., vol. 18, no. 11, pp. 5367-5379, 2019.

[50] Y. Chen, D. Ming, and X. Lv, "Superpixel based land cover classification of VHR satellite image combining multi-scale CNN and scale parameter estimation", Earth Sci. Inf., vol. 12, no. 3, pp. 341-363, 2019.

[51] U. Kanjir, H. Greidanus, and K. Oštir, "Vessel detection and classification from spaceborne optical images: A literature survey", Remote Sens. of Envir., vol. 207, pp. 1-26, 2018.

[52] A.G. Howard, M. Zhu, B. Chen, D. Kalenichenko, W. Wang, T. Weyand, and H. Adam,."Mobilenets: Efficient convolutional neural networks for mobile vision applications". arXiv:1704.04861 2017.

[53] J. Wu, C. Leng, Y. Wang, Q. Hu, and J. Cheng, "Quantized convolutional neural networks for mobile devices", IEEE Conf on Comp. Vision and Pattern Rec, pp. 4820-4828, 2016.

[54] P. Wang, and J. Cheng, "Accelerating convolutional neural networks for mobile applications", ACM Intl. Conf. on Multimedia, pp. 541-545, October 2016

[55] W.Yicong, and H. Xu, "Image analysis for user authentication", U.S. Patent No. 9,934,504. U.S. Patent and Trademark Office, 2018..

[56] L. Liu, H. Li, and M. Gruteser, "Edge assisted real-time object detection for mobile augmented reality", Annual Intl. Conf. on Mobile Comp. and Netw., pp. 1-16, August 2019.

[57] Z. Lu, S. Rallapalli, K. Chan, and T. La Porta, "Modeling the resource requirements of convolutional neural networks on mobile devices", ACM Intl. Conf. on Multimedia, pp. 1663-1671, October 2017.

[58] J. Hanhirova, T. Kämäräinen, S. Seppälä, M. Siekkinen, V. Hirvisalo, and A. Ylä-Jääski, "Latency and throughput characterization of convolutional neural networks for mobile computer vision". ACM Multimedia Syst. Conf., pp. 204-215, June 2018.

[59] Y. Chen, X. Jin, J. Sun, R. Zhang, and Y. Zhang, "Powerful Mobile app fingerprinting via power analysis", IEEE Conf. on Comp. Comm. pp. 1-9, May , 2017.

[60] W. Njima, I. Ahriz, R. Zayani, M. Terre, and R. Bouallegue, "Deep CNN for Indoor Localization in IoT-Sensor Systems", Sensors, vol. 19, no. 14, pp. 3127, 2019.

[61] M. Shin, W. Paik, B. Kim, and S. Hwang, "An IoT platform with monitoring robot applying CNN-based context-aware learning", Sensors, vol. 19, no. 11, pp. 2525, 2019.

[62] K. Muhammad, T. Hussain, M. Tanveer, G. Sannino, and V.H.C. de Albuquerque, "Cost-effective video summarization using deep CNN with hierarchical weighted fusion for IoT surveillance networks", IEEE Internet of Things Journl, vol. 7, no 5, pp. 4455-4463, 2019.

[63] S. Li, Y. Luo, K. Sun, N. Yadav, and K.K. Choi, "A Novel FPGA Accelerator Design for Real-Time and Ultra-Low Power Deep Convolutional Neural Networks Compared With Titan X GPU". IEEE Access, vol. 8, pp. 105455-105471, 2020.

[64] J. Redmon, S. Divvala, R. Girshick, and A. Farhadi, "You only look once: Unified, real-time object detection", In Proc. IEEE Conf. on Comp. Vision and Pattern Rec. pp. 779-788, 2016.

[65] J. Redmon, and A. Farhadi, "YOLO9000: better, faster, stronger", IEEE Conf. on Comp. Vision and Pattern Rec. pp. 7263-7271, 2017.

[66] R. Girshick, J. Donahue, T. Darrell, and J. Malik, "Rich feature hierarchies for accurate object detection and semantic segmentation" IEEE Conf. on Comp. Vision and Pattern Rec pp. 580-587, 2017.

[67] X. Zhang, X. Zhou, M. Lin, and J. Sun, "Shufflenet: An extremely efficient convolutional neural network for mobile devices". IEEE Conf. on Comp. Vision and Pattern Rec pp. 6848-6856, 2018.

[68] K. He, X. Zhang, S. Ren, and J. Sun, "Deep residual learning for image recognition". IEEE Conf. on Comp. Vision and Pattern Rec pp. 770-778, 2016.

[69] H. Mao, S. Yao, T. Tang, B. Li, J. Yao, and Y. Wang, "Towards real-time object detection on embedded systems". IEEE Trans. on Emerging Topics in Comp., 2016.

[70] J. Yu, K. Guo, Y. Hu, X. Ning, J. Qiu, H. Mao, and H. Yang, "Real-time object detection towards high power efficiency". Design, Autom. & Test in Europe Conf. & Exhib. pp. 704-708, IEEE, 2018.

[71] S. Ren, K. He, R. Girshick, and J. Sun, "Faster r-cnn: Towards real-time object detection with region proposal networks". In Adv. in Neural Inf. Proc. Syst. pp. 91-99, 2015.

[72] W. Liu, D. Anguelov, D., Erhan, et al. "Ssd: Single shot multibox detector". In Europ. Conf. on Comp. Vision pp. 21-37, 2016.

[73] Z. Huang, J. Wang, X. Fu, T. Yu, Y. Guo, and R. Wang, "DC-SPP-YOLO: Dense connection and spatial pyramid pooling based YOLO for object detection". Inf. Sci., 2020.

[74] "NVIDIA Turing GPU Architecture", NVIDIA whitepaper, WP-09183-001_v01, https://www.nvidia.com/content/dam/en-zz/Solutions/design-visualization/technologies/turing-architecture/NVIDIA-Turing-Architecture-Whitepaper.pdf, last visited on Sep. 2020

[75] T. Zheng, D. Nellans, A. Zulfiqar, M. Stephenson, and S.W. Keckler, "Towards high performance paged memory for GPUs", IEEE Intl. Symp. on High Perf. Comp. Arch. (HPCA), pp. 345-357, 2017.

[76] M.D. Hill, and M.R. Marty, "Retrospective on Amdahl's Law in the Multicore Era". IEEE Comp., vol. 50, no. 6, pp.12-14, 2016.

[77] Intel Movidius Vision Processing Units (VPUs), on https://www.intel.com/content/www/us/en/products/processors/movidius-vpu.html, last accessed Sep. 2020.

[78] X. Xu, S. Caulfield, J. Amaro, G. Falcao, and D. Moloney, "1.2 Watt Classification of 3D Voxel Based Point-clouds using a CNN on a Neural Compute Stick", Neurocomp., vol. 393, pp. 165-174, 2020.

[79] J.A. Cameron, P. Savoie, M.E. Kaye, and E.J. Scheme, "Design considerations for the processing system of a CNN-based automated surveillance system", Exp. Syst. with Appl., vol. 136, pp. 105-114, 2019.

[80] D. Pena, A. Forembski, X. Xu, and D. Moloney, "Benchmarking of CNNs for low-cost, low-power robotics applications". New Frontier for Deep Learning in Robotics, pp. 1-5, 2017.

[81] ARM Ethos-N Processor Series NPU, https://developer.arm.com/ip-products/processors/machine-learning/arm-ethos-n, last accessed Sept. 2020.

[82] NXP Announces Lead Partnership for Arm Ethos-U55 Neural Processing Unit for Machine Learning, Feb. 2020, https://media.nxp.com/news-releases/news-release-details/nxp-announces-lead-partnership-arm-ethos-u55-neural-processing, last accessed Sep. 2020.

[83] Cloud Tensor Processing Units, https://cloud.google.com/tpu/docs/tpus, last accessed Sept. 2020.

[84] Edge TPU Performance Benchmarks, Coral, Google Research, https://coral.ai/docs/edgetpu/benchmarks/, last accessed Sept. 2020.

[85] M.E. Ilas, and C. Ilas, "A New Method of Histogram Computation for Efficient Implementation of the HOG Algorithm", Computers, vol. 7, no.1, 2018, https://doi.org/10.3390/computers7010018

[86] X. Wei, C.H. Yu, P. Zhang, Y. et al "Automated systolic array architecture synthesis for high throughput CNN inference on FPGAs", Proc. of the 54th Ann. Design Autom. Conf. pp. 1-6, 2017.

[87] S. Akkas, S. Maini, and J. Qiu, "A Fast Video Image Detection using TensorFlow Mobile Networks for Racing Cars", Intl. Conf. on Big Data pp. 5667-5672, IEEE, 2019.

[88] S. Alyamkin, M. Ardi, et al "Low-power computer vision: Status, challenges, and opportunities", IEEE Journl on Emerg. and Select. Topics in Circ. and Syst. Vol 9, no 2, pp. 411-421, 2019.

[89] R.J. Wang, X. Li, and C.X. Ling, "Pelee: A real-time object detection system on mobile devices". In Adv. in Neural Inf. Proces. Syst. pp. 1963-1972, 2018.

Investigating the performance of MicroPython and C on ESP32 and STM32 microcontrollers

Valeriu Manuel Ionescu
Dept. Electronics, Communications and Computers
University of Pitesti
Romania
manuelcore@yahoo.com

Florentina Magda Enescu
Dept. Electronics, Communications and Computers
University of Pitesti
Romania
enescu_flor@yahoo.com

Abstract—**Python is a programming language that is used both by entry level programmers and advanced researchers. MicroPython is a software implementation of Python that runs on microcontrollers. This paper will investigate the MicroPython execution performance compared to similar C native code on low cost microcontrollers: STM32 and ESP32. The comparison will target: memory allocation speed; SHA-256 and CRC-32 performance and will present conclusions regarding the encountered problems and ways to improve the application performance.**

Keywords—MicroPython, ESP32, STM32, performance, hash, CRC

I   Introduction

Edge computing is a term associated to Internet of Things (IoT) devices that identifies the partial or complete processing of data captured from sensors on the system that makes the data acquisition. This is possible now because the processing power of the microcontrollers [9] has increased considerably while maintaining an acceptable price. While C code on these devices is preferable, having a larger codebase and a large number of programmers is also desirable, therefore being able to write code fast with as many coders as possible makes investigating the performance of Python code on microcontrollers a priority.

Python is a high level programming language that is often used in website development, GUI applications, machine learning and Internet of Things (IoT) applications. Some of its advantages are the ease of code writing and reading, running the same code (with no need to recompile) on multiple platforms as it is an interpreted programming language, the automatic memory management and the large number of libraries that allows developing prototype applications fast.

There are two major versions of Python that are used for development: 2 and 3. While the user base seemed to be split between the two versions, there is a rapid transition towards version 3 [1] [2] of more than 75% of the coders. Finally an announcement was made that version 2 will be retired at the end of 2020 [3] therefore all users will migrate to version 3.

MicroPython is an implementation of Python that runs on microcontrollers that was developed in 2013 by Damien George and drops backward compatibility by focusing on Python 3.5. MicroPython [4] implements a subset of Python functionality

and includes function and class libraries that are direct replacement for Python libraries but having as target the microcontroller market with their limitations: memory, hardware support and speed. For example the library cmath for mathematical functions for complex numbers is not available on the ESP8266 processor as it needs floating point support. Microcontrollers such as ESP32 or STM32 have floating point support and the functions in this library can be called [5]. As the multi core microcontrollers have started to appear for the low cost systems, the multithreading support in MicroPython is highly experimental and limited, while fully functional in Python.

Some variations of this software have appeared: CircuitPython in 2017 focused on simplicity and in the same year a version that runs on RISC-V processors [6] (that have seen an increase of use because they are open Instruction Set Architecture and offer trust and security).

The advantage of using MicroPython in the development of applications for microcontrollers is that there are many programmers that use Python as a programming language [7]. The transition to MicroPython is rather easy to do and involves mainly observing the limitations of the microcontrollers and optimizing the code in order to take advantage of its strengths and avoid the limitations (such as RAM speed and size, process of frequency and reduced number of cores).

Some of the disadvantages for developing MicroPython application include memory fragmentation and objects that grow in size (for example lists) [8].

This paper investigates the performance of microcontrollers running C and MicroPython for the ESP32 and STM32 devices in the following algorithms: computing Cyclic Redundancy Check (CRC) - used for error detection in data flows; floating point operations that involve memory allocation and management; computing SHA-256 (used in authentication, encryption algorithms and even cryptocurrencies).

II   Related literature

MicroPython performance was investigated before in many articles because the hardware is cheap and the performance is at a level where it can process data locally instead of simply sending it to a server.

In the article [10] investigates the usability of the two main solutions for running Python on microcontrollers: MicroPython and CircuitPython. MicroPython was the first of the two to appear and is targeted towards obtaining the most performance from the hardware by scarifying the ease of use of the software. CircuitPython is a fork MicroPython and is promoted by Adafruit Industries and many code examples are linked to the hardware products sold by this firm. CircuitPython focuses on the ease of use of by offering many libraries that directly perform the required tasks without need for much understanding of the microcontrollers where the code runs, targeting the beginner programmers.

In the book [11] the hardware focus is on the boards: PyBoard, micro:bit, Circuit Playground Express and ESP8266 and MicroPython. The investigation is thorough presenting many of the pitfalls that decrease performance (especially the memory related errors that a Python user will not encounter) but many advances have happened in hardware especially since ESP32 has appeared (late 2016).

The book [12] investigates the use of both Arduino C and MicroPyhton for neural network development with one hidden layer on the ESP32 board. The research aims to investigate neural networks data propagation speed.

At the conference PyCon held at Portland in 2017, Jake Edge [13] has presented a comprehensive investigation in the causes of performance problems due to the memory management specific to Python and the limitations of the MicroPython environment. CPython was focused on simplicity of use (while boasting a multitude of extensions) while MicroPython has performance in mind (for example losing backward Python version compatibility). The conclusion was to avoid the corner cases where MicroPython performs poorly (such as lists and the garbage collection mechanisms) in order to keep the performance level good.

Finally in [14] present the known general issues with MicroPython performance and give hints about program optimization aspects that anyone writing code with these libraries should follow. However the paper does not target any specific processor and the two main aspects for code optimization: using arrays instead of lists due to their fixed nature and paying attention with memory allocation as there will be a significant slow down if garbage collection is activated. Other articles like [15] give insights on how MicroPython code works in byte code, native code, and native code with native types and how it can further be improved to increase execution speed (such as calling for integer addition instead of the C function rt_binary_op, the machine instruction adds, which gives a significant performance boost).

III Used hardware

Two of the most used microcontrollers for the low cost projects are STM32 and ESP32. Both are Arduino compatible, being capable of using a large software library.

ESP32 is a 40 nm process microcontroller based on the Xtensa 32 bit processor, designed by Espressif (a company that had a rapid evolution surpassing in 2018 100 million IoT chips [16]). ESP32 has become an important platform for developing low cost projects due to the integration of many connectivity solutions at a very low price point, with applications ranging from simple sensor reading, and WiFi/Blutooth communication, to video streaming, machine learning and facial recognition systems [17]. ESP32 has a performance of up to 600 DMIPS (32bit, at the top speed of 240MHz) on Dhrystone 2.1 benchmark [18]. At a cost of 2$ for individual boards it is a prime dual core solution. The disadvantage is having less documentation and support when compared to STM32 devices. The board used for tests is ESP-WROOM-32.

STM32 is a 32-bit ARM Cortex microcontroller that has a very low price and is integrated in many boards used in low cost projects. The 90nm device (an older technology when compared to ESP32) used for this paper is STM32F401RE [20], with a frequency of 84 MHz, 105 DMIPS and was part of the first STM32 series to have DSP instructions. The unit also presents a CRC calculation unit, unlike ESP32 that has to implement it solution via software. Therefore in computing power we should expect 6 times lower when compared to ESP32 when comparing purely computing performance. As a comparison, a Raspberry Pi 3B+ has a 3039.87 DMIPS performance [21]. The cost is the same with ESP32. It is a very well documented microcontroller and the libraries available target many application domains. Many of the microcontrollers in this series are pin by pin compatible, making easy to upgrade a project if more processing power is necessary. The first version of MicroPython was developed [19] on STMicroelectronics' STM32F4 board called pyboard. This is why in this paper a board with microcontroller from this family will be used: STM32 Nucleo.

The C development environment used was ESP-IDF for ESP32 and Visual Studio with VisualGDB for STM32. For MicroPython development uPyCraft was used for both platforms.

IV Implementation and Results

The algorithms tested are: floating point operation performance; CRC computation; Hash computation. The tests we repeated multiple times for each device in order to avoid measurement errors and to investigate result consistency.

The implementation of these algorithms was simple to create and test in MicroPython because the code was the same.

The C implementation however was different for the two platforms because of the function that measures the execution time and other platform particularities. For ESP32 the used function was *esp_timer_get_time*() while for STM32 the *coreticks* [22] function was used that measures time based on the processor clock frequency and from that execution time was computed. The STM32 measurement method works for all ARM chips that support DWT, and the Cortex M4 has this support. For MicroPython, the function used to measure time was time.ticks_us().

The first algorithm implemented was floating point operation performance, necessary in areas that require a larger range for results with the trade off being the precision loss (truncation, roundoff, etc.). This is a known weakness of the MicroPython implementation mainly because of the garbage collection operation. The implementation consists of a 300 000 for loop that performs an addition operation between floating point numbers. The results are shown in Table I and in Fig. 1.

TABLE I FLOATING POINT MEMORY ALLOCATION AND COMPUTATION TIME IN MICROPYTHON AND C FOR ESP32 AND STM32

	ESP32 Micro Python (µs)	STM32 Micro Python (µs)	ESP32 C (µs)	STM32 C (µs)
Run 1	1021756	2408869	6258	127923.06
Run 2	1024445	2404664	6257	127923.14
Run 3	1024419	2404196	6257	127923.06
Run 4	1021713	2402580	6258	127923.06
Run 5	1021723	2404892	6258	127923.06

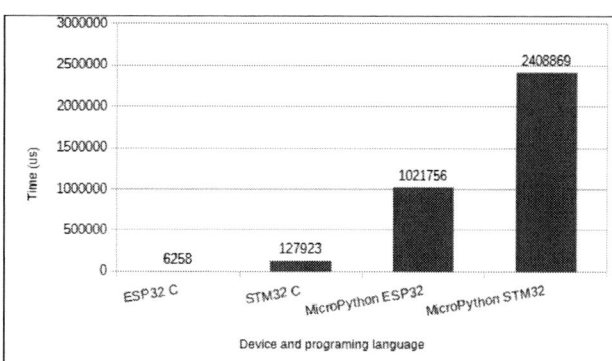

Fig. 1. Floating point computation results

As expected, the MicroPython is slower with an order of magnitude compared to the C implementation and ESP32 is faster than STM32 due to the clock and technology difference. The results also show that floating point operation is slower in MicroPython and should be avoided or replaced whenever possible with integer operations. As an additional observation, MicroPython has no problem with the allocation of a large number of floating point objects, as – for example – the use of: $a = a + 0.1$ in a loop will have to allocate the space for the object for every operation, instead of: $a = a + v$, where v is a constant value. This was noted as being a problem (double the time) in early versions of MicroPython [24]. This means that acceptable code can be written without paying attention to small details like frequency of variable allocation.

The second implementation was made for the CRC-32 algorithm computed for a 10 byte input. We have used the hardware CRC module for STM32 in C (offered by the TM_CRC_Calculate32 function) and the table implementation (same code) for ESP32 and in MicroPython. During research several aspects were observed: on Cortex M4 the CRC polynomial cannot be changed (Ethernet CRC-32 computation, while in higher modules like Cortex F7 the polynomial can be changed); the CRC hardware module in ESP32 is only used for calculating CRC of RTC fast memory; the C functions in ROM offered by ESP IDF use lookup tables to do computations in software. The results are presented in Table II and Fig. 2.

TABLE II CRC-32 COMPUTATION TIME IN MICROPYTHON AND C FOR ESP32 AND STM32

	ESP32 Micro Python (µs)	STM32 Micro Python (µs)	ESP32 C (µs)	STM32 C (µs)
Run 1	23056	15941	47	5
Run 2	23153	15952	47	5
Run 3	22780	16119	47	5
Run 4	23150	18961	47	6
Run 5	23602	19551	47	5

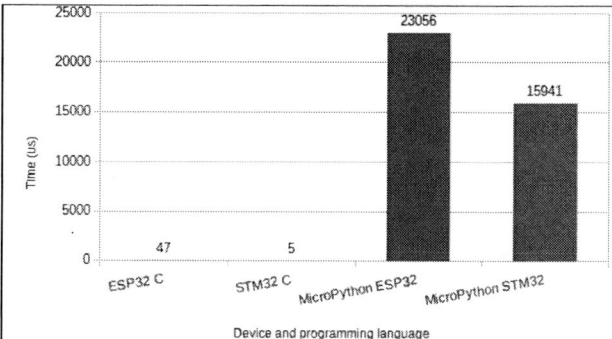

Fig. 2. CRC-32 computation results

The results show again that the C implementation is more efficient, in this case by three orders of magnitude. The hardware implementation used in the STM32 code allowed it to finish ahead of ESP32 even if the processor frequency favors the ESP32. However if other types of CRC need to be computed, the STM32 will perform worse than ESP32 as it will have to implement similar in memory lookup tables.

The final implementation is the SHA-256 hashing algorithm. The results are especially interesting to see because these micro platforms can be used to offload hash computation for blockchain transactions. The STM32F401RE does not have a hardware SHA-256 computation unit, this being available only in more expensive devices, like STM32F415. The implementation in this case was done in software, based on the algorithm implementation in [18]. ESP32 has hardware acceleration support for SHA and other cryptographic functions. It was easy to test the implementation using MicroPython using the hashlib library. The results for the ESP32 implementation are presented in Table III and Fig. 3.

TABLE III SHA-256 COMPUTATION TIME IN MICROPYTHON AND C FOR ESP32

	ESP32 Micro Python (µs)	ESP32 C (µs)
Run 1	295	163
Run 2	246	163
Run 3	244	163
Run 4	245	163
Run 5	245	163

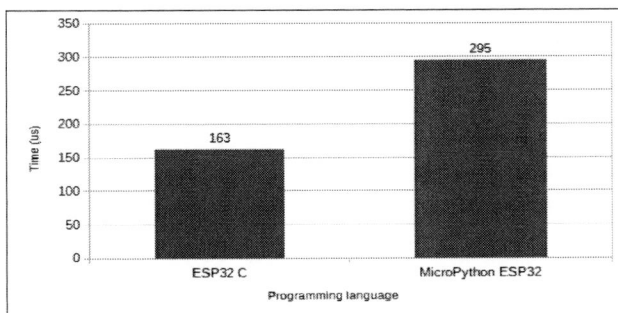

Fig. 3. SHA-256 computation results for ESP32

The C implementation is faster than the MicroPython one however only by 45%. Using hardware implementation for cryptographic functions gives better results then software implementations, however vulnerabilities found for the module in ESP32 [27] are harder to fix then a software implementation. The final observation will be made regarding the result consistency. The standard deviation of the results for of the MicroPython implementation is larger when compared to C, due to the memory operations that MicroPython is performing. Also, the standard deviation of the results for ESP32 is smaller when compared to STM32, in this case the motive is the much faster memory operations and processor speed in ESP32 that make memory management operations faster.

V CONCLUSIONS

In this paper the performance of MicroPython and C was tested for both STM32 and ESP32. The results show that the MicroPython performance level is lower than the C implementation as expected, but the difference varies depending of the task complexity and memory allocation. Writing code for MicroPython and comparing implementation results is simple because the code is portable, unlike C code where many platform specific changes needed to be performed. The result consistency is good only for the C execution. MicroPython shows large variations especially on the first runs. One of the possible applications of this study is teaching IoT programming [23] on low cost hardware using Python instead of C because of the code portability and ease of learning by students. In the future more test will be performed concerning network performance for MicroPython and C.

REFERENCES

[1] R. Olson, "Python usage survey 2014", January 30, 2015, web, http://www.randalolson.com/2015/01/30/python-usage-survey-2014/, Accessed 05.05.2020

[2] Jetbrains, "Python 3 vs Python 2", web, https://www.jetbrains.com/research/python-developers-survey-2017/, Accessed 05.05.2020

[3] Python Software Foundation, "Press Release 20-Dec-2019, PYTHON 2 SERIES TO BE RETIRED BY APRIL 2020", web, https://www.python.org/psf/press-release/pr20191220/, Accessed 05.05.2020

[4] D.P. George et al., "MicroPython Documentation Release 1.10 Jan 25, 2019", web, http://docs.micropython.org/en/v1.10/micropython-docs.pdf

[5] A. Schweizer, "ESP32 floating-point performance", 2016, web, https://blog.classycode.com/esp32-floating-point-performance-6e9f6f567a69, Accessed 05.05.2020

[6] B. Nilawar, "MicroPython Port for RISC-V soft Processor", 7th RISC-V Workshop Proceedings, November 28 – November 30, 2017, California

[7] L. Tung "Programming languages: Python developers now outnumber Java ones", 2019, web, https://www.zdnet.com/article/programming-languages-python-developers-now-outnumber-java-ones/, Accessed 05.05.2020

[8] D.P. George et al., "MicroPython on microcontrollers", 28 Jun 2020, web, https://docs.micropython.org/en/latest/reference/constrained.html, Accessed 05.05.2020

[9] J. Ivković, J. Ivković, "Analysis of the performance of the new generation of 32-bit Microcontrollers for IoT and Big Data Application", Proceedings of ICIST 2017, p330-336.

[10] -, "Profile: Scott shawcroft: This developer is squeezing python into microcontrollers," in IEEE Spectrum, vol. 56, no. 4, pp. 16-16, April 2019, doi: 10.1109/MSPEC.2019.8678507

[11] N.H. Tollervey, "Programming with MicroPython: Embedded Programming with Microcontrollers and Python", O'Reilly Media, Inc., Sep 25, 2017, ISBN 978=1=491-97273-1

[12] Dokic K., Radisic B., Cobovic M. "MicroPython or Arduino C for ESP32 - Efficiency for Neural Network Edge Devices", In: Brito-Loeza C., Espinosa-Romero A., Martin-Gonzalez A., Safi A. (eds) Intelligent Computing Systems. ISICS 2020. Communications in Computer and Information Science, vol 1187. Springer, Cham

[13] J. Edge, "Memory use in CPython and MicroPython", web, 2017, https://lwn.net/Articles/725508/, Accessed 05.05.2020

[14] D.P. George, P. Sokolovsky, "Maximising MicroPython speed", web, https://docs.micropython.org/en/latest/reference/speed_python.html, Accessed 05.05.202

[15] D. George, "The 3 different code emitters", 2013, web, https://www.kickstarter.com/projects/214379695/micro-python-python-for-microcontrollers/posts/664832, Accessed 05.05.202

[16] Espressif, "Espressif Achieves the 100-Million Target for IoT Chip Shipments", web, https://www.espressif.com/en/news/Espressif_Achieves_the_Hundredmillion_Target_for_IoT_Chip_Shipments, Accessed 05.05.2020

[17] M. Gravråk, "Machine Learning at the Edge with ESP32", web, https://www.bouvet.no/bouvet-deler/machine-learning-with-esp32, Accessed 05.05.2020

[18] B. Conte, "sha256.c", web, https://raw.githubusercontent.com/B-Con/crypto-algorithms/master/sha256.c, Accessed 05.05.2020

[19] J. Beningo, "Prototype to production: MicroPython under the hood". EDN Network. 2016, web, https://www.edn.com/prototype-to-production-micropython-under-the-hood/,

[20] STMicroelectronics, "STM32F401RE Datasheet", web: https://www.st.com/resource/en/datasheet/stm32f401re.pdf, Accessed 05.05.2020

[21] R. Longbottom, "Raspberry Pi 4B 64 Bit Benchmarks and Stress Tests". 10.13140/RG.2.2.33099.54562.

[22] Erich Styger, Cycle Counting on ARM Cortex-M with DWT, January 30, 2017, web, https://mcuoneclipse.com/2017/01/30/cycle-counting-on-arm-cortex-m-with-dwt/, Accessed 05.05.2020

[23] Pearson, Bryan & Luo, Lan & Zou, Cliff & Crain, Jacob & Jin, Yier and Fu, Xinwen. (2020). Building a Low-Cost and State-of-the-Art IoT Security Hands-On Laboratory. 10.1007/978-3-030-43605-6_17.

[24] Github. ""gc speedup" changes led to much increased heap usage #836", web, https://github.com/micropython/micropython/issues/836, Accessed 05.05.2020

[25] STMicroelectronics, "RM0368 Reference manual", web, https://www.st.com/resource/en/reference_manual/dm00096844-stm32f401xb-c-and-stm32f401xd-e-advanced-arm-based-32-bit-mcus-stmicroelectronics.pdf, Accessed 05.05.2020

[26] Limited Results, "Pwn the ESP32 crypto-core", August 2019 web, https://limitedresults.com/2019/08/pwn-the-esp32-crypto-core/, Accessed 05.05.2020

A Neuro Model for Weather Forecasting

T.G. Predună, V.A. Rusu, and D.I. Năstac

Faculty of Electronics, Telecommunications and Information Technology
POLITEHNICA University of Bucharest, Romania
nastac@ieee.org

Abstract—**Traditional weather forecasting, using physical methods, requires a lot of computational power and is generally slow. Modern developments in machine learning have led us to experiment with weather prediction using those newly available tools. In this article, we will present the capabilities, advantages and shortcomings of neural networks in the context of weather prediction. The tools consist of open source machine learning libraries, freely available data and two off the shelf computers for all the experiments, in order to show that weather forecasting can be done in a cost-efficient manner.**

Keywords—*weather forecasting; neural networks; machine learning*

I. INTRODUCTION

The problem we are trying to solve is to predict "weather" with the aid of machine learning.

But what is weather? Weather consists of many different phenomena that humans experience, such as wind, temperature, humidity, sunlight and other meteorological effects, linked together by the laws of physics. As an example, it's always cloudy when it's raining, but it's not always raining when it's cloudy. Therefore, weather can be characterized as a complex system of random interdependent variables, which are all time dependent. Everything weather event happens at the same time or after another one.

And what is machine learning? Machine learning is an application of artificial intelligence where a 'machine' learns by itself how to solve a problem. Most of the problems in the field are either classification or regression problems. This project started using three different algorithms, specifically, decision-trees, support vector machines and neural networks, but the first two were dropped due to bad results. Weather forecasting, by nature, is a regression problem, which is very well suited for neural networks. Traditionally, weather forecasting works by collecting as much data as possible about the atmosphere in the present moment and then inputting the data into computer programs that simulate the physical processes which take place in nature, in order to attempt to predict what the weather will be like at a certain location and time. Most resources and time are spent in those simulations, because you have to insert the current state of the atmosphere and then wait each time for the simulation to happen. The 'machine learning' approach is to collect as much 'past' data as possible about the weather at a location, train a neural network on a part of it, test the neural network on the remaining part and modify the network until a good enough accuracy is reached. Most of the resources and time are spent on the training and tuning part, but after this part,

every prediction is almost instantaneous, after the current state of the atmosphere is inserted.

Simple and obvious questions can then be asked: Is it possible that such a technique may predict accurately complex weather phenomena and events? What about just some of them? What is the most time efficient way to achieve accurate "enough" predictions in the context of machine learning? and Why should someone use machine learning and not the physical models that are already in use?

So, can complex phenomena be predicted with machine learning? As of now, no. Can at least "some" weather phenomena be predicted? Yes. What's the most efficient way? Multiple neural networks for each weather phenomena, all trained on the same data consisting of different phenomena. Why machine learning? It's very good for making localized predictions, as long as there is enough data for the point on the map for which the predictions are made.

But why should someone switch from physical models? Ideally, both physical models and neural networks would be used. An article that came out this June compares the two [1]. The authors used LSTMs (long short-term memory neural networks) trained on 5 months of data in a similar fashion to us and managed to get better results for up to twelve hours.

A more basic approach is also possible, using linear regression to achieve some degree of success [2] or using support vector machines [3]. Another option is the "Google Approach", which uses radar images and convolutional neural networks [4]. But the main advantages of neural networks and machine learning in general are their speed and ability to be tailored to a specific area. This flexibility makes them a better choice in places where using physical models is not justified, being too resource intensive or expensive. Neural networks can be trained in a similar fashion to how social media sites predict users' behavior, based on what they like and their interactions on the website, only this time the user is a specific area, the area where we want to forecast the weather. Traditional Numerical Weather Forecasting models require a large amount of computing power, which is expensive, whereas a neural network only needs to be trained once (or periodically), after which the predictions are instantaneous. They can also be linked to real time sensors to predict real time phenomena, as described in [5].

This article will therefore show just how easy and fast it is to train such a neural network, on inexpensive hardware, making accurate weather forecasting accessible even in remote locations.

II. METHODS

A. Data Preparation

The data must be prepared such that a neural network can be trained for each weather feature.

First, Neural Networks require a lot of data. Our dataset was obtained from [6], for the city of Basel, Switzerland. More specifically, it contains hourly data for the last 11 years for parameters like Temperature, Wind, Sunlight. The full list of parameters is presented in Fig. 1.

	Basel Temperature [2 m elevation corrected]	Basel Relative Humidity [2 m]	Basel Mean Sea Level Pressure [MSL]	Basel Precipitation Total	Cloud Cover Total	Basel Sunshine Duration	Basel Wind Speed [10 m]	Basel Wind Direction [10 m]	Basel Temperature
timestamp									
20090501T0000	12.016529	79.0	1018.6	0.0	100.0	0.0	2.280479	44.999985	11.72
20090501T0100	11.700529	83.0	1019.0	0.0	100.0	0.0	2.080661	39.805542	11.55
20090501T0200	11.420529	85.0	1019.1	0.0	100.0	0.0	1.787066	28.569033	11.33
20090501T0300	11.170529	87.0	1019.5	0.0	100.0	0.0	1.939421	15.945404	11.12
20090501T0400	10.540529	89.0	1019.5	0.0	100.0	0.0	1.787066	28.569033	10.97

Fig. 1. Sample from the dataset

In total, the dataset contains 96456 samples with 9 features. Missing or erroneous entries were removed from the dataset.

Looking back on it, we could have dropped some of the features, for example, there are two instances of 'temperature', which may have led to some degree of over-fitting. However, the effect is minimal.

Second, the weather features are signed values and have different orders of magnitude. Therefore, standardization was applied by subtracting the mean and dividing by the standard deviation. This also has the effect of making computations much easier.

Third, the data was split, 80% for training and 20% for testing the neural networks.

Next, the data is treated as a time series [7][8]. Therefore, separate delay vectors are created for each timestep in the future to be predicted. So, to create the delay vector for the temperature three hours into the future, the following steps were taken:

Firstly, we need training and testing vectors, let's call them X_train, Y_train, X_test, Y_test. X_train consists of the first 80% entries and X_test the other 20% of all features. This is simple enough. Y_train and Y_test consist only of the 'temperature' entry delayed by three, since it's hourly data and we want the temperature, one hour, two hours and three hours later. So, those vectors are made by selecting all entries, starting from the second up to the fourth, in order to 'move' three hours into the future. Keep in mind that the 'X' vectors are shorter than the 'Y' vectors by three, since the numbering started three entries later, therefore, the last three entries are dropped from the 'X' vectors in order for them to be the same length. This applies to any kind of delay we want to add, so you have to keep it in mind. For our paper, a 24 and 48 hour 'future' was used.

Finally, a sliding window [9] will be used for the training data. This is the way in which we use multiple hours to predict the future. In the situation above, it was assumed that only one hour will be used. The program will work such that it will use 48 and 192 hours of history for the prediction. So, if we want to predict the 'temperature' in three hours with a 48-hour history window, the data will be arranged such that we feed all the 48 hours of data in the neural net. Therefore, our X_train has a shape of (77093, 48, 9) and our Y_train_temp has a shape of (77093, 24, 1). The order of the items in parenthesis means entry size, sample size, feature size. Remember, there are 77093 entries on the training variables because only 80% of the initial set will be used.

B. Working Principles

As of now, we have X_train and X_test, which consist of entries of all the features, and different Y_train and Y_test for each feature we want to predict. For temperature specifically let us call them Y_train_temp and Y_test_temp.

In the following paragraphs, a very quick explanation of neural networks will be given. Let x_n be an input vector, containing numbers, let's assume natural numbers for simplicity. Let w_n be a vector of scalars, called the weight vector, which contains other natural numbers. Fig. 2 illustrates an extremely simple neural network, consisting of an input layer of n neurons, where all items in x_n are placed, and an output layer, consisting only of one neuron.

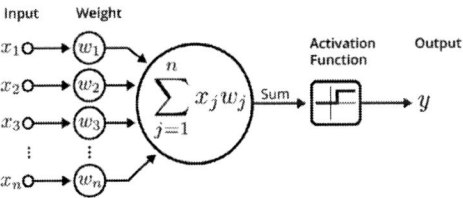

Fig. 2. Basic Neural Network Representation [10]

What is referred to as 'feeding forward' is the process by which, in this case, each value of the neurons of the input layer is multiplied by the corresponding value of the weight vector (x_1 is multiplied by w_1) and then, after this multiplication, all these values are added together. The number resulted after the summation is passed through an 'activation function' which has the sole purpose of translating the number in a limited domain of values, usually between –1 and 1 or 0 and one. After the activation function has done its job, the last neuron in the network represents what the network 'thinks' those inputs should output.

Consider a neural network which needs to learn how to predict the variable, or, vector of variables, Y in relation to the variable or vector of variables X. Datasets are usually given containing both X and Y with values for all of them. Some of the variables contained by our dataset (see the samples in Fig. 1) for example are Temperature [2 m elevation corrected], Relative Humidity [2 m] and Mean Sea Level Pressure. As it can be observed, the dataset has numerical entries for those variables for 10 years. If a toy neural network were to be trained to predict Temperature, the Y variable would become Temperature and the other variables would become the X variable. What is referred to as 'training' is the process where the X variables are 'fed forward' from the input of the neural network, the neural network then outputs a result, which represents its best estimation about what value the solution should have and then this result is compared to the actual Y value corresponding to the X values fed at the input (remember, the dataset is 'bijective' because all the columns are linked by a unique time of measurement'). To measure how different the neural network result is from the actual value, the 'error' function is used, and again, a multitude of error functions can be used. This error function computes the difference, and then, in order to adjust the weight vector, the error function is derived in regard to each weight in the weight vector. This derivative is later used to update the weights. This whole algorithm, in its most basic form is using Fermat's theorem, taking small steps to find the minimum of the error function. The most basic algorithm for this is called 'gradient descent' and the whole process of weight updating is called 'backpropagation'.

Lastly, the 'testing' process consists of using a portion of the dataset which was not used for training. Here, the inputs are fed through and again the outputs are compared to the actual values. This step exists to see if the neural network either 'generalized' the problem-solving or it just learned how to predict the training set very well, which is called 'overfitting'. We hope now it is clearer why we processed the data in the way that we did.

For the temperature example, X_train is fed at the input and Y_train at the output in order for it to train. Then the network is tested by 'feeding' it X_test and making it make predictions. We then compare those predictions with Y_test and see how close it was. Different error and activation functions [11] were tested. We settled on using the Mean Squared Error as our main error function, and the hyperbolic tangent and relu for our activation.

For the Sklearn implementation, the standard perceptron learning algorithm was used, better described here [12], and LSTMs [13] were used in Tensorflow. Keep in mind the different architectures, in Sklearn is feed-forward and in Tensorflow is recurrent.

Now, the best neural network architecture has to be found. In reality, neural networks are more complex than the example described above, with more neurons each layer and more layers, but the basic processes of training, backpropagation and testing are the same. Our project branches out in two directions, one using Sklearn and the other using Keras, both open-source and very well documented machine learning libraries that allowed us to quickly build neural nets. Both work on the same principle, that is, a separate network for each parameter. We decided to use every other input parameter to predict only one, for example, one network predicts 'temperature' using all other columns as inputs, one network predicts the 'humidity' using all the other columns all the other and so on. Why? Firstly, it's faster to train and secondly, it's harder for the model to over-fit.

C. Sklearn

On the Sklearn [14] part of the project, a simple grid-search type of program was implemented, so that it finds the 'best' neural net from a list of given hyperparameters. For the temperature example, the program functions as described in section B. It builds neural networks using different combinations of hyper parameters, trains them, tests them and picks the most accurate one. For the backpropagation, the program can either use the ADAM algorithm [15] or the stochastic gradient descent algorithm [16].

Regarding the number of layers and neurons per layer, here is a list of the neurons combinations (10, 10), (30, 10), (30, 15), (20, 20), (10, 15), all mixed with either an adaptive or constant learning rate, and a maximum number of 400 iterations, as this is a relatively small dataset. All the networks have the number of input neurons equal to the sample size multiplied by the feature size.

D. Tensorflow

Keras is the high-level API of Tensorflow 2.0, written in Python, designed with a focus on enabling fast experimentation [17]. Keras contains essential building blocks for creating complex neural network solutions. To enhance the speed at which different neural network architectures can be tested, Keras also supports GPU and TPU hardware acceleration.

We based our research on a previous multilayer perceptron network designed for energy usage forecasting. The results proved that reasonably accurate weather forecasting was possible and led us to continue our research in order to find networks better suited for this task. Various MLP, CNN and RNN network architectures were tested. We spent most of the time testing RNN networks with LSTM layers, as they're the optimal fit for this task.

The best results were achieved using a simple network with an LSTM input layer with 30 units and a fully connected output layer. We use 8 days of history (192 timesteps) to predict 4 days into the future (96 timesteps). The Adam optimiser [15] is used, with MSE as the loss function. To prevent overfitting, we used L1 and L2 kernel regularisation in the LSTM layer and implemented Early Stopping. The architecture is presented in Fig. 3.

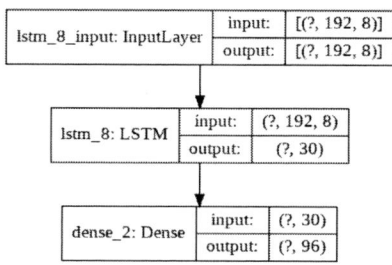

Fig. 3. Architecture of the Keras Network

III. RESULTS

The obtained results can be most easily visualized graphically. However, as there were no meteorologists involved in this project, the precision could only be measured for the Temperature. As none of the authors are familiar with the scale or measurement units of different meteorological phenomena, the only thing that could be judged was temperature, were even for 24 hour predictions, the error is mostly within a 0.5 margin, which is highlighted by the blue 'tube' around the red prediction line.

Fig. 4 represents the predictions for a 96-hour span made by the Tensorflow implementation. Figures 5 and 6 represent the predictions made by the Sklearn implementations on a 1 hour and 24 hours span. As it can be easily observed, the Tensorflow implementation not only offers more accurate results, but on a longer time span. The, Sklearn implementation, while marginally more inaccurate, is still very usable.

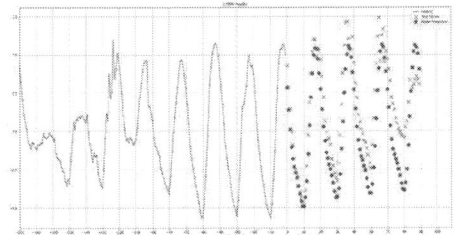

Fig. 4.   Tensorflow Results for 96 hour future

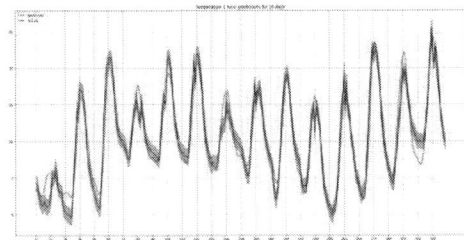

Fig. 5.   Sklearn Results for 1 hour future

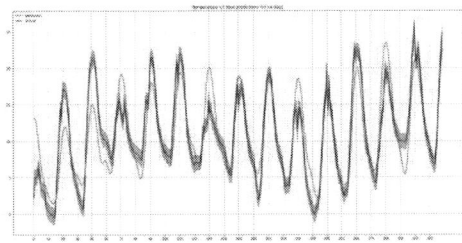

Fig. 6.   Sklearn Results for 24-hour future

IV. CONCLUSION

While it is obvious that the Tensorflow implementation provides more accurate overall results, it must be noted that it is harder to use, takes more time to train and is harder to implement. It is very flexible though.

The Sklearn implementation, while not as accurate and not able to make predictions that far into the future, it is a lot simpler to use, train, understand and reuse. It only provides the user to

select the number of basic neurons on each layer, it does not have other architectures or types of like Tensorflow, just dense layers.

Finally, for small, very localised applications of forecasting, we recommend the usage of Sklearn. For larger areas, where a lot more data is available and accuracy is needed, Tensorflow is a better choice. Overall, for local forecasting, both are adequate.

REFERENCES

[1] P. Hewage, M. Trovati, E. Pereira, A. Behera, "Deep learning-based effective fine-grained weather forecasting model," in Pattern Analysis and Applications, 2020.

[2] M. Holmstrom, D. Liu., "Machine Learning Applied to Weather Forecasting.", Stanford, 2016.

[3] J. Hickey, "Using Machine Learning to "Nowcast" Precipitation in High Resolution", Google, 2020.

[4] N. Singh, S. Chaturvedi and S. Akhter, "Weather Forecasting Using Machine Learning Algorithm," in 2019 International Conference on Signal Processing and Communication (ICSC), 2019, NOIDA, India.

[5] N.O. Jaiswal, "Weather Forecasting using Linear Regression In Machine Learning", 2020.

[6] MeteoBlue, https://www.meteoblue.com/en/weather/, 2020.

[7] E. Dobrescu, D.I. Nastac, and E. Pelinescu, "Short-term financial forecasting using ANN adaptive predictors in cascade" in International Journal of Process Management and Benchmarking, vol. 4, 2014, pp. 376-405.

[8] Tensorflow, "Time series forecasting", in Tensorflow Documentaion, 2020, https://www.tensorflow.org/tutorials/structured_data/time_series.

[9] J. Brownlee, "Time Series Forecasting as Supervised Learning", 2016.

[10] S. Saxena, "Artificial Neuron Networks (Basics) | Introduction to Neural Networks," in Becoming Human, https://becominghuman.ai/artificial-neuron-networks-basics-introduction-to-neural-networks-3082f1dcca8c, 2017.

[11] S. Bhardwaj, "Activation Function Basics," in Kaggle, https://www.kaggle.com/general/187778, 2020.

[12] Y. Freund, R.E. Schapire, "Large Margin Classification Using the Perceptron Algorithm". in Machine Learning, 1999, vol 37, pp.277–296.

[13] S. Hochreiter and J. Schmidhuber, "Long Short-term Memory. Neural computation," in MIT Press, 1997, vol. 9, pp. 1735–1780.

[14] Pedregosa et al., "Scikit-learn: Machine Learning in Python" in JMLR 12, 2011, pp. 2825-2830.

[15] D. Kingma and J. Ba, "Adam: A Method for Stochastic Optimization," in International Conference on Learning Representations, 2014.

[16] J. Kiefer, J. Wolfowitz, "Stochastic Estimation of the Maximum of a Regression Function," in Ann. Math. Statist., 1952, vol 23, pp. 462-466.

[17] F. Chollet, et al., "Keras", https://keras.io.

Intelligent Control for Dual-Boiler System with Digital Communication for Smart Buildings

V. Nicolau, M. Andrei, G. Petrea
Department of Electronics and Telecommunications
"Dunarea de Jos" University of Galati, Romania
Galati, Romania
viorel.nicolau{mihaela.andrei, george.petrea}@ugal.ro

E. Raducan
Department of Automation and Electrical Engineering
"Dunarea de Jos" University of Galati, Romania
Galati, Romania
elena.raducan@ugal.ro

Abstract—**Smart buildings have nowadays many integrated systems, like HVAC (Heating, Ventilation and Air Conditioning). In general, every component has its own control system, and these can be independent on each other or can collaborate for more complex control law. In this paper, an intelligent dual-boiler monitoring and control system (DBMCS) through digital communication using OpenTherm protocol is proposed. It is capable of controlling two independent boilers or implementing an optimal control of the dual-boiler system. DBMCS has a graphical user interface and it contains an interconnection device based on microcontroller with double function, gateway and switching device between boilers and coordinating system.**

Keywords—*HVAC, boiler, control, gateway, OpenTherm*

I. INTRODUCTION (*HEADING 1*)

Nowadays, most smart buildings have HVAC (Heating, Ventilation and Air Conditioning) systems. Its performance depends on the complexity of equipment installed, like sensors and actuators, and also on control system type [1]. Energy consumption is very important and the need for monitoring and control has increased over the years for saving purposes of energy and money [2].

The general structure of the HVAC control, monitoring and protection system can be centralized using one control unit (simple cases) or it can be distributed among multiple control units for every application of the system. In the second case, the control systems can be independent on each other or they can communicate each other, directly or through a supervisor system. A more flexible structure is one distributed on two levels, using a computer as coordinating system, such as desktop or PLC (Programmable Logic Controller), and several subordinate control systems, like embedded systems with microcontrollers [3].

An important operation of the HVAC system is the boiler control. It is important to implement real-time monitoring systems for energy efficiency and also to detect the performance degradation of condensing boilers [5]. In simple cases, boilers are controlled by room control systems, also known as thermostats, which send simple On-Off commands to boilers, through unidirectional wires. The trend in modern technologies towards high-efficiency appliances implies requirements of extensive communications capability between boilers and room controllers [5]. Hence, for energy efficiency, intelligent room controllers can be used.

For optimal control, the room controllers can be replaced by computers with graphical user interface. Computer is used as operating console, but also has specific functions regarding data management, high level data processing and communications. There are many applications in literature, by connecting computers with industrial or residential multiple boilers with predictive and optimal control laws [6], [7]. Boiler automation system using PLC was proposed in [8]. Remote boiler monitoring system based on SCADA is presented in [9] and with LabVIEW interface in [10].

Often, due to interconnected heterogeneous systems, communication devices are necessary between them to interface the communication protocols, such as gateway device. In general, a gateway connects networks with different communication protocols, changing the data packets structure and the type of signals on the physical communication channels. Furthermore, if several units are connected each other, a switching device is also necessary.

In this paper, an intelligent dual-boiler monitoring and control system (DBMCS) through digital communication using OpenTherm protocol is proposed. It consists of a distributed system with LabVIEW graphical user interface running on PC as coordinating system for up to 2 embedded boiler control systems, implemented with microcontroller. In addition, a Gateway device based on microcontroller is implemented, to assist the communications between the PC and the boiler. It has also switching capabilities, which permit to connect up 2 boilers for monitoring and control purposes.

The Gateway device was necessary because the boiler microcontroller communicates on an OpenTherm serial protocol, while the PC is connected to the RS232 asynchronous serial communication interface. The two interfaces have different electrical characteristics. Thus, in the RS232 standard, the signal is voltage type with NRZ line encoding, while for OpenTherm, the signal propagation is bidirectional on 2 wires, in current and voltage, respectively, using Manchester coding.

The paper is organized as follows. The structure and implementing aspects of boiler monitoring and control system are studied in section 2. In section 3, Gateway characteristics along with OpenTherm protocol specifications are presented. Section 4 describes the proposed boiler monitoring and control system with some experimental results. Conclusions are pointed out in section 5.

II. DUAL-BOILER MONITORING AND CONTROL SYSTEM

The structure of the HVAC control, monitoring and protection system can be centralized using one control unit or distributed among multiple control units for every application of the system. In general, every process can be divided in sub-processes which are controlled separately. The most flexible control structure is one distributed on two levels, using a computer coordinating system (CCS), and several subordinate control systems (SCS), like illustrated in Fig. 1.

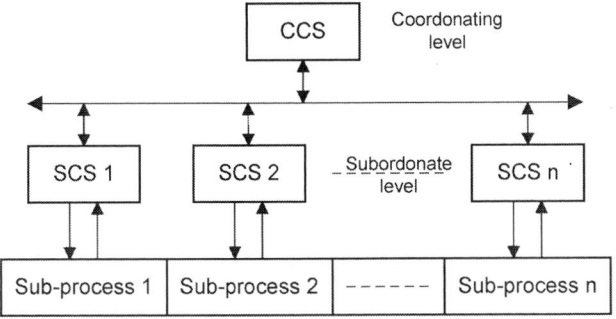

Fig. 1. Structure of HVAC control, monitoring and protection system

Regarding the boiler control, as sub-process of HVAC control, the proposed DBMCS was designed to operate in three modes of interconnection: connecting a single boiler as a slave unit with a PC as master unit, connecting a boiler, as a slave unit, and a programmable thermostat, as a master unit, with DBMCS for monitoring purposes, or connecting two boilers as slave units with a PC as master unit, in which case, the Gateway has also boiler switching functions (the solution presented in the paper).

The structure of the proposed dual-boiler monitoring and control system is illustrated in Fig. 2. It represents a distributed system with coordinate computer running LabVIEW graphical user interface and up to 2 embedded boiler control systems, implemented with microcontrollers. The Gateway device based on microcontroller assists the communications between the PC and boilers. It has also switching capabilities, permitting to connect up 2 boilers for monitoring and control purposes.

DBMCS performs the following functions: periodically acquires data from boiler control unit, visualizes the main parameters, and stores data in a file on the hard-disk, for further processing. The communication between master and slave is half-duplex, always initiated by the master.

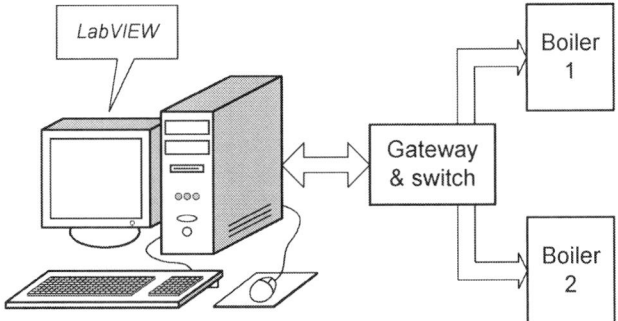

Fig. 2. Structure of dual-boiler control system

III. GATEWAY DEVICE WITH OPENTHERM PROTOCOL

The acquisition function is performed through a special serial communication between the PC and the boiler microcontroller, through Gateway device. It connects two different communication protocols, changing the data packets structure and the type of signals on the physical communication channels.

The Gateway contains two ports at OpenTherm side. At the second port it can be connected a second boiler, as slave unit, or a thermostat as master unit for boiler connected on first port. If the second port is unused, the system controls the boiler on first port as slave unit, with computer as master.

The Gateway device was implemented with PIC 16F628 microcontroller, and it has its own power source. The printed circuit board (PCB) with component places and physical implementation of the Gateway device are shown in Fig. 3, and Fig. 4, respectively.

The main program on microcontroller solves following tasks in bidirectional half-duplex communications, from master to slave and slave to master, respectively. The gateway tasks for master to slave communication are: receives on RS232 interface one packet of 10 bytes from PC, rearranges in packet of 4 bytes for OpenTherm protocol using Manchester line coding, and transmits the packet to boiler. In slave to master communication, the tasks are vice-versa, starting with receiving data from boiler based on interrupt system of the microcontroller, then Manchester decoding of the data.

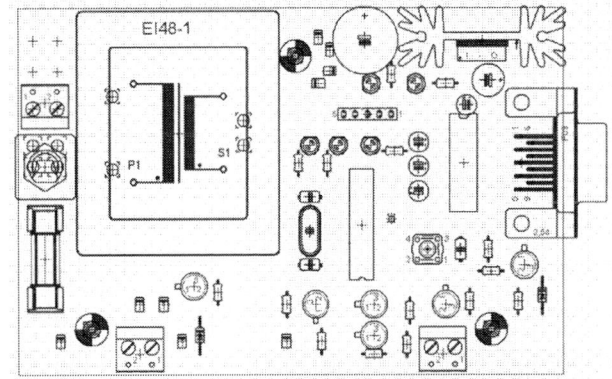

Fig. 3. PCB with component places of Gateway device

Fig. 4. Physical implementation of Gateway device

There are two communication protocols, between PC and Gateway using RS232 asynchronous serial communication interface, respectively between Gateway and Boiler with OpenTherm serial protocol. Both protocols are implemented on two levels of the OSI model: the physical level and the data link level. On physical layer, the two interfaces have different electrical characteristics: voltage type with NRZ line encoding for RS232 standard, and bidirectional on 2 wires, in current and voltage, respectively, using Manchester coding for OpenTherm protocol. On data link level, the structure of the data packets is different, with 4 bytes on the OpenTherm channel and 10 bytes on the RS232 channel.

OpenTherm is a point-to-point serial communication system that connects only two devices (connection without a bus). One device is master, and the other is slave. Communication is always initiated by the master. Used as such, it allows the direct connection of a programmable thermostat as master unit with a boiler as slave. The ports must have short-circuit feature, permitting also connection of old on/off boilers [5].

OpenTherm communication is low-speed half-duplex, with the nominal bit period of 1ms. After the dialog is initiated by the master, the slave unit must respond within a maximum time period of 800 ms, and a new communication can restart a minimum time period of 100ms.

There are two types of commands the master can request: *read data* and *write data*. Both commands have the same form: data-id and data-value. The slave response can be *read ack* or *write ack*, both with the same format. Data is divided in classes, regarding: control and status information, configuration information, and sensor and informational data.

The frame format is identical for both directions. It contains 4 bytes, with an added *Start* bit of '1' at the frame beginning and one *Stop* bit of '1' at the end, as represented in Fig. 5.

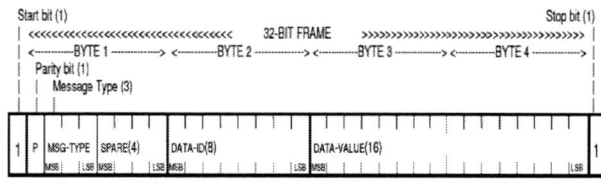

Fig. 5. Physical implementation of Gateway device

The frame bytes contain information about message type on 1 byte, data-id on 1 byte and data-value on 2 bytes. Every byte is serial transmitted starting with most significant bit (MSb). In serial communications, the communication lines have an idle state, with logical value '0' or '1' depending on standard. Hence, to signal the start of the communication, the starting bit must have opposite logical value to idle state of the lines. In this case, the idle state is '0', and hence, the starting bit is '1'.

The message type determines the contents and meaning of the frame, with seven possible defined values. Massages from master to slave can be: *Read-data*, *Write-data* or *Invalid-data*. Massages from slave can be: *Read-data*, *Write-data*, *Data-invalid* or *Unknown-dataID*.

IV. EXPERIMENTAL RESULTS

In this section, first mode of interconnection is considered, by connecting a single boiler as a slave unit with a PC as master unit, through Gateway device. In this case, the master has all functions of a programmable thermostat, but also has supplemental functions regarding data management, high level data processing and communications with other masters. The proposed DBMCS using one boiler is illustrated in Fig. 6.

The software program for DBMCS must solve all general task characterizing a monitoring system, but also specific tasks of boiler process. Data storage and visualization functions are implemented on a PC using the LabVIEW graphical programming environment.

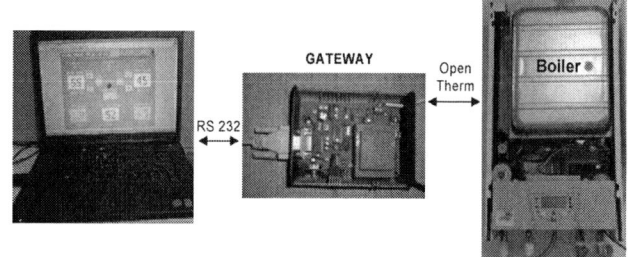

Fig. 6. Proposed DBMCS using one boiler

LabVIEW interface is organized in hierarchical manner with several software components running in multitasking mode, as shown in Fig. 7.

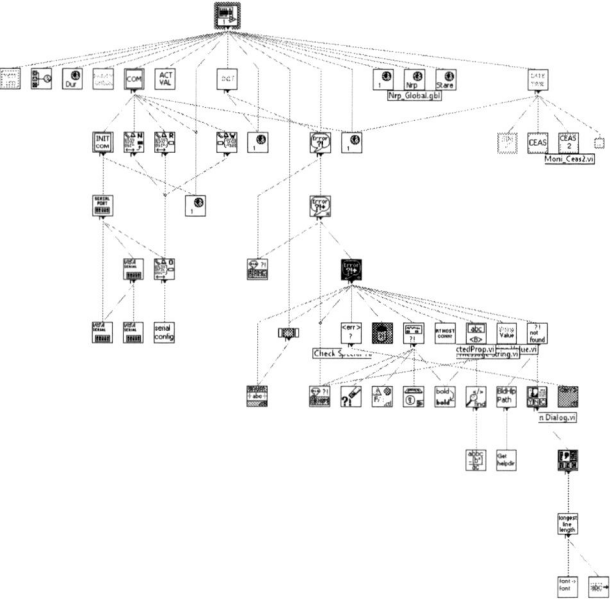

Fig. 7. Hierarchical structure of LabVIEW interface

It activates parallel processes with their own clocks, realizing following tasks:

- periodical commands generation to the boiler with appropriate line coding

- reception and decoding of packets from boiler

- error checking and error message display

- update instantaneous values received from slave

- compare instantaneous values with their setup points for normal operation and display of error messages

- update accumulated data vectors every minute and save them to hard disk

- update and display date and hour

- check and display the communication status and packets transfers.

The LabVIEW interface is organized in four Tabs for displaying: accumulated values over time in *Monitor* Tab, instantaneous values from the boiler in *Valori* Tab, and communication information about the two ports on the Gateway device, in Tabs denoted *OT L1* and *OT L2*, respectively. The packets on communication are signaled by a LED placed in upper-right corner.

The instantaneous values received from boiler are displayed by numerical and logical indicators, as illustrated in Fig. 8. The parameters refer to chilled water (CHW) and domestic hot water (DHW), and they are: temperature (TEMP) measured in (°C), pressure (PRESS) in (barr) and flow (FLOW) in (l/min). For example, DHW FLOW displays the instantaneous consumption of hot water. For every parameter, instantaneous values are displayed with big fonts. For temperatures, other values are also displayed: setup point, and threshold points, as maximum and minimum admissible values.

For example, CHW temperature has the following values displayed:

- CHW TEMP = instantaneous value of direct temperature of heating water = 54 °C

- CHW SETPOINT = preset direct temperature of heating water = 60 °C

- MAX CHW = maximum admissible value for CHW = 90 °C

- MIN CHW = minimum admissible value for CHW = 20 °C

At power up, DBMCS initiates communication parameters and start writing commands to the boiler, sending setup and threshold points.

The functioning state ON of CHW, Flame and DHW are indicated by LEDs: *CHW ON*, *FLAME ON*, and *DHW ON*, respectively. In addition, error state is signaling by a LED denoted *ERROR*, along with error numerical code, *ERR. CODE*.

The LabVIEW interface can be developed with new complex tasks, which can interact with other tasks by means of local and global variables. For example, new Tabs can be added in the LabVIEW interface, for optimal control laws, based on expert systems or artificial intelligence.

V. Conclusions

In this paper, an intelligent dual-boiler monitoring and control system through digital communication using OpenTherm protocol is proposed. It has a LabVIEW graphical user interface and it contains an interconnection device based on microcontroller with double function, gateway and switching device between boilers and coordinating system.

References

[1] L. Wang, S. Greenberg, J. Fiegel, A. Rubalcava, S. Earni, W. Pang, R. Yin, S. Woodworth, J. Hernandez-Maldonado, "Monitoring-based HVAC commissioning of an existing office building for energy efficiency", Journal of Applied Energy, Vol. 102, pp. 1382-1390, 2013.

[2] L. A. Mtungwa, J. C. Pretorius and S. P. Daniel Chowdhury, "Design of Energy Monitoring System for the Boiler and Chiller Plants in Netcare Private Hospital,"2018 IEEE PES/IAS PowerAfrica, Cape Town, 2018, pp. 817-820, doi: 10.1109/PowerAfrica.2018.8521124, 2018.

[3] S. Jalnekar and V. Gaikwad, "PIC16 based blowdown controller for industrial boilers," 2017 International Conference on Data Management, Analytics and Innovation (ICDMAI), Pune, 2017, pp. 58-63, doi: 10.1109/ICDMAI.2017.8073486.

[4] S. Baldi, T. Le Quang, O. Holub, P. Endel, "Real-time monitoring energy efficiency and performance degradation of condensing boilers", J. of Energy Conversion and Management, Vol. 136, pp. 329-339, 2017.

[5] OpenTherm Protocol Specification v2.2 - v3.0, OpenTherm Association.

[6] S. Wu and J. Li, "Intelligent and optimal control of energy saving of gas boiler group," 2010 2nd International Conference on Computer Engineering and Technology, Chengdu, 2010, pp. V3-50-V3-54, doi: 10.1109/ICCET.2010.5485766, 2010.

[7] M. A. Abbas, R. Naughton and J. M. Eklund, "System identification and predictive control of a building heating system with multiple boilers," 2011 24th Canadian Conference on Electrical and Computer Engineering(CCECE), Niagara Falls, ON, 2011, pp. 001483-001486, doi: 10.1109/CCECE.2011.6030710, 2011.

[8] S. M. Tahsin Labib, S. Ul Alam, S. Hossain, M. I. Hossain Patwary, R. Ahmed and M. A. Islam, "Design and Implementation of Boiler Automation System Using PLC," 1st Int. Conf. on Advances in Science, Engineering and Robotics Technology (ICASERT), Dhaka, Bangladesh, 2019, pp. 1-6, doi: 10.1109/ICASERT.2019.8934793, 2019.

[9] D. Wang and X. Su, "The boiler design of remote monitoring system based on the SCADA," 2013 Chinese Automation Congress, Changsha, 2013, pp. 864-869, doi: 10.1109/CAC.2013.6775854, 2013.

[10] A. Suresh, G. V. Krishna and B. Bhaskar, "Control level of boiler drum utilizing LabVIEW," 2016 IEEE International Conference on Computational Intelligence and Computing Research (ICCIC), Chennai, 2016, pp. 1-4, doi: 10.1109/ICCIC.2016.7919706, 2016.

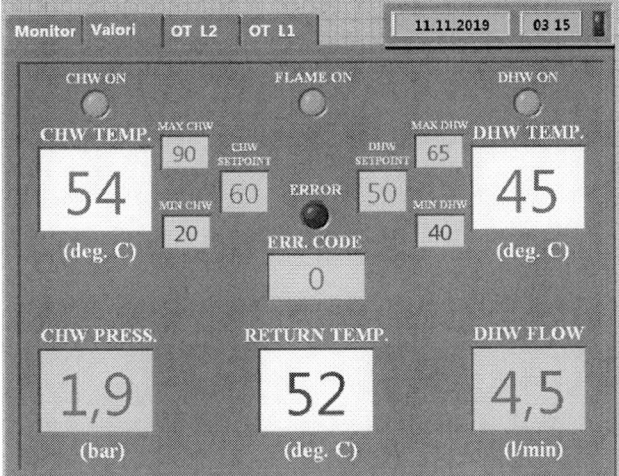

Fig. 8. LabVIEW Front Panel with monitoring values

On Image Processing System for Robot Control using DSK 6713 DSP Kit

G. Petrea, V. Nicolau, M. Andrei

Department of Electronics &Telecommunications,
"Dunărea de Jos" University of Galati, Romania.
george.petrea{mihaela.andrei, viorel.nicolau}@ugal.ro.

Abstract— **Real-time image processing techniques are used in many automation applications like in autonomous vehicle (AV) and robot control. In this paper, an image processing system for real-time video signal processing is studied, using DSK 6713 kit from Texas Instruments. The DSK kit is used to acquire frames from a camera module through HPI port and store them into external SDRAM memory. The DSP is used to extract the frames from memory through EMIF bus, and to run complex processing algorithms on 1D and 2D signals, like spectral analysis and artificial intelligence. Studying the system performances, three possible bottlenecks were identified and analyzed regarding image processing algorithms.**

Keywords— image processing, DSP, real time, FFT

I. INTRODUCTION

Image processing used in robot control is a very complex subject. When using video streams in robot control, the visual feedback needs to give specific properties of the extracted visual features. These features should be accurate, robust and stable in order to obtain a good control [1]. In this paper is considered a mechatronics line capable of processing a specific set of pieces, which is serviced by an autonomous vehicle with robotic arm. The visual features will be used in order to control this robotic arm to execute some specific tasks. This paper shows some aspects of real time grayscale image processing using FFT and a DSK 6713 kit from Texas Instruments.

Processing mechatronics lines are complex systems which require real-time control. This involves multiple tasks and conditions to be achieved for different kinds of operations. The processing time is a very critical issue for control of autonomous vehicle's tasks.

Visual feedback used in robot control needs to give enough and precise information extracted from video frame, in order to ensure task completion. This feedback is dependent on the features that are chosen to be extracted from the image. These features should not change when different operations such as scaling, translation or rotation are applied to the image.

In [4] and [5] are outlined the main feature types that are used for visual servoing. The quality of the features used gives the developer the ability to implement control laws suitable for each application. For efficiency reasons the control algorithm should be simple enough for working at the rate compatible with the desired bandwidth, but also robust, in order to deal

with the uncertainties that are unavoidable in a closed loop system[1]. This approach differs from the dynamic vision approaches which also exploit the motion of visual sensor or objects but do not control them.

Image processing techniques should ensure the maximum speed for real time control purposes but also should be accurate. A specialized DSP (Digital Signal Processor) is the best way for running complex processing algorithms on 1D and 2D signals, like: Discrete Fourier Transform (DFT), Fast Fourier Transform (FFT), autocorrelation functions, and complex image transformations. Depending on algorithm complexity and system performances, the real-time processing techniques are implemented on signal streaming with different parameters (e.g. sampling rate, image resolution, frame rate).

In this paper, aspects of real-time image acquisition and processing techniques used in robot control are studied using DSK 6713 kit from Texas Instruments, which is an evaluation platform using TMS320C6713 Digital Signal Processor (DSP). It has a 16 bit Host-Port Interface (HPI) for data transfer from external sources and a 32-bit External Memory Interface (EMIF) of 512 Mb total addressable memory space with the maximum speed of 100 MHz [12]. The video streaming is acquired through HPI port from OV7670 camera module, which has a maximum resolution of 640 x 480 pixels and rate up to 30 fps. The image processing results are send to a central computer which in turn gives commends to the mobile robot with manipulator.

The rest of the paper is organized as follows. Section 2 presents some elements about visual servoing systems. In section 3 are presented the main characteristics of the platform and the camera used for image processing and in section 3 are presented results after processing different types of images. Section 5 outlines conclusions of the work.

II. VISUAL CONTROL SYSTEMS CHARACTERISTICS

The main components of a visual control system are: a robotic arm, as the execution unit, a video camera for feature extraction of the work space and a regulator. This structure is represented in Fig. 1. The output signal of the regulator represents the input command for the manipulator. This signal is the reference speed of the camera and it is composed by the angular and linear speed.

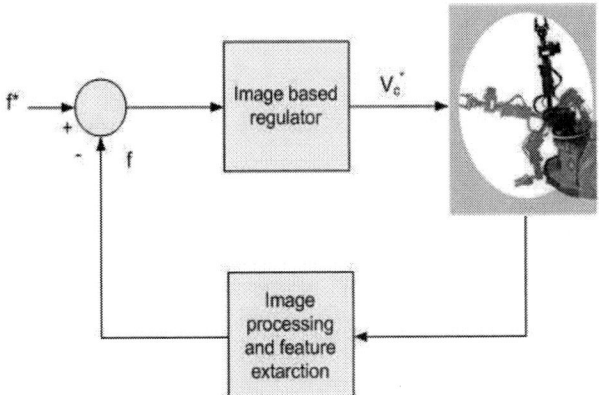

Fig. 1. Visual control system structure

The position of the camera regarding the robotic arm, gives the two possible control architectures for visual control. There is the eye-in-hand architecture, which has the visual sensor fixed on the last joint of the robotic arm and the eye-to-hand configuration in which the video sensor is mounted in a fix position in the workspace, as illustrated in Fig.2. For each of these configurations there are advantages and disadvantages and depending on application and the external conditioning, one of these or even both of the configurations could be chosen for controlling a robotic arm [2].

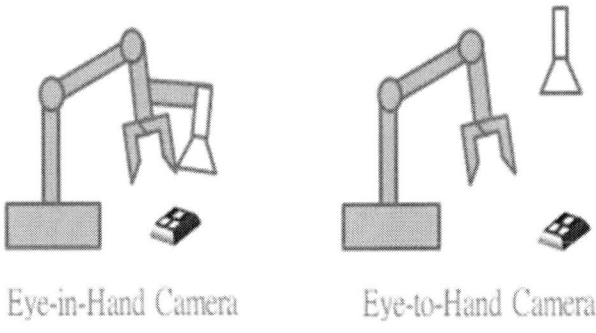

Fig. 2. Eye-in-hand and eye-to-hand visual control architectures

Talking about modeling a visual control system refers to minimizing the error between the real current features of the system extracted from the camera and the desired features of the workspace. In most of the applications it is preferred the use of image descriptors that are not changed if the image is scaled, translated or rotated. These kind of descriptors are called invariants. The visual features extracted will contain information about some properties of the objects in the field, such as area (total intensity), center of gravity etc.

A first step for modelling a visual control system is the analysis of the frame captured by the camera. After some preprocessing steps Fourier Transform is useful mostly for analyzing the texture of the images, but also for image filtering, image analysis, image compression or image reconstruction.

Generally speaking, Fourier Transform is a tool used for decomposing an image into sine and cosine components. The

result of this transformation will be the representation of the image in the frequency domain.

The algorithm for computing this transformation is called Discrete Fourier Transform (DFT) and for an $M \times N$ image is given by:

$$F(k,l) = \sum_{m=0}^{M-1} \sum_{n=0}^{N-1} f(m,n) e^{-i2\pi\left(\frac{km}{M} + \frac{ln}{N}\right)}, \quad (1)$$

where $f(a,b)$ is an image and each point of $F(k,l)$ is obtained by the multiplication of f with the corresponding base function and summing the result.

But the computing of this equation implies a complexity of $(M \times N)^2$ operations. In order to reduce processing time in practice is used Fast Fourier Transform (FFT).

One of the most used FFT methods is Cooley-Tukey algorithm. It applies a divide and conquer algorithm that will recursively break a DFT into many smaller DFT's along with multiplications by complex roots of unity.

In the case of images the algorithm means that it will be performed to all rows (respectively columns) and the resulting transformed rows (respectively columns) will be grouped together as another matrix. Then the FFT will be performed for each column (respectively row) of the new matrix and the results will be similarly grouped into the final matrix. The algorithm complexity will have a number of computations:

$$(M \times N) \log(M \times N) \quad (2)$$

Furthermore, the computational complexity can be reduced if the image is divided into smaller blocks, which usually are squared forms. This is useful for DSP which has limited internal memory L1D Cache to store an entirely data block.

III. 2D Signal Processing System for Visual Servoing

Different kind of solutions can be used for integrating video sensors in a visual control system. Cameras can be directly connected or by using an interface controller.

In figure 3 is illustrated a camera module connected directly to the system. In this case the host system is fast enough for frames transfer and processing or it is built with an dedicated interface.

The second case is when the system has a low speed processor and it is not built with a dedicated video interface. Thus, additional hardware should be added. Figure 4 shows this case. The frames are stored first in the RAM before the processor reads it.

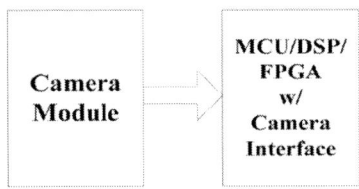

Fig. 3. Video sensor connected dirrectly to the system

2020 IEEE 26th International Symposium for Design and Technology in Electronic Packaging (SIITME)

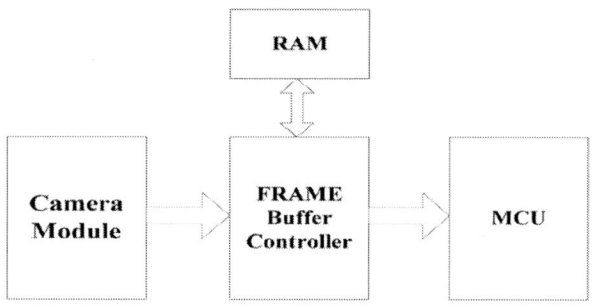

Fig. 4. Video sensor connected to system using interface controller

The camera module used in this paper is OV7670 with an 8-bit parallel interface. This video sensor is low power and with small dimensions. The OV7670 camera is based on a low voltage CMOS sensor interfaced with an integrated signal processor and FIFO memory AL422. The camera module can generate up to 30 frames per second (fps) with a maximum pixel resolution of 640x480. The camera module can be configured regarding data format, image settings or transmission mode. Using the image processor scaling or subsampling can be applied, and also parameters like hue, saturation, gamma and white balance can be adjusted.

In this paper, DSK6713 kit from Texas Instruments is used to interface the video sensor with the control system and also for different types of image processing, such as 2D Fast Fourier Transform. It is built in with TMS320C6713 DSP which is working at a 225 MHz clock speed, which can deliver up to 1350 million floating-point operations per second (MFLOPS), 1800 million instructions per second (MIPS), or with dual fixed/floating-point multipliers up to 600 million multiply-accumulate operations per second (MMACS). For simulations in this paper, P_s=1800 MIPS is considered as DSP computational speed [12].

In addition, 6713 DSP core benefits of a lot of additional features that help to interface with different external signals or systems. Some of these features are: 16 bit HPI (Host Port Interface), 32 bit EMIF (External Memory Interface) with 100 MHz maximum speed, and 16 MB external SDRAM [12].

The block diagram of the DSP core is shown in Fig. 5. It can be observed that DSP has an enhanced DMA controller with 16 individual channels, which can transfer information between peripheral systems, external and internal memory, without the need of DSP core.

The visual feedback system structure used in robot control for mobile robot with manipulator is illustrated in Fig.6.

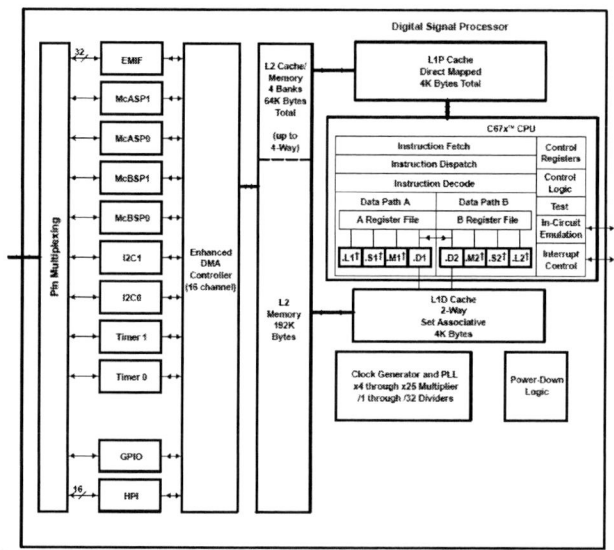

Fig. 5. Block diagram of 6713 DSP core

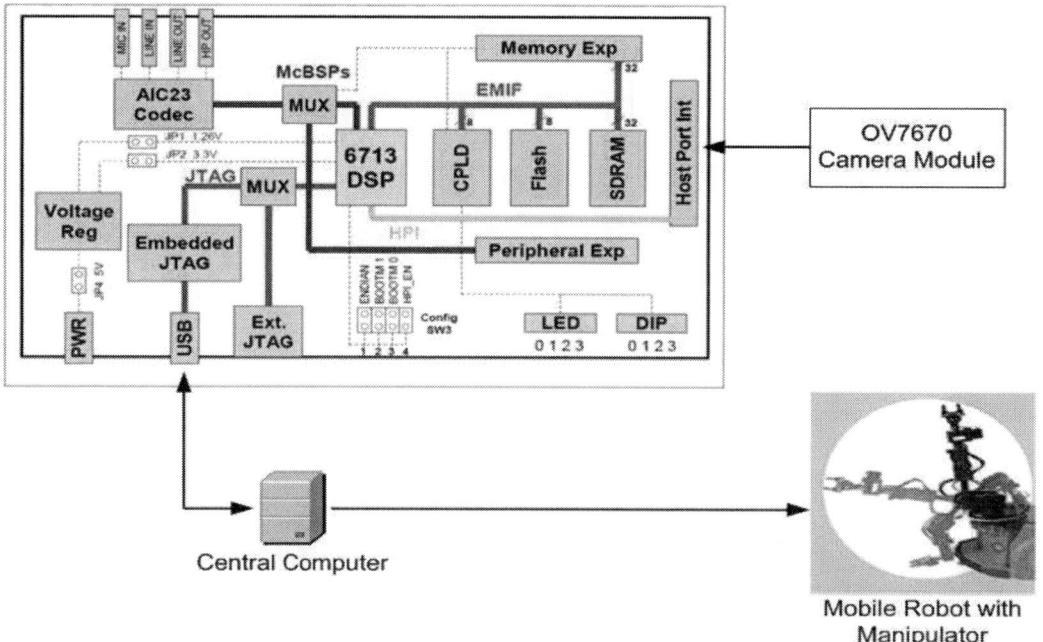

Fig. 6. Proposed visual control system

IV. SIMULATION RESULTS

Studying the system performances, three possible bottlenecks were identified and analyzed regarding image processing algorithms: image transfer from camera to memory, image transfer from memory to DSP, and image processing.

First limitation is imposed by the time of image transfer from the camera module through the HPI port into the external memory. The OV7670 camera module is used in order to achieve images from the working space. The camera module can be used in grayscale or color mode with different format, but for every format it transmits 2 bytes per pixel, in serial manner, byte after byte. Although the HPI port is 16 bit wide, the camera module transmits image information on 8 bit wide bus and, as a result, every transfer on HPI uses only one byte.

The maximum resolution of the camera is 640 x 480 pixels with frame rate F_{rate} = 30 fps. In this case, one frame has F_B = 640x480x2 = 614400 bytes. In addition, the byte transfer rate which is needed for transferring the images from the camera to the DSK kit using the HPI is:

$$r_{TB} = F_B \text{ x } F_{rate} = 640\text{x}480\text{x}2\text{x}30 = 18.432 \text{ MBps} \qquad (3)$$

Hence, the minimum time transfer of one byte from the camera module is: T_{TB} = $1/r_{TB}$ = 54.25 ns. With the DSP working at 225 MHz clock speed, every transfer on HPI bus can be done with a minimum transfer time: T_{HPI} = 44.6 ns < T_{TB}. It results that the HPI is fast enough to support real-time video transfer from camera module, even if every transfer on HPI uses only one byte.

The time of frame transfer on HPI bus is:

$$T_{FrHPI} = F_B \text{ x } T_{TB} = 33.33 \text{ ms} \qquad (4)$$

In addition, the time of frame transfer can be masked by processing time of previous frame.

Second possible issue is the time of transfer between external memory SDRAM and internal L1 and L2 cache memory of the DSP through EMIF bus. The EMIF bus has 100 MHz maximum speed of 32 bit, resulting a minimum transfer time: T_{EMIF} = 10 ns. Due to maximum speed of 32 bit EMIF bus, and DMA controller, this step is not a problem. For example, transferring of 4 bytes representing information of 2 pixels needs 4 transfers on HPI and only 1 transfer on EMIF bus: t_{4BHPI} = $4\text{x}T_{HPI}$ = 178.4 ns, and t_{4BEMIF} = $1\text{x}T_{EMIF}$ = 10 ns. Hence, the frame time transfer through EMIF bus for 640x480 image resolution is:

$$T_{FrEMIF} = F_B \text{ / } 4 \text{ x } T_{EMIF} = 1.536 \text{ ms} \qquad (5)$$

The third possible limitation is the processing time for FFT compution In the case of 640x480 image resolution, knowing the number of operations for 2D FFT which is N_{op}=1685736 and having a 225 MHz DSP with 8 instructions / cycle, it results 210717 cycles for processing algorithm, thus, 0.93 ms for processing a frame.

$$T_{FFT} = N_{op} \text{ / } P_s = 1685736/1800000 = 0.93 \text{ ms} \qquad (6)$$

In the case of artificial intelligence (AI), we take a neural network of N_h=10 hidden neurons with N_i=1024 input neurons and N_o=1024 output neurons and one hidden layer. The image will be decomposed into N_b=(MxN)/N_i=300 blocks and the processing time will be:

$$T_{block}=f(N_h, N_i, N_o) \text{ and } T_{AI} = N_b\text{x}T_{block}\text{/ } P_s = 6.65\text{ms} \qquad (7)$$

In Table 1 are shown some results for transferring and processing time of images of different resolutions.

TABLE I.   TRANSFER AND PROCESSING PERIODS PER FRAME

Image Resolution	Bytes	Transfer time per frame (30 fps)		Processing time per frame	
		T_{FrHPI} [ms]	EMIF, L1, L2 cache (DMA) T_{FrEMIF} [ms]	T_{FFT} [ms]	T_{AI} [ms]
640x480	307200	33.33	1.536	0.93	6.65
320x240	76800	8.33	0.384	0.21	1.66

V. CONCLUSIONS

This paper proposes a real time image processing system using DSK6713 lab kit in order to serve a visual control system for a robotic manipulator integrated into an automated processing line. Aspects of 2D FFT where pointed out and possible bottlenecks where studied, in order to know the system limitations. Also a comparison between using classic FFT algorithm versus neural network is approached.

REFERENCES

[1] B. Espiau, F. Chaumette, P. Rives. "A New Approach to Visual Servoing in Robotics", IEEE Trans.on Robotics and Automation, vol.8, no.3, pp. 313-326, June 1992.

[2] M.Vafadar, A.Behrad, S.Akbari, "Implementing a Visual Servoing System for Robot Controlling", Int. J. Electrical, Computer, Energetic, Electronic and Comm. Eng., 6(9), pp. 1022-1028, 2012.

[3] F. Chaumette, "Image moments: a general and useful set of features for visual servoing", IEEE Trans. on Robotics, 20(4), pp. 713-723, 2004.

[4] E. Marchand and F. Chaumette, "Feature tracking for visual purposes", In Robotics and Systems, 52(1), pag. 53-70, 2005.

[5] F. Chaumette, "Image moments: a general and useful set of features for visual servoing", IEEE Trans. on Robotics, 20(4), pp. 713-723, 2004.

[6] C. Solomon, T. Breckson, Fundamentals of Digital Image Processing: A Practical Approach with Examples in Matlab. Wiley-Blackwell, 2010.

[7] R.C. Gonzales, R.E. Woods, Digital Image Processing. Pearson Prentice Hall, 2008.

[8] S. Fang, E. Song, "A manufacturing process inf. model for design and process planning integration", J. of Manuf. System, vol22, pp.1-16, 2008

[9] A. Gasparetto, V. Zanotto, "A new method for smooth trajectory planning of robot manipulators", Mechanism and Machine Theory, vol. 42, pp. 455-471, 2007.

[10] Perez-Vidal, C., Garcia, L., Garcia, N. and Cervera, E., *Visual Control of Robots with Delayed Images*, Advanced Robotics, 2009, 725-745..

[11] Tzu-Chuen Lu, "A Survey of VQ Codebook Generation", J. of Inf. Hiding and Multimedia Signal Processing, Ubiquitous International, vol. 1, no. 3, July 2010.

[12] Texas Instruments Inc., "TMS329C6713B Floating-Point Digital Signal processor", June 2006.

Development and Test of a Data Framework for Prediction of Soldering Quality in Selective Wave Soldering Applying K-Nearest Neighbors

Reinhardt Seidel, Nils Thielen, Konstantin Schmidt, Christian Voigt, Jörg Franke

Institute for Factory Automation and Production Systems (FAPS),
Friedrich-Alexander University Erlangen-Nürnberg (FAU), Nürnberg, Germany
Reinhardt.Seidel@faps.fau.de

Abstract— **Machine Learning has been proven to be a powerful tool to model and predict complex applications. Selective wave soldering is a widely applied interconnection technology for THT components. It is mostly used if components are not substitutable by surface mount devices due to high thermal load or mechanical stress. Especially in power electronic circuit boards, large copper layers or high thermal mass components lead to critical soldering situations. This paper suggests a Machine Learning framework to identify thermally challenging solder joints. The hybrid approach consisting of an analytical thermal description of THT components and solder joints in the multilayer circuit board and the ML analysis allows the prediction of arbitrarily complex solder joint configurations. The data framework represents electronic components and solder joints. Utilizing solder joint, component and soldering process parameters as input, the K-Nearest Neighbors algorithm predicts the probable hole fill following IPC-A-610 with an overall accuracy of about 75%.**

Keywords—**Mini wave soldering; Solder rise; Grey box; Hybrid modeling; Data mining; Smart electronics production**

I. INTRODUCTION

In applications where components cannot be substituted by surface mount devices (SMD) due to high thermal load or necessary mechanical strength of the soldered connection through-hole devices (THDs) have to be soldered selectively. Selective wave soldering is a widely applied interconnection technology for THT components on mixed SMD and THD assembled boards. Especially in power electronic applications, thick copper layers or high thermal mass components lead to thermally critical solder joints due to increased heat dissipation. [1] In combination with elevated solidification temperatures of lead-free solder, the limited maximum core temperature of the component, and copper dissolution at high solder temperatures, the process window is limited. [2–5] According to J-STD-002D solderability is defined as equal to wettability, which is misleading when understood as manufacturability as it neglects other critical aspects. To provide a safe soft soldering process, design for manufacturing (DfM) in lead-free soft soldering is more important than ever. DfM in soft soldering according to [6] consists of four aspects

- Wettability

- Resistance against dissolution

- Resistance to soldering heat

- Thermal requirements of parts to be soldered

The successful identification of critical solder joints in the early design phase can help to reduce potential extra costs due to production problems and like necessary second design review and consequently the repetition of expensive electromagnetic compatibility tests after design release. However, the manual, experience or intuition-based estimation of the solderability of the solder joint comes with considerable risks of human error. Even a single solder joint that is difficult to solder leads to immense effects on cycle time or causes high rework costs. [5] Also the choice of suitable soldering parameters on the selective soldering machine for the successful capillary hole fill of the solder according to the IPC-A-610 acceptance standard is done experimentally or experience-based which potentially thermally damaging the component or the circuit board and risking early failure. [3, 7]

By predicting the manufacturability of a solder joint design, Machine Learning (ML) can contribute to mitigate the above-mentioned risks and help reduce costs by reducing the number of design reviews and time for new product introduction (NPI) on the shop floor. Additionally, experiential knowledge can be stored and therefore benefit later projects, even if the operator lacks the necessary information. Additionally, machine parameters can be recommended upfront based on the database to reduce time-consuming iterative optimization.

In this paper, the selective wave soldering process is modeled and predicted based on experimental data. The objective is to predict the capillary hole fill in dependence of the solder joint design, and THT-components. For that purpose, a suitable data framework for predicting the hole fill is developed and evaluated. The main innovation is to characterize and represent the complex solder joints and the components.

II. DATA BASED REPRESENTATION OF SOLDER JOINTS

Machine Learning (ML) has been proven to be a powerful tool in modeling and prediction of complex nonlinear problems and can considerably contribute to the solution of the above-described challenges. [8]

In literature, the main approaches for wave soldering quality prediction address the analysis of influence parameters. Besides traditional approaches with statistical design of experiments such as in [2], Liukkonen et al. analyze in [9] several ML-based approaches on wave soldering. The survey shows that data-based methods have considerable advantages over statistical approaches and show high potential in process parameter optimization and as an application in expert systems. In [10] self-organizing maps (SOM) are applied to the wave soldering process. By means of this unsupervised strategy, an impact of solder pump power of the solder bath and flux type on solder balling and bridging and was discovered. In [11] Coit et. al report a neural network (NN) approach on the prediction of a reliable soldering quality based on the circuit board surface temperature after the preheating phase wave soldering which comes close to the purpose of the paper. The objective was to suggest reliable process parameters for a wide range of circuit board designs such that parameters do not have to be adapted frequently. As circuit board characterization criteria and input parameters for the NN bare board weight, the mass of components, the thickness of the circuit board, and additional design information were used. However, the approach does not take into account that especially in multi-layered boards single solder joints can contain considerably different thermal conditions that lead to good or bad solderability. In our previous work in [12] an approach was worked out to identify critical solder joints and predict the capillary solder rise in selective wave soldering comparing a decision tree and a logistic regression model. Although, the number of data points and the variety of different solder joints in the data test set was low critical solder joints could successfully be identified. Therefore, the approach needs to be verified with a larger data set derived from more solder joints and components. A data framework was suggested that categorized various solder joints according to their layer stack. The decision tree showed a very good capability of differentiating very good and very poor hole filling. Logistic regression provides better overall results. Yet, some data samples were overestimated and predicted to be 100% fill although only below 75% was achieved.

Furthermore, the categorized representation approach can not cope with theoretically arbitrarily complex and geometrically varying solder joints is a very limited approach or causes high dimensional models. In this paper, the data framework is extended to prove the feasibility and transferability of the general concept on a broader basis.

III. DEVELOPMENT OF THE DATA FRAMEWORK

The key to successful data-based modeling is the data framework which contains features and labels as summarised in Fig. 1. To predict the process result based on a given solder joint in the design phase, the algorithm has to be trained with the process-related parameters. The data framework is the collection of the necessary features and labels. In order to choose relevant parameters for the data framework, a regression data analysis was carried out to figure out the statistically significant factors. The numerical representation of the soldering process can be provided by temperatures, times, contact diameters as those values are continuous. Discrete THT components and geometrically complex solder joint layer stack defined by its number of copper layers, layer thickness, and type of layer connection to the sleeve can not be easily represented by a number.

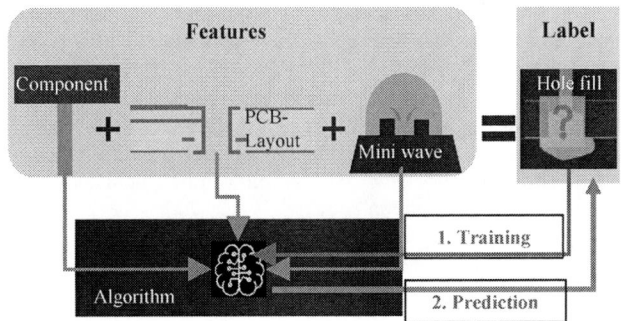

Fig. 1. Summary of the features and labels of the selective wave soldering process

Therefore, ultimately the thermal behavior of the solder joints and components has to be described using a constant set of features regardless of the number of layers and the complexity of the solder joint design. At the same time, the number of features has to be kept to a minimum to keep the dimension of the model low [13].

The vertical hole fill is the relevant label and is categorized into three classes according to IPC-A-610 as depicted in Fig. 2.

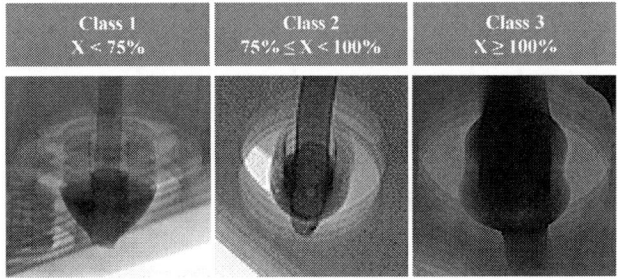

Fig. 2. Classification of the soldering result following IPC-A-610 [7, 12]

A. Thermal description of THT components

The applied model to describe any electronic component to its thermal behavior is done following the suggestion in [6] by the heat capacity of the component $C_{th,body}$ and the thermal resistance of the pin of the component $R_{th,pin}$ as shown in Fig. 3.

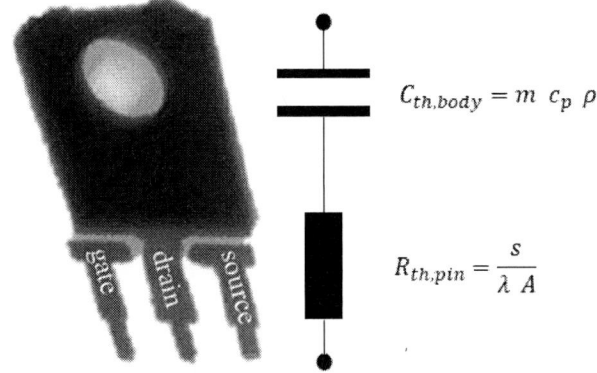

$$C_{th,body} = m \, c_p \, \rho$$

$$R_{th,pin} = \frac{s}{\lambda \, A}$$

Fig. 3. X-ray of a MOSFET with the massive bond from drain to case and wire-bonded source and gate; Thermal equivalent circuit description of a THT component [6]

The thermal heat capacity of the pin and the thermal resistance of the body is neglected. The observation that the product of material density and thermal capacity $c_p \rho$ for different materials used in electronics assemblies is relatively constant between 2.5 and 3.5 $10^6\ Jm^{-3}K^{-1}$, leads to the conclusion that the thermal capacity of a component highly correlates with its mass m [6]. The thermal resistance of the pin can be calculated with length s and cross-sectional area A.

Other than suggested in [6] previous soldering experiments showed a significant difference between single pins of one component. Fig. 3 shows a MOSFET with source, drain and gate pin. As source and gate are wire-bonded to the component body and drain is connected to the body with the entire cross-section of the pin, drain requires more soldering heat. Hence, the respective mass approached by the heat has to be estimated for every pin.

B. Thermal characterization approaches for solder joints in multilayer circuit boards

The description of the thermal behavior of solder joints with nearly arbitrary geometrical degrees of freedom can be done with an approach suggested in [6]. During soldering the copper layers are heated by the wave. Hence, the copper layer stack with high thermal conductivity heats the substrate with high thermal capacity which therefore acts as a heat sink. With the estimation of the ratios of thermal resistances of the heat dissipated to the substrate and the layer internal resistance, a characteristic thermal diffusion length $l_{thermal}$ can be calculated to:

$$l_{thermal} = \frac{1}{\sqrt{\alpha R_l}} = \sqrt{\frac{\lambda_{l\_spec}\, d_l\, d_s}{\lambda_s}} \tag{1}$$

The variable d corresponds to the thickness of the layer d_l and the substrate d_s.

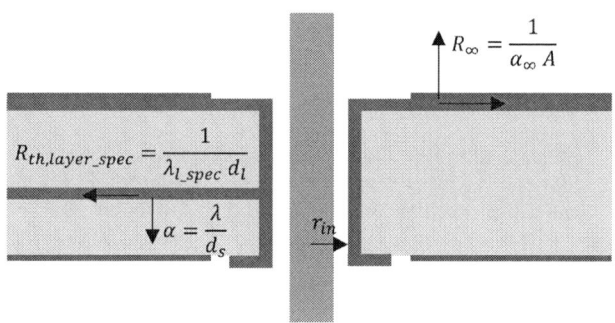

Fig. 4. Thermal characterization approach for solder joints [6]

From this thermal length the thermal resistance of the infinite layer can be calculated with:

$$R_{th,\infty} = \frac{1}{2\,r_{in}} \sqrt{\frac{d_s}{\lambda_s\, \lambda_l\, d_l}} \tag{2}$$

The thermal resistance of the finite layer is then:

$$R_{th,layer\ i} = R_{th,\infty} \coth\left(\frac{L}{l_{thermal}}\right) \tag{3}$$

When this approach is applied to every layer in the solder joint, the heated volume and hence, the thermal mass to be heated of each layer can be summed up for the solder joint.

By this means, the solder joint can be thermally characterized and represented numerically.

- $R_{th,\ joint} = \sum R_{th,layer\ i}$ (4)
- $C_{th,\ joint} = \sum C_{th,layer\ i}$ (5)
- $\tau_{th,\ joint} = R_{th,joint} * C_{th,joint}$ (6)
- Ratio of horizontal and vertical resistances in the solder joint $R_{th,joint}\big/R_{th,sleeve}$ (7)

This approach is referred to as thermal value representation in the following. A second approach is to represent every layer of the solder joint in the data framework with its above calculated characteristic thermal resistance $R_{th,\ joint}$, referred to as layer-wise representation.

C. Data framework

TABLE 1 shows the summarized sets for the numerical representation of the data framework consisting of naturally continuous process parameters and calculated, pre-evaluated numerical descriptions of discrete THT-components and solder joints.

TABLE 1. Summary of the data framework based on [12]

Feature		
Process parameters		
Solder time 1-10 s	Solder temperature 260 – 300°C	Preheat temperature 75 – 150°C
Thermal value solder joint representation		
$R_{th,\ joint}$	$C_{th,\ joint}$	$\tau_{th,\ joint}$ $R_{th,joint}\big/R_{th,sleeve}$
Layer-wise solder joint representation		
$l_{thermal}$ for 1, 2, and 6 layers		
THT-component representation		
$C_{th,\ body}$	$R_{th,\ pin}$	Component mass Annular gap width
Label		
Hole fill		Class 1 Class 2 Class 3

Using soldering experiments, 500 data points were generated. From the process parameter perspective, the entire process window for selective wave soldering is covered within the data. From the design perspective 1, 2 layered circuit boards with a copper layer thickness of 35 µm and 70 µm and a 6-layered circuit board with an inner copper layer thickness of 95 µm outer copper layer thickness of 140 µm were used. In each of the circuit boards solder joint designs with varying sleeve-layer connection are realized which lead to different thermal behavior during soldering. This behavior is attempted to describe as suggested above. Each board has a thickness of 1.6 mm. The inner diameter of the wettable nozzle used for the experiments is 3 mm and the solder alloy was SAC 305.

As part of the data preprocessing the min-max scaling for normalization [14] is used:

$$x' = \frac{x - min(x)}{max(x) - min(x)} \qquad (8)$$

To avoid unbalanced data sets, SMOTE random over-sampling is applied.

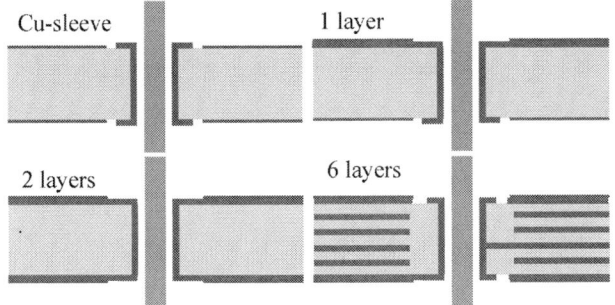

Fig. 5. Cu-sleeve without connected copper layers, single layer solder joints with and without heat trap; double layer solder joints with and without heat trap; 6-layered solder joints with and without heat traps and varying layer connection of sleeve and layer

D. Development of the algorithm

K-Nearest-Neighbor (KNN) algorithms belong to the family of instance-based learning concepts, i.e. test examples are compared with training examples that are temporarily stored in memory. KNN algorithms are supervised learning models that can handle regressive and categorical output but are mostly used in classification problems in the industrial context. The largest application areas are data mining and pattern recognition. [13] The KNN rule is one of the oldest and simplest methods of pattern categorization. [15] These algorithms achieve the best results when all features are expressed by numerical values. [16] This allows values to be interpreted as points and distances to be calculated based on standard definitions. A data point is then a feature-spanning vector in combination with a classification value. These can then be compared by calculating the Euclidean distance. [17] Then linear examples in the test set are classified by the majority category of the k times occurring neighbors from the training set. [18]

A prerequisite for KNN algorithms is a comprehensive scaling of the database. All features must be converted to the same range of values so that the later calculation of the distance is not weighted. Otherwise, this would distort the result. [13] Another important element of KNN is K-Fold-Cross-Validation. It is a standard method to avoid overfitting and to check the forecast quality. [19] Overfitting means that the model has been modeled too close to the training data, which harms the prediction performance for new data. [20] The argument K stands for the number of subsets into which the database is divided. Despite their simple implementation, KNN algorithms can be very meaningful and superior to other algorithms. The advantages here are good explainability, considerably shorter training time and a comparatively lower total number of necessary data points, especially in contrast to artificial neural networks. [21] The disadvantage of the KNN is a large loss in computational speed with increasing data volume. They are therefore not suitable for analyses based on a large database. [22] Another disadvantage is sensitivity to high-dimensional data. With increasing dimensionality, it becomes more difficult for the algorithm to calculate the distances to each other. [13]

In the context of wave soldering, this emphasizes the importance of dimensional reduction and numerical representation of the solder joints.

E. Benchmarking with auto-sklearn

Auto-sklearn is an automated supervised ML development toolkit that allows automated algorithm selection and optimization of hyperparameters for small and medium datasets. The system is based on sklearn and uses 15 classifiers and 14 feature preprocessing methods. [23, 24]

In [8] the outstanding performance of this automated approach was proven for a public data set. The data framework of this contribution is used to train the auto-sklearn model to figure out the best algorithm for the given data set. The model is trained with 75% of the data set and 25% is used for performance testing at default settings.

F. Evaluation measures

To evaluate the prediction quality of the approach accuracy, precision and recall are considered which can be derived from the confusion matrix. The measures are defined as follows:

$$Accuracy = \left(\frac{TP + \Sigma TN}{TP + \Sigma TN + \Sigma FP + \Sigma FN} \right) * 100 \quad (9)$$

$$Precision = \left(\frac{TP}{TP + \Sigma FP} \right) * 100 \qquad (10)$$

$$Recall = \left(\frac{TP}{TP + \Sigma FN} \right) * 100 \qquad (11)$$

In DfM of selective wave soldering the recall represents the correct assignment of a solder joint design to its true class of hole fill according to IPC-A-610.

Additionally, the overestimation of the solder rise has to be avoided under any circumstances, as this would lead to wrong implications in circuit board design, where the correct minimum rise shall be estimated before production. Therefore, the models are evaluated by a newly introduced measure, the risk of overestimation (ROE) defined as:

$$ROE_j = \frac{FN_{j,overestimation}}{TP_j} \qquad (12)$$

This measure counts all overestimated false negatives $FN_{j,overestimation}$ of class j and relates them to the true positives TP_j. Hence, this measure gives the model's specific risk of misclassifying thermally critical solder joints and therefore, the risk for unrecognized thermally critical solder joints during the design phase which would later lead to production problems.

IV. RESULTS AND DISCUSSION

The data framework described above is used to classify the data points within the experimental data set according to Fig. 2. The euclidian distance to five neighbors is calculated to identify the classification of the respective data point to be predicted in both models as this provided most robust results. The thermal value (eq. 4-7) and the layer-wise (eq. 3) characterization approach for solder joints are compared.

The confusion matrices TABLE 2 and TABLE 3 summarize the average results over the three folds of cross-validation for the thermal value solder joint representation. The overall accuracy is calculated to 77.2 % for the thermal value and 74.6 % for the layer-wise approach. The class-specific recall shows stable high classification correctness. Yet, the thermal value approach shows slightly better results in distinguishing class 2 as well as the novel ROE measure. Especially the overestimation of class 1 solder joints has to be taken seriously. The thermal value representation shows a significantly better performance in misclassification of class 2 solder joints as class 3.

TABLE 2. Data set with thermal value solder joint representation; average values of three-fold cross-validation of KNN algorithm

| | predicted | | | | | |
	class 1	class 2	class 3	precision	recall	ROE
actual class 1	48	10	1	0.73	0.813	0.231
actual class 2	14	43	2	0.68	0.727	0.039
actual class 3	3	10	45	0.94	0.767	0.000
avg				0.783	0.769	0.090

TABLE 3. Data set with layer-wise solder joint representation; average values of three-fold cross-validation of KNN algorithm

| | predicted | | | | | |
	class 1	class 2	class 3	precision	recall	ROE
actual class 1	48	9	2	0.71	0.82	0.222
actual class 2	15	38	6	0.69	0.65	0.148
actual class 3	5	9	45	0.85	0.77	0.000
avg				0.751	0.746	0.123

The auto-sklearn benchmark algorithm shows overall accuracy is calculated to be 80.9 % for the thermal value and 80.1 % for the layer-wise approach. TABLE 4 and TABLE 5 summarize the confusion matrix with the respectively calculated evaluation measures.

TABLE 4. Data set with thermal value solder joint representation; results from 25% test data set with an auto-sklearn algorithm

| | predicted | | | | | |
	class 1	class 2	class 3	precision	recall	ROE
actual class 1	34	6	2	0.756	0.810	0.235
actual class 2	7	26	3	0.765	0.722	0.115
actual class 3	4	2	42	0.894	0.875	0.000
avg				0.805	0.802	0.117

This shows that the automatically identified algorithm is superior especially in the area of misclassifying class 3 solder joints. However, the recall falls within a similar range to the

KNN. In this case, the data set with layer-wise solder joint representation performs slightly better ROE while the recall of class 2 lies within similar ranges in the case of both algorithms.

TABLE 5. Data set with layer-wise solder joint representation; results from 25% test data set with an auto-sklearn algorithm

| | predicted | | | | | |
	class 1	class 2	class 3	precision	recall	ROE
actual class 1	35	5	2	0.761	0.833	0.200
actual class 2	10	24	2	0.706	0.667	0.083
actual class 3	1	5	42	0.913	0.875	0.000
avg				0.793	0.792	0.094

The total number of predictions with the auto-sklearn model differ from the KNN the approaches of splitting up the data set for training and test vary.

Compared to the approach in [12] the degree of freedom of the proposed data frame is considerably higher in order to describe and predict arbitrary solder joints. This comes at the cost of classification quality especially considering ROE and overall accuracy.

Considering that the process variance of selective wave soldering lies between 5 and 10% according to our experiments, prediction accuracy by design can hardly reach values above 95%. Additionally, since the classified hole fill quality criterion is very coarse, the stated results can be seen as acceptable. Though further research has to be done on enhancing solder joint representation and enhancing the data basis. Furthermore, a thermal representation of solder joint designs with unknown layer stacks has to be developed to enable electronics manufacturing services without access to x-ray to reduce the time for NPI.

V. SUMMARY AND OUTLOOK

In this paper, a numerical data framework including numerical representation of solder joint, THT components, and process parameters to model the selective wave soldering process is developed and tested. The complex geometry of multilayer solder joints has to be represented. The thermal description of THT components and solder joints in multilayer circuit boards was realized using a mathematical model. This hybrid approach consisting of an analytical solder joint model as a basis for the ML analysis allows the description of arbitrarily complex solder joint configurations. By modeling with the K-Nearest-Neighbor algorithm and auto-sklearn as a benchmark, the performance of the data frameworks is tested. The prediction label was the hole fill according to IPC-A-610. Additionally, a new application-specific metric to measure the risk of overestimation (ROE) of the hole fill for a given solder joint configuration was introduced. Compared to the known literature approaches this is by now the most detailed approach for the use case of solder joint design-dependent prediction of the hole fill as the main quality criterion in industrial THT soldering. This allows for a more accurate design for manufacturing in selective wave soldering.

Yet, further studies are needed for the improvement of the prediction quality up to the process variance. As by design, classification comes along with information loss as a prediction on the upper and the lower limit of the hole fill range of a particular class is weighted equally, regression analysis has further improvement potential. Additionally, the data frameworks have to be tested on a broader basis of industrial solder joint designs.

ACKNOWLEDGMENT

The presented work is part of the IGF research project 19539N/1 accomplished by the Institute for Factory Automation and Production Systems (FAPS) and funded by the Federal Ministry for Economic Affairs and Energy (BMWi) on the basis of a resolution of the German Bundestag. Furthermore, the authors like to cordially thank all companies involved in the project for their support.

REFERENCES

[1] Z. Olah, M. Ruszinko, R. Batorfi, and Z. Illyefalvi-Vitez, "Process parameter optimization of selective soldering," *2012 IEEE 18th International Symposium for Design and Technology of Electronics Packages, SIITME 2012 - Conference Proceedings*, 2012, doi: 10.1109/SIITME.2012.6384359.

[2] P. Mach, P. Zeman, E. Kotrcova, and S. Barto, "Optimization of lead-free wave soldering process using taguchi orthogonal arrays," *3rd Electronics System Integration Technology Conference ESTC*, 2010, doi: 10.1109/ESTC.2010.5642946.

[3] Brian Czaplicki, "Advanced Through-Hole Rework of Thermally Challenging Components/Assemblies: An Evolutionary Process," Air-Vac Engineering Company, IncSeymour, CT. [Online]. Available: https://ww.airvacpumps.com/PDF%20Files/Advanced%20Through%20Hole%20R1.pdf

[4] Christopher Hunt and Davide Di Maio, "A Test Methodology for Copper Dissolution in Lead-Free Alloys," National Physical Laboratory Teddington, UK. [Online]. Available: http://www.circuitinsight.com/pdf/test_methodology_for_copper_dissolution_ipc.pdf

[5] R. Mendez, H. Lowe, I. Marin, and C. Monterrey, "Pb-Free Selective Wave Solder Guidelines for Thermally Challenging PCBs," in *IPC APEX EXPO Proceedings*. [Online]. Available: http://www.circuitinsight.com/pdf/lead_free_selective_solder_guide_ipc.pdf

[6] R. J. Klein Wassink, *Soldering in Electronics*, 2nd ed. Saulgau/Württ.: Leuze, 1991.

[7] *Acceptability of Electronic Assemblies*, IPC-A-610F.

[8] D. Kißkalt, A. Mayr, B. Lutz, A. Rögele, and J. Franke, "Streamlining the development of data-driven industrial applications by automated machine learning," *Procedia CIRP*, vol. 93, pp. 401–406, 2020, doi: 10.1016/j.procir.2020.04.009.

[9] M. Liukkonen, E. Havia, and Y. Hiltunen, "Computational intelligence in mass soldering of electronics – A survey," *Expert Systems with Applications*, vol. 39, no. 10, pp. 9928–9937, 2012, doi: 10.1016/j.eswa.2012.02.100.

[10] M. Liukkonen, E. Havia, H. Leinonen, and Y. Hiltunen, "Application of self-organizing maps in analysis of wave soldering process," *Expert Systems with Applications*, vol. 36, no. 3, pp. 4604–4609, 2009, doi: 10.1016/j.eswa.2008.05.016.

[11] D. W. Coit, B. T. Jackson, and A. E. Smith, "Neural network open loop control system for wave soldering," *Journal of Electronics Manufacturing*, vol. 11, no. 01, pp. 95–105, 2002, doi: 10.1142/S0960313102000217.

[12] R. Seidel, F. Leibold, N. Thielen, and J. Franke, "Prediction of the Solder Rise in Selective Wave Soldering Comparing Decision Tree and Logistic Regression," in *International Spring Seminar on Electronics Technology*.

[13] F. Pedregosa *et al.*, "Scikit-learn: Machine learning in Python," *Journal of machine learning research*, vol. 12, Oct, pp. 2825–2830, 2011.

[14] S. Patro and K. K. Sahu, "Normalization: A preprocessing stage," *arXiv preprint arXiv:1503.06462*, 2015.

[15] T. Cover and P. Hart, "Nearest neighbor pattern classification," *IEEE transactions on information theory*, vol. 13, no. 1, pp. 21–27, 1967.

[16] S. Cost and S. Salzberg, "A Weighted Nearest Neighbor Algorithm for Learning with Symbolic Features," vol. 78, pp. 57–78, 1993.

[17] J. M. Keller, M. R. Gray, and J. A. Givens, "A fuzzy k-nearest neighbor algorithm," *IEEE transactions on systems, man, and cybernetics*, no. 4, pp. 580–585, 1985.

[18] K. Q. Weinberger and L. K. Saul, "Distance Metric Learning for Large Margin Nearest Neighbor Classification," in *Journal of Machine Learning Research*, pp. 207–244. [Online]. Available: https://jmlr.csail.mit.edu/papers/volume10/weinberger09a/weinberger09a.pdf

[19] J. D. Rodriguez, A. Perez, and J. A. Lozano, "Sensitivity analysis of k-fold cross validation in prediction error estimation," *IEEE transactions on pattern analysis and machine intelligence*, vol. 32, no. 3, pp. 569–575, 2009.

[20] T. Dietterich, "Overfitting and undercomputing in machine learning," *ACM computing surveys (CSUR)*, vol. 27, no. 3, pp. 326–327, 1995.

[21] S. Belongie, J. Malik, and J. Puzicha, "Shape matching and object recognition using shape contexts," *IEEE transactions on pattern analysis and machine intelligence*, vol. 24, no. 4, pp. 509–522, 2002.

[22] M. Muja and D. G. Lowe, "Fast approximate nearest neighbors with automatic algorithm configuration," *VISAPP (1)*, vol. 2, 331-340, p. 2, 2009.

[23] M. Feurer, A. Klein, K. Eggensperger, J. Springenberg, M. Blum, and F. Hutter, "Efficient and Robust Automated Machine Learning," in *Advances in Neural Information Processing Systems 28*, C. Cortes, N. D. Lawrence, D. D. Lee, M. Sugiyama, and R. Garnett, Eds.: Curran Associates, Inc, 2015, pp. 2962–2970. [Online]. Available: http://papers.nips.cc/paper/5872-efficient-and-robust-automated-machine-learning.pdf

[24] *auto-sklearn*. [Online]. Available: https://automl.github.io/auto-sklearn/master/

Resource Utilization Comparison between Plain FPGA and SoC Combined with FPGA for Image Processing Applications Used by Robotic Arms

Roland Szabo

Applied Electronics Department
Fac. of ETcIT, Politehnica University Timisoara
Timisoara, Romania
roland.szabo@upt.ro

Aurel Gontean

Applied Electronics Department
Fac. of ETcIT, Politehnica University Timisoara
Timisoara, Romania
aurel.gontean@upt.ro

Abstract—**This paper presents a comparison of two FPGA implementations for controlling robotic arm with image processing. Image processing it's useful in robotic industry since robots can be made more autonomous this way to reduce as much as possible human intervention in production. Image processing it is also known to drain much resources from the control system. A comparison of more implementations can help chose the most suitable one. One implementation is the classic FPGA where everything is done from scratch and the other implementation is the more advanced one, where a microprocessor architecture is made from the FPGA, a graphical operating system is installed, and the control software is made in high level programming language.**

Keywords—*FPGA; image processing; resource utilization; robotic arm; SoC.*

I. Introduction

This paper shows the comparison of two embedded systems. One is the ATLYS development board with the Spartan-6 FPGA and the other is the ZYBO or ZedBoard development boards with the Zinq-7000 SoC (system on a chip). The two embedded systems are very powerful and in this experiment they are used for image processing to control robotic arms only by the information gathered by the video cameras. It is know that the control or robotic arms with image processing it's a task which requires high processing power and a high usage percentage from the control system [1]. A comparison of two systems which can control a robotic arm with image processing algorithms it has sense, due to the fact that can help in future developments which architecture to choose [2].

Controlling robotic arms with image processing it's a high demand task even in the industrial or other type of robotic systems, due to the fact that image processing can increase the robotic systems autonomous behavior [3].

The taken resources by a robotic system with image processing control is known to have a high demand in resources, due to this fact it is in continuous development and

The authors would like to thank Politehnica University Timisoara for the given support.

improvement [4]. Implementation on FPGA systems can solve the problem of high resource need, due to the fact that the FPGAs are very powerful and can make many operations in parallel [5], [6], [7]. The comparison of various FPGA implementations can show the utilization of devices and can demonstrate that this approach can solve a lot of high resource utilization problems [8], [9] [10]. This method has also another advance that it is embedded and portable [11].

II. Problem Formulation

There were made by us two robotic arms control systems with video cameras and the control system had image processing algorithms which could find the robotic arm position in space and also compute its future position in order to move close to the target [12], [13]. When creating these implementations the focus was on functionality and not on efficiency [14], [15]. Then came naturally to make a comparison of implementations in order to know which implementation is more suitable in a specific condition [16].

III. Problem Solution

There was made comparison of the two implementations base on the Xilinx ISE report.

The two implementations are somehow similar, yet very different. The first is plain FPGA, where all the image processing and control system is done from scratch and the second is the creation of a microcontroller architecture on the FPGA, over it is installed a graphical Linux operating system and the whole control is done in high level C programming language with the OpenCV image processing libraries. The only similarity is that both run on FGPA, but the first implementation is pure hardware and the second is more software.

On Table I. it can be observed the resources used by the Spartan-6 FPGA mounted on the ATYLS development board. It can be seen that not all the resources from the development board are used.

2020 IEEE 26th *International Symposium for Design and Technology in Electronic Packaging (SIITME)*

TABLE I. RESOURCES USED BY THE ATLYS FPGA BOARD FOR THE ROBOTIC ARM CONTROL SYSTEM WITH COLORED MARKER DETECTION PLACED AT THE JOINTS (SUMMARY)

Device Utilization Summary				[-]
Slice Logic Utilization	Used	Available	Utilization	Note(s)
Number of Slice Registers	1,103	54,576	2%	
Number used as Flip Flops	1,103			
Number of Slice LUTs	1,692	27,288	6%	
Number used as logic	1,602	27,288	5%	
Number using O6 output only	960			
Number using O5 output only	315			
Number using O5 and O6	327			
Number used as Memory	42	6,408	1%	
Number used as Single Port RAM	34			
Number using O6 output only	2			
Number using O5 and O6	32			
Number used as Shift Register	8			
Number using O6 output only	8			
Number used exclusively as route-thrus	48			
Number with same-slice register load	31			
Number with same-slice carry load	17			
Number of occupied Slices	591	6,822	8%	
Number of LUT Flip Flop pairs used	1,799			
Number with an unused Flip Flop	799	1,799	44%	
Number with an unused LUT	107	1,799	5%	
Number of fully used LUT-FF pairs	893	1,799	49%	
Number of unique control sets	98			
Number of slice register sites lost to control set restrictions	287	54,576	1%	
Number of bonded IOBs	108	218	49%	
Number of LOCed IOBs	108	108	100%	
IOB Flip Flops	2			
IOB Master Pads	4			
IOB Slave Pads	4			
Number of BUFIO2/BUFIO2_2CLKs	1	32	3%	
Number used as BUFIO2s	1			
Number of BUFG/BUFGMUXs	9	16	56%	
Number used as BUFGs	9			
Number of DCM/DCM_CLKGENs	2	8	25%	
Number used as DCMs	1			
Number used as DCM_CLKGENs	1			
Number of IODELAY2/IODRP2/IODRP2_MCBs	24	376	6%	
Number used as IODRP2s	2			
Number used as IODRP2_MCBs	22			
Number of OLOGIC2/OSERDES2s	55	376	14%	
Number used as OLOGIC2s	2			
Number used as OSERDES2s	53			
Number of BUFPLLs	1	8	12%	
Number of BUFPLL_MCBs	1	4	25%	
Number of MCBs	1	2	50%	
Number of PLL_ADVs	2	4	50%	
Average Fanout of Non-Clock Nets	3.66			

On Fig. 1 it can be observed the resource utilization chart by the Spartan-6 FPGA mounted on the ATLYS development board. It can be noticed that there are used almost 50% of the gates from the Spartan-6 FPGA mounted on the ATLYS development board. This result is due to the fact that the ATLYS development board with the Spartan-6 FPGA has far fewer gates than the ZYBO or ZedBoard development boards with the Zynq-7000 SoC.

2020 IEEE 26th International Symposium for Design and Technology in Electronic Packaging (SIITME)

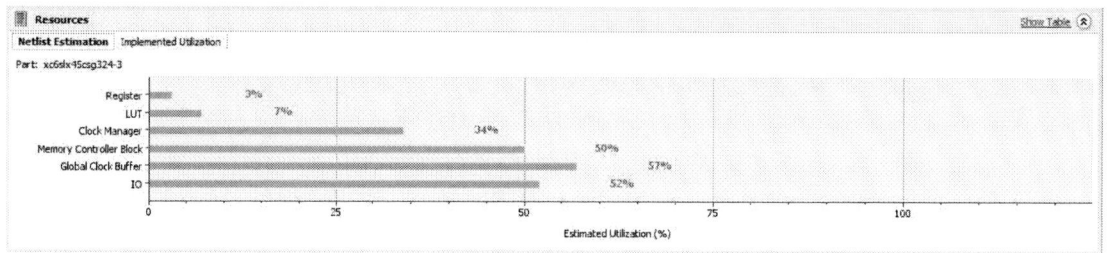

Fig. 1. Resource utilization chart by the ATLYS FPGA board for the robotic arm control system with colored marker detection placed at the joints.

On Table II. it can be seen the used resources to create the Zynq-7000 SoC on the ZYBO or ZedBoard development boards, these were used during the experiments. The table has been generated in Xilinx ISE. It can be seen that the used resources are minimal, despite the complexity of the system.

TABLE II.   FPGA RESOURCES USED FOR CREATING THE ZYNQ-7000 SoC (SUMMARY)

Device Utilization Summary				[-]
Slice Logic Utilization	Used	Available	Utilization	Note(s)
Number of Slice Registers	4,040	106,400	3%	
Number used as Flip Flops	4,026			
Number used as AND/OR logics	14			
Number of Slice LUTs	3,894	53,200	7%	
Number used as logic	3,513	53,200	6%	
Number using O6 output only	2,039			
Number using O5 output only	258			
Number using O5 and O6	1,216			
Number used as Memory	257	17,400	1%	
Number used as Dual Port RAM	232			
Number using O6 output only	180			
Number using O5 and O6	52			
Number used as Single Port RAM	20			
Number using O5 and O6	20			
Number used as Shift Register	5			
Number using O6 output only	5			
Number used exclusively as route-thrus	124			
Number with same-slice register load	100			
Number with same-slice carry load	24			
Number of occupied Slices	1,871	13,300	14%	
Number of LUT Flip Flop pairs used	5,271			
Number with an unused Flip Flop	1,643	5,271	31%	
Number with an unused LUT	1,377	5,271	26%	
Number of fully used LUT-FF pairs	2,251	5,271	42%	
Number of unique control sets	335			
Number of slice register sites lost to control set restrictions	1,445	106,400	1%	
Number of bonded IOBs	85	200	42%	
Number of LOCed IOBs	85	85	100%	
Number of bonded IOPAD	130	130	100%	
IOB Flip Flops	19			
Number of RAMB18E1/FIFO18E1s	9	280	3%	
Number using RAMB18E1 only	9			
Number of BUFG/BUFGCTRLs	3	32	9%	
Number used as BUFGs	3			
Number of ILOGICE2/ILOGICE3/ISERDESE2s	3	200	1%	
Number used as ILOGICE2s	3			
Number of OLOGICE2/OLOGICE3/OSERDESE2s	72	200	36%	
Number used as OLOGICE2s	72			

Number of PLLE2_ADVs		1	4	25%	
Number of PS7s		1	1	100%	
Average Fanout of Non-Clock Nets		3.49			

On Fig. 2 it is presented a chart generated by Xilinx ISE, where are shown the used resources to create the Zync-7000 SoC on the ZYBO or ZedBoard development boards, which are made by merging the ARM Cortex-A9 microprocessor and a 7th generation FPGA, usually an Artix 7. It can be seen a low usage percentage of the FPGA.

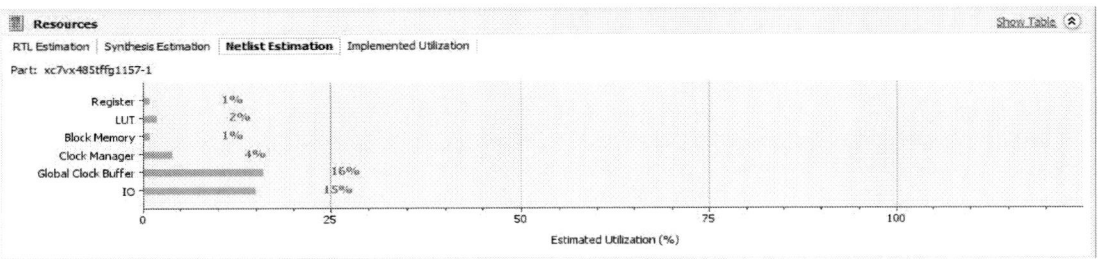

Fig. 2. Resources utilization chart by the Zync-7000 SoC when configured to run the graphical Linux OS.

IV. CONCLUSION

The number of the logic elements used by the Spartan-6 FPGA, it's higher than the logic elements used by the Zynq-7000 SoC (FPGA + μC) mounted on the ZYBO or ZedBoard development boards. This is usage is due to the fact that the Spartan-6 FPGA has fewer resources, than the Zynq-7000 SoC, so in terms of the number of logic elements used, the usage on the Spartan-6 FPGA may be higher, but the resource usage percentage in the two FPGA implementations (Spartan-6 and Zynq-7000) are similar.

ACKNOWLEDGMENT

The authors would like to thank Politehnica University Timisoara for the given support.

REFERENCES

[1] Shiuh-Jer Huang, Jian-Cheng Huang, "Vision guided dual arms robotic system with DSP and FPGA integrated system structure," Journal of Mechanical Science and Technology, vol. 25, issue 8, 2011, pp. 2067-2076.

[2] U. Meshram, P. Bande, P. A. Dwaramwar, R. R. Harkare, "Robot arm controller using FPGA," International Multimedia, Signal Processing and Communication Technologies, Aligarh, India, March 14-16, 2009, pp. 8-11.

[3] Jung Uk Cho, Quy Ngoc Le, Jae Wook Jeon, "An FPGA-Based Multiple-Axis Motion Control Chip," IEEE Transactions on Industrial Electronics, vol. 56, issue 3, 2009, pp. 856-870.

[4] R. Marin, G. Leon, R. Wirz, J. Sales, J. M. Claver, P. J. Sanz, " Remote control within the UJI Robotics Manufacturing Cell using FPGA-based vision," in Proc. European Control Conference, Kos, Greece, July 2-5, 2007, pp. 1378-1383.

[5] G. V. Persiano, S. Rapuano, F. Zoino, A. Morganella, G. Chiusolo, " Distance Learning in Digital Electronics: Laboratory Practice on FPGA," IEEE Instrumentation and Measurement Technology Conference, Warsaw, Poland, May 1-3, 2007, pp. 1-6.

[6] Jeong Seob Kim, Seul Jung, "Joint control of ROBOKER arm using a neural chip embedded on FPGA," IEEE International Symposium on Industrial Electronics, Seoul, South Korea, July 5-8, 2009, pp. 1007-1012.

[7] J. R. Guzman-Sepulveda, R. De Jesus Romero-Troncoso, "Digital System Control for Three-Degrees of Freedom Mechanical Arm with FPGA," Electronics, Robotics and Automotive Mechanics Conference, Morelos, Mexico, September 30-October 3, 2008, pp. 496-501.

[8] Jin Dang, Fenglei Ni, Yikun Gu, Minghe Jin, Hong Liu, "A highly integrated and flexible joint test system based on DSP/FPGA-FPGA," IEEE International Conference on Robotics and Biomimetics, Guilin, China, December 19-23, 2009, pp. 1877-1882.

[9] Zheng Yili, Sun Hanxu, Jia Qingxuan, Shi Guozhen, "Kinematics control for a 6-DOF space manipulator based on ARM processor and FPGA Co-processor," 6th IEEE International Conference on Industrial Informatics, Daejeon, South Korea, July 13-16, 2008, pp. 129-134.

[10] Min Xu, Wenzhang Zhu, Ying Zou, "Design of a Reconfigurable Robot Controller Based on FPGA," Fifth IEEE International Symposium on Embedded Computing, Beijing, China, October 6-8, 2008, pp. 216-222.

[11] V. Ramakrishnan, N. S. Gopal, R. Ashok, S. Moorthi, "FPGA based DC servo motor control for remote replication of movements of a surgical arm," IEEE Region 10 Conference, Bali, Indonesia, November 21-24, 2011, pp. 671-675.

[12] Woon Kyu Lee, Seul Jung, "FPGA Design for Controlling Humanoid Robot Arms by Exoskeleton Motion Capture System," IEEE International Conference on Robotics and Biomimetics, Kunming, China, December 17-20, 2006, pp. 1378-1383.

[13] J. Nikolic, J. Rehder, M. Burri, P. Gohl, S. Leutenegger, P. T. Furgale, R. Siegwart, "A synchronized visual-inertial sensor system with FPGA pre-processing for accurate real-time SLAM," IEEE International Conference on Robotics and Automation, Hong Kong, May 31-June 7, 2014, pp. 431-437.

[14] L. Zouari, M. Ben Ayed, M. Abid, "Embedded control of robot arm driven by Brushless DC motor on FPGA," Second World Conference on Complex Systems, Agadir, Morocco, November 10-12, 2014, pp. 722-727.

[15] S. Himavathi, D. Anitha, A. Muthuramalingam, "Feedforward Neural Network Implementation in FPGA Using Layer Multiplexing for Effective Resource Utilization," IEEE Transactions on Neural Networks, vol. 18 (3), 2007, pp. 880-888.

[16] Michael Hahnle, Frerk Saxen, Matthias Hisung, Ulrich Brunsmann, Konrad Doll, "FPGA-Based Real-Time Pedestrian Detection on High-Resolution Images," IEEE Conference on Computer Vision and Pattern Recognition Workshops, 2013, pp. 629-635.

Smart System for Incubating Eggs

L.A. Szolga, A. Bondric

Basis of Electronics
Technical University of Cluj-Napoca
Cluj-Napoca, Romania
Lorant.Szolga@bel.utcluj.ro, Anamaria.Oara@gmail.com

Abstract— **Egg incubation is a complex process that requires accuracy and precision in monitoring the essential factors that have direct influence on the embryonic development process. The key factors in the incubation process are temperature, humidity, ventilation and egg turning. Hence, the work presented here involves the design of an intelligent automated incubator system with LCD display, a stepper motor that ensures the eggs turning, high accuracy temperature and humidity sensors and a GSM module that inform the farmer about the status of incubator. The system keeps the last status (elapsed time/rotation angle) in the case of a power failure, thus permitting to continue the established processed after repowering it without starting it all over. The entire system is capable of continuously monitor and maintain the operating temperature (37°C) and a humidity (55%-66%) using a feedback control system.**

Keywords— incubation; temperature; humidity; GSM; stepper

I. INTRODUCTION

Electrical incubator is a device used for scientific incubation process in which temperature, humidity and other environmental variables can be maintained at desired temperature levels [1]. Eggs have been incubated by artificial means for thousands of years. Both the Chinese and the Egyptians were credited with artificial incubation procedures. The Chinese have developed a method by which they used coal as a source of heat while the Egyptians built brick incubators that heated them directly in the room where the eggs were placed [2]. Egg incubation is an area of interest today as the poultry industry has become one of the most efficient protein manufacturers for human consumption. There are several incubators on the market for either household or industrial use. However, they are continually working on improving the incubation process, being a field of technological development for engineers [3-8].

This paper aims to create an automated incubator used in the household.

II. HARDWARE

A. Essential Performance Parameters

During incubation four parameters must be properly monitored: temperature, humidity, air supply and egg turning. Oxygen is the key to life on embryonic development. The other parameter, humidity helps the eggs lose water during the incubation period. The eggs must be turned several times a day (every 5 hours). This will ensure that the embryo will not stick

to the shell. The other factor, the temperature, will be monitored continuously. Based on the studies on natural clotting and the analysis of various artificial incubation schemes, it is estimated that the limits between which the incubation temperature should oscillate are between +37.5°C and +37.9°C. If the temperature exceeds 39°C incubation can be compromised.

B. Components of the System

The system can imitate the natural environment for embryonic development. Humidity and temperature are monitored by two sensors. Temperature measurement is done by an NTC thermistor that will provide the required precision so that small temperature variations can be sensed and controlled so that the incubation is not compromised. Humidity measurement is done through a digital sensor.

The automatic return of the eggs is provided by a stepper motor. To ensure ventilation inside the incubator, a fan is used. The fan ensure that the temperature is uniform and at the same time maintains the required oxygen level inside the incubator. The heating system made of a filament lamp is operated via a control relay. The required humidity is controlled through a water pump that will disperse the vapor. The best embryonic development is achieved with limits between 56% and 65% for the humidity. The whole system can be remotely monitored so that at any time a current status of the parameters inside the incubator can be accessed via a GSM module. The whole process will be controlled by a Microchip microcontroller.

C. Design and Implementation

Fig. 1 presents the whole system for incubation including the main component parts.

The two sensors monitor the temperature and humidity inside the incubator and compare with the set reference values depending on the type of eggs to be introduced into the incubator and the values taken from the natural incubation environment. Depending on the difference between the prescribed values and the actual measured values of the sensors, the heating system or the humidity system is activated via the relays. The fan will be powered by 12VDC and will always work by providing the oxygen level required for empirical development and uniformity of temperature.

The egg return system is operated via the GSM module that receives the message from the farmer, the microcontroller interpreting the message that will trigger the mechanical action of the returned eggs. Also, through the GSM module, the farmer has a current state of parameters such as temperature and humidity inside the incubator.

With an LCD screen, temperature and humidity are displayed, and incubation temperature can be set to an optimal temperature using buttons.

The system was developed around the PIC16F1937 controller.

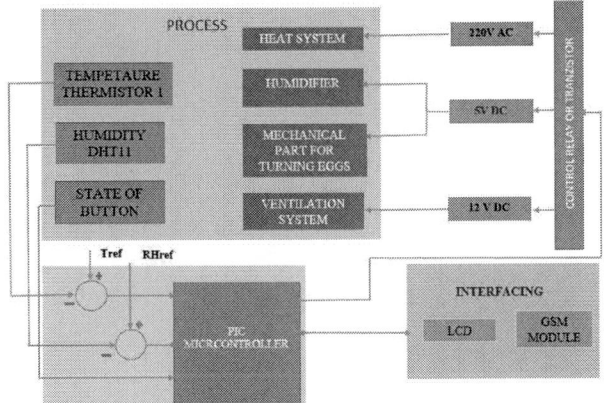

Fig. 1. Block diagram of the system.

The incubator chamber is made from a transparent plastic box (Fig.2). On the middle level of the box a horizontal rode axis is placed, that will hold the eggs support. The axis is rotated by a stepper motor in both ways, 45 degree related to vertical axis. The system can fit a maximum number of 25 eggs.

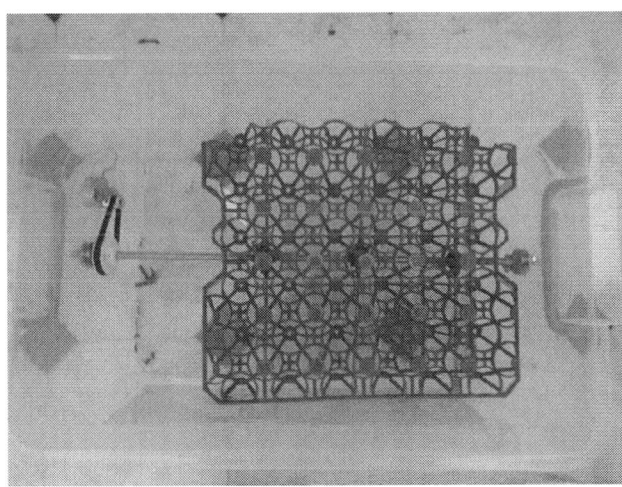

Fig. 2. Incubator box.

The electrical scheme of the system is presented in Fig.3 and the first prototype of board is shown in Fig.4 where the essential components are highlighted: 1- transformer and the bridge rectifier, 2- buck converter and IC LM350, 3- Pickit3 programmer, 4- reset circuit, 5- PIC16F1937 microcontroller, 6- LCD display, 7- circuit for temperature measurement, 8- humidity sensor, 9- buttons for menu interfacing, 10- heat

system, 11- stepper motor, 12- GSM module, 13- humidification system, 14- fan, 15- power fail security system.

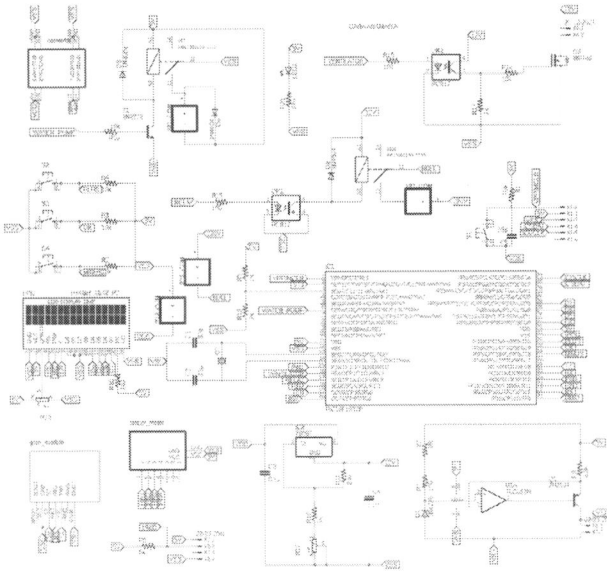

Fig. 3. Electrical scheme.

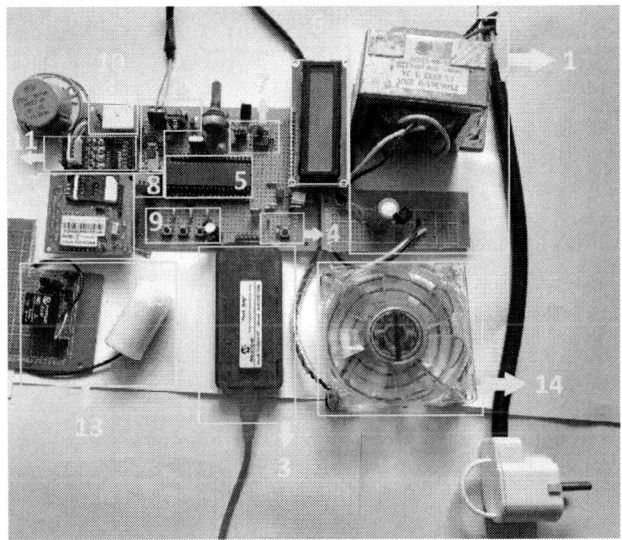

Fig. 4. Prototype of the electronic part of the incubator.

The transformer and the bridge rectifier convert 220VAC to 19VDC. Buck Converter converts 19VDC to 5VDC and will supply power for PIC16F1937, GSM module, water pump, stepper motor and a 5V relay. The IC LM350 is a tunable voltage stabilizer and it is set to have an output of 12V, which will be used to supply voltage for a DC fan and for a 12V relay. Reset circuit has been implemented so that the user can reset to an initial state the incubator. Fig.5 shows the power supply of the system.

For visual user interface an LCD 16x2 chars has been used to display different information for the user. Circuit for temperature measurement consists in an NTC thermistor in which a constant current is injected via a current source. The voltage across the thermistor is read by an internal ADC converter from PIC16F1937. Humidity sensor DHT11 is a digital sensor which provides both humidity and temperature. We use only humidity from this sensor because the NTC thermistor has a better accuracy in measuring temperature. Also, three buttons have been added so that the user can navigate in the menu.

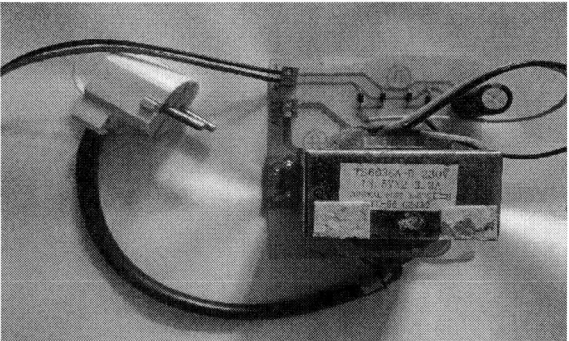

Fig. 5. Power supply.

For the heat system we had to choose between: the fiber heating cord, infrared incandescent light bulb, film for underfloor heating and ceramic resistances. Because the incubator is intended to be used in small households where the farmer can do and easy maintenance, we opted for the incandescent light bulb. It has an operation lifetime of 5000 hours. It can be replaced very quickly by the farmer and has the lowest price from all the above options. For a 25egg incubator a 100W infrared light bulb is more than suitable. Its consumption is less than a film for underfloor heating which presents a consumption of $150W/m^2$ to $200W/m^2$ consumption. A modern approach is by using fiber heating cord, but its replacement requires electronic measurements for the proper size of the cable which has a different electrical characteristic depending on the manufacturer. The light bulb is controlled by a 12V relay.

We chose to use a stepper motor to control as accurately as possible the angle of rotation of the support on which the eggs are placed. A 28BYJ-48 stepper motor was used to turn the eggs, which is powered by 5V. A ULN2003 driver was used to control the engine. Stepper motor will rotate the central axis with the eggs support (Fig.6). The activation of the egg return motor is done via the GSM module. The farmer will send a message signaling the microcontroller that it is the right time to turn the eggs at a 90° angle from the previous position.

Humidification system consists of a 5V relay which control a 5V water pump. The water pump will increase the humidity level. DC fan will make the temperature constant in all corners of the incubator and will maintain a correct level of oxygen.

D. Power Fail Security Circuit

To provide the best control for the egg turning we designed and implemented a power fail security circuit. An analog pin of the microcontroller was connected to a voltage divider and monitor all the time the voltage level given by this. The AD converter was configured to use a fixed internal voltage reference of 1.024V which is independent from the power supply. The voltage divider is dimensioned so that at a power supply of 5V, it will provide a voltage of 2.048V for the ADC. When the converter sense that the supplied voltage goes down, it will save some essential data in EEPROM which will be used to remember the last state of the incubator. To ensure that the power supply will have a smooth power decrease and to ensure that the PIC16F1937 will have enough time to save data in EEPROM we placed a high value capacitor which will store a reserve energy.

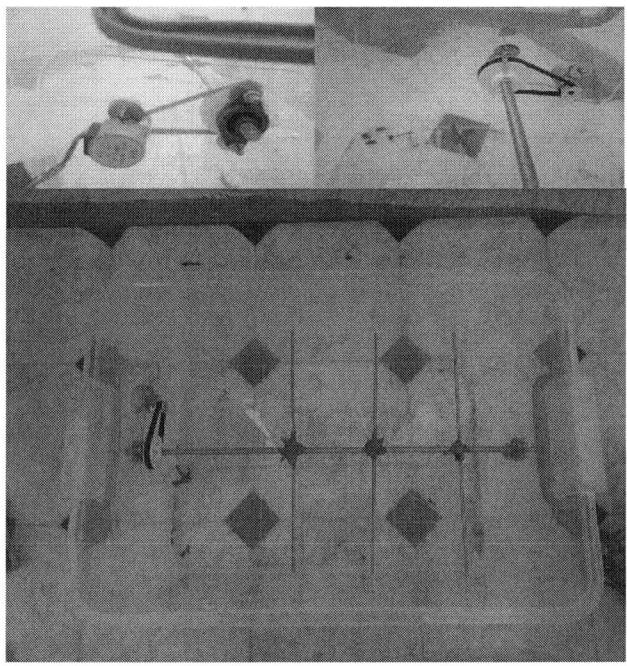

Fig. 6. Stepper motor connection to the central axis.

E. Remote Control Function

GSM module will have a remote-control function. The user can control the system via a phone with a SMS. Also, he can request information about the status of incubator or can order to rotate the eggs. As GSM module it is used M590E. In Fig.7 it is present this GSM module and the connection scheme.

Fig. 7. M590E GSM module.

The GSM module is controlled using AT commands. It is necessary to set a baud rate for initializing the communication between the microcontroller and module and ensure a current of 2A when the SMS is transmitted or received.

For transmission of messages to the farmer, we had to consider activating the transmission conditions. So, the farmer will send a message to the GSM module. The microcontroller will activate the routine to handle an interruption generated by the reception of a message and will trigger to send a SMS to the farmer with the current status of the parameters inside the incubator.

F. Final Assembly and testing

The final PCB of the system is presented in Fig.8.

Fig. 8. Final PCB.

First time it is necessary to initialize the system (Fig.9), which involves the configuration of pins, the ADC converter, timer, external oscillator and EUSART transmission. After that, the global and peripheral interrupts are enabled, and the LCD display and GSM module are initialized. We continuously monitor the signals from the temperature and humidity sensor and compare this value with the reference value. Depending on the comparison results, the heating or humidification of the system starts. Also, on the board there are the user interaction buttons that allow the users to navigate in the temperature setting menu or in the display of humidity and temperature. Using the GSM module, the farmer will have permanent access to essential parameters in incubation and can control the turning of the eggs by sending a specific SMS which the microcontroller interprets and acts on the stepper motor.

Several tests were done on chicken eggs to verify the functioning of the system, as presented in Table 1.

TABLE I.   INCUBATION TEST

Test #	Incubation parameters		
	No. of used eggs	Time	No. of chicks
1	25	21	23
2	20	21	17
3	25	21	22

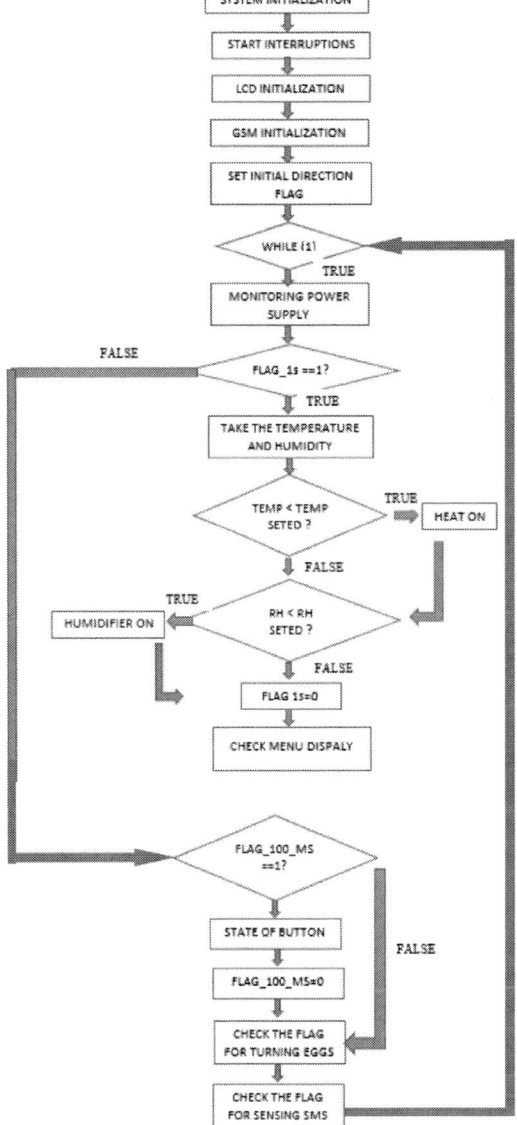

Fig. 9. Operation logic of the system.

III. CONCLUSIONS

The present work is aimed to develop a household egg incubator that offers accurate data to the farmer and remote control of the system.

Similar capacity egg incubators can be found on the market for a price ranging from 40 to 80 euro. Our system has a production value of 30 euro. The main differences between our device and those from the market is the real time information and control for the farmer by GSM connection. The mechanical turnaround of the eggs is done automatically by a stepper motor, compared to the manual activated mechanism found on the devices from the market. More than that, our device is capable to continue its turnaround process from where it stopped in case of a power failure. The elements used to build the system allows an easy maintenance for the farmer.

REFERENCES

[1] E.Idoke, G.O. Ogbeh, F. T. Ikule, "Design and Implementation of Automatic Fixed Factors Egg Incubator", IJIRMF, Vol.5(6), 2019;

[2] F. K. Sunyani, F. Peprah, B. Ahafo, "Design and Construction of an Arduino Microcontroller-based EGG Incubator" International Journal of Computer Application, Vol. 168(1), 2017.

[3] A. B. Umar, K. Lawal, M. Mukhtar, M. S. Adamu, "Construction of an Electrically-Operated Egg Incubator", International Journal of Modern Engineering Sciences, Vol.5(1), pp.1-18, 2016.

[4] P. E. Okpagu, A. W. Nwosu, "Development and Temperature Control of Smart Egg Incubator System for Various Types of Egg", European Journal of Engineering and Technology, Vol.4(2), 2016.

[5] A. Obidiwe, C. Ihekweaba, P. Aguodoh, "Design and Implementation of a Microcontroller Based Egg Incubator with Digital Temperature read out", Semnatic Scholar, ID 40378124, 2014.

[6] J.N. Olasunkanmi, O. Akintade, L.O. Kehinde "Development of a GSM Based DC Powered Bird Egg Incubator", International Journal of Engineering Research & Technology (IJERT), Vol. 4(11), 2015.

[7] A.S. Muhammad, M. Jimoh, A.S. Muhammad, J.N. Olasunkanmi, "Development of an Automatic Bird Egg Incubator", Rsearch & Reviews: A Journal of Embedded System & Applications, Vol. 5(1), 2017.

[8] S.K.K.S. Kumar et al."Novel Fully Automatic Solar Powered Poultry Incubator", In book: Emerging Trends in Computing and Expert Technology, Springer, pp. 1612-1618, 2020

Phosphor Based White LED Driver by Taking Advantage on the Remanence Effect

L.A. Szolga, R.G. Groza

Basis of Electronics
Technical University of Cluj-Napoca
Cluj-Napoca, Romania
Lorant.Szolga@bel.utcluj.ro, Robert.Groza@bel.utcluj.ro

Abstract—**This paper presents the development of a control circuit to enhance the performances of LED lamps. In this direction, a comparison between the luminous intensity of normal LED based lamps and mid-power ones, for both continuous and switching conditions has been made. The already well know control technologies were analyzed and a study was conducted to increase the lighting performances by rising the operating frequency and magnifying the contribution of remanence effect and thus increasing the efficiency of the light source. To achieve this, in the first stage of the project the power and control circuits have been modeled, related to desired parameters and tested in simulation software. In the second stage, the proposed circuit was implemented by functional blocks and in the last stage, tests were made on the circuit and on light sources in order to process the results. The power consumption has been decreased nearly to a half of it and the luminous flux raised with 15% due to overcurrent and remanence effect that we used.**

Keywords— LED; driver; remanence; PWM;

I. INTRODUCTION

Electronic science has known a spectacular evolution in the last few decades. Because this science side has proved more and more his utility among the „tools" we use every day, their majority contains at least one small electronic circuit.

On the other hand, the abundance of electronic systems in our homes has a disadvantage: power/dwelling consumption is increasing continuously. So that is why the necessity of optimization through consumption reduction is more and more often felt by anyone.

About 20 percent of dwelling power consumption is related to illumination systems. This percentage is slowly decreasing every year due to the electronic technology spreading among other domains, but also due to the benefits it brings to illumination systems [1].

WILA Company says that LED technology seems to follow the Moore's Law, while Light Emitting Diodes double their luminous flux every 18 months. In the last time, LED performances enhanced a lot. For instance, in 2008, the most powerful commercial LED came from South Korean company Seoul Semiconductor. Their LED, Z-Power P7 series could generate 900 Lumens for 10 Watts power consumed.

On 12th May 2010, Nexxus Lighting presented the most powerful LED lamp for that moment. Arraz LED PAR38 has a 50lm/W efficiency and its luminosity is similar to an 75 W incandescent light bulb.

This year, the efficiency reached an unpredictable level. So, in 2014 OSRAM Company revealed their LED light bulb model. It consumes 19 Watts and has a 3900lm luminous flux, reaching 205lm/watt efficiency.

Once again, the Moore's Law is confirmed by last released models. LED technology has its beginnings over 100 years ago, but the world didn't pay much attention to it until the first LED got commercial. It happened in 1962, when General Electric Company released their first light emitting diode model for 260$. It had scientific purposes.

First ever truly commercial LED has been made in 1962 by Mosanto. A piece had a price of 3$. It's odd, but the same model costs 18$ nowadays. So, even if only 5 decades passed from that moment, LED studies contrived to bring these diodes to new levels, making them more and more useful. Moreover, the technology got cheaper so that today a performant LED has the same price as 40 years ago, even if the performance is obviously bigger.

It's still considered a new technology. This fact and their efficiency lead to a high price compared to other technologies like HID (high intensity discharge), fluorescent and incandescent. Even so, people start to prefer LED for their multiple advantages, like: lifetime (35000–100000 hours), efficiency, resistance to temperature ripple, resistance to vibrations, resistance to mechanical shocks, reliability, spectrum (color is set by semiconductor material, not by filters), directional emission of light, working on low temperatures, no UV and IR emission, controllability, reduced environmental impact [2-5].

LEDs biggest disadvantage is their price/lumen ratio. Compared to other technologies, their efficiency is expensive.

As the Table 1 shows, LEDs don't lose too much heat through radiation (Rad.) and convection (Conv.), but they are losing a lot of it through the conduction (Cond.) phenomenon. So that's the reason a power LED needs additional cooling systems. Also, LEDs are the most efficient light source until now and the technology keeps getting more and more advanced while LED development's going on, following the tendency to double their luminous flux every two years.

TABLE I.   COMPARISON BETWEEN LIGHTING TECHNOLOGIES

| Light source | Eff. [lm/W] | Heat loss by | | | Cost [$/lm] | Usability |
		Rad. [%]	Conv. [%]	Cond. [%]		
Incandescent	10-20	>90	<5	<5	.001	Home
Fluorescent	75-90	40	40	20	.0005	Idustrial
HID	100-120	>90	<5	<5	.002	Outdoor
LED	150-200↑	<5	<5	>90	.02	Retail, Displays, General

From Table 1 it can be deducted that the most widespread field of use is LED technology. It is also the most efficient source of lighting, but also imposes the highest cost of efficiency. It is emphasized, moreover, that heat loss through radiation and convection is almost non-existent, hence the need for additional means of cooling, especially for power LEDs [6].

Another important aspect is the lifespan expressed in hours: incandescent bulb - 2000h; fluorescent lamp - 20000h; HID lamp - 15000; LED bulb - 100000.

Although they have a high price, today's LED bulbs are profitable even in the current acquisition costs, because, having the highest efficiency, the longest life, but also being very robust, their cost is amortized over time by low consumption energy and, in most cases, the lack of need to replace them as a result of damage [7-8].

Ordinary people find it harder to switch to new technology because of the cost, but in places where the power consumption associated with lighting sources is high (office buildings, industry, billboards, street lighting and decorative or architectural), LED bulbs are place getting easier.

In order to further reduce energy consumption, to ensure optimal brightness or simply to obtain play of lights, lighting control is used.

It can be a control module implemented directly in the bulb electronics, a stand-alone module or a network control module. The network module is used when it is necessary to control many sources in several rooms or in an entire building.

Generally used control equipment includes: DALI-bidirectional control and addressing of devices; DALI color - improved version, for color control; DMX (Digital MultipleX) - dedicated to theaters and stage lighting; Forward Phase Control - the most used method, the bulbs are powered in the second part of each half of the sinusoidal signal; Reverse phase-cut dimming - more expensive than the previous method, due to the complexity of electronic circuits, the bulbs are powered in the first part of each half of the sinusoidal signal.

The efficiency increases as new semiconductor materials are developed. But, in order to set efficiency higher, some LED drivers started to be implemented. The dimming effect can be reached by two basic methods: 1) controlling the luminous intensity by adjusting the current value - while for a good part of diode characteristic the luminous flux is proportional with the current intensity through P-N junction, the amount of light can be adjusted by changing deliberately the current value; not a good method when it comes to dimming the light under 20 percent of initial value; 2) controlling the luminous intensity by adjusting the duty cycle - with pulse width modulation (PWM), the LED is operated at rated current; to achieve the dimming effect, the light is cycled on and off at a high frequency, depending on the desired brightness; while the human eye can't detect these short pulses individually, instead, it notices only a lower total flux [9-11]. Based on these second method we designed a driving circuit in PWM for a row of 36 LEDs (Fig. 1). This circuit was used to test the power consumption and the light intensity of regular 5mm white LEDs and the SMD5630 LEDs.

II.   HARDWARE

A. Initial Conditions

We started this study from the idea of creating a light source of increased efficiency with the lowest cost possible. Being a research, the total anticipated cost was high anyway, but we wanted the final product to be as cheap as possible.

We started by studying the existing technologies, trends and methods usually chosen of the control circuit to understand the reason behind the design of such devices, as well as the lighting methods, possibly the configuration of the LEDs.

In order to obtain the highest possible efficiency, we opted for the use of medium power LEDs. Looking for efficient LEDs, we found the 5630 SMD model, which met the requirements since it has a luminous flux of 50-65lm/Watt. We took this LEDs from a commercial LED light bulk (Fig. 1) which contained a row of 36 LEDs and presented a power consumption of 8W.

Fig. 1.   8W LED light bulb with 36 LEDs of type 5730.

By opening the commercial light bulb, we could note down the scheme of the driver (Fig. 2) and calculate the electrical parameters used for the LEDs.

The measurement taken on the functioning of the light bulb revealed a filtered voltage at the input of the LED row of around 104.5V, which means that the voltage drop for each LED is 2.9V. Looking in the datasheet for the I-V characteristic of the 5630 (Fig. 3) at a voltage drop of 2.9V the current through the LED is 35mA.

2020 IEEE 26th International Symposium for Design and Technology in Electronic Packaging (SIITME)

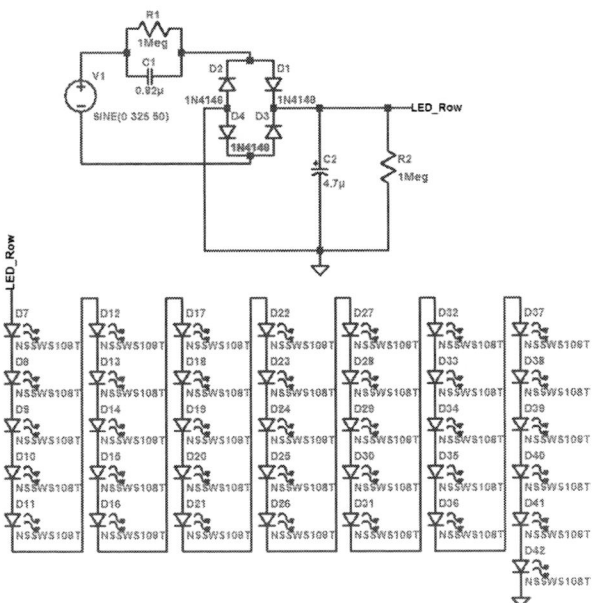

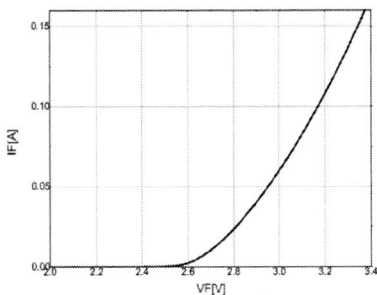

Fig. 2.   Commercial LED light bulb driver circuit.

Fig. 3.   I-V characteristic of the SMD5630 LED.

B. Operating Principle of the Proposed Driver

We imposed to design a circuit for the LEDs, to drive them in overcurrent conditions. Typically, the maximum pulse current is specified in the LED datasheet.

For the 5630, the maximum pulsed current is 150mA for a 1KHz signal, with a fill factor of 10%. Although, the LED lamp we tested does not have a control block, there are some bulbs that include circuits like this, in order to adjust the light intensity or increase the efficiency of the lighting source. Typically, such light bulbs have a control signal of 100 Hz, with a filling factor of 10-25%. Being an overcurrent, the luminous flux is high for a short time, after which it will suddenly attenuate to the level of 0lm during the LOW level of the control signal. At 100Hz the eye does not perceive the pulsation, instead it will notice a light intensity specific to a flux equal to the average flux over a period.

Through the driver proposed, we controlled the LEDs at a frequency of 1KHz and a fill factor of 10% (according to the specifications in the datasheet) in overcurrent, relying on the idea that the phosphor layer covering the diode has a large enough remanence to keep the radiating during the 0.9ms period, while the control signal is in the LOW state. Thus, the

remanence of phosphorus will contribute to a higher average value of the emitted flux. It is difficult to anticipate this effect, with an uncertain probability of occurrence due to lack of information. In all the consulted sources, only the idea is specified, that the phosphorus layer on the diode has a very small remanence, but no indication is given to know at least if it exceeds 1ms or not.

C. Driver Design and Implementation

Choosing the methods to implement the blocks needed for an LED power supply and control circuit involves making the best decisions to meet the project requirements, even if this requires some trade-offs. The proposed driver circuit is presented in Fig. 4 and its building blocks and operation is detailed in the next paragraphs.

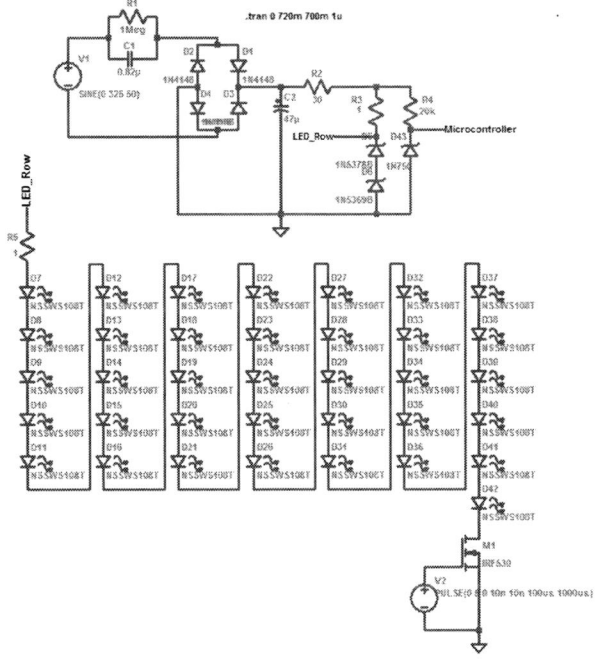

Fig. 4.   Proposed LED light bulb driver circuit.

To adapt the wall input signal to the LEDs necessary signal we used and RC filter on one of the inlets in the rectifier bridge. It is also called a capacitor-based power source. An ordinary capacitor will not work because the dielectric will break through due to large voltage variations of the main signal. A 250V or 400V metal propylene film capacitor will operate properly in this setup. In parallel, a high value resistor will be connected to discharge the capacitor when the circuit is not powered on. It has a high value (1MΩ in general) to greatly reduce the current and discharge speed of the capacitor, thus avoiding electric shocks. By choosing the value of the capacitor, the maximum current supplied to the assembly is limited. To take advantage of the LEDs specifications, we looked to obtain a value in the overcurrent as close as possible to 150mA. Thus, a 2.2µF capacitor can introduce a reactance of:

$$X_C = 1/2\pi fC = 1447\Omega \qquad (1)$$

which limit the current:

$$I = 230V / 1447\Omega = 159mA \quad (2)$$

We chose to use a full wave rectifier, to ensure enough voltage for the operation of the LEDs. Being a high consumption circuit, the rectifier bridge must withstand high voltages and currents. To work away from the operating limit, we chose to use the 2KBP06 bridge capable of 600V and 2A.

To smoothen the rectified signal an electrolytic capacitor was used with electrical specification of 250V and 47μF. The higher is the capacity, the voltage ripples are smaller.

From Fig. 3, it can be deducted that for a 150mA current the voltage drop on an LED is 3.3V, which means that for a row of 36 LEDs the minimum operation voltage should be around 120V.

To be able to control the switching frequency of the LEDs we used an ATmega8L which drives an IRF530 transistor. The microcontroller can be powered up between 2.7V and 5.5V for a proper operation. Although, the only function it will fulfill in carrying out this project is that of a pulse generator, we opted to use a microcontroller in order to implement a smart system which can include: voice detector, presence sensor, ambient light sensor, etc.

A regular 7805 type stabilizer is not suitable in our application, because of the maximum 20V accepted at its input compared to the 150V after the smoothening capacitor. Thus, to stabilize the voltage for the microcontroller with low costs we had the options of a resistive divider or a Zener diode to extract a voltage within the above-mentioned range. The drawback of the resistive divider is the excessive heat that could dissipate, so there are chances that their performance in the division structure will change due their value changes.

Thus, the Zener diode was preferred. For simplicity, we choose a 4.7V Zener diode to power up the microcontroller and two Zener diodes in series (100V and 51V) for the LED row.

The simulated power consumption of this designed driver is 4.25W (Fig. 5), which means that the power consumption has halved from the original driver of the light bulb.

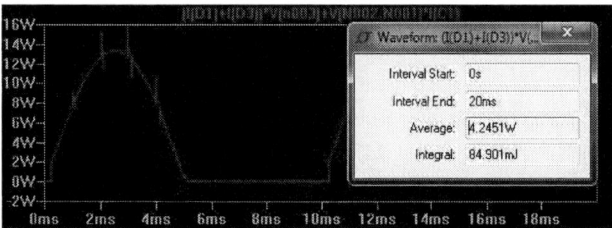

Fig. 5. Power consumption of the designed driver.

III. Experimental Measurements

To test out the flux efficiency of the PWM driver, we made tests by using the same driver with a row of 5mm white LEDs and with a row of 5630 LEDs (Fig. 6). The measurements were taken in a dark room by using a luxmeter and a professional photo camera with a fast shutter speed of 1/4000s.

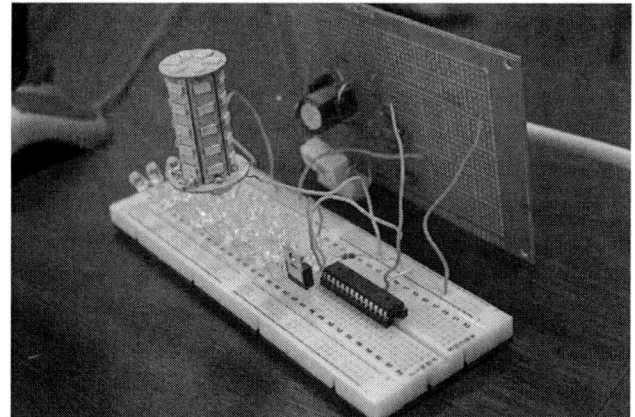

Fig. 6. Tested LED driver circuit.

The gathered result from Table 2, were obtained with a luxmeter at a 30cm distance from the top end of the LEDs. In the case of the 5mm LEDs the PWM control is not benefic if we want to obtain more flux. On the other hand, for the SMD LEDs this procedure increases the flux quantity with almost 15 percent compared with the continuous mode of operation.

TABLE II.   Loght Intensity Measurement Results

White LED type	Operating mode	Light intensity [lux]
5mm	Continous	272
	PWM	75
SMD5630	Continous	87
	PWM	100

In Fig. 7 it can be seen the high intensity of the continuous mode over the low intensity of the PWM mode operation for the 5mm LEDs.

Fig. 7. 5mm white LEDs powered up in continuous mode (left) and PWM (right).

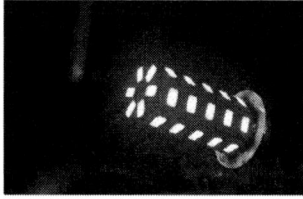

Fig. 8. SMD5630 white LEDs powered up in PWM mode (OFF state – left, ON state - right).

In Fig. 8 the ON/OFF PWM states of lighting of the SMD5630 shows that the remanence effect in this type of LED is benefic in keeping the LED to produce light in the OFF state of the PWM with almost the same intensity like in the ON state of the PWM.

IV. CONCLUSIONS

The aim of this work was to show the advantages of pulse LED control. Among the objectives of the research was the efficiency of the light source. We managed to halve the consumption while increasing the light intensity with 15 percent.

Another aspect that we wanted to underline was the advantage of the remanence effect of the phosphorus in the conditions of controlling the SMD LEDs in a higher frequency than those that are usually practiced.

The decision to use the microcontroller significantly increases the cost of the finished product but leaves room for further implementations.

There are still many ways to develop LED-based light sources, as the technology is still evolving.

REFERENCES

[1] O. Ayan, B.E. Turkay, "Comparison of lighting technologies in residential area for energy conservation" 2017 2nd International Conference Sustainable and Renewable Energy Engineering (ICSREE), pp. 116-120.

[2] M.H. Kane, N. Arefin, "Gallium nitride (GaN) on silicon substrates for LEDs in Nitride Semiconductor Light-Emitting Diodes (LEDs) (Second Edition)", Woodhead Publishing Series in Electronic and Optical Materials, 2018, pp. 79-121.

[3] J.H. Ryou, W. Lee, "GaN on sapphire substrates for visible light-emitting diodes in Nitride Semiconductor Light-Emitting Diodes (LEDs) (Second Edition)", Woodhead Publishing Series in Electronic and Optical Materials, 2018, pp. 43-78

[4] Z. Wu. Z. Xia, "hosphors for white LEDs in Nitride Semiconductor Light-Emitting Diodes (LEDs) (Second Edition)", Woodhead Publishing Series in Electronic and Optical Materials, 2018, pp. 123-208.

[5] T.H. KIM, W. WANG, Q. LI, "Advancement in materials for energy-saving lighting devices (Review Article)", Front. Chem. Sci. Eng. 2012, 6(1): 13–26.

[6] "Section 25 - Lighting/HVAC/Refrigeration in Handbook of Energy", Elsevier, 2013, pp. 827-838.

[7] Fereidoon P.Sioshansi, "Will Energy Efficiency make a Difference? in Energy Efficiency", Academic Press, 2013, pp. 3-50.

[8] C.R.B.S. Rodrigues, P.S. Almeida, G.M. Soares, J.M. Jorge, D.P. Pinto, H.A.C. Braga, "An experimental comparison .between different technologies arising for public lighting: LED luminaires replacing high pressure sodium lamps", 2011 IEEE International Symposium on Industrial Electronics, pp. 141-146.

[9] K. Szolusha, "Chapter 286 - 100V controller drives high power LED strings from just about any input in Analog Circuit Design", Newnes, 2015, pp. 615-616.

[10] Y. Wang, J.M. Alonso, X. Ruan, "A Review of LED Drivers and Related Technologies", IEEE Transactions on Industrial Elenics pp. (99):1-1, 2017.

[11] H.J. Chiu, Y. K. Lo, J. T. Chen, S. J. Cheng, C. Y. Lin, S. C. Mou, "A High-Efficiency Dimmable LED Driver for Low-Power Lighting Applications", IEEE Transactions on Industrial Electronics 57(2):735 - 743, 2010.

Integration of Internet of Things technology into a pill dispenser

1st Madalin Vasile Moise
Center for Electronics Technology
and Interconnection Techniques
Polytechnic University of Bucharest
Bucharest, Romania
madalin.moise@cetti.ro

2nd Ana-Maria Niculescu
Center for Electronics Technology
and Interconnection Techniques
Polytechnic University of Bucharest
Bucharest, Romania
ana_nic99@yahoo.ro

3rd Andreea Dumitrașcu
Center for Electronics Technology
and Interconnection Techniques
Polytechnic University of Bucharest
Bucharest, Romania
dumitrascu.andreea50@yahoo.com

Abstract—The main purpose of this paper is to present how a developing technology, such as the Internet of Things, can be used for the benefit of people. With the help of this technology, a system is implemented that can retrieve information received from several pill distributors, to store them in an online database, so that they can be analyzed later. It is also designed to send alerts in real time, if some pills were not taken by the patients at the prescribed interval. The system is designed to be used individually by users who need to check if a patient took its medication at home or in a hospital, but also to be used in creating a network of pill dispensers supervised by one administrator. In this way the exposure time to a virus, in case of a pandemic, is minimized. The embedded system is based on an ESP8266 microcontroller, that allows online remote access to the pill dispenser's status. The information about the pills is stored in a cloud storage, where it can be processed through a variety of applications. Several ESP8266 microcontroller can be connected, obtaining a network of pill dispensers, which will be monitored via IoT. The status of each pill dispenser can be viewed in real time using a website, this will contain valuable data on how patients receive treatment.

Keywords— IoT, ESP8266, microcontroller, pills

I. INTRODUCTION

Pills administration requires attention to details and focusing. For some people, it's difficult to do this on their own, which is why we believe that a pill dispenser can help them, being more than just a medicine storage space. Studies show that many errors can occur when administering medications [1]. The problem of medication errors, even made by medical staff, is very debated all over the world, so the health and even the life of patients is endangered.

There are many studies that have explored ways to improve the quality of prescribing in primary care [3]. A review of 10 randomized trials of computerized interventions found a reduction in medication errors in half of the studies. So, the adoption of electronic tools will be essential to improve safety of patients in many ways [3]. All the procedures regarding medication administration must assure the "Five rights" [2] of medication administration: right person, right medication, right dose, right time, right route and right documentation. Some of these rights can be more easily respected with the help of IoT

technology, more precisely, those related to: a specific patient, the dose of pills and the time of administration.

The main purpose of our system is to come to people's aid by using Internet connectivity to allow the caregivers to remotely manage the pill dispenser program and monitor whether or not a patient has taken the correct dose of pills at the right time. The integrated internet connection warns the caregiver whether the patient has taken a dose of the pills or not in a certain time.

Pill information is stored in a cloud storage and can be processed immediately of at a later time. What is interesting is that multiple ESP8266 microcontrollers can be linked to create a network of pill dispensers that are monitored via Internet of Things. The status of each pill dispenser can be viewed in real time via a website. From here, various statistics and research can be performed on how the treatments are being administered to different types of patients.

There aren't that many pill dispensers out there, and the ones available require a monthly subscription or are difficult to plan. With IoT technology, our system provides a user-friendly and secure environment, without monthly payment, where patients and caregivers can stay connected from anywhere.

II. PLATFORM DEVELOPING

In this project, the prototype is used for visualization the information from the pill dispensers network. The communication between the interface and the system which handle when pills are dispense is made through a serial communication.

Thanks to a reduced complexity and having a simple interface, people which are new to electronics and programming can use this prototype to put their own ideas in practice fast and easy.

A. System overview

The IoT is a sensor network of billions of smart devices that connect people, systems, and other applications for data collection and sharing. The entire IoT ecosystem consists of intelligent devices that use built-in processors, sensors and communication hardware to collect, send and process the data from their environment, which is essential in the development of this project [4].

With the help of an ESP8266 wireless (WI-FI) controller, the serial communication with the pill dispenser system is realized in order to allow access to monitoring and control remotely. The system uses an open-source platform that facilitate monitoring and controlling of IoT devices.

The ESP8266 microcontroller is a highly integrated Wireless System on Chip (SoC) solution that will meet user's continuous demand for efficient power usage; it has a compact design and reliable performance in the Internet of Things industry. It is already Internet-enabled, it takes care of all the Wireless, TCP/IP stack, and the overhead found in an 802.11 network. Thus, it is an extremely easy choice for anyone who wants to build an Internet of Things and start transmitting data to the Internet.

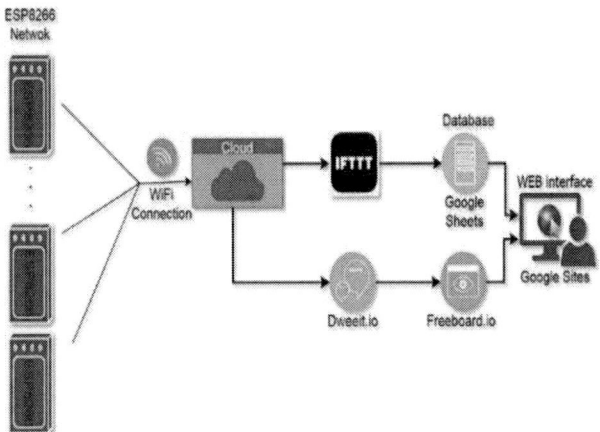

Fig. 1. *System block diagram*

The system has the following main components (Fig.1):
o   ESP8266 - wireless controller
o   Cloud server
o   IFTTT→Google Sheets/Dweet.io→Freeboard.io
o   Google Sites

B. About platforms used for communication

1) IFTTT

The main criteria considered in choosing the right platforms to use were ease of use, easy connection to the ESP8266 microcontroller, online data storage and viewing, and security of communications between server and devices.

Because communication between a device and the Google suite of applications requires a certificate, in order not to create unwanted delays in sending messages, a webhook platform was used, IFTTT, which takes messages sent by the microcontroller in HTTP format and provides fast integration with Google Sheets [5]. A webhook in web development is a method of increasing or changing the behavior of a page or web application with custom dialing. Because webhooks use HTTP, they can be integrated into web services without adding a new infrastructure.

2) Google Sheets

Although not an IoT platform in the true sense, it is a free platform, accessible from several different devices, being in an online environment and allows the creation of scripts using Google Apps Script in addition to the data storage function [6]. Apps Script is a fast application development platform, makes it possible quick and easy creation of business applications that integrate with the Google suite and run in the cloud. It offers increased security by using an identifier (an email address) and a password as a way of authentication, as well as a data processing environment. The biggest advantage is that it can be used as an online data storage base without the need to learn a new database. Security is also enhanced by using the HTTPS protocol. Google Sheets supports the Google Chart query language for sorting and filtering data so you can create charts that can be easily integrated into a WEB site.

3) DWEET.IO

The platform provides machine-to-machine (M2M) communication for the Internet of Things. It can also be used for free, but the messages sent can be viewed by all users of the service or a private name can be generated and at the same time the messages sent are kept for 30 days, but for which a monthly fee is paid [7].
Example of communication using a private name and a key:
https://dweet.io/dweet/for/{my_locked_thing}?key={my_key}&hello=world&foo=ba
r [http://dweet.io/faq].

4) FREEBOARD.IO

The Freeboard.io platform is intended exclusively for viewing information received from IoT nodes. These can be received directly from the microprocessor via the Dweet.io platform [8]. The use of the platform is ideal for viewing real-time information sent directly by the microprocessor. An intermediate platform, Dweet.io, is used to view the data received in real time from the IoT node, which provides the link to the Freeboard.io viewing environment. The data is available in the current HTTP connection, and the information sent to the server is in JSON format.
Example: *"POST / dweet / for / SUBIECT? Particle03 = 325"*

5) Google Sites

After implementing the working mode, the next step was to develop a WEB page, using Google Sites, to facilitate access to the way the data is viewed. The website is made using HTML, CSS and JavaScript and includes viewing from freeboard.io.

2020 IEEE 26th International Symposium for Design and Technology in Electronic Packaging (SIITME)

C. Printed circuit board developing

The printed circuit board (PCB) is capable of increase functionalities by attaching additional hardware. The role is to reduce the complexity of the work on new systems and to facilitate the immediate results.

We wanted to minimize the complexity of the circuit board and thereby increase its reliability, since large circuit boards are no longer so popular these days. The only additional component on the system board is the connector that carries the relevant signals to or from the adapter board (or programming board, debug expansion board, etc.); it incorporates test / debug functionality without significantly increasing the complexity or size of the application board (Fig. 2).

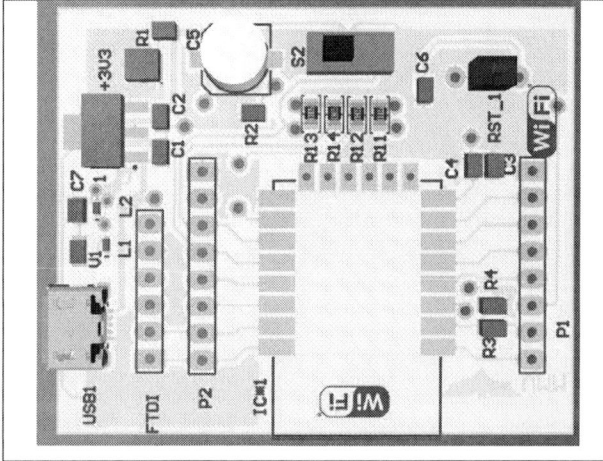

Fig. 2. *Printed Circuit Board*

III. SOFTWARE DEVELOPEMENT

The information received from one pill dispenser is transmitted via Internet using wireless connection to the cloud storage and to a real time visualization tool.

For this project, Google Sheets platform is used as an online database and Freeboard.io platform is used to visualize status from multiple pill dispensers in real time.

The connection between ESP8266 and Google Sheets platform is realized using IFTTT platform (Fig. 1), because a direct communication with Google application suite needs a certificate that ESP8266 could not use, or by using it, will create delays is sending the messages.

IFTTT platform takes over the messages sent by the microcontroller in HTTP format and offers fast integration of the sent data in Google Sheets.

An intermediate platform, Dweet.io, is used to provide a link to the Freeboard.io visualization environment, in order to view real-time data from ESP8266. The data is sent to the platform using POST request and the information sent to the server is in JSON format.

A website is created to show information's form both Google Sheets and Freeboard.io platforms, this way an administrator will have an easier overview of the pill dispensers' network. In this way, the medical staff will know exactly the current date and time, the set time interval, the time at which the pills were taken and if they were taken correctly and any other problems with the device (Fig. 3).

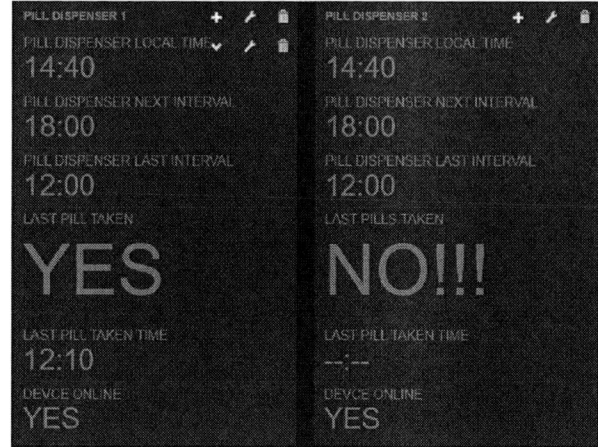

Fig. 3. *Visualization using Freeboard.io*

IV. RESULTS

The aim of this project is to integrate IoT technology into a pill dispenser, to make it easier for medical staff and patients to use it and to store information in a shared database. Possible ideas for developing the project include creating a smartphone application and then integrating it into an interconnected network of pill dispensers for use in hospitals.

It was calculated that, after 2000 records sent per day, a new document would be created in Google Sheets. In this way we have the possibility to send 83 recordings in one hour from several devices. Depending on the needs, the number of devices is limited only by the number of registrations per hour.

For security reasons, the data sent from the devices are concatenated into 3 data sets (Fig. 4) which are separated into Google sheets to extract the data (Fig. 5).

	A	B	C	D	E
1	Local time	Device ID	Value 1	Value 2	Value 3
2	July 21, 2020 at 02:17PM	ESP8266TEST1	07/08/2020_12:00;Yes	07/08/2020_08:00;No	07/08/2020_08:15
3	July 21, 2020 at 02:18PM	ESP8266TEST1	07/08/2020_12:00;Yes	07/08/2020_08:00;No	07/08/2020_08:15
4	July 21, 2020 at 02:17PM	ESP8266TEST1	07/08/2020_12:00;Yes	07/08/2020_08:00;No	07/08/2020_08:15
5	July 21, 2020 at 02:18PM	ESP8266TEST1	07/08/2020_12:00;Yes	07/08/2020_08:00;No	07/08/2020_08:15
6	July 21, 2020 at 02:19PM	ESP8266TEST1	07/08/2020_12:00;Yes	07/08/2020_08:00;No	07/08/2020_08:15
7	July 21, 2020 at 02:20PM	ESP8266TEST1	07/08/2020_12:00;Yes	07/08/2020_08:00;No	07/08/2020_08:15
8	July 21, 2020 at 02:21PM	ESP8266TEST1	07/08/2020_12:00;Yes	07/08/2020_08:00;No	07/08/2020_08:15
9	July 21, 2020 at 02:23PM	ESP8266TEST1	07/08/2020_12:00;Yes	07/08/2020_08:00;No	07/08/2020_08:15
10	July 21, 2020 at 02:24PM	ESP8266TEST1	07/08/2020_12:00;Yes	07/08/2020_08:00;No	07/08/2020_08:15
11	July 21, 2020 at 02:25PM	ESP8266TEST1	07/08/2020_12:00;Yes	07/08/2020_08:00;No	07/08/2020_08:15
12	July 21, 2020 at 02:26PM	ESP8266TEST1	07/08/2020_12:00;Yes	07/08/2020_08:00;No	07/08/2020_08:15

Fig. 4. *Data sent to Google Sheets*

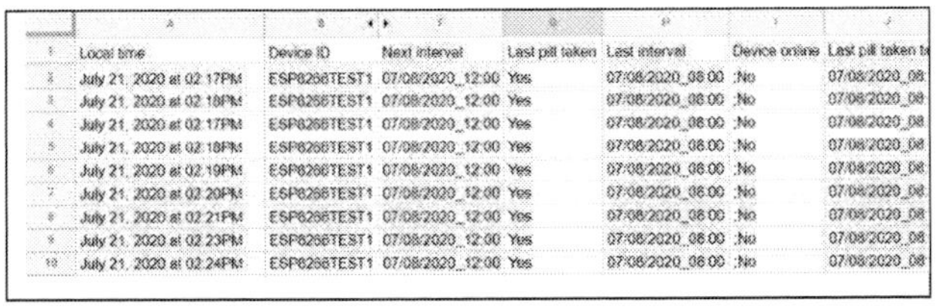

	Local time	Device ID	Next interval	Last pill taken	Last interval	Device online	Last pill taken t
2	July 21, 2020 at 02:17PM	ESP8266TEST1	07/08/2020_12:00	Yes	07/08/2020_08:00	;No	07/08/2020_08
3	July 21, 2020 at 02:18PM	ESP8266TEST1	07/08/2020_12:00	Yes	07/08/2020_08:00	;No	07/08/2020_08
4	July 21, 2020 at 02:17PM	ESP8266TEST1	07/08/2020_12:00	Yes	07/08/2020_08:00	;No	07/08/2020_08
5	July 21, 2020 at 02:18PM	ESP8266TEST1	07/08/2020_12:00	Yes	07/08/2020_08:00	;No	07/08/2020_08
6	July 21, 2020 at 02:19PM	ESP8266TEST1	07/08/2020_12:00	Yes	07/08/2020_08:00	;No	07/08/2020_08
7	July 21, 2020 at 02:20PM	ESP8266TEST1	07/08/2020_12:00	Yes	07/08/2020_08:00	;No	07/08/2020_08
8	July 21, 2020 at 02:21PM	ESP8266TEST1	07/08/2020_12:00	Yes	07/08/2020_08:00	;No	07/08/2020_08
9	July 21, 2020 at 02:23PM	ESP8266TEST1	07/08/2020_12:00	Yes	07/08/2020_08:00	;No	07/08/2020_08
10	July 21, 2020 at 02:24PM	ESP8266TEST1	07/08/2020_12:00	Yes	07/08/2020_08:00	;No	07/08/2020_08

Fig. 5. *Data decoded in Google Sheets*

Once the information is available on the Google Sheets platform, it is automatically imported by the site created in Google Sites (Fig. 6).

	Local Time	Next Interval	Last Interval	Last Pill Taken	Last Pill Taken Time	Device Online
2	07/08/2020 10:35	07/08/2020 12:00	07/08/2020 08:00	YES	7/8/2020 6:15:00	YES
3	07/08/2020 16:30					
4	07/08/2020 16:36					

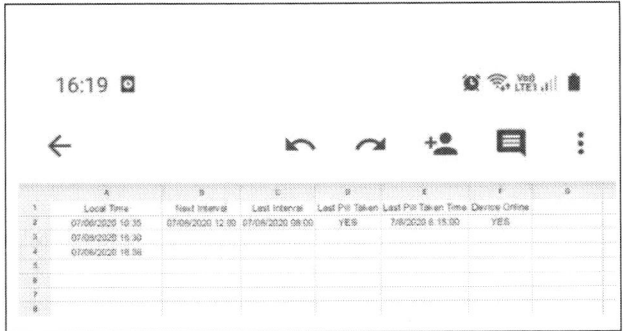

Fig. 6. *Visualization received data using Googel Sites*

REFERENCES

[1] JONA: The Journal of Nursing Administration, June 2009, 39(5):204-10;

[2] Medication Administration Curriculum, Department of Human Services Seniors and People with Disabilities, 2010;

[3] Medication Errors: Technical Series on Safer Primary Care, World Health Organization, 2016, 978-92-4-151164-3;

[4] Mattern, Friedemann; Floerkemeier, Christian. "From the Internet of Computers to the Internet of Things". ETH Zurich. pp. 1-6 Retrieved 23 October 2012;

[5] https://ifttt.com/;

[6] https://docs.google.com/;

[7] http://dweet.io/;

[8] https://freeboard.io/.

Design of a command and control system for an automatic pill dispenser

1st Madalin Vasile Moise
Center for Electronics Technology
and Interconnection Techniques
Polytechnic University of Bucharest
Bucharest, Romania
madalin.moise@cetti.ro

2nd Daniela-Mihaela Pavel
Center for Electronics Technology
and Interconnection Techniques
Polytechnic University of Bucharest
Bucharest, Romania
danielamihaelapavel@gmail.com

3rd Nicolae Elisei
Center for Electronics Technology
and Interconnection Techniques
Polytechnic University of Bucharest
Bucharest, Romania
elisei.nicolae08@gmail.com

Abstract—**The main purpose of this paper is to present a system designed to facilitate the distribution of pills to people in need and the work of medical staff with the help of technology. The project consists of several parts, it includes the design of a system which will use a programming interface with a keyboard and an LCD and design of a mechanical unit which will store the pill boxes and will deliver pills at a specific time and day. Our system is developed using an ATmega328PB microcontroller, interfaced with two solenoids, a stepper motor, two microswitches, one buzzer, a real time clock module and an LCD with 16 lines by 2 columns display. We have developed an algorithm that consists in setting the number of pills specified in a prescription and setting the interval to deliver those pills. The system alerts the user when it is time to take the pills, the alert will intensify if the pills are not removed from the tray. In order to keep the microcontroller peripheral use to minimum, the system benefits from a 5-button keyboard based on the principle of a voltage divider which will only use one ADC pin. The keyboard is used to program the number of pills and the specified intervals.**

Keywords— keyboard, stepper, pill, microcontroller

I. INTRODUCTION

With aging, some categories of people need help when it comes to taking medication, in hospitals doctors and nurses have to regularly administrate the treatment to patients. We wish to help patients prone to following their treatments incorrectly but also to provide a solution for hospitals to administrate the prescription without the need of regularly entering the hospital ward especially in case of a pandemic, minimizing the exposure time to a virus.

Several studies reveal that only about 50% of people who need medication follow their treatments as prescribed. [1] Out of those who do not take their medicine correctly, approximately half do so intentionally, while the other half do not realise that they are making a mistake or their treatment regimens are too complex for them to follow. Our device could solve the problem for patients who are unaware that they are not following their treatment correctly and it would reduce the percentage of people who do so intentionally with the help of audio alerts.

This decrease in incorrect treatment administration would also reduce consequences such as waste of medicine, disease progression, a lower quality of life, hospital visits and hospital admissions as a result of incorrect treatment administration. [2]

Incorrect treatment administration has negative effects not only for the patient, but also for the doctor or even the medical system the patient is a part of. This phenomenon can lead to higher risks of severe relapses, antibiotic resistant germs and even preventable hospitalizations, which makes it an important public health concern. [3]. The pill dispenser is designed to be waterproof and lockable to avoid overdose, so we can be sure that the person taking the drug has access only to the prescribed dose.

II. PLATFORM DEVELOPING

The system was created in the following steps. In the first stage, we developed a modular platform printed circuit board (PCB) with a microprocessor which allows us to balance power consumption and processing speed. In addition, we can control all sensors and peripherals.

The second stage consists in designing the body of the pill dispenser. We also developed a 5-button keyboard, based on the working principle of a resistive network.

For the last stage, we 3D printed the pill distributor, its housing and manufactured the PCB board.

A. System overview

We have developed the embedded system on an ATmega382P microprocessor that has a fast response and is easy to understand, thus allowing further development. The user has the possibility to control the system using the keyboard. Any information will be displayed on a LCD screen.

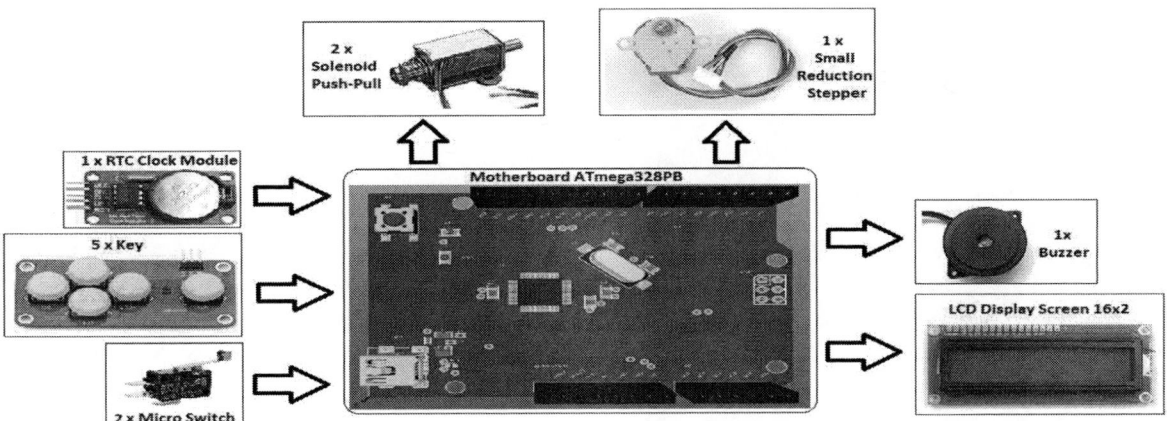

Fig. 1. System Block Diagram

Various alerts can be set in order to notify the user about the pills that are placed in the tray.

System components as seen in the System Block Diagram (Fig. 1):

o ATmega328PB - microprocessor

o Micro Switch – to detect the removal of the pill tray

o LCD display – displays various information about pills, time intervals, instructions

o Fig. 3. Printed Circuit Board

o Five button keypad – used for programming

o Small Reduction Stepper motor – rotates the shaft of the dispenser

o Solenoid Push-Pull – two of them, one for every level; opens the lid of every slot

o Buzzer - signaling device

o Real time clock module - used for keeping track of the current time

B. Printed circuit board design

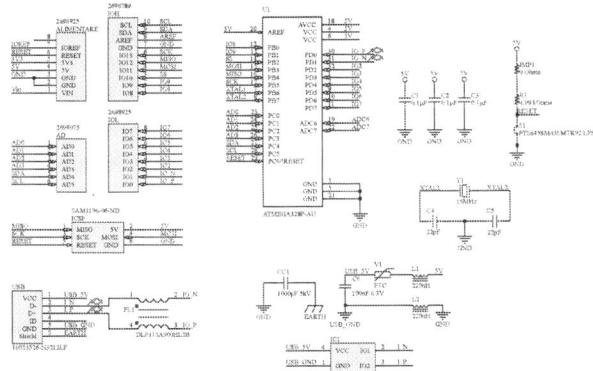

Fig. 2. Circuit Schematic

The ATmega328PB microprocessor will meet user's continuous demand for efficient power usage, being one of the most popular AVM microprocessor from Atmel. It has a maximum operating frequency of 16 MHz, a memory of 32KB, an EEPROM memory of 1 KB, a 2 KB SRAM memory and various analog and digital peripherals.

We chose to minimize the complexity of our board, facilitating immediate results. The system was designed in an user friendly manner, making this an exceptionally easy choice for anyone who does not have experience in this field.

We developed a printed circuit board (PCB) which is capable of increase functionalities by attaching additional hardware.

The existing connectors provide access to the input / output pins of the microcontroller.

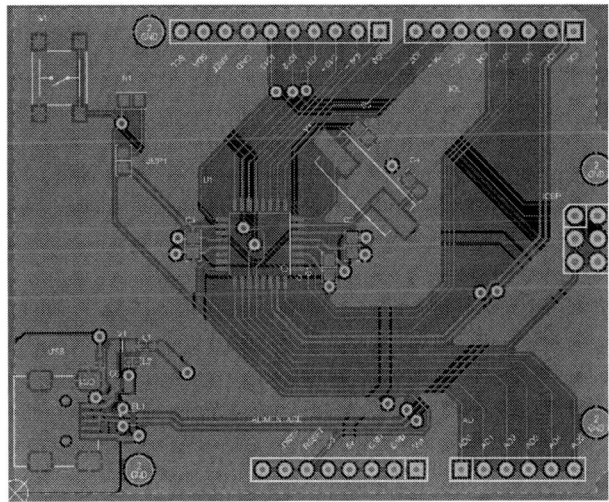

Fig. 3. Printed Circuit Board design

C. Pill dispenser housing design

Fig. 4. Illustrates a rendering of the design of the pill dispenser with transparent housing.

Fig. 4. Pill dispenser design

The dispenser is designed to accommodate 16 pill boxes on two levels of 8 boxes each. Each box of pills is designed so that it can only be introduced in the correct position, removing possibilities for human errors.

In Fig. 5 the red lid covering the slot at the bottom of the box is operated by a solenoid to dispense pills. Each pill box has an interior diameter of 2.8 cm and a height of 7.4 cm, so the volume is 182.262 cubic centimeters.

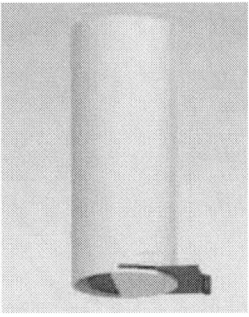

Fig. 5. Pill box design

III. SOFTWARE DEVELOPMENT

The system continuously reads the current time and date and checks for active intervals during that time. When it finds a match, the system notifies the user that it is starting to prepare the dosage by displaying a message on the LCD and generating a sound with the buzzer. The organigram of the main function of the system can be seen in Fig. 6.

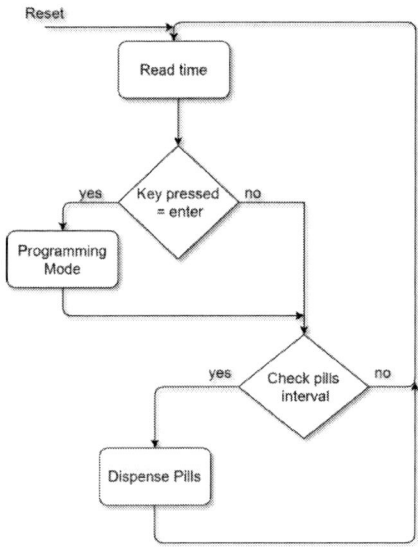

Fig. 6. System organigram

IV. RESULTS

The experimental results were obtained by developing the system and using it to dispense pills at certain predefined intervals. Different time intervals and different dosages were set for different days using the 5-button keyboard.

A 3D printer was used to create the housing and internal elements of the pill dispenser (Fig. 7).

Fig. 7. 3D printed Pill Dispenser

In idle, the system displays a message similar to the one shown in Fig. 8.

Fig. 8. Idle message

When it is time to administer pills, the system displays a message, informing the user it is preparing the dose (Fig. 9).

Fig. 9. Message informing the user that the dose is being prepared

The system uses the stepper motor to select the correct pill box by making a 45° rotation for each consecutive box (16 pill boxes distributed in 2 rows stacked on top of each other arranged in a circle, see Fig. 4.), which means that the motor needs to execute 256 steps to move between consecutive pill boxes (2,048 steps for a full rotation times 45 degrees for consecutive pill boxes). Once the box is in the correct location, a solenoid operates the lid of one of the two boxes to dispense pills from the correct one into the collector below.

If the pills are not removed from the collector in a programmed amount of time, the system changes the sound tone to a more irritating one and displays a message on the screen (Fig. 10).

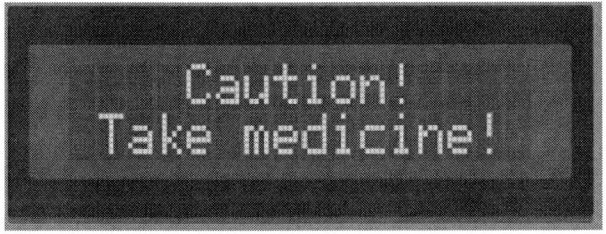

Fig. 10. Warning message shown if the pills were not removed from the tray

If the user wishes to change the schedule for administering the pills, he can use the keypad. By pressing enter, the system enters programming mode, where the user will select the day, the number of intervals in a day and the number of pills from each pill box, in this order. The system is programmed to admit a maximum of 9 intervals per day and up to 10 pills from each box. An example of schedule for one day (i.e. Thursday) is illustrated in Fig. 11.

D a y	Interval (0-9)	INTERVAL TIME	PILL BOX (1-16)															
			1	2	3	4	5	6	7	8	9	10	11	12	13	14	15	16
T h u r s d a y	1	00:00	4	0	0	3	1	2	0	0	0	4	0	0	0	0	0	0
	2	08:00	1	1	1	4	2	3	1	1	7	0	0	0	0	1	4	0
	3	12:00	4	0	0	3	1	2	0	0	0	4	0	0	0	0	0	0
	4	16:00	0	3	2	0	5	0	4	0	0	2	0	1	0	4	0	0
	5	18:00	4	0	0	3	0	1	0	1	0	0	0	0	1	0	0	2
	6	-	0	0	0	0	0	0	0	0	0	0	0	0	0	0	0	0
	7	-	0	0	0	0	0	0	0	0	0	0	0	0	0	0	0	0
	8	-	0	0	0	0	0	0	0	0	0	0	0	0	0	0	0	0
	9	-	0	0	0	0	0	0	0	0	0	0	0	0	0	0	0	0

Fig. 11. Example of pills schedule for one day

The LCD display paired with the five button keypad makes creating a schedule for dispensing pills easy and intuitive for the user. In order to set the number of intervals, the corresponding time and the pill box and number of pills for a day the user begins by choosing the day he wishes to programme (Fig. 12), then setting the number of intervals for that day (Fig. 13), using the up and down keys to increment or decrement this number. The right key is used to go to the next step of the programming mode, while enter quits programming mode, saving the changes.

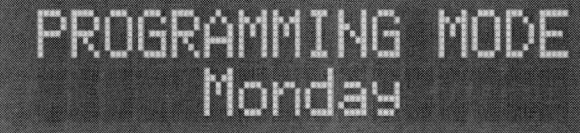

Fig. 12. Choosing the day to programme

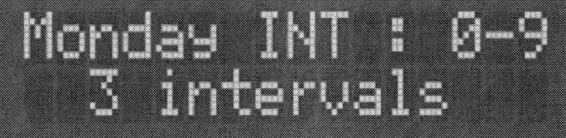

Fig. 13. Setting the number of intervals

After setting the number of intervals, the system will ask for the time of the intervals (Fig. 14).

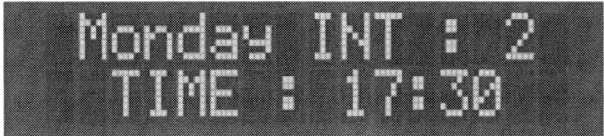

Fig. 14. Setting the time for an interval

Once the time of the intervals is known, the last step is to set the correct pill box and the number of pills to be dispensed (Fig. 15).

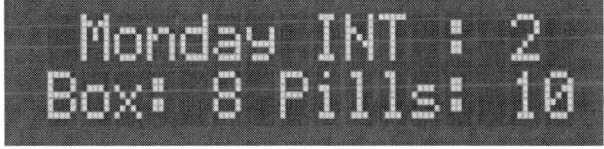

Fig. 15. Choosing the pill box and the number of pills for an interval

REFERENCES

[1] World Health Organization. 2003. Adherence to long term therapies: evidence for action [online]. Available at http://www.who.int/chronic_conditions/en/adherence_report.pdf;

[2] Sullivan S, Kreling D, Hazlet T. Noncompliance with medication regimens and subsequent hospitalizations: A literature analysis and cost of hospitalization estimate. J Res Pharmaceut Econ 1990;2:19-33;

[3] Sokol MC, McGuigan KA, Verbrugge RR, Epstein RS. Impact of medication adherence on hospitalization risk and healthcare cost. Med Care 2005. Jun;43(6):521-530 10.1097/01.mlr.0000163641.86870.af

Design of touch ECG detection system based on STM32 and Android mobile phone

Junzhuo Zhou[1]
[1] School of Microelectronics
Southern University of Science and Technology
Shenzhen, China
zaozao12138@163.com

Ang Li[2]
[2] School of Information Science and Technology
University of Science and Technology of China
Hefei, China
ansonlee@mail.ustc.edu.cn

Zeying Tian[3]
[3] School of Electrical Engineering
Xi'an Jiaotong University
Xi'an, China
3127233223@qq.com

Corresponding Author: *Zeying Tian*
Email:3127233223@qq.com

Abstract—Cardiovascular disease is one of the main diseases that endanger human health, especially because of the sudden and transient characteristics of cardiovascular disease. For patients who have been discharged from the hospital, the abnormality of ECG cannot be detected and recorded in time, which will make the doctor lose The most valuable medical data makes it impossible to make accurate judgments on subsequent treatment. This paper proposes a touch ECG detection system based on STM32 and Android mobile phones. The ECG detection system can obtain the human body's electrocardiogram only by touching the electrode pads with both hands and quickly analyze and diagnose, avoiding the traditional ECG After a series of complicated operations such as wearing electrodes, the timely capture of abnormal ECG is of great significance to the development of medical electronics.

Keywords—Cardiovascular; touch ECG detection system; medical electronics

I. INTRODUCTION

In recent years, with the continuous deepening of urbanization and population aging, cardiovascular disease has become the main killer of human health. The incidence of cardiovascular disease is increasing year by year, and the affected population is gradually younger. This makes people have to pay attention to the prevention and treatment of cardiovascular diseases. According to the research report [1], the prevention and treatment of cardiovascular diseases in my country has achieved initial results, but it still faces many severe challenges. The social burden brought by cardiovascular diseases in China has become a major public health problem.

Electrocardiograph (ECG) is a graph that uses an electrocardiograph to record the potential changes of the heart during each cardiac cycle from the human body surface. A high-quality electrocardiogram can reflect the health of the human heart and is an important basis for the diagnosis of abnormal heart rhythms, atrial hypertrophy, and myocardial ischemia. Traditional electrocardiographs require the patient to lie flat on a bed and wear electrode pads on the upper limbs, lower limbs and chest when measuring the human body's

electrocardiogram signal. After a period of time, the electrocardiogram will be printed on the electrocardiograph. Due to the suddenness of cardiovascular disease, the large size of the equipment and the complicated operation, although this measurement method is powerful, but the cumbersome measurement process makes the electrocardiograph unable to capture the abnormal electrocardiogram in time, thus wasting the gold of detection time.

With the continuous progress of society, portable medical equipment has also achieved great development in the field of medical electronics. Especially with the rapid development and widespread use of smart phones, the research on portable medical devices using smart phones as a platform has become a hot spot in the field of electrocardiography. At the same time, data shows that by the end of 2017 [2], the market share of mobile medical has reached 20.09 billion yuan, a year-on-year increase of 90.2%. It is estimated that by the end of 2020, the market size of mobile healthcare will exceed RMB 40 billion. It is foreseeable that the combination of wearable medical devices and smart phones will be the main direction of the development of medical devices in the future. Among many smart phone systems, the Android system is favored by researchers due to its open source and free nature. The design of this article is also based on the Android platform.

This paper designs a touch ECG detection system based on STM32 and Android mobile phones. It solves the troublesome problem of traditional ECG equipment with electrodes and enables patients to measure their ECG anytime and anywhere. The system can collect the electrocardiogram of the human body only by touching the electrodes on the circuit board with both hands, and can display the electrocardiogram and heart rate of the human body in real time. The design of this article is a real-time ECG detection system composed of a touch ECG acquisition module, Bluetooth module and Android smart phone. The system uses STM32F107 as the control core. The hardware circuit of the system consists of a data acquisition part and a core processor. The function of the system is to amplify and filter the ECG signal, and then convert it into a digital signal through an ADC analog-to-digital converter. At the same time, the SG and IIR digital filtering of the ECG signal and the calculation of the heart rate are realized on the STM32. Finally, data such as ECG and heart rate are sent to the smart phone via Bluetooth, and ECG and heart rate are displayed on the mobile terminal (APP).

II. INTRODUCTION TO ECG SIGNAL

The heart is composed of a large number of cardiomyocytes. When the cardiomyocytes are stimulated, the cardiomyocytes will change the potential on both sides of the cell membrane, from the initial "internal negative and external positive" to "internal positive and external negative" process, it is called the process of depolarization, the process of restoring the initial potential after the excitement is conducted is called repolarization. The depolarization and repolarization of multiple cardiomyocytes in the human body form the ECG signal [3].

ECG signal is relatively weak and easily affected by the environment[4]. It mainly has the following characteristics: (1) Weakness: its normal amplitude range is usually 0.05~5mV, and the typical value is 1mV. (2) Low frequency: The frequency range is 0.05Hz~100Hz, and the energy is mainly concentrated in 0.25Hz~33Hz. (3) Randomness and instability: ECG signals will change with the human body's environment, and there are random and unstable differences [5]. Although the ECG waveform recorded by different leads is different, it basically includes P wave, QRS wave and T wave [6]. Sometimes a small U wave will appear after the T wave. The ECG waveform is shown in Fig.1:

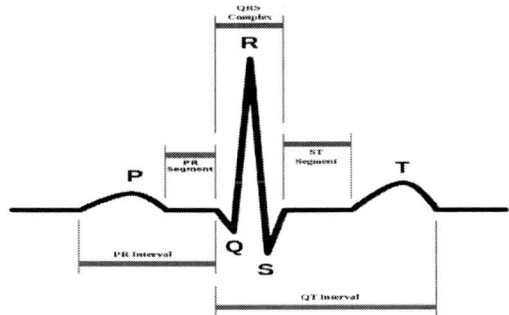

Fig. 1. ECG signal waveform.

- P wave: shows the potential change process of the atrium. Normal people's P wave duration is generally 0.06~0.11s, and the amplitude of the wave is generally 0.05~0.25mV [7].

- P-R interval: represents the time from atrium to ventricle to start depolarization. Normal time is generally: 0.12~0.20s [8].

- QRS complex: represents the depolarization process of the left and right ventricles. The width of the entire QRS complex is 0.06~0.12s, and the amplitude does not exceed 5mV. The first wave is the downward Q wave, then the upward R wave, and finally the downward S wave. The QRS wave is the most important part of the electrocardiogram, reflecting the

health of the heart, and is the focus of medical research [9].

- S-T band: the band between the end of QRS wave and the beginning of T wave. It represents the state from the end of ventricular depolarization to the beginning of repolarization. At this time, the potential difference is very small, the rising amplitude does not exceed the baseline 0.1mV, and the decrease is not less than 0.05mV. Therefore, the ST segment is generally flush with the baseline. If the ST segment amplitude deviates too much, it means cardiac ischemia and myocardial cell damage [10].

- T wave: represents the potential change during the repolarization of the ventricle. The duration is generally 0.05~0.25s. Its waveform is relatively round and blunt. Generally, the main wave direction of T wave and QRS wave is the same [11].

- U wave: represents the subsequent potential of the ventricle. This low-width wave generally occurs 0.02~0.04s after the T wave, and the direction is consistent with the T wave. The duration is generally 0.1~0.3s, and the amplitude is generally not greater than 0.05mV [12].

III. SYSTEM OVERALL DESIGN

The overall framework of the ECG detection system is shown in Fig.2 acquisition circuit, ECG signal processing module with STM32 as the core, Bluetooth module, and the upper computer Android mobile phone client.

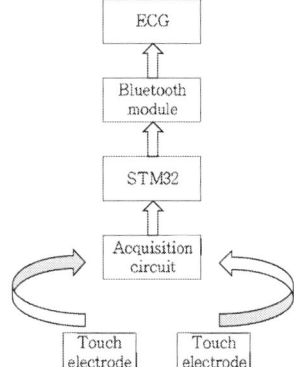

Fig. 2. System overall structure frame diagram.

- Touch electrode: it is composed of two common PCB copper-plated pads. The purpose is to increase the contact area when the left and right hands touch the electrodes and reduce the impedance when touching with both hands.

- Acquisition circuit: the body fluids and tissues around the heart can conduct electricity and generate weak electrical signals and reflect them on the body surface. The ECG signals can be introduced by touching the electrodes with both hands, and then the weak signals are amplified by the corresponding amplifier circuit, and finally filtered through simple analog filtering. The signal is transmitted to the ADC for further processing.

- STM32 core processing module: the pre-processed analog signal is collected, held, quantized and encoded into a digital signal by ADC, and then filtered through the designed IIR digital notch filter to filter out 50H and low frequency interference, and then further smooth the signal to obtain a clean ECG, the algorithm that realizes heart rate calculation. Finally, ECG, heart rate and other data are sent to the mobile phone.

- Bluetooth module: it can send ECG data and heart rate data to the mobile phone for analysis through a certain protocol.

- Android mobile phone client: Using multi-threading technology on the Android Studio platform to analyze the data transmitted from the lower computer to the mobile phone and display the ECG and heart rate.

A. Design of ECG analog acquisition circuit module

The human body's ECG signal is a low-frequency and weak signal, its frequency range is 0.05-100 Hz, the energy concentration frequency is between 0.25-30 Hz, the normal amplitude range is 0.05-5 mV, and the typical value is 1mV.

In addition, the ECG signal has instability and randomness, which will change with the environment and behavior of the human body, resulting in differences, resulting in a lot of interference and noise in the ECG signal, including myoelectric interference, engineering Frequency interference, etc. Therefore, in order to effectively extract the ECG signal of the human body, the analog acquisition circuit module needs to fully consider the low frequency, weakness, and instability of the ECG signal, and filter out the influence of various noises of the ECG signal.

This system uses the electrode connection of the left and right hands to obtain the body surface potential of the human body, and uses the potential difference to obtain the electrocardiogram of the human body. In order to amplify the ECG signal, a differential input method is designed to amplify the left and right hand electrode signals. The specific ECG analog acquisition circuit is shown in Fig.3:

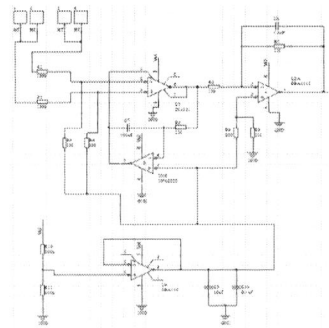

Fig. 3. ECG analog acquisition circuit diagram.

Considering that the ECG signals collected by the left and right fingers are relatively weak, the acquisition circuit uses four positive and negative electrodes, HT1, HT2, HT3, and

HT4, to measure the electrical signals of the two fingers on the left and right. First, the front-end amplifier INA321 is used for 5 times amplification , And then use the main operational amplifier chip OPA2325 (U2A) to further amplify by about 100 times, input into the STM32 main chip from the port ECG_IN10 to filter the ECG signal.

B. Software Design of Digital Filter

Before designing the filter, we must first find out the source of noise in the ECG signal. ECG noise sources are divided into the following categories:

- Power frequency interference: Power frequency interference is a kind of interference caused by the power system, which is composed of 50hz and its harmonics, and the amplitude is about 50% of ECG

- Baseline drift: It is caused by respiration. It is caused by the thermal noise amplified by the impedance instrument between the motor and the human skin when the object is breathing during the test. The frequency is about 0.15-0.3hz. There is a 15% change in ECG amplitude during respiration.

- Electromyographic interference: Muscle tremors originate from the human body. Electromyographic interference generates millivolt potential, which can be regarded as instantaneous Gaussian zero-mean band-limited noise .

Power frequency interference can be processed with a 50hz notch filter, and baseline drift can be eliminated by a high-pass filter at frequencies below 0.5hz. Commonly used fixed frequency designs include IIR filters and FIR filters, among which FIR filters have a good linear phase. But under the same performance conditions, the order is higher than that of the IIR filter, and the amount of calculation is large. Therefore, the power frequency notch and high-pass filtering of the original design are designed with IIR filters.

With the help of MATLAB's FDATOOL toolkit, the relevant parameters of the direct type II 50HZIIR notch filter can be directly generated, as shown in the figure.

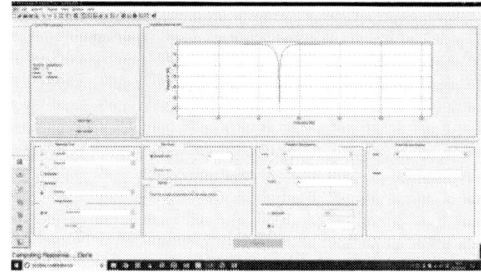

Fig. 4. Example of a figure caption. *(figure caption)*

Get the relevant parameters of the transfer function through the automatic header file generation function of the toolkit. The transfer function is as follows:

$$H_{(z)} = \frac{0.902398 - 0.5577128Z^{-1} + 0.90239778Z^{-2}}{1 - 0.5577128Z^{-1} + 0.8047958Z^{-2}} \qquad (1)$$

Also using Matlab's Fdatool toolkit to design a 0.5hz second-order IIR high-pass filter, derive the filter coefficients, and get the transfer function as:

$$H_{(z)} = \frac{0.991153*(1-2*Z^{-1}-Z^{-2})}{1-1.982289*Z^{-1}+0.982385*Z^{-2}} \quad (2)$$

Although it has gone through 0.5HZ high-pass filtering and 50HZ notch, the signal still has certain glitches. In order to obtain a smooth signal, the signal needs to be smoothed and filtered. In this design, SG smoothing filter with length 11 and order 5 is used.

C. Heart rate calculation

Heart rate is an important data that can be derived from the human body's ECG signal [13]. A dynamic threshold method is used in the calculation of heart rate. A segment of sampled data is used to obtain the maximum and minimum values before taking the data. The data obtained after averaging is the average extreme value of the heart rate, and then 3/5 of the average extreme value is taken as the threshold value of the sampled data, which are the minimum threshold and the maximum threshold respectively, and then the maximum and minimum values of the data are taken respectively, using the threshold and The maximum and minimum data comparison, when the maximum value is greater than the maximum threshold, continue to determine if the minimum value is less than the minimum threshold count +1 at this time, and the maximum and minimum values are cleared at the same time, the judgment continues until the entire segment of data is traversed [14]. In this way, the number of heartbeats of the data can be obtained.

D. System Simulation

In order to verify the effect of the ECG acquisition system in actual operation, this design uses TINA simulation software to verify whether the output waveform, magnification and bandwidth of the acquisition system meet the requirements by building an analog acquisition circuit model, as shown in Fig.5.

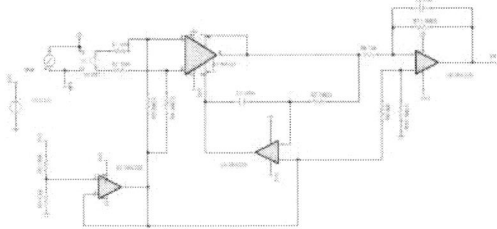

Fig. 5. ECG analog acquisition circuit model.

Since the frequency of the ECG signal is between 0.05Hz-100Hz, the typical waveform peak value is 1mV, so set the signal source of the ECG acquisition circuit module to a sine waveform with an amplitude of 1mV and a frequency of 30Hz. The following figure 2-2 shows the ECG The setting interface of the signal source of the circuit module, after running the circuit in TINA, the amplified signal waveform is obtained at the output terminal Vo, as shown in Fig.6.

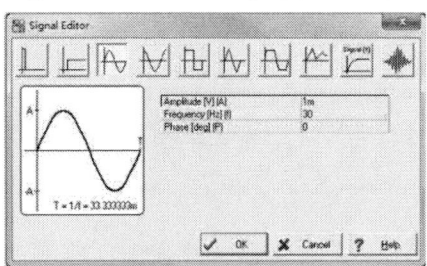

Fig. 6. Signal source design interface.

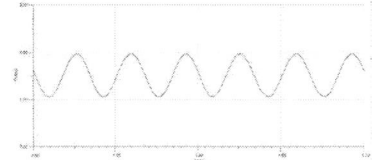

Fig. 7. Waveform diagram of circuit output.

As shown in Fig.7, this sine wave is a waveform amplified by the signal acquisition system. The peak voltage is as high as 1.98V, which is between 0-3.3V. It can be collected and input by the STM32 main control chip to meet the set requirements. In addition, due to the existence of the voltage follower OPA330 in the acquisition circuit module, the overall signal source voltage is increased by 1.52V. According to the magnification formula, the acquisition system gain is A=(1.98-1.52)V/1mV=460. It is impossible to fully work in an ideal state. There is resistance in the filter capacitor in the op amp module, which will affect the signal amplification to a certain extent, so the gain A=460 is close to 500 in the ideal state, which is acceptable in this design.

Fig.8 is the amplitude-frequency characteristic curve of the simulated acquisition system at the time of operation obtained by simulation. From the set point a and point b, it can be seen that the frequency of point a is 1.55 Hz when the system maximum gain is near, b The frequency of the point is 102.24Hz, and the frequency of the ECG signal is between 0.05-100Hz, so the bandwidth setting of the acquisition circuit can be verified to meet the requirements.

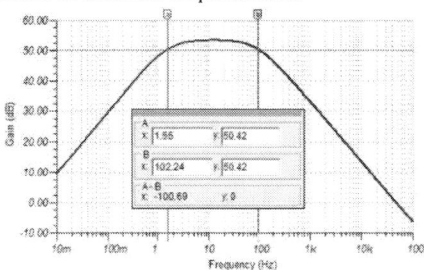

Fig. 8. Amplitude-frequency characteristic curve of analog acquisition circuit.

Virtually create two analog serial ports on the computer through the serial port simulator to connect the host computer with the serial debugging assistant, and use the serial debugging assistant to generate a set of random data and send it to the host computer to verify the effect of SG smoothing and filtering. The effect is satisfactory. It can be seen from

Fig.9 that the waveform after smoothing filtering is much smoother than the original waveform.

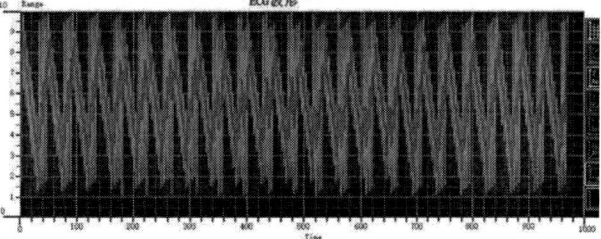

Fig. 9. SG smooth filtering effect diagram.

IV. RESULTS

In order to verify the availability of the system, first test each module and its function of the entire system. First measure the unprocessed ECG signals of the human body, as shown in Fig.10 below.

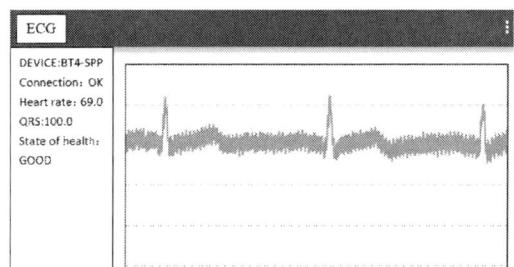

Fig. 10. Unprocessed human ECG signals.

After software filtering, the final ECG is shown in Fig.11 below.

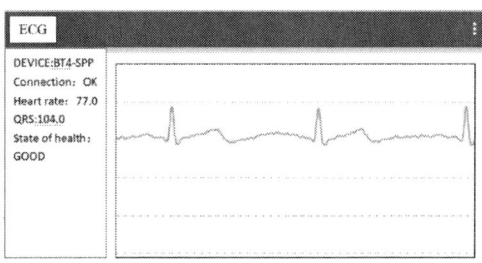

Fig. 11. The processed electrocardiogram of human body.

As can be seen from the above figure, this system can accurately reflect the user's electrocardiogram and physical health status on the mobile phone app, which is of great significance for the prevention of cardiovascular and cerebrovascular patients.

V. CONCLUSION

This article designs and develops a touch ECG detection system based on STM32 and Android mobile phones based on the system design goals and using the architecture of STM32 and Android mobile phones working together. The system consists of an ECG acquisition and processing module and a mobile client application. It can measure the human ECG anytime and anywhere through external electrodes, and can display and analyze the ECG in real time through the mobile client.

REFERENCES

[1] Chen Weiwei, Gao Runlin, Liu Lisheng, et al. Summary of "Chinese Cardiovascular Disease Report 2017"[J].Chinese Journal of Circulation,2018, 33(01):1-8.

[2] Mobile medical market scale[EB/OL].http://www.chinabgao.com/,2018-03-05.

[3] Abdolrahman Peimankar,Sadasivan Puthusserypady. DENS-ECG: A deep learning approach for ECG signal delineation[J]. Expert Systems With Applications,2021,165.

[4] Aboli N. Londhe,Mithilesh Atulkar. Semantic segmentation of ECG waves using hybrid channel-mix convolutional and bidirectional LSTM[J]. Biomedical Signal Processing and Control,2021,63.

[5] Navdeep Prashar,Meenakshi Sood,Shruti Jain. Design and implementation of a robust noise removal system in ECG signals using dual-tree complex wavelet transform[J]. Biomedical Signal Processing and Control,2021,63.

[6] Peter Magnusson,Adam Lyren,Gustav Mattsson. Diagnostic yield of chest and thumb ECG after cryptogenic stroke, Transient ECG Assessment in Stroke Evaluation (TEASE): an observational trial[J]. BMJ Open,2020,10(9).

[7] Xiaolong Zhai,Zhanhong Zhou,Chung Tin. Semi-supervised learning for ECG classification without patient-specific labeled data[J]. Expert Systems With Applications,2020,158.

[8] Georgios Petmezas,Kostas Haris,Leandros Stefanopoulos,Vassilis Kilintzis,Andreas Tzavelis,John A Rogers,Aggelos K Katsaggelos,Nicos Maglaveras. Automated Atrial Fibrillation Detection using a Hybrid CNN-LSTM Network on Imbalanced ECG Datasets[J]. Biomedical Signal Processing and Control,2021,63.

[9] Mathis K. Stokke,Anna I. Castrini,Meriam Åström Aneq,Henrik Kjærulf Jensen,Trine Madsen,Jim Hansen,Henning Bundgaard,Thomas Gilljam,Pyotr G. Platonov,Jesper Hastrup Svendsen,Thor Edvardsen,Kristina H. Haugaa. Absence of ECG Task Force Criteria does not rule out structural changes in genotype positive ARVC patients[J]. International Journal of Cardiology,2020,317.

[10] Zhu Zeyang,Li Jianhua,Zhang Shuang,Geng Ning,Xu Lisheng,Greenwald Steve E. Quality evaluation of signals collected by portable ECG devices using dimensionality reduction and flexible model integration.[J]. Physiological measurement,2020.

[11] Francesco Ancona. The dynamic of ECG in Takotsubo Syndrome and myocardial infarction: the long quest for an intriguing non-invasive differential diagnosis between ischemic syndromes.[J]. International journal of cardiology,2020.

[12] Dural Muhammet,van Stipdonk Antonius M W,Salden Floor C W M,Ter Horst Iris,Crijns Harry J G M,Meine Mathias,Maass Alexander H,Kloosterman Mariëlle,Vernooy Kevin. Association of ECG characteristics with clinical and echocardiographic outcome to CRT in a non-LBBB patient population.[J]. Journal of interventional cardiac electrophysiology : an international journal of arrhythmias and pacing,2020.

[13] Satti Afraiz Tariq,Park Jinsoo,Park Jangwoong,Kim Hansang,Cho Sungbo. Fabrication of Parylene-Coated Microneedle Array Electrode for Wearable ECG Device.[J]. Sensors (Basel, Switzerland),2020,20(18).

[14] . ZOLL Medical Corporation; Patent Application Titled "Ecg And Defibrillator Electrode Detection And Tracking System And Method" Published Online (USPTO 20200261712)[J]. Medical Patent Business Week,2020.

Sensorless BLDC Control Method

A. Zîrnea[1], G. Bărbulescu[2], M. Păunoiu[2], C. Pop[2], N. Codreanu[1] and M. Enăchescu[1]

[1] Faculty of Electronics, Telecommunications and Information Technology, University "Politehnica" of Bucharest, Romania

[2] Microchip Technology S.R.L., Bucharest, Romania

alexzirnea@gmail.com

Abstract—**This paper proposes a new control method for BLDC motors using sensorless trapezoidal control capable of regenerative braking, designed to be used on low-power and moderate computational power microcontrollers. It exhibits good performance on systems where the load does not have abrupt changes over time and the requirements for efficiency are not very demanding. Trapezoidal control targets mainly sub-kilowatt systems. Due to increased power dissipation, going beyond this figure is not feasible.**

Keywords—*BLDC motor; Bipolar Switching; Regenerative Braking; Trapezoidal Control*

I. INTRODUCTION

During the past years, multiple power-efficient drive systems were developed to target not only the industrial applications, but also mass-market by the increasing number of electrical vehicles which require more and more range. Most of these implementations depend of additional feedback to detect the position of the rotor, which is an absolute requirement in brushless motor control, as, unlike DC motors, each of the three phase of a brushless motor needs a different potential according to the above-mentioned rotor position. These sensors take the form of Hall sensors which are placed inside the motor and detect the magnetic field of the permanent magnets. The present solution does not use any sensor, eliminating the extra cost associated with sensored designs.

II. CONTROL METHOD

A. Trapezoidal control (Six-step)

Trapezoidal control represents an approximation of sinusoidal control. It has a significant amount of harmonics which gets dissipated inside the motor's stator coils as heat.

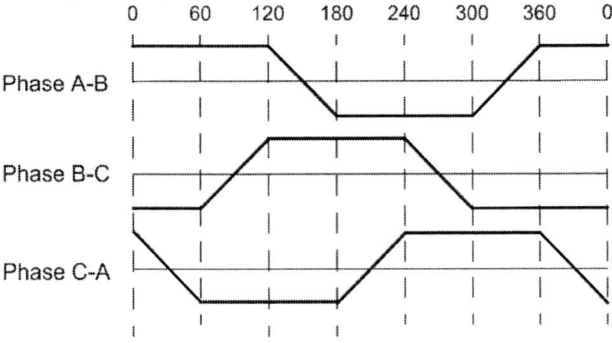

Fig. 1. Trapezoidal waveform [1]

It is also known as six-step control as it requires precisely six PWM sequences to complete one electrical cycle. The main characteristic of the trapezoidal control is that two of the motor

phases are driven and the third is left floating. Thus, if the motor is spinning, then, according to Faraday's law, a back-electromotive voltage (BEMF) is induced in the floating phase. By monitoring the voltage on the phase, the position of the motor can be determined with enough accuracy. The point of interest in this control scheme is the zero-cross. That is, for a split-supply system, the point where the voltage crosses the zero level. In this case, the system runs on a single DC supply, so the entire measurement reference has an offset of VBUS/2.

B. Thirty-degrees timing

After detecting the zero-cross point, the system must wait for 30 electrical degrees in order to perform commutation to the next step. Unlike sinusoidal control, where the control is much more robust in terms of maximum torque, trapezoidal control lacks torque in high efficiency applications. Thus, a compromise must be done between the efficiency of the overall system and the torque delivered by the motor. This tradeoff is chosen in accordance with the desired behavior of the system. Any alteration to the thirty-degrees timing is called advance (when the commutation happens earlier than it should) and retard (when the commutation happens later).

A side effect of adding an advance angle to the commutation is that the motor has a higher speed since the applied magnetic field is constantly ahead of that of the motor.

C. Stall detection

In a normal operating environment, it may appear an unwanted condition where the motor can not deliver the required power to the load (that is, either the load is over the capacity of the motor, or a mechanical failure occurred). It is mandatory to stop the motor as quickly as possible to avoid any damage to the power stage, motor, or the supply system. This can be accomplished in a few ways:

1. Have a maximum time frame for the zero-cross to be detected (timeout mode). In this case, the zero-crossed fails to be detected due to an insufficient BEMF amplitude or other factors.

2. Detect the absolute variation (delta) of the current time of the detected zero-cross point relative to the previous one. Too high of a value could indicate that the detected zero-cross point is a false one (one of the factors could be electromagnetic noise), or desynchronization to the BEMF

3. During a desynchronization or a stall condition, the current through the windings will rise dramatically and it can be used as a trigger to stop the PWM to the power stage.

D. BEMF Amplitude

The BEMF amplitude is highly dependent on the angular speed of the rotor. The equivalent circuit for a driven 3 phase motor looks like in the figure below. Phase A is undriven, and thus, the voltage of interest is for that specific phase. For a perfectly symmetrical motor, the two resistances and inductances of the driven coils are equal, thus, the common voltage (COM) is half of the supply voltage. As mentioned above, the voltage swings from 0 to VBUS, thus, the BEMF induced in the "A" coil will be contained in the interval [-VBUS/2, VBUS/2], in normal operating conditions.

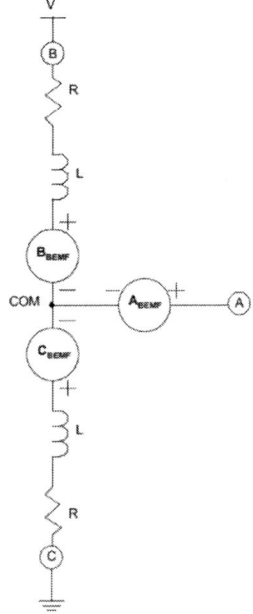

Fig. 2. Equivalent three-phase motor circuit [2]

The actual voltage measured on terminal "A" will be expressed by the formula:

$$BEMF = \frac{RPM}{K_e} \qquad (1)$$

Where RPM is the mechanical rotations-per-minute of the rotor with no load, and K_e is a motor-characteristic parameter called back-EMF constant.

As the zero-cross depends on the BEMF, too low of an amplitude will be comparable to captured electromagnetic noise which, will not be discernable to decide if a zero-cross happened, and will also require high-performance hardware. Thus, at zero speed and low speeds the position of the rotor is not available. This calls for an open-loop ramp-up sequence which brings up the motor RPM to a minimum value where the BEMF amplitude is enough to be used in the zero-cross detection.

E. BEMF filtering

In a usual application, the captured waveforms are noisy due to the magnetic coupling between the coils as well as the control method used. Using unipolar switching and a high time-constant filter, the filter will behave as expected in low RPMs, but at high speeds, the introduced delay will introduce phase shifts which will lead to the desynchronization of the motor. The solution to this problem is to avoid using unipolar switching, as is induces the PWM into the free phase, and switch instead to bipolar switching, as is does not suffer from this problem. Instead, the induced voltage on the free phase is cancelled by the complementary state, thus, only high-frequencies spikes are present, comparable to the dead-time used, which are easily filtered. This eliminates the need for using heavy filtering, and the BEMF is as noise-free as possible, regardless of the applied duty-cycle.

F. Four-quadrant Operation

By using the bipolar switching method, the system can run in a "four-quadrant" system.

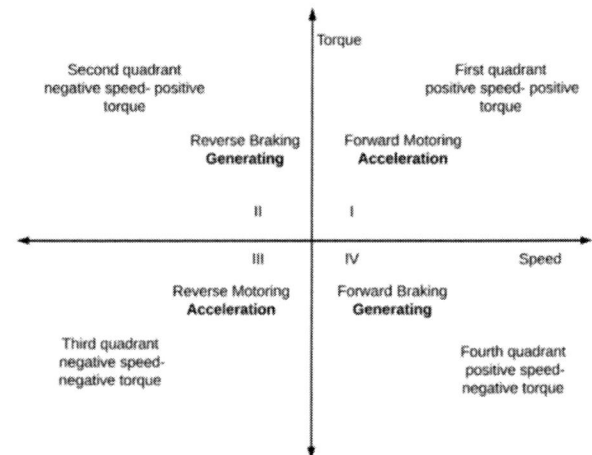

Fig. 3. Four quadrant operation [3]

This allows the motor to brake when the applied duty cycle is lower than the corresponding duty cycle associated to the current RPM. This makes the system run in quadrants I and IV when the duty cycle is between 51% and 100% and, correspondingly, for duty cycle between 0 and 40% when the system runs in quadrants II and III. The maximum braking torque depends on the resistance of the coils, the absolute difference in the set RPM and actual RPM, and in the ability of the power supply to sink the current. As the braking is regenerative, the system behaves like a boost circuit, effectively increasing the voltage on the bus. This change of voltage makes the voltage higher than the supply and the current through the system reverses and is pushed back into the supply. Therefore, for this implementation a battery is needed, or an additional monitoring circuit to make sure that the voltage on the bus is not higher than the maximum input voltage, and dissipate it as heat, if storing is not an option.

III. HARDWARE IMPLEMENTATION

A. System overview

The system is built around a microcontroller which handles all the required control signals going to the inverter. The inverter is made-up of three half-bridges driven by three MOSFET drivers. The microcontroller gets its input from the motor's phases which voltage levels are reduced and filtered. Additionally, the system has a current measurement to avoid damaging the inverter or other parts. The interface with the user will be the speed reference, which can be indirectly set by directly changing the duty cycle of the output PWM, or, it can be fed in a PI controller in software and maintain that specific speed of the motor. The block diagram of the system can be seen in Fig. 4.

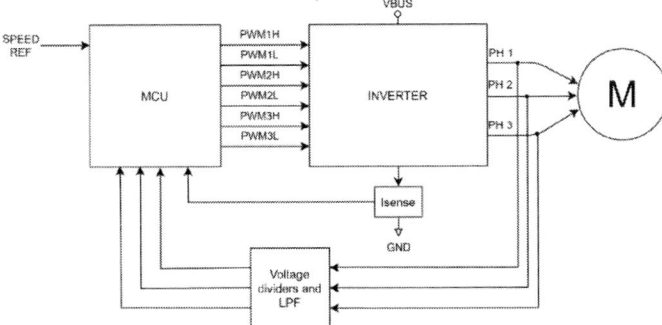

Fig. 4. System block diagram

B. Zero-cross sensing

To achieve the zero-cross sensing, there are a few ways. One is to use an ADC which measures the voltage on the phases, and by using a simple comparator. By default, motors do not have the neutral point (in the case of star-connected coils). Thus, a reference is needed, which comes under the form of a sum of the three phases, called "virtual neutral" or "virtual ground". This point typically has the VBUS/2 voltage, but it varies due to motor construction. A fixed reference could not have been used as it would have introduced phase shifts.

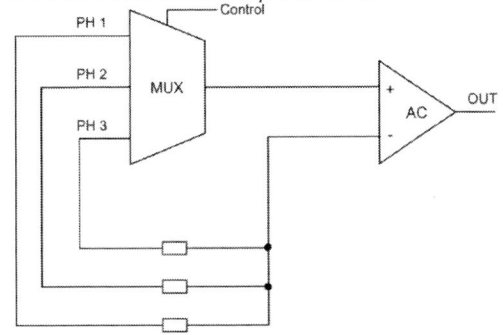

Fig. 5. Comparator zero-cross sensing method

In Fig.5, it can be observed how the implementation of the zero-cross detection can take place. In this solution, the comparator is an internal peripheral of the microcontroller, and a single unit is used. As in trapezoidal control only a single phase should be monitored (the undriven phase), a single comparator

will be enough. The analogue signal is multiplexed, and each phase is selected according to the current trapezoidal step.

C. Bipolar waveforms generation

The bipolar waveforms are generated using a single peripheral. It generates the characteristic signal, and then it generates the complementary by negation and subtracts the deadtime zones to avoid shoot-through. Each waveform is then routed to the necessary transistor, according to the current trapezoidal step. By using this method, the core is free to process the user layer leaving plenty of headroom. The novelty is contained within the method of waveforms generation itself which, at the current moment is not public and falls under the NDA contract and cannot be described further.

D. Electronic modules/assemblies

The system is divided in two electronic modules/assemblies based on two printed circuit boards (PCB).

The first one is the baseboard which contains all the power regulation, the inverter (the three half-bridges and drivers), voltage dividers, and a bidirectional hall current sensor.

The second PCB has the logic part of the circuit, which contains the microcontroller, inputs, the LPF and other miscellaneous parts. It plugs into the baseboard using a 2.54mm header. This allows flexibility in the logic part, which requires only a logic board to be designed, and not the whole system.

Fig. 6. The two electronic modules/assemblies which compose the system

Careful layout design was needed because of the following considerations:

1. The board has only 2 layers due to manufacturing costs;

2. Good thermal performance was needed, so the transistors have thermal vias under them in order to transfer the heat to the bottom layer which can be dissipated using a heatsink;

3. The board must handle high currents in excess of 50A, and the copper thickness is only 17μm. Thus, the high current traces are as wide as possible and there is no solder-mask, allowing during the assembling stage to fill the exposed areas with solder, which increases the current capability;

4. The high-current traces must be as short as possible to reduce the EMI, power connectors and bulk capacitors

978-1-7281-7507-2/20 $31.00 © 2020 IEEE   285   21-24 October, Pitesti, Romania

being placed as close as possible to the switching transistors for the same reason;

5. The analogue part must be as isolated as possible from any possible disturbances, which include the power switching part, as well as the baseboard buck converter. The current sensor zone, which is one of the most noise sensitive areas, is placed as far as possible from the buck converter and is placed in the lower-side of the layout.

IV. CONCLUSIONS

Having the perspective of a trapezoidal drive system, the results are above the expectations. Because of the bipolar switching, the RPM range is broader than unipolar switching, and the resource usage is minimal. It lacks torque in the lower RPMs, which is characteristic to all sensorless trapezoidal drives, but this can be compensated by increasing the advance angle of the applied magnetic field and thus, decreasing the overall system efficiency. The startup sequence must be tailored to match the motor and load characteristics and is tolerant to some extent to certain variations. The system self-adapts by altering certain parameters according to the current RPM.

The system is mainly suited for high-speed applications where the efficiency is not critical, and the trapezoidal drive does approach sinusoidal drive characteristic in the upper range of RPM. A field-oriented control designed for such high speeds requires very expensive hardware, but with this approach, it can be implemented using an 8-bit microcontroller in sub-$2 range price point.

This implementation is part of a much more complex system, which integrates all the concepts presented here, adds novel ways to generate all the waveforms and adopts multiple protections, in the same package, with reduced costs.

ACKNOWLEDGEMENTS

The author thanks to his coordinators, Prof. Norocel Codreanu and Lect. Marius Enachescu, for their continuous support during the development of the diploma project and of the present paper and to all his colleagues who offered valuable ideas and solutions during the design and manufacturing of the electronic system.

REFERENCES

[1] Padmaraja Yedamale, "Brushless DC (BLDC) Motor Fundamentals", Microchip Technology Inc., AN885, pp. 2;

[2] Ward Brown, "Brushless DC Motor Control Made Easy", Microchip Technology Inc., AN857, pp. 9;

[3] Shivam Tiwari, S. Rajendran, "Four Quadrant Operation and Control of Three Phase BLDC Motor for Electric Vehicles", Department of Electrical Engineering, Indian Institute of Technology Gandhinagar, Gandhinagar, India, pp. 3, May 16, 2019.

2020 IEEE 26th International Symposium for Design and Technology in Electronic Packaging (SIITME)

Increasing Students' Motivation Using Project-Based Learning on the Topic of Electrical Filters

Adriana N. Borodzhieva

Department of Telecommunications
University of Ruse "Angel Kanchev"
Ruse, Bulgaria
aborodzhieva@uni-ruse.bg

Abstract—**The paper presents an approach for increasing students' motivation using project-based learning on the topic of electrical filters by creating bilingual multimedia interactive applications in English and Bulgarian. The applications are used in the educational process in the course "Communication Circuits" for Bachelors in the specialty "Internet and Mobile Communications" in our University, and for Erasmus students studying this course.**

Keywords—active learning; electrical filters; interactive and multimedia applications.

I. INTRODUCTION

Project-Based Learning (PBL) is a student-centered pedagogical method involving a dynamic classroom approach, where students are believed to acquire more in-depth knowledge using active research of real-world problems. Students learn the material by working for a long period of time to research and find an answer of a complex question, challenge, problem or scenario [1]. Undoubtedly, PBL can provide our students with knowledge, skills and abilities that help them achieve future success. Many scientific publications are devoted to the benefits of project-based learning, summarizing only some of these benefits: collaboration, problem solving, creativity, in-depth understanding, self-confidence, critical thinking, perseverance, project management, curiosity and empowerment. Precisely in order to develop these skills in our students, project-based training has been introduced in the course "Communication Circuit".

The paper describes an approach for increasing students' motivation using project-based learning on the topic of electrical filters by creating bilingual multimedia interactive applications in English and Bulgarian. The applications created using the platform LearningApps [2, 3] are used in the training process in the course "Communication Circuits" for students-bachelors in the specialty "Internet and Mobile Communications" in the University of Ruse, and for Erasmus students attending the course. Creating the applications with various types of testing exercises, allows students to learn basic concepts in the theory of electrical filters, i.e. the main principle applied in the educational process is learning by doing while having fun.

The work presented in this paper is completed as partial fulfilment of Project 2020 – FEEA - 03 "Design and development of a multifunctional robot for implementation and evaluation of autonomous navigation algorithms", financed under the Scientific and Research Fund of the University of Ruse "Angel Kanchev".

II. PROJECT-BASED LEARNING ON THE TOPIC OF ELECTRICAL FILTERS

A. Education 4.0 and Project-Based Learning

The Industrial Revolution 4.0 with its technologies like Artificial Intelligence (AI), Big Data and Internet of Things (IoT) is changing the world. It will not only affect the major industries but the roles for which students will be prepared, resulting in the evolution of education. Fig. 1 shows the main characteristics of education during this evolution of education – from Education 1.0 to Education 4.0 [4].

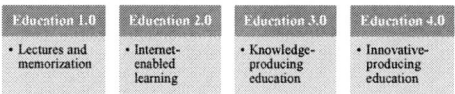

Fig. 1. Evolution of education – from Education 1.0 to Education 4.0

The Industrial Revolution 4.0 will require "educational institutions to produce a workforce for working in this technologically transformed era and the current workforce to upgrade their skills and knowledge to match these newly created job roles" making the revolution in education essential [4].

The future of education, Education 4.0, will change the teaching-learning methodologies to prepare the students for the future progressive, intellectual and knowledge-driven world. A few trends in this evolution are shown in Fig. 2.

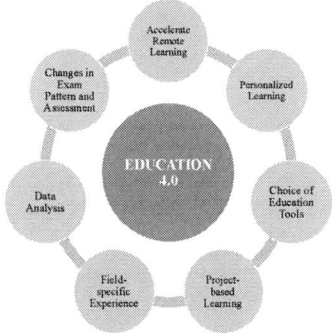

Fig. 2. Trends in Education 4.0

One of these trends is Project-Based Learning (PBL), defined as "a teaching method in which students gain knowledge and skills by working for an extended period of time to investigate and respond to an authentic, engaging, and complex question, problem, or challenge" [5, 6]. Project-Based Learning includes the core components, presented in Fig. 3 [6].

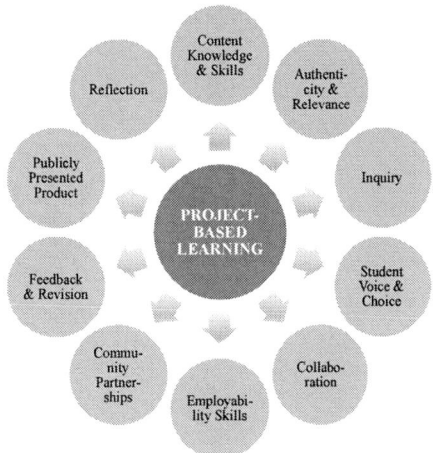

Fig. 3. Core components of Project-Based Learning

B. Project-Based Learning on the Topic "Electrical Filters

At the beginning of the semester, students get acquainted with the concepts of Education 4.0 and project-based learning (paragraph II A, Fig. 1…Fig. 3). The purpose of this introductory lesson is to indicate trends in future education and what to expect from the university as an educational institution, as well as the need from applying the project-based training in the course "Communication Circuits".

Each student receives an individual assignment from the theory of electrical filters, one of the topics studied in the course (Fig. 4). The course "Communication Circuits" covers three sections: Resonant Circuits, Electrical Filters and Modulation. Due to the wide scope of the section "Electrical Filters", only topics from this section were given to the students this academic year.

LECTURES		
№	Topic	Classes
1.	Electrical filters. Types of filters depending on the component basis used. Approximations. Passive electrical filters. Analysis and design of reactive chain filters.	2
2.	Chain LC filters of type "K" and type "m" – serial and parallel derivatives.	2
3.	Passive RC filters. Crystal and ceramic filters.	2
4.	Active filters. Cascading implementations of active filters. Second-order active filters with low and mediate quality factor – sections with one operational amplifier with single-loop and multi-loop negative feedback.	2
5.	Second-order active filters with low and mediate quality factor – sections with one voltage controlled voltage source, biquadratic sections with one operational amplifier.	2
6.	Second-order active filters with high quality factor – filter sections with two operational amplifiers using the model with generalized impedance converter	2
7.	Second-order active filters with high quality factor – universal biquadratic sections with three and four operational amplifiers. Active filters with switchable capacitors (SC filters).	2
	Total	14

Fig. 4. Topics in the section "Electrical Filters"

The assignment includes the development of a multimedia interactive application. With this aim, students review existing free platforms that allow the development of such applications. One of these platforms is LearningApps, a Web 2.0 application, helping the teaching and learning processes using small interactive modules for self-studying or directly embedded in learning materials. These reusable building blocks (the Apps) are collected and made available to everyone. The most important functions in LearningApps are explained in details in [2, 3]. Students with good programming knowledge and skills can create their own applications, Web- or Windows-based. Students without programming knowledge and skills are got acquainted with the platform, and based on the presented examples with many details about creating such applications (in different courses), can easily cope with the task. When choosing another environment, the student must argue for his/her choice.

Then, students must look for suitable literature sources (in English or Bulgarian) on the specified topic, select suitable templates for the testing exercises, formulate questions and answers and look for illustrative material (pictures, animated images, video-clips, etc.) to use in creating this application. Students should cite the sources used to formulate the questions and answers, as well as the illustrative material used, in order to protect copyright. Before creating the tests, the lecturer checks all the materials prepared by the students. Students are expected to solve this task in the first half of the semester.

For solving the problem according to the individual assignment, students apply screenshots of stages of creating the application (Fig. 5) describing in details the steps for creating it (Fig. 6). Sometimes it is necessary to make an image based on different pictures in Internet, for example for using it as a background image. Such an example is shown in Fig. 7, a background image with different circuits of RC filters where the students must select one of the components (circuits) for each type of passive and active filters. Some additional explanations about creating the marks in different colors on the image and annotating with different content are shown also in the student's solution, Fig. 8. When creating the applications, students are encouraged to make different experiments with the template chosen and to describe them in the documentation. The next task in the individual assignment is solving the testing exercises and applying a description of the tests done by the students (Fig. 9) and screenshots of its functioning with different possible options for answers (right or wrong) (Fig. 10). For creating the test, presented in Fig. 5…Fig. 10, the references [7-12] are used.

Students are expected to solve these tasks in the second half of the semester.

Implementing these tasks, students are taught to:

1) use specialized literature (in English and Bulgarian) related to their field of interests and select the useful facts and illustrative material for the tests improving their skills for analyzing, synthesizing and evaluating scientific information;

2) create a technical documentation – a soft skill indisputably necessary for software professionals today;

3) communicate effectively with their colleagues and lecturers in case of difficulties when reading and understanding the material taught, as well as developing their applications.

2020 IEEE 26th International Symposium for Design and Technology in Electronic Packaging (SIITME)

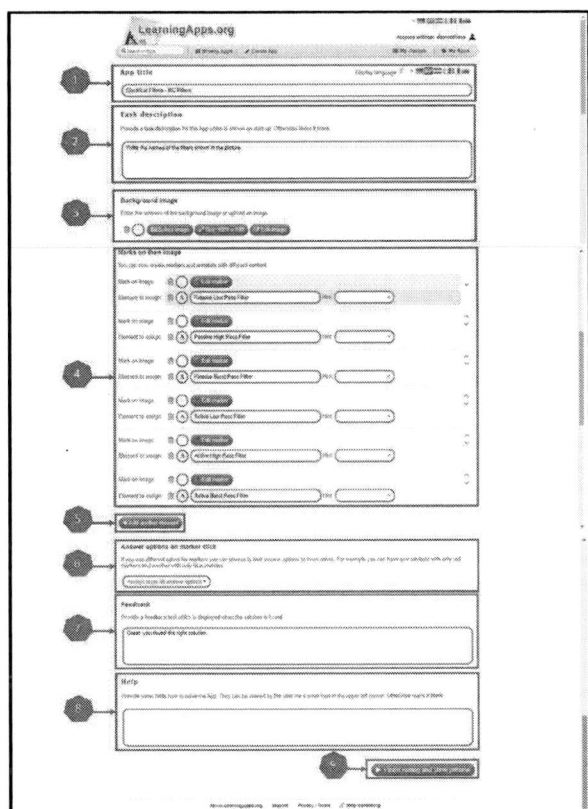

Fig. 5.   Task 1: Applying screenshots when creating the Apps

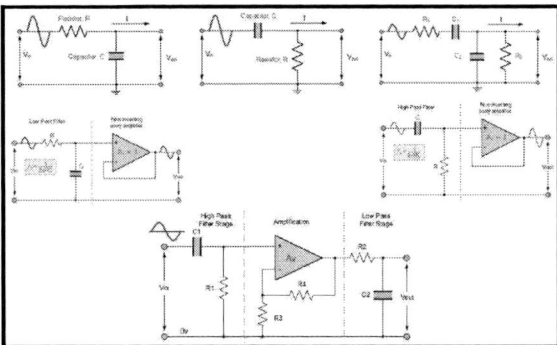

Fig. 7.   Background image with different circuits of RC filters

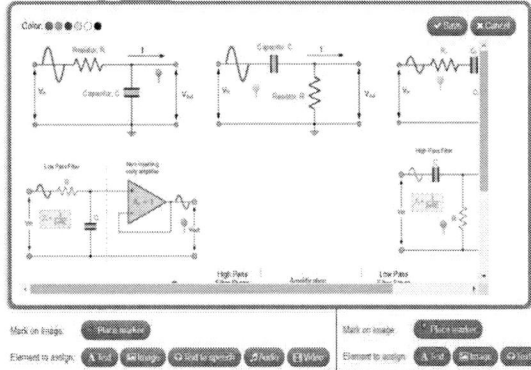

Fig. 8.   Creating the marks in different colors on the image and annotating with different content

For building a testing exercise on the topic "Electrical Filters – RC Filters", the template "Matching Pairs on Images" is selected. A new App based on this template will be created by clicking "Create new App". At the beginning of the form, a title (1) and a task description (2) are specified. The following settings are specific for this template. For example: selecting a background image (3) by entering the address of the background image or uploading an image; this is the image on which the marks will be created; creating the marks in different colors (red, green, blue, yellow, white, black), on the image and annotating with different content; here, different element (text, image, text to speech, audio, video), might be assigned (4); in this case, the following texts are assigned to the marks: "Passive Low Pass Filter", "Passive High Pass Filter", "Passive Band Pass Filter", "Active Low Pass Filter", "Active High Pass Filter", "Active Band Pass Filter"; other elements might be added (5); determining the answer options on marker click – there are two options: "Always show all answer options" (the selected one) and "Only show answer options with matching color"; if different colors for markers are used, then answer options can be chosen to limit to their colors; for example, one attribute with only red markers and another with only blue markers (6); providing a feedback: the text which is displayed when the solution is found; in this case, the text "Great, you found the right solution." is displayed (7); providing some hints how to solve the App that can be viewed by the user via a small icon in the upper left corner (8). After making the changes for creating the new App, the user may click on "Finish editing and show preview" (9) at the end of the form and then the user can customize the App again or save it to his App collection.

Fig. 6.   Task 2: Describing the steps in the process of creating the Apps

The initial state of the exercise and the formulation of the task are given in (1). A partial solution, i.e. finding one pair, after clicking on a pin for determining the type of the filter marked with this pin, is shown in (2). Here, the selected answer options are marked with darker background. The final state where all the answers are right (marked in green background) and the corresponding greeting message are shown in (3). The situation with a wrong choice, marked in red, not in green, is presented in (4). As a hint, passive filters are marked with green pins and active filters – with red pins.

Fig. 9.   Task 3: Applying a description of the tests done by the students of its functioning

In this academic year, due to COVID-19 crisis, an experiment for applying project-based learning was conducted with 36 students-bachelors of the specialty "Internet and Mobile Communications" studying the course "Communication Circuits". The students had to implement such interactive and multimedia applications and describe the steps in the process of creating and testing. All of them chose the platform LearningApps due to its interface in Bulgarian, detailed instructions and many available examples developed with this platform.

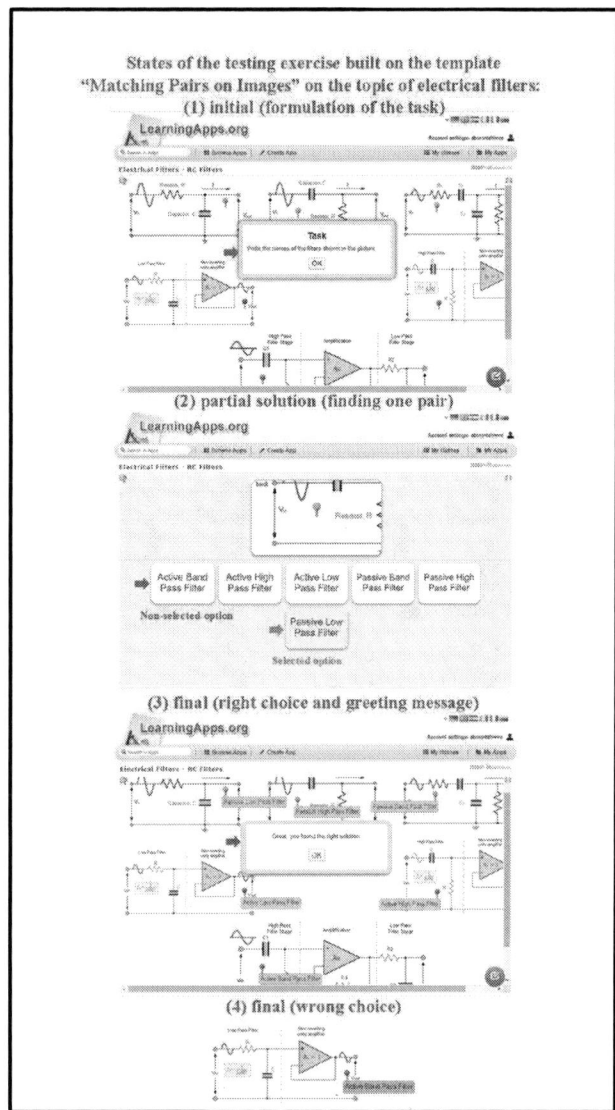

Fig. 10. Task 4: Applying screenshots of Apps functioning with different possible options for answers (right or wrong)

The students did their assignments and demonstrated great interest in the material studied in the course and the platform. The most frequently used templates were "Matching Pairs" (the most preferred by the students), "Matching Pairs on Images", "Group assignment", "Cloze text", "Free-text input", "Multiple-Choice Quiz", "Crossword", "Word grid", "Guess the word".

At the end of the semester the students had to present their applications to their colleagues, but due to Covid-19 crisis, the presentation could not take place in person, but using the system BigBlueButton [13], Fig. 11, preferred for synchronous distant learning by most of the teachers and students at our University.

At the end of the semester, a survey was conducted to study the opinion of students about the applied project-based approach to learning. 97% of the respondents approve this way of training and share that they have learned a lot of the topics while looking for relevant information, developing the application and testing the applications of their colleagues, and they have studied the material with a lot of interest and curiosity. Only 3% of the respondents appreciate the method as less attractive.

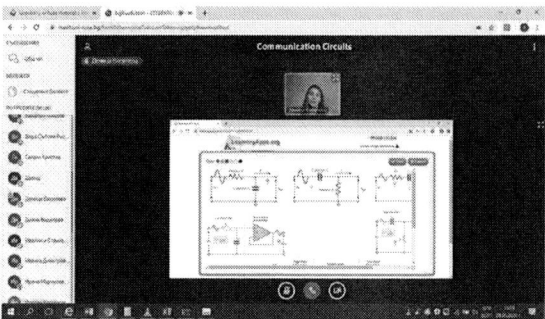

Fig. 11. Presenting the students'projects using BigBlueButton

III. CONCLUSIONS

The paper presents an approach for increasing students' motivation using project-based learning on the topic of electrical filters by creating bilingual multimedia interactive applications in English and Bulgarian, are used in the educational process in the course "Communication Circuits". Using PBL, students are taught to use specialized literature, create a technical documentation, communicate effectively with their colleagues and lecturers.

REFERENCES

[1] D. Yasseri, P.M. Finley, P. M. Mayfield, E. Blayne, D.W. Davis, P. Thompson, J. S.Vogler, "The hard work of soft skills: augmenting the project-based learning experience with interdisciplinary teamwork". Instructional Science, 46 (3): pp. 457-488, doi:10.1007/s11251-017-9438-9, ISSN 1573-1952, June, 2018.

[2] https://learningapps.org (Last visited: June 2020)

[3] M. I. Mazola Ortega, J. Wysowski, A. Borodzhieva, "Designing an Interactive Multimedia Bilingual Application for the Course "Pulse and Digital Devices"". 58th Science Conference of Ruse University, Bulgaria, 2019, Ruse, 24-26 October 2019, pp. 37-44, ISSN 1311-3321.

[4] Education 4.0. https://www.futurereadyedu.com/what-is-education-4-0-how-you-can-adapt-this-in-the-learning-environment/

[5] What is PBL? https://www.pblworks.org/what-is-pbl

[6] Project-based learning. https://www.magnifylearningin.org/what-is-project-based-learning

[7] Passive Low Pass Filter. https://www.electronics-tutorials.ws/filter/filter_2.html

[8] Passive High Pass Filter. https://www.electronics-tutorials.ws/filter/filter_3.html

[9] Passive Band Pass Filter. https://www.electronics-tutorials.ws/filter/filter_4.html

[10] Active Low Pass Filter. https://www.electronics-tutorials.ws/filter/filter_5.html

[11] Active High Pass Filter. https://www.electronics-tutorials.ws/filter/filter_6.html

[12] Active Band Pass Filter. https://www.electronics-tutorials.ws/filter/filter_7.html

[13] Big Blue Button, https://bigbluebutton.org/

Computer-Based Education for Teaching the Topic "Galois Linear Feedback Shift Registers"

Adriana N. Borodzhieva
Department of Telecommunications
University of Ruse "Angel Kanchev"
Ruse, Bulgaria
aborodzhieva@uni-ruse.bg

Abstract—**The paper presents different computer-based tools, such as MS Excel, Logisim and ISE Project Navigator, for teaching the topic of Linear Feedback Shift Registers (LFSRs) and different types of LFSRs (Fibonacci, Galois), as well as building cryptosystems based on them, in the course "Telecommunication Security". The course introduces the students-bachelors of the specialty "Internet and Mobile Communications" in the University of Ruse "Angel Kanchev" to the main issues of cryptography and information protection.**

Keywords—computer-based education; Galois linear feedback shift registers; cryptography.

I. INTRODUCTION

During the last years, computers have changed the manner people teach and learn, live and work. There is no area where computers do not play a very essential role. Education has also gone beyond textbooks.

Computer-Based Education (CBE) refers to teaching and learning methodologies using computers as a main ingredient of information transfer. For example, computers might be used to present the content of the lesson to students in a more attractive manner or used in test-assessment to make grading and evaluation easier. Computer-based learning is mainly used in the following areas: 1) knowledge-based training and assessment; 2) simulation-based learning and training; 3) creative and instructional games; 4) problem-solving training [1].

The paper presents an approach for teaching the topic "Galois Linear Feedback Shift Registers" in the course "Telecommunication Security" for Bachelors of the specialty "Internet and Mobile Communications" in the University of Ruse, using different computer-based tools, such as MS Excel [2], Logisim [3], FPGA design, etc. The course introduces the students to the main issues of cryptography and info-protection, but it might be used in other courses covering these issues.

Applying active and interactive teaching methods during the classes, based on the student-centered approach, such as learning by doing, learning by teaching, one-minute papers, class debates, think-pair-share, collaborative learning group, etc., increases the interest and success of students in the course.

The work presented in this paper is completed as partial fulfilment of Project 2020 – FEEA - 03 "Design and development of a multifunctional robot for implementation and evaluation of autonomous navigation algorithms", financed under the Scientific and Research Fund of the University of Ruse "Angel Kanchev".

II. SYNTHESIS AND ANALYSIS OF GALOIS LINEAR FEEDBACK SHIFT REGISTERS

A. Galois Linear Feedback Shift Registers

The *Linear Feedback Shift Register* (LFSR) is a standard shift register with feedback from two or more points (taps), in the register chain. The input to the LFSR is generated by using XOR or XNOR logic gates covering the tap bits. The sequence of values generated by LFSRs is determined by its feedback function and tap selection. They are simple for constructing and useful for a wide range of applications, but often neglected by designers [4, 5].

The *Galois Linear Feedback Shift Register* (GLFSR), named after Évariste Galois (1811-1832), a French mathematician, also known as *internal XORs*, *modular*, or *one-to-many LFSR*, is an alternative structure generating "the same output stream as a conventional LFSR", but with offset in time. In GLFSRs, when the register is clocked, "bits that are not taps are shifted one position to the right unchanged, the taps are XORed with the output bit before they are stored in the next positions and the new output bit is the next input bit". When the output bit is 0, the input bit will be 0 and all the register's bits are right-shifted unchanged. When the output bit is 1, the input bit will be 1, the bits in the tap positions all flip (0s become 1s, and 1s become 0s) and then all the register's bits are right-shifted [4, 5]. If the taps' order of GLFSR is in reverse of the taps' order for the conventional LFSR, then the same output stream (but not in reverse) is generated.

GLFSR have the following advantages: **1)** they do not concatenate every tap for producing the new input bit, i.e. the operations XOR are done, not in serial, within the GLFSR, making possible XORs to be computed in parallel for each tap, and therefore the propagation time is reduced to that of one XOR gate and the speed of execution is increased; **2)** in software implementations, the Galois configuration is more effective, as the XOR operations might be implemented as a word at a time: only the output bit must be examined individually [4, 5].

2020 IEEE 26th International Symposium for Design and Technology in Electronic Packaging (SIITME)

B. Applying Active and Interactive Teaching Methods

In recent years, in connection with the idea of Education 4.0 [6], the term "flipped classroom" [7] has been increasingly used. Fig. 1 illustrates the idea of the "flipped classroom", giving a definition of the term, a comparison between the "traditional classroom" and the "flipped classroom", as well as the activities of the teacher and the students at home and at school when the "flipped classroom" is applied.

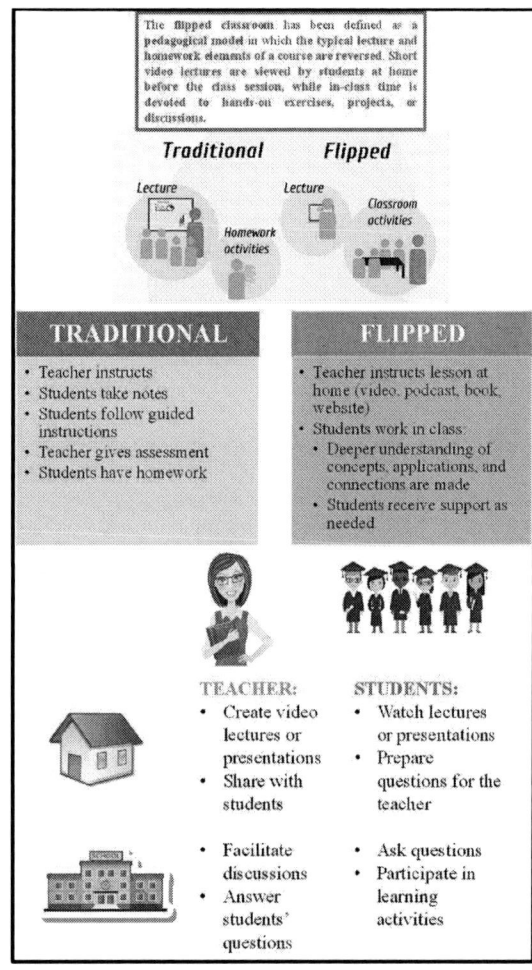

Fig. 1. "Flipped classroom" concept

The "flipped classroom" method [7] is actively applied in the course "Telecommunication Security". It motivates students and improves quality of interaction between students, as well as inspires changes in teaching methods. The main points in the "flipped classroom" are as follows: **1)** learning content at home; **2)** coming to class to apply what students learnt at home; **3)** students learn for and by themselves; **4)** more interaction and face-to-face time; **5)** higher students' engagement; **6)** small group work. Of course, there are some risks, such as: **1)** the "flipped classroom" relies highly on motivated learners; **2)** "flipped homework" is still homework; **3)** not all students

learn well using visual learning; **4)** Internet accessibility and modern technologies are required [7].

In the course "Telecommunication Security", based on the idea of the "flipped classroom", students must get acquainted in advance at home with the theoretical formulation of the topic – a definition of linear feedback shift registers, their mathematical foundations and their application in the field of computer and communication technologies, by searching appropriate references (recommended textbooks) or Internet sources. During lectures, they had the opportunity to discuss issues with the teacher and their colleagues. Students must consider the three types of LFSRs (conventional, Fibonacci and Galois) in detail, as well as the examples solved, the development of applications in MS Excel, the circuits built in Logisim, including the implementation of the devices with integrated circuits from the library 74xx, and the circuits built in ISE Project Navigator.

MS Excel-based applications "implementing and graphically illustrating the processes of encryption and decryption of 3-symbol words in English using cryptosystems" based on 8-bit LFSR and 16-bit Fibonacci LFSR with details about the functions used for the implementation and describing the problems that arise when encrypting and decrypting texts are presented in [8] and [9]. FPGA implementation of crypto-systems based on 8-bit LFSRs for educational purposes is presented in [10]. Based on these numerous examples obtained by the applications and the circuits, students must solve an individual assignment during the classes – first, manually, and then implementing it using different computer-based tools. Within the first lesson (90 minutes classes) students work on Fibonacci LFSR and within the second lesson (90 minutes classes) they work on Galois LFSR (Fig. 2), and at the end of the classes students have to compare the results of encryption/decryption of English texts with the two cryptosystems. Different primitive polynomials are given to the students in the individual assignments (Fig. 2). Also, different length of the registers might be given to the students, depending on their knowledge and skills. The table in Fig. 2 lists maximal-length polynomials for shift registers with lengths up to 20 [4]. For any given length of shift registers, more than one maximal-length polynomial may exist and a list of alternative maximal-length polynomials for 16-bit shift registers are presented in Fig. 2, necessary for the individual assignments. A list of alternative maximal-length polynomials for shift registers with lengths 4–32 are available in [11].

In the paper, a 16-bit Galois LFSR and a cryptosystem built on it are considered (Fig. 3 and Fig. 4) using maximal-length polynomial $802D_{HEX}$. During the classes (the second lesson), students develop an application in MS Excel [2] illustrating the principle of operation of a 16-bit Galois LFSR and a cryptosystem built on it (Fig. 3) using different maximal-length polynomials.

Then the students implement the cryptosystem using Logisim [3], "an educational tool for synthesis and analysis of digital logic circuits", allowing them to investigate the principle of operation of the real device changing the input signals (Fig. 4). After building the circuit in Logisim, the students implement it with integrated circuits from the library 74xx of Texas Instruments, choosing appropriate integrated circuits.

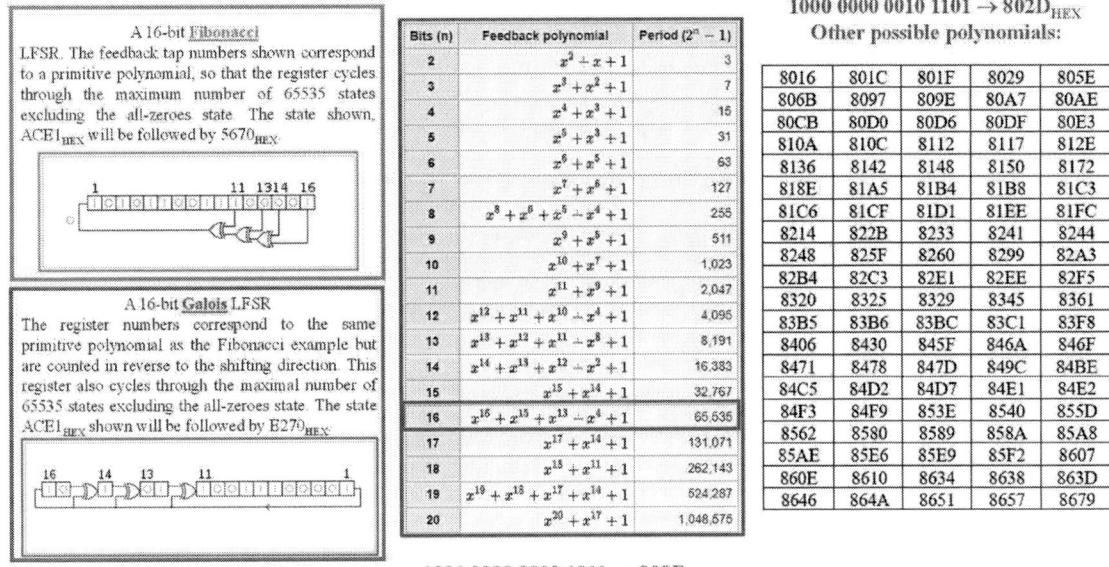

Fig. 2. 16-bit Fibonacci and Galois linear feedback shift registers and primitive polynomials used for building maximum-length LFSR

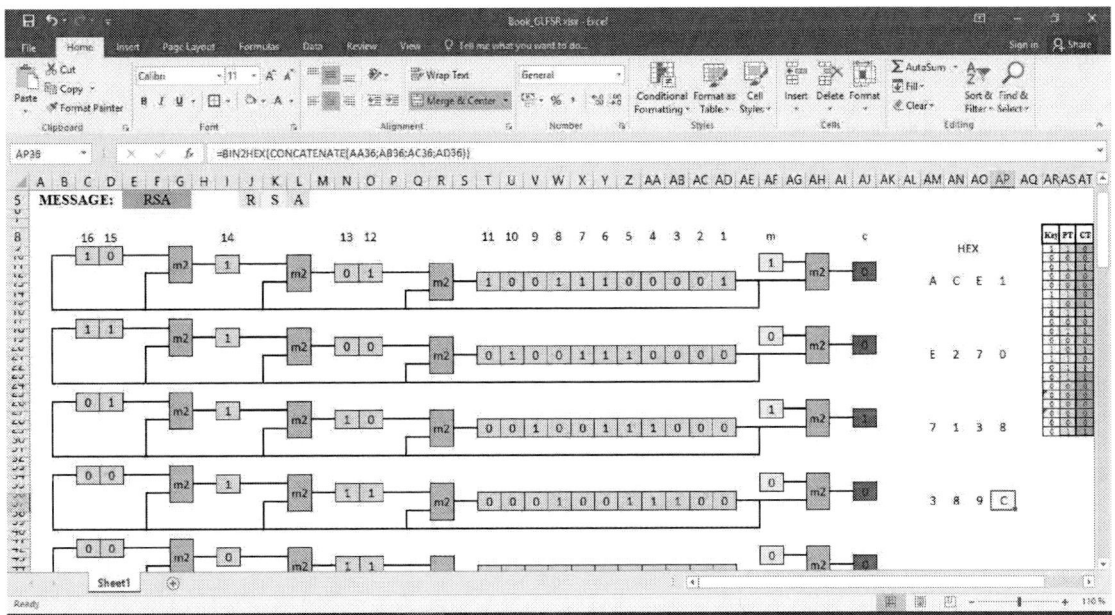

Fig. 3. Principle of operation of a 16-bit Galois linear feedback shift register and a cryptosystem built on it – the initial state (ACE1$_{HEX}$) and the next three states (E270$_{HEX}$, 7138$_{HEX}$ and 389C$_{HEX}$) of the register, plaintext and ciphertext, presented in ASCII code; MS Excel implementation

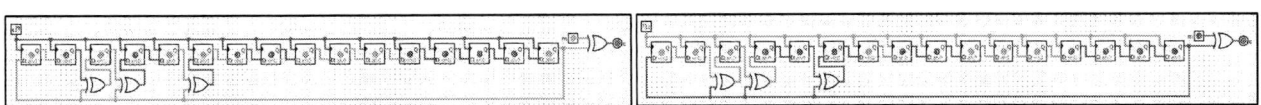

Fig. 4. Principle of operation of a 16-bit Galois linear feedback shift register and a cryptosystem built on it – the initial state (ACE1$_{HEX}$) and the next state (E270$_{HEX}$) of the register; implementation in Logisim

978-1-7281-7507-2/20 $31.00 © 2020 IEEE 293a 21-24 October, Pitesti, Romania

Finally, using the example presented in [10], the students might draw the circuit of the 16-bit GLFSR in the environment of ISE Project Navigator for FPGA design of the cryptosystem on a laboratory board, developed at the University of Ruse for the purposes of the courses studying different digital devices.

The individual assignments given to the students include as intermediate tasks: 1) determining the states of the 16-bit FLFSR/GLFSR for a given initial state (seed) and a given maximal-length polynomial (Fig. 2); 2) encrypting a given plaintext (3-symbol English text, abbreviations in the field of computer and communication technologies) using the 16-bit FLFSR/GLFSR; 3) decrypting a given cipher-text using the 16-bit FLFSR/GLFSR. The texts (plaintext or cipher-text) might contain Latin letters, decimal digits, punctuation signs, arithmetic and logical operators, control symbols, etc. During the practical exercises, the students must solve the tasks manually and then using different computer-based tools. There are opportunities for extra work for the students, depending on their knowledge and skills: 1) decrypting a given cipher-text when the initial state of the 16-bit FLFSR/GLFSR is not known using the application in MS Excel or the circuit built in Logisim; 2) implementation and testing a device for encryption and decryption on the laboratory board – in this case, students have to consider how to organize the outputs of the device due to the need to visualize 16 outputs (of the register) and one bit of the encrypted/decrypted text in the presence of only 10 outputs on the output module of the laboratory board, requiring the use of output multiplexing.

Due to time constraints, the teacher divides the students into groups of two students. The first student implements the application in MS Excel (Fig. 3) and the other implements it in Logisim (Fig. 4). The first student encrypts the plaintext of his/her choice with the application in MS Excel (using 7-bit ASCII code or 8-bit EBCDIC code) and says the encrypted text to the second student, who decrypts it using the circuit he/she built in Logisim, and vice versa. The following week, when another device is studied, students change their places. At the moment, the time during the classes has not been enough to realize the device considered on the laboratory module.

In the academic 2019-2020 year, an experiment was conducted with 45 students of the specialty "Internet and Mobile Communications" studying the course "Telecommunication Security". At the end of the classes, a survey was conducted to analyze the opinion of students about the way of teaching the material. 83 % of the respondents approve this way of training (the green sector in Fig. 5). Only 17 % of the respondents do not approve it (the red sector in Fig. 5). All of these respondents (17 %) had to implement the device in MS Excel – the more difficult task in the case.

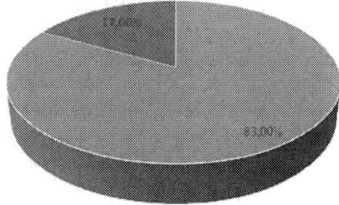

Fig. 5. Students' approval of using computer-based tools (YES:NO = 83:17)

III. CONCLUSION

The applications in MS Excel and Logisim described in the paper are used in the educational process in the course "Telecommunication Security", compulsory in the curriculum of the specialty "Internet and Mobile Communications" for Bachelors at the University of Ruse but they can be applied in other courses covering topics of digital devices or cryptography and information protection, such as "Cryptography and Data Security" (Bachelor Program "Computer Systems and Technologies"), "Security in the Telecommunication Networks" (Master Program "Internet and Multimedia Communications"), "Coding and Security in Telecommunication Networks" (Master Program "Telecommunication Networks"), etc.

The applications in MS Excel and the circuits in Logisim allow easily understanding the material studied by the students thanks to visualizing the steps at each stage of encryption and decryption in details.

For more effective learning, the "flipped classroom" and active learning methods are applied in the course "Telecommunication Security". Using some of the most important pointers describing the future of education, known as Education 4.0, such as accelerate remote learning, personalized learning, choice of education tools, project-based learning, field-specific experience, data analysis, changes in exam pattern and assessment, motivates the students in the learning process and increases the final grades of the students in the course.

REFERENCES

[1] J. Evans, "Importance of Computers in Education", 2017, https://www.clntraining.net/importance-computers-education/

[2] Microsoft Excel, https://products.office.com/en/excel

[3] Logisim, http://www.cburch.com/logisim/

[4] Linear-feedback shift register. https://en.wikipedia.org/wiki/Linear-feedback_shift_register

[5] W. Press, S. Teukolsky, W. Vetterling, B. Flannery, "Numerical Recipes: The Art of Scientific Computing", Third Edition. Cambridge University Press, 2007, ISBN 978-0-521-88407-5, http://e-maxx.ru/bookz/files/numerical_recipes.pdf.

[6] Education 4.0, https://www.futurereadyedu.com/what-is-education-4-0-how-you-can-adapt-this-in-the-learning-environment/

[7] Europass Teacher Academy, Flipped classroom; 2020, https://www.teacheracademy.eu/course/flipped-classroom/

[8] A. Borodzhieva, "Modeling of Cryptosystems Based on Linear Feedback Shift Registers Using Spreadsheets". 42nd International Convention on Information and Communication Technology, Electronics and Microelectronics, MIPRO 2019, 20 – 24 May 2019, Opatija, Croatia, Engineering Education, Proceedings, pp. 1713 – 1718, ISBN: 1847-3946.

[9] A. Brodzhieva, "MS Excel-Based Application for Encryption and Decryption of English Texts Using Fibonacci Linear Feedback Shift Registers". 13th Annual International Technology, Education and Development Conference, Valencia (Spain), 11 – 13 March 2019, Proceedings of INTED2019 Conference, pp. 4328-4337, ISBN: 978-84-09-08619-1.

[10] A. Borodzhieva, I. Stoev, V. Mutkov, "FPGA Implementation of Cryptosystems Based on Linear Feedback Shift Registers for Educational Purposes". 29th Annual Conference of the European Association for Education in Electrical and Information Engineering (EAEEIE) 2019, 4-6 September 2019, University of Ruse, Ruse, Bulgaria, Conference Proceedings, pp. 306 – 309, ISBN: 978-1-7281-3221-1.

[11] A list of alternative maximal-length polynomials for shift registers with lengths 4–32, http://www.ece.cmu.edu/~koopman/lfsr/index.html

This page intentionally left blank.

2020 IEEE 26th International Symposium for Design and Technology in Electronic Packaging (SIITME)

Computer-Aided Tools for Synthesis and Analysis of Pseudorandom Number Generators

Adriana Borodzhieva[1], Iordan Stoev[2], Snezhinka Zaharieva[2] and Valentin Mutkov[2]

[1] Department of Telecommunications, [2] Department of Electronics

University of Ruse "Angel Kanchev"

Ruse, Bulgaria

aborodzhieva@uni-ruse.bg

Abstract—**The paper presents the usage of various computer-aided tools, such as MS Excel, Logisim and ISE Design Suite, for synthesis and analysis of pseudorandom number generators studied in the course "Digital Electronics" for students-bachelors in the specialties "Computer Systems and Technologies" and "Electronics" in the University of Ruse. Using active teaching methods during the classes increases the motivation for learning, students' interest and grades in the course.**

Keywords—*computer-aided tools; MS Excel; Logisim; ISE Design Suite; pseudorandom number generators.*

I. Introduction

Today, it is impossible to find a place in the world where there is no digital electronics. Various applications such as computers, digital video recorders, microwaves, etc. would be impossible without digital electronic controls.

Digital electronics is an important subject in the curriculum of electrical and computer engineering programs in our universities. It covers the topics of synthesis and analysis of combinational and sequential circuits. In our university, the classes in the course "Digital Electronics" consist of lectures and labs. Various active learning methods during the laboratory exercises in addition to the components of the traditional, presentation-based lecture are used, for example: "think, pair, share", minute papers, decision-making activities, case-based learning, etc. [1, 2]. Tasks related with modelling, simulation and FPGA design allow the students to solve problems and understand the material studied. Applying the research approach to the training process using labs develops communication, collaboration, and problem solving skills, adaptive, analytical and critical thinking, personal management, creativity and innovation [1, 2, 3, 4].

Today, in the information age, computers play a significant role in many fields. Doubtless, in the field of education computers improve the quality of learning and teaching based on various technologies and resources. Lecturers at the universities must be aware about the effect of computers in the area of education as well as their subject area to make effective learning and teaching processes.

The work presented in this paper is completed as partial fulfilment of Project 2020 – FEEA - 03 "Design and development of a multifunctional robot for implementation and evaluation of autonomous navigation algorithms", financed under the Scientific and Research Fund of the University of Ruse "Angel Kanchev".

The paper will discuss the usage of various computer-aided tools for synthesis and analysis of pseudorandom number generators studied in the course "Digital Electronics" for students-bachelors in the specialties "Computer Systems and Technologies" and "Electronics" in the University of Ruse, for example MS Excel [5], Logisim [6, 7], ISE Design Suite [8] for FPGA design, etc. Active teaching methods during the classes, such as decision-making activities, "think, pair, share", case-based learning, minute papers, etc. are used to ensure the understanding of the course contents, to increase the motivation for learning, interest and grades of students in the course.

II. Active Learning in the Course "Digital Electronics"

A. Generator of pseudorandom sequences of binary numbers

The generator of pseudorandom sequences of binary numbers is a cyclic shift register, in the feedback of which "a sum modulo 2" ("exclusive OR", XOR) of the outputs of two of the memory elements (flip-flops) of the register is used. The idea of such a five-bit circuit, with a shift to the left and a feedback to the flip-flop of the least significant bit of the signals from the outputs of the third and fourth flip-flops is illustrated in Fig. 1.

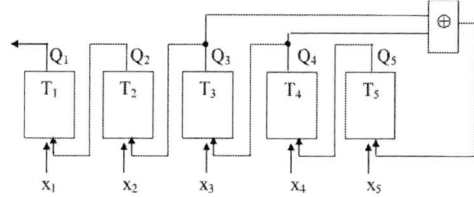

Fig. 1. An example of a five-bit pseudorandom number generator

B. Describing the experiment in the academic 2019-2020 year

During the practical exercises in the course "Digital Electronics", the students must solve the following tasks:

The students should study the topic of pseudorandom number generators – the theory and the practice, as well as their applications in the area of computer and communication technologies, in advance, at home. For this purpose, the students use Internet or references given by the teacher in the beginning of the semester. For clarifying the material, the students can discuss the topic with their colleagues, as well as with the lecturer by e-mail. At the beginning of the class, within 2-3 minutes, they answer the question "What did I learn about this?" After that, each student must solve a task, such as the example below.

Example. Build a pseudo-random number generator based on a 6-bit right-shift register using (1) *JK* flip-flops with a feedback from the outputs of flip-flops numbered (2) 5 and (3) 6. Find the pseudo-random sequence of binary numbers starting from (4) 110111, for 6 cycles (clocks) of the abstract automatic time.

The possible options are: (1) *JK* or *RS*; (2) a number from 1 to 6; (3) a number from 1 to 6, without repetitions with position (2); (4) any binary sequence of 6 bits. The solution of the task follows the steps presented below.

The block diagram of the synthesized pseudorandom number generator with the purpose of determining the excitation functions of the memory elements is presented in Fig. 2.

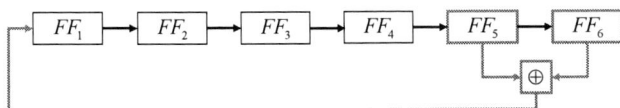

Fig. 2. Block diagram of the synthesized pseudorandom number generator

The excitation functions of the memory elements and the output functions of the generator are determined as follows (Table I):

TABLE I.   EXCITATION FUNCTIONS OF THE FLIP-FLOPS AND OUTPUT FUNCTIONS OF THE GENERATOR

Excitation functions of the memory elements	
At SET inputs (*S* or *J*)	At RESET inputs (*R* or *K*)
$J_1 = Q_5 \oplus Q_6;\ J_2 = Q_1;$ $J_3 = Q_2;\ J_4 = Q_3;$ $J_5 = Q_4;\ J_6 = Q_5$	$K_1 = \overline{Q_5 \oplus Q_6};\ K_2 = \overline{Q_1};$ $K_3 = \overline{Q_2};\ K_4 = \overline{Q_3};$ $K_5 = \overline{Q_4};\ K_6 = \overline{Q_5}$
Output functions of the pseudorandom number generator	
$z_1 = Q_1$ $z_2 = Q_2$ $z_3 = Q_3$ $z_4 = Q_4$ $z_5 = Q_5$ $z_6 = Q_6$	

The structural diagram of the pseudorandom number generator is shown in Fig. 3. The determination of the pseudorandom series of binary numbers, starting from the state 110111, for 6 clocks of the abstract automatic time is presented more clearly in Fig. 4.

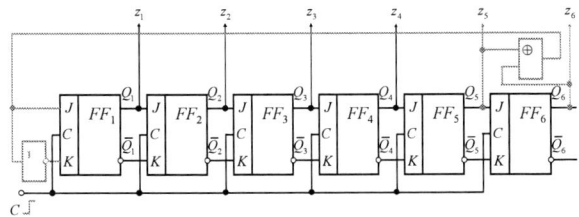

Fig. 3. Structural diagram of the pseudo-random numbers generator

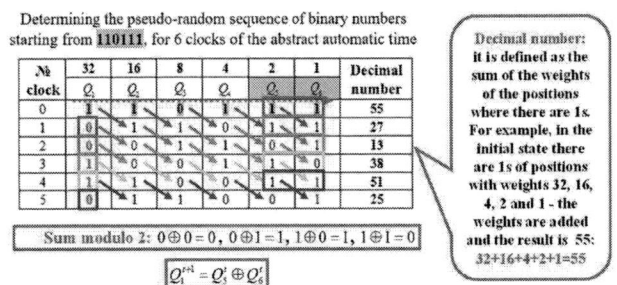

Fig. 4. Illustrative method for determining the pseudorandom series of binary numbers for the example

In the table of Fig. 4, the sequence of binary numbers obtained at the outputs of the generator is found. The first column shows the clock number of the abstract automatic time, and the last column – the decimal number, corresponding to the information stored in the generator. The arrows show the information movement (shifting). The output Q_1 of the first flip-flop at a given moment is determined by the sum of modulo 2 of the outputs Q_5 and Q_6 at the previous moment. The last column shows that, at least at a first glance, the original sequence of numbers is "random". If this analysis is continued, after a certain number of cycles (clocks) the resulting sequence will begin to repeat, i.e. it is pseudorandom.

During the classes, first the students must solve their assignment manually and then use different computer-based tools for illustrating the principle of operation of the pseudo-random number generators. The activities of the students are shown in Fig. 5 ... Fig. 9.

First, they develop an application in MS Excel, as shown in Fig. 5. For this purpose, students use the following tools: validating the data, inserting formulas, conditional formatting.

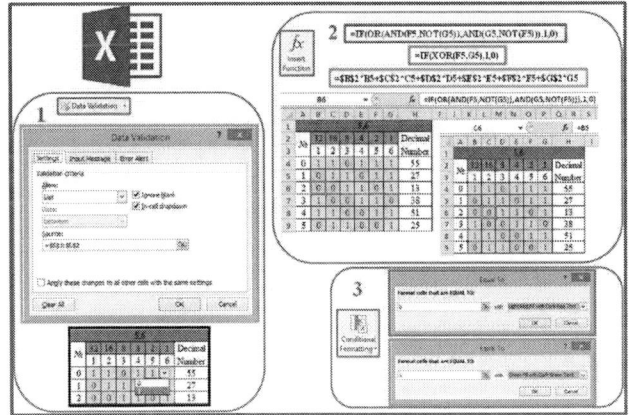

Fig. 5. Students' activities for creating the application in MS Excel

The application in MS Excel allows the student to select the bits of the initial state (seed) of the generator using a drop-down menu to select one of the two possible values 0 or 1 (Fig. 5, block 1). The next states of the generator are calculated using the formulas: for example, C6=B5 and B6=IF(XOR(F5,G5),1,0), if the function XOR is available in MS Excel or B6=IF(OR(AND(F5,NOT(G5)),AND(G5,NOT(F5))),1,0),   if the function XOR is not available in MS Excel (for versions lower than 2013). In the application, functions embedded in MS Excel allowing the usage of Boolean functions (NOT, AND, OR, and XOR), and a modification of IF function for correct visualization of the results (0s or 1s instead of False and True) (Fig. 5, block 2) are used. For calculating the decimal number (of the generator's state), the following formula is used: H6=$B$2*B6+$C$2*C6+$D$2*D6+$E$2*E6+$F$2*F6+$G$2*G6 (in the last column). Finally, conditional formatting is used in the information presentation for displaying 0s and 1s with different colours of fonts and backgrounds (Fig. 5, block 3).

During the classes, according to the second task, the students draw the circuit of the pseudorandom number generator (Fig. 6, block 1) in Logisim based on the excitation functions of the flip-flops and the output functions of the generator previously determined by the students (Table I). For this purpose, the students use *RS* or *JK* flip-flops and logic gates of type NOT and XOR. The circuits of 4-bit and 6-bit generators are presented (Fig. 6, block 1).

According to the third task, the students look for information in Internet about the integrated circuits used for building the generator's circuit (Fig. 6, block 1), the integrated circuits SN7476 (Dual Master-Slave *JK* Flip-Flops) [9] and SN7486 (Quad 2-Input Exclusive-OR Gate) [10] (Fig. 6, block 2), and draw the circuit of the pseudorandom number generator (Fig. 6, block 3) according to the assignment using TTL integrated circuits of the library 74xx mentioned above. For eliminating the NOT logic gate and the necessity of using an individual integrated circuit with inverters, the logic gate XOR is used for implementation of the inverse using the relationship $\bar{a} = 1 \oplus a$.

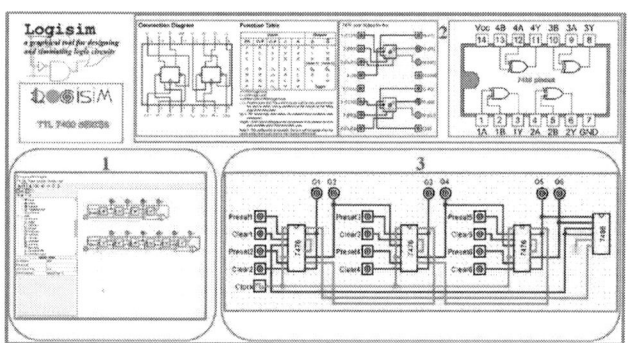

Fig. 6.   Students' activities for creating the circuits in Logisim

After implementing the circuits in Logisim (Fig. 6, block 1, and Fig. 6, block 3), the students must test the circuits and apply the simulation results to the solution of their assignments. The simulation results are shown correspondingly in Fig. 7 (without integrated circuits) and Fig. 8 (with integrated circuits). The circuit of the generator is drawn in Logisim, its initial state is set to 110111 and tested for 6 clocks. It can be seen that the results of the simulation study presented in Fig. 7 and Fig. 8 confirm those in the table in Fig. 4. Fig. 8 a presents the signals necessary to be given to the inputs PRESET and CLEAR of the flip-flops for setting the initial state of the generator – 110111. The next figures (Fig. 8 b… Fig. 8 g) presents its next states, when PRESET = 0 and CLEAR = 0 for all flip-flops.

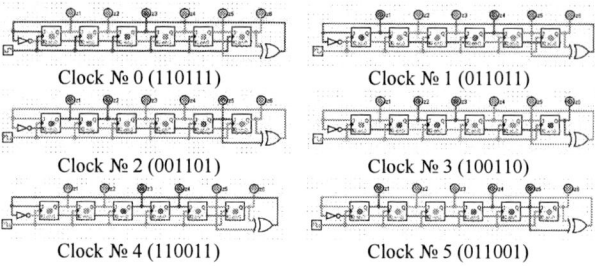

Fig. 7.   Results of the simulation study of the pseudorandom number generator in the example, implementation without integrated circuits

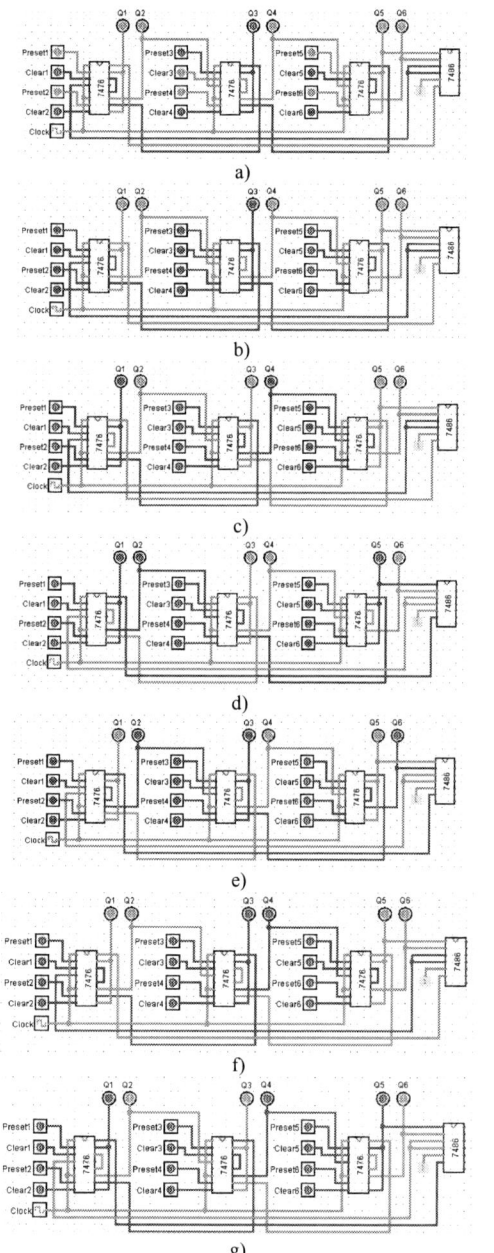

Fig. 8.   Results of the simulation study of the pseudorandom number generator in the example, implementation with integrated circuits

According to the fourth (last) task, the students draw the circuit in ISE Design Suite for FPGA design (Fig. 9 a) and examine it (Fig. 9 b and Fig. 9 c). For building the circuit, the students use XOR logic gates and flip-flops of the specified type, based on the solution of the problem. After drawing the circuit of the device, the students generate the corresponding bit file and with its help, they program the FPGA-based laboratory board, based on Spartan-6 FPGA Family and developed at the University of Ruse. Then, the students must compile an algorithm for testing the device regarding the order of entering the input combinations in order to observe the principle of operation of the synthesized device, to test the operation and confirm the results of its simulation investigation in Logisim. If the students do not have enough time for drawing the circuit, it is given to them drawn in advance and the students look for a match between the inputs (x0…x9) and outputs (Y0…Y9) of the lab-board and the inputs and outputs of the device. The following activity of the students is testing the device with the lab-board, for the initial state given by the teacher using the "bit" file, generated in the environment of ISE Design Suite (Fig. 9 a).

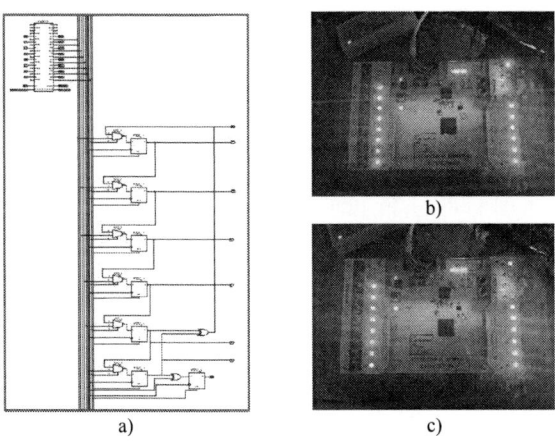

Fig. 9. Principle of operation of a pseudorandom number generator: implementation in ISE Design Suite (a) and its testing – the initial state (b) and the next state (c)

In the academic 2019-2020 year, an experiment was conducted with 69 students of the specialty "Computer Systems and Technologies" and 17 students of the specialty "Electronics" studying the course "Digital Electronics". At the end of the classes, students were asked about their opinion about the way of teaching the material using different computer-aided tools. Fig. 10 presents the distribution of students' approval for the use of computer-based tools based on the survey conducted with students.

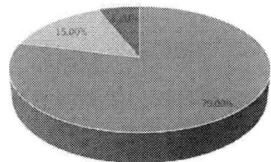

Fig. 10. Distribution of students' approval for the use of computer-based tools

79 % of students approve this way of learning and they have received excellent marks (the green group). 15 % of students were satisfied of this way of learning and they have received good marks (the yellow group). Only 6 % of students do not approve it and they have received bad marks (the red group).

III. CONCLUSIONS

The paper presents the use of various computer-aided tools, such as MS Excel, Logisim and ISE Design Suite, for designing pseudorandom number generators studied in the course "Digital Electronics" for students-bachelors at the University of Ruse.

During the classes, students develop an application in MS Excel illustrating the principle of operation of a pseudorandom number generator. Then the students implement the generator using Logisim, where they can see the operation of the real device when changing the input signals. In the end, the students draw the circuit of the generator in the environment of ISE Design Suite for FPGA design and test it on the laboratory board.

Through the laboratory exercises, students are expected to develop design experience and problem-solving skills that are important for the engineering profession. They have gained skills for analysing design problems, synthesizing and evaluating design information. Using active teaching methods increases the motivation for learning, students' interest and grades.

REFERENCES

[1] B. Balamuralithara, P. C. Woods, "Virtual laboratories in engineering education: The simulation lab and remote lab." Computer Applications in Engineering Education, pp.108-118, 2009.

[2] J. Pang, "Active Learning in the Introduction to Digital Logic Design Laboratory Course". ASEE Zone III Conference. 2015. https://www.asee.org/documents/zones/zone3/2015/Active-Learning-in-the-Introduction-to-Digital-Logic-Design-Laboratory-Course.pdf.

[3] A. Borodzhieva, "Computer-Based Tools Applied in the Course "Telecommunication Security"". 16th International Scientific Conference "eLearning and Software for Education" (eLSE 2020), 23-24 April 2020, Bucharest, Romania, Proceedings, Volume 2, pp. 42-52, Publisher: Carol I National Defence University Publishing House, ISSN 2360 - 2198, DOI: 10.12753/2066-026X-20-091.

[4] A. Borodzhieva, I. Stoev, V. Mutkov, "Active learning methods applied in the course "Digital electronics" on the topic "Arithmetic circuits using FPGA design"". 29th Annual Conference of the European Association for Education in Electrical and Information Engineering (EAEEIE) 2019, 4-6 September 2019, University of Ruse, Ruse, Bulgaria, Conference Proceedings, pp. 310-313, ISBN: 978-1-7281-3221-1.

[5] Microsoft Excel, https://products.office.com/en/excel

[6] Logisim, http://www.cburch.com/logisim/

[7] Related links, Libraries, 7400 series Logisim library from Ben Oztalay, http://www.cburch.com/logisim/links.html

[8] Xilinx documentation, https://www.xilinx.com/content/xilinx/en/downloadNav/design-tools/v2012_4---14_7.htm

[9] SN7476, Dual Master-Slave J-K Flip-Flops, https://datasheetspdf.com/datasheet/7476.html

[10] SN7486, Quad 2-Input Exclusive-OR Gate, https://datasheetspdf.com/datasheet/7486.html

Education 4.0: An Adaptive Framework with Artificial Intelligence, Raspberry Pi and Wearables - Innovation for Creating Value

Monica Ionita Ciolacu [1) 2)], Ali Fallah Tehrani[2)], Paul Svasta[1)], Ioan Tache[1)], and Dan Stoichescu[1)]

[1)] Faculty of Electronic, Telecommunications and Information Technology, CETTI, UPB, Bucharest, Romania

[2)] Faculty of Computer Science, Deggendorf Institute of Technology, Deggendorf, Germany

monica.ciolacu@th-deg.de

Abstract—**The Education 4.0 process can be used to foster students' performance, to motivate them by means of adaptive and personalized learning, to automate answering routine questions and to improve the quality of online examinations. It releases teachers from routine tasks, and it enables them to be more involved in the individualization of the educational process and in innovation. We apply artificial intelligence methodology to motivate students to improve their performance, and to identify early in course which students are at-risk of dropping out. In the adaptive Education 4.0 IoT framework, the environmental and embedded sensors from wearables examine biosignals with biofeedback, such as pulse (HR) and heart rate variability (HRV), thereby increasing students' self-reflection, and allowing for a more personalized learning experience. The first experiments corroborate meaningful correlations between the data-points reached in the self-assessments, and values of the biosignals.**

Keywords— Biosignals, Internet of Things (IoT), Technology Enhanced Learning, Education 4.0 Learning Lab, self-regulate learning, Biofeedback

I. Introduction

The COVID-19 pandemic caused a global crisis in education that has led to an exponential growth in the number of participants in online assisted education. Academia have provided students with as much support as possible, either by employing the use of MOOCs (Massive Open Online Courses), or by providing university specific digital platforms in order to substitute traditional teaching methods during quarantine period. The importance of the Education 4.0 topic has gained considerable attention. It is accelerating the 4th revolution in education with the paradigms: personalized and adaptive materials for self-regulate learning, biometrical authentication, online lectures, and new examination forms. Education 4.0 enhances the quality of teaching and learning using the Internet of Things (IoT) and Artificial Intelligence (AI) [1].

The insights of the BCG survey [2], which collected data from 12000 professionals before and during COVID-19 reveals that remote and flexible work will have significant implications for the way we organize our work and learning. BCG reported that a significant number (75% of employees) improved or maintained their level of productivity when working on their individual tasks, but just half of the employees reported an improvement when working on collaborative tasks in virtual, hybrid or completely remote settings. The new world of work includes: social connectivity, creating tools for mental and physical health, building capabilities to use the technologies that enable workstream collaboration, creating awareness with workplace analytics, and employees digital experience monitoring. Under these conditions, the difficulties associated with Technology Enhanced Learning have become obvious and the solution more important than ever.

The Education 4.0 process [4], [9], [11], [12] offers concrete and innovative solutions, theoretically and empirically well-founded, with an interdisciplinary character, but deeply rooted in the field of Electronic Engineering. The rest of this paper is organized as follows: in Section II we review some existing concepts. In Section III we describe the methodology and the development of an adaptive framework. In Section IV we present our experiments and the corresponding results. Finally, Section V concludes this paper.

II. Purpose and Background

A. Intelligent Agents

The concept of intelligent agents is crucial to our AI approach. Figure 1 illustrates an intelligent agent that interacts with the environment through sensors and acts upon an environment using actuators [1].

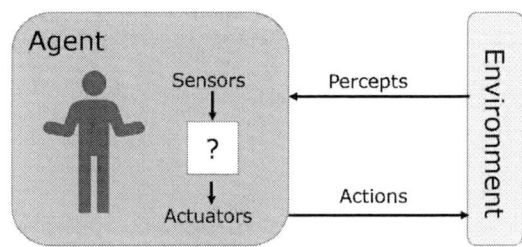

Fig. 1. A general intelligent agent adapted from [1].

A human agent has sensory organs, such as eyes, nose, ears, tongue, and skin with which he can perceive the environment, and actuators such as legs, hands, and mouth. The agent behavior is described by the agent function that maps any given percept sequence to an action. The agent function is implemented by an agent program. For designing an intelligent agent, we had to specify the performance (P) measurement, the environment (E), actuators (A) and sensors (S) – PEAS description. Table I illustrates an example of agent type interactive teletutor in higher education.

TABLE I.   AGENT TYPE TELETUTOR

Agent Type	PEAS Description			
	Perfor-mance	*Environ-ment*	*Actuators*	*Sensors*
Intelligent Teletutor[a]	Student's score on test.	A group of students; Teacher.	Display of exercises, recommenda-tions, rewards, corrections. Text to speech technology.	Key-board, speaker, microph one.

[a.] The Nature of Environments [1]

B. Motivating Learning Performance

In recent years, efforts have been made towards the development of Generalized Intelligent Framework for Tutoring for an effective adaptive instruction tailored to the needs of each learner [13]; adaptive instruction of a collective of independent individuals to allow communication and interaction with each other in order to perform assigned tasks [17]; sensor-based effect sensitive tutoring systems for interaction with the online learning environment [6], self-regulated learning in the context of learning from multimedia [11], [12], [14], and the further developments are still on going. Vansteenkiste et al. [3] tested the self-determination theory and hypotheses such as, intrinsic goals and autonomy-supportive learning climates improve students learning performance, and persistence.

Smart eyeglasses, such as WISEglass [25] or other head-mounted devices allow users to enhance and improve their subjective perceptions of reality. Scheiter et al. [14] uses eye movement modelling to support students in choosing and regulating their cognitive process. Lawson et al. [16] investigates the emotional tone displayed by both human or virtual instructor during a lesson that could affect learning outcomes. Le and Pinkwart [7] have suggested adopting a system-analytic approach for gifted students. This approach helps students during learning and makes them familiar with new technology rapidly by employing different applications, various types of underlying concepts can be learned.

C. Biofeedback

Biofeedback is a method that involves using electronic devices for visual and auditory feedback in order to gain awareness of physiological functions. Heart Rate Variability (HRV) is an indicator for biofeedback with parameters such as:

- Pulse or heart rate (HR).

- Tension and energy level (LF) – "fight or flight" – an indication of stress, i.e., higher sympathetic arousal.

- Rest and regeneration or fatigue level (HF) – "rest and digest" – an indicator of low motivation and tiredness, i.e., parasympathetic nervous system), and

- Recovery level after effort (rMSSD).

With biofeedback methods, using wearable devices both participant (student) and observer (researcher) can become aware of physiological responses of the body. It can be very helpful since it can identify unseen and unfelt physiological changes in real-time such as muscle tension [5]. These physical responses can be viewed on a screen as they occur, so the feedback is clear and objective.

D. Education 4.0 Process

The Education 4.0 adaptive framework [4], [12] combines the benefits of information gained with smart devices, biofeedback [5] of physiological data, AI algorithms for improving student's self-regulated learning, teaching efficacy and health. Taking advantages of AI in the Education 4.0 process, makes it possible to improve student's well-being and health [8], to recognize clusters of strategic importance for student's success, and to alert when students are at risk [9]. Traditional education can no longer meet these requirements, and there is a growing discrepancy between Academia and Industry requirements.

According to Ciolacu et al. [4], [12], [19] the Education 4.0 process assisted by AI technology and wearable devices consists of the following seven phases:

1) Orientation: an entrance test that assesses students previous knowledge about the subject (passive adaptivity) [4], [15]; an overview of the course content, motivation with an activity diagram that displays the grades achieved in examination, and strategic planning with intrinsic learning goals; Biometrics registration and authentication methods.

2) Digital preparation: personalized content according to learning types (interactive book or video), learning control for students self-monitoring with adaptive quizzes and self-assessments (continuous adaptivity) [9], [19]; Biofeedback.

3) Interactive presence: the teacher discusses case studies and acts as a coach; AR/VR experience; students work in groups; hands-on experiments and makerspaces, for example Education 4.0 Learning Lab (E4LL).

4) Collaboration: "Communities of Practice" with material from students for students [9], students work on short projects so called assignments – with continuously increasing difficulty and enhancing problem-based tasks, such as Learning Lab for Digital Technologies (LL4DT) [20].

5) Follow-up and performance: Feedback – self-assessments where the correct solutions and answers are explained and studied; the level of covered learning material is recognized and the evolution of knowledge is assessed. An "intelligent" teletutor (chatbot) answers simple questions with expert knowledge from the lecture manuscript, such as using an ontology [11], [12]; Natural Language Processing (NLP) and Text Mining; Biofeedback.

6) Reflection and motivation: "early recognition system" applies neural networks for self-monitoring and self-observation to continue the educational process. Future scenarios can be an extension of the Learning Analytics Cockpit based-on data from online and physiological activities [19].

7) Evaluation and examination: E-assessment - parts of the exams are automatically evaluated competence tests. We performed experiments with the latent Semantic Analysis (LSA) and Word2Vec [11]. Teachers are relieved of their correction work. Biometrics will play an important role, for example the two factors authentication in online examinations.

Students usually progress cyclically. After the follow-up and the reflection they go back to the digital preparation phase.

E. Well-being in Higher Education

The new 24/7 connectivity to working teams and to social networks means that there is little time left to reflect, to concentrate, or to integrate our thoughts. Information is constantly flooded through our visual and auditory world.

For example, Peper et al. [5] discuss "Ways to reduce Tech Stress" associated with technology overuse and offer practical tools such as biofeedback to increase learning, namely, how to stay healthy, to enhance productivity in front of screens, and to reduce physical stress. Awada and Mocanu [8] introduce a student's multilingual platform that aims to encourage students to have a more active life, suggesting specific physical exercises, realized through an avatar and a multimodal interface that is controlled by the student in a real-life game. Peper et al. [22] explain with biofeedback monitoring that the body posture (collapsed or upright) and position affects recall of emotional (positive or negative) memories, and recommend to sit more upright as a strategy to increase positive effects. Behavior that is more autonomous has been associated with more creative learning and engagement, greater energy and vitality, lower stress and a higher level of well-being [10].

As challenges in Higher Education we can mention electronic technology, personalized services and Lab mindset, creation of free innovation laboratories for development and testing of key enabling technologies (KET), where teachers and students from different disciplines are involved [4].

III. EXPERIMENTS METHODOLOGY

The use of embedded sensors from wearable devices extract more information about a student, such as physiological signals and learning environment. Initial monitoring and coaching of the students occurred in an Education 4.0 workshop. In our case, the biofeedback was provided using wearable devices (such as a smartwatch and a smartphone) worn all day by students .

A. Key Performance Indicators (KPI) in Education 4.0

To understand key performance indicators (KPI) of the Education 4.0 framework, the dimensions and indicators of the three IoT devices, students attended a workshop. The first dimension was a smartphone, the second a smartwatch, and the third dimension an embedded computer with Raspberry Pi, Grove Pi and Groove Pi sensors. Fig. 2 illustrates the three different KPI dimensions (D_i) with indicators for each of these dimensions ($I_{i,j}$).

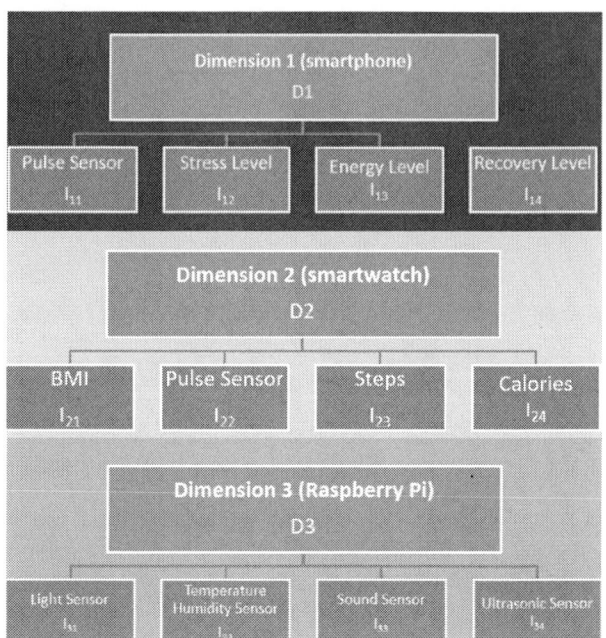

Fig. 2. Three dimensions of Education 4.0 IoT System

Adaptive IoT with Raspberry Pi

The IoT in Education 4.0 architecture provides us real-time data about the learning environment from Raspberry Pi 3.0 [21], Grove Pi, intelligent Grove Pi (ultrasonic range, sound, light, humidity, temperature) sensors, Bluetooth, I2C and Wi-Fi protocols. We use the I2C (Inter-Integrated Communication) serial interface - ideal for applications with micro-controllers. The I2C protocol allows the interconnection of 128 different devices, using only two bidirectional lines, one for the clock signal (SCK) and the other for the data (SDA). We select Grove Pi [18] digital and analog sensors available on the market. Fig. 3 illustrates the plug and play Grove Pi sensors. The hardware implementation of our prototype is illustrated in Fig.4.

Fig. 3. Example of plug and play Grove Pi sensors - LED and sound

Fig. 4. Education 4.0 prototype consist of a Raspberry Pi, Grove Pi - board and sensors, backlight display

In the Education 4.0 Learning Lab (E4LL) we experimented with the use of IoT in higher education. We developed an AI agent-based software with Python 3.7.3 with Raspberry Pi for analyzing student's environment. Table II presents the Education 4.0 Lab Agent.

TABLE II.   AGENT EDUCATION 4.0 LAB

Agent Type	PEAS Description			
	Perfor-mance	*Environ-ment*	*Actuators*	*Sensors*
Edu. 4.0 Lab	Students score with maximum gains on tests; well-being.	Pleasant learning; quiet; distraction free; non invasive.	Displays exercises; recommend actions; rewards; corrections. LCD display; LEDs; monitor; speakers.	Temperature; ultra-sonic ranger; humidity; light; barometer; sound.

One AI area is Text-to-Speech (TTS) technology that converts text into spoken audio output with a computer-generated voice that can read text. The recent progress of text-to-speech synthesis (TTS) technology has allowed computers to synthesize speech with an AI voice [26]. Google "Text-to-Speech" technology module for Python is a command line that coverts text into speech and saves as a .mp3 audio file. Google cloud TTS synthesizes AI voices available in various languages. Regarding specific actions, we mention that the adaptive IoT system greets in three different languages (Romanian, English and German) and sends out warning messages using Google cloud "Text-to-Speech" technology such as whether room temperature is too cold, the light intensity is too low, or environment is too loud. Those values can be than controlled by speech messages such as turn heater on/off; turn light on /off; make room silent. Fig. 5 illustrates the algorithm flowchart with the Grove Pi humidity and temperature sensor the values displayed on the Grove Pi "LCD RGB backlight" screen and as spoken alert message.

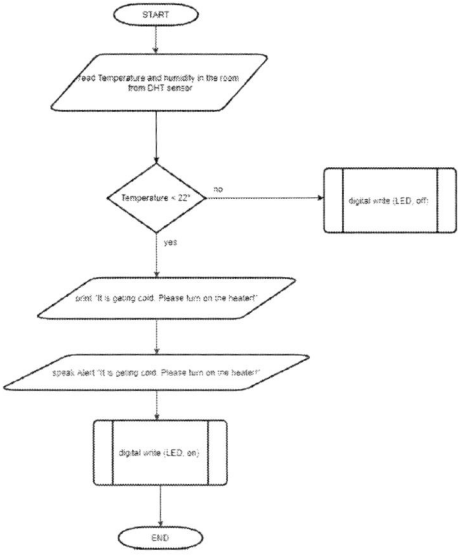

Fig. 5.   Exemplification of algorithm flowchart with Grove Pi temperature and humidity senor

B. Case Studies Biofeedback

For the biofeedback experiments we looked at the activities of a heterogeneous group of four healthy undergraduate (mean age 21, two women and two men) students. In addition to [9], [11], [12], in the first biofeedback case study students wore a wearable device (smartwatch Samsung Gear Fit) during the day for twelve weeks. In the second biofeedback case study participant students performed measurements during learning, and data such as heart rate variability was gathered for a three weeks period by means of a smartphone camera (Samsung S8) and ECG for everybody software [23].

IV. RESULTS AND OUTLOOK

The proposed solution for higher education can be implemented on any kind of platform. Such an architecture can be utilized for other age groups or profiles, as well as to support programs in human resources development. The first case studies highlight that an adaptive framework considers the heterogeneity of learners in a better manner.

In a first biofeedback case study we analyzed pulse, number of steps, active minutes, daily calories intake during learning for twelve weeks in 2019/2020 winter term. Fig. 6 illustrates real time data obtained with wearable devices by the embedded sensors of a smartwatch.

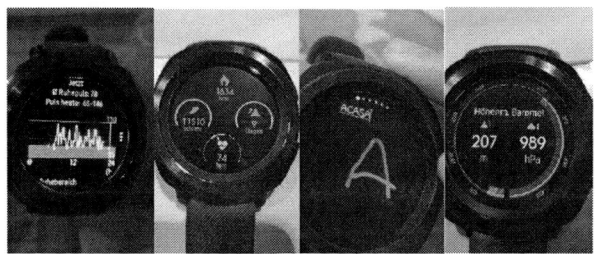

Fig. 6.   Experiment with smartwatch embedded sensors and AI features such as text or voice recognition

Fig. 7 depicts the number of steps in the first week compared to the number after twelve weeks. In the first week our students had a median value of 2419 steps, and the median increased to 4013 steps at the end of the experiment. The size of the boxplot shows the variation in the number of the steps. A large boxplot means a large deviation (see left) over time, and a small one implies that the number of steps varies slightly (see right). We noticed that at the beginning of the experiments the number of steps was between 1583 and 3986 with a large deviation. At the end of the experiment the number of steps increased significantly between 3076 and 4623 with a smaller deviation, meaning the student recognized the usefulness of being active while learning and studying.

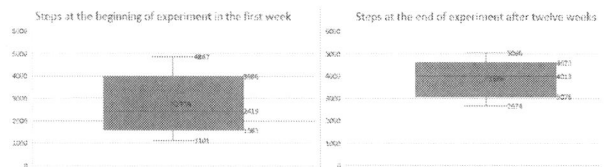

Fig. 7.   Boxplot with the number of steps for participant

Fig. 8 illustrates the measured pulse (HR) values for a healthy student. Legend: blue means average pulse, orange means minimum pulse, and grey means maximum pulse. After twelve weeks, participants reported a decrease in pulse (HR) also during the exam period in the winter term (24.01-13.02).

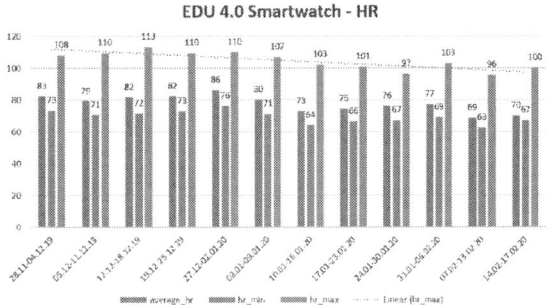

Fig. 8. The participant pulse decrease during examination period in an Education 4.0 experiment after twelve weeks

The second biofeedback case study we analyzed the physiological such as heart rate variability (HRV) parameters such as HR, LF, HF, rMSSD, active minutes, and the number of steps with a smartphone camera for three weeks during a learning period. We use the Photopletismography (PPG) technology [24] available on smartphones for measuring the HRV. As shown in Fig. 9, the top illustrates a strong positive correlation between HF (fatigue – "rest and digest" level) and recovery level (rMSSD), while the bottom figure provides a negative correlation between the pulse (HR) and recovery level (rMSSD). After 12 weeks the following conclusions are drawn: there exists a strong positive correlation between the level of fatigue (HF) fatigue level (HF), steps, active minutes, and pulse; the minimum pulse decreased while the number of steps increased.

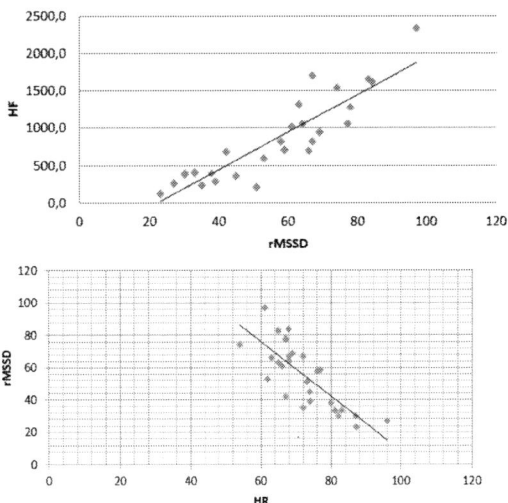

Fig. 9. The figure top provides example of correlation between the fatigue level (HF) and the recovery level (rMSSD), where the bottom figure provides a strong negative correlation between the pulse (HR) and recovery level (rMSSD).

In addition, pulse (HR) decreased as the number of steps increased. Roughly speaking, there is a direct relationship between the level of stress and pulse. In this regard, Fig. 8 and Fig. 9 confirm that while the pulse decreased, the recovery level increased. Students learned through biofeedback methods to be more active. In our experiments, we have shown that when students became aware of tension, they learned to reduce stress. They identified activity as being helpful in dealing with stress while learning.

As an outlook, the development of the adaptive framework with AI methods can take advantage of multimodal data of students. In the "reflection and motivation" phase of the Education 4.0 process, we achieved promising results with the "early recognition system" [4], [19]. This classified at-risk students after the first 2 months of the semester for the Mathematics program of the Faculty of Economic Sciences ("Business Administration") and after the first 2.5 months of the semester for the Business Information Technology program of the Faculty of Computer Sciences ("Business Information Technology") at the Deggendorf Institute of Technology (DIT).

Fig. 10 illustrates the block diagram of the system, which can use the multimodal data of both online courses, but also the data of non-invasive sensors, which are obtained by smart devices (smartphone, smartwatch, smart glasses) worn during the day. So far, we have analyzed the sensor data in a descriptive manner. Although another off-shoot of this work is to study the dependencies between sensor data and learning quality. This is in fact a supervised learning problem, and the goal is to model and recognize the statistical dependencies between input factors and learning quality. Basically, modeling such dependencies can be accomplished in a regression framework, yet, since the learning quality is linked to scores, one can instead use ordinal regression.

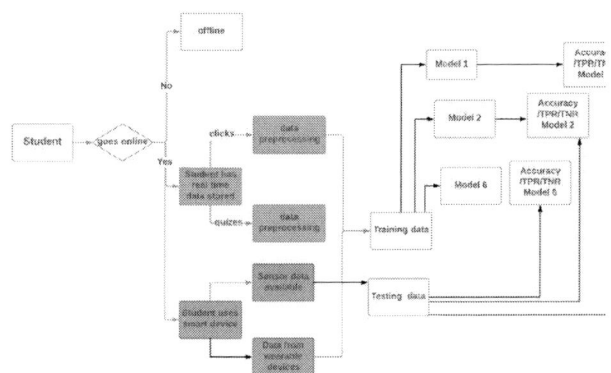

Fig. 10. Future development with AI methods and data from wearables.

In this regard, the scores can be treated as some ordinal classes, i.e., the better the score is, the higher is the ordinal class. Once the model has been trained, it is capable of predicting the scores given sensor data. This allows us to make an early prediction of at-risk students. Apart from the prediction, explaining the ability of the multinomial logistic regression model [27] is a key feature which allows us to recognize important factors and meaningful interactions among features derived from sensor data.

V. Conclusion

This paper is a first step in creating this basis, leading to the development of an exemplary prototypical system that illustrates the possibilities of using a systematic and engineering approach. The first results are promising and encourage us to pursue and conduct further studies, specifically, in terms of an adaptive framework for self-regulated learning. The Education 4.0 framework with wearables and AI models improves students' experience, performance, and motivates teaching and learning.

Acknowledgment

The authors would like to acknowledge Prof. Dr. Dr. Heribert Popp, Prof. Dr. Cezar Ionescu for their valuable support. I would like to express my gratitude to M.Sc. Leon Binder, M.Sc. Charan Singh and all participant DIT students. Furthermore, the project "Bayern Digital - Digital Knowledge Management" of the Bavarian Ministry of Culture and the "smart" project of the "Virtual University of Bavaria" for funding this research.

References

[1] S. Russell, P. Norvig, Artificial Intelligence: A Modern Approach, 3th ed., Pearson, Boston, USA, 2010.

[2] A. Dahik, D. Lovich, C. Kreafle, A. Bailey, J. Kilmann, D. Kenedy, P. Roongta, F. Schuler, L. Tomlin, J. Wenstrup, "What 12,000 employees have to say about the future of remote work", BCG, "Leading in the new reality", August 2020.

[3] M. Vansteenkiste, J. Simons, W. Lens, K.M. Sheldon, E. L. Deci, "Motivating learning, performance, and persistence: the synergistic effects of intrinsic goal contents and autonomy-supportive contexts," Journal of Personality and Social Psychology, vol. 87, pp. 246–260, 2004.

[4] M. I. Ciolacu, "An Adaptive Framework for Computer-Based Learning Technology," Doctoral Thesis, University Politehnica of Bucharest, Faculty of Electronics, Telecommunications and Information Technology, July 2020.

[5] E. Peper, R. Harvey, N. Faass, "Tech Stress: How Technology is Hijacking Our Lives, Strategies for Coping, and Pragmatic Ergonomics," North Atlantic Books, Berkley, California, United States, 2020.

[6] J. A. DeFalco, J.P. Rowe, L. Paquette, V. Georgoulas-Sherry, K. Brawner, B. W. Mott, R. S. Baker, J. C. Lester, "Detecting and Addressing Frustration in a Serious Game for Military Training," International Journal of Artificial Intelligence in Education, vol. 28, pp. 152–193, September 2018.

[7] N. Le, N. Pinkwart, "The system-analytic approach for gifted high school students to develop computational thinking," Proceedings of International Conference on Computational Thinking Education (CTE 2019), Hong Kong, pp. 2-7, 2019.

[8] I. A. Awada, I. Mocanu, "A platform to promote a more active lifestyle between students," Procedings of the 16th International eLearning and Software for Education Conference (eLSE), Bucharest, Romania, vol. 3, pp. 355-362, April 2020.

[9] M. I. Ciolacu, P. Svasta, D. Hartl, S. Görzen, "Education 4.0: smart blended learning assited by Artificial Intelligence, biofeedback and sensors," Proceedings of the 14th IEEE International Symposium on Electronics and Telecommunications (ISETC 2020), Timisoara, Romania.

[10] N. Weinstien, A. K Przybylski, R. M. Ryan, "The index of authonomous functioning: development of a scale of human autonomy," Journal of Research in Personality, vol. 46, pp. 397-413, 2012.

[11] M. I. Ciolacu, L. Binder, H. Popp, "Enabling IoT in Education 4.0 with biosensors from wearables and artificial intelligence," Proceedings of the 2019 IEEE 25th International Symposium for Design and Technology in Electronic Packaging (SIITME), Cluj-Napoca, Romania, pp. 17-24, 2019.

[12] M. I. Ciolacu, L. Binder, P. Svasta, I. Tache and D. Stoichescu, "Education 4.0 – jump to innovation with IoT in higher education," Proceeding of the 2019 IEEE 25th International Symposium for Design and Technology in Electronic Packaging (SIITME), Cluj-Napoca, Romania, pp. 135-141, 2019.

[13] R. A. Sottilare, R. S. Baker, A. C. Graesser, J. C. Lester, "Special issue on the generalized intelligent framework for tutoring (GIFT): creating a stable and flexible platform for innovations in AIED research," Interarntional Journal of Artificial Intelligence in Education, vol. 28, pp. 139–151 , 2018.

[14] K. Scheiter, C. Schubert, A. Schüler, "Self-regulated learning from illustrated text: eye movement modelling to support use and regulation of cognitive processes during learning from multimedia," British Journal of Educational Psyhology, vol. 88, Special Issue: The intersection between depth and the regulation of strategy use, pp. 80-94, 2018.

[15] B. Motyl, G. Baronio, D. Speranza, S. Filippi, "TDT-L0 a test-based method for assessing students' prior knowledge in engineering graphic courses," Advances in Design Engineering. INGEGRAF 2019. Lecture Notes in Mechanical Engineering. Springer, pp. 454-463, 2020.

[16] A. P. Lawson, R. E. Mayer, N. Adamo-Villani, B. Benes, X. Lei, J. Cheng, "Recognizing the emotional state of human and virtual instuctors," Computers in Human Behavoir, Elsevier ltd., vol. 114, 2021, in press.

[17] J. D. Fletcher, R. A. Sottilare, "Shared mental models in support adaptive instruction for teams using GIFT tutoring architecture," International Journal of Artificial Intelligence in Education, vol. 28, pp. 265-285, 2018.

[18] Grove Pi, https://www.seeedstudio.com/GrovePi-p-1672.html.

[19] M. I Ciolacu, A. F. Tehrani, L. Binder, P. Svasta, "Education 4.0 – Artificial Intelligence assited higher education – early recognition system to support students' success," Proceedings of the 2018 IEEE 24th International Symposium for Design and Technology in Electronic Packaging (SIITME), Iasi, Romania, pp. 23-30, 2018.

[20] A. Humpe, L. Brehm, "Problem-based learning for teaching new technologies", 2020 IEEE Global Engineering Education (EDUCON), Porto, Portugal, pp. 493-496, April 2020.

[21] E. Upton, W. Archer, D. Crookes, P. J. Evans, G. Halfacree, R. Hattersley, N. King, B. Nuttall, M. Scott, D. Staple, M. Vanstone, The official Rapberry Pi Projects book, The MagPi, vol. 5, Seymour, London, 2019.

[22] E. Peper, R. I.-M. Lin, R. Harvey, J. Perez, "How posture affects memory and mood," Biofeedback, vol. 45(2), USA, pp. 36-41, 2017.

[23] S. Jokic, I. Jokic, S. Krco, V. Delic, "ECG for Everybody: Mobile Based Telemedical Healthcare System", International Conference on ICT Innovations 2015, Advances in Intelligent Systems and Computing (AISC), vol. 399, Springer, pp. 89-98, 2015.

[24] N. Pinheiro, R. Couceiro, J. Henriques, J. Muehlsteff, I. Quintal, L. Gonçalves, "Can PPG be used for HRV analysis?," 38th Annual International Conference of the IEEE Engineering in Medicine and Biology Society (EMBC), Orlando, USA, pp. 2945-2949, 2016.

[25] F. Wahl, "Methods for monitoring the human circadian rhythm in free-living", Doctoral Thesis, University Passau, Germany, 2019

[26] Y. Shiga, J. Ni, K. Tachibana, T. Okamoto, "Text-to-Speech synthesis." In: Y. Kidawara, E. Sumita, H. Kawai (eds) Speech-to-Speech Translation. Springer Briefs in Computer Science, Springer, Singapore, pp.39-52, 2020.

[27] D. W. Hosmer, S. Lemeshow, Applied Logistic Regression, Wiley Series in Probabilty and Statistics, 2nd ed., 2001.

Adaptation of Electrical Engineering Education to the COVID-19 Situation: Method and Results

B. I. Evstatiev
Department of Electronics
University of Ruse Angel Kanchev,
Ruse, Bulgaria
bevstatiev@uni-ruse.bg

T. V. Hristova
Department of Electrical Engineering
University of Mining and Geology "St. Ivan Rilski"
Sofia, Bulgaria
teodora@mgu.bg

Abstract—**In this paper are presented a new extended methodology and tools for providing electrical engineering education in a distant form. It was used by two Bulgarian universities, in order to adapt their Electrical engineering classes to the COVID-19 situation. The methodology includes several phases: needs analysis, preparation of educational materials, selection of teaching methods, increasing competencies and selection of assessment methods. In the results of the paper is presented the implementation of the methodology based on the EVEEE environment for electrical engineering equipment. At the end of the semester a questionnaire was conducted among the students. The results about the students' opinion clearly indicate that engineering education should be implemented in a distant form only during emergency situations.**

Keywords—electrical engineering, distant education, virtual lab, COVID-19

I. Introduction

In March 2020 Bulgaria officially entered emergency situation aimed at preventing the spread of the COVID-19 virus. As a result, all non-critical spheres of the economy were either closed or certain restrictions were implemented, mainly aimed at isolation. Therefore, all educational institutions in Bulgaria were forced to switch to distant forms of education. The implementation of distant education has several aspects – preparation of the teachers, of the administrative staff and of the students. This study is mainly focused on the first group.

Engineering education relies on several learning activities – lectures, tutorials, practical and laboratory exercises [2]. The possibilities for implementing distant lab exercises are generally limited to circuit simulation software, such as SPICE, Multisim, Microcap, etc. and virtual and remote laboratories [3]. Numerous approaches exist, which allow virtual electrical engineering education at different levels of abstraction. They vary from simplified 2D labs [8] to realistic 3D virtual reality [9]. National instruments have developed a combination of hardware and software tools, which allow the remote control of virtual equipment [10]. Other well-known virtual laboratories are supported by Masachusets Institute of Technology (iLab) [11], Tecnológico de Monterrey University (eLab) [12], the VISIR project [13], and many others.

According to [1], the quality of distant education is ensured with several steps during its implementation: needs analysis,

guidelines and implementation for preparing instructional materials, ensuring learning and practical experience for the teachers and following the syllabus. Numerous studies were aimed at improving the quality during distant forms of education. In the ROLE project [4], the goal was to use web technologies in order to provide personalized and adaptive learning environment. In another study, a dashboard was integrated into an online learning system [5]. Furthermore, a tool was developed that guides students in their work, traces their search for reference materials on the web and tracks their interaction with the system.

Another important factors in distant education are the potential cybersecurity problems and the cybersecurity expertise of the participants [6,7]. This aspect concerns both teachers and student and should be taken care of during design time.

This study presents a methodology and results from the implementation of distant education during the spring of 2020 as part of the Electrical engineering courses in two Bulgarian universities – University of Ruse Angel Kanchev (URAK) and University of Mining and Geology "St. Ivan Rilski" (UMG). It concerns a wide range of problems, aimed at ensuring the necessary quality of education.

II. Materials and Methods

In this section is described the methodology used for preparation of the educational process for distant learning.

A. Needs analysis

The courses that are object of this study are "Electrical engineering" and "Electric measurements". The first was toughed to electrical and computer engineering students in URAK and non-electrical engineering students in UMG. The second course was taught to electrical engineering students in UMG. According to their syllabuses, the courses require the use of lectures, laboratory and tutorial exercises. The laboratory exercises are a fundamental requirement for these courses as during them the students acquire practical knowledge and experience on working with electrical equipment.

In the period 2018-2019, a team from RUAK developed the so-called Engine for Virtual Electrical Engineering Equipment (EVEEE), which is a 2D virtual laboratory, which represents a 3D virtual reality. Therefore, this tool was selected for the implementation of the laboratory exercises, i.e. a shared infrastructure was used by both URAK and UMG students.

B. Development of training materials

Next, several types of educational materials have to be developed, in order to meet the requirements of the target courses syllabus. This includes preparation of:

1. New virtual laboratory exercises, corresponding to the course syllabus;

2. Guidelines for implementation of the virtual labs;

3. Electronic reports, where the student can fill in and summarize the results from their virtual experiments.

Initially, the necessary laboratory exercises have to be identified and selected. However, this should be done in accordance with the available virtual equipment, supported by the EVEEE environment. Once the topics are selected and the necessary equipment is identified, the virtual lab is created using the engine's system for automated design of labs.

Next, methodological instructions are developed for each virtual lab. They could vary in a wide range, such as pdf files (text + screenshots), PowerPoint presentations and recorded video instructions.

In traditional education, students are commonly required to summarize the experimental results in a report. Furthermore, they often require the help of their tutors to summarize the lab results in graphical form and to proceed with further analysis. In a distant form of education, such help would be hard to implement, therefore a different approach is selected. For each virtual lab an electronic report should be developed, which can provide the necessary guidelines, including graphical representation of the experimental results.

C. Choosing the educational methods

Considering the wide variety of specialties, involved in this study, the level of students' IT competence were significantly different. This means that if the students spend too much time on understanding how to work with the virtual tools, this would prevent them from understanding the idea behind the labs and the electrical engineering basics. According to [14], in order to provide gradual progress in the educational process, it is necessary to provide step-by-step instructions during the preparation, implementation and reporting phases. According to the principles of segmentation and adaptation [15], in this case either synchronous introductory virtual laboratory exercise or asynchronous recorded video instructions are extremely necessary. With the help of these tools, teachers demonstrate how to work with the virtual environment. Therefore, the following educational methods should be considered, in order to provide the necessary teaching quality:

1. Sharing materials, such as recorded video instructions, pdf files, etc., using e-Learning websites;

2. Synchronous communication based on video conferencing, including presentations, screen sharing (by both teacher and student), etc.;

3. Asynchronous communication using e-mail.

D. Increasing competencies

According to Bloom's taxonomy [16], the development of competencies is divided into two sub-stages, the first covering the skills of analysis and the second including the skills of synthesis and evaluation. In order to increase the competencies of the students, discussions should be held at the same time to improve the competencies in the chat rooms. The synthesis and assessment skills of advanced students have to be enhanced by executing complex tasks, which is demonstrated to the others using screen sharing.

E. Assessment of the results

To assess the real level of progress of students and the correct segmentation of the material [15] it is necessary to introduce another step in the methodology used so far - assessment of the result. The assessment of the education results has two aspects: the obtained knowledge by the students and the efficiency of the educational process.

The assessment of the acquired knowledge by the students could be implemented in several forms:

1. Assessment of the electronic reports of each student;

2. Discussion of the obtained results.

The first approach could be implemented by assigning points to each section of the report, which take part in the forming of the final mark of the course. The second one could be implemented in synchronous form through a video conference session. The discussion could also be achieved asynchronously, using appropriate questions at the end of each laboratory exercise, which the students should answer in their reports.

Finally, in order to assess the efficacy of the provided distant education, a short survey is prepared (Table 1).

A summary of the described methodology in this section is presented using a Use-case diagram (Fig. 1). The role of the Professor is to provide the educational process with the necessary materials and means. The preparation of the labs includes creating the necessary virtual labs, instructions for them and templates for the lab reports. Different forms of instructions are used, such as e-Learning websites, instructions in text form, recorded video instructions. Furthermore, synchronous and asynchronous instructions were provided using video conferencing and messengers/e-mail communication. In order to learn the course material, the Student actor has to execute the

TABLE I.   QUESTIONS AND ANSWERS OF THE SHORT SURVEY.

Question	Answers
Which from of laboratory exercises do you prefer - virtual or real?	- I prefer real labs - I have no preferences - I prefer virtual labs
Which of the two forms is more interesting for you?	- Real labs - There is no difference - Virtual labs
Which of the two forms is easier to execute?	- Real labs - There is no difference - Virtual labs
Which of the two forms has higher impact on your education?	- Real labs - There is no difference - Virtual labs

2020 IEEE 26ᵗʰ International Symposium for Design and Technology in Electronic Packaging (SIITME)

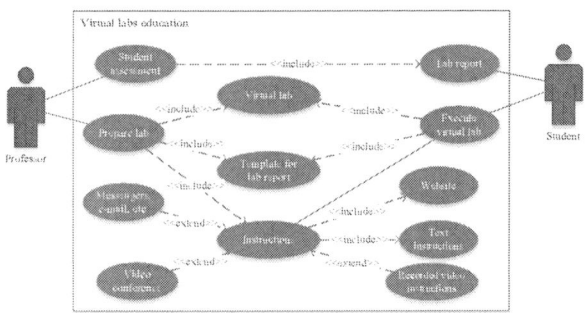

Fig. 1. Use-Case diagram of the adopted methodology

virtual labs by following the instructions and to summarize their results in reports, which are assessed by the Professor.

III. RESULTS AND DISCUSSION

A. Implementation of the methodology

Following the developed methodology, a list of virtual laboratory exercises has been selected, which can be implemented in the EVEEE environment (Table 2).

An example from the virtual lab for investigation of series resonance is presented in Fig. 2. It includes a function generator, a multimeter, a breadboard and three passive elements – resistor, capacitor and inductor. In the other laboratory exercises are also used AC power source, DC power source, power meter, a "black box" two-port device and others devices used in real laboratories.

Next, for each lab were developed implementation guidelines, which have the form presented in Fig. 3. They were structure into "steps" for easier implementation by the students according to segmentation e-learning.

The final stage from the preparation of the learning materials was the development of electronic report templates for each lab. This was achieved online using Google Docs and more precisely Google Sheets. An example report is presented in Fig. 4. According to the developed guidelines were structured the

TABLE II. LIST OF THE SELECTED VIRTUAL LABS

Topic	Course
Investigation of Kirchhoff's laws	EE
Investigation of nonlinear elements in DC circuits	EE
Investigation of series RLC in sinusoidal steady state	EE
Investigation of series resonance	EE
Obtaining the parameters of a two-prot network	EE
Measurement of electric resistance	EM
Measurement of the power factor	EM
Investigation of Thevenin's equivalent circuit	EM
Measurement of electric capacitance	EM

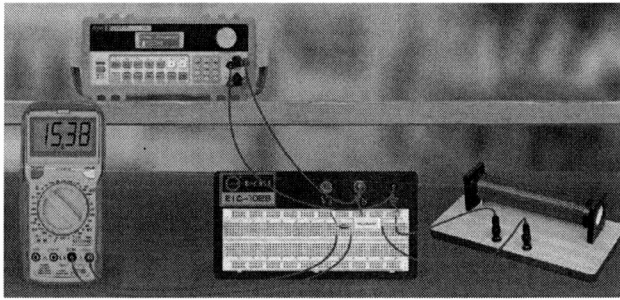

Fig. 2. A virtual lab for investigation of series resonance

Task 1. Investigate the series resonance for the circuit in Fig. 3, using the following elements: $C = 8,4\ \mu F$, $L = 300\ mH$ and $R = 30\ \Omega$.

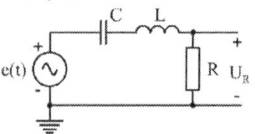

Fig. 3. Circuit schematics for Task 1

Step 1. Connect the circuit from Fig. 3 on the breadboard. Power it from the function generator with input voltage $E = 20\ Vpp$.

Step 2. For each of the frequencies in the report do:
- Set up the frequency from the function generator;
- Measure the RMS value of the voltage on the inductor U_L:

Fig. 3. A fragment from the guidelines for implementing the series resonance virtual lab

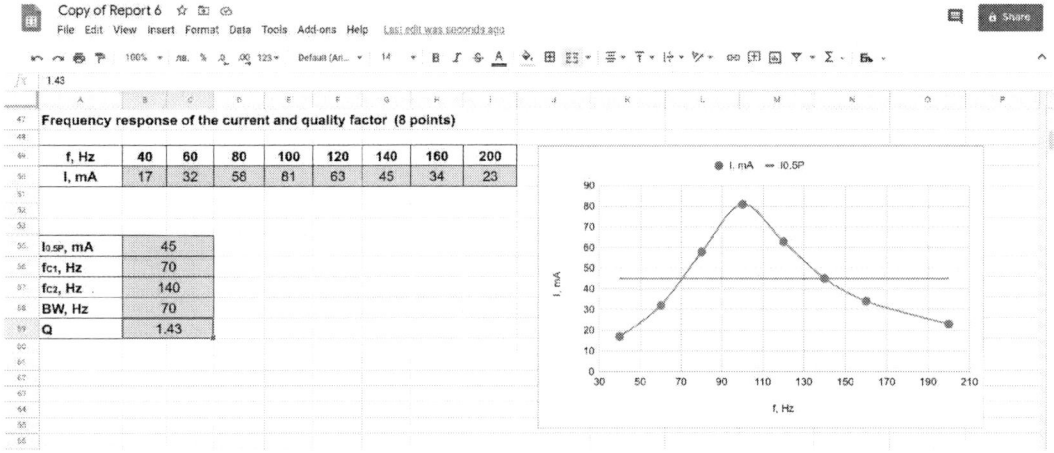

Fig. 4. A lab report implemented in Google Docs

978-1-7281-7507-2/20 $31.00 © 2020 IEEE   306   21-24 October, Pitesti, Romania

necessary tables. Furthermore, charts were pre-created for automatic visualization of the necessary results when the table data is filled in. This was necessary because the tutor do not have the opportunity to provide online help with the drawing of the graphs. From Fig. 4 can also be noticed that the different sections of the lab report are assigned points, which are later used in the assessment process.

The links to the templates were shared with the students allowing them to make a personal copy of each document. The only disadvantage to this approach is that each student should have a Google account.

Next, the developed materials were shared with the students using e-learning websites. A screenshot from the URAK's e-Learning Shell is presented in Fig. 5. On the presented page are available all materials, necessary for the implementation of one of the labs:

- Instructions in PDF format;

- Recorded video instructions;

- Link to the virtual lab;

- Link to the electronic report and instructions for working with it.

Additionally, synchronous and asynchronous form of communication with the students were used in order to provide them with the necessary instructions during the implementation of the virtual labs.

The discussion with the students was implemented in two ways – synchronously, using online video conferencing and asynchronously, by providing guiding questions, which should be answered in the summary of the reports. For example, the questions at the end of the "Series resonance" lab are presented in Table 3.

B. Results from the performed survey

In order to improve and adapt the training process, at the end of the semester all students were asked to fill in a short

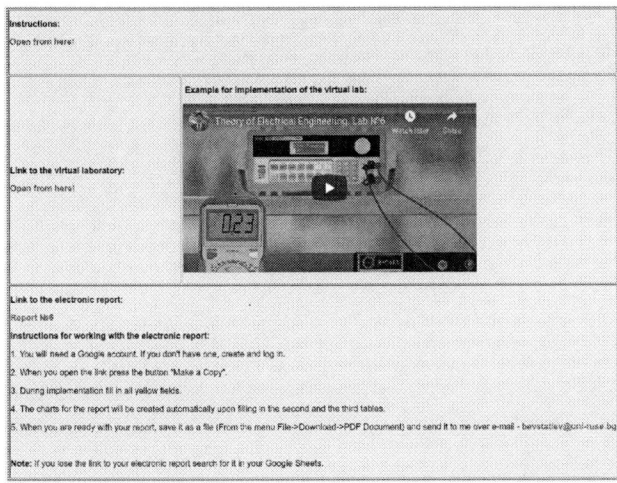

Fig. 5. Screenshot from the E-Learning Shell website of the University of Ruse, providing instructions (in PDF and video format) and links to the virtual lab and electronic report

anonymous questionnaire. It was aimed at obtaining their opinion on learning electrical engineering classes in a distant form. The obtained results are presented in Fig. 6-9. A dominating majority of the participants stated that "they prefer real labs", "real labs are more interesting" and "real labs have higher impact on their education". Therefore, we can make the conclusion that students are fully aware that virtual experiments cannot be a full substitute for real experiments.

TABLE III. DISCUSSION QUESTIONS FROM THE LAB FOR INVESTIGATION OF SERIES RESONANCE.

№	Question
1	What are the voltage drops on the inductor and the capacitor during series resonance?
2	What is the reactance of the inductor and capacitor during series resonance?
3	What is the current in the circuit during series resonance?

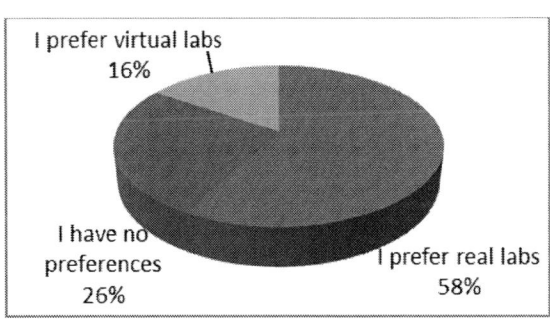

Fig. 6. Answers to the quesiton "Which form of laboratory exercises do you prefer?"

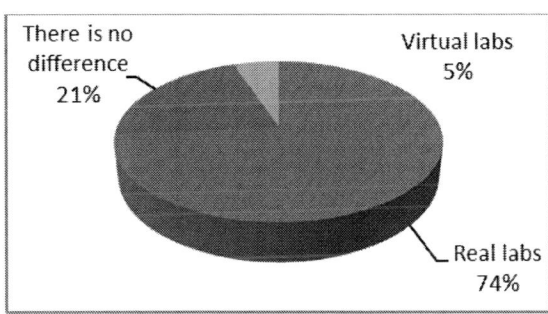

Fig. 7. Answers to the question "Which of the two forms is more interesting for you?"

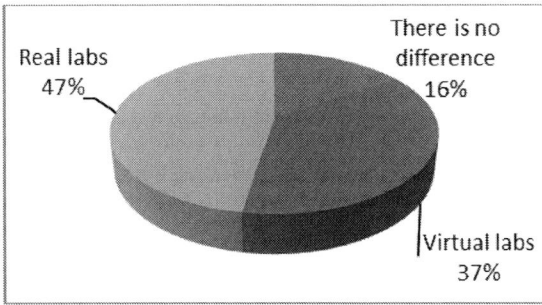

Fig. 8. Answers to the question "Which of the two forms is easier to execute?"

2020 IEEE 26th International Symposium for Design and Technology in Electronic Packaging (SIITME)

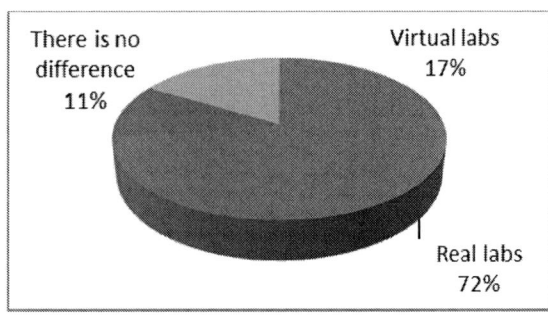

Fig. 9. Answers to the question "Which of the two froms has higher impact on your education?"

The only question in which the votes were divided relatively equally was "Which of the two forms is easier to execute" (47% and 37%, respectively for real and virtual labs). This is probably caused by the difference in the expertise of students on using virtual technologies and the combination of asynchronous and synchronous forms of teaching.

Determining appropriate forms of teaching and methods for adapting the training process will be the next goal of the team.

IV. CONCLUSIONS

This study presents the implementation of distant education in Electrical engineering courses during the summer semester of 2020 for two Bulgarian universities – University of Ruse Angel Kanchev and University of Mining and Geology "St. Ivan Rilski". The students involved were from electrical and non-electrical engineering specialties, studying "Electrical engineering" and "Electrical measurements".

Initially a methodology was presented, which includes the following phases: needs analysis, aimed at identifying the key requirements of the education; learning material development; selection of teaching methods; increasing competencies; assessing students and assessing the education methodology.

Next, the required virtual labs were selected and implemented in the EVEEE environment. For each one was also created methodological instructions and templates of electronic reports. All materials were shared with the students using different e-Learning websites. Both synchronous and asynchronous forms of communication were used to implement the communication with the student. This was achieved with the use of video conferencing, screen sharing, preparation of recorded video instructions, e-mail communication, etc.

At the end of the semester a short survey was perform amongst the students in order to assess their attitude towards distant education in electrical engineering courses. The results were predominantly in favor of real labs. Approximately 70% of the student stated that they prefer real labs, and that real labs are more important for their future. This allows us to make the conclusion that students are quite aware of what is important for them and pure distant education should only be used during emergencies, such as the COVID-19 lockdown in 2020.

Virtual laboratories are suitable when it is necessary to refresh knowledge or in preparation for tests. Therefore, they give opportunities for testing and experimenting. In our opinion, the virtual laboratories can be established as an accompanying training in technical specialties but not as a substitution for real practical training.

ACKNOWLEDGMENT

This work was supported by the University of Ruse Research Fund under contract no 2020-FNI-FEEA-02.

REFERENCES

[1] M. Bušelić , Distance Learning – concepts and contributions, Prethodno priopćenje, Oeconomica Jadertina 1/2012, ISSN 1848-1035.

[2] V. Potkonjak, M. Vukobratovic, K. Jovanovic and M Medenica, "Virtual mechatronic/robotic laboratory - a step further in distance learning", Computers & Education, vol. 55, pp 465-475, 2010;

[3] L. Gomes and S. Bogosyan, "Current trends in remote laboratories", IEEE Transactions on Industrial Electronics, vol. 56(12), pp. 4744-4756, 2009;

[4] Website of project "Responsive Open Learning Environments (ROLE)", FP7, European Commission: https://cordis.europa.eu/project/id/231396 .

[5] D. Taibi, F. Bianchi, P. Kemkes, and I. Marenzi, "A learning analytics dashboard to analyse learning activities in interpreter training courses", Communications in Computer and Information Science, pp. 268-286, Springer International Publishing, 2019.

[6] W. Dimitrov, B. Jekov, E. Kovatcheva, L. Petkova, "An analysis of the new challenges facing cyber security expertise", 12th International Conference on Education and New Learning Technologies, 6-7 July, 2020.

[7] W. Dimitrov, B. Jekov, E. Kovatcheva, T. Ostrovska, "A High-Efficient Low Budgetary Approach to Cyber-Security Exercises", 12th International Conference on Education and New Learning Technologies, 6-7 July, 2020.

[8] J. Fuertes, et al., "Virtual and Remote Laboratory of a DC Motor", Proceedings of the 9th IFAC Symposium Advances in Control Education, Nizhny Novgorod, Russia, June 19-21, pp. 288-293, 2012.

[9] A. Carpenoa, et al., "3D virtual world remote laboratory to assist in designing advanced user defined DAQ systems based on FlexRIO and EPICS", Fusion Engineering and Design, Vol. 112, pp. 1059-1062, 2016.

[10] "Using myDAQ with the NI ELVISmx Function Generator Soft Front Panel": http://www.ni.com/tutorial/11503/en

[11] G. Viedma, I. J. Dancy, and K. H. Lundberg, "A Web-based linear-systems iLab", Proc. the 2005 American Control Conference Proceedings, Portland, USA, June 8-10, pp.5139–5144, 2005.

[12] M. E. Macias, and I. Mendez, "elab - Remote Electronics Lab In Real Time", Frontiers In Education Conference - Global Engineering: Knowledge Without Borders, Opportunities Without Passports, 2007. FIE '07. 37th Annual, pp. S3G-12 -S3G-17, 2007.

[13] L. Claesson, L. Hakansson, "Using an Online Remote Laboratory for Electrical Experiments in Upper Secondary Education", International Journal of Online Engineering (iJOE), vol. 8, pp. 24-30, 2012.

[14] D. Izvorska, G. Velev, M. Avdeeva, "Good practices in language training at the Technical University of Gabrovo", Journal of Mining and Geological science, section Humanities And Economics, vol. 62, p.113-117, 2019.

[15] Sv. Toncheva-Pencheva and Yo. Anastasova, Personalization of distance and e-learning for learning content, Journal of Mining and Geological Sciences, Vol. 62, ISSN 2683-0027 (online) n. 4, p.108-111, 2019

[16] B. S. Bloom, "Taxonomy of educational objectives", Vol.1: The cognitive domain. New York, NY: McKay, 1956.

Higher Education with Distance Learning during COVID-19 Pandemic – a Transitional Semester from the Viewpoint of Teachers

Attila Géczy, Olivér Krammer

Department of Electronics Technology
Faculty of Electrical Engineering and Informatics
Budapest University of Technology and Economics
Budapest, Hungary
gattila@ett.bme.hu

László Sujbert

Department of Measurement and Information Systems
Faculty of Electrical Engineering and Informatics
Dean's Offices
Budapest University of Technology and Economics
Budapest, Hungary

Abstract—**2020 presented unprecedented challenges in the higher education system. The first wave and the still-lasting COVID-19 pandemic brought the need for a new mindset, both from educators-teachers and students around the world. In this paper, we focus on our country Hungary and present the general feedback on distance education at Budapest University of Technology, Faculty of Electrical Engineering, and Informatics from the viewpoint of teachers. After a survey within the course coordinators at the faculty, we present the general results and the lessons learned from the obtained data. The feedback form questions focused on the aspect of knowledge transfer and the examination methodologies at the faculty. The results and the responses give an overall outlook on the success of the transitional semester and might offer directions for future problem-solving in remote education.**

Keywords— Remote education, distance learning, teacher, COVID-19

I. INTRODUCTION

The topic of COVID-19 affected most relatable academic fields recently, ranging from basic research to applied studies and even education. Most of the published studies are in the preliminary phase after an approximately half-year period of experience with exciting and noteworthy initial results. Our university also faced challenges during the semester, when the first wave of the pandemic arrived. During a fast and efficient transition, the Faculty of Electronic Engineering and Informatics lead a pioneering role in the transformation of higher education in our country and elected a remote teaching system, which leads to the successful completion of the given semester. The paper focuses on the results of this transitional period and the following months of the semester from the viewpoint of the teachers and course coordinators. The paper includes the feedback of our colleagues and presents the lessons learned during this period, also highlighting that the unfortunate pandemic can be a catalyst for digitalized education.

II. REMOTE TEACHING IN THE LITERATURE

Distance education was investigated years before the pandemic arrived, from the aspect of different perspectives, the catalysts of a positive outcome, and experience during such teaching methodology [1]. Student satisfaction was also investigated recently during such courses with relatively low numbers (~100); however, the analysis was thorough, with frequency, mean, std. deviation, min-max, one-way ANOVA, and T-Tests to investigate the results in in-depth details [2].

The recent literature of remote teaching focuses on technical methodologies and the effects on academic life. The example of Laplante points to a clear tendency in our profession: restaurants and their "contactless delivery" approach must be imported to higher education too, and in this world, computer science professionals can be the advocates for the rest of the education field [3]. Online education is a commonly discussed approach (either on computers or mobiles [4]), where online teaching, or blended (live broadcasting + MOOC - massive open online courses) may be a solution of the upcoming semesters during crisis times, such as a global pandemic. Chen et al. [5] highlighted the possibility to adapt to BOPPPS (Bridge-in; Objective/Outcome; Pre-assessment; Participatory Learning; Post-assessment; Summary) methodology as a reforming approach during COVID-19 periods [6]. Others focused on their country related aspects. Feng et al. discussed the situation in the universities of China [7]. Khattar et al. investigated the activities and mental health of students in India with machine learning techniques [8]. In a most recent survey, it was found that Arabs welcomed the social distancing initiatives during distance education, despite their usual social closeness [9]. A general remark can be concluded from the works: teaching is not merely technical - the content is an equally fundamental aspect. As the distance between students and teachers increased in the traditional meaning, the personal distance became smaller everywhere.

Many of the works focus on tackling the practical laboratories for engineering education with remote access [10-12], mostly for technology-based education. Focus on information technology and electronics engineering (such as in our case) is also vital; usually, this field yields the fastest advances of the remote education of laboratory practices and computer-aided design education [13-15].

The latest results also focus on the premise of telepresence robots, which can improve student engagement and experience, compared to distance learning tools (DLT) [16]. However, this approach is not entirely applicable to the given higher education concept.

III. METHODOLOGY

In our paper, we focus on the remote education introduced to Budapest University of Technology, Faculty of Electrical Engineering and Informatics, where the teachers (generally experienced in computer science and IT tools) were asked about their participation and connected experience during the transitional semester.

The faculty adapted to remote education in the March of 2020 on the platform of MS Teams accompanied by Webex for massive colloquium sessions and Moodle for further administration and remote examinations. The courses were directed to an MS Teams group as a central gathering point for students and teachers.

Briefings were held by the management time to time with the given tools, focusing on real-time streams and connecting students from different years and curricula for the effective flow of information. A new weekly newsletter was initialized for the colleagues to keep every worker informed at the faculty about recent changes in the directives.

General lectures were held in online classes, where real-time feedback, recording, interactivity and Q+A options were offered to the students. Available laboratory classes were tackled with proprietary online tools, where video and software-based materials were presented to the students, with tasks to complete at home with the provided tools. Then the reports were handed in by the students, then online consultation and feedback were delivered from the teachers. Examinations were performed in Moodle, with various possibilities offered by the platform.

According to the faculty, the following active numbers were reported under the given transitional semester. [17]

- 5200 students (international: 13 %),
- 369 teachers or educators,
- 873 courses,
- 873 Teams groups,
- 550 internal courses (the rest is handled by other faculties).
- 545 active WebEx accounts.

A survey was performed at the end of the semester in the circle of ~190 course coordinators in both electronics engineering and computer engineering programs. The coordinators teach in BSc, MSc, and PhD levels. The fourteen questions spanned from the knowledge transfer aspect to the examination methodologies. Most of the questions were based on a 0-10 Likert scale; four questions focused on explanatory responses. The number of the faculty's students is practically one-quarter of the total student count of the university, so we can say that the numbers show strong feedback on the transitional semester.

IV. RESULTS

In this chapter, the following questions were evaluated. Table 1 summarizes the questions regarding knowledge transfer (Q1-4) and examinations (Q5-9)

TABLE I.   QUESTION SUMMARY

Questions regarding knowledge transfer and examinations.	
Q1	Overall rating of additional workload, caused by remote Education Transition.
Q2	Overall satisfaction with the attendance and attitude of the students.
Q3	Overall satisfaction with the regulations and instructions coming from the University or the Faculty.
Q4	Overall satisfaction with the newly introduced informatics systems?
Q5	Opinion about fair assessment, given the fact that students might used non-acceptable tools during examination.
Q6	Overall rating of additional workload of examination virtualization.
Q7	Opinion about fair assessment, and evaluation of obtained knowledge, given the fact that students were behaving ethical during examinations.
Q8	Overall satisfaction with the regulations and instructions coming from the University or the Faculty regarding examinations.
Q9	Overall satisfaction with the newly introduced informatics systems regarding examinations.

A. Knowledge Transfer

Figure 1 presents four questions and their evaluation regarding knowledge transfer given by the percentage of answers on the scale.

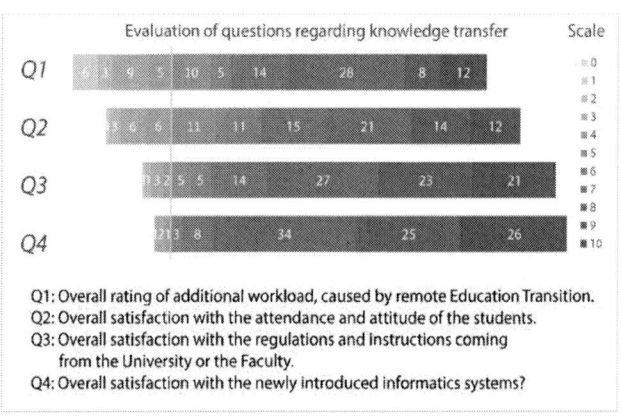

Q1: Overall rating of additional workload, caused by remote Education Transition.
Q2: Overall satisfaction with the attendance and attitude of the students.
Q3: Overall satisfaction with the regulations and instructions coming from the University or the Faculty.
Q4: Overall satisfaction with the newly introduced informatics systems?

Fig. 1. Knowledge transfer questions evaluation

The overall additional workload was not found to be significant by the colleagues. Most of the answers were located on the positive side of the scale. However, a few colleagues found it problematic to transition to online from offline education. From the aspect of students, the attendance was a bit mixed too. Still, it is easy to see that general satisfaction can be outlined from the results. The colleagues found satisfactory tools for the given tasks in the system, which was set up by the faculty during the transitional time. The regulations and instructions were mostly based on general regulations of the Hungarian Government. Due to the fast reaction, the effective communication on the faculty, it was found that the colleagues were satisfied with the final, distilled regulations and the communication coming from the university or the faculty itself. It was found that most of the course coordinators were generally positive about their personal experience regarding the knowledge transfer capabilities of the proposed remote learning systems.

It can also be concluded that the percentage of negative feedback is slightly increased when asked about the attitude of the students and the additional workload caused by the transitions. Still, the overall results could be stated as overwhelmingly positive, with the occasional critical mindset. These criticisms were taken into account on a personal basis, where the will for improving the processes helped the colleagues to work together on new solutions for the following autumn semester.

B. Examinations

Figure 2 presents five questions and their evaluation regarding examinations given by the percentage of answers on the scale.

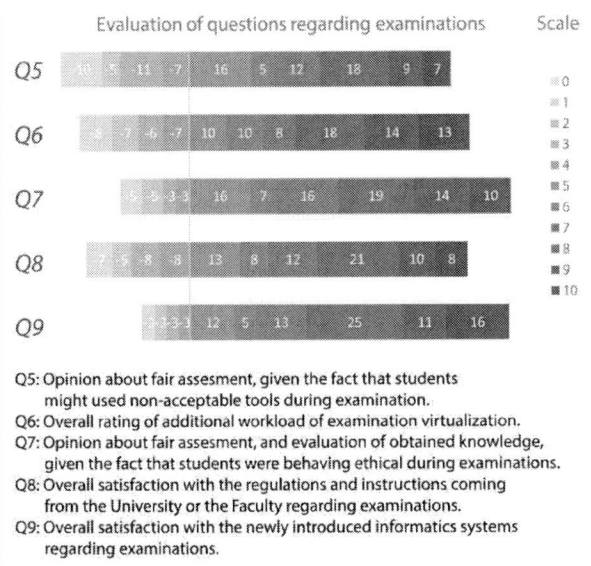

Q5: Opinion about fair assesment, given the fact that students might used non-acceptable tools during examination.
Q6: Overall rating of additional workload of examination virtualization.
Q7: Opinion about fair assesment, and evaluation of obtained knowledge, given the fact that students were behaving ethical during examinations.
Q8: Overall satisfaction with the regulations and instructions coming from the University or the Faculty regarding examinations.
Q9: Overall satisfaction with the newly introduced informatics systems regarding examinations.

Fig. 2. Examination questions evaluation

As Figure 2 shows, the results are more mixed than that in the case of previous questions. Examinations showed more significant challenges, but this part of the questionnaire also presented mostly positive comments.

The colleagues were the most dissatisfied about the lack of fair assessment – the online examination could result in the use of non-acceptable, unfair tools during the tests. It was an overall additional reaction that the method of examination should be revised and configured so that the students could rely more on their critical thinking and less on facts and lexical knowledge. The results were slightly better regarding the additional workload of virtualized exams. The negative score comes from the fact that many teachers tried to revise their previous examination habits to fit the new challenges of remote assessment. If the teachers suggested an ethical approach from the students, the results pointed to more positive responses regarding student behavior during exams. The examination

regulations were not welcomed as positively as the general regulations regarding knowledge transfer; however, most of the reactions were still showing a positive attitude from the educators' side. The colleagues mostly praised the informatics systems regarding examinations.

C. Selected feedback from colleagues

Four questions focused on the written opinion, description of teachers, course coordinators about the distance education, and remote examination. Some selected feedbacks from colleagues are organized below along these questions.

*Q10 – Please describe your **positive** experience with remote teaching!*

"The captured videos about the lectures can be used and also viewed later, thus resulting in more flexible time-schedule for teachers and students."

"Prompt feedback from students during the lecture via the chat channel; higher interactivity; the students answered each other's questions too."

"Higher student attendance in lectures; students could join in a more convenient way to the lectures."

*Q11 – Please describe your **negative** experiences about remote teaching!*

"Less non-verbal feedbacks from students, which makes the teacher hard to decide whether the explanations are clear and was followed or not."

"Less control over the progress of the students in the semester (flexibility); some students were failed because of their inappropriate time-schedule."

"Remote teaching is inappropriate for lab courses or for practicing seminars; lack of tools like intelligent/digital boards in the university as a whole."

"A lot of effort was needed from the teachers to transfer their teaching materials into distance learning form; the ratio reached 6 to 1 between preparing materials and delivering the specific lecture."

*Q12 – Please describe your **positive** experiences about remote examination!*

"The possibility of easy/automatic evaluation of well-prepared examination."

"The question banks can be used for further examinations; huge work at the beginning to prepare the question, but much less effort is needed in the preparation of further exam sheets."

"Remote examination provided tools for the creative assessment of students' knowledge."

*Q13 – Please describe your **negative** experiences about remote examination*

"Finding out the correct methodology for the distance examination, and the preparation of questions were required much more work to avoid the unfairness during the examinations."

"The distance examination can hardly assess the synthesizing knowledge of students; the solution of engineering problems cannot be expected via distance examinations."

"In some cases, the submission of home-works was late, maybe because of the lack of self-motivation of the students."

V. CONCLUSION

This paper shows the repercussions of COVID-19 pandemic induced remote education and the rapid transition during the 2020 spring semester at Budapest University of Technology, Faculty of Electrical Engineering and Informatics from the side of teachers and educators. Overall, the responses were positive, showing that the almost 200 course coordinator colleagues were satisfied with the technical and ethical possibilities presented by the university and the faculty. Knowledge transfer and technical questions related to the topic were scoring better than the examination possibilities, where the most concern was recorded in relation to the availability of non-accepted, unfair tools during examinations from the side of students, the resulting additional workload, and additional directives coming from the university. Overall, all questions scored mostly positively; however, the future tasks must be focused on unfolding the negative aspects and comments to improve the remote education efficiency, quality, and experience for both students and teachers.

The presented results were taken into account during the preparations for the second wave of the pandemic, which is currently tackled during the submission of this paper.

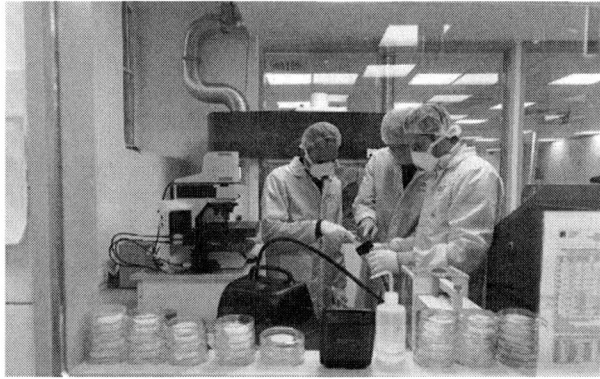

Fig. 3. Mask filtration material analysis at BME VIK (Photo source: MTI)

During the first wave of the COVID-19 pandemic, the faculty was also involved in other state-of-art research, such as evaluating masks and filtration materials [18] (as seen in Figure 3.) or proprietary respiratory system development.

ACKNOWLEDGMENT

The help and time of the course coordinator colleagues are highly appreciated.

REFERENCES

[1] E.C.Boling, M.Hough, H.Krinsky, H.Saleem, M.Stevens, Cutting the distance in distance education: Perspectives on what promotes positive, online learning experiences, The Internet and Higher Education, Volume 15, Issue 2, March 2012, Pages 118-126

[2] Semih Caliskan, Sibel Suzek, Deniz Ozcan, Determining student satisfaction in distance education courses, Procedia Computer Science Volume 120, 2017, Pages 529-538

[3] Phil Laplante, Contactless U: Higher Education in the Postcoronavirus World, Computer, 2020. July, pp. 76-79, DOI: 10.1109/MC.2020.2990360

[4] J. Romero-Rodríguez, I. Aznar-Díaz, F. Hinojo-Lucena and G. Gómez-García, "Mobile Learning in Higher Education: Structural Equation Model for Good Teaching Practices," in IEEE Access, vol. 8, pp. 91761-91769, 2020.

[5] Y. Chen, Y. Zheng and T. Yu, "Construction and Implementation of Blended Online Teaching Mode Based on Live Broadcasting and MOOC," 2020 IEEE 2nd International Conference on Computer Science and Educational Informatization (CSEI), Xinxiang, China, 2020, pp. 260-263,

[6] W. Peng, X. Li and L. Fan, "Research on Information-based Teaching and its Influence on Future Education under the Background of Epidemic Situation," 2020 IEEE 2nd International Conference on Computer Science and Educational Informatization (CSEI), Xinxiang, China, 2020, pp. 340-343

[7] X. Feng, X. Hu, K. Fan and T. Yu, "A Brief Discussion About the Impact of Coronavirus Disease 2019 on Teaching in Colleges and Universities of China," 2020 International Conference on E-Commerce and Internet Technology (ECIT), Zhangjiajie, China, 2020, pp. 167-170

[8] A. Khattar, P. R. Jain and S. M. K. Quadri, "Effects of the Disastrous Pandemic COVID 19 on Learning Styles, Activities and Mental Health of Young Indian Students - A Machine Learning Approach," 2020 4th International Conference on Intelligent Computing and Control Systems (ICICCS), Madurai, India, 2020, pp. 1190-1195, doi: 10.1109/ICICCS48265.2020.9120955.

[9] Abdulrahman Essa Al Lily, Abdelrahim Fathy Ismail, Fathi Mohammed Abunasser, Rafdan Hassan, Alhajhoj Alqahtani, Distance education as a response to pandemics: Coronavirus and Arab culture, Technology in Society, Volume 63, November 2020, 101317

[10] Mykhailo Poliakov, Ibrahim Rida, Remote laboratories for engineering education: status and prospects, 2020 Advances in Science and Engineering Technology International Conferences (ASET), 4 Feb.-9 April 2020, 10.1109/ASET48392.2020.9118221

[11] Wânderson de Oliveira Assis, Alessandra Dutra Coelho, Hugo da Silva Bernardes Gonçalves, WebLabs: Remote Access Experiments for Teaching Process Control in Engineering Courses, 2020 XIV Technologies Applied to Electronics Teaching Conference (TAEE), 8-10 July 2020, 10.1109/TAEE46915.2020.9163667

[12] Ricardo Costa, Paulo Bastos, Gustavo Alves, Manuel Carlos Felgueiras, André Fidalgo, An educational remote laboratory for controlling a signal conditioning circuit with an LDR sensor, 2020 XIV Technologies Applied to Electronics Teaching Conference (TAEE), 8-10 July 2020, 10.1109/TAEE46915.2020.9163688

[13] Dušan Gleich, Andrej Sarjaš, Marko Malajner, Poliksena Miteva, Jelena Stojanovic Josifovska, Natasha Bozinovska, Zivko Kokolanski, Bodan Velkovski, Srečko Simović, Matic Podobnik, Matjaž Šegula, Zlatko Ruščić, Marijan Pavosević, CORELA Collaborative Learning Environment for Electrical Engineering Education, 2020 International Conference on Systems, Signals and Image Processing (IWSSIP), 1-3 July 2020, 10.1109/IWSSIP48289.2020.9145082

[14] Aranzazu D. Martin, Juan M. Cano, Jesus R. Vazquez, Diego A. López-García, A Low-Cost Remote Laboratory for Photovoltaic Systems to Explore the Acceptance of the Students, 2020 IEEE Global Engineering Education Conference (EDUCON), 27-30 April 2020, 10.1109/EDUCON45650.2020.9125211

[15] Ricardo Martin Fernandez, Felix Garcia-Loro, Clara Perez, Manuel Castro, Work-in-Progress: Matrix Analyser and Circuit Design Automator: a Software Tool, 2020 IEEE Global Engineering Education Conference (EDUCON), 27-30 April 2020, 10.1109/EDUCON45650.2020.9125181

[16] Naomi T. Fitter, Nisha Raghunath, Elizabeth Cha, Christopher A. Sanchez, Leila Takayama, Maja J. Matarić, Are We There Yet? Comparing Remote Learning Technologies in the University Classroom, IEEE Robotics and Automation Letters, 5(2), April 2020, pp. 2706-2713

[17] Attila Schopp, Átállás a távoktatásra a BME VIK-en (Transition to remote education at BME VIK) ITBusiness.hu 26-04-2020. Link: https://itbusiness.hu/human/behaviour/munkaero-fejlesztes/tiz-nap-alatt-tiz-evnyi-fejlodes-zajlott-le (accessed at 28-09-2020.).

[18] Haijun He, Min Gao, Balázs Illés, Kolos Molnar, 3D Printed and Electrospun, Transparent, Hierarchical Polylactic Acid Mask Nanoporous Filter, International Journal of Bioprinting, Vol 6, Issue 4, 2020, Article identifier:278

Integrated topics approach for teamwork students projects

1st Madalin Vasile Moise
Center for Electronics Technology
and Interconnection Techniques
Polytechnic University of Bucharest
Bucharest, Romania
madalin.moise@cetti.ro

2nd Paul Mugur Svasta
Center for Electronics Technology
and Interconnection Techniques
Polytechnic University of Bucharest
Bucharest, Romania
paul.svasta@cetti.ro

3rd Elena Valentina Dumitrascu
Center for Electronics Technology
and Interconnection Techniques
Polytechnic University of Bucharest
Bucharest, Romania
valentina.dumitrascu@cetti.ro

Abstract—The choice for continuous solutions that demand connectivity is spreading globally, this paper presents a university program aimed to create an environment where students can develop professional competences through practice. We have designed a pedagogical program able to evaluate, train and generate mechanisms to support learning for students even if they need to work remotely. In this way, we managed to expand digitalization, hybridization and ubiquitous learning; and we promoted internal reflection on the renewal of teaching and learning model. The program is designed to run throughout the 6th semester of bachelor student's courses at electronics universities. Students receive the project theme from teaching stuff, it consist from a complex embedded system, at the beginning of the semester; this is divided and assigned to teams, each consisting of two students. This way the division into smaller projects is promoted, the union of which actually represents the final project. This approach has multiple advantages: each team is provided with the individualization of the theme, but at the same time the teams are working on a complex project, each contributing to it. When solving the projects, the fact that the topics interact means that the students need to cooperate with one another in order to find optimal solutions. These subsystems are assembled at the end of the semester and interconnected using a standard interface without facing problems like the lack of wires or that other teams are using other protocols.

Keywords— embedded, project, teamwork, students

I. INTRODUCTION

Since its foundation, universities, like any other social institution, have had to face devastating epidemics that have affected their daily functioning. However, they need to survive and continue their mission even with their doors closed; pandemic adds a further degree of complexity to higher education globally. Inevitably, the loss of social contact and socialization routines that are part of the daily experience of a higher education student will have consequences [1].

Students must strive to adapt to what for many of them are new formulas for teaching and learning [2]. The demand for an almost instant digital transformation of educational institutions requires not only the inclusion of technologies, but also the creation or change of processes. Our approach helps them by providing content for distance learning and trying to make up for the lack of face-to-face classes and hardware components with regular courses and simulators using virtual platforms that can provide the necessary information's and tools for them to finish their project while keeping in touch with colleagues. Application of technology in remote teaching and learning are still in the development phase [3], benefits are evident, and students can use their own device to access an environment where they can develop, program and test the project before the final physical implementation.

II. OVERVIEW

As per now, in science education, the most common learning method use is "on-site learning", this is the classic way of education and it provide high efficiency due to direct interaction between the teacher and student.

After the outbreak of the COVID pandemic, universities were forced to adopt a distance learning system using on-line platforms. The main disadvantage of this system is that it is not the most efficient method, mainly due to the latency in teacher-student communication.

We have implemented a program aimed to create an environment where student can work much easier and in a team, where they are not tight to a fixed schedule while maintaining the interaction with the teacher for making the learning process faster and efficient.

The flexibility of the schedule is achieved by using remote laboratory or virtual laboratory instead of real laboratory.

The scenario focuses on the activities related to laboratory and to maximize the efficiency of the learning process while reducing the cost of activities.

The emphasis was on practical achievement, looking to familiarize students with the specific aspects of building, testing and troubleshooting a prototype, as well as gaining experience inherent in overcoming problems that arise.

The program is an integrative project that uses algorithms, notions of microprocessor architecture, microcontrollers, and basic knowledge of analog and digital electronics to achieve an optimal learning solution,

Is designed to run throughout the 6th semester of bachelor student's courses at electronics universities. Students receive the project theme from teaching stuff, it consist from a complex embedded system, at the beginning of the semester; this is divided and assigned to teams, each consisting of two students. This will encourage teamwork, the results of which will actually constitute the final project. (Fig.1).

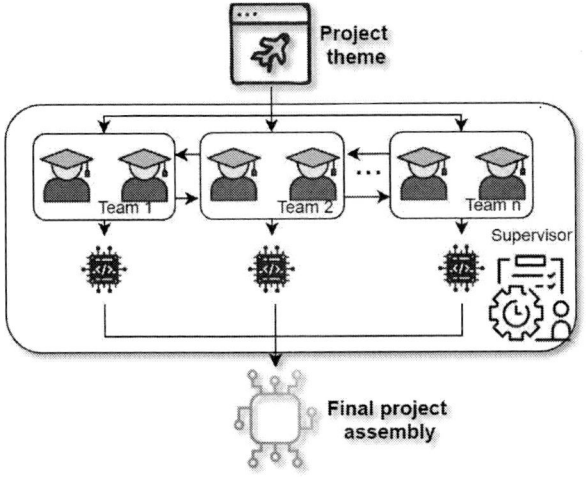

Fig. 1. Program diagram

III. IMPLEMENTATION

After each team has received their subsystem as part of the project, they will also receive the hardware necessary for its physical realization, even without them having access to a laboratory (i.e. breadboard, microcontroller, wires, sensors).

This approach has the advantage that each team has an individual topic, which contributes to the realization of a complex project, this means that teams will have to work together to find optimal solutions.

The program flow:

- Each team perform circuit simulations in an electronic simulation environment (i.e. PSpice, Proteus, Tinkercad)

- Using a desktop utility (i.e. Microsoft Teams, AnyDesk, Google Teams) for regular meeting to monitor and quide the students work

- Running experiments in a virtual laboratory

- Running experiments in a real laboratory, if necessary

The subsystems are assembled at the end of the semester and interconnected using a standard interface (i.e. DC-BUS transceiver [4] Fig.2) for all teams, without reaching the situation in which connections cannot be established due to the use of other protocols.

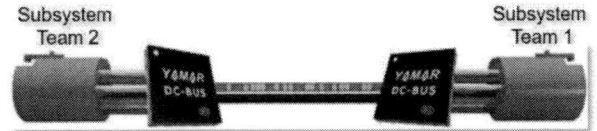

Fig. 2. Communication using Yamar's DC-BUS transceiver

Complex embedded systems can range from partial or full electronic architecture of a car to the infrastructure of a smart home or factory. Each subsystem assigned to a team can perform basic functions such as signal acquisition, data processing, peripheral control, user interface and data storage using the hardware chosen by them and approved by the teaching stuff.

Resources and technological solutions are shared between the students via a website created by each team; it contains all the steps that were carried out in realizing the project (Fig.3).

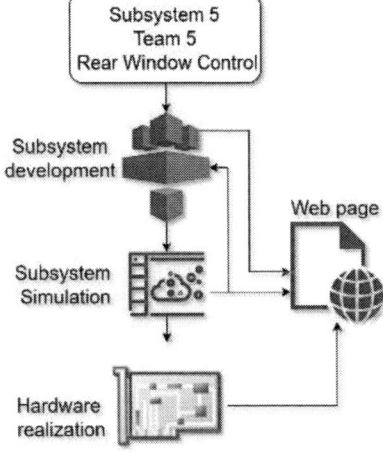

Fig. 3. Subsystem development steps

IV. RESULTS

As an initial program we chose as the theme the realization of a structure of several existing modules inside a car (i.e. door control module, seat control module) which are controlled by a main module (Fig.4.-BCM).

Communication with the embedded systems of other teams is done using a device that uses DC-BUS technology developed by Yamar [4] that can transmit different communication protocols using the DC power line at a maximum speed of 1.4 Mbit/s.

2020 IEEE 26th International Symposium for Design and Technology in Electronic Packaging (SIITME)

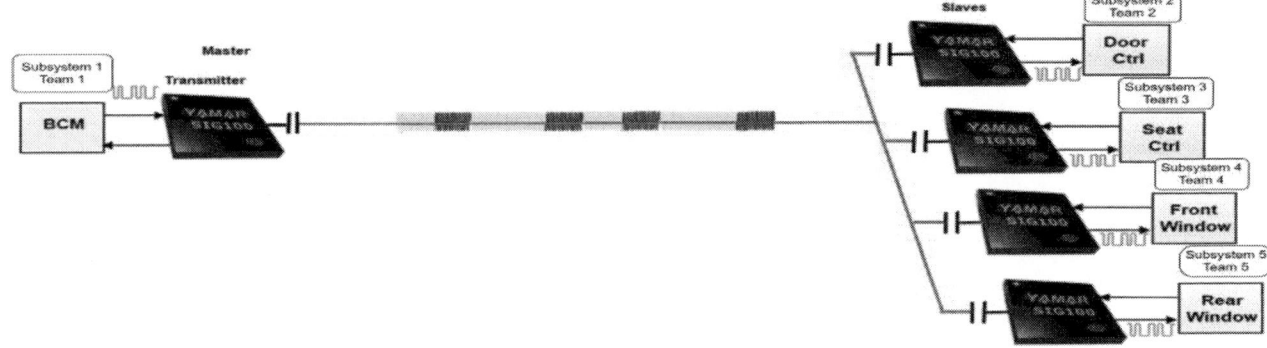

Fig. 4. Example of complex embedded system using Yamar's devices

In this way, the number of electrical wires that can have a large volume is reduced and the risk of problems in the development phases is also reduced becoming a simple plug & play solution. The communication standard differs depending on the destination of the project, for example, if it will be a car architecture then CAN, UART or LIN will be used.

Each team was able to finish the project, the results of the examination at the end of the program: from the group of 14 students, 48% of them obtained a grade of 10, 30% of them obtained a grade of nine and 22% of them obtained a grade of seven. Grades were given from 1 to 10, 10 being the highest grade (Fig.5). Compared to the regular program, out of the group of 60 students, only 3% managed to obtain a grade of 10, the most persistent grades being 9 and 5 (Fig.6).

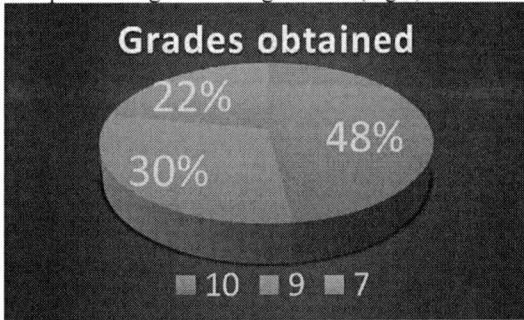

Fig. 5. Grades obtained during program

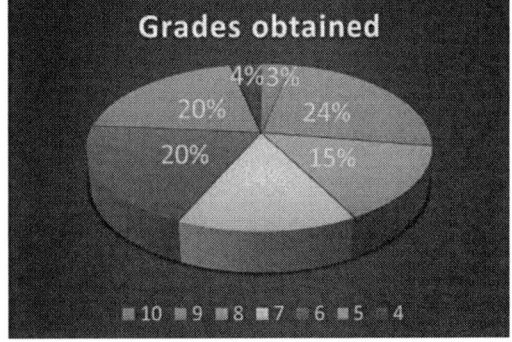

Fig. 6. Grades obtained during normal program

At the end of the project, a questionnaire was completed that targeted students from several groups, including those involved in the program. We have conducted the survey with different questions in order to have a feedback and to be able to compare the students opinions with those from other programs so that we can decide the efficiency of the program not only according to the grades obtained.

Most students in the target group categorized the approach of the project as effective (Fig. 7) compared with students from the other group where it was mostly satisfactorily (Fig.8).

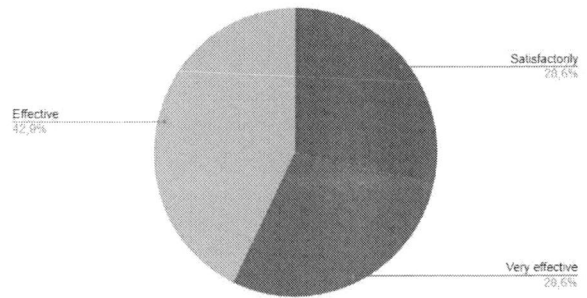

Fig. 7. Target group feedback on the program approach

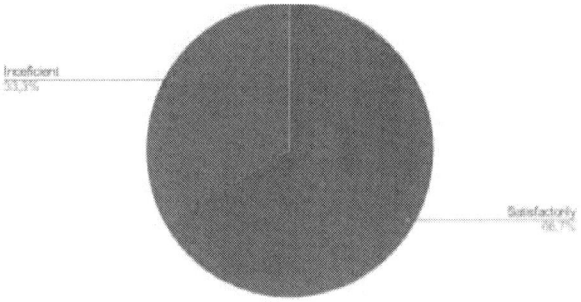

Fig. 8. Other group feedback about their program

All participating students responded that they think the approach to the program helped them in their professional development, compared to the students from the other group who answered negatively.

To the question "Was the subject as difficult as you expected it to be?", 87.5% of the students in the target group answered "Yes" (Fig. 9), compared to only 33.3% in the other group.

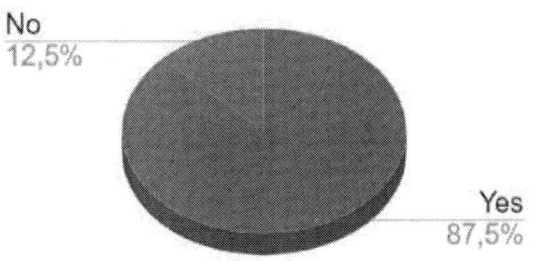

Fig. 9. Answers from tacget group students to "Was the subject as difficult as you expected it to be?"

When asked if "Would you recommend this type of project to other students?" 87.5% of the responses of the students in the target group answered "Most likely", compared to only 33.3% of the other group. From a scale from 1 to 5, 5 beeing the highest, at "How efficient do you consider the matter to have been taught?" the majority of students in the target group answered 4, the students in the other group answered the majority 2.

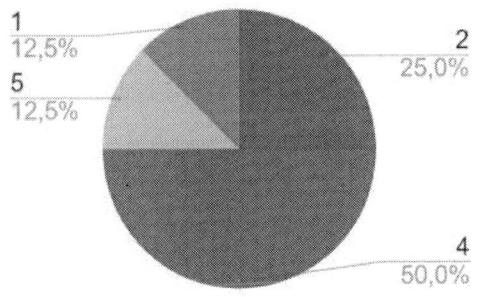

Fig. 10. The opinion of the target group students on how effective the program was

Fig. 11. The opinion of the other group students on how effective the program was

Asked if they would start a project in the same way, 62.5% of students in the program answered "likely" (Fig.12), compared to only 33.3% from the other group (Fig.13).

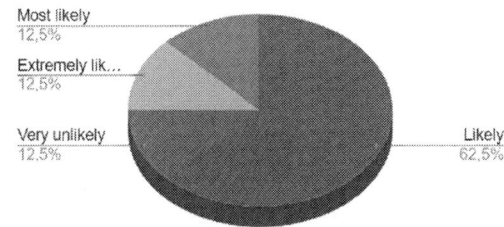

Fig. 12. Opinion of the target group students, ask if they will start a project in the same way

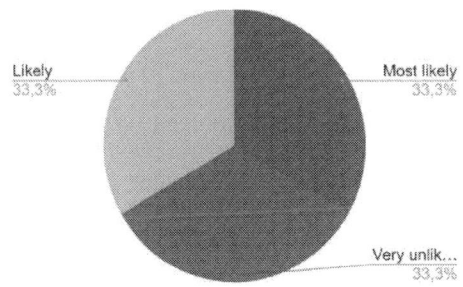

Fig. 13. Opinion of the other group students, ask if they will start a project in the same way

V. CONCLUSIONS:

Given the widespread use of electronic technologies in industry and state-of-the-art electronic products, education in electronics should be given more attention and importance. The limited resources available in the provision of laboratory hardware have increasingly marginalized the quality of education in electronic education. The introduction of e-learning in electronics, as a complement to the face-to-face learning process, fills the gap between theory and experiments when face-to-face learning is not possible.

REFERENCES

[1] Instituto Internationl para la Education Superior en America Latina y el Caraibe, COVID-19 and higher education: Today and tomorrow, Impact analysis, policy responses and recommendations, April 9, 2020

[2] Anshari, M., Alas, Y. & Guan, L.S. Developing online learning resources: Big data, social networks, and cloud computing to support pervasive knowledge. Educ Inf Technol 21, 1663–1677 (2016). https://doi.org/10.1007/s10639-015-9407-3

[3] Benson, Vladlena (2008) Unlocking the Potential of Wireless Learning. Learning and Teaching in Higher Education (2). pp. 42-56. ISSN 1742-240X

[4] https://yamar.com/

Optimization of Silver-PDMS and Gold-PDMS Surface Nanocomposite Fabrication Technologies Considering LSPR and SERS applications

Alexandra Borók, Zsanett Izsold, Shireen Zangana, Attila Bonyár

Department of Electronics Technology
Budapest University of Technology and Economics
Budapest, Hungary
borok@ett.bme.hu

Abstract—Silver and gold nanoparticles embedded into the surface of PDMS (polydimethylsiloxane) were investigated. The surface nanocomposites were synthesized both on PDMS membranes and inside PDSM based microfluidic chips. The fabrication methods were optimized in order to maximize the bulk refractive index sensitivity of the nanocomposites. The sensors are intended for biosensor applications, where a bioreceptor layer would be formed on the surface of the nanoparticles to bind target molecules. The produced nanolayers were characterized by Raman-spectroscopy and optical spectroscopy based on their localized surface plasmon resonance (LSPR) properties. These cost-effectively produced nanocomposites can be successfully utilized for LSPR and SERS (surface-enhanced Raman-spectroscopy) purposes.

Keywords— PDMS-Ag; PDMS-Au; nanocomposite; Surface-enhanced Raman spectroscopy; SERS; Localized Surface Plasmon Resonance; LSPR.

I. INTRODUCTION

With the help of POC (point-of-care) and LOC (lab-on-a-chip) systems, biological samples could be successfully analyzed in a rapid and personalized way. It can also be stated that with the use of LOC systems, a shorter amount of time is needed to perform diagnosis, thus the medical attendance can be started sooner, the monitoring of patients can be ongoing and more effective, and modifications of the used treatment can be adapted in a short amount of time. Affinity type biosensors (e.g., nucleotide sensors) integrated into microfluidic systems can be regarded as LOC systems, as their main goal is to detect DNA of bacteria or viruses from biological fluids, like blood or saliva [1].

Poly(dimethylsiloxane) (PDMS) is a widely used polymer material for fabricating microfluidic chips due to its favorable qualities, like transparency, outstanding elasticity, good thermal stability, ease of fabrication, and capability to form strong bonds with other silicon-based material like glass [2]. Gold and silver nanoparticles were used to form a nanocomposite layer on the surface of PDMS. During the synthesis process, the residual curing agent in the PDMS matrix is used to reduce $HAuCl_4$ to

gold and $AgNO_3$ to silver nanoparticles [3]. These embedded nanoparticles can be functionalized with bioreceptor molecules, and they can act as optical (plasmonic) transducers for signal generation. The optimization of the nanolayer fabrication to maximize the plasmonic sensor's sensitivity is crucial for a working biosensor [4].

Localized surface plasmon resonance (LSPR) based sensing utilizes collective electron oscillations, which are incited on the nanoparticles for refractive index sensing and promises the possibility of high throughput biomolecule sensing integrated into miniaturized point-of-care (PoC) devices [5]. Surface-enhanced Raman spectroscopy (SERS) is a promising ultrasensitive method in which the Raman scattering signal strength of molecules, absorbed on the surface of metallic nanoparticles with intensive near-fields, is enhanced with several orders of magnitude [6,7]

The current work aims to investigate and optimize the technological parameters of PDMS-Au/Ag nanocomposite membranes for SERS applications and fabricate cheap but sensitive microfluidic biosensors for LSPR applications.

II. MATERIALS AND METHODS

A. Preparation of PDMS membranes and microfluidic chips

PDMS was used to create a membrane with a thickness of 5 mm and microfluidic chips. As a first step, SYLGARD® 184 silicone elastomer was mixed with its corresponding curing agent in 1:10 and 1:5 mass ratio (η). The created mixture was degassed in a vacuum exicator then poured into a molding form. In the case of the membranes, the form is a simple glass Petri dish, while in the case of the microfluidic chip, the used form is a specifically designed 3D printed case. To increase the speed of polymerization, the molding forms were placed in a ceramic oven for 45 min at 80 oC. Before the subsequent nanocomposite preparation, the PDMS membrane was cut into circa 1.5 cm × 1.5 cm blocks. The polymerized microfluidic chips were finalized by bonding the PDMS cell part to a corona discharge treated glass underplate (Fig. 1)

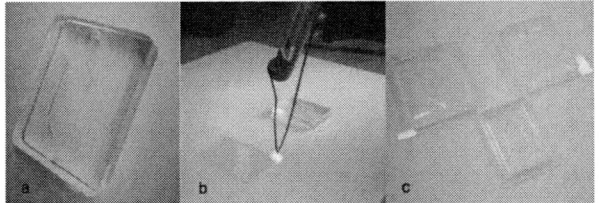

Fig. 1. Polymerized PDMS microfluidic chip without underplate (a). Bonding process (b). The final form of microfluidic chips (c).

B. PDMS-Au and PDMS-Ag nanocomposite preparation

For the preparation of PDMS-Au/Ag nanocomposite films on the surface of PDMS 100 µL of 2% (m/m) chloroauric acid ($HAuCl_4$, from Sigma Aldrich (Saint Louis, MO, USA)) or 2% (m/m) silver nitrate solution ($AgNO_3$, from Sigma Aldrich (Saint Louis, MO, USA)) was pipetted on top of the PDMS blocks. These samples were put into a hermetically sealed Petri dish, along with drops of water during the nanocomposite preparation to avoid evaporation (Fig. 2). In the case of the microfluidic cells, these precursor solutions were injected into the cell via the opening ports. These ports were hermetically sealed later with a small cylinder of Parafilm to hinder evaporation and to keep the pressure inside the cell on a constant level (Fig. 2).

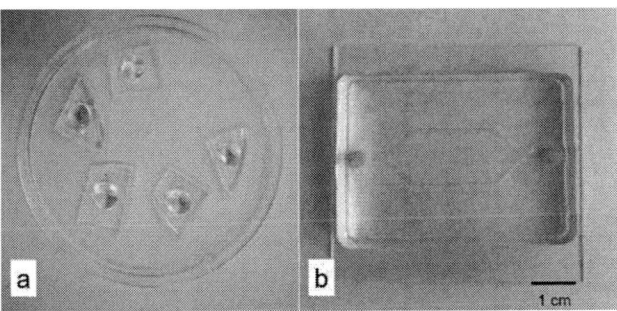

Fig. 2. a) Preparation of the PDMS-Au and PDMS-Ag membrane samples in a Petri dish. b) A finished PDMS microfluidic cell with Ag nanoparticles covering the surface of the reaction chamber.

During the study, first the nanocomposite layers were prepared at room temperature (24 °C) for 48 hours. As the preparation technique developed, the membrane with the droplet solution and the microfluidic chips filled with precursor solution were put into the oven for 90 minutes of incubation time at 80°C of temperature. After removal, the samples were cleaned with deionized water, dried with airflow, and subjected to optical spectroscopy.

C. Spectrophotometry

To obtain the LSPR spectra of the samples, optical spectroscopy measurements were performed. Avantes Avaspec 2048-4DT spectrometer and an Avantes Avalight DHS halogen light source were used between 350 nm and 750 nm. For the evaluation of the obtained spectra, a custom Matlab program was written and used.

D. SERS Measurements

SERS measurements were performed with a Renishaw 1000 micro-Raman spectrometer. Different laser sources were used for the PDMS-Ag and PDMS-Au samples to match the excitation with the nanocomposites' plasmon absorbance (488, 532, and 785 nm). The diameter of the excitation spot was around 1µm, which was monitored with a 50× objective. The spectra were recorded with 10 s integration time.

The SERS performance of the samples was tested with two target molecules:

1) A 1 mM benzophenone-isopropyl alcohol solution was drop coated onto the PDMS nanocomposite membranes. The two characteristic peaks of benzophenone at 1590 cm^{-1} and 1660 cm^{-1} were used to evaluate sample performance (for more details, see our previous paper [4]).

2) Nucleotide detection was also tested by using 20 bases long probe- and target-DNA molecules that form a specific sequence from the parasite Giardia lamblia (the β-giardin gene). We used the same DNA functionalization protocols, which were tested in our previous work [8].

III. RESULTS AND DISCUSSIONS.

A. LSPR Measurements

In one of our previous papers, the effect of the technological parameters of PDMS-Au/Ag nanocomposite membrane formation was already investigated [9]. These essential parameters are the concentration of the precursor solution, the mass ratio of curing agent/PDMS base material (η), the incubation temperature, and time. The main drawbacks of nanocomposite membrane synthesis are the solvent evaporation and bubble formation inside the solution (on the membrane surface), which frequently occurs at elevated temperatures, even when the Petri dish containing the membranes during synthesis (see Fig. 2) is hermetically sealed. To increase the reliability of synthesis, we now investigate the preparation of nanocomposites inside a microfluidic channel. After filling the microfluidic channel with the precursor solution, the channel is hermetically sealed to avoid evaporation and bubble formation.

Our general aim is to optimize the fabrication parameters to maximize the bulk refractive index sensitivity (S) of the composites. This can easily be measured by changing the medium surrounding the sensors, e.g., between air and water. In the first set of experiments, the samples were prepared at room temperature with 1:10 (curing agent/PDMS base material) mass-ratio (η), but these sensors showed only minor bulk refractive index sensitivities (a couple of nm/RIU, spectra not shown). In order to increase the number of nanoparticles synthesized on the surface of PDMS, this mass ratio was increased from 1:10 to 1:5. The presence of two times more curing agent catalyzed the reduction of more silver/gold nanoparticles on the surface while keeping the other parameters of the synthesis constant.

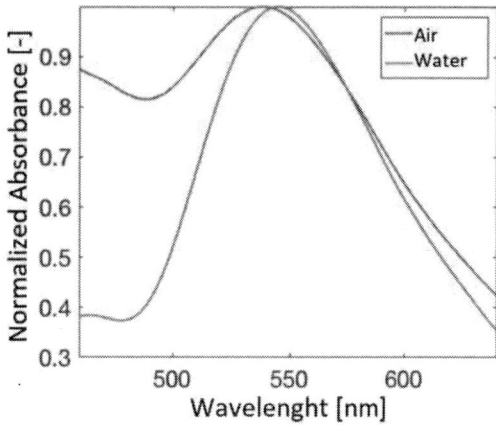

Fig. 3. Absorbance spectra of PDMS-Au nanocomposites measured in air and water, respectively. Parameters of the fabrication technology: η =1:5, T = 24°C, t = 1 day, S = 30 nm/RIU.

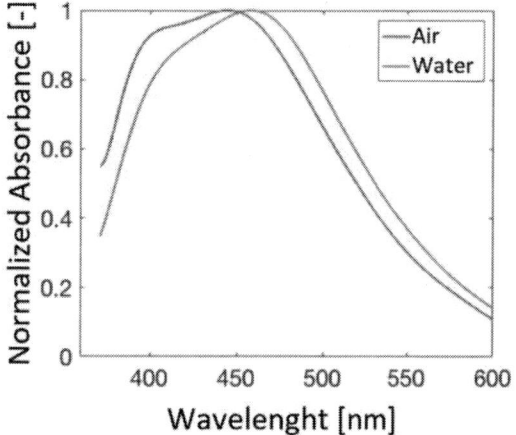

Fig. 4. Absorbance spectra of PDMS-Ag nanocomposites measured in air and water, respectively. Parameters of the fabrication technology: η =1:5, T = 24°C, t = 1 day, S = 39 nm/RIU

Fig. 3 and Fig. 4 show that increasing the curing agent's mass ratio provided measurable sensitivity of the samples (30, 39 nm/RIU for the PDMS-Au and PDMS-Ag nanocomposites, respectively).

As the next step, the effect of elevated temperature on the synthesis was investigated. The best sensor performance (90 nm/RIU) was reached with samples prepared at 60 °C temperature for 1 h, with 1:10 mass-ratio (η), as can be seen in Fig. 5. (Sensors fabricated on elevated temperatures and increased, 1:5 mass-ratio were not yet tested by the time of submission of this paper.)

As a general conclusion, we can state that increasing the curing agent's mass-ratio and increasing the temperature of the synthesis both improve the fabricated sensors' performance. Also, the time required for synthesis can be significantly reduced by using higher temperatures.

Unfortunately, the sensitivity of the sensors was not stable in time. After only days of sensor fabrication, the sensitivity of the nanocomposites dropped to nearly zero, for both PDMS-Au and PDMS-Ag materials. Thermally annealing the PDMS composite after synthesis (as suggested by [3]) did not solve this instability issue for us. The possible cause of this loss of sensitivity will be discussed later in Section III.C.

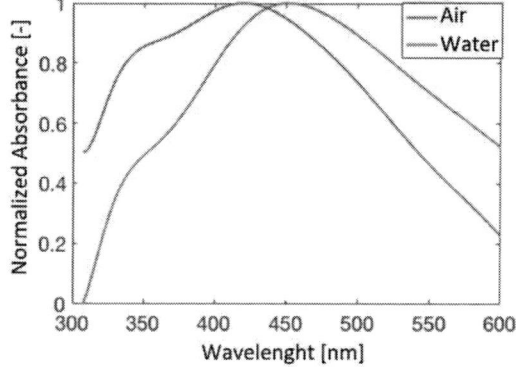

Fig. 5. Absorbance spectra of PDMS-Ag nanocomposites measured in air and water, respectively. Parameters of the fabrication technology: η =1:10, T = 60°C, t = 1 h, S = 90 nm/RIU

B. SERS Results

To test the potential applicability of the PDMA-Au/Ag nanocomposites as SERS substrates, two target molecules were investigated. First, a benzophenone-isopropyl alcohol solution was used to test the response of the nanocomposite membranes. A reference Raman signal was obtained on a polished Si substrate. Compared to the Raman signal of the reference Si wafer, our SERS substrates showed a 30×–60× increase in the signal at the same experimental conditions. Both PDMS-Au and PDMS-Ag nanocomposites were investigated, Fig. 5 presents sample results from the latter. More detailed information on these measurements can be found in [4].

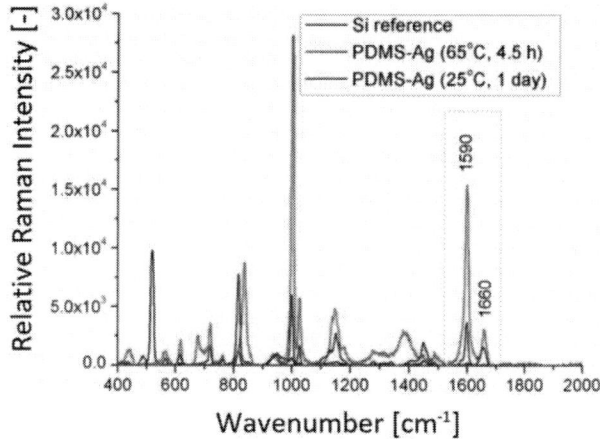

Fig. 6. Raman (Si reference) and SERS (PDMS-Ag nanocomposites) spectra of the of the benzophenone-isopropyl alcohol test solution measured with 785 nm excitation [4].

PDMS-Au membranes were also tested for nucleic acid detection with SERS. The protocols for DNA sensor fabrication are described in detail in our previous work [8]. A 20 bp long probe-DNA was bound to the surface of gold nanoparticles with thiol chemistry and the corresponding complementary target-DNA was hybridized with the probe. Fig. 7 presents the Raman spectra for the bare PDMS-Au composite (As reference) and the sensors after DNA hybridization. It is clear that all peaks are originating from the PDMS substrate, the presence of peaks corresponding to the bases in the DNA cannot be seen.

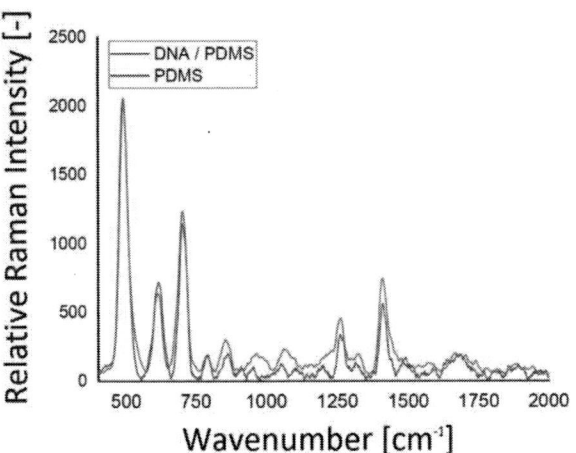

Fig. 7. SERS spectra of the of PDMS-Au nanocomposite membrane (fabrication technologies were 1:5 PDMS, 60°C, 1h) and SERS spectrum of 20 base pair long ds-DNA bound on the gold nanoparticles.

C. Discussion, explanation of SERS performance

A contradiction can be observed in the applicability of the PDMS-Au/Ag composites as SERS substrates for benzophenone and DNA detection. While they were demonstrated to be suitable to detect the former, no signal was obtained with DNA molecules. This contradiction is connected to the previously mentioned instability of the samples.

A significant loss in the bulk refractive index sensitivity was observed when the prepared samples were examined for a few days after preparation. This loss of sensitivity was not due to surface contamination, since cleaning the sensors with O_2 plasma did not restore their lost sensitivity. This suggests that the nanoparticles, which were originally synthesized on the surface of the PDMS slowly "sank" into the bulk of the PDMS. In other words, unbound chains from the polymer diffuse from the bulk and slowly cover the surface of the particles. This can account for the loss of sensitivity and also why O_2 plasma could not clean the surface.

This hypothesis – a few nm thick PDMS layer on top of the nanoparticles (Fig. 8) – could also explain the observed SERS results. For good SERS performance an intensive plasmon field (near field) is required. If the PDMS layer, which covers the particles, is only a few nm thick, the plasmon field can still penetrate into the investigated sample solution. Thus the small benzophenone molecules can yield a SERS signal if they are inside the plasmon field. However, the DNA molecules are meant to bind covalently to the gold particles with thiol groups.

If the particles are covered with a small layer of PDMS, the DNA cannot bind to the gold and they will be washed away from the surface during the rinsing steps after probe immobilization and target hybridization. Since the SERS measurements were performed after these rinsing steps, the lack of DNA signal is accounted for. It has to be emphasized that this explanation for the observed time-dependent degradation and SERS performance is currently a theory, in the future, surface analytical techniques will be used to confirm the presence of the thin PDMS layer on top of the gold particles.

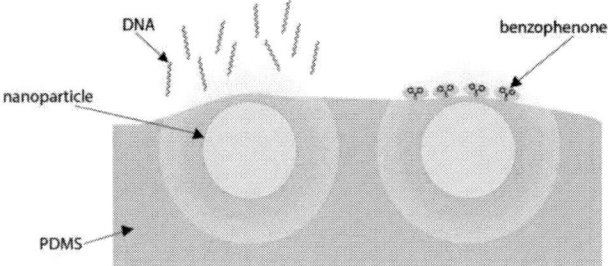

Fig. 8. Schematic illustration and explanation for the PDMS-Au/Ag nanocomposites samples' SERS performance. Due to a time-dependent degradation, a thin PDMS layer covers the surface of the particles.

CONCLUSIONS

PDMS-gold/silver nanoparticle composite films were prepared on PDMS membranes and in PDMS based microfluidic chips by the direct reduction of chloroauric acid and silver nitrate by the residual curing agent of the polymer. The test of the nanocomposites as LSPR sensor elements showed that by optimizing the parameters of the fabrication technology (curing agent mass-ratio, incubation time and temperature) the refractive index sensitivity of the sensors could be increased from 30 nm/RIU to 90 nm/RIU.

The applicability of the nanocomposite membranes as SERS substrates was also demonstrated. The substrates were suitable to detect benzophenone molecules in a solution dropped onto the surface of the nanocomposite; however, the presence of covalently bound DNA could not be confirmed while performing nucleic acid sensing.

As an interpretation of the controversial SERS performance, a hypothesis was presented, which explains the unsuccessful DNA detection with a thin PDMS layer which covers the surface of the metallic nanoparticles, and which develops with time. This development of the thin PDMS layer can also be observed through a loss of sensitivity. This hypothesis will be put to more investigation in the future.

ACKNOWLEDGMENT

The research reported in this paper and carried out at the Budapest University of Technology and Economics has been supported by the National Research Development and Innovation Fund (TKP2020) based on the charter of bolster issued by the National Research Development and Innovation Office under the auspices of the Ministry for Innovation and Technology.

REFERENCES

[1] Attila Bonyár, "Label-Free Nucleic Acid Biosensing Using Nanomaterial-Based Localized Surface Plasmon Resonance Imaging: A Review", ACS Appl. Nano Mater. 2020, 3, 9, 8506–8521.

[2] Ren, Kangning; Zhou, Jianhua; Wu, Hongkai "Materials for Microfluidic Chip Fabrication, Accounts of Chemical Research" Volume: 46, Issue: 11, Pages: 2396-2406; Published: NOV 19 2013

[3] SadAbadi, H.; Badilescu, S.; Packirisamy, M.; Wüthrich, R. "Integration of Gold Nanoparticles in PDMS Microfluidics for Lab-on-a-Chip Plasmonic Biosensing of Growth Hormones". Biosens. Bioelectron. 2013, 44 (1), 77–84.

[4] Bonyár, Attila ; Izsold, Zsanett ; Borók, Alexandra ; Csarnovics, István ; Himics, László ; Veres, Miklós ; Harsányi, Gábor "PDMS-Au/Ag Nanocomposite Films as Highly Sensitive SERS Substrates", Proceedings 2:13 Paper: 1060 , 4 p. (2018)

[5] Daiki Kawasaki, Hirotaka Yamada, Kenichi Maeno, Kenji Sueyoshi, Hideaki Hisamoto, and Tatsuro Endo. "Core−Shell-Structured Gold Nanocone Array for Label-Free DNA Sensing". ACS Applied Nano Materials 2019 2 (8), 4983-4990

[6] I Rigó, M Veres, T Váczi, E Holczer, O Hakkel, A Deák, P Fürjes, "Preparation and Characterization of Perforated SERS Active Array for Particle Trapping and Sensitive Molecular Analysis" Biosensors 9 (3), 93, 2019.

[7] Shinki and Subhendu Sarkar, "Au0.5Ag$_{0.5}$ Alloy Nanolayer Deposited on Pyramidal Si Arrays as Substrates for Surface-Enhanced Raman Spectroscopy" ACS Appl. Nano Mater. 2020, 3, 7, 7088–7095.

[8] Tomáš Lednický and Attila Bonyár, "Large Scale Fabrication of Ordered Gold Nanoparticle–Epoxy Surface Nanocomposites and Their Application as Label-Free Plasmonic DNA Biosensors". ACS Appl. Mater. Interfaces, 12(4), 4804-4814, 2020.

[9] Bonyar, A.; Izsold, Z.; Himics, L.; Veres, M.; Csarnovics, I., "Investigation of PDMS-gold nanoparticle composite films for plasmonic sensors." In Proceedings of the 23rd IEEE International Symposium for Design and Technology in Electronic Packaging (IEEE-SIITME), Constanta, Romania, 26–29 October 2017; pp. 25–28.

Solar Cell Types and Technologies with Applications in Energy Harvesting

Andrei Drăgulinescu
Electronics Technology and Reliability Department
University Politehnica of Bucharest
Bucharest, Romania
andrei.dragulinescu@upb.ro

Ana-Maria Claudia Drăgulinescu
Telecommunications Department
University Politehnica of Bucharest
Bucharest, Romania
ana.dragulinescu@upb.ro

Abstract— Energy harvesting using solar cells is a domain that has received a lot of interest and improvement in the recent years. Various techniques and new materials have been proposed in order to attain this purpose. Our paper has the goal to present the main types of solar cells, comparing their advantages and shortcomings, together with the most recent technologies proposed for energy harvesting by using solar cells. It reviews state-of-the-art solutions in this domain and may be a helpful reference for the researcher in order to obtain a thorough view of the present stage of the studies in this field and also as a starting point for new solutions towards increasing the efficiency of such devices. We also performed simulations of some solar cell structures in order to determine their performance parameters and to compare them with the ones mentioned in the literature.

Keywords— energy harvesting, solar cells, photovoltaics (PV)

I. SOLAR CELL OPERATION AND LIMITATIONS ON THE SOLAR CELL EFFICIENCY

A solar cell is a solid-state device which converts sunlight (as a stream of photons) into electrical power [1]. Mainly, solar energy is used in outdoor applications for producing large power. As materials, the most used for this purpose is crystalline silicon, but there are also other candidates with promising properties. For indoor applications, for powering electronic devices, solar cells are increasingly used in order to obtain microenergy harvesting. In this domain there is still a lack of standard procedures (for indoor conditions) by which photovoltaic devices could be measured. There are, however, several studies in which indoor perfomances and materials used for solar cells were compared [2-3]. In this paper we review recent studies that proposed solutions for energy harvesting with solar cells, proposing various technologies and materials towards achieving this goal. We also simulated two types of solar cells (CdTe and CIGS) in order to compare their performance parameters with the ones of a DSSC solar cell proposed in literature (for the CdTe solar cell) and to evaluate the effect of the CIGS layer thickness on the efficiency and other parameters (for the CIGS solar cell). The main parameters that should be taken into account for evaluating the performance of a solar cell are: the short-circuit current I_{SC}, the open-circuit voltage V_{OC}, the fill factor FF, the energy-conversion efficiency η. The first two parameters can be determined from the I-V characteristic of the solar cell [1, 4].

The maximum theoretical efficiency of a solar cell (with a single p-n junction) is given by the Shockley-Queisser limit. It was estimated initially at about 29-30%, but a more recent calculus indicated the value of 33.7%, for a semiconductor with the bandgap of 1.34 eV, taking into account only the limitation given by the radiative recombination. If Auger recombination is also considered, the theoretical efficiency decreases from 33.7% to 29.4%. This limit can be outperformed by cells with multiple layers and by using concentrated sunlight. In these conditions, in the assumption of an infinite number of layers, the limit reaches 86.8% (and, in unconcentrated light, 68%) [4-7].

II. SOLAR CELLS TYPES

The most used type (80-90% of the solar cell market) is made of crystalline silicon. The best efficiency obtained for crystalline Si solar cells has reached 25.6% for research prototypes [1-2]. Newer technologies that have received a huge amount of interest are thin-film solar cells. The best types of thin-film solar cells are considered to be those based on CdTe and CIGS materials, respectively. The best efficiency for CdTe solar cells is 21.0% (July 2017), whereas for CIGS solar cells it is 21.7% (2017) [1-2, 5]. Thin-film solar cells can also be fabricated using amorphous Si, with the benefits of a much lower cost than CdTe and CIGS. The best efficiency of amorphous Si solar cells is 13.6%. Another thin-film alternative to crystalline Si is presented by GaAs-based solar cells. The best efficiency for GaAs solar cells is 28.8% (Alta Devices) [1-2, 4-5].

Newer and promising solutions for thin film solar cells are the ones that use organic materials. They have the advantage of being easier to manipulate, flexible, light-weighted, with low material and manufacturing costs. However, they have the drawbacks of a diffusion length of about 10 nm, a low efficiency, unstable materials and limited solar cell lifetime. The best efficiencies achieved by organic solar cells are of the order of 12-13% [2, 8].

Dye-sensitized solar cells (DSSC) are another thin-film solution that undergoes a lot of research in the last years. These materials present the advantages of being semi-transparent and requiring a simple technology. However, the electrolyte and the price of the dye are still limiting factors in the path towards a widespread use. The best efficiencies obtained by DSSC are of the order of 12-13%, close to the ones for organic solar cells [2, 9].

Finally, another type of thin-film solar cell is represented by the ones using as materials perovskites (ABX$_3$, with X being Br, I or Cl). Advantages of these materials are their high absorption, high diffusion length, high efficiency and the fact that they are easy to fabricate. On the other hand, they are unstable, not reproducible on a large area and, moreover, most perovskite solar cells contain Pb, which is toxic [2] (although lead-free perovskite PV cells have recently been demonstrated). The best efficiency (as of June 2018) was achieved by a structure that combines silicon and perovskite, in a tandem solar cell (see below): 27.3%.

In order to increase the efficiency of PV, scientists succeeded in developing multi-junction (MJ) solar cells (also named tandem solar cells). The fabrication method consists in stacking different bandgap junctions and connecting them in series. The materials with a wider bandgap are placed on top of the structure; the bandgap decreases from top to bottom. MJ solar cells are able to reach higher efficiencies than single-junction solar cells, with even higher values when light concentration is used; therefore, they can be used in space (where efficiency is the most important aspect, regardless of price). Other advantages are the decreased thermalisation and decreased transmission of light. The main disadvantage is that they are expensive to manufacture. The most efficient ones are the ones with III-V compounds, but crystalline Si, thin-film and organic MJ cells are under study. The best efficiency for an MJ solar cell is 46% (reached in 2014, for a 4 junction solar cell) [1-2, 5].

Table 1 summarizes and compares the best efficiencies reached by solar cells fabricated using various materials.

Table 1 Comparison of the best efficiencies of various types of solar cells (adapted from [2] and enhanced)

Solar Cell Type	Crystalline Si	CdTe	CIGS
Best efficiency (%)	25.6	21.0	21.7
Solar Cell Type	a-Si	GaAs	Organic
Best efficiency (%)	13.6	28.8	12-13
Solar Cell Type	DSSC	Perovskite (tandem, with Si)	MJ (tandem)
Best efficiency (%)	12-13	27.3	46

III. SOLAR CELLS FOR ENERGY HARVESTING

Taking into consideration the disadvantages of fossil fuels: their limited amount that will be exhausted in the future and the pollution they create (by producing gases such as carbon monoxide, sulphur dioxide and carbon dioxide), also contributing to the global warming, a clean alternative that has gained a lot of interest in recent years is represented by the solar cell [9].

"Microenergy harvesting" is a new concept that designates the reduction of the demand of electric power up to the range of microwatts, for small electronic devices. This eliminates the need of batteries and wiring and may lead to new applications, mainly for indoor photovoltaics (IPV) (such as for WSN – wireless sensor nodes – in automated buildings [10]). Such IPV devices can be used for prolonging the life of batteries for mobile phones and tablet PCs (which require a 10 mW, respectively 10-100 W power) [3]. PV can be used for energy harvesting mainly in indoor applications (such as: homes, offices etc.), where the light level is lower than outdoors. The factors that influence the amount of harvested power can be classified in two categories: light characteristics (such as: intensity, spectral content, incident angle) and solar cell characteristics (type, size, sensitivity, temperature) [2].

For indoor energy harvesting purposes, we must take into account the fact that white LEDs and fluorescent lamps emit mainly in the 400-500 nm wavelength domain (i.e. 45% of the white LED power and 33% of that of the fluorescent lamp). Thus, in order to protect the human eye while also generating enough electrical power, a proposed solution consists in the use of OPV (organic photovoltaics) with optimized absorption of blue-light, with a material having a bandgap such as not to block the injection and extraction of carriers [11].

Another type of solar cells used for energy harvesting is the dye-sensitized one (DSSC), due to its unique properties, as it is able to maintain high values of the photovoltage even when ambient light is diffuse [12]. DSSC also possess advantages such as: low cost; environmental friendly; ease of manufacturing; suitable for production on a large scale; relatively high efficiency [9].

Multi-junction solar cells (MJSC) have proved to be the most efficient devices for converting solar energy into electrical power. An ideal MJSC would be able to harvest the whole solar spectrum, also extending to mid-infrared. Such a cell would achieve a theoretical maximum efficiency of 86%, when a 45900 suns full solar concentration is assumed [13]. As a consequence of the filtering of the solar spectrum by the atmosphere of the Earth, the spectral band between 300-2500 nm contains ~99% of the power corresponding to the direct-beam airmass 1.5 (AM 1.5) reference spectrum. This has an impact on the optimal value of the bandgap for MJSC: for a 4-7 junctions cell, the optimum lowest energy was calculated to be ~0.5 eV (2500 nm), in order to obtain practically an energy harvesting for the entire spectrum. However, the diffuse part of the irradiation cannot be captured by concentrator photovoltaic (CPV) cells, thus leading to significant losses, as this diffuse portion, in terrestrial applications, is a large fraction of the total irradiation. In order to also capture the diffuse light, hybrid approaches were proposed, combining large area solar cells and CPV cells, on the back-plane of the module, thus increasing the efficiency of the device [13].

In order to produce MJSCs that are able to capture all the solar spectrum, no direct-bandgap lattice-matched (LM) III-V alloys have yet been found. The maximum wavelength in conventional solar cells for concentrator applications is ~1800 nm, thus leading to a waste of a significant part of the power available from the Sun, which is transmitted through the cell, instead of being absorbed. Thus, currently, techniques for extending the bandgap range of MJSC are under study. For example, the semiconductor alloys that can be lattice-matched to GaSb offer direct bandgap, ranging from 0.27 to about 1 eV [13].

IV. MATERIALS FOR ENERGY HARVESTING

For harvesting the solar energy, methods have proposed, that strive to emulate natural photosynthesis. In this process, RC (reaction center) and LH (light harvesting) pigment-protein complexes perform a separation of charges, photochemically, generating an electron for almost each absorbed photon [14].

In order to fabricate photo-bioelectrochemical cells, frequently chosen materials are RCs obtained from Rhodobacter sphaeroides (a purple photosynthetic bacteria) and also complexes formed by RC and LH1 protein. With these proteins, the values of the photocurrent were enhanced in cells with three electrodes by improving the electron transfer process, but still very few attempts were made to obtain high values of the photocurrent by using the natural property of photosynthetic bacteria concerning light harvesting. Problems faced in fabricating such proteins, partially synthetic, for solar cells include: cost, complicated device fabrication, use of rare or environmental-unfriendly materials [14].

As materials for DSSC cells, as early as 1991 a group of researchers obtained a dye based on TiO₂ and ruthenium, achieving a record efficiency of 13% with a low cost. Nowadays, ruthenium and osmium metal-organic complexes are preferred, that are the most effective and stable DSSC dyes, but on the other hand they necessitate procedures of preparation involving many steps and also a very careful chromatographic purification. These difficulties triggered scientists to try using instead natural dyes for solar energy conversion, that have the advantages of being cheaper and environmentally friendly [9].

Organic photovoltaic cells (OPV) also raised a sustained interest in finding the most suitable materials. Donor polymers were the first materials studied for this purpose, chosen because of their narrow bandgaps matched with the energy levels of the fullerene derivatives, considered at that time the best energy acceptors. More recently, small-molecule (SM) organic acceptors have emerged, based on non-fullerene compounds. Thus, perspectives appeared for improving the power conversion efficiency (PCE) (~12% for OPVs based on polymeric donors), using the possibility offered by the well-defined molecular structures to reduce the morphological complexity in organic solar cells. Concerning the donor materials, recently derivatives of porphyrin have been proposed, that are SM donors having the advantages of very good properties of charge transport and a broad optical absorption, from visible to NIR range. Devices with porphyrin donors and fullerene acceptors achieved PCEs of more than 8%, whereas those with non-fullerene acceptors were proposed for the first time in 2018, with a PCE of 6.13% [15].

As concerns the optimum material or the maximum theoretical efficiency for indoor photovoltaics (IPV), no systematic investigations have been performed. However,

several works treated some materials suitable for IPVs, such as CdTe, CIS, CIGS, GaAs, GaInP, GaAs/GaInP for MJSCs [3].

Table 2 compares various materials used in solar cells and, according to their bandgap values, it can be noticed which ones are preferable for outdoor and indoor conditions, respectively.

TABLE 2 COMPARISON OF THE BANDGAP ENERGY EG FOR VARIOUS MATERIALS USED IN SOLAR CELLS (ADAPTED FROM [2]). THE MATERIALS ARE ORDERED FROM THE ONE WITH THE LOWEST TO THE MATERIAL WITH THE HIGHEST VALUE OF THE BANDGAP

Material	Si	GaAs	CdTe	Perovskites
E_g (eV)	1.12	1.42	1.44	1.5 – 2.2
Observations	Ideal bandgap for outdoor: ~1.3 - 1.4 eV			
Material	Dye	a-Si	GaInP	Organic (P3HT:PCBM)
E_g (eV)	1.62	1.6 – 1.8	1.88	1.9
Observations	Ideal bandgap for indoor: ~1.9 eV			

V. SIMULATION RESULTS FOR SEVERAL TYPES OF SOLAR CELLS

In order to compare the performance of several types of solar cells, we first simulated a Cd-Te solar cell using SCAPS software (developed by Marc Burgelman, from University of Gent, Belgium). We considered a value of the bandgap energy of CdTe of 1.44 eV (see Table 2 above), which we considered to be more accurate than the one provided by the software (1.5 eV). By using a value of the incident power of 100 mW/cm² (equal to the maximum one in the DSSC simulations in [9]), we were able to compare our obtained values for the CdTe solar cell with the ones provided for the DSSC cell in [9]. Thus, we obtained the following results: V_{OC} = 0.9979 V; J_{SC} = 24.185 mA/cm²; FF = 66.45%; PCE = 16.04%. We can observe that the values of the first three quantities are significantly higher than the ones for the DSSC; however, the efficiency is lower (16.04%, as compared to 24.6%). In these simulations we did not take into account the limitations imposed by the Auger recombination. However, by considering an approximate value of the Auger coefficient of 10⁻³¹ cm⁶/s, we obtained new values: V_{OC} = 0.8336 V; J_{SC} = 24.183 mA/cm²; FF = 77.11%; PCE = 15.14%. The value of the efficiency is lower than above, which is a normal behaviour, because Auger recombination is a limiting factor on the efficiency of a solar cell.

TABLE 3 INFLUENCE OF THICKNESS OF THE CIGS LAYER ON SEVERAL IMPORTANT PARAMETERS OF A CIGS SOLAR CELL

Thickness of CIGS layer (µm)	Power conversion efficiency η (%)	Open-circuit voltage V_{OC} (V)	Short-circuit current density J_{SC} (mA/cm²)	Fill factor FF (%)
1	14.41	0.59	31.1	78.61
3	16.5	0.61	33.775	79.81
5	16.75	0.615	34.12	79.8
10	16.76	0.6155	34.146	79.76
20	16.76	0.6155	34.176	79.7

In another group of simulations, we evaluated the effect of the CIGS-layer thickness on the main parameters of a CIGS solar cell. The device consisted in a p-CIGS layer, deposited on an n-CdS and an n-ZnO layer. The thicknesses of the CIGS, CdS and ZnO layers are 3 µm, 50 nm, 50 nm, respectively. We changed the value of the CIGS layer thickness to a lower value (1 µm) and to higher values (5, 10 and 20 µm, respectively). The results are shown in Table 4. We can observe that for a smaller thickness the value of the efficiency decreases, whereas for higher values of the thickness the efficiency is higher than for 3 µm, remaining almost constant when the thickness increased.

VI. CONCLUSIONS

For outdoor applications, mature solar cell technologies are already developed, reaching relatively high values of the efficiency at a low cost. The most used is crystalline silicon technology, but thin film materials continue to gain an increased interest, mainly due to their promising advantages and to the possibility of developing new materials and fabrication techniques [2].

As concerns the indoor applications, there is still no international measurement standard developed. However, under low levels of indoor light, several implemented technologies demonstrate good results. For indoor applications, the need for optimizing solar cell performance harvesting is still needed [2]. In this paper, we presented some state-of-the-art technologies proposed for indoor energy harvesting using solar cells. Our review shows that there are multiple solutions and materials that can be used towards this goal and that there are promising perspectives concerning future research in the domain of indoor energy harvesting technologies and applications using solar cells. We also simulated two types of solar cells (CdTe and CIGS). We compared the performance parameters of the CdTe solar cell with the ones of a DSSC solar cell proposed in literature and evaluated, for a CIGS solar cell, the effect of the CIGS layer thickness on the efficiency and other parameters of the cell.

ACKNOWLEDGMENT

This research was funded by the Operational Programme Human Capital of the Ministry of European Funds through Financial Agreement 51675/09.07.2019, SMIS Code 125125 and under Project PN - III - P1-1.2-PCCDI-2017-0560, contract no. 41PCCDI/2018, and under PN-III-P1-1.2-PCCDI-2017-0419, contract no. 71PCCDI/2018.

REFERENCES

[1] C. J. Chen, "Physics of solar energy", John Wiley & Sons, 2011.

[2] A. Kaminski-Cachopo, "Solar cells for energy harvesting", IMEP-LAHC, Grenoble, France, 2016.

[3] Monika Freunek (Müller), Michael Freunek, L. M. Reindl, "Maximum Efficiencies of Indoor Photovoltaic Devices", IEEE Journal of Photovoltaics, Vol. 3, No. 1, pp. 59-63, 2013.

[4] M. Green, "Solar Cells: Operating Principles, Technology and System Applications", Prentice-Hall, 1981.

[5] J. Nelson, "The Physics of Solar Cells", Imperial College Press, 2003.

[6] S. Rühle, "Tabulated values of the Shockley–Queisser limit for single junction solar cells", Solar Energy, Vol. 130, pp. 139–147, 2016.

[7] A. De Vos, "Detailed balance limit of the efficiency of tandem solar cells", Journal of Physics D: Applied Physics, Vol. 13, No. 5, pp. 839-846, 1980.

[8] A. M. Bagher, "Introduction to Organic Solar Cells", Sustainable Energy, Vol. 2, No. 3, pp. 85-90, 2014.

[9] H. S. Hafez, S. S. Shenouda, M. Fadel, "Photovoltaic characteristics of natural light harvesting dye sensitized solar cells", Spectrochimica Acta Part A: Molecular and Biomolecular Spectroscopy, Vol. 192, pp. 23-26, 2018.

[10] A. Drumea, R. A. Dobre, "Modelling, simulation and testing of an autonomous embedded system supplied by a photovoltaic panel", IEEE 20th International Symposium for Design and Technology in Electronic Packaging (SIITME), pp. 309-312, 2014.

[11] M. C. Jung, H. Kojima, I. Matsumura, H. Benten, M. Nakamura, "Diffusion and influence on photovoltaic characteristics of p-type dopants in organic photovoltaics for energy harvesting from blue-light", Organic Electronics, Vol. 52, pp. 17-21, 2018.

[12] M. Freitag, J. Teuscher, Y. Saygili, X. Zhang, F. Giordano, P. Liska, J. Hua, S. M. Zakeeruddin, J.-E. Moser, M. Grätzel, A. Hagfeldt, "Dye-sensitized solar cells for efficient power generation under ambient lighting", Nature Photonics, Vol. 11, No. 6, pp. 372-378, 2017.

[13] M. P. Lumb, S. Mack, K. J. Schmieder, M. González, M. F. Bennett, D. Scheiman, M. Meitl, B. Fisher, S. Burroughs, K.-T. Lee, "GaSb-Based Solar Cells for Full Solar Spectrum Energy Harvesting", Advanced Energy Materials, Vol. 7, No. 20, pp. 1700345 (1-9), 2017.

[14] S. K. Ravi, Z. Yu, D. J. K. Swainsbury, J. Ouyang, M. R. Jones, S. C. Tan, "Enhanced Output from Biohybrid Photoelectrochemical Transparent Tandem Cells Integrating Photosynthetic Proteins Genetically Modified for Expanded Solar Energy Harvesting", Advanced Energy Materials, Vol. 7, No. 7, pp. 1601821 (1-7), 2017.

[15] W. T. Hadmojo, D. Yim, S. Sinaga, W. Lee, D. Y. Ryu, W. D. Jang, I. H. Jung, S. Y. Jang, "Near Infrared Harvesting Fullerene-Free All-Small-Molecule Organic Solar Cells Based on Porphyrin Donors", ACS Sustainable Chemistry & Engineering, Vol. 6, No. 4, pp. 5306-5313, 2018.

Characteristics of a Dilute Nitride InGaAsN Double Quantum Well Laser at 1047 nm

Andrei Drăgulinescu
Electronics Technology and Reliability Department
University Politehnica of Bucharest
Bucharest, Romania
andrei.dragulinescu@upb.ro

Mihail Dumitrescu
Optoelectronics Research Center
Tampere University of Technology
Tampere, Finland
mihail.dumitrescu@tut.fi

Abstract—**Quantum well (QW) semiconductor lasers working at various wavelengths have been extensively used in the last decade due to their advantages in a wide range of application domains. One of the material systems used for fabricating these laser structures is InGaAsN with a very small percent of nitrogen, having useful applications for lasers working at long wavelengths and for high-efficiency multi-junction solar cells. We simulated the performance characteristics of a 1047 nm InGaAsN QW laser, specifically the I-V and L-I characteristics, energy band diagram, wave intensity distribution, threshold current, slope efficiency and external differential quantum efficiency, with the purpose of gaining a wide perspective on the structure performance in terms of some of its most important parameters and to constitute a useful reference for applications**

Keywords— laser; quantum well; dilute nitride; InGaAsN

I. INTRODUCTION

InGaAsN material system has gained a lot of interest since the discovery that it presents a very useful property, which consists in obtaining a significant reduction of the bandgap energy, with 30%, due to an inclusion to InGaAs of a very percent, of only 2%, of nitrogen [1].

This is the principle on which dilute nitride QW lasers based on InGaAsN are working, making this alloy system particularly useful for lasers functioning at longer wavelengths, photodetectors in infrared and solar cells with multi-junction, of high efficiency.

Among the QW lasers grown on substrates of GaAs, the only ones that enable laser emission at long wavelenths are those based on InGaAsN(Sb) [2]. Recently, this alloy was employed in injection microdisk lasers. These devices, as well as the microring lasers, possess the advantages of being easy to fabricate, at small sizes (thus, having the potential to be used in applications where radiation sources of very small dimensions are necessary). Moreover, they benefit from low values of the threshold current. Probably, the application where they will be most widely used in the near future is in transmission of optical data on very short distances, such as, e.g., between boards or even between the chips in those boards [2]. For such injection microdisk lasers, a comparison was recently performed by a research group [2] between these devices, based on InGaAsN QW grown on GaAs substrates and on InAs/InGaAs/GaAs quantum dots, in terms of their threshold characteristics.

Microdisk lasers with InGaAsN/GaAs QWs, with diameters in the range between 11 and 31 μm, emitting at wavelengths higher than 1200 nm, at room temperature, were found to perform better than the lasers with comparable sizes, but based on quantum dots and with other materials than InGaAsN, as mentioned above, in terms of several spectral characteristics [2].

InGaAsN QW heterostructures were also analyzed recently, on two types of substrate (GaAs and InP, respectively) [3]. The paper assessed the way in which the characteristics of the optical gain for the heterostructures were influenced by the choice of the substrate. The carrier confinement in the QW and the optical gain were computed by using the so-called k·p method (from quantum mechanics). The authors concluded that the characteristics of the optical gain were significantly influenced by the choice of the substrate. Moreover, the effect of the cladding in the heterostructures based on InGaAsN/InP was also analyzed. The values obtained for two parameters (optical gain and wavelength), for the InGaAsN QW heterostructure, grown on GaAs substrate, with claddings, and on InP substrate, with and without claddings, respectively, are presented in Table 1.

TABLE 1 COMPARISON BETWEEN OPTICAL GAIN AND WAVELENGTH FOR INGAASN QW HETEROSTRUCTURES, WITH TWO TYPES OF SUBSTRATE, WITH AND WITHOUT CLADDING. ADAPTED FROM [3].

Substrate	Presence of claddings in the InGaAsN QW heterostructure	Optical gain (cm^{-1})	Wavelength (nm)
GaAs	Yes	2100	1300
InP	Yes	2353	1100
InP	No	1675	1140

The results showed that not only the optical gain was influenced by the presence of the claddings, but also the wavelength and (not shown in the table) the photonic energy. The optical gain was significantly increased in the InGaAsN/InP QW heterostructure, when claddings were present. For InGaAsN on the other substrate (GaAs), both peak values for the optical gain and for the modal gain (the product between the optical gain and the confinement factor) were obtained at approximately the same wavelength, namely 1300 nm, whereas for the ones grown on InP substrate, the wavelength was much lower, the presence of the cladding lowering it even more, but not significantly (from 1140 nm to 1100 nm). Thus, from the simulation results, it was concluded that the influence of the substrate on optical gain characteristics is very important and, together with the cladding effect, it should be taken into account for the fabrication of such devices, in various applications [3].

Another device where InGaAsN has proved a suitable material, with high performance and promising perspectives, is the photodetector, particularly e-SWIR ones (extended short wavelength infrared photodetectors). A recent proposal [4] included InGaAsN/AlGaAs QWs in e-SWIR photodetectors, benefiting from the shift of a technology already well developed. This technology, named QWIP (used for fabricating GaAs quantum-well infrared photodetectors) was shifted from LWIR (long-wave infrared) and MWIR (mid-wave IR) to shorter IR wavelengths, in the e-SWIR spectral band (that covers wavelengths in the range between 1000 and 2500 nm). This e-SWIR domain has two main advantages: the images obtained have a high-quality resolution close to the one for the visible domain; e-SWIR photons are very well transmitted through smoke particles and fog. Moreover, e-SWIR enables both active imaging, using photons from e-SWIR lasers and LEDs, and passive imaging, using as light source the natural night glow of the atmosphere in this spectrum. Additionally, InGaAsN/AlGaAs QW was demonstrated in [4] to avoid the performance reduction observed, at cutoff wavelengths higher than 1700 nm, for the InGaAs/InP technology, and the high costs, crystal growth and fabrication problems, for MCT (mercury-cadmium telluride), these two being the technologies currently on the market of e-SWIR photodetectors. The authors included nitrogen into the QW material, in order to decrease the bandgap and to increase the CBO (conduction band offset) up to approximately 1 eV, large enough for the structure to be suited for an efficient e-SWIR detection at room temperature, with the added benefit of a significant reduction of the dark current. Thus, two structures were devised, both with InGaAsN QWs and AlGaAs barriers, with the same parameters, except that one included 1% nitrogen and the other one 2%. The results showed that the structure with 2% N presented, at room temperature, a lower value of the dark current, together with a stronger photoresponse, as compared to the 1% N structure, where only at temperatures below 150 K the results were comparable [4].

Besides lasers and photodetectors, a third major type of device that employs successfully InGaAsN materials is the solar cell. Recently, a multi-junction solar cell, based on InGaP/InGaAs-GaAsP/InGaAsN QW, was studied, using APSYS numerical device simulator, in terms of deep level defects [5]. The efficiency obtained had a record value of 38% at 1-sun and 44% at 500-sun AM0 spectrum. When considering more realistic values for two parameters (1016 cm-3 for the trap concentration, and 104 cm·s-1 for the electron/hole surface recombination velocity, SRV), the values of the efficiency obtained were with 4% lower. However, these values are still

higher than the limit of the current MJSCs (31% at 1-sun AM0 spectrum). The high efficiency, together with the tolerance to radiation, make the proposed cells good candidates for applications in space satellites [5].

Another research group published recently a characterization of layers made of dilute nitride InGaAs(Sb)N, with a thickness of 2-3 μm, grown under near-equilibrium conditions, by low-temperature LPE (liquid phase epitaxy) [6]. Their structural properties and local microstructure were investigated. X-ray based methods were employed in order to determine the composition and crystal quality of the layers. The characteristics of the obtained layers make them suitable for being used in solar cell applications [6]. More recently, the same group reported results on growing both dilute nitride InGaAsN and GaAsSbN layers, using the same low-temperature LPE method [7]. Moreover, X-ray photoelectron spectroscopy (XPS) was used for studying the local bonding of N in the InGaAs and GaAsSb alloys and, for assessing the electronic band structure and optical quality of these grown alloys, PL (photoluminescence) and SPV (surface photo-voltage) spectroscopy, at room temperature, were employed. Most of the challenges in the epitaxial growth of such structures were overcome, making the resulting materials very good candidates for the inclusion in solar cell devices [7].

In this paper we simulated, using LASTIP software, a structure of InGaAsN QW laser, in terms of the I-V and L-I characteristics, energy band diagram, wave intensity distribution, threshold current, slope efficiency and external differential quantum efficiency. The paper is organised as follows: Section II explains the simulation setup and provides the parameters of the structure and the parameters used for simulation. Section III depicts the simulation results and, finally, Section IV concludes the paper.

II. SIMULATION SETUP

A. QW Laser Structure

The structure of the 1047 nm InGaAsN QW laser that we characterized is presented in Table I.

The structure consisted of an n-GaAs buffer layer, graded-index (GRIN), cladding and GRIN-waveguide (GRIN-WG) Al_xGa_yAs layers, constant-composition waveguide (CC-WG) GaAs layers, two $Ga_{0.64}In_{0.36}As_{0.988}N_{0.012}$ quantum well (QW) layers, separated by a GaAs barrier layer, and a GaAs contact layer.

TABLE I.   LAYER STRUCTURE FOR THE SIMULATED 1047 NM DILUTE NITRIDE INGAASN DOUBLE QUANTUM WELL (DQW) LASER

Layer type	Material	Thickness (nm)	Doping
Contact	GaAs	100	p-type doped
Graded index (GRIN)	$Al_xGa_{1-x}As$	200	p-type doped
Cladding	$Al_yGa_{1-y}As$	1000	p-type doped

Graded index waveguide (GRIN-WG)	$Al_xGa_{1-x}As$	160	undoped
Constant composition waveguide (CC-WG)	GaAs	20	undoped
Quantum well	$In_{0.36}Ga_{0.64}As_{0.988}N_{0.012}$	7	undoped
Barrier	GaAs	20	undoped
Quantum well	$In_{0.36}Ga_{0.64}As_{0.988}N_{0.012}$	7	undoped
CC-WG	GaAs	20	undoped
GRIN-WG	$Al_xGa_{1-x}As$	160	undoped
Cladding	$Al_yGa_{1-y}As$	1000	n-type doped
GRIN	$Al_xGa_{1-x}As$	200	n-type doped
Buffer	GaAs	300	n-type doped

B. Simulation Parameters

For the simulation of this structure, we used, instead of the old default parameters provided by LASTIP, the parameters obtained by calculations from more accurate interpolation formulae for the bandgap energy, effective masses of electrons, Luttinger parameters, elastic stiffness constants, hydrostatic deformation, shear deformation potential and lattice constant, using values for the parameters of GaAs, InAs and GaN provided in one of our previous works [8].

III. SIMULATION RESULTS

With the considerations in Section II, we simulated the contact total current as a function of the contact voltage (i.e. the I-V characteristics) and the laser output power as a function of the applied current (i.e. the luminescence-current curve, or the L-I characteristics) and obtained the result shown in Fig. 1, respectively, Fig. 2.

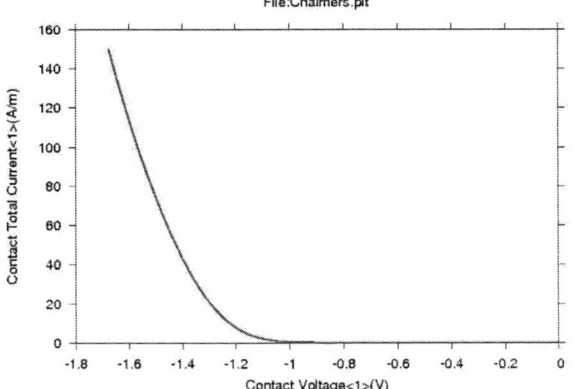

Fig. 1. I-V characteristics for the InGaAsN QW laser structure.

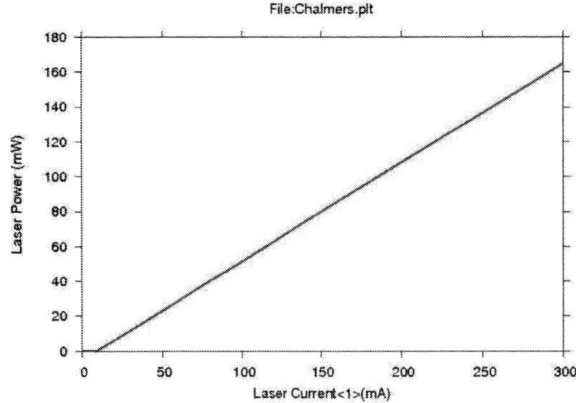

Fig. 2. L-I characteristics for the InGaAsN QW laser structure.

In Fig. 3 we represented the energy band diagram of the simulated InGaAsN laser structure. The dashed lines in the figure are the quasi-Fermi levels.

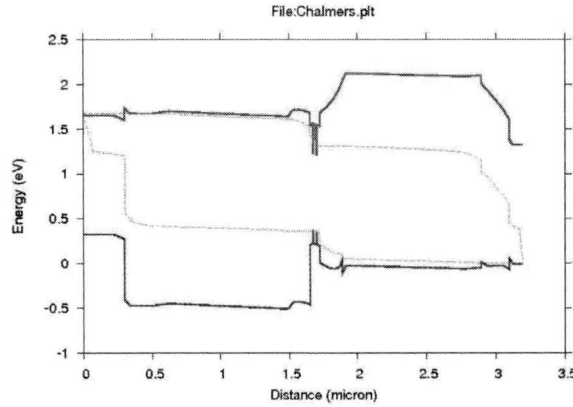

Fig. 3. Energy band diagram of the InGaAsN QW laser structure.

The wave intensity profile is represented in Fig. 4 for the studied structure as a function of the distance along the direction of crystal growth. One can observe the parabolic shape of the obtained profile, with a maximum value of the relative intensity (expressed in arbitrary units) of 1 at about 1.65 µm, in the active region of the laser.

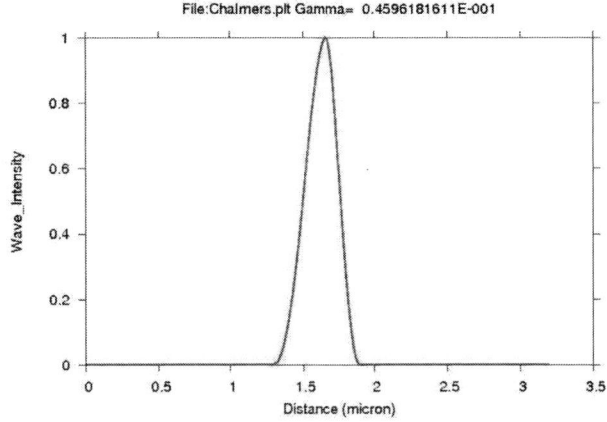

Fig. 4. Wave intensity profile for the InGaN QW laser structure.

We also determined three important performance parameters for this structure, namely the threshold current, slope efficiency and external differential quantum efficiency, which are given in Table II.

TABLE II. RESULTS OBTAINED FOR THREE MAIN PERFORMANCE PARAMETERS OF THE InGaAsN QW LASER STRUCTURE

Parameter	Obtained value
Threshold current	9.17 mA
Slope efficiency	564 mW/A
External differential quantum efficiency	20.96%

IV. CONCLUSIONS

As mentioned throughout the paper, there are three main optoelectronic devices that use InGaAsN material: QW lasers, photodetectors and solar cells. In this paper, we chose to characterise by simulation QW lasers. As explained in the introductive section, the need for a good assessment of such structure is essential in countless domains, therefore our approach would be very useful in near-infrared applications of quantum well lasers.

In conclusion, we simulated some of the main performance characteristics for a structure of 1047 nm $In_{0.36}Ga_{0.64}As_{0.988}N_{0.012}$ QW laser, with the purpose of gaining a wider perspective on the performance of this device. In other papers we shall focus on improving these performance characteristics for dilute-nitride InGaAsN QW laser in order to obtain an enhanced structure best suitable in applications in various domains.

ACKNOWLEDGMENT

This work was funded by ENI project, contract no. 41PCCDI/2018 and by SENSIS project, contract no. 71PCCDI/2018.

REFERENCES

[1] A. Allerman, S. Kurtz, E. Jones, N. Modine, J. Gee, D. Follstaedt, "InGaAsN: A novel material for high-efficiency solar cells and advanced photonic devices", Sandia Report SAND 99-0922, Sandia National Laboratories, Albuquerque, USA, July 1999, pp. 1-35.

[2] E. I. Moiseev, M. V. Maximov, N. V. Kryzhanovskaya, O. I. Simchuk, M. M. Kulagina, S. A. Kadinskaya, M. Guina, A. E. Zhukov, "Comparative Analysis of Injection Microdisk Lasers Based on InGaAsN Quantum Wells and InAs/InGaAs Quantum Dots", Semiconductors, vol. 54, no. 2, Mar. 2020, pp. 263–267.

[3] M. I. Khan, K. Sandhya, A. M. Khan, P. A. Alvi, "Gain Characteristics of InGaAsN Quantum Well Heterostructures with GaAs and InP Substrates", IOP Conference Series: Materials Science and Engineering, vol. 594, no. 1, August 2019, pp. 012044.

[4] A. Albo, D. Fekete, G. Bahir, "The opportunity of using InGaAsN/AlGaAs quantum wells for extended short-wavelength infrared photodetection", Infrared Physics and Technology, vol. 96, Jan. 2019, pp. 68-76.

[5] M. Sukeerthi, S. Kotamraju, S. E. Puthanveettil, "Study of deep level defects in InGaP/InGaAs-GaAsP/InGaAsN quantum well based multi-junction solar cell using finite element analysis", Superlattices and Microstructures, vol. 130, June 2019, pp. 28-37.

[6] M. Milanova, V. Donchev, P. Terziyska, E. Valcheva, S. Georgiev, K. Kirilov, I. Asenova, N. Shtinkov, Y. Karmakov, I. G. Ivanov, "Investigation of LPE grown dilute nitride InGaAs(Sb)N layers for photovoltaic applications", AIP Conference Proceedings, vol. 2075, Feb. 2019, pp. 140004.

[7] V. Donchev, M. Milanova, S. Georgiev, "Dilute nitride InGaAsN and GaAsSbN layers grown by liquid-phase epitaxy for photovoltaic applications", Journal of Physics: Conference Series, 21st International Summer School on Vacuum, Electron and Ion Technologies (Sozopol, Bulgaria), vol. 1492, 2020, pp. 012049.

[8] A. Drăgulinescu, A. Laakso, M. Dumitrescu, M. Guină, "Simulations of dilute nitride quantum well InGaAsN semiconductor lasers", *Proc. of SPIE*, vol. 7297, Advanced Topics in Optoelectronics, Microelectronics, and Nanotechnologies IV (ATOM-N 2008), Jan. 2009, pp. 729707 - 729707-4.

Dependence of Shear Strength of Adhesive Conductive Joints on Adhesive Modification with the Silver Nanoparticles and Climatic Aging

Pavel Mach

Czech Technical University in Prague, Faculty of Electrical Engineering
Department of Electrotechnology
Prague, Czech Republic
Pavel.Mach@fel.cvut.cz

Abstract—**Electrically conductive adhesive was modified by the addition of silver nanoparticles in two levels of concentration. Homogenization of the adhesive was carried out by rotary agitation for two different long times. The modified adhesive was used for the formation of adhesive joints by mounting jumpers on the test board. The joints were climatic aged at high temperature and high humidity. The analysis of the influence of individual parameters and their interactions on the shear strength of adhesive joints was carried out using DOE. It was found that the homogeneity of the distribution of nanoparticles in the adhesive is more significant than their concentration and that the treatment in high humidity influences the shear strength more than thermal treatment.**

Keywords—*conductive adhesive; silver nanoparticle; adhesive assembly; climatic treatment*

I. INTRODUCTION

Soldering assembly is based on using different types of lead-free solders. The soldering temperature of these alloys is near to 255 °C usually. However, there are different types of electronic devices, which can be damaged due to too the high temperature requested for this operation. Therefore for a montage of such devices must be used other materials, which will make conductive joining possible. It should be that the electrical, mechanical, and other properties of the conductive joint formed by solder and this material comparable. Suitable replacement soldering where this technology can not be, for various reasons, used is adhesive joining using electrically conductive adhesives [1-3].

The main problem of electrically conductive adhesives is that their electrical, as well as mechanical properties, are worse in comparison with the lead-free solders [4]. Therefore, various methods are tested to improve the properties of these materials [5-7]. One of them is the modification of the adhesives with conductive, mostly silver, nanoparticles. The concentration of nanoparticles must be low because they increase the number of contacts in the conductive network primarily made by microparticles of conductive filler in the adhesive.

The goal of the work was to inspect the shear strength of adhesive joints formed from adhesive affected with nanoparticles in the two different concentrations. The influence of the homogeneity of the distribution of nanoparticles in the adhesive on the shear strength of the joints was also monitored. The nanoparticles were added to the adhesives by rotary agitation. A mechanical stirring tool was used for mixing. It was necessary to take care of the appropriate viscosity of the adhesive during this process. The last monitored parameter whose influence on the shear strength of adhesive joints was observed was climatic aging [8].

II. MATERIALS AND METHODS

Adhesive under test

Jumpers (resistors with "zero" resistance) of the type 1206 were mounted using adhesive assembly. The adhesive ECO SOLDER AX 20 (AMEPOX, Poland) [9]. was applied by stencil printing. The concentration of filler particles is 75 wt%. The adhesive is cured at the temperature 150 °C for 10 min.

Modification of the adhesive was performed by silver nanoparticles with dimensions of 5-55 nm (Sigma Aldrich) at a concentration of 3.8 and 7.4 wt %. Homogenization nanoparticles in the adhesive was carried out by mixing using a special stirring tool at 1300 rpm for two different periods: 10 minutes and 30 minutes.

The adhesive was cured using the curing schedule recommended by the manufacturer of adhesive. The curing was carried out in an air-circulated oven at the relative humidity 54 %.

Climatic aging

Climatic aging was performed under the following conditions: at 125 ° C and relative humidity of 65% RH, and at 24 ° C and relative humidity of 98 % in the climatic chamber WTB Binder. The aging time in both cases was 700 hours. The aging was carried out in a climatic chamber B

Application of adhesive and test boards

The adhesive was applied by screen printing. The Cu pads had no special surface finish.

Seven jumpers were mounted on one test PCB, it means that fourteen adhesive joints were formed on one board. Three test PCBs were prepared for each combination of test conditions used. Thus, it was obtained twenty-one values of shear force for every combination of test conditions.

Measurement of shear strength

The shear strength was measured using a digital force meter KERN MH10K10. The PCB with mounted jumpers was placed vertically and the force was applied to the jumper from the bottom upwards.

III. MATHEMATICAL PROCESSING OF MEASURED RESULTS

Modeling of different types of relationships between the parameters of the adhesive and technological factors of their preparation is frequently used for the optimization of these parameters [10, 11].

As already noted, a total of twenty-one values of shear strength were measured for each combination parameters studied. The maximum and minimum values in every group were deleted and a cut mathematic average was calculated from the remaining values. These averages were then used for the processing using a method of design of experiments (DOE).

The shear strength was examined depending on the following factors:

A ... the time of homogenization (stirring) of nanoparticles in adhesive,
B ... the time of thermal treatment,
C ... the concentration of nanoparticles in adhesive.

AC, AB, BC, and ABC are the interactions of these parameters. The table of factorial experiments is shown in Tab. 1.

TABLE I: DIAGRAM OF FACTORIAL EXPERIMENTS 2^3

A_1				A_2			
B_1		B_2		B_1		B_2	
C_1	C_2	C_1	C_2	C_1	C_2	C_1	C_2
$A_1B_1C_1$	$A_1B_1C_2$	$A_1B_2C_1$	$A_1B_2C_2$	$A_2B_1C_1$	$A_2B_1C_2$	$A_2B_2C_1$	$A_2B_2C_2$
(1)	c	b	bc	a	ac	ab	abc
$y_{1,1}$	$y_{2,1}$	$y_{3,1}$	$y_{4,1}$	$y_{5,1}$	$y_{6,1}$	$y_{7,1}$	$y_{8,1}$
$y_{1,2}$	$y_{2,2}$	$y_{3,2}$	$y_{4,2}$	$y_{5,2}$	$y_{6,2}$	$y_{7,2}$	$y_{8,2}$
.	.	.	.	.	.	.	.
.	.	.	.	.	.	.	.
$y_{1,r}$	$y_{2,r}$	$y_{3,r}$	$y_{4,r}$	$y_{5,r}$	$y_{6,r}$	$y_{7,r}$	$y_{8,r}$
R_1	R_2	R_3	R_4	R_5	R_6	R_7	R_8

Here factors and their interactions are in a green part of the table and a symbolic expression of interactions is in the orange part. Measured values of the shear strength $y_{i,j}$ are in the white part as well as R_i totals measured values in each column.

At first, the influence of changes of factors A, B, C, and their interactions from the lower to the upper level on the shear strength of the joints must be inspected. These influences are also named contrasts. The contrasts are calculated using the following formulas:

$$Z_A = a + ac + ab + abc - (1 + b + c + bc) \quad (1)$$

$$Z_B = b + bc + ab + abc - (1 + a + c + ac) \quad (2)$$

$$Z_C = c + bc + ac + abc - (1 + b + a + ab) \quad (3)$$

$$Z_{AB} = (1) + c + ab + abc - (a + b + ac + bc) \quad (4)$$

$$Z_{AC} = (1) + b + ac + abc - (a + c + ab + bc) \quad (5)$$

$$Z_{BC} = (1) + a + bc + abc - (b + c + ab + ac) \quad (6)$$

$$Z_{ABC} = a + b + c + abc - (1 + ab + ac + bc) \quad (7)$$

Note that all relationships are given in symbolic variables (see the orange line in table 1).

When the contrasts are calculated, it is inspected if the influence of individual technological factors and their interactions on the shear strength of the joint is or is not statistically significant. Such information can be calculated by statistical testing. In the case of factorial experiments, an F-test is mostly used for such testing.

First, the test characteristics are calculated using the following formulas:

$$F_A = S_A/(S_R/v) \quad (8)$$

$$F_B = S_B/(S_R/v) \quad (9)$$

$$F_C = S_C/(S_R/v) \quad (10)$$

$$F_{AB} = S_{AB}/(S_R/v) \quad (11)$$

$$F_{AC} = S_{AC}/(S_R/v) \quad (12)$$

$$F_{BC} = S_{BC}/(S_R/v) \quad (13)$$

$$F_{ABC} = S_{ABC}/(S_R/v) \quad (14)$$

Where

$$S_i = Z_i^2/(8r) \quad (15)$$

and residuum sum of squares S_r

$$S_r = \sum_{i=1}^{d} \sum_{j=1}^{r} \left(y_{i,j} - \frac{\sum_{j=1}^{r} y_{i,j}}{r} \right)^2 \quad (16)$$

Degrees of freedom v are calculated using a formula:

$$v = 8 \cdot (r - 1) \quad (17)$$

Measured and calculated values are presented in Tab. II and Tab. III, where Tab II is for experiments joined with the thermal treatment of the test samples and the Tab. III for experiments joined with the humidity treatment. The following labeling is used in the tables: A ... concentration of nanoparticles used for modification of adhesive, B ... time of stirring representing the homogeneity of mixing nanoparticles into adhesive, C ... time of thermal or humidity treatment, n ...the number of technological factors, r ... the number of repetitions of experiments under the same conditions, d ... number of columns of the table of experiments ($d = 2^n$), N ... total number of experiments ($N = r.d$), v ... degrees of freedom, R_i ... column sum of column i, Z_j ... contrast of the factor or interaction j, S_j ... sum of squares of deviations for the factor or interaction j, F_j ... test characteristic for the factor or interaction j.

TABLE II: Measured and Calculated Values for Samples After Thermal Treatment

A_1	3.80	R_4	137.80	S_B	2566.80
A_2	7.40	R_5	296.10	S_C	505.08
B_1	10.00	R_6	229.90	S_{AB}	143.08
B_2	30.00	R_7	117.30	S_{BC}	32.90
C_1	0.00	R_8	101.90	S_{AC}	33.84
C_2	700.00	Z_A	-128.80	S_r	6392.41
n	3.00	Z_B	-496.40	S_0	9868.39
r	12.00	Z_C	-220.20	F_A	2.38
d	8.00	Z_{AB}	-117.20	F_B	35.34
N	96.00	Z_{BC}	56.20	F_C	6.95
v	88.00	Z_{AC}	57.00	F_{AB}	1.97
R_1	301.90	Z_{ABC}	45.40	F_{BC}	0.45
R_2	229.90	S_A	172.81	F_{AC}	0.47
R_3	204.40	S_{ABC}	21.47	F_{ABC}	0.30

$F_{0.025}(1,88) = 5.20$

TABLE II: Measured and Calculated Values for Samples After Humidity Treatment

A_1	3.80	R_4	299.80	S_B	3277.18
A_2	7.40	R_5	231.60	S_C	4217.48
B_1	10.00	R_6	393.00	S_{AB}	40.95
B_2	30.00	R_7	121.30	S_{BC}	299.98
C_1	0.00	R_8	154.20	S_{AC}	23.70
C_2	700.00	Z_A	-310.10	S_r	13084.54
n	3.00	Z_B	-560.90	S_0	22077.82
r	12.00	Z_C	636.30	F_A	6.74
d	8.00	Z_{AB}	62.70	F_B	22.04
N	96.00	Z_{BC}	-169.70	F_C	28.36
v	88.00	Z_{AC}	-47.70	F_{AB}	0.28
R_1	284.70	Z_{ABC}	112.70	F_{BC}	2.02
R_2	526.30	S_A	1001.69	F_{AC}	0.16
R_3	199.40	S_{ABC}	132.31	F_{ABC}	0.89

$F_{0.025}(1,88) = 5.20$

IV. Results and Discussion

The results are presented in Fig. 1 and Fig. 2.

It was found that the concentration of the nanoparticles used for modification of adhesive was not the most significant factor that influences the shear strength of adhesive. More significant seems to be the quality of the mixing of nanoparticles in adhesive. This information was surprising.

The orange columns present the situation before the climatic treatment. The strongest influence on the shear strength has the quality of mixing followed by the concentration of the nanoparticles and the interaction of both these factors. When the nanoparticles are not dislocated in the adhesive with

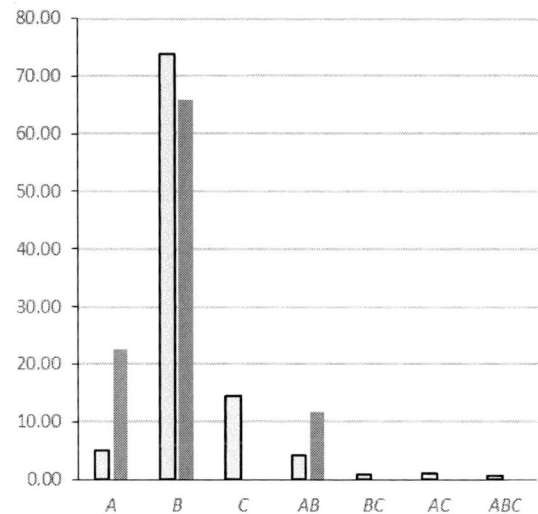

Fig. 1. Percentage expression of the influence of the individual factors and their combinations on the shear strength of adhesive joints. Grey columns represent the situation after the thermal treatment of the joints, orange ones before the treatment.

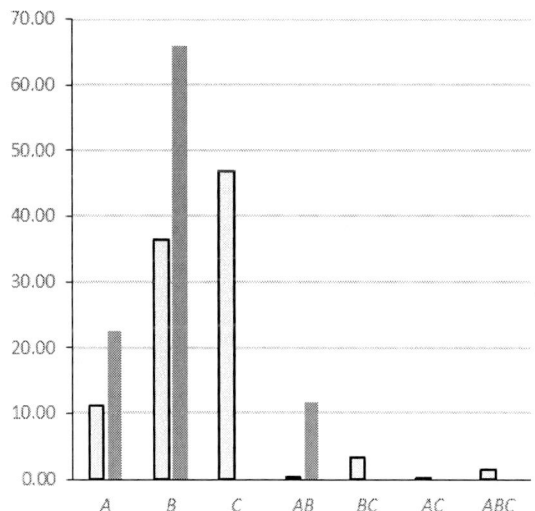

Fig. 2. Percentage expression of the influence of the individual factors and their combinations on the shear strength of adhesive joints. Grey columns represent the situation after the humidity treatment of the joints, orange ones before the treatment.

sufficient homogeneity, they form clusters and therefore can not contribute to the overall shear strength of the joint at such level as the particles uniformly distributed in the bulk adhesive. The significance of the distribution of particles in adhesive was confirmed also in samples thermally aged and the samples aged in the higher humidity.

Whereas for samples aged at a higher temperature and normal humidity it was found that the quality of mixing is the most significant parameter to obtain the joint with the maximum shear strength, for samples treated at the high humidity the mixing is also significant for the shear strength but more significant is the treatment in humidity.

This fact can be explained by the properties of epoxy resin that forms the insulating matrix of adhesive. This resin has a high susceptibility to moisture absorption. Wetting of the matrix causes the formation of AgOH thin films on the filler particles and so changes the adhesion mechanism between the resin and the filler. This change will influence the mechanical properties of the adhesive joint.

V. CONCLUSIONS

Electrically conductive adhesive with isotropic electrical conductivity was modified by silver nanoparticles. Then it was used for the manufacturing of adhesive joints. The shear strength of the joints was measured.

Then the first half of the adhesive joints was aged at the temperature of 125 ºC and the relative humidity of 65 % for 700 hours, and the second half at the temperature of 24 ºC and the relative humidity of 98 % for also 700 hours. Then the measurement of the shear strength was repeated.

The results were processed using DOE to find how strong is the relationship between the concentration of the nanoparticles in adhesive, between the quality of their mixing in the adhesive, and between the climatic aging and the shear strength of the adhesive joints. It was found that a very strong is a feedback between the shear strength and the quality of mixing of the nanoparticles in adhesive and, in the case of the joints aged at the high humidity, also this aging.

The work showed interesting relationships between the shear strength and the parameters of the preparation of the modified

adhesive as well as the shear strength and the different types of climatic aging.

It was also found the effectiveness of using of DOE in the case of processing of results of such the type.

REFERENCES

[1] Morris, J.E., Liu, J., "Electrically Conductive Adhesives: A Research Status Review. Micro- and Opto-Electronic Materials and Structures: Physics, Mechanics, Design, Reliability, Packaging." Springer US, (2007), pp. B527-B570.

[2] Yim, M. J.: Review of Electrically Conductive Adhesive Technologies for Electronic Packaging, Electronic Materials Letters, Vol. 2 No. 3 (2006), pp. 183-194.

[3] Kristiansen, H. and J. Liu, Overview of conductive adhesive interconnection technologies for LCD's. IEEE Transactions on Components, Packaging, and Manufacturing Technology Part A, 21 (2), (1998), p. 208-214.

[4] Li, Y.; Moon, K.; Li, H.; Wong, C.P.: Development of isotropic conductive adhesives with improved conductivity, Proceedings of the ninth international symposium on advanced packaging materials (2004), pp.1–6.

[5] Lu, D.; Wong, C.P.: Isotropic conductive adhesives filled with low melting-point alloy filler, IEEE Trans Electron Pack Manuf (2000) 23(1), (2000), pp.185–90.

[6] Fan, L.; Su, B.; Qu, J.; Wong, C.P.: Effects of nano-sized particles on electrical and thermal conductivities of polymer composites, Proceedings of the ninth international symposium on advanced packaging materials (2004), pp. 193–9.

[7] Lee, H-H., Chou, K-S., Shih Z-W.: Effect of nano-sized silver particles on the resistivity of polymeric conductive adhesives. International Journal of Adhesion and Adhesives, 25 (5), (2005); pp. 437-441.

[8] Kim, H.K.; Shi, F.G.: Electrical reliability of electrically conductive adhesive joints: dependence on curing condition and current density, Microelectronics Journal, 32, (2001), pp. 315-321.

[9] Amepox Microeletronics Ltd., „ELPOX AX 20 - Technical sheet," 2018. [Online]. Available: http://www.amepox-mc.com/files/E-S_AX20.pdf Accessed September 15, 2020;

[10] Li, L., Morris J. E.: Electrical Conduction Models for Isotropically Conductive Adhesive Joints, IEEE Trans. on Electronics Packaging Manufacturing, 22, (1997), pp.3-8.

[11] Kim, D.O., Jin, J.H., Shon, W.I., Oh, S.H.: Observation for mechanical property variations of single polymer. Journal of Applied Polymer Science, 105 (2), (2007), pp. 585 - 592.

Effect of different analyte solutions on the SERS process examined on gold nanoisland samples

Petra Pál, István Csarnovics
Department of Experimental Physics
Faculty of Science and Technology
University of Debrecen
Debrecen, Hungary

Miklós Veres
Institute for Solid State Physics and
Optics
Wigner Research Centre for Physics
Budapest, Hungary

Attila Bonyár
Department of Electronics
Technology
Budapest University of Technology
and Economics
Budapest, Hungary

Abstract—**In this work, we are examining the effect of the analyte on the process of Surface-Enhanced Raman Scattering (SERS). Benzophenone and riboflavin solutions were used as analytes for the measurements. The enhancement was measured on the same gold nanoisland substrates for both analytes at two different excitation wavelengths (532 and 633 nm). The enhancement factor was obtained for both analytes, and the influence of the analytes was studied.**

Keywords— gold nanoisland substrate, surface-enhancement Raman scattering, plasmonics, photonic devices, sensors

I. INTRODUCTION

Raman spectroscopy is a vibrational spectroscopic method that allows the analysis of the composition and structure of materials in solid, liquid, or gaseous states. [1] The main disadvantage of this method is that the Raman scattering intensity is very weak. Several solutions have been proposed to address this, including surface-enhanced Raman scattering. Plasmonic enhancement is one of the most efficient chemical analysis methods and optical sensors, which is performed on nanostructured plasmonic materials, most in the form of colloids or substrates. [2] SERS substrates can be made of gold, silver, or copper. [1] Today, the most common SERS substrates include metallic nanoparticles in suspension, nanostructures formed on a solid substrate, and nanostructures produced by nanolithography. [3] The efficiency of SERS depends on a number of factors, such as the geometric properties of the nanoparticles (shape, size, interparticle distance), the wavelength of the excitation source, or even the analyte being tested. [4,5] In our previous work, we investigated the geometric properties of glass surfaces coated with gold and silver nanoparticles to maximize SERS enhancement. The results showed that the larger and closer nanoislands resulted in greater enhancement. [6] However, the effect of the analyte has not been investigated yet. The aim of this work is to study two different analytes at two different excitation wavelengths on gold nanoisland substrates.

II. MATERIALS AND METHODS

A. Preparation of SERS substrates

Gold nanoislands formed on a glass substrate were used as SERS substrates. The basis of the structure, i.e. the thin layer, was prepared by thermal vacuum evaporation. Thin layers of different thicknesses (6-12 nm) were prepared. The layers were then placed in an oven where the heat treatment was performed at different temperatures (450°C, 500°C, 550°C), and for different times (15, 30, 60, and 120 minutes). Thus, nanoislands with different parameters (e.g., particle size, the distance between particles) were obtained.

B. Sample preparation for SERS measurements

Two different analyte solutions were used for Raman measurements. The first analyte is 50 mM benzophenone dissolved in isopropyl alcohol. The application of the analyte to the surface of the substrate is done using a spin coater. The other analyte was an aqueous solution of 10^{-5} M riboflavin. In this case, the substrate was immersed in the solution for about 20 hours, and after removal, the excess analyte solution remaining on the surface was washed with distilled water. [7]

C. SERS measurements

Renishaw inVia and Horiba LabRam Raman spectrometers were used for the Raman measurements on substrates. In our work, we used excitation sources of 532 nm and 633 nm to examine both analyte solutions. For both analytes, the beam was focused on the sample surface with a 50x objective. The measurement parameters are summarized in Table I. and Table II.

TABLE I. Measurement parameters for 532 nm excitation

532 nm	Benzophenone	Riboflavin
Laser intensity	1 %	1 %
Measurement time	10 s	50 s
Accumulation	1	20

TABLE II.Measurement parameters for 633 nm excitation

633 nm	Benzophenone	Riboflavin
Laser intensity	5 %	3.2 %
Measurement time	10 s	40 s
Accumulation	1	7

2020 IEEE 26th International Symposium for Design and Technology in Electronic Packaging (SIITME)

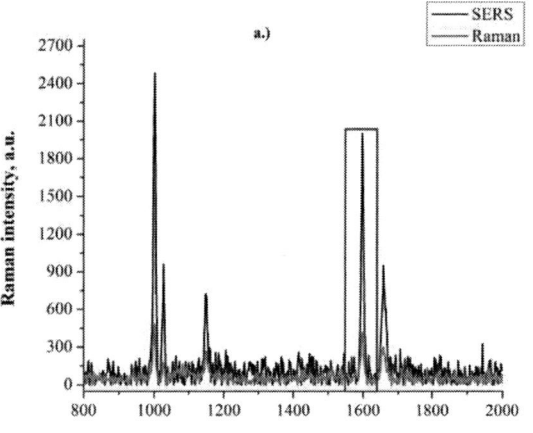

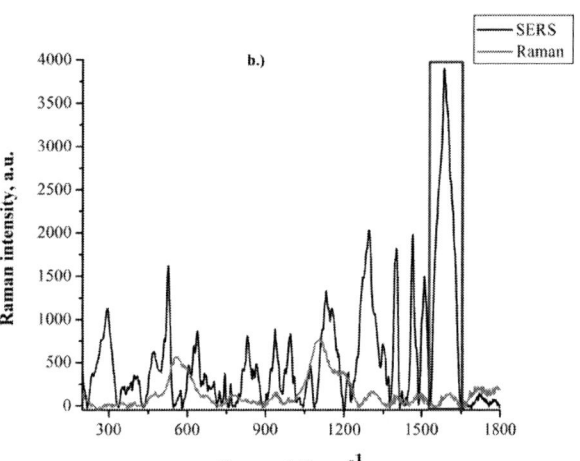

Fig. 1. *a.) The measured Raman and SERS spectra of benzophenone. b.) The measured Raman and SERS spectra of riboflavin*

The characteristic spectra of benzophenone and riboflavin were examined on plain glass and metal nanoisland samples. Figure 1 shows how the intensities of the major bands of the two analytes change in the presence of nanoislands.

Based on these, the SERS enhancement was calculated from the intensity of the 1596 cm⁻¹ peak for benzophenone, which is related to the vibration of the C-C bonds of the phenyl ring, and the 1576 cm⁻¹ peak for riboflavin, which can be assigned to the vibration of the C-N bond. [8,9]

In this work, the SERS enhancement factor (EF) [10] was calculated based on the following equation:

$$EF = \frac{I_{SERS}/N_{SERS}}{I_{RS}/N_{RS}} \qquad (1)$$

III. Results and discussion

In our previous article, we reported how the value of EF is affected by the size of the nanoislands, the distance between them, and the effect of the excitation laser wavelength. Our measurements showed that we obtained the best results for a gold nanoisland substrate using an excitation source with a wavelength of 532 nm. [6]

In addition to the benzophenone a new analyte, riboflavin solution was also tested on a gold nanoisland substrate with two excitation lasers of different wavelengths (532 nm and 633 nm) to study whether the analyte affects the enhancement.

Since we used a gold nanoisland substrate, we would expect greater enhancement using 532 nm excitation. Indeed, our results show that in the case of the benzophenone analyte, a higher degree of enhancement was achieved with this light source, since the plasmon resonance of these nanoislands is in the 500–600 nm range. Figure 2. shows the SERS spectra measured with two different excitation sources, it is clear that the 532 nm laser has a higher enhancement. The EF value for benzophenone and 532 nm excitation showed 2-6-fold enhancement.

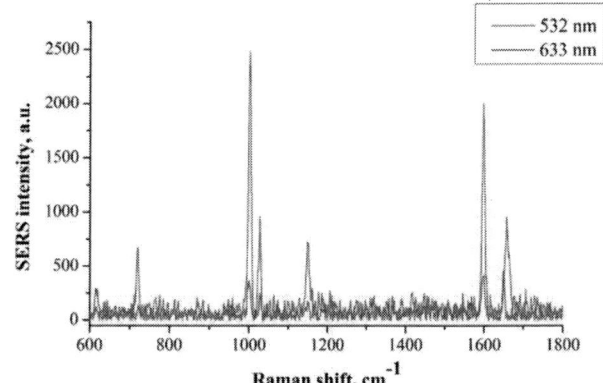

Fig 2. *Enhancement of characteristic peaks of benzophenone on a gold nanoisland substrate with different excitation wavelengths*

The same gold nanoisland substrates were also used for SERS enhancement analysis with riboflavin. For this analyte, the results show that the 633 nm excitation source proves to be better in terms of enhancement. As shown in Figure 3, 633 nm laser has a higher degree of enhancement with well distinguishable characteristic peaks of riboflavin. In contrast, the spectrum obtained with 532 nm laser hardly shows a peak. This can be explained by the fact that, due to its electronic structure, the resonant Raman scattering conditions for riboflavin occur for excitation photon energies closer to 633 nm than to 532 nm.[11] Partially this also can be explained by the fact that riboflavin is a fluorescent compound with an emission peak between 520 and 530 nm so that fluorescence suppresses the characteristic peaks of riboflavin and also reduces the degree of enhancement. [12] Therefore, the spectra obtained with 633 nm excitation were evaluated to calculate the value of the enhancement, which turned out to be almost 100-fold.

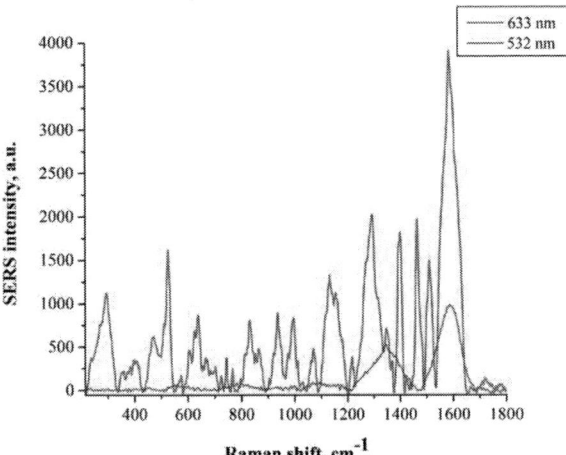

Fig.3. *Enhancement of characteristic peaks of riboflavin on a gold nanoisland substrate with different excitation wavelengths*

The study of the optimization of SERS substrates is a particularly important area of research on the enhancement of Raman scattering. In our previous work we showed how this is affected by the size of the nanoislands, the distance between them, and the wavelength of the excitation laser, and we determined which wavelengths are suitable for gold and silver substrates. [6] However, in addition to the parameters listed above, the properties of the analyte studied by the Raman experiment has an effect on the SERS process as well. So, during the SERS measurements, the analyte behavior should be considered as well. Also, the EF values obtained during the measurement of riboflavin analyte with a 633 nm laser may vary depending on the size of the nanoislands and the distance between them, which will be investigated later.

IV. CONCLUSIONS

In this work, we examined the effect of the analyte on the SERS process. Benzophenone and riboflavin solutions were used as probe molecules for the measurements. The enhancement was measured on the same gold nanoisland substrates for both analytes at two different excitation wavelengths (532 and 633 nm).

Using benzophenone, a higher enhancement (2-6-fold) was obtained with the 532 nm excitation source, while for riboflavin the SERS EF was higher for 633 nm excitation. This can be explained by the fact that the resonance Raman scattering wavelength of riboflavin is closer to 633 nm than to 532 nm. Furthermore, it can also be explained by the fluorescent property of riboflavin. During the evaluation of the results obtained with the 633 nm excitation source, an increase of about 100-fold was found for this analyte.

Based on our results, it can be said that for surface-enhanced Raman measurements of different samples/materials, the appropriate excitation source must first be found, and then the substrate material must be selected and optimized for the analyte and the excitation source, thus maximizing SERS enhancement and detection efficiency.

ACKNOWLEDGMENT

This work was financially supported by the grant GINOP-2.3.2-15-2016-00041. The projects are co-financed by the European Union and the European Regional Development Fund. This work was supported by the VEKOP-2.3.2-16-2016-00011 grant, which is co-financed by the European Union and European Social Fund.

Istvan Csarnovics is grateful for the support of the János Bólyai Research Scholarship of the Hungarian Academy of Sciences and the support through the New National Excellence Program of the Ministry of Human Capacities.

REFERENCES

[1] S. Tokonami, Y. Yamamoto, H. Shiigi, T. Nagaoka," Synthesis and bioanalytical applications of specific-shaped metallic nanostructures: A review" Analytica Chimica Acta, vol. 716, pp. 76–91. 2012

[2] L.T. Hoang, H.V. Pham, M.T.T. Nguyen," Investigation of the Factors Influencing the Surface-Enhanced Raman Scattering Activity of Silver Nanoparticles" Journal of ELECTRONIC MATERIALS, vol. 49, pp. 1864-1871. 2020

[3] P. A. Moiser-Boss," Review of SERS Substrates for Chemical Sensing," Nanomaterials, vol. 7. pp.1-30. 2017

[4] M. Fan, G. F. S. Andrade, A. G. Brolo, "A review on the fabrication of substrates for surface enhanced Raman spectroscopy and their applications in analytical chemistry" Anal. Chim. Acta. vol. 693 pp. 7–25. 2011

[5] R.A. Álvarez-Puebla, "Effects of the Excitation Wavelenght on the SERS Spectrum" J. Phys. Chem. Lett. vol.3 pp. 857-866, 2012

[6] P.Pál, A. Bonyár, M. Veres, L. Himics, L. Balázs, L. Juhász, I. Csarnovics," A generalized exponential relationship between the surface-enhanced Raman scattering (SERS) efficiency of gold/silver nanoisland arrangements and their non-dimensional interparticle distance/particle diameter ratio" Sensors and Actuators A: Physical, 2020

[7] A. Kokaislová, P. Matějka," Surface-enhanced vibrational spectroscopy of B vitamins: what is the effect of SERS-active metals used?" Anal Bioanal Chem, vol. 403, pp. 985-993, 2012

[8] L. Babkov, J. Baran, N.A. Davydova, V.I. Mel'nik, K.E. Uspenskiy," Raman spectra of metastable phase of benzophenone." Journal of Molecular Structure, vol. 792-793, pp. 73–77 2006

[9] F. Liu, H. Gu, Y. Lin, Y. Qi, X. Dong, J. Gao, T. Cai," Surface-enhanced Raman scattering study of riboflavin on borohydride-reduced silver colloids: Dependence of concentration, halide anions and pH values" Spectrochimica Acta Part A, vol. 85, pp. 111-119, 2012

[10] H. K Lee, Y. H. Lee, C. S. L. Koh, G. C. Phan-Quang, X. Han, C. L. Lay, X. Y. Ling, H. Y.F. Sim, Y-C. Kao, Q. An, Designing surface-enhanced Raman scattering (SERS) platforms beyond hot-spot engineering: emerging opportunities in analyte manipulations and hybrid materials, Chemical Society Reviews. vol. 48 pp. 731-756, 2019

[11] M. Šubr, A. Kuzminova, O. Kylián, M. Procházka, " Surface-enhanced Raman scattering (SERS) of riboflavin on nanostructured Ag surfaces: The role of excitation wavelength, plasmon resonance and molecular resonance." Spectrochimica Acta Part A: Molecular and Biomolecular Spectroscopy, vol. 197, pp. 202–207. 2018

[12] D.M. Gore, A. Margineanu, P. French, D. O`Brart, C. Dunsby, B.D. Allan, "Two-Photon Fluorescence Micriscopy of Corneal Riboflavin Absorption" Investigative Opthalmology & Visual Science, vol. 55, pp. 2476-2481, 2014

2020 IEEE 26th International Symposium for Design and Technology in Electronic Packaging (SIITME)

Optical properties of core-shell Ag@Au and Au@Ag nanoparticles

Géza Szántó, István Csarnovics
Department of Experimental Physics
University of Debrecen
Debrecen, Hungary

Attila Bonyár
Department of Electronics Technology
Budapest University of Technology and Economics
Budapest, Hungary

Abstract— **The optical properties of spherical, bimetallic: (core-shell) plasmonic nanostructures made of gold and silver were studied. The simulations were performed using the boundary element method. The refractive index sensitivity (RIS) of the nanoparticles was calculated from the peak shift in the extinction spectrum due to the change of the surrounding medium's refractive index. The effect of the particle's material, the size of the core, and the shell on the extinction spectrum and the RIS were investigated and they were compared with the data of the pure spherical nanoparticles.**

Keywords— extinction spectra; bulk refractive index sensitivity; localized surface plasmon resonance; LSPR; core-shell spheres;

I. INTRODUCTION

Research in plasmonic nanomaterials is in the focus of scientific interest today. One of the most important future applications of localized surface plasmon resonance (LSPR) is the fabrication of biosensors based on this phenomenon. Such sensors can be used for several measurements in parallel, so they can play a significant role in medical diagnosis in the future.

Gold and silver nanoparticles are often used as sensors because of their favorable optical properties. By applying a shell to a nanoparticle synthesized from one of the metals, the two materials can be easily combined [1-3]. Using simulation, it can be investigated how the optical properties of a core-shell nanoparticle evolve by changing the core radius and shell thickness parameters.

Sensitivity, a measure often used to characterize LSPR sensors, shows the degree of peak shift caused by a unit refractive index change. Sensitivity alone does not give a complete characterization of a plasmonic structure, so it is important to evaluate the extinction spectra of different structures as well.

The optical properties of LSPR in nanoparticle systems are influenced by the material and shape of the particles and the distance between them, as we described in our previous article [4]. We have not previously studied core-shell type nanoparticles, so in the present research, we investigate the changes in the spectrum and RIS of these nanostructures in the function of material type and core, shell sizes.

II. METHODS

The bulk refractive index sensitivity [1] of core-shell plasmonic nanostructures, namely Ag@Au and Au@Ag, were simulated with the MNPBEM MATLAB toolbox [5]. The inner radius of the nanoparticles and the shell thickness were the running parameters. The boundary element method was used with retarded, i.e. full, Maxwell's equations solver, and the surfaces of both the inner and the outer sphere were approximated by triangulation with 256 vertices. Such systems were excited by plane waves to obtain the extinction cross-section. The wavelengths of the plane waves were in the range of visible and near-UV light.

The result of the simulation is influenced by the frequency-dependent permittivity used for the metals, which are experimental data. For both gold and silver, the results of Johnson and Christy [6] were used.

The results obtained from the simulations are extinction cross-section data, which are the sum of the scattering and absorption cross-sections. Peak or peaks appeared on the resulting extinction spectrum.

The RIS of the investigated nanoparticles was calculated by Eq 1. Changing the medium surrounding the nanoparticles between $n = 1.33$ RIU and $n = 1.35$ RIU (so the refractive index change was $\Delta n = 0.02$) results in a $\Delta\lambda$ peak shift. (see Fig. 1).

$$S = \frac{\Delta\lambda}{\Delta n} \qquad (1)$$

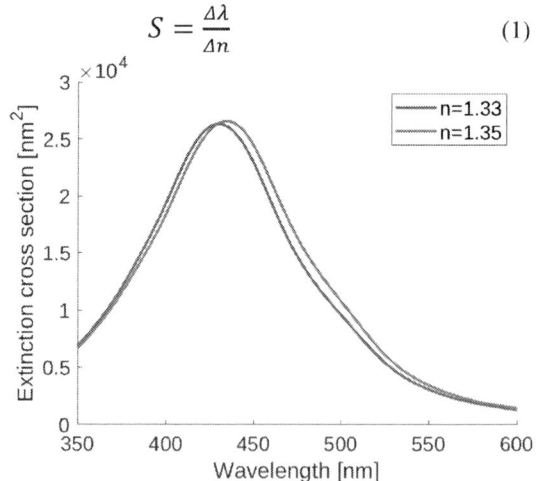

Fig. 1. Illustration of the extinction peak shift due to refractive index change

III. RESULTS AND DISCUSSION

In both Ag@Au and Au@Ag cases, the effect of a shell applied at 2 nm pitch intervals between 0 nm and 8 nm was demonstrated at 10 nm, 20 nm, 30 nm radii cores. Extinction spectra are shown up to 8 nm shells because no radical change in the spectra is observed in thicker shells, except that the spectrum is more and more like a sphere made of the shell material.

The applied refractive index change is small, so the shapes of the spectra corresponding to the two values are very similar. Thus only the spectrum corresponding to the $n = 1.33$ RIU case will be presented.

A. Ag@Au

The extinction spectrum of a 10 nm silver sphere is characterized by a single thin peak, as can be seen in Fig. 2. By applying a 2 nm gold shell, it causes a significant broadening of the spectrum. In the case of a 4 nm shell, the peak of the gold shell appears, while the peak of the core widens further. If the shell thickness is increased to 6 nm, the gold peak becomes dominant, and the shell peak is redshifted due to the increase in particle size. At the thickest 8 nm shell in the figure, the trend observed in the previous case continues, with the only dominant peak already coming from the shell.

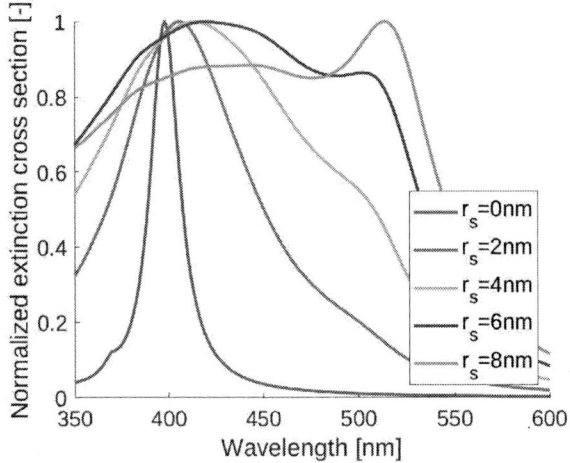

Fig. 3. Normalized extinction of r_c =20nm Ag@Au nanoparticles

In the case of a core with a radius of 30 nm (Fig. 4), increasing the thickness of gold to a thickness of 4 nm, the widening of the peak is the dominant phenomenon. At 6 nm, the peaks of the shell and the core are of similar height, merged peaks are formed, and the local minimum between the peaks is barely observable. At a shell thickness of 8 nm, the gold peak becomes dominant.

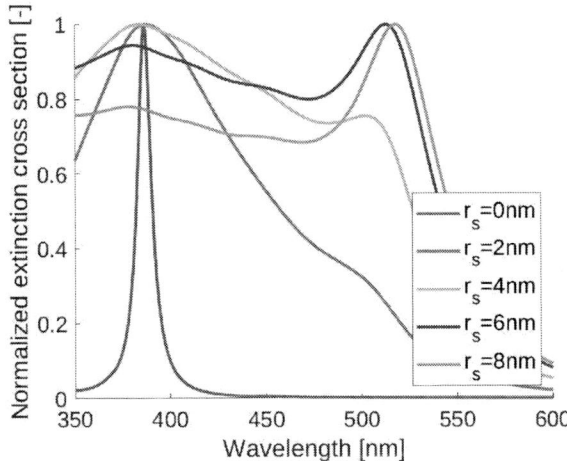

Fig. 2. Normalized extinction of r_c =10 nm Ag@Au nanoparticles

The peak of a silver sphere with a radius of 20 nm is slightly wider than the 10 nm sphere (Fig. 3). In the case of a 2 nm gold shell, a noticeable redshift occurs in addition to the widening of the peak. If the Au shell is 4 nm, a single very wide peak is present. By applying an additional layer of gold, the silver peak is completely flattened, and the peak from the gold shell appears at the same time. With an 8 nm thick shell, the peak of the shell is already the dominant one.

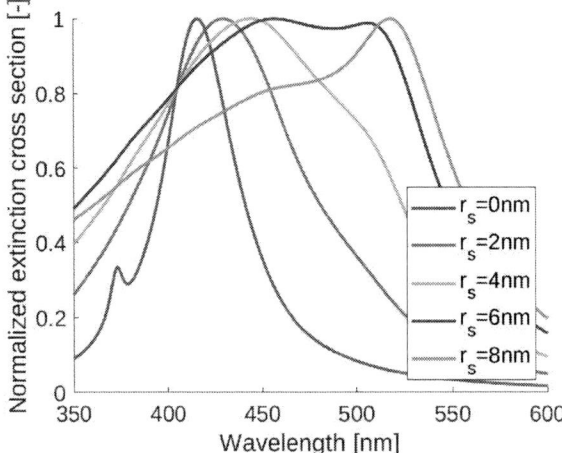

Fig. 4. Normalized extinction of r_c =30 nm Ag@Au nanoparticles

B. Au@Ag

A 10 nm gold particle has a peak at approximately 540 nm (Fig. 5), which does not drop down completely at lower wavelengths. Applying 2 nm silver to the sphere causes a huge change in the general shape of the spectrum. The peak of the gold sphere is significantly blueshifted (to around 500 nm), the dominant peak forming at a wavelength lower than the minimum wavelength in the figure, in the UV region. As additional layers of silver are applied, a redshift occurs as the size increases, while the spectrum becomes increasingly silver dominated. As the role of the gold core becomes insignificant, the width of the peak decreases.

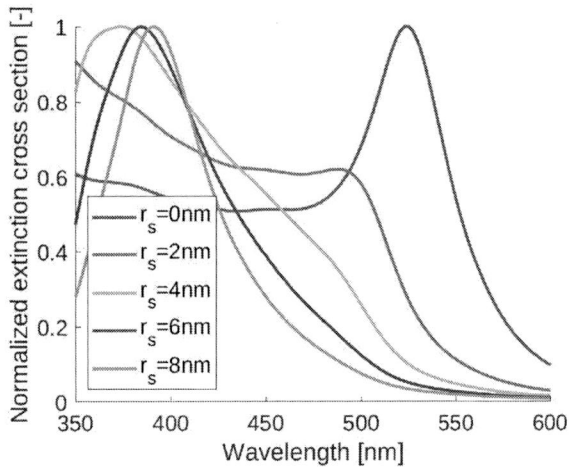

Fig. 5. Normalized extinction of r_c =10 nm Au@Ag nanoparticles

In terms of shape, the 20 nm gold sphere (presented in Fig. 6) is slightly different from the previously presented 10 nm, only some redshift can be observed and the height of the maximal extinction has increased, although this is not shown in the figure due to normalization. Due to the large gold core, the 2 nm Ag layer does not change the shape of the spectrum much, apart from the blueshift. In the case of a 4 nm thick shell, the silver's flattened peak is already dominant. At 6 nm, a complete fusion of the peaks is observed. At an 8 nm shell, a single extremely wide peak is formed, which is shifted towards a larger wavelength compared to a less thick shell.

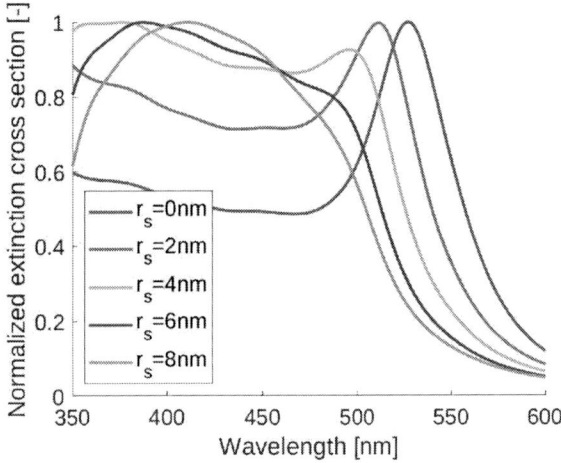

Fig. 6. Normalized extinction of r_c =20 nm Au@Ag nanoparticles

In the case of a 30 nm gold core (Fig. 7), the addition of silver layers causes a blueshift of the gold peak. At the 8 nm shell, the peak belonging to the core can still be observed, however, moving from shorter wavelengths towards the peak, the rise of the extinction is less steep. The trend persists even at larger shell sizes; no sudden formation of a new peak is observed with increasing shell, only a single merged peak is formed.

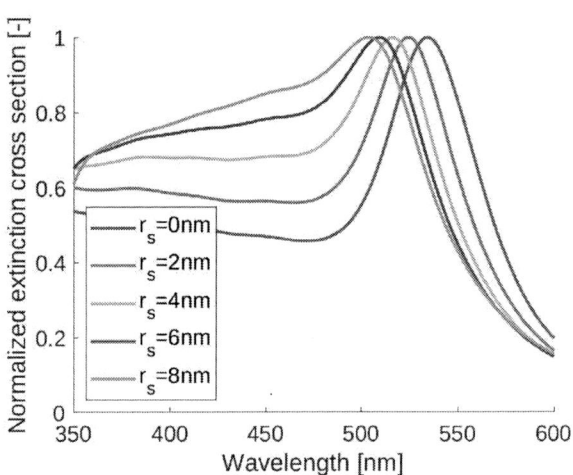

Fig. 7. Normalized extinction of r_c =30 nm Au@Ag nanoparticles

C. RIS of the investigated samples

Despite the differences in the extinction spectrum, similar sensitivity trends can be observed for both Au@Ag and Ag@Au systems. The value of RIS can be increased in the case of both compositions by adding a shell, however, in cases of higher sensitivity, the merging and broadening of the peaks can also be observed. Tables 1 and 2 collects the RIS of the investigated Ag@Au and Au@Ag core-shell nanoparticles, respectively.

TABLE I.   RIS OF THE INVESTIGATED AG@AU NANOPARTICLES

Ag@Au					
r_c [nm]	r_s [nm]	Wavelength of peak1 [nm]	Sensitivity of peak1 [nm/RIU]	Wavelength of peak2 [nm]	Sensitivity of peak2 [nm/RIU]
10	0	387	112	-	-
10	2	388	120	-	-
10	4	384	72	502	89
10	6	381	51	513	63
10	8	379	48	518	61
20	0	398	136	-	-
20	2	406	167	-	-
20	4	413	195	-	-
20	6	419	354	502	147
20	8	443	185	513	88
30	0	-	-	416	185
30	2	429	223	-	-
30	4	444	211	-	-
30	6	457	296	506	186
30	8	-	-	518	109

TABLE II.   RIS OF THE INVESTIGATED Au@Ag NANOPARTICLES

Au@Ag					
r_c [nm]	r_s [nm]	Wavelength of peak1 [nm]	Sensitivity of peak1 [nm/RIU]	Wavelength of peak2 [nm]	Sensitivity of peak2 [nm/RIU]
10	0	-	-	523	55
10	2	338	26	487	101
10	4	374	171	-	-
10	6	384	133	-	-
10	8	391	166	-	-
20	0	-	-	527	68
20	2	336	10	511	77
20	4	376	146	495	110
20	6	387	183	-	-
20	8	411	281	-	-
30	0	-	-	534	97
30	2	-	-	524	90
30	4	-	-	516	99
30	6	-	-	509	120
30	8	-	-	503	144

IV. CONCLUSIONS

Examining both Ag@Au and Au@Au core-shell nanoparticles, it was found that the size of the core and the shell has a significant effect on the shape of their spectra, and also on their sensitivity. By adding a shell structure to a core material, the optical spectra will be changed, it getting wider with increasing of the shell size. In some cases, several peaks can be observed at the same time. A typical process is the merging of different peaks into a single broad peak. Also the sensitivity could be enhanced, but usually only in a weak size region.

ACKNOWLEDGMENTS

We acknowledge KIFÜ for awarding us access to resource-based in Hungary.

This work was supported by the GINOP- 2.3.2-15-2016-00041 Project, which is co-financed by the European Union and the European Regional Development Fund. Istvan Csarnovics is grateful for the support of the János Bólyai Research Scholarship of the Hungarian Academy of Sciences and the support through the New National Excellence Program of the Ministry of Human Capacities.

The research reported in this paper and partially carried out at the Budapest University of Technology and Economics has been supported by the National Research Development and Innovation Fund (TKP2020) based on the charter of bolster issued by the National Research Development and Innovation Office under the auspices of the Ministry for Innovation and Technology.

REFERENCES

[1] T. Ghodselahi, S. Arsalani, and T. Neishaboorynejad, "Synthesis and biosensor application of Ag@Au bimetallic nanoparticles based on localized surface plasmon resonance," *Applied Surface Science*, vol. 301, pp. 230–234, May 2014, doi: 10.1016/j.apsusc.2014.02.050.

[2] A.-M. Hada *et al.*, "Fabrication of gold–silver core–shell nanoparticles for performing as ultrabright SERS-nanotags inside human ovarian cancer cells," *Nanotechnology*, vol. 30, no. 31, p. 315701, Aug. 2019, doi: 10.1088/1361-6528/ab1857.

[3] Daiki Kawasaki, Hirotaka Yamada, Kenichi Maeno, Kenji Sueyoshi, Hideaki Hisamoto, and Tatsuro Endo. "Core−Shell-Structured Gold Nanocone Array for Label-Free DNA Sensing". ACS Applied Nano Materials 2019 2 (8), 4983-4990

[4] A. Bonyár, I. Csarnovics, and G. Szántó, "Simulation and characterization of the bulk refractive index sensitivity of coupled plasmonic nanostructures with the enhancement factor," *Photonics and Nanostructures - Fundamentals and Applications*, vol. 31, pp. 1–7, Sep. 2018, doi: 10.1016/j.photonics.2018.05.004.

[5] U. Hohenester and A. Trügler, "MNPBEM – A Matlab toolbox for the simulation of plasmonic nanoparticles," *Computer Physics Communications*, vol. 183, no. 2, pp. 370–381, Feb. 2012, doi: 10.1016/j.cpc.2011.09.009.

[6] P. B. Johnson and R. W. Christy, "Optical Constants of the Noble Metals," *Phys. Rev. B*, vol. 6, no. 12, pp. 4370–4379, Dec. 1972, doi: 10.1103/PhysRevB.6.4370.

Modelling Thermally-Induced Mechanical Faults in Power Integrated Circuits Assemblies

A. Bojiță, M. Purcar, V. Țopa
Department of Electrotechnics and Measurements
Technical University of Cluj-Napoca
Cluj-Napoca
Marius.Purcar@ethm.utcluj.ro

R. Oneț, M. Neag
Bases of Electronics Department
Technical University of Cluj-Napoca
Cluj-Napoca

Abstract— **The structure of a Double Diffused Metal-Oxide-Semiconductor (DMOS) device is complex and consists of multiple routing metallization layers interconnected through vias embedded in Silico Dioxide (SiO2). The predominant failure mechanisms that occur after a large number (e.g. 10^6) of active cycles are the crack formation between two adjacent signal metal lines and/or delamination of power metal plates from the SiO2 layer. Numerical simulation of power IC's based on finite element analysis is one of the most used simulation tools for the defect risk assessment. A novel subdivision method combined with the homogenization and nonconformal mesh of computational structure is applied to the model to efficiently simulate the complex IC structures. Results are assessed on a commonly DMOS layout substructure.**

Keywords—FE model simplification, failure mechanism. Power DMOS devices, metallization system.

I. INTRODUCTION

DMOS technologies, predominantly use the routing metallization system arranged on multiple layers interconnected through vias sets and contacts, [1] and [2]. Many papers, i.e. [3]-[5], extensively analyzed by numerical simulation the predominant failure mechanisms as, crack formation between two adjacent signal metal lines and/ or delamination of the power metal plates from the dielectric layer in DMOS power devices subjected to repetitive induced thermal stress. Such thermo-mechanical processes can be studied with the help of the Finite Element Method (FEM). To assess the reliability of the power IC's, efficient computational structures, that can easily highlight the accumulation of plastic deformations inside the DMOS metallization structures, are required. A more efficient simulation process can be achieved only if some simplifications are performed, especially at the level of the metallization system. One of the used simplification methods replaces the high detailed heterogenous regions with one homogenous region [6]. Moreover, the FE model mesh density can be reduced by using a nonconformal mesh approach at subdomain interfaces, as presented in [7]. Thus, the mesh is easily refined only inside the interest region. In addition, this paper proposes an additional simplification, by subdividing the regions occupied by the metallization system function of the thermal induced elastic or plastic behavior and introduces a plasticity indicator-based selection method. Hence, the degrees of freedom (DOF) in the matrix system considerably reduces because the elastic material

properties model operates only 1 unknown compared to plastic material properties one which operates 8 unknowns.

The simplification procedure of computational model starts already with the definition of geometrical model. It is followed by homogenization, meshing and the typical boundary conditions. Section III presents the thermal field distribution who justifies and introduces the implementation of subdivision algorithm of metallization function of elastic or plastic material properties behaviors. Results denotes a CPU time gain up to 10.3% while the accuracy remains unchanged.

II. FINITE ELEMENT MODEL OF POWER INTEGRATED CIRCUITS

A. Model configuration

In order to highlight the failure mechanism, a computational structure commonly found in Bipolar CMOS-DMOS (BCD) technologies is proposed. The structure consists of a 400μm thickness silicon (Si) substrate covered by a 4 μm thick insulating (dielectric) SiO_2 layer. The SiO_2 encapsulates the routing metallization (signal metallization) - composed of multiple metal line levels, interconnected by via groups in the vertical direction. Above the SiO_2 substrate, two power metal plates of 10μm thickness are placed and further connected with the signal metallization by via groups [5].

The active area of analyzed structure is more or less symmetrical along the longitudinal and transversal axis in BCD technologies. Therefore, only one quarter of the entire active area is extracted for further simulation (see Fig. 1). Adiabatic Boundary Conditions (BC) and symmetry BC's are added to the thermal and respectively mechanical simulation model, as shown in Fig. 1.

The accumulation of plastic deformation is more pronounced at the upper level of the routing metallization system [2] and [3]. Thus, only the topmost thin signal metal layer of metallization is practically necessary in order to assess the crack formation risk between two adjacent signal metal lines.

The high geometric aspect ratio of the metallization lines and the number of the interconnection paths (very long and thin structures) leads to computational meshes in the range of 10^8-10^{10} nodes and elements. In order to further simplify the computational model, only the topmost thin routing metal lines are considered, while the Power Vias and the lower routing metal system are usually excluded from the computational model, [2] and [3]. The high detailed heterogenous regions are replaced with one homogenous region. The equivalent material properties of these regions are extracted using a FEM approach applied to a Representative Volume Element, as described in [6]. The extracted material properties are linear-elastic and orthotropic due to the distribution of the geometrical details inside the homogenization regions. Hence, most of small geometrical details are simplified for a faster FE mesh generation with a smaller number of nodes and elements (in the range of 10^6 nodes and elements) compared to a detailed simulation structure.

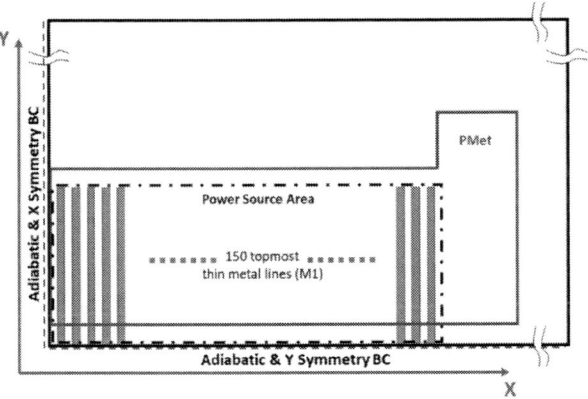

Fig. 1. Schematic representation of reduced computational model for thermo-mechanical analysis (dimensions are not to scale).

B. FE mesh of the computational model

Although simplified, the computational structure still requires a special (nonconformal mesh at the interfaces of the computational subdomains [7]) in order to further reduce the FE mesh size. Hence, the computational model is divided into an interest region and the others. The interest region incorporates all topmost thin signal metal lines (M1). A high-density hexahedral mesh is generated inside M1 metal lines. Two Viscous Layers mesh elements are defined inside the SiO$_2$ region around the M1 lines, Fig. 2.

Fig. 2. Clipping view through the final nonconformal finite elements mesh distribution with zoom focused on the nonconformal interface.

The number of computational nodes reduces more than half compared to the conformal mesh, 1.25×10^6 nodes for the nonconformal compared to 2.5×10^6 for the conformal mesh.

C. Boundary conditions

Thermally induced deformations inside the model are computed by coupling two field problems: a thermal with a mechanical one. The temperature distribution inside the model is emulated by a uniform power source placed at the Si/SiO$_2$ interface (dashed doted rectangle in Fig. 1 b). For a 10 ms power pulse duration, the temperature profile over time will evolve according to Fig. 3 inside the model. This applied heating-cooling cycle is repeated at 1 Hz frequency.

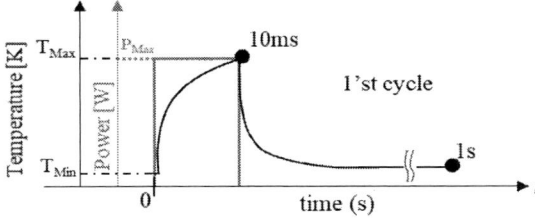

Fig. 3. Power pulse applied to the source and the evolution of temperature profile over time.

For a fixed temperature condition, imposed at the bottom face of Si die, a temperature gradient arises across the model, which will further develop a mechanical displacement with stress accumulation withing the IC structure. The X and Y symmetry planes shown in Fig. 1 b), mechanically constrain the movement of the model along oX and respectively oY axis. The bottom surface of the Si die structure is set fixed along oZ axis.

D. Material properties

Temperature dependent material properties are considered for the DMOS sandwich structure according to [5]. The Si and SiO$_2$, being a brittle material, are modeled with an elastic behavior model, while Al metal regions are modeled with the elastoplastic non-linear kinematic hardening law of Chaboche [8]. For the homogenized regions orthotropic temperature dependent material properties are defined according to [6].

III. RESULTS AND DISCUSSION

A. Thermal study

The uniform power source from the computational model (see Fig. 1 b) heats up the model according to the evolution of the power pulse described in Fig. 3.

Fig. 4. Thermal field distribution at 10ms for 623-423K temperature interval.

Thus, the temperature reaches the maximum value after 10ms. Two temperature intervals are studied in this paper: (i) 623-423K and (ii) 500-423K. Both temperature intervals (i) and (ii) were chosen such that the deformation at the level of the metallization system enters predominantly in the plastic respectively elastic domain. Thus, the efficiency of the proposed simplification algorithm can be assessed. For the studied case (i), the maximum temperature value is 623K, see Fig. 4. These are typical temperature variation for inside power IC's during the active cycling. The temperature variation between 623K and 423K leads to a temperature gradient (the source of thermally induced stress). For the studied case (ii), the power level is decreased such that the temperature is cycled between 500 and 423K, while the other conditions are the same.

B. Mechanical study

In order to generate the thermal-induced deformation conditions, the thermal field distribution is further interpolated on the simulation mesh of the analyzed structure to compute the thermal induced stress.

Modelling all computational subdomains with elastic material properties leads to simple and very fast simulations. Nevertheless, such study is only suitable for static structural cursory analysis. In case of transient analysis, the materials with elastic behavior do not account the deformation accumulation and the remnant thermally induced stress at the end of each temperature cycle. The Von Misses stress distribution at the end of first thermal cycle (1s) when metal regions are modeled with elastic behavior is presented in Fig. 6 a). The stress values are under 1 Pa. For capturing the deformation accumulation and the remanent thermally induced stress, appropriate mathematical models to account the plastic behavior of metals, are required. The nonlinear elastoplastic kinematic hardening processes in metallization system are approached with Chaboche model [8]. Although this model allows an accurate simulation of internal stress state [5], at the end of each temperature cycle (see Fig. 6 b) it also leads to a higher computational time. Therefore, an extra simplification stage is required in order to further accelerate the simulation of thermally induced stress problem. As an elastic behavior material model is described only by one internal variable, at the opposite end, a plastic behavior material model [8], is described by 8 internal variables. For reducing the number of unknowns in the system matrix, an algorithm based on the element plasticity indicator will be used to sort the mesh elements of metallization regions in two groups: one with

elements of elastic and another one with elements of plastic material properties. Hence, the computational time and memory used by simulation are reduced too.

The partition algorithm starts assigning plastic material properties to all mesh elements of metallization region. Further, when the model reaches the highest temperature (at 10ms), the thermally induced stress is computed. A sorting loop iterates over the metallization region mesh elements. Based on the plasticity indicator computed earlier, the sorting loop appends the mesh elements IDs into two lists for the plastic respectively elastic material region. At the end of the selection algorithm two mesh groups are generated containing the plastic and elastic mesh elements from the metallization system. A top view of subdivided metallization systems is presented in Fig. 5 for both temperature intervals, (i) and (ii).

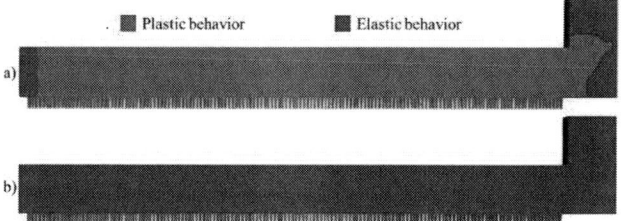

Fig. 5. Power IC metallization system subdivided into plastic and elastic material behaviors: a) study case (i) and b) study case (ii).

The mesh elements assigned with plastic material properties depends on temperature level reached by the metallization system during the active thermal cycling. Higher, temperatures lead to a larger number of mesh elements of metallization system, enriched with plastic material properties behavior. After the subdivision stage of metallization system in plastic or elastic material properties regions function of plasticity indicator, the accumulated thermally induced stress is computed at each temperature cycle. The gain of the computational effort after simplification is expressed in terms of CPU time and RAM memory usage for one temperature cycle during the mechanical simulation stage.

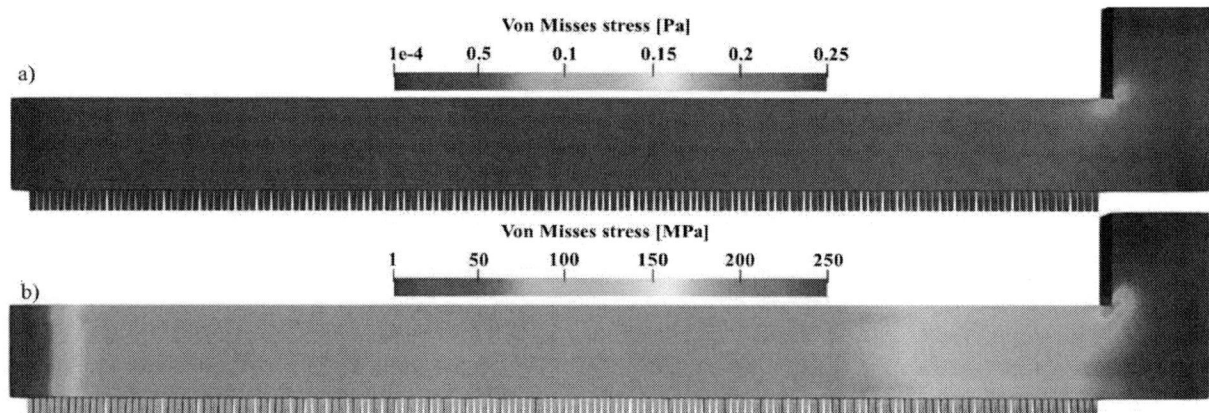

Fig. 6. Distribution of the Von Misses stress inside the metallization system for case (i): a) fully elastic material properties and b) plastic material properties.

2020 IEEE 26th International Symposium for Design and Technology in Electronic Packaging (SIITME)

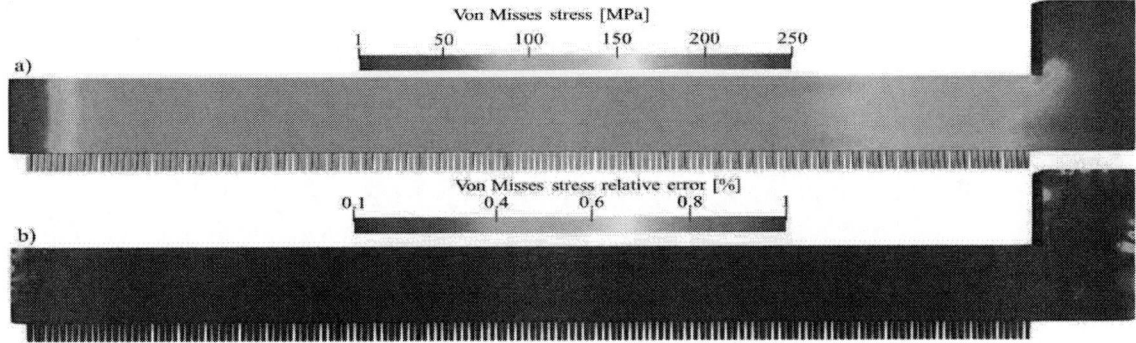

Fig. 7. Study case (i): a) Distribution of Von Misses stress inside the metallization system; b) Von Misses stress relative error distribution between the unpartitioned (fully plastic material properties behaviors) and partitioned (function of elastic and plastic material behaviors) models.

In case of first temperature interval (i), 95% of mesh elements of metal region were assigned with plastic material behavior. The CPU time was 32min, for 28.3GB RAM usage. These represented 4% gain in CPU time and 2% gain in RAM usage, compared to the case where the complete metallization system was modeled as material with plastic behavior. In case of second temperature interval (ii), although only 32% of mesh elements of metal region were enriched with plastic material behavior, the CPU time and RAM memory usage, did not linearly decrease because the matrix factorization has a larger weight in the final CPU time. The CPU time was 24min, for 26.1GB RAM usage. These represented 10.3% gain in CPU time and 12% gain in RAM usage, compared to the case where the complete metallization system was modeled as material with plastic behavior.

The Von Misses stress field inside the metallization system is presented in Fig. 7 a) at the end of cooling phase (after 1s). The results are in line with the case where the complete metallization system was modeled as an material with plastic behavior, Fig. 6 b). The relative error distribution of Von Misses stress field between the unpartitioned (fully plastic material properties behavior) and partitioned (function of elastic and plastic material properties behavior) models is presented in Fig. 7 b), showing a good accuracy, less than 1%.

IV. Conclusions

A methodology for efficient FEM simulation of thermally induced mechanical faults inside power IC devices is presented. Techniques as simplification and homogenization, nonconformal mesh generation and subdivision of the metallization system are applied to accelerate the CPU time, reduce the RAM requirements, but keep the solution accuracy of detailed BCD IC models. The simplification and homogenization with nonconformal mesh generation methodology, reduces the number of computational nodes from c.a. 10^8 to $x10^6$ nodes. The material properties assignment to the mesh elements of metallization region function of elasticity/plasticity indicator, further reduces the number of the unknowns inside the system matrix. The gain in CPU time is 4% and in RAM usage is 2%, for the large temperature interval case (i). The gain in CPU time is 10.3% and in RAM usage is 12%, for the smaller temperature interval case (ii). For both analyzed cases the accuracy of results remains unchanged.

The assignment of mesh elements with elastic or plastic material properties function of temperature distribution further improves the CPU time and RAM usage. The partition algorithm also allows a simpler homogenization approach to elastic regions.

Acknowledgment

This work was fully supported by the project iDev40. The project iDev40 has received funding from the ECSEL Joint Undertaking under grant agreement No. 783163. The JU receives support from the European Union's Horizon 2020 research and innovation program. It is co-funded by the consortium members, grants from Austria, Germany, Belgium, Italy, Spain and Romania. The information and results set out in this publication are those of the authors and do not necessarily reflect the opinion of the ECSEL Joint Undertaking.

References

[1] R. Rudolf et al., "Automotive 130 nm smart-power-technology including embedded flash functionality," in Proc. Int. Symp. Power Semicond. Devices ICs, San Diego, CA, USA, May 2011, pp. 20–23.

[2] G. Pham, M. Ritter and M. Pfost, "Influence of metallization layout on aging detector lifetime under cyclic thermo-mechanical stress," 2016 IEEE International Reliability Physics Symposium (IRPS), Pasadena, CA, 2016, pp. 5B-5-1-5B-5-6, doi: 10.1109/IRPS.2016.7574551.

[3] T. Smorodin, J. Wilde, P. Alpern and M. Stecher, "A Temperature-Gradient-Induced Failure Mechanism in Metallization Under Fast Thermal Cycling," Transactions on Device and Materials Reliability, vol. 8, no. 3, pp. 590-599, 2008.

[4] G. Kravchenko, B. Karunamurthy, M. Nelhiebel and H. E. Pettermann, "Finite element analysis of fatigue cracks formation in power metallisation of a semiconductor device subjected to active cycling" EuroSimE 2013.

[5] A. Bojita, C. Boianceanu,M. Purcar,C. Florea,D. Simon and C. Plesa, "A simple metal-semiconductor substructure for the advanced thermo-mechanical numerical modeling of the power integrated circuits", Journal of Microelectronics Reliability, Volume 87, pages 142-150, 2018.

[6] C. Florea, C. Bostan, D. Simon, V. Topa and M. Purcar, "Extraction of Equivalent Mechanical Properties for Power ICs Metallization," 25th International Workshop on Thermal Investigations of ICs and Systems, Lecco, Italy, 2019, pp. 1-4, doi: 10.1109/THERMINIC.2019.8923452.

[7] A. Bojita, M. Purcar, C. Boianceanu and V. Topa, "Efficient Computational Methodology of Thermo-Mechanical Phenomena in the Metal System of Power ICs," 25th International Workshop on Thermal Investigations of ICs and Systems, Lecco, Italy, 2019, pp. 1-4, doi: 10.1109/THERMINIC.2019.8923502.

[8] Chaboche, J.-L., 2008. A review of some plasticity and viscoplsticity constitutive theories. International Journal of Plasticity, 24(10), pp. 1642.

DC/DC Converter Output Capacitor Bank's Reliability Comparison using Prediction Standard MIL-HDBK-217F and IEC 61709

Dan Butnicu
Technical University of Iasi
Iasi, Romania
dbutnicu@etti.tuiasi.ro

Luminiţa-Camelia Lazăr
Institute of Computer Science,
Romanian Academy-Iasi Branch
Iasi, Romania

Abstract—When it comes of reliability many studies has shown that the output capacitor bank is demonstrated to be the most critical component [7,9,10,13]. A higher level of reliability became a compulsory demand in modern DCDC converters [7]. In this work, the failure rate of an output capacitor bank used within a high current low voltage buck converter is calculated with both the latest prediction standard IEC 61709-2017 and the older standard MIL-HDBK-217, providing a comparison between the two which is a helpful tool for the output capacitor selection in the early stage design. This newer standard is trying to make compensation of the lack of the newest component technology in older standards .The components' environmental conditions were defined by a standard buck converter POL used for both calculation methodologies [4,5]. Results are compared by taking account of the influence of component's temperature and application of specific concepts like references conditions, operating conditions, ripple capability, internal selfheating.

Keywords—reliability; MTBF; MIL-HDBK-217F; IEC 61709; DC/DC converter; POL; polymer electrolytic capacito; MLCC.

I. INTRODUCTION

Modern power electronics needs for a proper functioning of DC/DC converters to ensure higher output quality, less energy consumption and longer lifespan. This paper is focused on the last requirement by analyzing the reliability of the equipment from a comparison point of view. The United States Navy's failure rate prediction of electronic components standard Military Handbook 217 published in 1965 was widely accepted for decades in order to reliability prediction even on industrial electronics and is still used today under critical manner because no more update after its latest version MIL-HDBK-217F - Notice 2 released in 1995[1]. Since then, newer failure rate prediction standards appear on the electronic systems reliability market. Over the time, military standard approach was mostly used but it does not accurately model the reliability because lack of taking account of the *mission profile*. Therefore IEC (International Electrotechnical Commission) provided a newer standard named TR 62380 (Technical Report) which take account of the temperature cycling for the failure rate predictions by a means of annual

mission profile [2]. Being not updated with data source for the new device technologies, IEC released in 2017 [3] a new standard IEC 61709 providing prediction by taking account of *References conditions* and *Oprating conditions* concepts. Since than several other standard were developed especially for telecommunication systems. None of the ulterior proposed standards has accomplished to become well accepted, all of them being criticized or defended.

II. BRIEF DESCRIPTION OF THE RELIABILITY CONCEPT AND IT'S KEY INDICATOR MTBF

Reliability is the ability of an item to perform a specific function under given conditions but in a specific period of time, often expressed by the failure rate λ

$$R(t) = e^{-\lambda t} \qquad (1)$$

A more practical and commercial factor of reliability is mean time between failure (MTBF) which is the average length of time before the first failure appears after it starts to work, then the item no longer able to continue functioning in normal operation. It is expressed by the integral of equation (1)

$$MTBF = \int_0^{+\infty} R(t)dt \qquad (2)$$

A simpler form is resulting in:

$$MTBF = 1/\lambda \qquad (4)$$

with λ signifying intrinsic the failure rate, excluding early failures and wear-out failures which is assumed constant during the *Life time* period in bath tub curve (Fig.1) and expressed in [F/10^6h] i.e. failures per one million component hours or FIT i.e. failures in time or one failure per one billion hours. Therefore all the standards provide a constant failure rate for system's components during useful lifetime. More details on the reliability prediction are provided by the reliability standards [1,2,3].

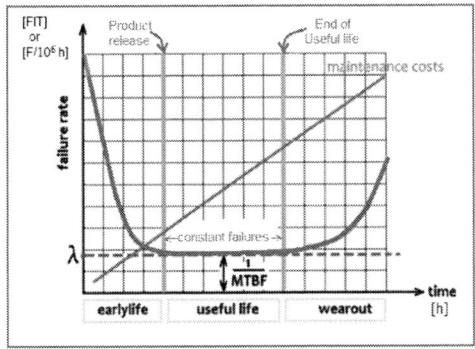

Fig.1 Failure rate and his bathtub curve shaped format.

III. THE METHODOLOGY OF RELIABILITY PREDICTION USING IEC 61709-2017 STANDARD

The reliability methodology for IEC 61709-2017 Standard uses the concept of *References conditions* and *Operating conditions* which are the typical values of stress that are observed by the components in majority of applications, with environmental conditions, operating and non-operating conditions, and stress models. The component failure rate under operating conditions is calculated as follows:

$$\lambda_{61709\text{-}2017} = \lambda_{ref} \times \pi_U \times \pi_T \times \pi_I \times \pi_E \times \pi_S \times \pi_{ES} \qquad (4)$$

where π_I is current dependence factor, π_E is environmental application factor, π_S is switching rate dependence factor, π_{ES} is electrical stress dependence factor.

$$\pi_U = \exp\{C_3 \times [(U_{op}/U_{rat})^{C_2} - (U_{ref}/U_{rat})^{C_2}]\} \qquad (5)$$

where U_{op} is operating current, U_{ref} is reference current, U_{rat} is rated current, C_2 and C_3 is a constant depending of capacitor technology provided by specific table. where π_I is current dependence factor:

$$\pi_I = \exp\{C_4 \times [(U_{op}/U_{rat})^{C_5} - (U_{ref}/U_{rat})^{C_5}]\} \qquad (6)$$

where Iop is operating voltage, I_{ref} is reference voltage, I_{rat} is rated voltage, C_2 and C_3 is a constant depending of capacitor technology provided by specific table.

Stress factor for temperature dependence π_T is :

$$\pi_U = \exp[(E_{a1}/k_0) \times (1/T_{ref}) - (1/T_{op})] \qquad (7)$$

which derives from Arrhenius equation and it describes the temperature dependence of the failure rate , where : $k_0 = 8,616 \times 10^{-5}$ [eV/°K], $T_{ref} = \theta_{Uref} + 273$ [°K] , $T_{op} = \theta_{op} + 273$ [°K], *and Ea_1-* are activation energy in [eV] ia a constant, we can find these within specific table standard.

π_E is environmental application factor and for stationary use at weather protected location environment (E1) is equal to 1 (from standard's table);

π_S is switching rate dependence factor and for a number of operating cycles per hour S < 0.01 is equal to 1.

π_{ES} is electrical stress dependence factor it is only applicable to certain devices and is explained in detail in the related clauses within standard.

For the case of capacitors the failure rate is reduced at :

$$\lambda_{Capacitor\ 61709\text{-}2017} = \lambda_{ref} \times \pi_U \times \pi_T \qquad (8)$$

π_T is obtained (from table 40 within standard) as a function of the actual capacitor temperature: $\theta_{op} = \theta_{op} + \Delta T$ in [°C] where ΔT is the temperature change due to operating conditions and as a function of the capacitor temperature under reference conditions (see table 36 within standard): $\theta_{ref} = 40°C$.

λ_{ref} is calculated by doing conversion to reference condition using the λ values from [4], and π_U, π_T, U_{ref}/U_{rat} values from Table 38 , 40, 36 within IEC 61709 standard.

IV. CASE STUDY ; CALCULATION OF PARAMETERS RELIABILITY COMPARISON

A. Input data for calculus with MIL-HDBK-217standard.

This paper continues the reliability calculus within [4] which was focused on output capacitor bank's failure rate calculation according to *MIL-HDBK-217 rev.2 cap.10.1 – Capacitors* , thus we use the same schematic for the converter from [5] where SPICE simulation was used for multiple-constraint choice of capacitor bank (Fig.2) with the same parameters (Table I).

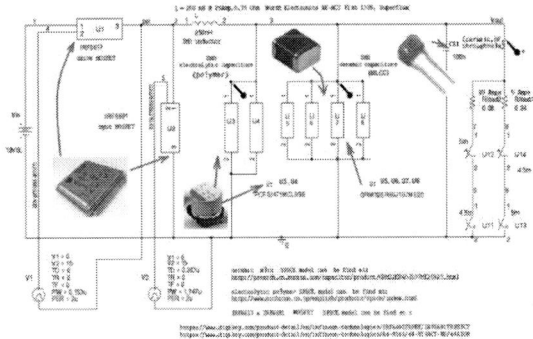

Fig.2 Converter under investigation. High fidelity dedicated SPICE model provided by manufacturer was used for MOSFETs, polymer electrolytic and MLCC capacitors [5].

Because of lack of space we show here only the equation for failure rate of either electrolytic or ceramic capacitors stated by above mentioned standard :

$$\lambda_{capacitor\text{-}MIL\text{-}HDBK\text{-}217} = \lambda_{base} \times \pi_T \times \pi_Q \times \pi_V \times \pi_{SR} \times \pi_E \times \pi_C \qquad (8)$$

It is well knoun that all reliability standards assume that there is no redundancy built in the system, meaning that if one component fails, then the entire system fails. Therefore, system failure rate results in calculation of the sum of all components' failure rates. Because our investigated dc/dc converter is a buck converter who has a series structure from reliability calculation point of view, we apply the parts count approach. Thus, the overall system failure rate can be written as being the sum of all components' failure rate.

$$\lambda_{system} = \sum_{i=1}^{N} \lambda_i \qquad (9)$$

Equation (9) illustrates the total failure rate [4] with N = total number and λ_i = the failure rate for i^{th} component.

The calculus in [4] had has the following results for the three types of capacitors used in converter (according with Table I):

- $\lambda_{polymer\ electrolytic\ capacitor\text{-}MIL\text{-}HDBK\text{-}217}$ = 0.015333864 [F/10^6 h] or 15.334[FIT] (for two pieces in parallel)

- $\lambda_{MLCC\ capacitor\text{-}MIL\text{-}HDBK\text{-}217}$ = 0.52585624 [F/10^6 h] or 525.9 [FIT] (four pieces in parallel)

- $\lambda_{ceramic\ HF\ capacitor\text{-}MIL\text{-}HDBK\text{-}217}$ = 0.007666932 [F/10^6 h] or 7.666 FIT (only one pieces)

and for entire bank:

- $\lambda_{capacitor\ bank\ \text{-}MIL\text{-}HDBK\text{-}217}$ = the sume of the above three values (eq.8) = 0.548861812 [F/10^6 h] or 548.9 [FIT] (2× polymer + 4×MLCC + one ceramic HF throgh hole)

- $MTBF_{capacitor\ bank\ \text{-}MIL\text{-}HDBK\text{-}217}$ = 1,821,952 hours

TABLE I.

Parameters of convertor	Value
Rated output active power, P_o	max. 0.06 Ω ×100 Amp = 600 W
Nominal load current	I_{out} = 25 A (5A@transition step)
Input voltage, V_{in}	12 V DC with 5% tolerance
Output voltage, V_{out}	1.2 V DC ± 50 mV [7]
Switching frequency, f_{sw}	500 khz @ Duty cycle = 7.65 %
Inductor, L	250 nH
Output capacitor bank, C	Electrolytic: 2 pcs. × 470 µF Ceramics: 4 pcs. × 100 µF HF ceramic: 1 pcs. × 100 nF
Transient load step	Current variation between : I down = 5A, I up = 25A

B. Input data for calculus with IEC 61709 standard.

IEC 61709 calculation mode of failure rates are valid for operating stress conditions that are typical in industrial environment pretty similar to IEC 60654-1 class C requirements – means a sheltered location who have an average temperature of 40°C during a long period of time. θ_1 =

40°C from standard table means 25°C from ambient temperature plus the internal selfheating due to ripple current through capacitor. The polymer and MLCC capacitors used in schematic adopted for buck converter (Fig. 2) have a very low ESR (equivalent series resistor) ensuring a maximum ripple capability and longer operating time life. θ_2 is the actual capacitor's temperature that was obtained from SPICE simulation [5] and is shown in Table II [11,12].

TABLE II.

Capacitor	Temperature
PCF0J471MCL6GS – Polymer electrolytic SMD Can-type – from Nichicon	28 °C
GRM32ER60J107ME20 – MLCC , SMD - Class II (X7R) - from Murata	32 °C

Taking account of Section III. and after doing the extracting work of data from IEC 61709 standard's tables and making the conversion from reference to operating conditions we found that :

- For polymer capacitor we have : C_2 = 1.9; C_3 =3; U_{rat} = 6.3 V; U_{ref}/U_{rat} = 0.8; Uop = 1.2V ; Uop/U_{rat} =0.19048; Ea_1=0.14 ; θ_{ref} = 40°C; θ_1 = 40°C; θ_2 = 28°C (see Table 36-40.); θ_{op} = θ_{amb} +ΔT=53°C ; T_{ref} = 313[°K], T_1 =313[°K], T_2 = 301 [°K]; λ_{ref} =3.898043136 ; π_U = 1,6 ; π_T = 1.229225438;

- For MLCC capacitor we have :C_2 = 1; C_3 =4; U_{rat} = 6.3 V; U_{ref}/U_{rat} = 0.5; Uop = 1.2V; Uop/U_{rat}=0.19048; Ea_1= 0,35; θ_{ref} = 40°C; θ_1 = 40°C; θ_2 = 32°C (see Table 36-40.); θ_{op} = θ_{amb} +ΔT=57°C; T_{ref} = 298[°K], T_1 =313[°K], T_2 = 305 [°K]; λ_{ref} =199.2045455 ; π_U = 0.3; π_T = 1 ;

- For ceramic through hole HF capacitor we have :C_2 = 1; C_3 =4; U_{rat} = 25 V; Uop = 1.2V ; U_{ref}/U_{rat} = 0.5; Uop/U_{rat} =0.048; Ea_1= ; θ_{ref} = 40°C; θ_1 = 40°C; θ_2 = 32°C (see Table36-40.); T_{ref} = 313[°K], T_1 =313[°K], T_2 = 301 [°K]; λ_{ref} = 11.60606061

- ; π_U = 0.3; π_T = 1;

V. RESULTS

And after doing the math we have:

- $\lambda_{polymer\ electrolytic\ capacitor\text{-}\ IEC\ 61709}$ = 0.007,666 [F/10^6 h] or 7.666 [FIT](one piece)

- $\lambda_{MLCC\ capacitor\text{-}\ IEC\ 61709}$ = 0.059,761,363 [F/10^6 h] or 59.761 [FIT] (one piece)

- $\lambda_{ceramic\ HF\ capacitor\text{-}\ IEC\ 61709}$ = 0.003,481,8 [F/10^6 h] or 3.481,8 [FIT](one piece)

And for the entire capacitor bank we have:

- $\lambda_{capacitor\ bank\ \text{-}\ IEC\ 61709}$ = 0.257,854 [F/10^6 h] or 257.854[FIT]

(2× polymer + 4×MLCC + one ceramic HF through hole)

- $MTBF_{capacitor\ bank\ -\ IEC\ 61709}$ = 3,878,151.3 hours

These data are synthesised within Table III.

TABLE III.

Capacitor's technology	$\lambda_{MIL-217}$ $MTBF_{MIL-217}$	$\lambda_{IEC\ 61709}$ $MTBF_{IEC\ 61709}$
2×polymer electrolytic SMD	15.3 [FIT] 65,215,134 hours	15.333,036,1 [FIT] 65,218,655.53hours
4×MLCC SMD	525.8 [FIT] 1,901,660 hours	239.045,454,5 [FIT] 4,183,304.811hours
1×ceramic through hole HF	7.6 [FIT] 130,430,268 hours	3.481,818,182 [FIT] 287,206,266.3hours
Entire caps bank	548.9 [FIT] 1,821,825 hours	257.854 [FIT] 3,878,151.3 hours

A diagram comparison of the two failure rates for the entire capacitor bank is shown in Fig.2. After applying the newer standard's failure rate calculations, result about half of the old one. A less difference is obtained by calculus with International Electrotechnical Commission's IEC TR 62380 standard and for the very same schematic and capacitors [9].

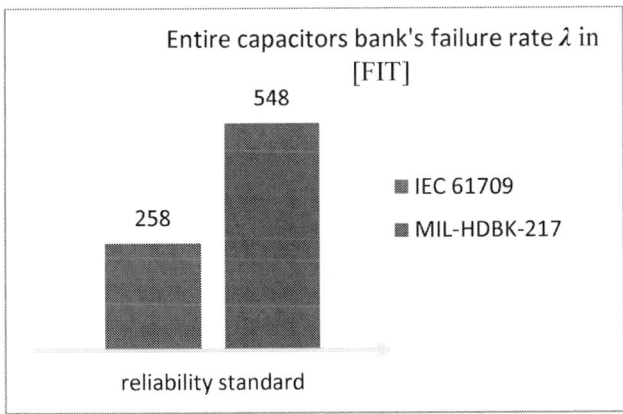

Fig.2. Comparison of the failure rate's value for the two standards.

An embedded cause of it appear to be the values for stress factors within Mil-HDBK-217 standard's tables which was stated for the capacitor's technology at that time. An excerpt of basic failure rate „ λ_b" from standard shows values for ceramic capacitors described as „*Capacitor, Chip, Multiple Layer, Fixed, Ceramic Dielectric, Established reliability*" in table „ *10.1 Capacitors* " is equal to 0.020 and for polymer capacitors described as „*Capacitor, Fixed Electrolytic (DC, Aluminum, Dry Electrolyte, Polarized)*" is equal to 0.00012 . If we have to discuss this (with no intention to criticized the wide world accepted and used standard above mentioned) it must have taking account that λ_b = 0.020 value for MLCC vs. λ_b = 0.00012 value for polymer capacitors seems to be reversed regarding today technology of the two types of capacitors (nowadays MLCCc are characterized by a longer useful life than polymer capacitors [12,13])

VI. CONCLUSION

This paper provide a comparison calculation using two reliability standard: a world wide used one (being the very first appeared on the reliability prediction market) and a newer one who takes account of the newer capacitor's technology like polymer. According to calculus using the newer standard a better MTBF results. But is necessary to notice that due to a λ_b = 0.020 value for MLCCs in older standard a greater failure rate result. Another key point when using the newer standard is taking account that it not specifying the MLCCs technology when calculating stress factors but the older one is doing .

REFERENCES

[1] Anon., Military Handbook - Reliability Prediction of Electronic Equipment, MIL-HDBK-217F, Notice 2, Feb 28, 1995.

[2] "IEC TR 62380: Reliability Data Handbook-Universal Model for Reliability Prediction of Electronics Components, PCBs and Equipment," 2006.

[3] "IEC 61709 (2017): Electric Components - Reliability - Reference Conditions for Failure Rates and Stress Models for Conversion," 2017.

[4] D. Butnicu and D. O. Neacsu, "Using SPICE for reliability based design of capacitor bank for telecom power supplies," *2017 IEEE 23rd International Symposium for Design and Technology in Electronic Packaging (SIITME)*, pp. 423-426, Constanta, 2017.

[5] Dan Butnicu, Dorin O. Neacsu, "Using SPICE for multiple-constraint choice of capacitor bank for telecom power supplies" , 2017 IEEE 23rd International Symposium for Design and Technology in Electronic Packaging (SIITME)

[6] Dorin O. Neacsu , "Telecom Power Systems ", ISBN 9781138099302 Published December 8, 2017 by CRC Press

[7] Dorin O. Neacsu, Dan Butnicu, "A review and ultimate solution for output filters for high-power low-voltage DC/DC converters", 2017 International Symposium on Signals, Circuits and Systems (ISSCS)

[8] M. G. Pecht and F. R. Nash, "Predicting the reliability of electronic equipment," in Proceedings of the IEEE, vol. 82, no. 7, pp. 992-1004, July 1994. doi: 10.1109/5.293157.

[9] Dan Butnicu, D.O. Neacsu, Cristian M. Neacsu, "Reliability Calculation Method for Output Capacitor Bank used in Telecom Power Supplies" October 2019 DOI: 10.1109/SIITME47687.2019.8990892 Conference: 2019 IEEE 25th International Symposium for Design and Technology in Electronic Packaging (SIITME)

[10] D. Zhou, H. Wang, and F. Blaabjerg, "Mission Profile Based System-Level Reliability Analysis of DC/DC Converters for a Backup Power Application," IEEE Trans. Power Electron., vol. 33, no. 9, pp. 8030–8039, 2018

[11] Anon., http://www.nichicon.co.jp/english/products/spice/index.html

[12] Anon.,http://psearch.en.murata.com/capacitor/product/GRM32ER60J10 7ME

[13] M.Askari and B.Abdi, Reliability investigation of PC's SMPSs,WSEAS, CSECS'10, 2010.

An Efficiency Comparative Workbench Study of eGaN and Silicon Discrete Transistor based Buck Converters

Dan Butnicu
Faculty of Electronics
Technical University of Iasi
Iasi, Romania
dbutnicu@etti.tuiasi.ro

Luminiţa-Camelia Lazăr
Institute of Computer Science,
Romanian Academy-Iasi Branch
Iasi, Romania
camelia.lazar@iit.academiaromana-is.ro

Abstract—**This paper presents a comparative efficiency experimental study of eGaN and conventional Si MOSFET power transistors within similar power ratings. The aim is to prove that the use of eGaN-FET power devices can be beneficial for 12V/1.2V POL buck converters required for microprocessors, FPGA, and other such power supply- demand architectures. Two evaluation boards (a Silicon MOSFET based one and an eGaN FET based one) are tested from an efficiency point of view on the work-bench; for a higher precision, the output current delivered in load was measured by mean of a current probe, avoiding in this way the adding extra resistance in series with the load which has very small values. Because the temperature highly impacts the efficiency performance and reliability of converters within power supplies, making the environment a critical factor in their selection, a thermal scanning of the power devices' package was taken in order to show how is the heat dissipated and giving a suggestive picture of converter's reliability.**

Keywords—buck converter;POL;eGaN; efficiency;WBG

I. INTRODUCTION

73,000,000,000 kWh of consumed electric energy-this huge amount is the estimate of energy that will be taked by U.S. data centers in 2020, according to a 2016 study by Lawrence Berkeley National Laboratory [1]. As long as our appetite for consumer electronics and telecommunication services continues to increase, so will the need to deliver more energy in less space to run these as efficiently as possible. Also in industrial automation or automotive is more encountered the need to supply high-density power systems. Recently, many studies have stated that eGaN devices (-e-stands for enhancement mode), due to its intrinsically superior properties (which are closely related to switching performance when compared to traditional silicon solutions) when implemented in power supply alow higher levels of efficiency. than was possible in silicon MOSFET based technology [2.3]. Rising converter's efficiency by implementing the WBG technologies help to improve the key performance parameters like weight, size, battery life in the portable electronics field.

This generate for many end users energy saving, lower BOM costs and lowered quantities of carbon which they release in the ambient. Figure 1 show the proposed two POL converters using two different technologies for the comparative study of efficiency .

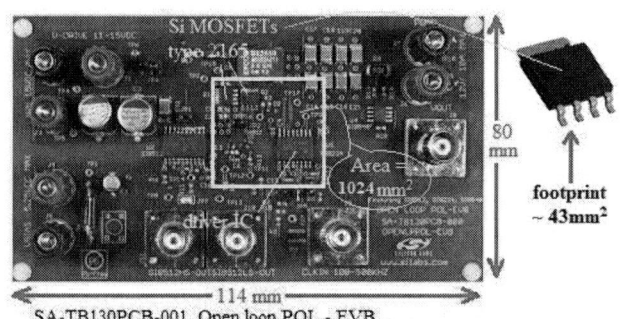

SA-TB130PCB-001 Open loop POL - EVB
based on Silicon MOSFETs - HAT2165H-EL-E MOSFET
30V, 55A, LFPAK, N-Channel, Surface Mount LFPAK
Rds On (Max) 3.3mOhm @Id= 27.5A, Vgs= 10V

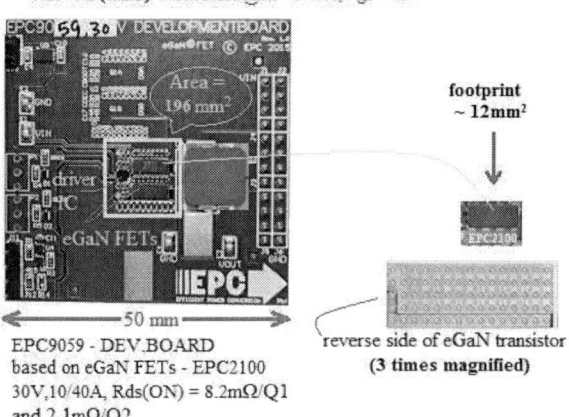

EPC9059 - DEV.BOARD
based on eGaN FETs - EPC2100
30V,10/40A, Rds(ON) = 8.2mΩ/Q1
and 2.1mΩ/Q2

Fig. 1: The yellow outlines are the footprint area of the actual POL converter.

II. POWER LOSSES IN POL CONVERTERS DETERMINE EFFICIENCY

According to figure 2, these power losses were identified.

- Conduction losses due to *Rds(ON)*

$$Ids2 \times (Rds(ON)\_HS + Rds(ON)\_LS) \qquad (1)$$

- Drain charge switching loss

$$0.5 \times (C_{ds\_HS} + C_{ds\_LS}) V_{ds}^2 \times f_{sw} \qquad (2)$$

- Gate charge switching loss

$$(Cg\_HS + Cg\_LS)Vds \times fsw \qquad (3)$$

- Reverse recovery losses due to the diode formed by the drain and the transistor body

$$Q_{rr} \times Vin \times f_{sw} \qquad (4)$$

Note that gallium nitride products are free of recovery losses because unlike silicon MOSFETs, have no minority carriers to be stored in a junction hence there is no Q_{rr} [4].

- Overlap losses due to $L_{parasitic}$: these refer to situations in which current is flowing through the transistor, but in the same time the drain voltage is high, in which case the dissipated power, calculated as V×I, can be quite substantial. One challenge in the design of a DC/DC converter is that large currents are switched rapidly between flow paths, which causes parasitic inductors to charge and discharge their stored magnetic energy rapidly, resulting in ringing of various voltage terminals. This ringing energy is dissipated generating so called overlap loss.

In this paper the following formulae were used for determination of the efficiency (η).

$$P_{LOSS} = P_{IN} (1 - \eta) \text{ and } \eta(\%) = P_{OUT} / P_{IN} \times 100\% \qquad (5)$$

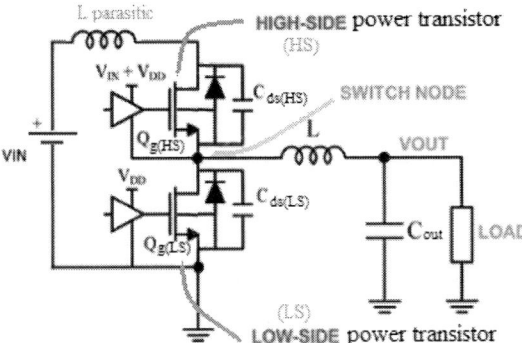

Synchronous buck-converter

Fig. 2: Typical synchronous buck converter with real parasitic inductances and capacities usually founded in most POL converters.

Note that gallium nitride products are free of recovery losses because unlike silicon MOSFETs, have no minority carriers to be stored in a junction hence there is no Q_{rr}.

III. MODE OF OPERATION, EXPERIMENTAL SETUP AND RESULTS

The proposed two evaluation board converters [5,6] who are the subject of efficiency comparison require to be capable to deliver power in buck mode, with no forced air dissipation (i.e. natural convection) in all its power range (figure 1). From the technological point of view, the extreme conditions of high current and low voltage operation in above-mentioned applications are well assumed by the use of eGaN power devices, which ensures a good efficiency. An intensely used converter in these supplying applications is the synchronous buck converter [7]. This topology becomes well suited for these applications due to having only two active switches, two storage elements, and a very simple control approaches. In order to estimate the reliability of the two converters technologies a thermal scan was taken which is depicted in figure 3. A natural convection thermal image of the converter is shown (at room temperature) for 12V input voltage at 6A output current. The IC is the hottest component at 59.7°C. The 1.2V output ripple is shown in red below, DC coupled with offset 6A. It can be seen that a far better thermal behaviour for the eGaN transistors pair was found: about 49 °C for Silicon technologies and about 35 °C for eGaN technologies. The two tested converters achieve efficiencies around 95% vs.79% when delivers about 6 A current load at 250 kHz switching frequency (figure 4). Both converters were operated at a switching frequency of 250kHz and with 12 V as the input voltage. Because as the switching frequency increase converters efficiency tend to achieve lower values, a neutral value of 250 kHz was adopted for testing. The MOSFETs [8] selected were based on physical fit into the evaluation board and having a similar gate threshold voltage as the eGaN FET transistors . Both converters were tested using the same output inductor, which is EPCOS - B82559 SMD0921 package size with an inductance value of 3μH, 15mΩ, and one-piece 470 μF Nichicon polymer electrolytic 1010 size as output capacitors. The devices compared were the EPC2100 Enhancement-Mode GaN Power Transistor Half-Bridge rated 30V, 8.2/2.1mΩ, and HAT2165H-MOSFET n-channel rated 30V at 55A, power dissipation (Max) 30W, surface mounted LFPAK packaging SOT-669 case, Rds(ON) = 3.3mΩ. Gate drive voltages were chosen 5 V for the ON state, and 0 V for the OFF state. Tests were performed for a load range from zero to 10 A and the converters measurements are taken for open-loop operating mode and the duty cycle was adjusted for the appropriate output voltage.

The switching voltage node waveforms for each of the buck converters are shown in figure 5 where the rise time and dv/dt values for the two technologies were determined from these. It has been also proved that the efficiency of the eGaN based converter is better due to smaller footprint and capsule

2020 IEEE 26th International Symposium for Design and Technology in Electronic Packaging (SIITME)

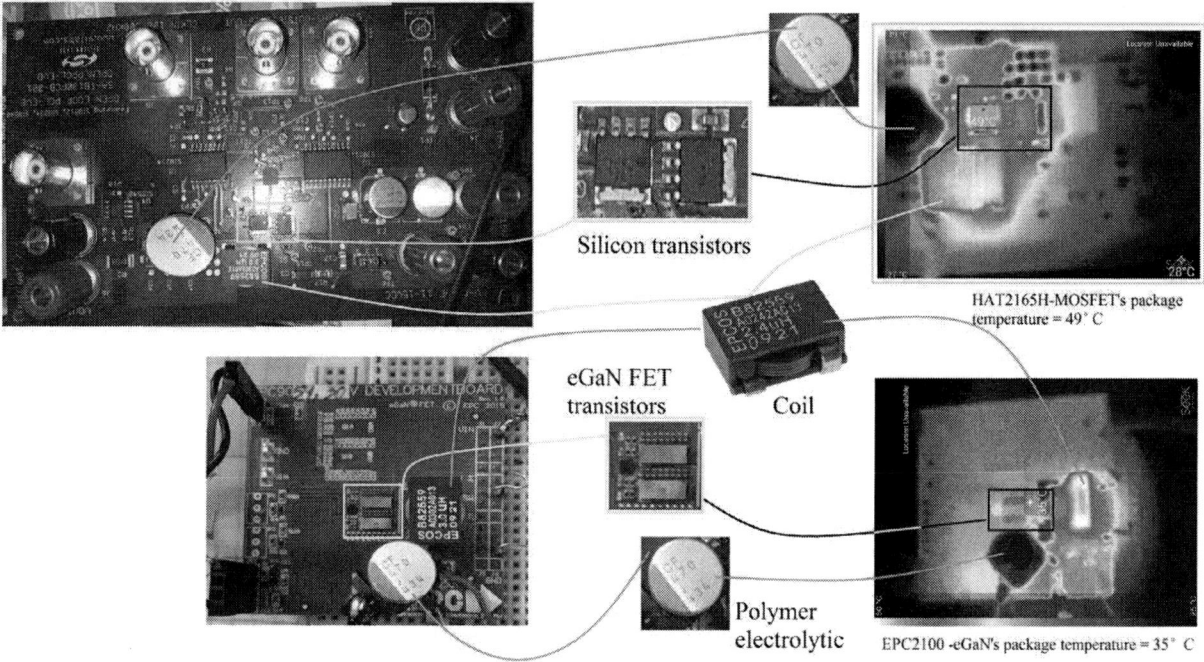

Fig. 3: Experimental thermal comparison between eGaN FET vs. Si MOSFET based buck converter with $f_{sw} = 250$ kHz, $I_{load} = 6A$. It is shown the graphic correspondence between the EVB tested plates and thermal image of the scanned area of the plate revealing a 35°C package temperature for eGaN way smaller than 49°C package temperature for Silicon (with natural convection).

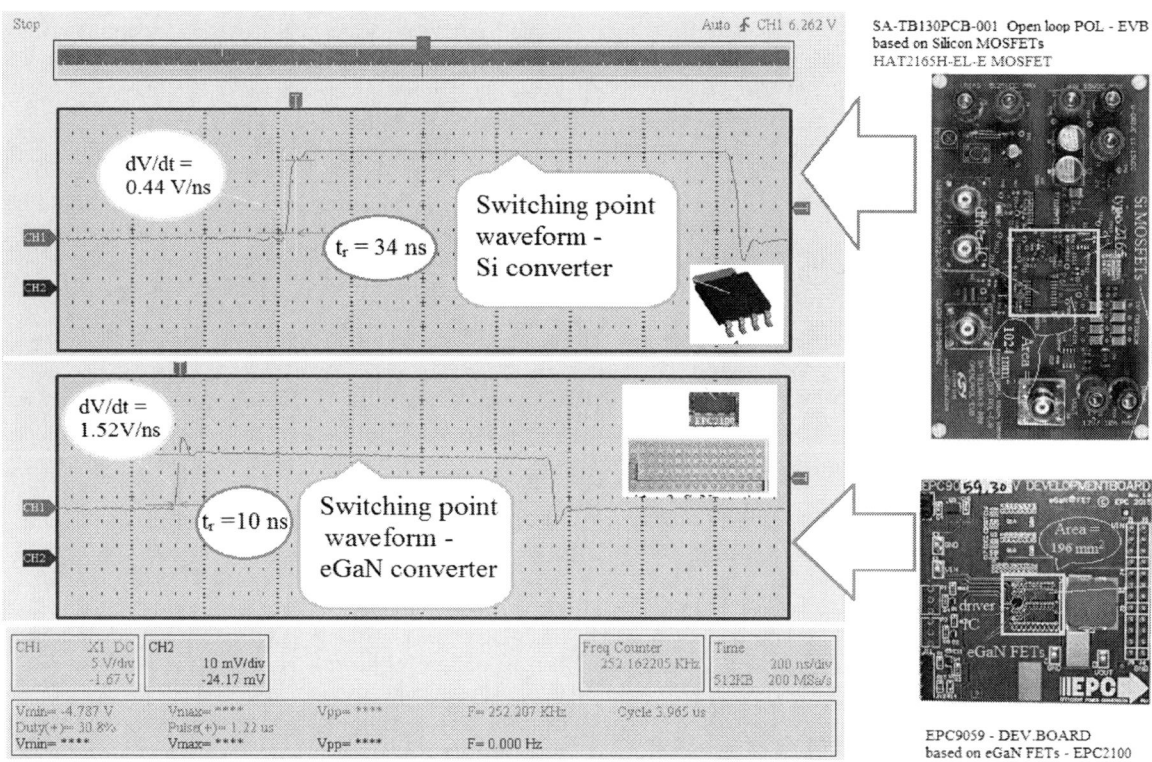

Fig. 4: Experimental measurement of t_r (rise time) for the two converters at switching point with $f_{sw} = 250$ kHz, $I_{load} = 6A$ revealing about three times better behavior of the eGan vs.Silicon (with natural convection)

978-1-7281-7507-2/20 $31.00 © 2020 IEEE   352   21-24 October, Pitesti, Romania

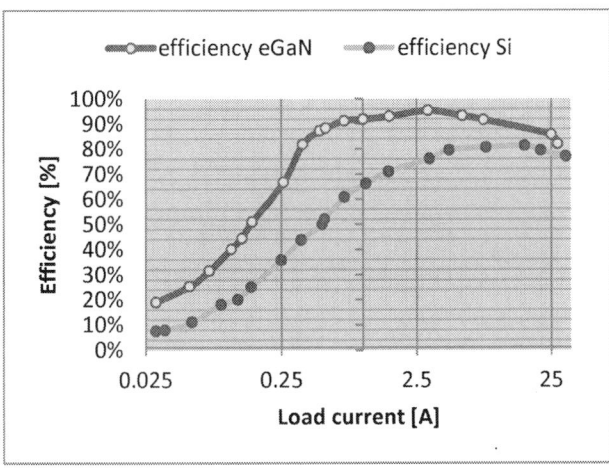

Fig. 5: The efficiency measurements for the two converters over a wide load current range (using a current probe for preventing errors due to lower values of the load). The graphic representation shows a better efficiency performance of the eGaN based converter vs. the MOSFET one. The peak efficiency of 95% is encountered for the WBG converter.

temperatures thus outperforming the Silicon based converter approach. The main-reason for this efficiency's amelioration the implementation of the POL by means of an eGaN - FETs instead of Si - MOSFETs because the GaN device has a lower ON resistance, which leads to a decrease of the conduction losses. In addition, the smaller values of parasitic capacitances, which lead to an increased transconductance and faster transitions between ON/OFF states, allowing smaller switching losses. It is also important to add that choosing an appropriate capacitor and inductor technology, is important for DC/DC switching POL converters efficiency with high levels of input and output currents.The main features of the two transistor technologies incorporated within evaluation boards used for efficiency investigation are grouped in Table I. Workbench setup for efficiency investigation of the two converters as an overview is depicted in figure 6.

IV. CONCLUSION

The efficiency of a GaN-based and Si MOSFET-based DC/DC converters has been investigated. The results of the

TABLE I

	Transistor part number	V_{ds} [V]	I_d [A]	Rds (ON) [mΩ]	Qg [nC]	Qgd [nC]	Package	PCB area [mm²]
Si tech.	HAT216 5H [9]	30	55	3.3	33	-	LFPAK SOT-669	36
eGaN tech.	EPC 2100[10]	30	10 40	8 2.1	3.6 15	0.6 2.7	solder bumps	13.9

Fig. 6: Workbench setup for efficiency investigation of the two converters.

mesurements show that the figures of merit of eGaN devices compared to best-in-class silicon-based MOSFETs indicate an improvement in performance so this WBG (wide bandgap materials) technology open the way for designers to reduce the space occupied by the DC/DC converters in portable applications exceeding the efficiency of converters based on conventional silicon power MOSFETs. Higher efficiency is obtained when implement POL converters with eGaN-FETs instead of Si - MOSFETs because of the lower R_{ds}(ON) resistance of the WBG device, which conducts to a decrease in the conduction losses. Also the MOSFET efficiency is reduced by the inherent body diode conduction. Due to all these facts it is understandable why nowadays power designers migrating from good old fashioned silicon to promising WBG materials.

REFERENCES

[1] ***https://www.comsoc.org/publications/tcn/2019-nov/energy-efficiency-data-centers

[2] ***https://epc-co.com/epc/GaNTalk/Post/15421/Design-Efficient-High-Density-Power-Solutions-with-GaN

[3] Alex Lidow, Michael de Rooij, Johan Strydom, David Reusch John Glaser, "GaN Transistors for Efficient Power Conversion'', 2019 Print ISBN:9781119594147, DOI:10.1002/9781119594406, John Wiley & Sons Ltd.

[4] ***https://www.ti.com/lit/wp/slyy071/slyy071.pdf?ts=1600348596166&ref_url=https%253A%252F%252Fwww.google.com%252F

[5] ***https://www.silabs.com/documents/public/user-guides/SATB130PCB-001-evb.pdf

[6] ***https://epc-co.com/epc/Products/DemoBoards/EPC9059.aspx

[7] Dorin O. Neacsu, "Telecom Power Systems ", ISBN 9781138099302 Published December 8, 2017 by CRC Press

[8] D. Butnicu, "A Reliability Comparison between Disrupting eGaN-FET and Cutting Edge Silicon MOSFET Devices in POL Buck Converters" , 2019 International Symposium on Signals, Circuits and Systems, ISSCS, Iasi, July 2019

[9] ***https://www.alldatasheet.com/datasheet-pdf/pdf/248999/RENESAS/HAT2165H-EL-E.html

[10] ***https://epc-co.com/epc/Products/eGaNFETsandICs/EPC2100.aspx

Modelling of the Thermal Conditions of a LED Driver

N. L. Evstatieva, B. I. Evstatiev
Department of Electronics
University of Ruse Angel Kanchev
Ruse, Bulgaria
{nevstatieva,bevstatiev}@uni-ruse.bg

Abstract—**Light Emitting Diodes offer an efficient way to provide lighting. In order to ensure the stability of their operation and performance, it is important to ensure a stable thermal regime and electronic control. This study presents the development of a model of the thermal conditions of LED drivers based on physical dependencies and the Finite difference method. The temperature response of several parts of the driver are investigated – the PCB, the non-SMD components, the casing and the air within the casing. The model is verified and its parameters are obtained by comparing the modelled data with experimental one. The developed model could be useful for investigating the stability of the LED drivers under different environmental conditions, the influence of its structural components on the temperature response, etc.**

Keywords—LED driver, temperature response, model

I. INTRODUCTION

Light Emitting Diodes (LED) have proven themselves as an efficient mean to provide lighting. In order to provide the stable and reliable performance, it is important to be aware of the thermal regime of its components. The application of LED ensures high energy efficiency and lifetime for lighting procedures. However, it is important to provide the necessary electronic control and thermal management, in order to stabilize their performance [1]. The reliability of LED luminaires depends highly on the operation of their drivers, providing the necessary power supply. Commonly the volume of the drivers' casing should be small, which limits the size of the PCB and the possibilities for positioning the components. Therefore, it is critical to ensure appropriate thermal regime of the driver for providing stable power parameters. Furthermore, the environmental conditions could differ significantly for the different spheres of application, such as indoor application, street lighting, etc. [2].

Different approaches are used to model the thermal conditions LED application. In [3,4] is performed a computer simulation of double pulse flash-lamp pump laser electrical system. In another study is presented precise thermal modelling of a LED module using CFD software [5]. In [6] the Deconvolution method was adopted in order to model a LED element. In a similar study, multi-domain compact models of LEDs and compact thermal models of their thermal environment was investigated [7]. Thermal modelling of some critical for the reliability components of a LED driver were simulated using FEM simulations [8]. Another approach for thermal modelling is the Finite difference method, which allows to perform algorithmic modelling [9].

While the available studies cover different aspects of thermal modelling, they do not account for all constructional parameters of the investigated objects and the conditions in the environment. This study aims to develop and verify a model for thermal simulation of a LED driver that allows assessing the influence of the driver's physical, dimensional and environmental parameters.

II. MATERIALS AND METHODS

A. Objet of the investigation

The object of the investigation is the thermal regime of a LED driver. The elements of such drivers are commonly placed on a PCB, mounted in a plastic box (Fig. 1). Part of the elements are implemented with SMD installations, therefore most of the heat power is dissipated towards the PCB. The rest of the elements are DIP or PDIP installed, therefore their pins are mounted to holes in the PCV, and the body of the element is within the volume of the box, surrounding the PCB. Such elements are the transformers, capacitors, etc. Commonly, the most heat is dissipated by one or two components. In order to ensure even distribution of the thermal energy within the box, sometimes the elements are filled with silicon or resin. Radiators are also used for better energy exchange with the environment.

In order to model the temperature regime of a LED driver, we consider it contains the following structural parts: PCB (including the SMD elements), non-SMD components, the casing, the air within the casing and the environment. Furthermore, the following energy flows are considered:

- Convective heat transfer between the non-SMD elements and the air in the casing - Q_{com}^{air};

Fig. 1. A common LED driver from the outside and its elements mounted on the PCB

- Convective heat transfer between the PCB and the air in the casing - Q_{pcb}^{air};

- Convective heat transfer between the air and the casing - Q_{air}^{box};

- Conductive heat transfer between the PCB and the casing - Q_{pcb}^{box};

- Convective heat transfer between the casing and the environment - Q_{box}^{out}.

The energy flows can be expressed using known dependencies for convective and conductive heat transfer:

$$E_1 = h.A.\Delta T.\Delta\tau, J \qquad (1)$$

$$E_2 = \frac{A}{\frac{d}{k}}.\Delta T.\Delta\tau, J \qquad (2)$$

where h is the heat transfer coefficient, $W.m^{-2}.K^{-1}$;

A – the heat transfer surface, m^2;

ΔT – the temperature difference between the surface and the fluid, K;

$\Delta\tau$ – the time interval, s;

d – the width of the layer, m;

k – the thermal conductivity of the material, $W.m^{-1}.K^{-1}$.

The heat transfer coefficient from equation (1) is estimated according to [10,11]:

$$h = 5.6 + 4.v, W.m^{-2}.K^{-1} \qquad (3)$$

where v is the fluid velocity I $m.s^{-1}$ and is obtained experimentally. The dynamics of the heat transfer can be obtained using the following calorimetric equation:

$$E_3 = m.c.\Delta T, J \qquad (4)$$

where m is the mass of the body in kg, c is the specific heat capacity of the body in $J.kg^{-1}.K^{-1}$ and ΔT is the change in the temperature of the body in K when the energy E_3 is absorbed.

The power dissipated in the LED driver is estimated using the electrical power P_{el} in W and the efficiency of the driver η:

$$Q_{el} = P_{el}.(1 - \eta) \qquad (5)$$

The modelling of the temperature regime of a LED driver is implemented using the above physical dependencies and the finite difference method. In the present study, the following assumptions are made:

- The air withing the casing of the driver has the same parameters in its whole volume;

- The temperature of the casing and the temperature of the PCB have the same parameters in their whole volume;

- The plastic casing is attached to a radiator, which has the same temperature as the casing.

B. Algorithm of the model

The algorithm for modelling the temperature regime of the LED driver is presented in Fig. 2.

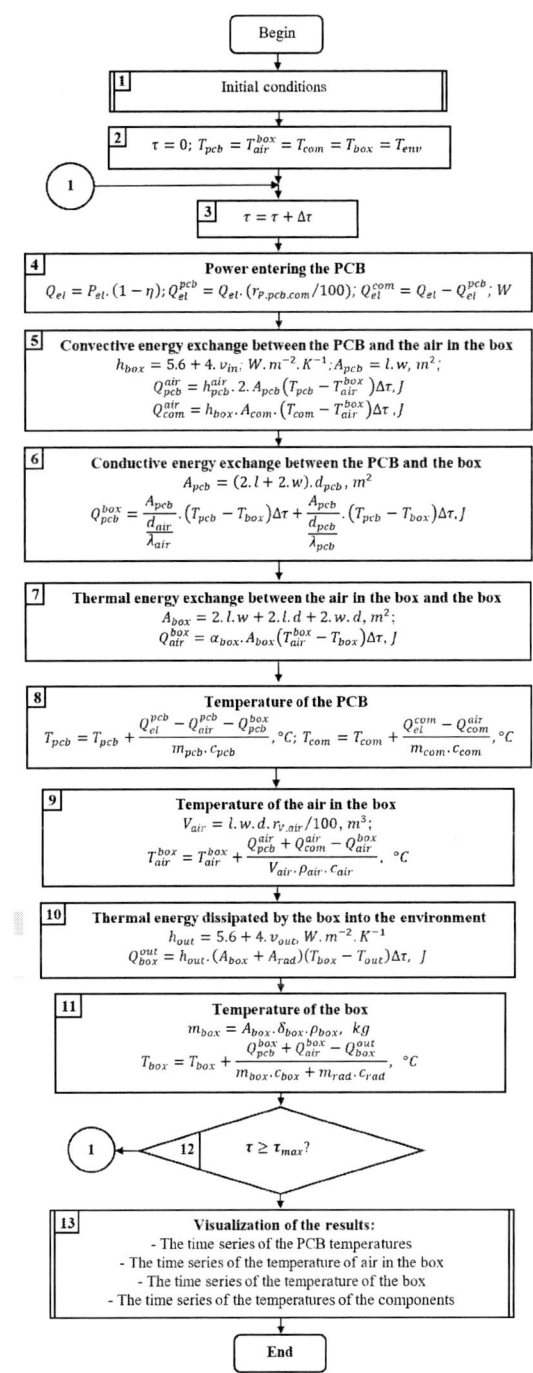

Fig. 2. Algorithm for modelling of the thermal regime of a LED driver

In block 1 are set the initial conditions and the variables are initialized in block 2. Next, in blocks 3 and 12 is organized the time cycle.

The heat power dissipated into the PCB Q_{el}^{pcb} and heat power dissipated into the non-SMD elements Q_{el}^{com} is estimated in block 4, using the ratio $r_{P.pcb.com}$, which can be obtained experimentally.

The convective heat transfer between the PCB/elements and the air within the casing is estimated in block 5, the conductive heat transfer between the PCB and the casing - in block 6 and the convective heat transfer between the casing air and the box is estimated in block 7. Next, in blocks 8 and 9 are obtained the new temperatures of the PCB and of the air in the casing. After the convective heat transfer between the casing and the environment is obtained (block 10), in block 11 is estimated the new temperatures of the casing. Finally, in block 13 are visualized the results of the simulation.

III. RESULTS AND DISCUSSION

The object of the investigation is a LED driver with power 25 W and dimensions of the box 107 x 37 x 30 mm. In order to verify the model and obtain its properties, an experimental study has been conducted. The temperature measurements were done using DS18B20 sensors, which have resolution 0.1 °C. The sensors were initially calibrated for the temperature of the environment – 17.5 °C. Five sensors were used to measure the temperatures of the driver components, three of which are visible in Fig. 3. The first sensor was attached to the PCB. Considering the good thermal conductivity of textolite and copper, it can be assumed that the temperature is relatively equal everywhere. The second sensor is attached to the radiator of the voltage regulator, which is one of the main heat producers. Other two floating sensors were used to measure the temperature of the air in the casing and of the environment (near the LED driver). The fifth sensor is used to measure the temperature of the casing and is attached to its outer surface.

All sensors are connected to a monitoring system using a One-Wire network, which has been setup to store their records automatically on a 1 minute interval. A measurement cycle has been prepared and carried out, the results from which are presented in Fig. 4.

Next, in order to verify the model, a software tool has been developed implementing it. Numerous simulations have been

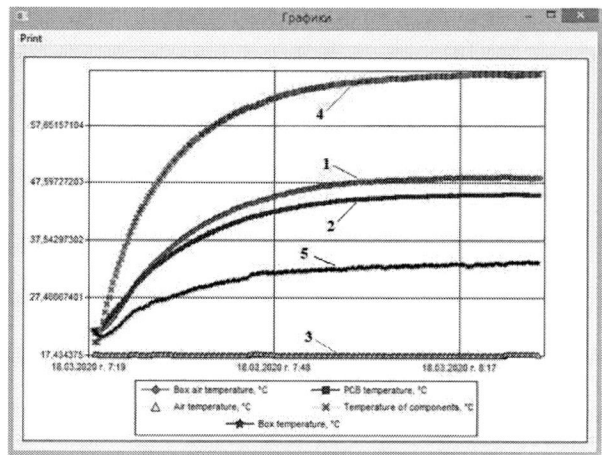

Fig. 3. Experimental results about the measured temperatures: temperature of the air in the box (1); PCB temperature (2); temperature of the environment (3); temperature of the radiator (4); and temperature of the box (5)

performed in order to estimate the optimal model parameters. The predefined parameters of the simulations are presented in Table 1. The best-fit values of the remaining parameters of the model are presented in Table 2. A graphical comparison between the experimental and the modelled values is presented in Fig. 5. In order to improve the visibility, the experimental values are presented with 5 minutes time step. During the first 1500 seconds, a small delay in the measured air temperature could be noticed, compared to the modelled one. This can be explained with the delayed response of the sensor, caused by its own specific heat capacity and purely convective heat transfer.

TABLE I. PREDEFINED PARAMETERS OF THE SIMULATIONS

Parameter	Value
T_{air}	$17.5, °C$
d_{pcb}	$1, mm$
d_{box}	$0.8, mm$
d_{air}	$4, mm$
ρ_{air}	$1.2, kg.m^{-3}$
C_{air}	$1005, J.kg^{-1}.K^{-1}$
k_{air}	$0.0288, J.m^{-1}.K^{-1}$
$\Delta\tau$	$0.05, s$
τ_{max}	$4200, s$
λ_{air}	$0,0288, W.m^{-1}.K^{-1}$

TABLE II. OBTAINED PARAMETERS OF THE MODEL

Parameter	Value
$m_{pcb}.C_{pcb}$	$6, J.K^{-1}$
$m_{com}.C_{com}$	$26, J.K^{-1}$
$m_{rad}.C_{rad}$	$0.15, J.K^{-1}$
$\rho_{box}.C_{box}$	$910000, J.K^{-1}.m^{-3}$
λ_{pcb}	$0,133, W.m^{-1}.K^{-1}$
A_{com}	$0.0148, m^2$
A_{rad}	$0.00023, m^2$
v_{in}	$0.65, m.s^1$
v_{out}	$0.41, m.s^1$
η	0.88
$r_{V.air}$ (volumetric share of air in the box)	$80, \%$
$r_{P.pcb.com}$	$27, \%$

Fig. 4. LED driver with installed sensors for measuring the temperature of the PCB (1), the air in the box (2) and the radiator of the voltage regulator (3)

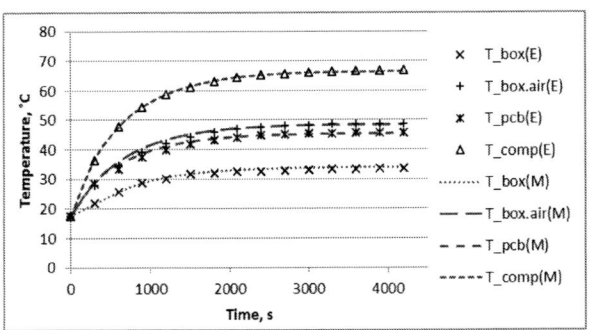

Fig. 5. Experimental and modelled temperatures of the LED driver components

As can be seen from Fig. 5, the developed and verified model gives a very accurate representation of the experimental values. This includes both the steady-state values and the transient response.

IV. CONCLUSION

In this study is developed and verified a model, describing the thermal conditions of a LED driver. It is based on the Finite difference method and well-known physical dependencies for thermal convection and thermal conduction. The temperatures of several elements of the driver are modelled, such as the driver's casing, the PCB, the non-SMD components and the air within the casing.

In order to verify the model and acquire its unknown parameters, an experimental study has been conducted with a real driver, where the above temperatures are measured by a monitoring system. A best-fit approach is used to obtain them.

The developed and verified model could be used to perform virtual investigations of the temperature regime of the LED driver under different environmental conditions. Furthermore, it could be used to assess the influence of the different constructional parameters of the driver on the thermal regime and find out the most optimal means for its optimization.

ACKNOWLEDGMENT

This work was supported by the University of Ruse Research Fund under contract no 2020-FNI-FEEA-02.

REFERENCES

[1] M. Fathi, A. Aissat, M. Abderrazak, "Optimization of the electronic Driver and thermal management of LEDs lighting powered by solar PV", Energy Procedia, Vol. 18, pp. 291 - 299, 2012.

[2] Y. Zhou and N. Narendran, "Photovoltaic-powered light-emitting diode lighting systems", Optical Engineering, Vol. 44, No. 11, 111311, November, 2005.

[3] Sv. Ivanov, Y. Ivanova, "Analisys and simulation investigation of Double pulse flash-lamp pump laser electrical system", Journal of the Technical University – Sofia, Plovdiv branch, Bulgaria, Fundamental Sciences and Applications, vol. 19, 2013, pp. 247-250.

[4] Sv. Ivanov, Y. Ivanova, "Theoretical and graphical representation of transition process in the scheme for obtaining high-energy pulses for excitation of laser with discharge lamps", Journal of the Technical University – Sofia, Plovdiv branch, Bulgaria, Fundamental sciences and applications, Volume 20, 2014, pp. 11-16.

[5] K. Baran, M. Leśko, H. Wachta, and A. Rózowicz, "Thermal modeling and simulation of high power LED module", AIP Conference Proceedings, 2018, 020048 (2019).

[6] Marcin Janicki, Tomasz Torzewicz, Agnieszka Samson, Tomasz Raszkowski, Andrzej Napieralski. "Experimental identification of LED compact thermal model element values", Elsevier, Microelectronics Reliability, 86 (2018), 20–26, Journal homepage: www.elsevier.com/locate/microrel .

[7] András Poppe, "Simulation of LED based luminaires by using multi-domain compact models of LEDs and compact thermal models of their thermal environment", Elsevier, Microelectronics Reliability, 72 (2017), 65–74, Journal homepage: www.elsevier.com/locate/microrel .

[8] H. Niu, H. Wang, X. Ye, S. Wang, F. Blaabjerg, "Converter-level FEM simulation for lifetime prediction of an LED driver with improved thermal modelling", Elsevier, Microelectronics Reliability, 76-77 (2017), 117–122, Journal homepage: www.elsevier.com/locate/microrel .

[9] I. Evstatiev, "Application of modelling in the electronic systems for control of agricultural processes", EE&AE'2009-International Scientific Conference, 2009, Ruse, Bulgaria, pp. 754-758.

[10] Kuchling H., Reference book on physics, Moscow, Mir, 1982, 520 p., (in Russian)

[11] S. Rudobashta, „Thermal Engineering", Pero, Moscow, 2015, 672 p., (in Russian).

Spent Battery Classification by Electrical Characterization

A. Fazakas, M. Purcar, A. C. Vonsza

Technical University of Cluj-Napoca

Cluj-Napoca, Romania

Albert.Fazakas@bel.utcluj.ro, Marius.Purcar@ethm.utcluj.ro, andavonsza@gmail.com

Abstract— Recovering raw minerals from spent batteries represents a challenge worldwide. To obtain clear, uncontaminated minerals, one of the most important tasks is to separate batteries of same size (AA or AAA) but different chemistries. This paper presents a study of separation methods and proposes a simple, fast, and cost-effective analysis to classify alkaline and zinc-carbon batteries, based on battery impedance.

Keywords—battery, alcaline, zinc-carbon, impedance

I. INTRODUCTION

In the early 90s the main objectives for recycling were, in order, Waste reduction, Environmental protection and Preservation of finite resources. At that time technology for recycling Lead/Acid and Nickel/Cadmium batteries was existent and implemented, while technologies for Zinc/Carbon and alkaline batteries were only in pilot stage. At that time batteries in the household waste were considered that are not causing pollution of water, air, or soil [1].

Nevertheless, the directives changed over time drastically when later researches proven much higher environmental impact of household batteries. Therefore, environmental protection moved to the highest priority level. Manufacturing and selling very low mercury and cadmium content batteries were not enough. Nowadays all batteries are selectively collected for special disposal in EU countries.

On a resource management level, batteries can be considered as a source of secondary raw materials. Several recycling processes, both pyrometallurgical and hydrometallurgical ones are proposed and applied for recover Mn and Zn from zinc-carbon and alkaline batteries mainly in Europe [2]

In an alkaline battery, the negative electrode is zinc - Zn, and the positive one is manganese dioxide $-MnO_2$, similar to a zinc-carbon battery, where the negative electrode is also Zn and the positive electrode is a mixture of graphite and MnO_2.

The major difference between alkaline and zinc-carbon batteries is the type of the electrolyte. While the alkaline uses potassium hydroxide – KOH for electrolyte, the zinc-carbon uses acidic ammonium chloride – NH_4Cl or zinc chloride - $ZnCl_2$ as electrolyte.

Most of the recycling processes start with battery decomposition, either by mechanical operations - milling or chemical – leaching. Mixing batteries with different electrolytes, especially KOH with NH4Cl in the decomposition process will result in results in a massive ammoniac emission leading not only to environmental and health issues but also to a nonefficient recycling process.

It results that before any milling or leaching process applied to the disposed batteries, a separation must be done based on the battery chemical composition. The aim of this work is to find a cost-effective battery chemistry classification method using electrical measurements for battery characterization and chemistry detection

The first thing that comes in mind for spent battery separation is optical inspection. However, this type of sorting comes with significant disadvantages:

- waste sorting systems based on optical inspection usually detect the shape of the battery versus other garbage shapes such as bottles or cans [3], being unable to recognize details such as branding, or battery type form the case.

- Many manufacturers produce batteries with both alkaline and zinc-carbon chemistry, therefore knowing a battery brand is not a proof of its chemistry e.g. not all the alkaline battery cases hold the inscription "alkaline"

- Note that the commonly used battery cases such as AA, AAA, C etc. can be any of alkaline, zinc-carbon, NI-MH and even older NI-CD composition. All these types must be separated from each other based on composition, not the size

- Battery case can be damaged from electrolyte leaking. In this case its label becomes unreadable

The well-known method for chemistry determination is the Electrochemical Spectroscopy Impedance estimation (EIS) [4], [5], [6]. The first goal of the research is to find an equivalent circuit model that fits to the EIS curve. The circuit models found are either composed by R-C cells or R-L series and R-C parallel cell that model the battery impedance. Note that the chemistry of the batteries studied were already known. Variations in the equivalent circuit models were found for the same type of battery at different State of Charge (SOC) or State of Health (SOH) values [4].

However, the research mainly focuses on testing the consistency of a battery pack cells [5]. Further research focuses on determining the dynamic behavior of the battery to extend its run time [6]. On the other hand, EIS equipment can be very expensive and testing time can raise up to tens of minutes, being inefficient in a large-scale battery separation.

II. IMPEDANCE ANALYSIS OF SPENT BATTERIES

Keeping in mind the drawbacks of the EIS method, we investigated an impedance analysis method, using the Digilent Analog Discovery device Impedance Analyzer feature [7].

To perform the impedance analysis, the batteries are initially connected with a 100Ω reference resistance in a divider configuration, the battery having the role of the numerator component.

Note that most of the spent batteries disposed are not completely depleted. They usually provide 0.8V...1.2V open-circuit voltage and about 0.2A...1.1A short-circuit current. It means that the batteries were at different SOC. A DC current flowing through the battery can affect impedance measurements. Initially, to minimize impedance variations due to internal electrochemical reactions:

- For each battery, its open-circuit voltage was measured

- The applied sine wave signal offset was adjusted to the battery open-circuit voltage, compensating the battery voltage

- The sine wave amplitude was only up to 20% of the battery open circuit voltage

During the testing procedure it was found that DC current flowing through the battery and larger sine wave amplitude emphasizes the differences between alkaline and zinc-carbon battery impedance characteristics.

The impedance characteristics resulted are represented in Fig. 1 a) and b), respectively, for a group of six alkaline and two zinc carbon batteries. The Fig. 1 a) characteristics were taken by applying a 200mV sine wave with the battery voltage compensating offset method described above. For the Fig. 1 b) characteristics, the applied sine wave had zero offset voltage and 800mV amplitude, allowing DC current to flow from the battery while testing. The reactance characteristics show a series LC behavior, presenting a series resonance. As expected, the resulted internal impedances of the alkaline batteries are much lower comparing to zinc-carbon ones. However, the difference between the observed resonance frequencies of the alkaline is about a half decade in the case of Fig. 1 a) and a decade in the case of Fig. 1 b).

The obtained results open the opportunity to differentiate between the two battery chemistries using impedance measurements. In Fig. 1 b) is also visible that larger amplitudes and DC current flowing decreases and spreads the resonance frequency values, but also largens the difference between alkaline and zinc-carbon resonance frequencies, allowing a comfortable measurably difference to classify batteries with different compositions.

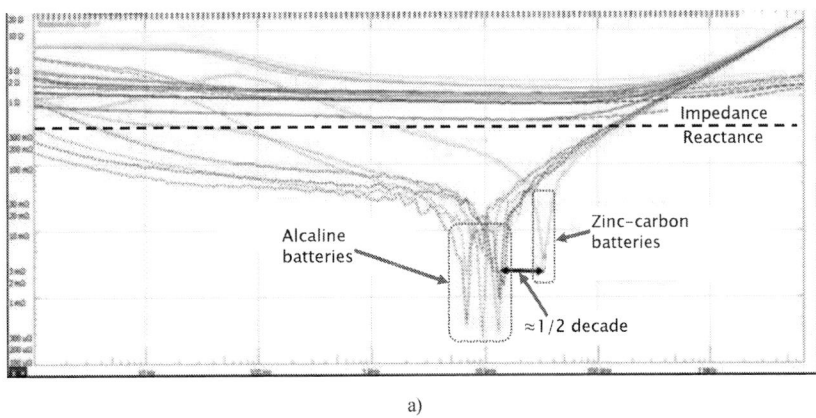

a)

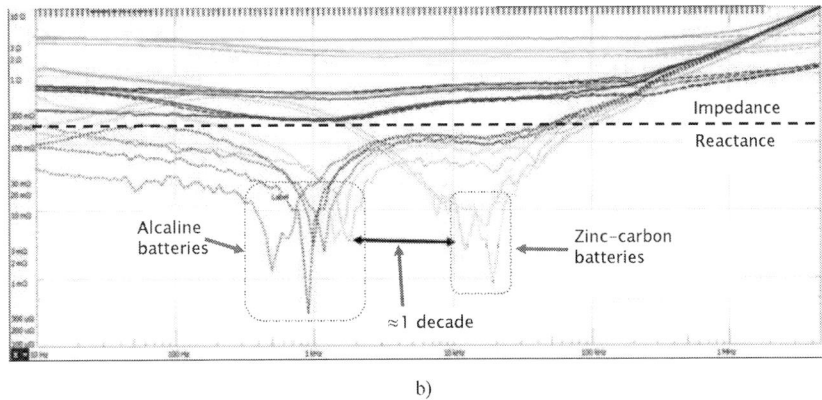

b)

Fig. 1. Impedance characteristics for 6 alcaline and 2 zinc-carbon batteries a) Applying a 200mV sine wave with battery voltage compensation offset b) Applying a 800mV sine wave and 0V offset voltage

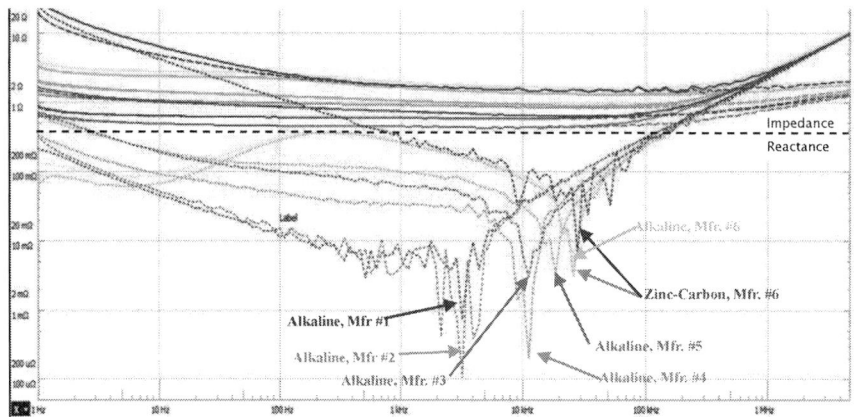

Fig. 2. Impedance characteristics for 6 alcaline and 2 zinc-carbon batteries from different manufacturers

However, various manufacturers use different materials and material quality for battery production. This fact clearly affects impedance response in frequency. Fig. 2 shows the impedance characteristics from six alkaline batteries from different manufacturers, compared to the impedance characteristics of two zinc-carbon batteries. It is visible from Fig. 2 that some of the alkaline battery resonant frequencies overlap with the zinc-carbon ones, resulting in impossibility of battery separation

III. HIGHLIGHTING SECONDARY IMPEDANCE EFFECTS

A. Bode diagram

Consequently, secondary impedance effects were searched in both alkaline and zinc-carbon battery impedance behavior. The first approach was to represent the Bode diagram of an RC lowpass filter, where the battery takes the role of the capacitor.

Fig. 3 shows the resulted amplitude and phase characteristics. There are no easily measurable differences between the alkaline and zinc-carbon characteristics, but the characteristic shapes suggest the presence of different secondary effects. For instance, the capacitive effect at low frequency of the zinc-carbon batteries is more visible than for alkaline batteries. On the other hand, zinc-carbon battery phase characteristics show a small lag at lower frequencies. These differences lead to the idea of involving a different impedance representation.

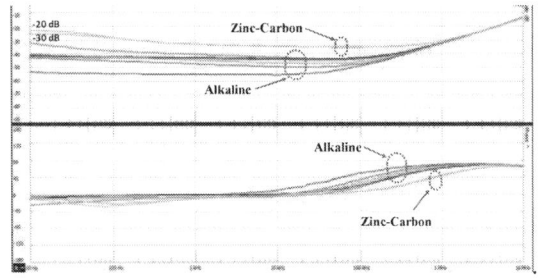

Fig. 3. Bode diagram for 6 alkaline and 2 zinc-carbon bateries in RC lowpass configuration, where C was replaced by the battery

B. Modifying impedance representation

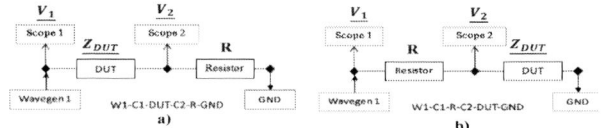

Fig. 4. Impedance Analyzer circuit connections. DUT = Device Under Test [8]

The Analog Discovery Impedance Analyzer circuit can be constructed in two different ways [8], as Fig. 4 shows:

In the case of Fig. 4 a) configuration, the represented $\underline{Z}_{DUT}$ impedance is calculated at each frequency point as

$$\underline{Z}_{DUT} = \frac{V_{Z_{DUT}}}{I_{Z_{DUT}}} = \frac{(V_1 - V_2) \cdot R}{V_2} \tag{1}$$

while, in the case of Fig. 4 b), $\underline{Z}_{DUT}$ is calculated as

$$\underline{Z}_{DUT} = \frac{V_{Z_{DUT}}}{I_{Z_{DUT}}} = \frac{V_2 \cdot R}{V_1 - V_2} \tag{2}$$

The Digilent Analog Discovery Impedance Adapter uses Fig. 4 a) connection for impedance determination. Therefore, when in the Digilent Waveforms software [8] the Impedance Analyzer Adapter configuration is selected, Equation (1) is applied.

To highlight alkaline and zinc-carbon battery secondary impedance effects, one comes in mind to apply Equation (2) for the circuit in Fig. 4 a). That is, the represented impedance will be

$$\underline{Z}_{DUT}^{*} = \frac{V_{Z_{DUT}}}{I_{Z_{DUT}}} = \frac{V_2 \cdot R}{V_1 - V_2} \tag{3}$$

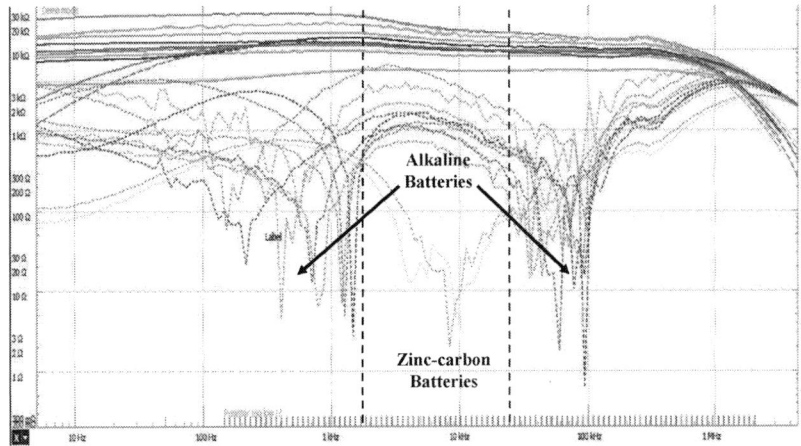

Fig. 5. Impedance characteristics for 6 alcaline and 2 zinc-carbon batteries from different manufacturers, using modified impedance representation

Keeping in mind that the circuit used is the one in Fig. 4 a), results

$$I = \frac{V_2}{R} = \frac{V_1 - V_2}{Z_{DUT}} \Rightarrow V_1 - V_2 = \frac{V_2}{R} \cdot Z_{DUT} \qquad (4)$$

From here, results that the represented impedance will be, in fact, the complex conjugate of the Z_{DUT}-impedance.

$$Z_{DUT}^{\ *} = \frac{V_2 \cdot R}{V_1 - V_2} = \frac{R^2}{Z_{DUT}} \qquad (5)$$

Fig. 5 shows the impedance characteristics of six alkaline batteries from different producers and two zinc-carbon batteries. The results show a clear difference between alkaline and zinc-carbon batteries: Only the alkaline batteries present double resonance frequencies.

IV. CONCLUSIONS

Spent batteries represent a valuable source for raw minerals. Battery recycling process is often battery chemistry dependent. In this paper a classification method for spent batteries was presented, targeting alkaline and zinc-carbon battery separation. The method is based on battery impedance analysis under various conditions. Measurements concluded that battery impedance depending on its chemical composition can be easier highlighted while current flows from the battery. Also applying a sine wave excitation with an amplitude close to the battery open circuit voltage helps finding the differences between various batteries.

However, the high diversity of the materials used by various alkaline battery manufacturers lead to a wide resonance frequency range, making impossible battery separation.

The problem was solved by changing the battery impedance representation in the impedance analysis procedure by changing the equivalent circuit in the Waveforms software. With this approach, a clear difference was found between alkaline and zinc-carbon batteries. The former presents two resonant frequencies while the latter only a single resonant frequency.

The impedance analysis performed by Analog Discovery, in the 10Hz...10MHz frequency domain, with a granularity of 151 samples i.e. 21 samples/decade takes only about 2...3 seconds before displaying the result. Therefore, the proposed method is fast, cost-effective, and simple, offering a valuable solution for battery classification.

ACKNOWLEDGMENT

This work was supported within the research program PN-III-P1-1.2-PCCDI-2017-0652, project NR. 84PCCDI - 01/03/2018 TRADE-IT.

REFERENCES

[1] H.-.Kiehne, "Batteries and environmental requirements," Thirteenth International Telecommunications Energy Conference - INTELEC 91, Kyoto, Japan, 1991, pp. 209-213, doi: 10.1109/INTLEC.1991.172398.

[2] Francesco Ferella∗, Ida De Michelis, Francesco Veglio, "Process for the recycling of alkaline and zinc–carbon spent batteries", Journal of Power Sources 183(2):805-811, September 2008

[3] Hongyi Luo, Jiming Sa, Ruibin Li, Jin Li, "Regionalization Intelligent Garbage Sorting Machine for Municipal Solid Waste Treatment", 6th International Conference on Systems and Informatics (ICSAI), Shanghai, China, 2019, pp. 103-108, doi: 10.1109/ICSAI48974.2019.9010575.

[4] M. Abedi Varnosfaderani and D. Strickland, "A Comparison of Online Electrochemical Spectroscopy Impedance Estimation of Batteries," in IEEE Access, vol. 6, pp. 23668-23677, 2018, doi: 10.1109/ACCESS.2018.2808412.

[5] Jee-Hwan Jang et al., "Method for grouping Unit Cells Using Pattern Maching Technology of Impedance Spectrum," (US. Patent No. 6674287), January 6, 2004

[6] M. K. Hossain and S. M. R. Islam, "Battery Impedance Measurement Using Electrochemical Impedance Spectroscopy Board," 2017 2nd International Conference on Electrical & Electronic Engineering (ICEEE), Rajshahi, 2017, pp. 1-4, doi: 10.1109/CEEE.2017.8412902.

[7] Digilent, Inc., Analog Discovery™ Technical Reference Manual, Copyright Digilent, Inc. Revised March 18, 2015

[8] Digilent, Inc., Waveforms Software Reference Manual, Copyright Digilent, Inc. Revised February 23, 2017

Converter topologies for MVDC traction transformers

Izsák Ferencz, Dorin Petreuş
Department of Applied Electronics
Technical University of Cluj-Napoca
Cluj-Napoca, Romania
izsak.ferencz@ael.utcluj.ro

Pietro Tricoli
Department of Electronic, Electrical and Systems Engineering
University of Birmingham
Birmingham, United Kingdom
p.tricoli@bham.ac.uk

Abstract— **The paper compares the conventional railway traction systems with new traction systems proposed and developed in the last decade that are also suitable for the medium voltage DC (MVDC) railway electrification concept presented in this project. Differences and requirements of the MVDC traction system will be considered while investigating converter topologies for MVDC transformers (Power Electronic Traction Transformers – PETTs or Solid State Transformers – SSTs). Then, the paper will focus on presenting the most suitable DC-DC converters for this application, defining an example of optimal configuration and requirements of control, which in the future can be further developed for a novel MVDC railway electrification's traction systems on-board.**

Keywords – phase-shift converter, RDC snubber, active snubber, battery charger

I. INTRODUCTION

The reason behind railway electrification (RE) was the reduction of operating costs, CO_2 emissions and an improvement in energy efficiency. The newly developed electric locomotives (EL) achieved more power than diesel engines, while showing better reliability. High power EL can also pull heavier freight at higher speed over slopes, thus increasing capacity in mixed traffic conditions. Having no local emissions, electric propulsion has a great advantage over diesel in urban areas as well. In the past decade, RE constantly increased and the account of electrified tracks almost reached one third of the total tracks globally by the year 2012. According to a report of the International Energy Agency, between 1990 and 2015 due to the increase of RE both the energy consumption per transport unit decreased and the CO_2 emissions per transport unit decreased by 35.8% and 31.6% respectively. Moreover, half of these reductions were achieved in the decade of 2005 to 2015: rail energy consumption per passenger-km decreased by 27.8% and energy consumption per freight tonne-km decreased by 18.1%; rail CO_2 emissions per passenger-km decreased by 21.7% and CO_2 emissions per freight tonne-km decreased by 19.0%. In this time, the share of oil products decreased from 62.2% to 56% in the global railway fuel mix, while the share of electricity increased, whereof electricity generated by renewables has shown an increase of 65% [1]. The data presented in these reports points to the future tendency of integration of renewables in the railway electrification system (RES) and to the necessity of innovation and better compatibility between them.

However, electrification of new and existing railway lines have required a substantial investment for the railway infrastructure. This is because RE uses AC single-phase power that requires connection to high-voltage transmission lines, which are not always available in places where the railway feeder stations should be located and usually require complicated and extremely expensive modifications of the existing layouts, i.e. tap and looped connections of the substations. Moreover, the typical power level of heavy trains or even high-speed trains is compatible with the typical capabilities of medium-voltage (MV) distribution systems. On the other hand, the connection of railway feeder stations to the power distribution network (PDN) would be possible only with the condition not to introduce any imbalance into the system. In contrast to the single-phase AC electrification system, DC systems satisfy this requirement, however, the level of the DC voltage is limited to around 3kV because of the limitation on the maximum short-circuit breaking current of circuit breakers, which in turn limits the maximum power of the railway. Additionally, a higher voltage of the power supply would pose problems for the traction system of the trains, which operates at voltage levels of few kV. Besides, the traditional concept of DC railways does not fit very well with the future vision of electric railway better integrated with the PDNs. The most promising concept is a smart interoperable electric railway grid including green energy plants. The aim of MVDC Electric Railway Systems (MVDC-ERS) project is to propose a new type of MVDC traction power supply based on controlled bidirectional converters to improve the connectivity of the railway to the grid and to integrate renewable power sources to the RES. This would not only improve the efficiency of the railway supply, but it will give additional capacity to the power distribution grid, as railway electrification lines could be used to provide extra capacity between the nodes where the substations are connected. This would be especially important for future scenarios where a higher proportion of renewable energy sources will be introduced in the power system and the control of the power flow will be vital to maintain the correct functionality of the power system. [2], [3], [4].

This paper has four main sections. This first section presents the context and the issues to be solved. The second section contains a detailed comparison between MVDC and MVAC RESs including traction. The third part analyses different PETT topologies to define the suitable one for MVDC traction. The penultimate section represents the simulation model and results for the proposed MVDC traction configuration, while the last section includes observations and conclusions.

978-1-7281-7507-2/20 $31.00 © 2020 IEEE
21-24 October, Pitesti, Romania

II. MVDC-ERS COMPARED TO MVAC-ERS

At the moment, modern RESs use AC to produce higher voltages using transformers. For the same amount of power, the higher the voltage the lower the current. Having lower currents, the line losses are reduced, and higher power can be delivered. The earliest systems choose DC because at that time, AC was not understood well and good insulator materials for such high voltages were not available. However, the DC equipment was massive high currents being implied to obtain enough power for the low voltage locomotives (first at 600 V and then 1500 VDC). These high currents lead to large transmission losses. Areas like Eastern Europe, where catenaries operate at 3kV DC, two 1500VDC motors in series are used, but even at 3kV to power a heavy train the currents needed can be excessive. Later AC motors became predominant as they developed, used on longer routes. The higher voltages of tens of thousands allowed the use of low currents and losses could be minimized meaning cheaper wires. Such high voltages could not be used with DC locomotives due to the difficulty of the voltage/current transformation in a so efficient way as AC transformers. Now better semiconductor devices being available, DC lines are still used and under development. Both RESs converts and transports high-voltage AC from the grid to lower voltage DC in the locomotive, the difference between the two RESs is the location where the conversion from AC to DC is done: at the feeding substation (in case of DC) or on the locomotive (AC). The choice of which one to be used, often depends on the already existing RES in the respective country or area and the costs of a new infrastructure.

PETTs are popular in applications where power density and high efficiency are targeted, therefore it is highly researched for traction applications in electric railways and ships. Fig. 1 illustrates the concept of a PETT, as the LFT is replaced by an MFT as part of the chosen topology. With a higher operating frequency, MFTs achieve a reduced volume and weight at the same winding current density and maximum magnetic field strength, as the induced voltage is proportional to frequency. Additional features of PETTs include control of input and output voltages and currents, the flow of power and load protection.

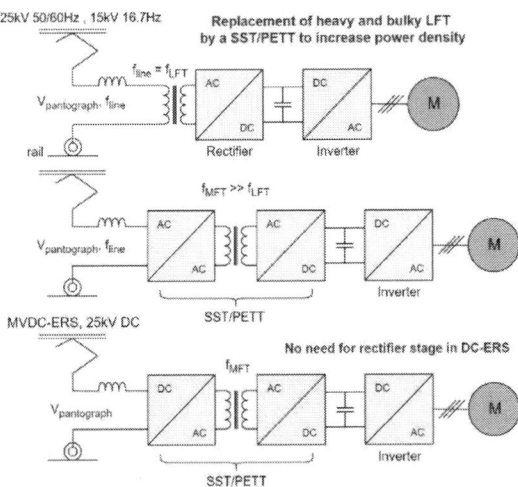

Fig. 1. PETT replacing traditional LFTs and the difference between currently used MVAC-ERS and the MVDC–ERS.

To illustrate this, let's take the example of 15kV/16.7Hz ERS: ABB reported a system weight and volume reduction of 50% and 20% respectively applying only a 400Hz PETT instead of the LFT system [5]. With the appearance of low-floor vehicles or roof mounted traction equipment as well as higher power demand in the case of high-speed trains, the features offered by PETT technology are highly attractive. Fig. 2 summarizes the advantages of MFT technology used in PETTs as the new traction transformer tendency.

	LFT	MFT
Power density	low	high
Efficiency	limited and lower	high
Transformer design complexity	low	high, moreover different applications need different and specific design
Operating/switching frequency	line frequency (low)	hundreds of Hz to tens of kHz
Power quality	fair	good, due to more control options
Technical maturity	reached its maturity	not yet mature, some topologies and configurations are reaching their potential faster than others
Fault current limitation	low	good
Fault isolation capability	poor	good, also redundant configuration is available
Control complexity	low	high and in some applications can be difficult, but rewarding
Switch and drives count	low	high number of devices, due to modular/multi-level structure, however WBG high-voltage devices can lower it
Flexibility	low	high, offers additional functionalities like fault limitation and isolation, voltage flicker compensation
Controllability	low, no control over transmitted power	high, good control over power flow
Availability	high	fair, difficult to design and manufacture
Reliability	high	lower; under research and development, different configurations, like redundancy can bring improvements
Costs	low cost compared to state of the art technologies, much better kW/cost value	due to multilevel/multi-stage and/or multi-modular structure they have a higher cost (still low kW/cost value)
Losses	higher losses	lower losses

Fig. 2. Advantages of MFT technology over LFT in current RESs.

In table I, the most widely used AC RES – 25kV, 50/60Hz ([6]) – will be analyzed in comparison with the novel MVDC RES, in terms of: power supply system technology, number of connections to utility grid, substation interaction, current feeding back to the grid, overhead lines, current transportation, rolling stock, power fed back to the overhead line through braking and current return, corrosion and leaks. Regarding DC RES, most of the drawbacks of LVDC systems are caused by having low voltage, implying higher number of substations, heavier overhead lines and higher traction losses. Due to the higher current, corrosion should also be considered. Because of these, current DC systems are not economical regarding overhead lines – implying higher investments and operational costs (tear and wear) – and regarding substations – higher number meaning more expensive connections to the grid and higher maintenance costs.

2020 IEEE 26th International Symposium for Design and Technology in Electronic Packaging (SIITME)

TABLE I.   MVAC AND MVDC RAILWAY ELECTRIFICATION SYSTEM – COMPARISON WITH ADVANTAGES AND DISADVANTAGES

	25 kV 50/60Hz AC	MVDC-ERS
Utility/main grid (power supply)	− possible unbalance on the utility grid − strong electric connections needed + medium to low number of connections (depending on substation technology if it is transformer or converter based)	+ low impact and no unbalances on the grid + low number of connections to the grid + possible connection to weaker parts of the utility grid + possibility to develop smart grids
Substation	+ low number of substations, meaning lower investments and maintenance costs + simple circuit breakers and switching devices + simple fault detection − in case of using converters, two conversion stages AC/DC/AC to solve unbalanced loading, larger substation (need of land)	+ fewer substations (no inductive voltage drop, allows more distance between substations) meaning lower investments and maintenance costs + bilateral supply, substations can be paralleled to share the load + Possibility of controlling DC short circuit currents by substation converters and using low-load or no-load DC circuit breakers + only one conversion stage, thus improved efficiency and smaller substation
Interactions in substations	− complex power supply diagram due to phase separation − less flexibility in case of substation incident	+ simple power supply diagram since there is no phase separation, beneficial in dense areas of traffic + substations in parallel flexible in case of incident
Current fed back to the utility grid	+ basic transformers needed to feed back currents to overhead line, or the two stage AC-DC-AC converters could also be used	+/− inverters needed with harmonics generated, but power factor of AC-DC converter and harmonics injected to the grid can be controlled to meet standards
Overhead line and current transportation	− high insulation distances, thus difficult implementation in urban areas and tunnels − complex impedance jωL, therefore presence of inductive voltage drops + low losses due to high voltage in traction circuit + light overhead line due to lower current: lower costs and higher speeds + low tear & wear of contact wire + one contact wire − neutral zones	− high insulation distances, thus difficult implementation in urban areas and tunnels + absence of jωL part, thus no inductive voltage drops and reactive power consumption + low losses (high voltage) and light overhead line due to lower current + no skin effect, thus smaller cross-sections; light overhead line due to lower current: lower investments + low tear & wear of contact wire, low maintenance costs + no neutral sections, avoiding power transfer interruptions and speed loss as well as mechanical and electrical stresses in locomotive circuit breakers
Rolling stock	− large and heavy transformers on-board, thus heavy rolling stock − need of rectifiers on-board + simple circuit breakers − converter complexity and reliability	+ smaller PETTs on-board, thus lighter rolling stock + no rectifier on-board, thus lighter and more reliable rolling stock − converter complexity and reliability, complex circuit breakers, current has to be controlled and limited in faults in on-board PETTs − need of rolling stock development
Regenerative braking	− necessity to adjust the phase of the current with overhead line current	+ no adjustments of the phase of the feedback current is needed
Current return	+ low levels of current returning to substations due to high voltage	+ lower levels of current returning to substations due to high voltage, but the new system must be able to mitigate stray currents
Corrosion and leaks	+ low risk of corrosion due to low current leaks	+ limited corrosion due to lower return currents
Interferences	− ground currents may interfere with communication devices near the railway installations and when power electronics are used − large filters and compensators needed to improve power quality	+/- possible interference with signaling systems, no induced voltages in adjacent lines − high power converters may produce high order harmonics − EMI, EMC noise emissions have to be investigated
Conclusions	Allows more powerful traffic if well dimensioned.	Will combine advantages of current AC and DC systems, (most drawbacks in current DC systems are due to low voltage), however new operation procedures and regulations are needed.

As Table I summarize, a new MVDC-ERS is a promising project, since it combines the advantages of the current MVAC and LVDC ERSs and at the same time opening new opportunities for the design of future smart grids. This will imply new areas of study and some aspects presented in table 1 need more research and investigations. Some examples are the faults detection in real time, new circuit breakers for HVDC in substations as well as in PETTs on-board in rolling stocks, insulating materials, overhead line design, flexible power-supply diagrams, necessary modifications in rolling stocks for compatibility, the impact of high DC voltage on current collection. Several studies like [7] have shown environmental, and system stability benefits of High Voltage DC (HDVC) transmission lines. In the case of DC train systems potential cost savings, complexity of infrastructure and more friendly integration into the grid are highlighted as further advantages in [8], [9], [3],[10].

III. MVDC TRACTION

A. Topological Overview

The new wide band-gap semiconductor materials like silicon carbide (SiC) encourages PETT development, especially when the 6.5kV and 10kV and later 15kV SiC components will be ready to be commercialized. SiC semiconductors allow switching frequencies as high as tens of kilohertz. Due to this advantage the switching frequency of MFTs could be increased and when SiC devices will appear at higher voltages it will also allow to use fewer converter modules and/or stages. A new ERS such as a flexible MVDC-ERS requires high-performance, novel PETT structures to handle new challenges such as fault handling, protection circuits and smart-grid compatibility. In the MVDC-ERS concept the setup looks different in comparison to the main topological families defined in state of the art literature, since a rectifier stage is not needed in MVDC traction topologies. This will further improve efficiency and power density. However, to improve the new MVDC line-based traction devices voltage balancing stages (VBS) could be used instead of the rectifier stages.

This section will present an overview of different PETT topologies developed in the last three decades, presenting the trends. However, the present study will focus only on MV PETTs (15kV and 25kV AC ERSs), since currently all state of the art MV PETTs are developed for AC ERSs.

The topology on Fig. 3 was developed by Weiss in '85 as the first ever PETT for railway applications. It consisted of a matrix converter and 400Hz MFT. Later the concept was further studied and the newly available IGBTs of high voltage were employed. Currently it is efficiently applicable to LV systems. Its advantages are: higher switching frequency, lower losses, less modules and costs with a future potential, when the 10-15kV SiC transistors will appear. However the design for reliability is challenging (not having redundancy) or increases complexity.

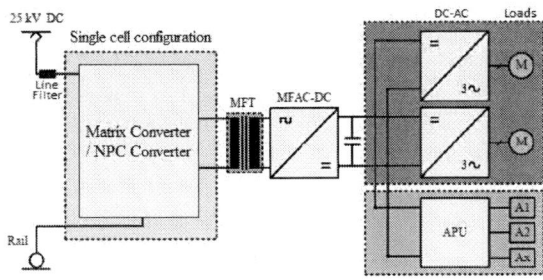

Fig. 3. Single cell matrix/NPC based PETT for MVDC traction. (APU means Auxiliary Power Unit and M is Motor).

In 2001 researches demonstrated the necessity of series connection of converters in the front end. Right after, in 2002, the multi-cell concept was presented, see Fig. 4. During 2003-2005 Siemens also developed such a system of 2MVA power. This topology is scalable to higher voltages, it is reliable (redundant cells), and it has dynamic voltage sharing capability. A single MFT can be an advantage and a disadvantage too in some situations, moreover having many stages and levels increases costs and control can be more difficult.

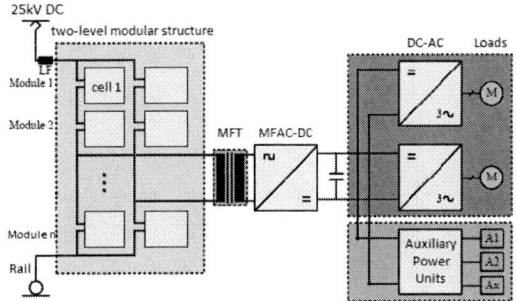

Fig. 4. Multi-cell (modular multi-level) PETT topology.

Currently the most commonly used converter configuration is the input-series output-parallel (ISOP). In 2003 Alstom developed a converter with semi-separated multi-winding transformer, as in Fig. 5, usable for electric-multiple-units (EMU) setup with independent output DC links in secondary. The advantages of this topology was the balanced power distribution between modules, it became mature and popular and the modular design is fully controllable. However, the joint multi-winding MFT is difficult to make and it has weaker fault-handling capability. Regarding the more advantageous ISOP setup, it is a compromise, since increases control complexity.

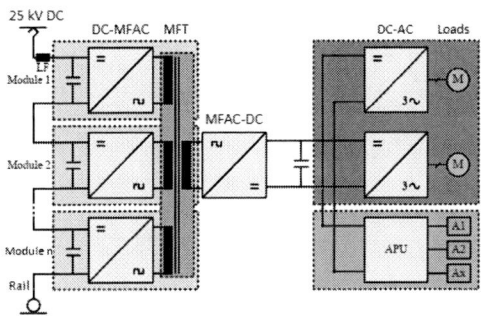

Fig. 5. Cascaded ISOP setup with semi separated multi-winding isolation (SSMW).

In 2014 a similar configuration as the previous one was developed in China, but with multi-port configuration in the secondary as a novelty. It is usable in 25kV ERSs. Similar to the previous topology, voltage-balancing control is achievable and as a plus more ports are available. Its disadvantages is the joint multi-winding MFT, which is difficult to make and has weaker fault-handling capability.

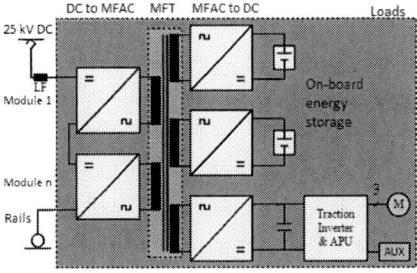

Fig. 6. Multi port multi-winding PETT configuration.

Currently the most preferred topological family is the one on Fig. 7. It is mostly proposed for 15kV ERS, and since 25kV is a higher voltage by 66% than 15kV, it implies more cascaded modules (high number of power devices) and higher costs, when applied to a 25kV ERS. The whole system is ISOP structure, fully controllable and with improved reliability due to separated windings, currently is the most popular and mature topology, the transformer is less difficult to produce and has better fault handling capability.

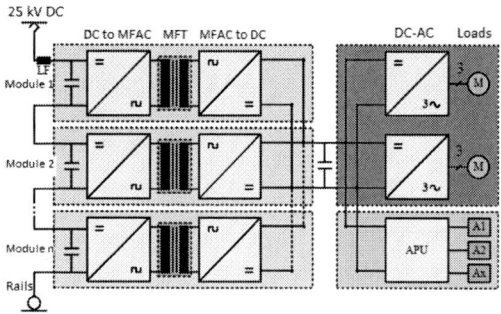

Fig. 7. Cascaded ISOP setup with separated multi-winding (SMW) isolation.

ABB developed the first ever PETT implemented into an actual locomotive and then tested in 2011 on the Swiss Federal Railways [11]. It was a half-bridge topology. Asymmetrical active H bridge topologies include different LLCs, Phase-Shifts and other suitable configurations for modular applications. They work well in different projects depending on each application requirement. Other projects presented full-bridge topologies also, up to 3MVA. In [12] it is highlighted, that CHB configurations are more mature and can reach higher voltages than other state of the art multilevel topologies, including diode-clamped configurations, which can have higher switching losses and unbalanced voltage.

The CHB and the matrix converters are the two preferred candidates for front-end converters, with CHB systems more mature and popular. In terms of the transformer, the multi-separated MFTs configuration is better than joint multi-winding configuration, because it is easier to manufacture and has better fault handling performance. In terms of the system configuration, cascaded front-end (CFE) converters with fully controllable power electronic devices must be used, in order to withstand the input voltage from the catenary. ISOP is the most popular, as it uses a modular design that is good for redundancy. However, they have the disadvantages of high cost and low-power density necessity, thus topologies with reduced number of CFE converters would ultimately be preferred when new semiconductors with higher blocking voltage will become available.

As a conclusion for the MVDC-ERS project the most suitable topology is the one in Fig. 7 with an H Bridge as the converter stage using the latest SiC technology and MFT above 20 kHz. The example presented in this paper is a MVDC Dual Active Bridge (DAB) traction converter with.

B. Specifications

The specifications of the MVDC RES catenary voltage are: Lowest permanent voltage 19kVDC, Nominal voltage 25kVDC, and highest permanent voltage 27.5kVDC. On this range (19-27.5kV) the total power factor λ (active power /apparent power) must be:

$$\begin{cases} \lambda \geq 0.95, & if\ P_{pant} > 2MW \\ \lambda > 0.85, & if\ P_{pant} < 2MW \end{cases} \quad (1)$$

,where P_{pant} is the instantaneous power at the pantograph. In the cases when this power is below 2 MW, the overall (traction and auxiliaries) average power factor must be greater than 0.85 over a complete timetabled journey [13].

$$\lambda = \sqrt{\frac{1}{1 + \left(\dfrac{W_Q}{W_P}\right)^2}} \quad (2)$$

Equation 2 presents the calculation of the overall average λ for a train journey, including stops as a function of active (WP in MWh) and reactive energy (WQ in MAh).

Inside yards and depots, when traction power is switched off but all auxiliaries are still running and the power drawn is greater than 200 kW, the power factor should be ≥ 0.8. During regenerative braking the power factor can decrease freely to keep the voltage within limits. Capacitive power factor of a train is not limited, since a train should not behave as a capacitor, however during regenerative mode if there is capacitive power, it shall be limited to 150kvar on the range 19-27.5kV.

Each train should have automatic regulation device, which in case of weaker network or abnormal operation adapt the level of maximum power consumption depending on the contact line voltage in steady state. Therefore, power selection devices must be installed, which can limit the power demand to the electrical capacity of the line. According to standard EN 50388 from 2012, the maximum allowable current on classical lines is 800A and 500 to 1500A on HS TSI (High-Speed Technical Specifications for Interoperability) lines, in the case of the 25kV AC ERS.

IV. SIMULATIONS AND RESULTS

The simulation model of the proposed MVDC PETT was implemented in Matlab/Simulink and Powersim too. Fig. 9 on the left shows one DAB module and on the right 8 modules in ISOP setup. In Fig. 8 the PETT module's waveforms can be seen. The values of maximum primary current and voltage depends on how the converter is designed.

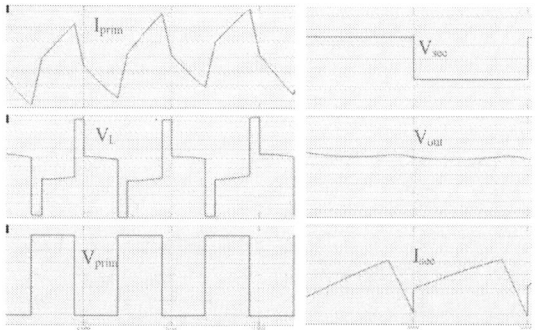

Fig. 8. DAB module results. Primary and secondary waveforms.

The design parameters depends on the maximum train current allowed and the maximum voltage and current of the SiC devices used. This can vary the number of modules too.

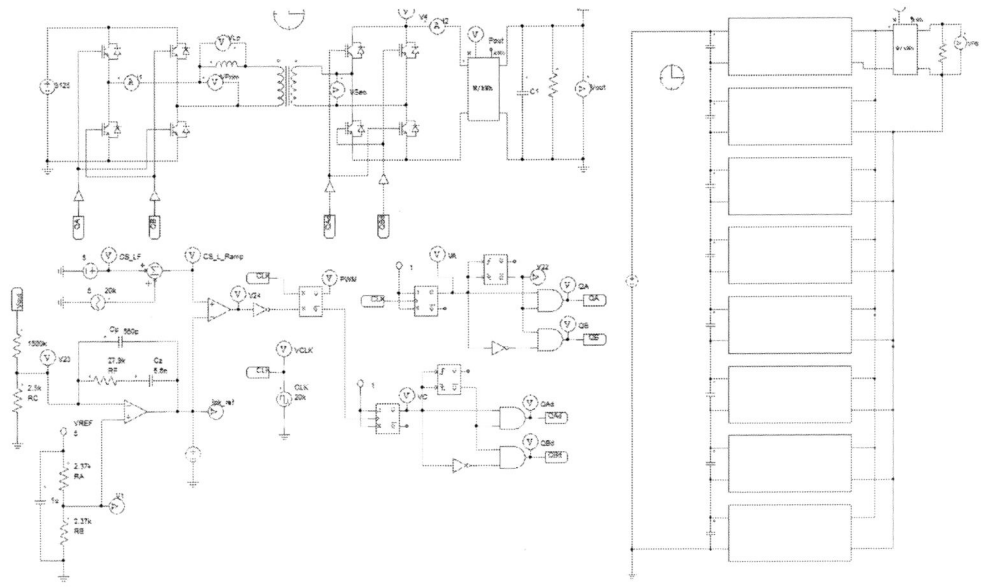

Fig. 9. The schematic of DAB converter. A single module and 8 modules in ISOP configuration.

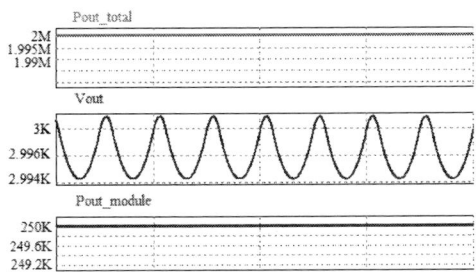

Fig. 10. MVDC PETT System example – 8 modules waveforms.

In this example, the following parameters were used: 20kHz switching frequency, 25kV input voltage, 3kV output voltage, 2MW maximum total and 250kW module power.

V. CONCLUSIONS

As a conclusion, it is important to notice that the benefits of modern PETTs are evident - firstly, the improved efficiency and power quality, secondly a redundant design, which improves availability, and thirdly the increased power density, while most drawbacks are technology and material dependent and the development of power devices and materials, as well as investigation of topologies and control methods will probably mitigate most of them. MVDC-ERS presents a concept based on various new technology that makes possible its implementation. Such a novel system will open new opportunities and functionalities of an interoperable smart DC grid. At the same time the new system will combine the advantages of current ERSs. The on-board PETTs will have to be redefined also for the new system and its needs.

ACKNOWLEDGMENT

This project has received funding from the Shift2Rail Joint Undertaking (JU) under grant agreement No 826238. The JU receives support from the European Union's Horizon 2020 research and innovation programme and the Shift2Rail JU members other than the Union.

REFERENCES

[1] International Energy Agency, "Railway Handbook 2017."

[2] M. Brenna, F. Foiadelli, and D. Zaninelli, *Electrical railway transportation systems*. Wiley, 2018.

[3] A. Verdicchio, P. Ladoux, H. Caron, and S. Sanchez, "Future DC Railway Electrification System - Go for 9 kV," *2018 IEEE Int. Conf. Electr. Syst. Aircraft, Railw. Sh. Propuls. Road Veh. Int. Transp. Electrif. Conf.*, pp. 1–5, 2019, doi: 10.1109/esars-itec.2018.8607304.

[4] L. Peng, S. Wang, L. Xu, Z. Zheng, and Y. LI, "Onboard DC Solid State Transformer Based on Series Resonant Dual Active Bridge Converter," *2018 IEEE Int. Conf. Electr. Syst. Aircraft, Railw. Sh. Propuls. Road Veh. Int. Transp. Electrif. Conf.*, pp. 1–6, 2019, doi: 10.1109/esars-itec.2018.8607334.

[5] N. Hugo, P. Stefanutti, M. Pellerin, and R. Sablières, "Power Electronics Traction Transformer," *Proc. Eur. Conf. Power Electron. Appl.*, pp. 1–10, 2007.

[6] *EN 50163: Railway applications. Supply voltages of traction systems.* 2007.

[7] K. Meah and S. Ula, "Comparative evaluation of HVDC and HVAC," *Proc. IEEE Power Eng. Soc. Gen. Meet.*, pp. 1–5, 2007.

[8] A. Gomez-Exposito, J. M. Mauricio, and J. M. Maza-Ortega, "VSC-Based MVDC railway electrification system," *IEEE Trans. Power Deliv.*, vol. 29, no. 1, pp. 422–431, 2014, doi: 10.1109/TPWRD.2013.2268692.

[9] A. Verdicchio, P. Ladoux, H. Caron, and C. Courtois, "New Medium-Voltage DC Railway Electrification System," *IEEE Trans. Transp. Electrif.*, vol. 4, no. 2, pp. 591–604, 2018, doi: 10.1109/TTE.2018.2826780.

[10] H. Shigeeda, H. Morimoto, K. Ito, T. Fujii, and N. Morishima, "Feeding-loss Reduction by Higher-voltage DC Railway Feeding System with DC-to-DC Converter," *2018 Int. Power Electron. Conf. (IPEC-Niigata 2018 -ECCE Asia)*, pp. 2540–2546, 2018, doi: 10.23919/IPEC.2018.8507567.

[11] M. Claesens, D. Dujic, F. Canales, J. K. Steinke, P. Stefanutti, and C. Veterli, "Traction transformation: A power-electronic traction transformer (PETT)," *ABB Rev.*, no. 1, pp. 11–17, 2012.

[12] L. Heinemann, "An actively cooled high power, high frequency transformer with high insulation capability," *Conf. Proc. - IEEE Appl. Power Electron. Conf. Expo. - APEC*, vol. 1, no. c, pp. 352–357, 2002, doi: 10.1109/apec.2002.989270.

[13] European Committee for Electrotechnical Standardization, "Railway Applications - Power supply and rolling stock - Technical criteria for the coordination between power supply (substation) and rolling stock to achieve interoperability," 2013.

2020 IEEE 26th International Symposium for Design and Technology in Electronic Packaging (SIITME)

Electro-Thermal Simulation of Power DMOS Devices Operating under Fast Thermal Cycling

Ciprian Florea, Vasile Țopa
Department of Electrotechnics and Measurements
Technical University of Cluj-Napoca (UTCN)
Cluj, Romania
Ciprian.Florea@ethm.utcluj.ro

Dan Simon
Infineon Technologies Romania & Co. SCS
Bucharest, Romania

Abstract—**An electro-thermal simulation setup using a FEM-based simulator is being calibrated for adequate modelling of the thermal behavior of a DMOS power transistor operating under fast thermal cycling. The computational domain is defined by selecting the essential geometry details of the chip. A method of homogenization is used to reduce the complexity of the chip metallization. Equivalent anisotropic thermal properties and their temperature dependencies are extracted for the homogenized regions. The simulation stimuli are adjusted so that they replicate the testing conditions. Transient electro-thermal simulations are performed and their accuracy is assessed by comparison with the experimental data. A second comparison is performed between a highly detailed model and the homogenized one in order to assess the reduction in simulation time and possible accuracy loss.**

Keywords—*electro-thermal simulation; DMOS power transistors; homogenization; anisotropic thermal properties*

I. INTRODUCTION

Modern double-diffused metal oxide semiconductor (DMOS) transistors which act as switches in automotive applications dissipate high power levels for short periods of time. Power dissipation leads to device self-heating and junction temperatures up to 400°C can be reached. The electro-thermal phenomena reduce the safe operating area (SOA) of the DMOS due to formation of local hotspots which cause the device to slip into thermal runaway [1]. Accurate temperature estimation including electro-thermal effects is needed for chip lifetime estimation, but is difficult to achieve due to the complexity of the integrated circuit. Therefore, 3D multi-physics simulators based on the finite element method (FEM) can be used to accurately estimate the chip temperature distribution. This requires good modelling of the device, its metallization and parts of its packaging.

The scope of this work is to calibrate a simulation setup using the simulator described in [1] so that the simulated peak temperature fits to the temperature measured by the embedded temperature sensor. Another objective is the reduction of the geometrical complexity of the 3D model through homogenization. Equivalent anisotropic thermal parameters are extracted for the homogenized region using the method described in [2]. In the end, a comparison between the detailed model and the simplified model is done in order to assess the impact of homogenization on simulation time and accuracy.

The article is structured as follows: In Section II the simulation setup is described. In Section III equivalent thermal properties and their temperature dependencies are extracted. In Section IV simulation results are discussed. In Section V conclusions regarding the entire activity and possible future applications are presented.

II. THE SIMULATION SETUP

A. The computational domain

The first step in every simulation is the definition of the domain which is subjected to analysis. The DMOS device is manufactured in a standard automotive technology, similar to the one described in [3], which contains multiple layers of routing metallization and plates of thick metal on top, known as power metal. The chip is packaged in a 24-pin dual-in-line ceramic package (CDIP24). The 3D model is extracted from the actual 2D layout of the real device in accordance with the manufacturing technology and the adopted package. This results in a fairly large structure with a lot of chip details which render the simulation extremely time consuming. However, considering that for short power pulses (in the millisecond range) it is most likely that several parts of the packaging are not heated, the adopted 3D model can be reduced, as presented in Fig. 1. The model contains in full size: (1) the DMOS device, (2) its metallization, (3) the silicon die and (4) the die attach. The leadframe and the ceramic packaging are partially represented, while other parts (package cap, pins, sockets, PCB) are ignored. A perfect heatsink is placed under the package. Its temperature is equal to the ambient temperature. All other faces are adiabatic.

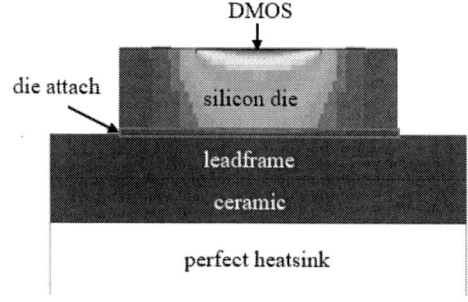

Fig. 1. The computational domain

2020 IEEE 26th International Symposium for Design and Technology in Electronic Packaging (SIITME)

B. Setting up the simulation

The relevant geometry is extracted from the 2D layout and extruded to 3D. Then, material thermal properties, boundary conditions and simulation stimuli are specified. Some thermal properties for thin-film materials as well as their nonlinear temperature dependence are provided in [1], [4] and [5]. The perfect heatsink is used to force a constant temperature of 125°C at the interface with the bottom ceramic layer, in this way replicating the temperature forced inside the oven during testing. All other outer surfaces are set adiabatic, as the surrounding environment is air (no mold compound present, because a ceramic package is used). At the surface of the silicon die lies the DMOS transistor which acts as a heat source. The DMOS is provided with a dedicated electro-thermal model similar to those described in [1] and [5]. The electro-thermal model accounts for current density variation with temperature for both the intrinsic MOS and the parasitic bipolar transistor, for various drain to source voltages (V_{DS}). The power pulse has a magnitude of approximately 23W and a pulse length of 1.6ms (rectangular power pulse with constant V_{DS} and I_{DS}). The adopted V_{DS} is 30V, the one used in the measurements. The power pulse obtained from measurements is sampled and introduced as a piecewise linear (PWL) input to the simulator which solves the heat equation and delivers a temperature response.

III. HOMOGENIZATION

Due to the high amount of contacts and vias ($10^5 \div 10^6$) in the routing metallization a full-chip-full-detail simulation is impossible. The computational domain is simplified through homogenization. Equivalent thermal properties are extracted. Although the properties of the constituents are isotropic, due to the alternation of metals and oxides the metallization, a certain degree of anisotropy is expected for the equivalent thermal conductivity. The method of extraction is detailed in [2] and [6]. Three regions selected for this study are: (1) the arrays of contacts and the oxide between them, (2) the arrays of vias and the oxide between them and (3) a metal stack starting from the contact layer up to the i-th layer of routing metallization. Two analysis are carried out: (1) an assessment of the degree to which the equivalent thermal conductivity varies with the scale length of the of the elementary cell and (2) an assessment of the variation of equivalent thermal conductivity with temperature. The first analysis is done by extracting the equivalent thermal conductivities for the considered basic cell, and for a bigger slice obtained by multiplying the basic cell by 10x10 on X and Y directions. The variation is expressed as a percentage of the equivalent thermal conductivity of the basic cell, according to the formula:

$$\Delta k_{norm} = (k_{10 \times 10} - k_{basic}) / k_{basic} \times 100 \qquad (1)$$

, where Δk_{norm} is the thermal conductivity variation relative to that of the basic cell, $k_{10 \times 10}$ is the equivalent thermal conductivity of the bigger slice and k_{basic} is the equivalent thermal conductivity of the basic cell. The second analysis consists in sweeping the ambient temperature in the 25°C ÷ 400°C range and extracting the equivalent thermal conductivity at each 25°C step, thus deriving a nonlinear temperature variation model for the equivalent anisotropic thermal conductivity. The extracted equivalent properties and their nonlinear temperature models are added to the simulation setup. No model was extracted for the

equivalent specific heat variation with temperature as the specific heat variation of the constituents is negligible.

IV. RESULTS

The first batch of results are obtained from homogenization. In Table I. are listed the calculated Δk_{norm} values in all three directions, for the considered subdomains of the metallization. The variation in thermal conductivity barely exceeds 2% with the increase in the scale length of the basic cell, therefore it can be stated that the homogenization method is independent on the size of the considered sample, as long as the it is representative for the entire metallization, which in fact is the case, given that geometrical details are distributed periodically in X and Y directions. The extracted nonlinear temperature dependence of the equivalent thermal conductivity is presented in Fig. 2. and Fig. 3. The traces $k(T)/k_0$ represent the thermal conductivity for the temperature T expressed in Kelvin normalized to k_0, the conductivity at T=300K. The variation of thermal conductivity over the temperature range of interest is not negligible at all, decreasing with 22% for high conductivity directions and increasing with 30% for low conductivity directions, therefore it is very important to model this variation as well, especially the decrease in conductivity on Z direction, as it limits the heat transfer in the thermal capacity of the power metal, leading to higher junction temperatures. Not modelling this variation would lead to a serious underestimation of the real temperature.

TABLE I.   THERMAL CONDUCTIVITY SENSITY TO THE SCALE OF THE CELL

Region	Δk_{norm}		
	X direction	Y direction	Z direction
contacts + oxide	-1.033%	-1.033%	-0.003%
vias + oxide	2.04%	2.04%	0.02%
contacts up to i-th metal layer	-2.18%	-0.19%	-0.02%

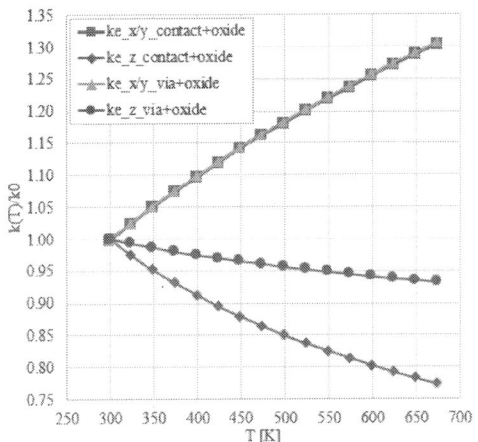

Fig. 2. The equivalent thermal conductivity variation with temperature for contacts and vias

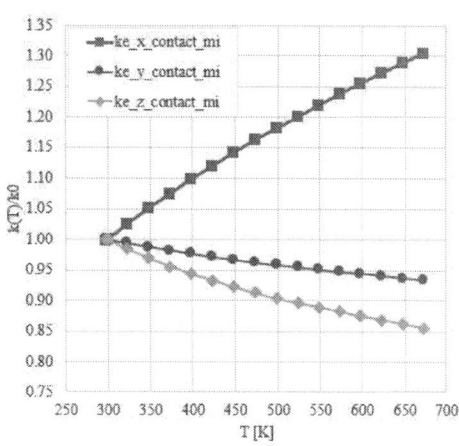

Fig. 3. The equivalent thermal conductivity variation with temperature for a metal stack ranging from the contact layer up to the i-th level of metal

With the geometry defined, nonlinear material properties annotated, boundary conditions and input stimuli set, transient simulations are carried out. There are two simulation setups: (1) one setup where only the contacts and vias were homogenized while all other regions (routing metal lines, power vias) are kept detailed, referred to as the detailed model and (2) one setup where all layers of metallization starting from the contact layer and up to the i-th layer are homogenized, referred to as the simplified model. For accuracy reasons, the first simulations and the calibration are performed on the detailed model. The temperature and power density distributions at the surface of the DMOS device at the end of the short pulse are plotted in Fig. 4. The hotspot reaches about 380°C. From the power density distribution, one might argue that the DMOS device operates below the thermal compensation point (TCP), because the power density increases with temperature. However, given the high junction temperature it is very likely that the parasitic bipolar junction transistor is active as well, as the electro-thermal model accounts for that effect as well, so there is no way to distinguish which phenomenon is predominant. One way to verify this is by dropping the ambient temperature down to 25°C and extend the

pulse length, maintaining the same values for power and V_{DS}, therefore, the same operating point. In Fig. 5. can be seen that for the specified biasing conditions the device actually operates slightly above TCP (the power density decreases with temperature) yet due to the high junction temperature, the leakage current through the parasitic bipolar transistor is significant, but lower than in the case where the ambient temperature is 125°C. This result is in accordance with the one in [5], where a similar electro-thermal model is used. In Fig. 6. is shown a vertical cross-section through the chip assembly in the vicinity of the die attach. It can be observed that the heat does not propagate beyond the die attach. The leadframe and the ceramic layer underneath remain at the ambient temperature. This result justifies the reduced version of the computational domain adopted in Section II. In fact, the ceramic package can be eliminated and the leadframe thickness reduced in order to save simulation time.

In order to assess the accuracy of the simulation, the simulation results on both the detailed model and the simplified one are compared to the measurements. In Fig. 7. are plotted the measured and the simulated peak temperature rises ($\Delta T = T_{peak} - T_{amb}$), during the power pulse. All traces are overlapping almost perfectly so in Fig. 8. is a close-up around $\Delta T = 250°C$ for a better view. The blue spurious trace represents the measured temperature rise, the trace with square markers represents the simulated temperature rise on the detailed model and the trace with triangle markers represents the simulated temperature rise on the simplified model. The simulations are in good agreement with the measurements and the peak temperature is underestimated by less than 2%. There is virtually no difference in accuracy between the simulation result obtained with the

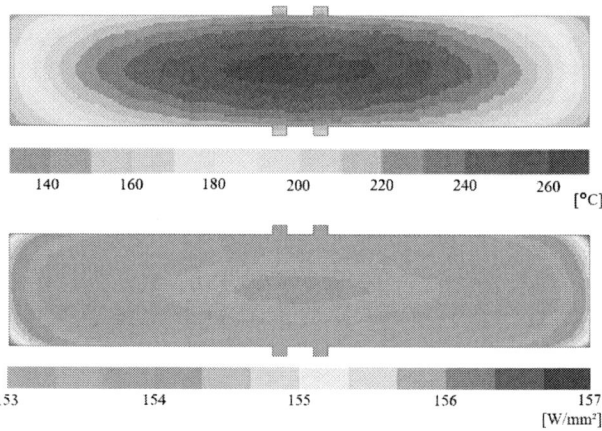

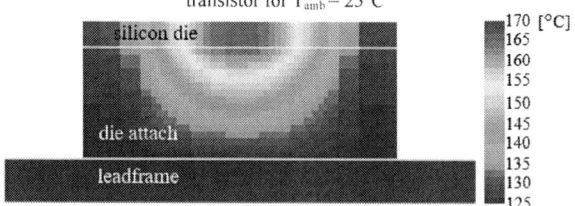

Fig. 5. The temperature distribution (top) and power density distribution (bottom) at the end of the power pulse at the surface of the DMOS transistor for $T_{amb} = 25°C$

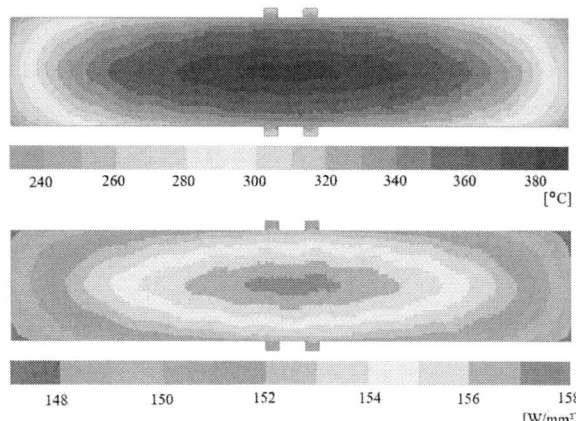

Fig. 4. The temperature distribution (top) and power density distribution (bottom) at the surface of the DMOS transistor at the end of the power pulse for $T_{amb} = 125°C$

Fig. 6. The temperature distribution on a vertical cross-section in the die attach underneath the center of the DMOS

978-1-7281-7507-2/20 $31.00 © 2020 IEEE   370   21-24 October, Pitesti, Romania

detailed model and the one obtained with the simplified model, which indicates that the modelling of the anisotropic nonlinear equivalent thermal conductivity was carried out properly. In Table II. is presented a comparison in terms of simulation time between the detailed model and the simplified model. The "a.u." stands for arbitrary units used for quantifying the simulation time, being proportional with the real simulation time, as the latter cannot be disclosed. However, the reduction in simulation time is the same and it is proportional to the reduction in the number of mesh elements. A reduction of approximately 30% in simulation time with virtually no accuracy loss is a very good result which can be used for future optimizations.

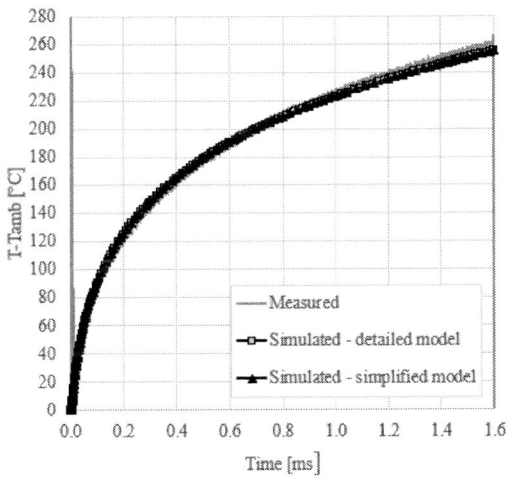

Fig. 7. The measured and simulated temperature rise during the short pulse

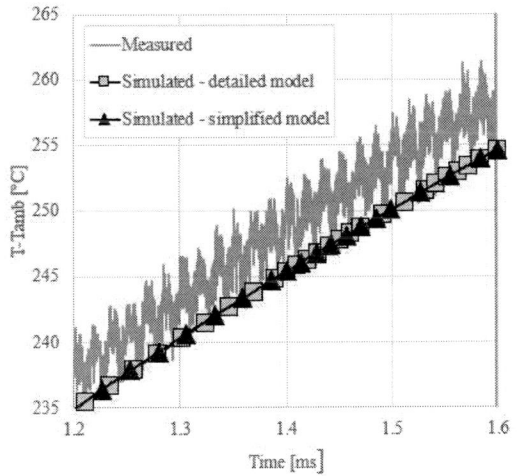

Fig. 8. The measured and simulated temperature rise during the short pulse – a close-up around $\Delta T = 250°C$

TABLE II.   COMPARISON BETWEEN THE DETAILED AND SIMPLIFIED MODEL

Model	Number of elements	Simulaton time [a.u.]
Detailed	7925237	77576
Simplified	5372077	50996

V. CONCLUSIONS

In this paper an electro-thermal simulation setup was calibrated in order to replicate the behavior of a DMOS transistor operating under fast thermal cycling.

The computational model was generated from the 2D layout, material properties were adjusted according to the technology, the simulation inputs were set so that they replicate the testing conditions as accurately as possible.

Equivalent thermal properties and nonlinear models were extracted for different portions of the metallization and were added to the setup.

Transient electro-thermal simulations show a good matching between the simulated peak temperature rise and the measured one (relative error < 2%).

The simplified model obtained through partial homogenization of the metallization offers 30% reduction in simulation time with virtually no accuracy loss.

One possible application of such high-accuracy thermal simulation consists in using the temperature distribution as input for a thermo-mechanical simulation in order to identify major stress points in the chip metallization.

ACKNOWLEDGMENT

This work was fully supported by the project iDev40. The project iDev40 has received funding from the ECSEL Joint Undertaking under grant agreement No. 783163. The JU receives support from the European Union's Horizon 2020 research and innovation program. It is co-funded by the consortium members, grants from Austria, Germany, Belgium, Italy, Spain and Romania. The information and results set out in this publication are those of the authors and do not necessarily reflect the opinion of the ECSEL Joint Undertaking.

REFERENCES

[1] M. Pfost, C. Boianceanu, H. Lohmeyer, M. Stecher, "Eletro-Thermal Simulation of Self-Heating in DMOS Transistors up to Thermal Runaway", IEEE TRANSACTIONS ON ELECTRON DEVICES, Vol. 60, Issue 2, Feb 2013, pp. 699-707.

[2] S. de Filippis, H. Köck, M. Nelhiebel, V. Košel, S. Decker, M. Glavanovics, A. Irace, "Modeling of highly anisotropic microstructures for electro-thermal simulations of power semiconductor devices", Elsevier Microelectronics Reliability 52 (2012) 2374-2379, August 2012.

[3] R. Rudolf, C. Wagner, L. O'Riain, K.H. Gebhardt, B. Kuhn-Heinrich, B. von Ehrenwall, A. von Ehrenwall, M. Strasser, M. Stecher, U. Glaser, S. Aresu, P. Kuepper, A. Mayerhofer, "Automotive 130 nm Smart-Power-Technology including embedded Flash Functionality", Proc. ISPSD, Issue 23, May 2011, pp. 20-23.

[4] M. von Arx, O. Paul, H. Baltes, "Process-Dependent Thin-Film Thermal Conductivities for Thermal CMOS MEMS", JOURNAL OF MICROELECTROMECHANICAL SYSTEMS, VOL. 9, NO. 1, MARCH 2000, pp. 136-145.

[5] D. Dibra, "Single Pulse Safe Operating Area of Trench Power MOSFETs in Automotive Power Integrated Circuits", PhD Thesis, Otto-von-Guericke-Universit, Magdeburg, Nov. 2011.

[6] D. Simon, "RELIABILITY IMPROVEMENT OF DMOS POWER SWITCHES WHICH OPERATE UNDER REPETITIVE THERMAL CYCLING", PhD Thesis, Technical University of Cluj-Napoca, 2016

Estimating Power Dissipation through Thermal Measurements in Power Circuits

A. Fodor, G. Chindris
Applied Electronics Department,
Technical University of Cluj-Napoca
Cluj-Napoca, Romania
Alexandra.Fodor@ael.utcluj.ro

Abstract— **In systems working in strongly distorted regimes, electrical measurements are insufficient for determining the total power dissipation. The study proposes an alternative method for estimating power dissipation in an electronic system in which thermal measurements are correlated with thermal modelling and simulations for a power circuit, to estimate power dissipation in the assembly. It is shown that modelling the component packages and measuring the temperature rise in a power system, the total power dissipation can be estimated.**

Keywords— *thermal modelling; power dissipation; power circuit.*

I. INTRODUCTION

Motion control is a key factor in automation. There is an increasing demand in reliable and efficient motion control devices, for applications such as electric vehicles [1], machine tools, handheld devices, pumps, or even aerospace [2]. In such systems, temperature and power consumption have become key parameters in designing modern, high power, high performance systems. Simulation tools are indispensable in the design of coherent power and thermal models, that need to be validated. This paper proposes a thermal study on a motor controller board that incorporates the high-power MOSFET drivers and the current sense resistor, for which power consumption is estimated via thermal measurements. The thermal study is done in stages – firstly thermal models are developed for the high-power components, secondly, they are validated with thermal measurements done with an infrared camera and thirdly, the measurement results are included in the simulations, to have a complete thermal and power consumption model.

The paper is structured as follows: chapter II presents the state-of-the-art results related to estimating power consumption through thermal measurements, chapter III shows the thermal measurements using an IR camera, followed by the thermal simulation setup and models in which the measurement data was entered. Chapter IV covers the thermal analysis conclusions and future work.

II. RELATED WORK

Several studies can be found in the literature that characterize power estimation via thermal measurements. In [3], a setup is presented in which power consumption is estimated through thermal measurements, done directly on the silicon die. The setup provides a strategy to obtain the leakage and dynamic components of the power dissipation of modern processors,

based on thermal measurements. To increase the quality of the thermal image, coolant and oil are added to the exposed die, so most heated parts are isolated and thermal resistances lowered. In [4], the thermal load is measured for an IGBT module, more focused on the lifetime estimation, based on the thermal cycling and corresponding strength models of power devices. Thermal imagery is used to measure the case temperature of the opened IGBT module, in different operation scenarios. The measurements are used to obtain thermal profiles to validate the simulation models. At a larger scale, in [5], thermography is applied for monitoring the losses in power transformers. The temperature distribution on the surface of a single device was examined, the images being used to characterize the losses throughout the system.

III. THERMAL ANALYSIS AND MODELLING

A. Initial thermal measurements

As stated before, the paper proposes estimating the power dissipation in a motor control system that uses power MOSFETs to drive the motor. In order to validate the test setup, system has a known power dissipation of 44.81 W. The physical module is shown in Fig. 1.

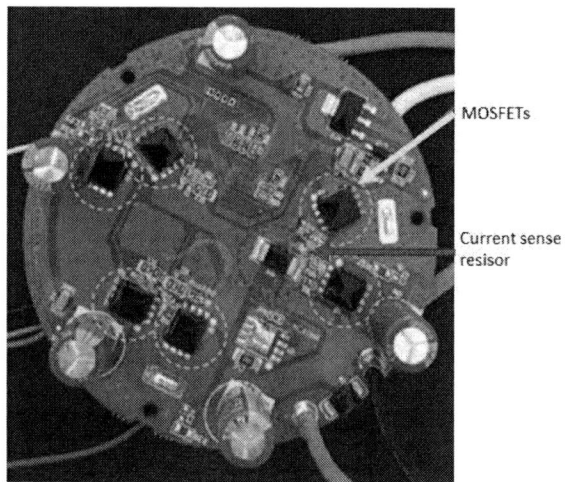

Fig. 1. Bottom side of the driver PCB, with power components exposed only to passive cooling.

In order to characterize the thermal profile of losses, a thermal imaging camera was used for the following test setup (Fig. 2):

- The PCB of the driver circuit (driver microcontroller, transistor bridge and all related electronic components) was separated from the motor assembly.
- The only cooling method the PCB was exposed to is passive cooling.
- The motor driver was controlled with 80% control factor. (percentage of full operating mode).
- Initial temperature was 20°C and the motor was running continuously until thermal shutdown was activated.

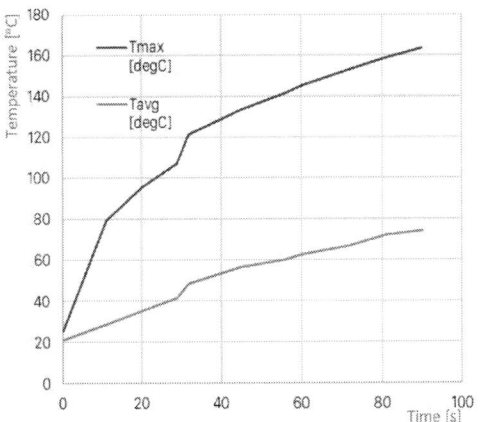

Fig. 2. Initial step of the experiment at 20°C.

The test shows that after 90 seconds of continuous operation, at 80% of the maximum power, in continuous operation with passive cooling, the systems enters the thermal shutdown state due to the high temperature rise (up to approximately 160°C for the current sense resistor) as presented in the next images.

Fig. 3. Final step of the experiment just before the thermal shutdown, with measurement points for each component.

Fig. 4. Temperature rise for PCB during the experiment.

B. Modelling power losses

A general method for computing the thermal behaviour of various geometries [6], including electronic components, is the use of FEM (Finite Element Method). The model of the driver was developed using SolidWorks, where each power component was modelled by its own electronic package, since the package carries most of the thermal characteristics, their material structure being presented in Fig. 5 and Fig. 6.

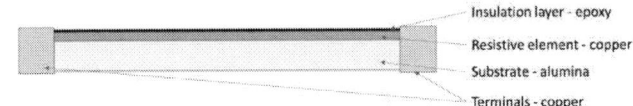

Fig. 5. Thermal model of the current sense resistor.

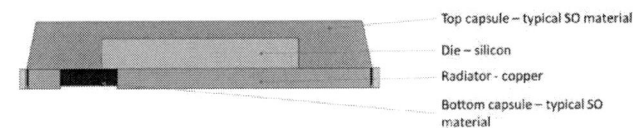

Fig. 6. Thermal model of the MOSFET driver.

The PCB was modelled as a homogenous FR4 board, with top and bottom copper planes, fitted with six power MOSFET drivers, one current sense resistor and a voltage regulator, as seen in Fig. 7 (top view) and 8 (side view).

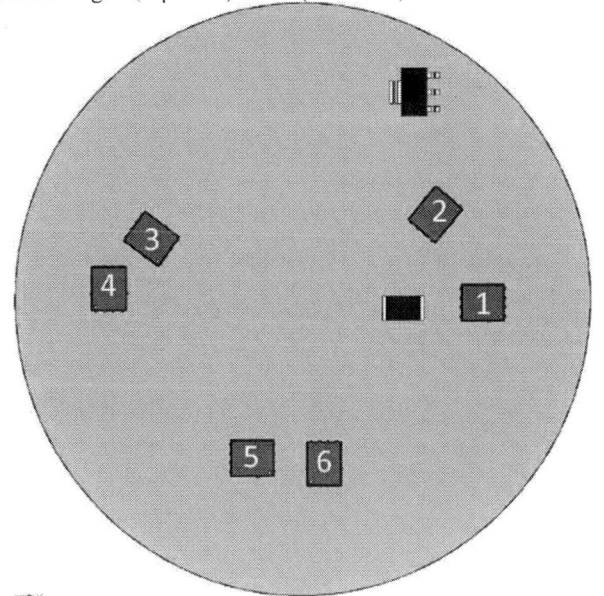

Fig. 7. Thermal model of the motor driver system. Numbered from 1 to 7 are the power MOSFETs.

Regarding the simulation boundary conditions, the only cooling method set in the simulation profile was passive cooling, in an ambient temperature of 20°C.

2020 IEEE 26th International Symposium for Design and Technology in Electronic Packaging (SIITME)

Fig. 8. Thermal model of the motor driver system- side view.

The data from the thermal measurements presented above is used as input parameters for the power distribution of the system. Thus, for each component a thermal sweep analysis was carried out: for the same time duration (90 seconds) the thermal power was swept from 0W to 40W, in order to obtain the same temperature rise of the majorly heat dissipating components. From the obtained sweep analysis results, the power values producing the same temperature rise as thermal imaging results were kept.

C. Thermal analysis results

The parametric sweep is presented for the current sense resistor. Similar sweeps were performed also for each power MOSFET. From the graph in Fig. 9, values for the power dissipation on the current sense resistor can be extracted and used as the input data for analyzing the temperature rise in the whole assembly.

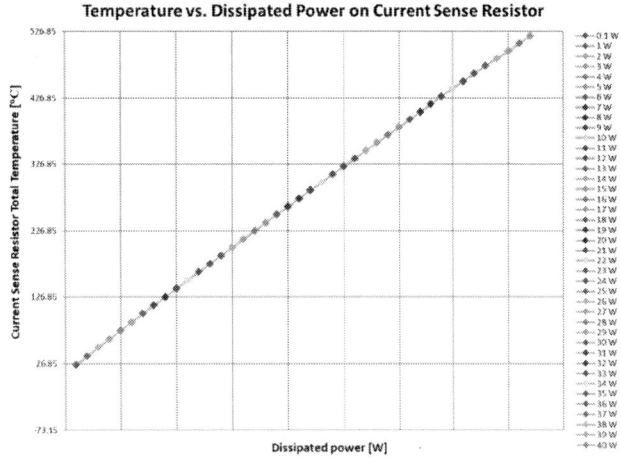

Fig. 9. Thermal sweep on current sense resistor showing the temperature-rise depending on the dissipated power.

From the sweep analysis the following results were obtained: 8W for the current sense resistor and 5.8W for the transistors resulting a total of 42.8W dissipated in the system. Setting the obtained power dissipation from the parametric sweep as input data for a thermal steady-state analysis, all components had the temperature rise in Fig. 9 (graph) and Fig 10 (contour plot).

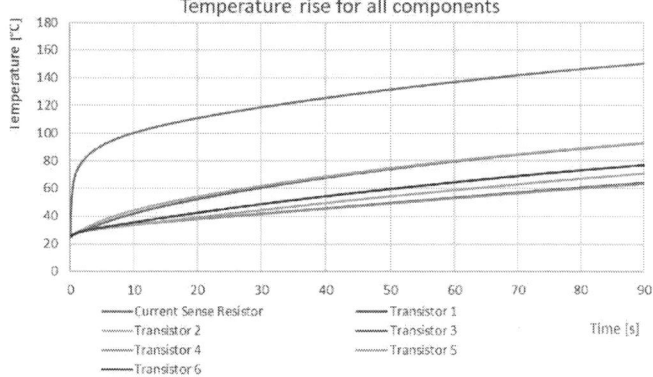

Fig. 10. Temperature-rise produced by 8W dissipated on the current sense resistor and 5.8W on the power MOSFETs.

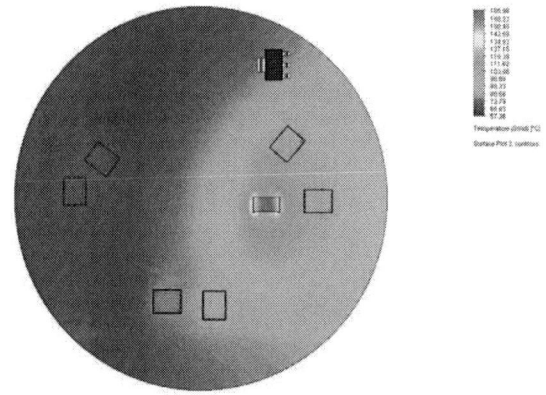

Fig. 11. Contour plot for temperature distribution of the power in the analyzed system, in case 8W were dissipated on the current sense resistor and 5.8W for the power MOSFETs.

From the first set of results, the temperature rise for current sense resistor matches the thermal measurements done with the IR camera, thus for a second run, the 8W set to be dissipated by the resistor can be kept. The power dissipated by the power MOSFETs can be slightly increased for their temperature rise to be more correct in the simulations. Thus, for the second run, a power dissipation of 6.1W was set on each transistor.

It is also seen that connecting all components with one copper plane eases the temperature transfer between one component and the others, the transistors in the immediate proximity of the current sense resistor have a higher temperature rise than the others, even if they are dissipating the same amount of power. By modelling more accurately the top and bottom layers, with correct interconnections and with thermal vias for the high-power components, a more accurate estimation of the total power dissipation for each component can be obtained.

As a comparison, the simulation results in Fig. 12 shows approximately 6.1W for the power MOSFETs and 8W for current sense resistor, at the end of the 90s simulation time.

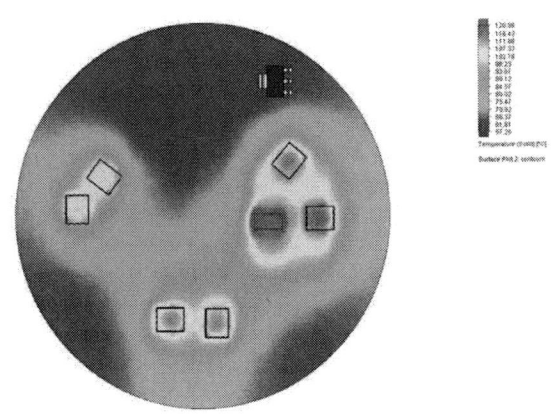

Fig. 12. Contour plot for temperature distribution of the power in the analyzed system, in case 8W were dissipated on the current sense resistor and 6.1W for the power MOSFETs.

IV. CONCLUSIONS AND REMARKS

The scope of the simulations was to obtain the thermal power dissipation values by matching experimental and simulation results. Only thermal relevant components were modelled. However, all the other passive and active components can act, at least, as cooling devices, if their thermal contribution is not known. A more detailed PCB modelling, containing traces, holes and copper planes can change the simulation results, as the current model acts as a better passive radiator than the actual PCB. The obtained results in the second run (44.6W) are consistent with the measured values (44.81W) for the entire system, validating the proposed thermal model.

REFERENCES

[1] M. F. Bhuiyan, N. Sakib, M. R. Uddin and K. M. Salim, "Experimental Results of a locally developed BLDC Motor Controller for electric tricycle," 2019 1st International Conference on Advances in Science, Engineering and Robotics Technology (ICASERT), Dhaka, Bangladesh, 2019, pp. 1-4, doi: 10.1109/ICASERT.2019.8934491.

[2] V. S. Veena and R. P. Praveen, "Mathematical modeling of advanced PMBLDC motor drive for aerospace application," 2014 Annual International Conference on Emerging Research Areas: Magnetics, Machines and Drives (AICERA/iCMMD), Kottayam, 2014, pp. 1-5, doi: 10.1109/AICERA.2014.6908267.

[3] F. Javier Mesa-Martinez, J. Nayfach-Battilana, J. Renau, Power model validation through thermal measurements, 34th International Symposium on Computer Architecture (ISCA 2007), June 9-13, 2007, San Diego, California, USA, DOI: 10.1145/1273440.1250700

[4] K. Ma, M. Liserre, F. Blaabjerg and T. Kerekes, "Thermal Loading and Lifetime Estimation for Power Device Considering Mission Profiles in Wind Power Converter," in *IEEE Transactions on Power Electronics*, vol. 30, no. 2, pp. 590-602, Feb. 2015, doi: 10.1109/TPEL.2014.2312335.

[5] Lisowska, A. Thermographic monitoring of the power transformers.Meas. Autom. Monit.2017,63, 154–157

[6] Guo, Yong-feng et al. "Power Losses and Temperature Variations in a Power Converter for an Electronic Power Steering System Considering Steering Profiles." (2014).

A Comparison between State of Charge Estimation Methods: Extended Kalman Filter and Unscented Kalman Filter

Adelina Ioana Ilieş, Gabriel Chindriş and Dan Pitică
Applied Electronics Department, Technical University of Cluj-Napoca,
Cluj-Napoca, Romania
Adelina.Ilies@ael.utcluj.ro

Abstract—The Battery Management System (BMS) plays an essential role in the optimal and safe operation of a battery. One task performed by the BMS is the battery parameters monitoring. State of charge is a critical parameter that indicates the amount of charge contained in a battery. An accurate estimation of the state of charge of the battery is important not only for informing the user but also in establishing a control strategy for keeping the battery parameters within the safe limits in order to maximize its lifespan. In this paper, a comparison in terms of performance between two variations of the Kalman filter (the Extended Kalman filter and the Unscented Kalman filter) for state of charge estimation is presented.

Keywords—State-of-Charge, Lithium-Ion Battery, Extended Kalman Filter, Unscented Kalman Filter.

I. INTRODUCTION

Even if the development of new and better battery chemistry is in full swing, aiming to propose a more efficient and safe battery, the Lithium-ion batteries are still enjoying a great success being used in more and more applications like energy storage systems, portable devices or electric vehicles. Optimal battery monitoring and control is handled by the battery management system. State of charge (SOC) is a critical parameter that characterizes the amount of charge stored in the battery relative to its total capacity. The SOC cannot be directly measured and must be estimated using other parameter of the battery. Knowing the SOC of a battery as real as possible is vital not only to inform the user that the device/system needs to be recharged and to avoid unpredictable interrupts of the system caused by a discharged battery but also to avoid overcharging or over-discharging of the battery, which can permanently damage the battery's internal structure. Therefore, estimation of state of charge is an extremely researched subject [1]. Various estimation methods have been developed in the past years and they can be categorized into four groups: estimation of SOC by direct measurements (open circuit voltage method, internal resistance method, terminal voltage method), book-keeping methods (Coulomb counting), methods based on adaptive algorithms (Kalman filter, Neural network, Particle filter, Fuzzy logic) and combined methods.

One of the most used and simple methods for estimating the SOC of a battery is the Coulomb counting method. It is based on current measurements and current integration and therefore, in order to have a precise information about the state of charge, the current measurement must be precise and the initial value of the

SOC must be well estimated. A more accurate estimation can be done by using a Kalman filter algorithm. This is a model-based method that provides real-time predictions by using an error correction mechanism. This method needs an accurate model and identification of parameters for the battery. The Extended Kalman filter and the Unscented Kalman filter are extensions of the basic Kalman filter that can be used for nonlinear systems [2].

To determine and compare the performances of the two estimation methods, a battery model was used in this paper, based on already available models found in literature [3]. The battery model parameters (open-circuit voltage, terminal resistance, RC pairs) are estimated based on experimental measurements applied on the same type of battery for which the estimation is performed. The estimation of the battery SOC is made by using the Extended Kalman filter and Unscented Kalman filter algorithms, by comparison, determining also the errors compared to the real SOC. Different charge and discharge scenarios are applied to the battery to observe the behavior and performance of the estimations in various conditions.

II. THEORETICAL BACKGROUND

The state of charge of a battery represents the energy quantity available at a certain moment of time with respect to the total amount of energy. State of charge cannot be directly measured so it must be estimated by means of other measured parameters of a battery. In order for the battery, or the battery-powered device, to operate at full capacity, and for the battery itself not to be damaged due to improper practices caused by erroneous data about the battery, an accurate estimation of the state of charge is essential.

The SOC indication require battery measurements and modelling. For example, the battery voltage (V) can be measured and the relationship between voltage and SOC is stored in a look-up table function in a microcontroller. The size and accuracy of the look-up tables depend on the number of stored values, i.e. the number of stored V-SOC data points [4].

Battery models are often used in a Battery Management System to indirectly estimate and monitor battery's functioning by voltage, current and temperature measurements. When it comes to characterizing batteries by modeling, a balance between model accuracy and complexity is essential, so that it can be easily integrated into a simple and inexpensive processor, similar to those found in electric car BMS hardware. The battery models presented in the current literature can be split into the following three categories: electrochemical models, empirical or data-based models and equivalent circuit models [5].

A. Kalman Filter method

Kalman filter (KF) method is used to estimate a system state, when it cannot be directly measured. KF can be applied by viewing each battery cell as a dynamic system whose inputs are the current and temperature and the output is the terminal voltage.

The algorithm consists of two stages: prediction and update. In the first stage, the prediction, the algorithm produces a priori estimation of the state variable along with the error covariance. In the update stage, new estimations are made based on current measurements along with the priori estimation and error covariance.

KF advantage is that it eliminates the error of the Ah integral method which can be accumulated over time. At the same time, it does not have a high requirement for the accuracy of the initial SOC, so even if its initial value has a certain deviation it can well converge to the real value [1].

B. Extended Kalman Filter (EKF)

Recently, EKF has attracted more and more attention and has become one of the most widely used methods of estimating battery SOC even when the initial SOC is unknown. KF is the optimal state estimator for linear systems. If the system is nonlinear, a linearization process can be used at every time step to approximate the nonlinear system with a linear time varying system which is then utilized in the KF, resulting an EKF on the real, nonlinear system [6].

C. Unscented Kalman Filter (UKF)

Unscented Kalman Filter deals with non-Gaussian noises as well and the Jacobian matrix is not calculated, it makes the method more suitable for SOC estimation [7]. The unscented Kalman filter uses a deterministic sampling technique known as the unscented transformation (UT) to pick a minimal set of sample points (called sigma points) around the mean. The sigma points are then propagated through the nonlinear functions, from which a new mean and covariance estimate are then formed.

III. METHODOLOGY

The battery model used for this research is an equivalent circuit model adapted from [3]. The battery model parameters (Em, R0, R1, τ1, R2, τ2, R3, τ3) are estimated based on measurement data used from [8]-[10], using Simulink Design Optimization which will tune the model parameters to meet the required measured data. It consists of current and voltage measurements, representing a series of discharge pulses at constant current followed by longer periods of rest. The battery that was used is a Lithium-ion cell, INR 18650-20R, which has a 2000mAh capacity rating.

First, a parameter estimation was performed by considering three time constants (three RC pairs that simulate the dynamic behavior of the battery cell). The model's estimated parameters are then loaded into the model used for the SOC estimation application, together with the physical characteristics of the battery cell (cell wight, length, thermal mass etc.). The result of

the parametrization is a vector of equivalent circuit parameters with different values for different levels of SOC.

For the SOC estimation experiments performed in this paper, different charge / discharge scenarios, with different current pulses, are applied to the battery model to observe and compare the behavior and efficiency of the estimations.

The used estimation techniques are the Extended Kalman filter and the Unscented Kalman filter. For each experiment, different initial SOC (SOC0) are considered in the estimation process, in comparison, to check the performances of the two methods in converging to the SOC real value. Five charge-discharge scenarios were taken into consideration and are described in Table 1. Waveforms of applied current sequences are shown in Fig. 1, Fig. 2, Fig. 3, Fig. 4 and Fig. 5. For all experiments, a 20°C temperature is considered.

TABLE I.   CHARGE-DISCHARGE SCENARIOS DESCRIPTION

Scenario no.	Charge	Discharge
1	1A (standard charge)	0.4 A (used for nominal discharge capacity test)
2	2A	2A
3	4A (rapid charge)	22A (max. discharge)
4	1A (standard charge)	1A (pulsed discharge – 60% duty cycle)
5	1A (standard charge)	0.4 – 4A (random pulsed discharge – 60% duty cycle)

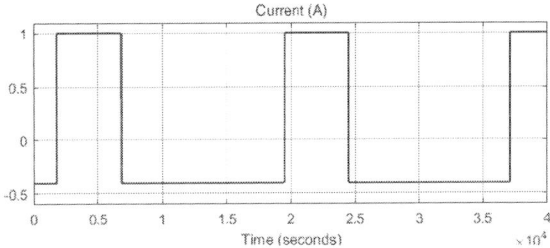

Fig. 1. Applied current sequence (1st scenario)

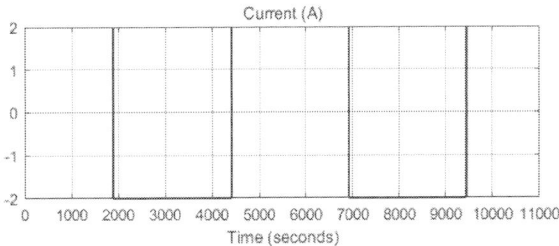

Fig. 2. Applied current sequence (2nd scenario)

2020 IEEE 26th International Symposium for Design and Technology in Electronic Packaging (SIITME)

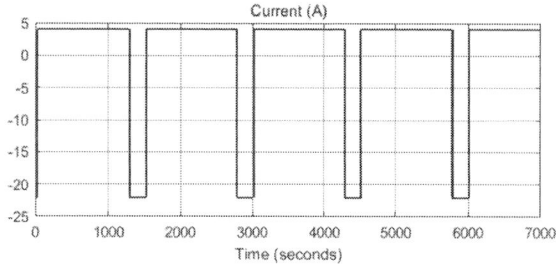

Fig. 3. Applied current sequence (3rd scenario)

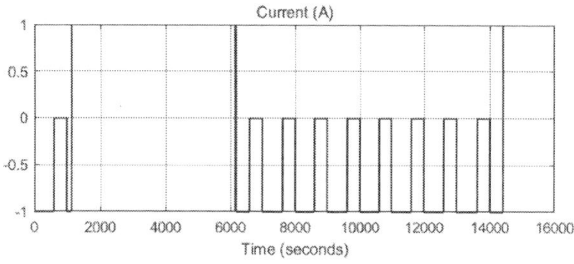

Fig. 4. Applied current sequence (4th scenario)

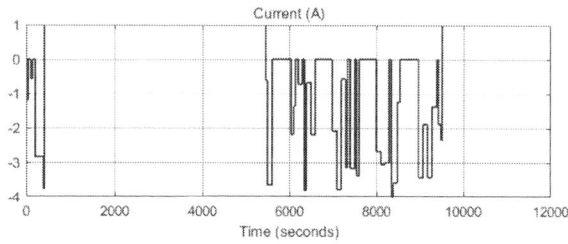

Fig. 5. Applied current sequence (5th scenario)

IV. EXPERIMENTAL RESULTS

For each method and each test scenario, graphs representing the estimated SOC and the real SOC are presented, as well as the evolution of the estimation error over time. The obtained results are presented below.

A. Extended Kalman Filter method

1) Scenario no. 1

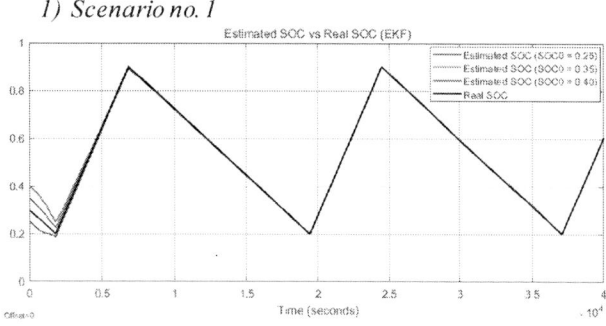

Fig. 6. Estimated SOC vs Real SOC (EKF method) considering three different initial states for the SOC and the 1st charge-discharge scenario

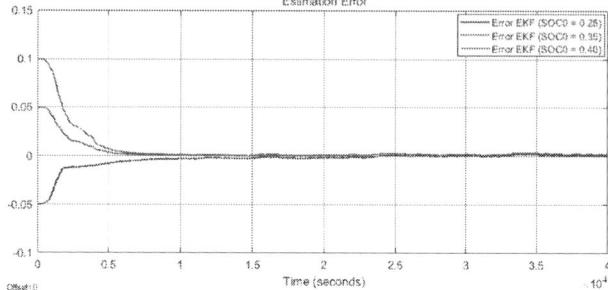

Fig. 7. Estimation error (EKF method) for the 1st charge-discharge scenario

2) Scenario no. 2

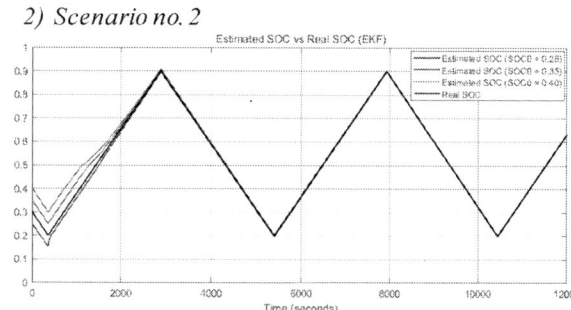

Fig. 8. Estimated SOC vs Real SOC (EKF method) considering three different initial states for the SOC and the 2nd charge-discharge scenario

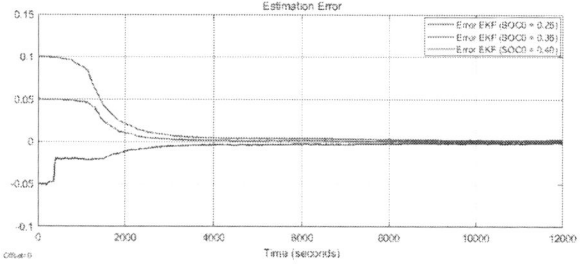

Fig. 9. Estimation error (EKF method) for the 2nd charge-discharge scenario

3) Scenario no. 3

In this scenario, it can be observed that the errors have some small spikes during the discharge periods, due to the very high discharge current.

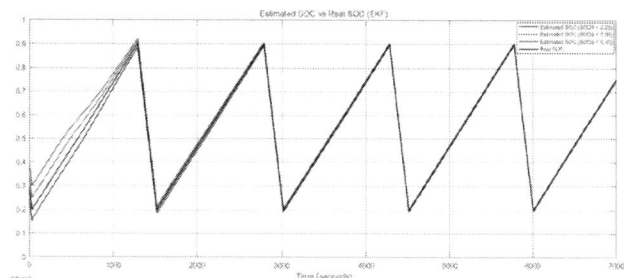

Fig. 10. Estimated SOC vs Real SOC (EKF method) considering three different initial states for the SOC and the 3rd charge-discharge scenario

978-1-7281-7507-2/20 $31.00 © 2020 IEEE   378   21-24 October, Pitesti, Romania

2020 IEEE 26th International Symposium for Design and Technology in Electronic Packaging (SIITME)

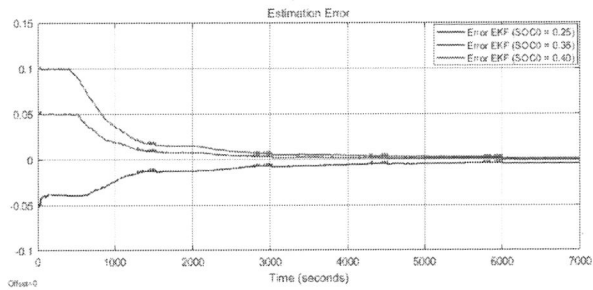

Fig. 11. Estimation error (EKF method) for the 3rd charge-discharge scenario

4) Scenario no. 4

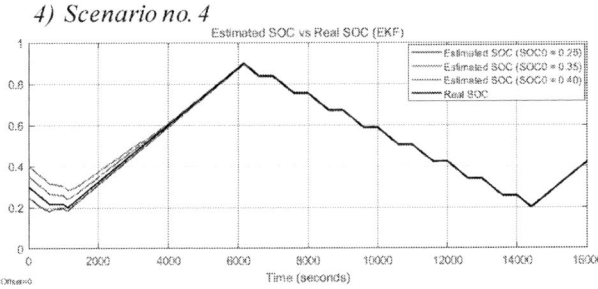

Fig. 12. Estimated SOC vs Real SOC (EKF method) considering three different initial states for the SOC and the 4th charge-discharge scenario

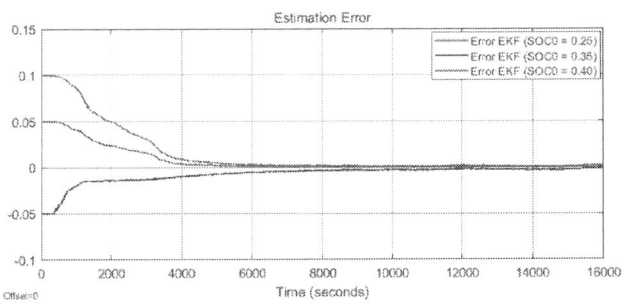

Fig. 13. Estimation error (EKF method) for the 4th charge-discharge scenario

5) Scenario no. 5

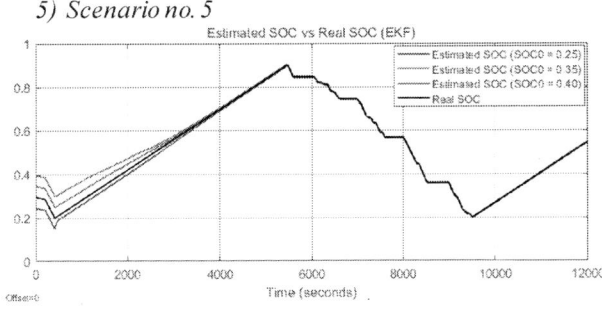

Fig. 14. Estimated SOC vs Real SOC (EKF method) considering three different initial states for the SOC and the 5th charge-discharge scenario

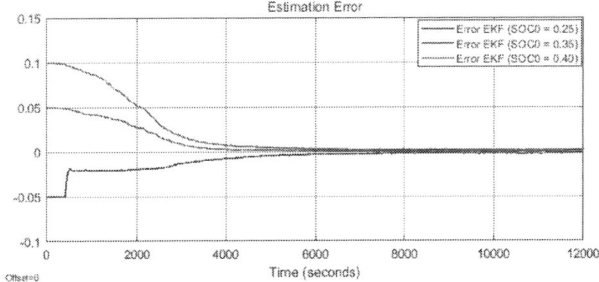

Fig. 15. Estimation error (EKF method) for the 5th charge-discharge scenario

B. Unscented Kalman Filter method

1) Scenario no. 1

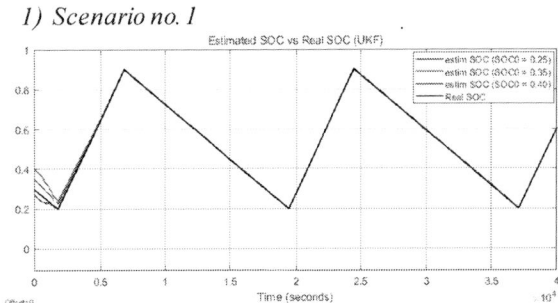

Fig. 16. Estimated SOC vs Real SOC (UKF method) considering three different initial states for the SOC and the 1st charge-discharge scenario

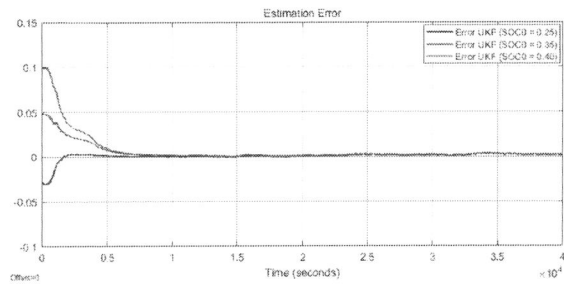

Fig. 17. Estimation error (UKF method) for the 1st charge-discharge scenario

2) Scenario no. 2

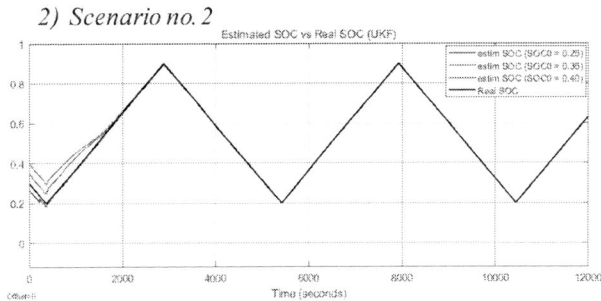

Fig. 18. Estimated SOC vs Real SOC (UKF method) considering three different initial states for the SOC and the 2nd charge-discharge scenario

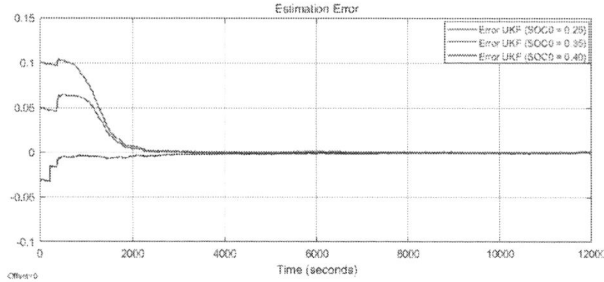

Fig. 19. Estimation error (UKF method) for the 2nd charge-discharge scenario

3) Scenario no. 3

In this scenario, as in the EKF method, it can be observed that the error values have some small spikes during the discharge periods, due to the very high discharge current.

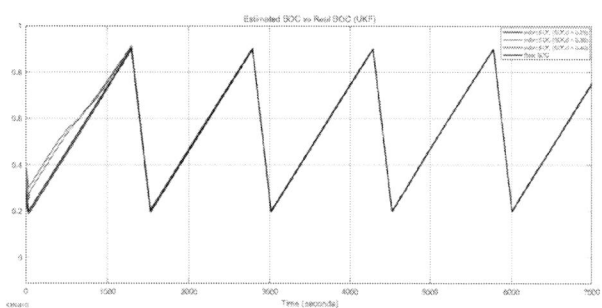

Fig. 20. Estimated SOC vs Real SOC (UKF method) considering three different initial states for the SOC and the 3rd charge-discharge scenario

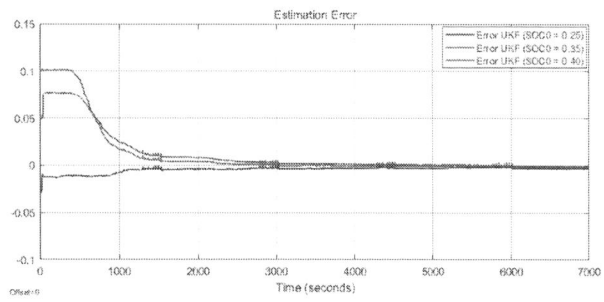

Fig. 21. Estimation error (UKF method) for the 3rd charge-discharge scenario

4) Scenario no. 4

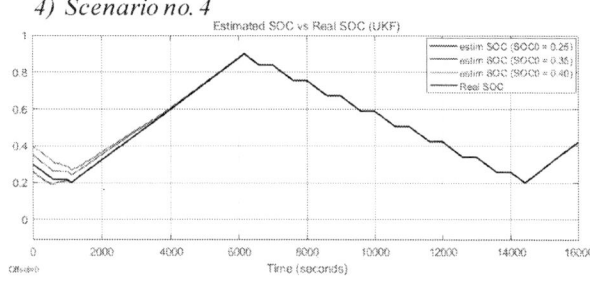

Fig. 22. Estimated SOC vs Real SOC (UKF method) considering three different initial states for the SOC and the 4th charge-discharge scenario

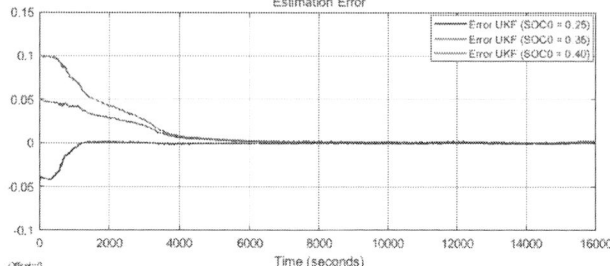

Fig. 23. Estimation error (UKF method) for the 4th charge-discharge scenario

5) Scenario no. 5

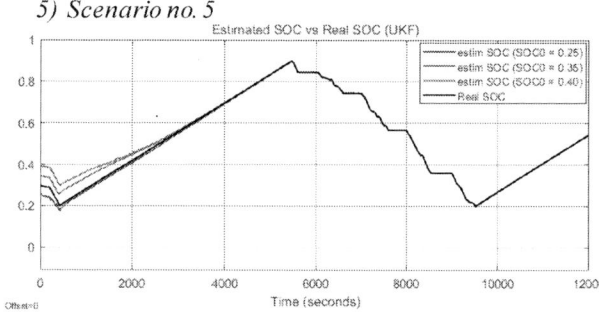

Fig. 24. Estimated SOC vs Real SOC (UKF method) considering three different initial states for the SOC and the 5th charge-discharge scenario

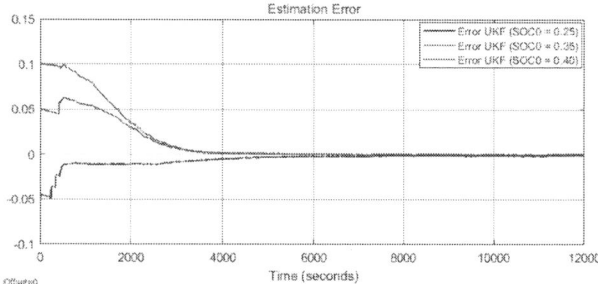

Fig. 25. Estimation error (UKF method) for the 5th charge-discharge scenario

V. CONCLUSIONS

Accurate SOC estimation is a key element for a Battery Management System. When thinking about electric vehicles, for example, one must know how long it will take until they must recharge the vehicle's battery. Unlike a conventional car's fuel tank, which can be measured, an electric vehicle battery state of charge can only be determined indirectly.

When we use indirect methods for SOC estimation, a battery model is usually used. In order to have high accuracy estimation, the model's behavior must resemble as close as possible to that of the real battery. In this research, an equivalent circuit model is used to model a 2000mAh lithium-ion battery. The model's parameters are estimated based on experimental data and using the Simulink Design Optimization. The obtained parameters and then loaded into the model used in the SOC estimation application.

For the SOC estimation process, two methods are used and compared: Extended Kalman filter method and Unscented Kalman filter method. The performances of each estimation technique are compared based on different charge-discharge scenarios (using different current sequences) and considering different initial SOC values. For each method, the estimated SOC waveforms (considering three different initial SOC values) together with the real SOC waveforms and the errors for each estimation are plotted.

From the performed experiments can be observed that both methods, being adaptive methods, manage to bring, in time, the estimated SOC value to match the real SOC value, even if the initial SOC is not correctly estimated. The Extended Kalman filter technique and the Unscented Kalman filter technique are compared to each other in terms of estimation performances.

Summarizing the obtained results considering both methods and all five experimental scenarios, the convergence time is determined for each particular case and the results are presented in Fig. 26 and Fig. 27. By comparing the figures, it is noticeable that the Unscented Kalman filter has higher performances because the estimated value converges faster in all the studied cases.

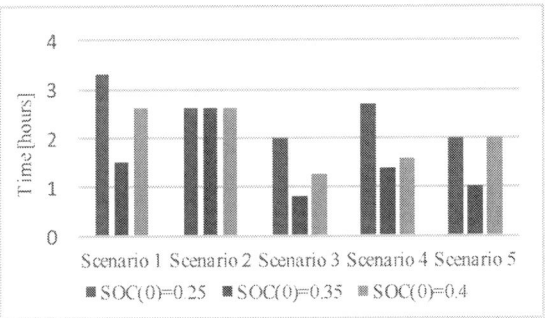

Fig. 26. Correspondence between initial SOC and convergence time
(Extended Kalman filter)

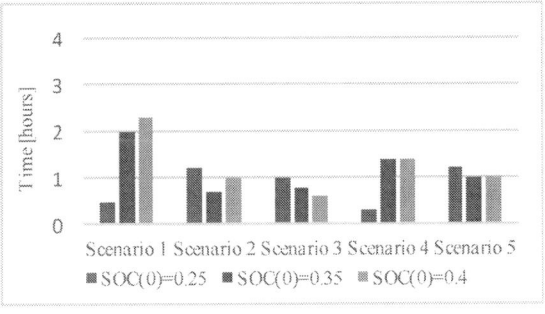

Fig. 27. Correspondence between initial SOC and convergence time
(Unscented Kalman filter)

ACKNOWLEDGMENT

This paper was supported by the Project "Entrepreneurial competences and excellence research in doctoral and postdoctoral programs - ANTREDOC", project co-funded by the European Social Fund.

REFERENCES

[1] Zhang, M.; Fan, X., "Review on the State of Charge Estimation Methods for Electric Vehicle Battery", World Electr. Veh. J. 2020, 11, 23.

[2] A.M.S.M.H.S.Attanayaka, J.P.Karunadasa, K.T.M.U.Hemapala, "Estimation of state of charge for lithium-ion batteries - A Review", AIMS Energy, 2019, 7(2): 186-210.

[3] Javier Gazzarri (2020). "Battery Modeling" (https://www.mathworks.com/matlabcentral/fileexchange/36019-battery-modeling), MATLAB Central File Exchange. Retrieved July 15, 2020.

[4] V. B. H. D. D. R. P. N. P. Pop, "Battery Management Systems. Accurate State-of-Charge Indication for Battery-Powered Applications", Springer, 2008.

[5] D. G. D. S. S. Nejad, "A systematic review of lumped-parameter equivalent circuit models for real-time estimation of lithium-ion battery states", Journal of Power Sources, vol. 316, pp. 183-196, 2016.

[6] N. M. H. O. S. Juan Pablo Rivera-Barrera, "SoC Estimation for Lithium-ion Batteries: Review and Future Challenges", Electronics, vol. 6, 2017.

[7] M. G. C. W. e. a. Zhiwei He, "Adaptive State of Charge Estimation for Li-Ion Batteries Based on an Unscented Kalman Filter with an Enhanced Battery Model", Energies , vol. 6, no. 8, pp. 4134-4151, 2013.

[8] Fangdan Zheng, Yinjiao Xing, Jiuchun Jiang, Bingxiang Sun, Jonghoon Kim, Michael Pecht, "Influence of different open circuit voltage tests on state of charge online estimation for lithium-ion batteries", Applied Energy, 183, pp.513–525, 2016.

[9] Yinjiao Xing, Wei He, Michael Pecht and Kwok Leung Tsui, "State of Charge Estimation of Lithium-Ion Batteries Using the Open-Circuit Voltage at Various Ambient Temperature", Applied Energy, 113, pp.106-115, 2014.

[10] Wei He, Nicholas Williard, Chaochao Chen, Michael Pech, "State of Charge Estimation for Li-Ion Batteries Using Neural Network Modeling and Unscented Kalman Filter-based Error Cancellation", International Journal of Electrical Power & Energy Systems, 62, pp.783-791, 2014.

2020 IEEE 26th International Symposium for Design and Technology in Electronic Packaging (SIITME)

Cooling Techniques for M.2 to PCI(e) Adapters

Rajmond Jánó, Alexandra Fodor

Applied Electronics Department
Technical University of Cluj-Napoca
Cluj-Napoca, Romania
Rajmond.Jano@ael.utcluj.ro

Abstract—**This paper studies different cooling techniques used in the design and manufacturing process of PCI and PCIe adapter cards used to convert and install NVMe M.2 solid state drives on legacy motherboards. Since modern, high-speed solid-state drives dissipate well over 5 W during high workloads, the design of these adapters needs to take into account thermal considerations. Several cooling techniques built into the PCB are simulated and the most efficient one is determined for best overall performance**

Keywords—SSD; cooling; thermal; simulation; M.2, PCIe, adapter

I. INTRODUCTION

Solid state drives (SSDs) are becoming ever more popular for use both in desktop as well as in mobile computing systems, as they offer higher significant transfer speeds in smaller form factors compared to classical hard disk drives (HDDs). However, SSDs themselves come in various form factors, such as the classical 2.5-inch, but also the mSATA and the 2280 M.2 variants.

Fig. 1. SSD form factors

The manufacturing of these various form factors means that compatibility between a desired SSD to be installed in a given system and the motherboard used for the same system is not always ensured. Therefore, there are several adapter card designs which are able to convert the interface of the SSD to expansion slots available on the motherboard. Such is the case of M.2 to PCI or PCIe adapters, which can ensure the insertion of an M.2 SSD into any PCI(e) slot available on the motherboard while also maintaining transfer speeds.

II. COOLING SSDs

A. Necessity

The first question that needs to be answered is whether keeping SSDs cool is a real necessity. This topic is highly

debated. It is true that some SSDs can dissipate up to 10W and as a result they do heat up, however it needs to be determined if this heating has any effect on the health, lifetime and data retention of the SSD drive itself.

Most commonly an SSD drive is comprised of the following components: NAND flash for data storage, a DRAM cache for fast data access and indexing, a memory controller responsible for managing the read/write and store operations and additional passive components such as resistors and capacitors. It is therefore important to be determined how these components are affected by high temperatures [2][3].

TABLE I.   NAND FLASH AGING ACCELERATION FACTOR RELATIVE TO 55°C [1]

Operating temperature	Acceleration factor relative to 55°C
125°C	939
85°C	26
70°C	5
55°C	1
25°C	0.01988

NAND flash has two main characteristics that are affected by temperature. Thermal stress can cause logical errors in the device, such as charge de-trapping and bit-flips, corrupting the data, while operating at high temperatures for prolonged periods also accelerates the aging of the device [1].

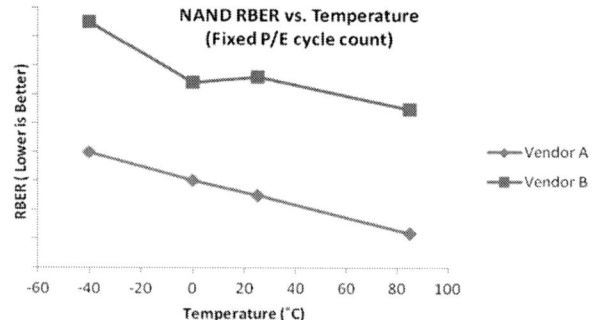

Fig. 2. Raw bit error rate (RBER) vs. temperature in NAND flash chips [1]

As seen in Table 1, the aging of NAND flash is exponentially accelerated with the increase in temperature of over 55°C. On

978-1-7281-7507-2/20 $31.00 © 2020 IEEE

the other hand, data error rate typically drops, and data retention rates increase at higher temperatures [4] (as seen in Figure 2). Therefore, we can find an optimum balance between these factors and determine that the recommended operating temperature of the NAND flash chips hovers between 25-55°C.

Another component found in the design of an SSD which is impacted by high temperature conditions is the memory controller. It has been shown in several studies that at high temperatures, controllers can suffer from clock drift, slow down in performance, and if exposed to such temperatures for prolonged periods of time, even be irreparably damaged [5][6]. As a result, the operating temperature of these components should also be kept as close to possible to room temperature.

Fig. 3. M.2 SSD cooler and heat sink

In addition to this, passive components can also be undesirably influenced by excessive heat: resistance values may drift, capacitance values may decrease, and these variations can have a significant impact on the performances of the SSD.

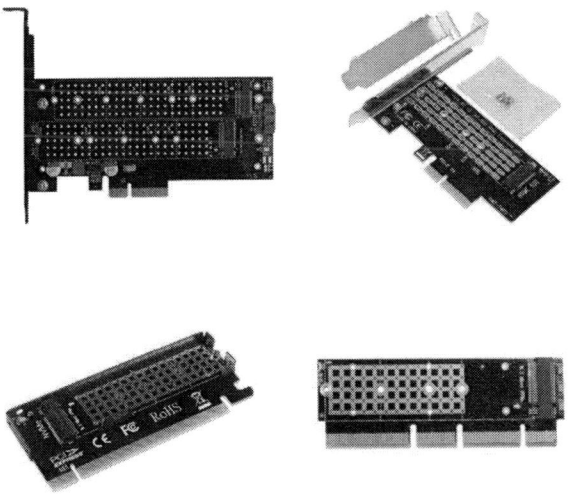

Fig. 4. Different M.2 to PCIe adapters

In consequence, most high-end motherboard manufacturers provide some sort of heat sink or heat dissipating device for M.2 mounted SSDs (Figure 3). However, when using a PCIe adapter card, very often these heat sinks do not fit or cannot be attached as the adapter is missing the mounting mechanism form them.

B. Design considerations

For all the reasons mentioned in the previous section, when designing an M.2 to PCIe adapter, thermal considerations need to be taken into account. This means that the adapter should not only not prevent the efficient heat dissipation from the SSD drive, but is should also in some way contribute to keeping it cool.

Manufactures employ different techniques to achieve this goal. As seen in Figure 4, the most common techniques are either to place metal-plated cooling holes throughout the PCB or run copper cooling traces on the boards, below the SSD. However, it can be observed that the shape of the holes varies between the manufacturers, and so does the thickness and density of the cooling traces.

III. SIMULATION SETUP

The aim of the simulations performed was to determine the optimal cooling methodology for an M.2 to PCIe adapter from the several different configurations. These cooling techniques were analyzed using SolidWorks Flow Simulation, to determine which is the most efficient. First of all, a FEM simulation was run with no cooling at all (Figure 5), with the SSD mounted directly to the PCB. Then, the following cooling setups were used:

- copper thermal tracks between the SSD and PCB (Figure 6),

- round copper plated cooling holes through the PCB (Figure 7),

- square copper plated holes through the PCB (Figure 8),

- oval copper plated holes (Figure 9) and

- oval copper filled holes through the PCB (Figure 10).

The dimensions of the simulated components as well as the materials used are presented in Table.

Component	Dimensions	Material
Adapter PCB	90 mm x 25 mm x 1.60 mm with PCIe keys	FR-4
Solid State Drive (SSD)	80 mm x 23 mm x 2.4mm	Aluminium
Cooling tracks	2 mm wide with 1.75 mm spacing	0.2 mm thich copper
Round holes	2 mm diameter	Plated with 0.2 mm copper
Square holes	2 mm x 2 mm	Plated with 0.2 mm copper
Oval holes	2 mm x 4 mm	Plated with 0.2 mm copper

*2020 IEEE 26th International Symposium for Design and Technology in Electronic Packaging (**SIITME**)*

Fig. 5. M2 to PCIe adapter with no cooling

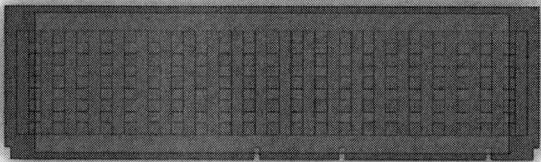

Fig. 6. Copper traces for cooling

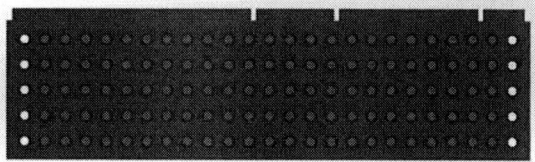

Fig. 7. Copper plated round holes for cooling

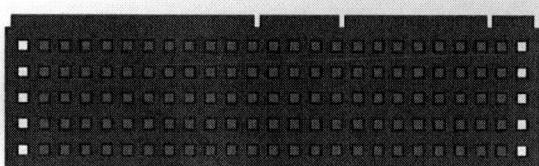

Fig. 8. Copper plated square holes for cooling

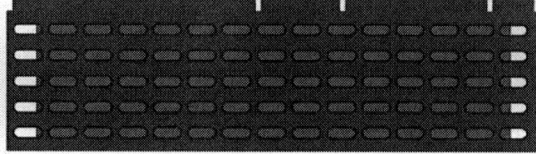

Fig. 9. Copper plated oval holes for cooling

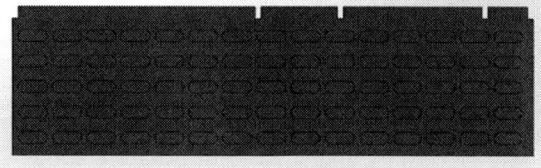

Fig. 10. Copper filled oval holes for cooling

Finally, all assemblies were inserted inside an aluminum case, at 20 mm from the bottom. The case has an air inlet at the bottom, which has a Papst 252N external inlet fan, circulating 25°C air and 25 (in a 5-by-5 grid) square air outlet holes on the back side, each 5.14 mm x 5.14 mm (Figure).

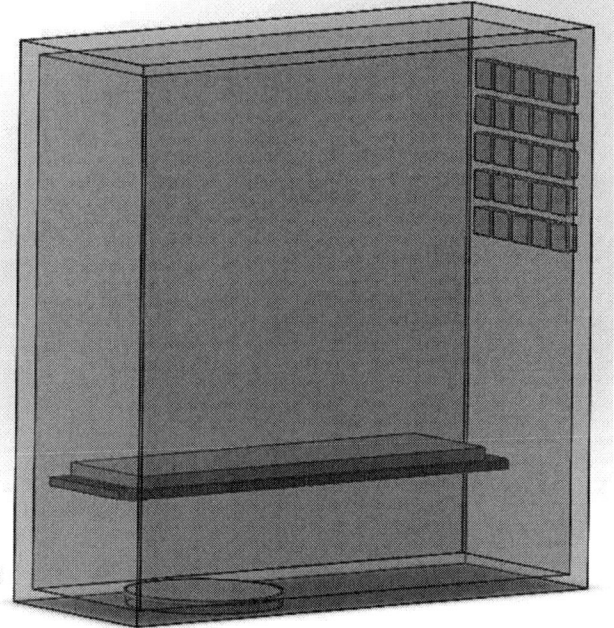

Fig. 11. Simulation arrangement with the M2. to PCIe adapter inside a cooled computer case

Then, the SSD component was set as a volumetric heat genage generetion rate of 5W and the average temperature of the SSD was set as a simulation goal.

IV. SIMULATION RESULTS

The evolution of the temperature of the SSD until convergence (i.e. thermal equilibrium) are presented in Figure 12, while the maximum temperature reached over a period of 1000 seconds are presented in Figure 11.

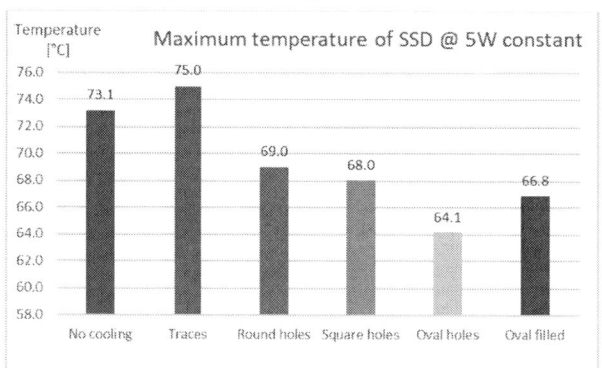

Fig. 12. Maximum SSD temperatures using the cooling methods presented

As it can be seen, with no cooling method implemented on the adapter PCB, the temperature of the SSD reaches 73.1°C when dissipating a sustained 5W. When cooling traces are implemented on the PCB, the situation is made even worse, as the maximum temperature of the SSD reached 75°C. This may be due to the fact that the cooling traces are on the opposite side of the PCB to the side that is directly being cooled by the airflow from the bottom mounted fan. So rather than aiding in heat dissipation, in this case, the traces actually enable heat retention, due to the fact that air (which is a thermal insulator) is trapped in the gaps between which are formed between the SSD and the PCB where no traces are routed.

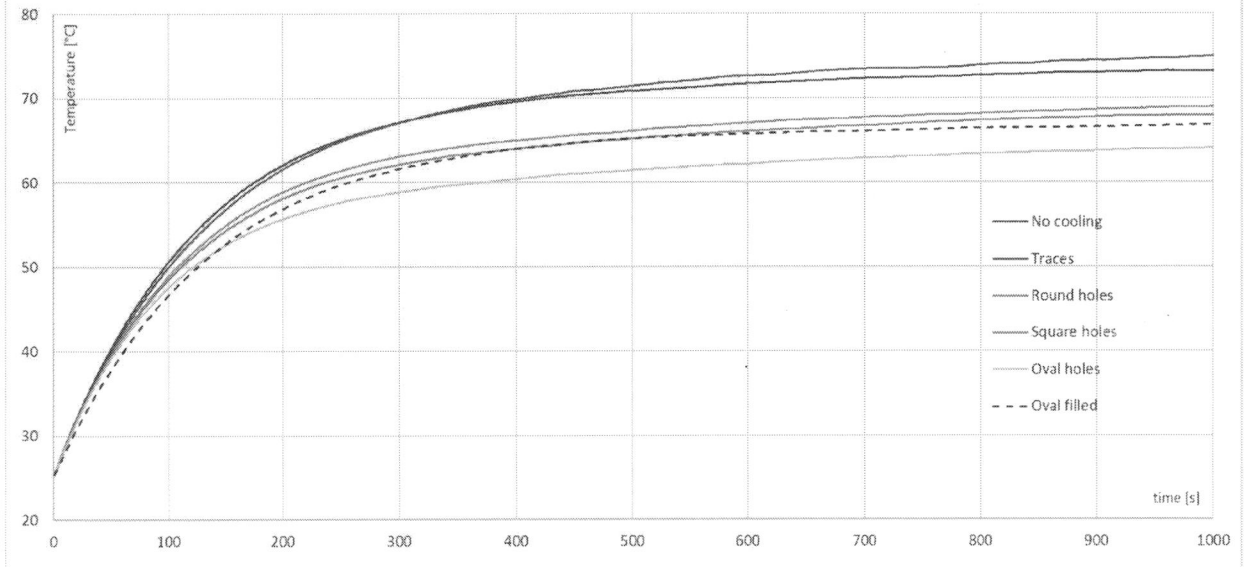

Fig. 13. SSD temperature evolution when using the cooling methods analyzed

The results also show that there is no significant difference between using round holes versus square holes with only 1°C of difference between the two setups, in favor of the square holes. This minor difference can be explained by the slightly higher are exposed to airflow in case of the square holes in comparison to the round holes (Figure 14).

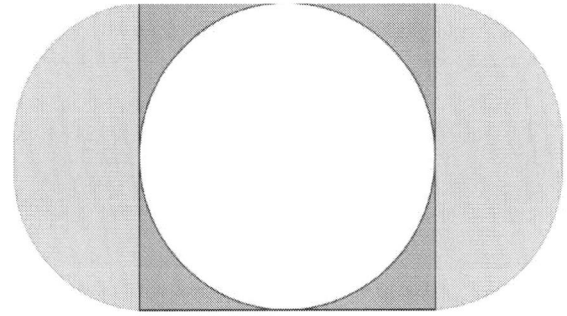

Fig. 14. Cooling area gained (in green) by using square shaped holes versus round holes and are gained (in yellow) by using oval holes as opposed to square holes

The most efficient cooling method is using oval shaped holes. These ensure the largest surface area exposed to airflow and therefore provied the lowest obtained temperatures during the simulations (64.1°C), a gain of 9°C as opposed to no cooling used at all. Taking into account Arrhenius' law regarding component aging, according to which the lifetime of a component halves with every increase of 10°C, a gain of 9°C can result in almost doubling the lifetime of the SSD.

Filling the oval holes with copper did not make heat dissipation better, and therefore leaving the holes directly exposed to the circulating air is the most efficient cooling solution.

V. CONCLUSIONS

In this paper it has been shown that significant gains can be achieved regarding the expected lifetime of an M.2 form factor solid state drive, when mounted on a PCIe adapter, if the adapter is designed with thermal considerations in mind.

It has been shown that adding cooling holes to the PCB of the adapter provides more efficient cooling, however making the holes oval shaped offers the most gains. Therefore, wherever possible, these types of cooling holes should be used when designing M.2 to PCIe adapters.

REFERENCES

[1] EEWeb, "Industrial Temperature and NAND Flash in SSD Products", https://www.eeweb.com/industrial-temperature-and-nand-flash-in-ssd-products/, retrived October 2020

[2] C. Zambelli, R. Micheloni and P. Olivo, "Reliability challenges in 3D NAND Flash memories", 2019 IEEE 11th International Memory Workshop (IMW), Monterey, CA, USA, 2019, pp. 1-4

[3] D. Resnati, A. Goda, G. Nicosia, C. Miccoli, A. S. Spinelli and C. Monzio Compagnoni, "Temperature Effects in NAND Flash Memories: A Comparison Between 2-D and 3-D Arrays", in IEEE Electron Device Letters, vol. 38, no. 4, pp. 461-464, April 2017

[4] C. Zambelli, L. Crippa, R. Micheloni and P. Olivo, "Cross-Temperature Effects of Program and Read Operations in 2D and 3D NAND Flash Memories", 2018 International Integrated Reliability Workshop (IIRW), South Lake Tahoe, CA, USA, 2018, pp. 1-4

[5] A. Fodor, R. Jano and D. Pitica, "Investigations of thermal influences on clock speed of embedded devices", 2017 40th International Spring Seminar on Electronics Technology (ISSE), Sofia, 2017, pp. 1-4

[6] J. F. Dawson, I. D. Flintoft, A. P. Duffy, A. C. Marvin and M. P. Robinson, "Effect of high temperature ageing on electromagnetic emissions from a PIC microcontroller", 2014 International Symposium on Electromagnetic Compatibility, Gothenburg, 2014, pp. 1139-1143

A Generalized Model for Stacked Boost Single-Switch Converters

Septimiu Lica, Vlad Vătău, Dan Lascu, Mircea Tomoroga

Applied Electronics Department
Politehnica University Timișoara
Timișoara, Romania
septimiu.lica@upt.ro, vlad.vatau@student.upt.ro, dan.lascu@upt.ro, mircea.tomoroga@upt.ro

Abstract— **In this paper a generalisation for an *n* stages Boost topology is proposed. The starting point is the two stages stacked boost converter. The dc conversion ratio and a set of properties are theoretically derived. The results are analysed using MATLAB™ and the simulations performed in CASPOC have validated the theoretical considerations. The converter is useful in applications with high step-up dc gain, where it can successfully replace a cascade of several classical Boost converters.**

Keywords— *Stacked converter; DC conversion ratio; duty cycle control; step-up topologies*

I. INTRODUCTION

Step-up converters with three or more stages have been reported in the literature [1]. In [2] the two stages stacked boost converter was presented. In this paper a generalization is presented by adding *n* stages of boost cells, in order to achieve a higher static conversion ratio.

The paper is structured as follows: in section II the two stages stacked boost converter is shortly revised. A three stages stacked boost converter is introduced in section III, while the four stages counterpart is analysed in section IV. The generalised model is presented in section V. The simulation results for validating the theoretical considerations are performed in section VI. section VII is devoted to the conclusions.

II. THE TWO STAGES STACKED BOOST CONVERTER

The stacked DC-DC boost converter with two stages was published in [2] and is depicted in Fig. 1. It consists of an active switch and two passive switches, together with two inductors and two capacitors suppling energy to a resistive load in parallel with a capacitive output filter. The converter can be considered as two boost topologies stacked together, sharing the same

transistor, while D_1 and D_2 act as freewheeling diodes for the inductive currents.

This DC-DC converter operates with a pulse width modulated (PWM) signal, of frequency f_s and the duty cycle D. For continuous conduction mode (CCM) operation with respect to diodes D_1 and D_2 the converter switches between two topological states.

When S is ON, the two diodes are reversely biased thus turned OFF, the converter being in the first topological state. The inductor currents are increasing, and two current loops appear. The first one includes the voltage source V_g and inductor L_1, while the second consists of voltage supply V_g, capacitor C_2, coil L_2, and capacitor C_1. At the output, the filter capacitor C_o discharges on the load resistor R.

In the second topological state, the transistor S is OFF, and freewheeling is achieved through diodes D_1 and D_2. Consequently, inductor currents split into capacitors C_1 and C_2, and the sum of inductor currents is injected into the output group C_o-R.

By writing the volt-second balance equations [3] and Kirchhoff voltage law (KVL) in the loop containing the two capacitors C_1 and C_2, the DC voltages across capacitors result in:

$$V_{C1} = V_{C2} = \frac{D}{1-D} \cdot V_g \qquad (1)$$

and

$$V_{Co} = \frac{1+D}{1-D} \cdot V_g \qquad (2)$$

From the charge balance equation [3] and Kirchhoff current law (KCL) written for the nodes adjacent to coils L_1 and L_2, the DC inductor currents are obtained:

$$I_{L1} = \frac{(1+D)^2}{(1-D)^2} \cdot \frac{V_g}{R} \qquad (3)$$

$$I_{L2} = \frac{1+D}{1-D} \cdot \frac{V_g}{R} \qquad (4)$$

The static conversion ratio of this converter is:

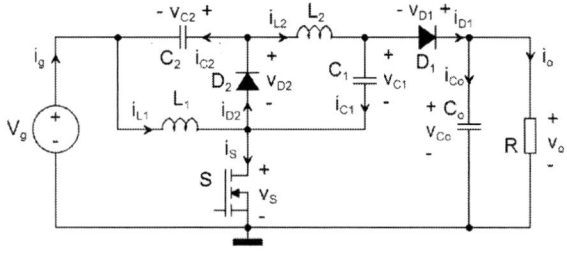

Fig. 1. The stacked boost converter [1].

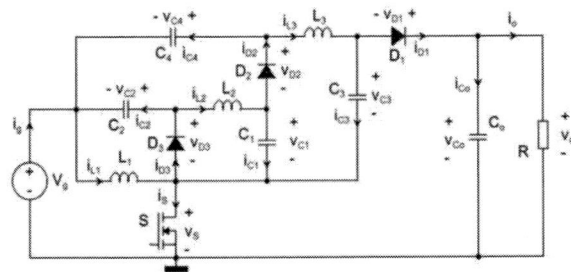

Fig. 2.   The stacked boost converter for $n = 3$

$$M_2 = \frac{1+D}{1-D} \qquad (5)$$

Where index 2 in the DC conversion ratio denotes the no. of stages.

Other formulas, as current and voltage stresses of components, can be easily derived based on previous equations.

III.   THE THREE STAGES STACKED BOOST CONVERTER

By inserting an additional boost stage to the previous presented converter, the structure from Fig. 2 is obtained.

By applying the volt-second balance principle to the inductors [3], the following equations can be written:

$$V_{L1} = D \cdot V_g + (1-D) \cdot (-V_{C2}) = 0 \qquad (6)$$

$$V_{L2} = D \cdot (V_{C2} - V_g - V_{C1}) + (1-D) \cdot (V_g - V_{C3} - V_{C4} - V_o) = 0 \quad (7)$$

$$V_{L3} = D \cdot (V_g + V_{C3} - V_{C4}) + (1-D) \cdot (V_g - V_{C3} - V_o) = 0 \quad (8)$$

As can be easily observed there are 5 unknowns, namely the capacitive voltages, and only 3 equations. Two additional equations should be written. Applying KVL in DC for the loops $L_1 - C_1 - L_2 - C_2$ and $C_3 - L_3 - C_4 - L_1$ and taking into account that the coils are short-circuits in DC, one obtains:

$$V_{C1} = V_{C2} \qquad (9)$$

$$V_{C3} = V_{C4} \qquad (10)$$

Now, the system consisting of (6) – (10) may be solved, resulting in:

$$V_{C1} = V_{C2} = \frac{D}{1-D} \cdot V_g \qquad (11)$$

$$V_{C3} = V_{C4} = \frac{2 \cdot D}{1-D} \cdot V_g \qquad (12)$$

$$V_o = \frac{1 + 2 \cdot D}{1-D} \cdot V_g \qquad (13)$$

The static conversion ratio for $n = 3$ stages is recognized as:

$$M_3 = \frac{1 + 2 \cdot D}{1-D} \qquad (14)$$

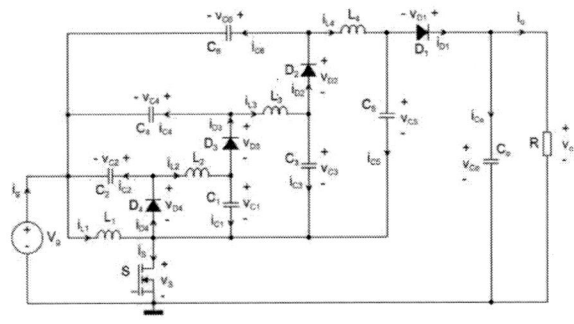

Fig. 3.   The stacked boost converter for $n = 4$

IV.   THE THREE STAGES STACKED BOOST CONVERTER

To better observe the evolution, another stage is added to the converter in section III, obtaining $n = 4$ stages, as in Fig. 3.

The equations resulting from the volt-second balance principle [3] are:

$$V_{L1} = D \cdot V_g + (1-D) \cdot V_{C2} = 0 \qquad (15)$$

$$V_{L2} = D \cdot (V_g - V_{C2} + V_{C1}) + (1-D) \cdot (-V_{C1}) = 0 \qquad (16)$$

$$V_{L3} = D \cdot (V_g - V_{C3} + V_{C4}) + (1-D) \cdot (V_{C4} - V_{C1}) = 0 \qquad (17)$$

$$V_{L4} = D \cdot (V_g - V_{C6} + V_{C5}) + (1-D) \cdot (V_{C5} - V_{C4}) = 0 \qquad (18)$$

The same issue obviously appears in connection to the no. of equations. In these converters there are more capacitors that inductors, thus additional equations from KVL must be added. In this case the resulting equalities of voltages from the loops containing coils are:

$$V_{C1} = V_{C2} \qquad (19)$$

$$V_{C3} = V_{C4} \qquad (20)$$

$$V_{C5} = V_{C6} \qquad (21)$$

Then the capacitive voltages are the following:

$$V_{C1} = V_{C2} = \frac{D}{1-D} \cdot V_g \qquad (22)$$

$$V_{C3} = V_{C4} = \frac{2 \cdot D}{1-D} \cdot V_g \qquad (23)$$

$$V_{C5} = V_{C6} = \frac{3 \cdot D}{1-D} \cdot V_g \qquad (24)$$

$$V_o = \frac{1 + 3 \cdot D}{1-D} \cdot V_g \qquad (25)$$

Finally, the DC conversion ratio is:

$$M_4 = \frac{1 + 3 \cdot D}{1-D} \qquad (26)$$

The procedure can be extended to any no. of stages n.

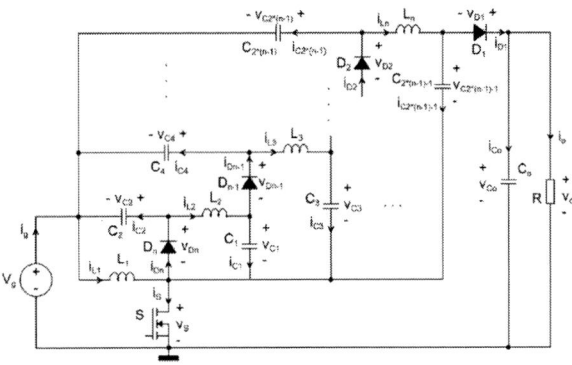

Fig. 4. The generalized stacked boost converter.

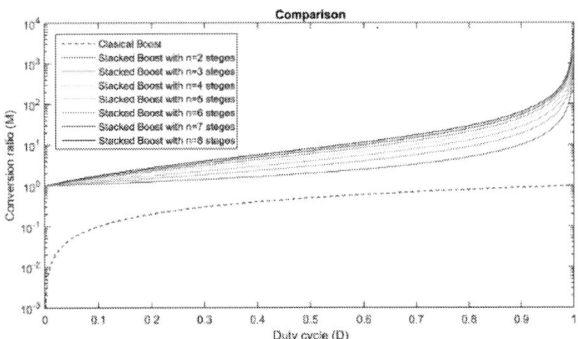

Fig. 5. The conversion ratio curves for converters with different no. of stages. (log y scale).

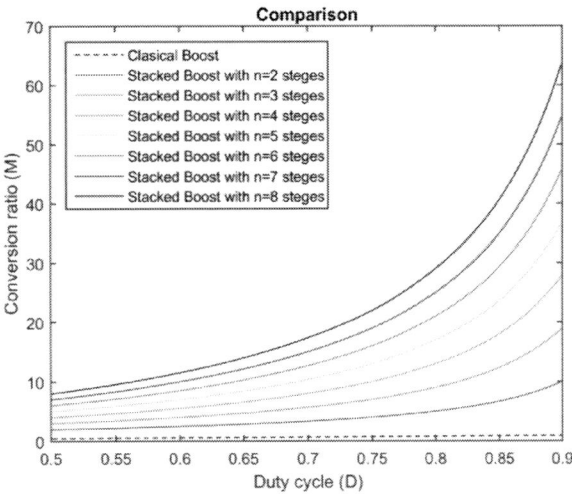

Fig. 6. Detail on the static conversion ratio comparison.

V. THE GENERALISED CONVERTER MODEL

The proposed generalised structure of the stacked boost DC-DC converter is depicted in Fig. 4. The DC conversion ratio of this converter can be written as:

$$M_n = \frac{1 + (n-1) \cdot D}{1 - D} \tag{27}$$

The recurrence equation for the static conversion ratio was demonstrated through mathematical induction and is defined by the formula below:

$$M_{n+1} = M_n + \frac{D}{1-D} \tag{28}$$

As a first remark it can be seen that this formula is an arithmetic sequence with the common difference:

$$r = \frac{D}{1-D} \tag{29}$$

A representation of the static conversion ratio M_n against the duty cycle D is depicted in Fig. 5, for a couple of stages no. The y axis has a logarithmic scale in order to easily distinguish between characteristics, since a high boost is produced at high duty cycles D. The classical boost converter is obtained for $n = 1$.

A detailed plot for duty cycle D between 0.5 and 0.9, this time in linear axes, is shown in Fig. 6. In high-step-up applications this is the domain of interest and it is clear that big differences occur at higher no. of stages and higher duty cycle.

It is predictable, that for a high number of stages results:

$$\lim_{n \to \infty} M_n = \infty \tag{30}$$

An increase in the number of stages n will cause the DC conversion ratio to become very high at moderate duty cycles D, in the vicinity of 50%. The equation $M_n = M_{n+1}$ has the unique solution $D=0$. Then the only common point of the static conversion ratio family curves will be the origin. In the far end the curves will asymptotically approach infinity.

The capacitive voltages will result in:

$$V_{C1} = V_{C2} = \frac{1 \cdot D}{1 - D} \cdot V_g \tag{31}$$

$$V_{C3} = V_{C4} = \frac{2 \cdot D}{1 - D} \cdot V_g \tag{32}$$

$$\dots$$

$$V_{C2 \cdot (n-1)-1} = V_{C2 \cdot (n-1)} = \frac{(n-1) \cdot D}{1 - D} \cdot V_g \tag{33}$$

The voltages across the inner capacitors resemble to the classical buck-boost converter output voltage.

VI. SIMULATION RESULTS

Simulations for the converter with $n = 2$ were already presented in [2] with more details. In this section, three converters having $n = 2$, $n = 3$, $n = 4$ are compared. They are designed to operate at the same input voltage $V_g = 10$V, load resistor $R = 82\Omega$, switching frequency $f_s = 100$kHz and duty cycle $D = 0.6$, in order to prove the correctness of the mathematical model. All the simulations are performed using CASPOC software [4].

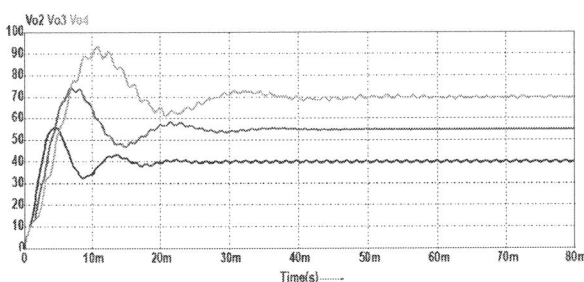

Fig. 7. Output voltage V_o for different no. of stages n: 2 stages with navy-blue, 3 stages with red and 4 stages with cyan.

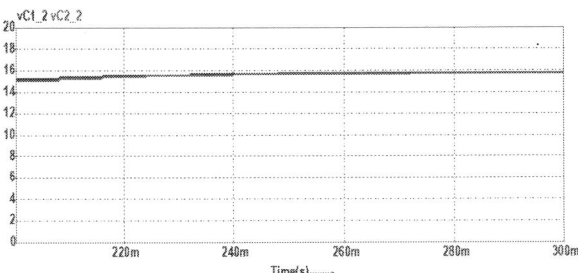

Fig. 8. Capacitive voltages for $n = 2 - V_{C1}$ with navy blue superimposed on V_{C2} with red.

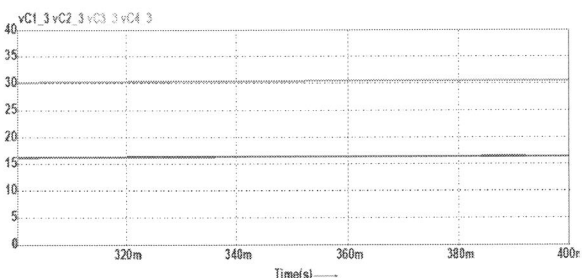

Fig. 9. Capacitive voltages for $n = 3 - V_{C1}$ with navy blue superimposed on V_{C2} with red, V_{C3} with cyan superimposed on V_{C4} with green.

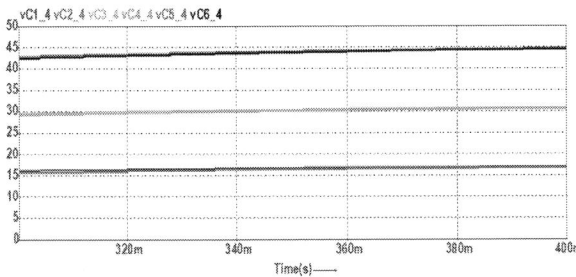

Fig. 10. Capacitive voltages for $n = 4 - V_{C1}$ with navy blue superimposed on V_{C2} with red, V_{C3} with cyan superimposed on V_{C4} with green, V_{C5} with magenta superimposed on V_{C6} with black.

The stacked DC-DC converter with $n = 2$ has $C_1 = C_2 = C_o = 22\mu F$ and $L_1 = L_2 = 3.3mH$.

The converter with $n = 3$ consists of the following components: $C_1 = C_2 = C_3 = C_4 = C_o = 22\mu F$ and $L_1 = L_2 = L_3 = 3.3mH$.

The stacked boost converter having $n = 4$ stages employs these values for the passive components: $C_1 = C_2 = C_3 = C_4 = C_5 = C_6 = C_o = 22\mu F$ and $L_1 = L_2 = L_3 = L_4 = 3.3mH$.

In Fig. 7, the output voltages for the three simulated converters are depicted. As may be easily observed, the distance between the steady state voltages is equal, as predicted by the theory, according to the arithmetic progression. The output voltage is directly proportional with the no. of stages.

Also, the capacitive voltages are shown in Figs. 8, 9 and 10. The values are over-imposed as the capacitors that are adjacent to the same coils, has the same voltage, because of the DC loops they are included in, as explained in the theoretical part, in section V.

VII. CONCLUSIONS

The new family of converters introduced is suited for applications with high step-up voltage, replacing the traditional cascade of several classical boost converters that have independent active switches with their own controllers. The proposed converters overlap the domain of quadratic converters and hybrid converters. They may feature high dc gain without the use of a transformer with a certain turns' ratio, such as in hybrid converters with coupled inductors. A transformer may be a custom-made component and opposed to it the proposed converters use "from the shelf" electronic components available on the market. A specific application is a photovoltaic system that provides relatively low voltage and high current which needs to be connected to an inverter, as the inverter needs a high voltage at the input, equal to the maximum power line standardised voltage.

Future work will focus on using these converters for maximum power point tracking and renewable energies in general. Another aspect under study is to develop small signal models for this family, also in generalised form.

REFERENCES

[1] M. Forouzesh, Y. P. Siwakoti, S. A. Gorji, F. Blaabjerg, B. Lehman, "Step-Up DC–DC Converters: A Comprehensive Review of Voltage Boosting Techniques, Topologies, and Applications", IEEE Transactions on Power Electronics, Vol. 13, No. 12, pp. 9143-9178, December 2017.

[2] S. Lica, I. M. Pop-Călimanu, D. Lascu, A. Cireşan, and M. Gurbină, "A New Stacked Step-Up Converter", 40th International Conference on Telecommunications and Signal Processing (TSP), Barcelona, Spain, July 5-7, pp. 315-319, 2017.

[3] R. W. Erickson and D. Maksimovic, Fundamentals of Power Electronics, 2nd Ed., Kluwer Academic, 2001.

[4] Simulation Research, Caspoc, online user manual, http://www.caspoc.com/support/manuals/, April 2013.

Vector Control Of Permanent Magnet Synchronous Machine

Ana-Maria Petri, Dorin Petreuș

Faculty of Electronics, Telecommunications and Information Technology
Technical University of Cluj-Napoca,
Cluj-Napoca, Romania,
Email: Ana.Petri@ael.utcluj.ro

Abstract—**This paper presents the implementation of a vector control for a permanent magnet synchronous machine. MATLAB-Simulink simulations were used to obtain preliminary results for three implementations of the vector control block. For a maximum torque the current i_d was set to zero and the q component of the current was used to control the torque. Three implementations of vector control were done, based on how the two components of the current are controlled (open loop or close loop). The results obtained prove the advantages and disadvantages of each implementation.**

Keywords — *Permanent Magnet Synchronous Machine, Vector Control*

I. INTRODUCTION

In the last decades the trend in electric systems is to reduce size and energy consummation, using more efficient devices, with reduced size. Permanent magnet synchronous machine (PMSM) is a widely used electric machine in all kind of applications, from home appliances to industrial applications, and even to electric vehicles [1].

The permanent magnets assure a high-power density, eliminate the filed copper losses and the construction of the rotor is robust. Compared to an induction machine a PMSM has higher efficiency, smaller size for the same power, but for a bigger cost [1]. The disadvantage of PMSM is the possibility of demagnetization of the permanent magnets, which can be affected by temperature, large stator currents and aging of the magnet [2].

The control of the torque or the speed of the machine is an important task in all PMSM drive system. Two control methods are well known for PMSM control: direct torque control (DTC) and vector control (VC). DTC is a good choice in applications where a high torque ripple is not a problem [3]. But if the application requires high accuracy VC is a better choice [4].

Depending on the construction of the rotor there are two types of PMSMs: surface mounted magnets and interior magnets. The surface mounted magnets machines are known as non-salient pole, characterized by $L_{dm} = L_{qm}$. In the case of interior magnets machines $L_{dm} < L_{qm}$ and are known as salient pole machines. In this paper a non-salient pole machine will be used for the implementation of VC [5].

This paper presents the implementation of a vector control drive system for a PMSM. In section 2 the mathematical model of a PMSM is presented and the principles of vector control for PMSM are illustrated in section 3. Then the implementation of the VC drive system with the corresponding results are presented in section 4. In the end, based on the results from section 4 some conclusions are given in section 5.

II. MATHEMATICL MODEL OF PERMANENT MAGNET SYNCHRONOUS MACHINE

The implementation of the vector control requires the dynamic model of the PMSM in dq rotating frame. The expressions of the machine's voltages in dq representation are presented in (1) and (2):

$$v_{ds} = R_s i_{ds} - \omega \Psi_{qs} + \frac{d}{dt} \Psi_{ds} \tag{1}$$

$$v_{qs} = R_s i_{qs} + \omega \Psi'_{ds} + \omega \Psi_f + \frac{d}{dt} \Psi_{qs} \tag{2}$$

Where R_s is the stator resistance; ω is the synchronous speed of the machine; i_{ds}, i_{qs}, Ψ_{ds}, Ψ_{qs} are the stator currents, respectively the stator flux linkages and Ψ_f is the flux linkage generated by the permanent magnets.

Based on the stator currents and stator inductances the flux linkages expressions are illustrated in (3), (4) and (5) as follow:

$$\Psi_{ds} = \Psi'_{ds} + \Psi_f \tag{3}$$

$$\Psi'_{ds} = i_{ds}(L_{ls} + L_{dm}) = i_{ds}L_{ds} \tag{4}$$

$$\Psi_{qs} = i_{qs}(L_{ls} + L_{qm}) = i_{qs}L_{qs} \tag{5}$$

Also, the electromagnetic torque expression is the following in (6):

$$T_e = \frac{3}{2}\left(\frac{P}{2}\right)(\Psi_{ds}i_{qs} - \Psi_{qs}i_{ds}) \tag{6}$$

Where P is the number of poles of the PMSM.

Because a non-salient pole machine is used and substituting (3), (4) and (5) in (6) a simpler expression of the torque is obtained in (7):

$$T_e = \frac{3}{2}\left(\frac{P}{2}\right)\Psi_f i_{qs} \tag{7}$$

Knowing that Ψ_f the flux linkage generated by the permanent magnets is constant, it can be seen in (7) that the electromagnetic torque can be control by controlling the q component of the stator current.

Equation (6) shows the electrical expression of the torque, but the electromagnetic torque developed by PMSM is influenced also by the load torque (T_L), the acceleration, respectively deceleration and the inertia (J) of the machine, as following in (8):

$$T_e = T_L + \frac{2}{P} J \frac{d\omega}{dt} \qquad (8)$$

III. PRINCIPALES OF VECTOR CONTROL

Vector control of permanent magnet synchronous machine has the same principals as in the case of induction machine. The stator current is represented in the dq frame and the two orthogonal components of the current are used to obtain the decoupling effect between the torque and the flux.

In the case of the induction machine the torque is aligned the direction of the q component, and the flux is oriented in the direction of d component of the stator current.

For the PMSM it is in the same way, because from (7) the q component is proportional with the torque, so i_{qs} will be aligned with the torque.

The permanent magnets from the rotor generate the main flux linkage of the machine. So, the stator current will be used only for torque. In order to maximize the sensibility of the torque with the stator current, it is needed to maximize the q component of the current. Also, because the flux is assured by the permanent magnets, the d component of the current can be set to zero.

For a good alignment of the q component of the current the synchronous rotating angle is needed. For the induction machine the slip frequency is used, but for synchronous machine the slip frequency is equal to zero, because the rotor speed is equal to the synchronous speed. Also, the synchronous machine has fixed positions of the poles and the position of the rotor can be determined using an encoder. In Fig. 1 is presented a block diagram of a vector control PMSM drive system:

Vector control of PMSM is similar to the one of induction machine, but it is simpler because:

- The slip frequency is zero, because the rotor speed is always equal to the synchronous speed.

- The reference value of d component of the current is set to zero, because the flux is generated by the permanent magnets and to maximize the torque with smallest stator current.

- The synchronous rotating angle is obtained using an encoder.

- The parameters of the machine are not required for the implementation of the vector control block.

An important element for the implementation of a vector control block for a PMSM is obtaining the correct rotating angle. A high accuracy encoder assures a good efficiency of a vector control block. In order to reduce the cost and to improve the reliability of a vector control block there are sensorless algorithms, which can be used to estimate the rotor position and the speed of the motor. In this paper is used an encoder, but for future work a sensorless algorithm will be proposed.

It must be noted that vector control of PMSM is valid only for speed below the rated speed of the motor. The rated speed of the motor is the limit of the constant torque region. In this region the stator voltage increases proportionally with the increase of the speed, until the voltage saturates at the edge of constant torque region. If higher speed is required field weakening it can be used.

In order to operate with a speed above the rated speed, because the voltage is saturated and $V_s = \omega \cdot \Psi_s$, the flux must be reduced, so current control remains applicable. This method is known as field weakening control and in general is achieved by reducing the magnetizing current.

For PMSM the flux is generated by the permanent magnets, but the air gap flux can be weakened by a negative value of i_{ds}. So, in order to control the speed of a PMSM beyond the rated speed the d component of the stator current must have negative values and it cannot be kept at zero.

For PMSM the speed range for field weakening region is small because at high negative values of the d component of the current exist the danger of permanent demagnetization the magnets.

IV. IMPLEMENTATION AND RESULTS

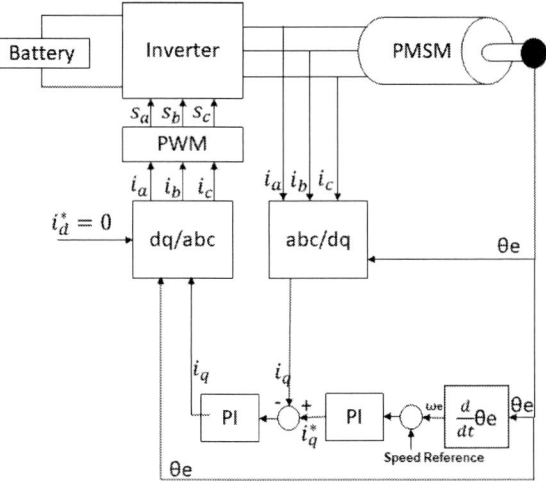

Fig. 1. PMSM Vector Control drive system

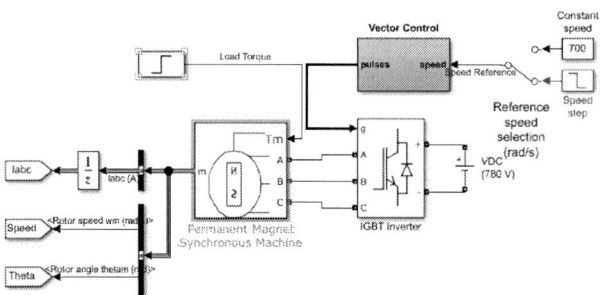

Fig. 2. MATLAB\Simulink PMSM Vector Control drive system

2020 IEEE 26th International Symposium for Design and Technology in Electronic Packaging (SIITME)

Using MATLAB\Simulink a PMSM vector control drive system was implemented. Using a DC source and an inverter the PMSM is supplied. The IGBTs from the inverter are driven by PWM signals generated by the vector control block. In Fig. 2 the drive system is illustrated:

Based on the block diagram from Fig. 1 three vector control implementations were done.

The simplest way to implement a vector control block for a PMSM is to set the d component of the current to zero and the q component based on the speed error. Both currents are controlled in open loop. Fig. 3 presents the implementation:

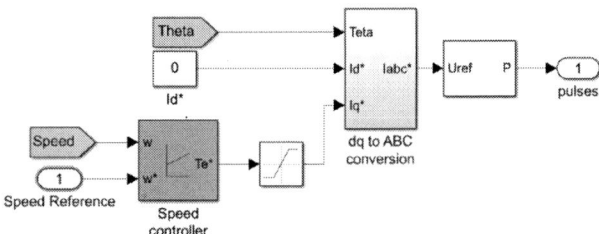

Fig. 3. PMSM Vector Control block with open loop current control

This implementation has a dq-abc transformation block, a PWM generator and only one PI controller for speed error, in order to calculate the corresponding torque, which is proportional with the q component of the current. In this case the vector control block requires a small amount of computational resources, but the response of the block has a small efficiency. In Fig. 4 and Fig 5 are presented the speed and torque response for speed and load torque step chances:

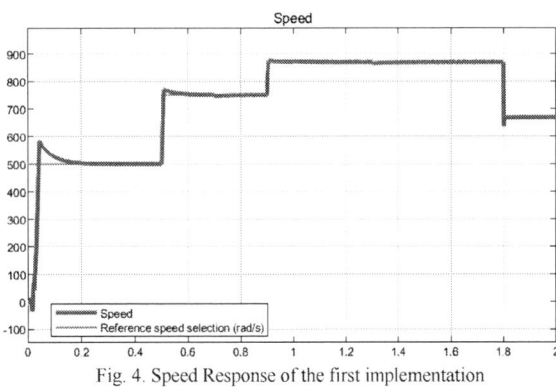

Fig. 4. Speed Response of the first implementation

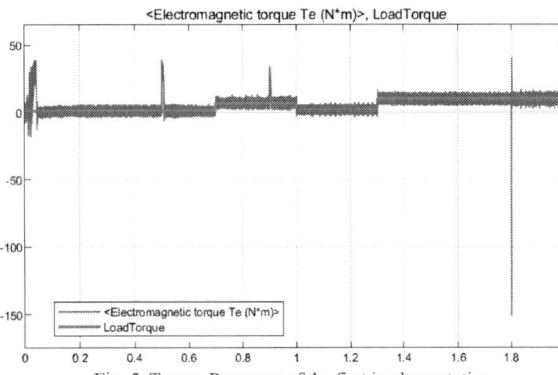

Fig. 5. Torque Response of the first implementation

The second implementation is done according the block diagram from Fig. 1, with a close loop control for i_{qs} and an open loop control for i_{ds}, which is set to zero. Fig. 6 presents the second implementation:

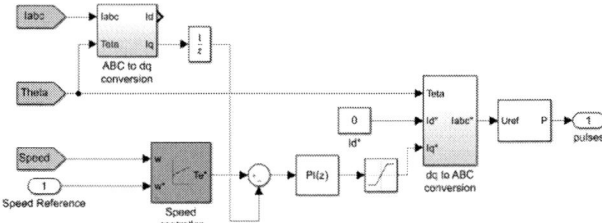

Fig. 6. PMSM Vector Control block with i_{qs} close loop control

The second implementation of the vector control block, beside the dq-abc transformation block, PWM generator block and PI speed error controller, has a block of abc-dq transformation, needed for the feedback signal of i_{qs}, and a second PI controller for the close loop control of q component of the stator current. The d component of the current is still set to zero in an open loop control.

This implementation required bigger computational resources than the previous one, but with higher efficiency.

Fig. 7 and Fig. 8 illustrate the speed and torque response:

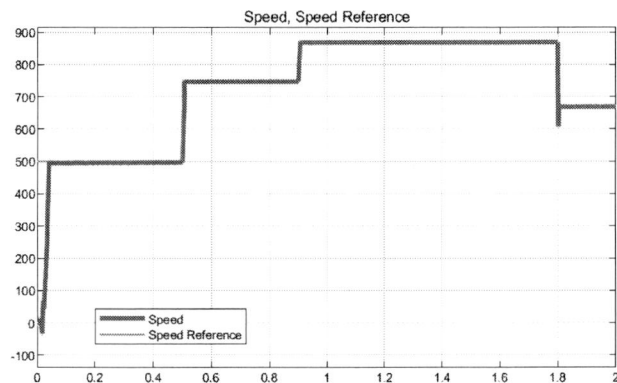

Fig. 7. Speed Response of the second implementation

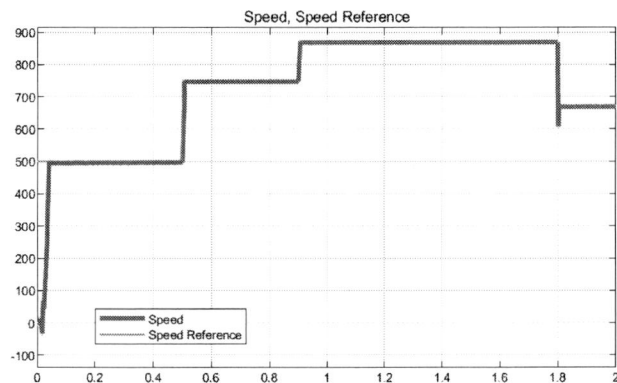

Fig. 8. Torque Response of the second implementation

In the third implementation a close loop control was used for i_{ds} too, with the reference value set to zero. This approach can be further improved, adding a field weakening control, so the reference value of the d component of the current can be set to zero for speed below the rated speed and to negative values in case of speed beyond the rated speed value.

Fig. 9 shows the third implementation:

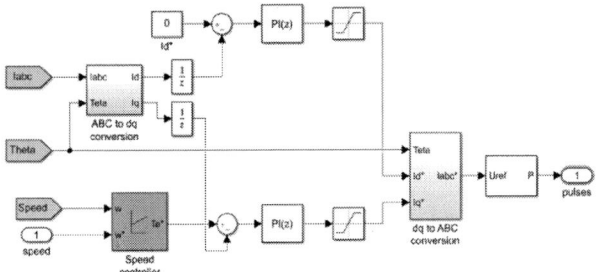

Fig. 9. PMSM Vector Control block with i_{qs} and i_{ds} close loops control

This implementation has three PI controllers, one for speed error, another for i_{qs} error and the third one is for i_{ds} error.

The responses are presented in Fig. 10 and Fig. 11.

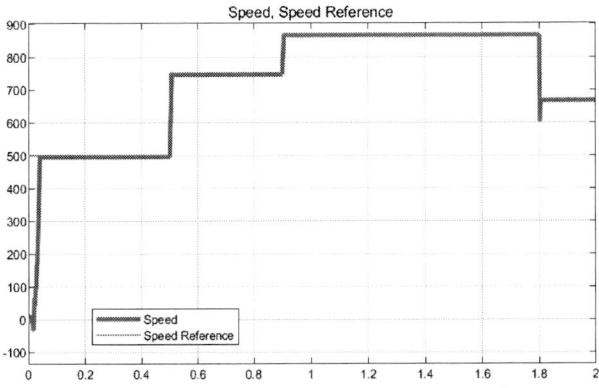

Fig. 10. Speed Response of the third implementation

Analyzing the speed responses of the three implementations of the vector control blocks, the first implementation has a small efficiency in tracking the reference speed value, in comparison with the second and third implementation, where the response has higher accuracy.

The torque response of the second implementation has smaller rippler that the other two vector control blocks.

Every project has its specific requirements and one of the three implementations can be better than the others. But, based on the results obtained from the three implementation the second implementation, with open loop control for i_{ds} and close loop control for i_{qs}, it can be the best choice. Because it has a good

speed response, low torque ripple and requires medium computational resources.

This is applicable only if the operation of the PMSM is required only in the constant torque region. But, if the speed range must exceed the rated value, then field weakening control must be added in the control block.

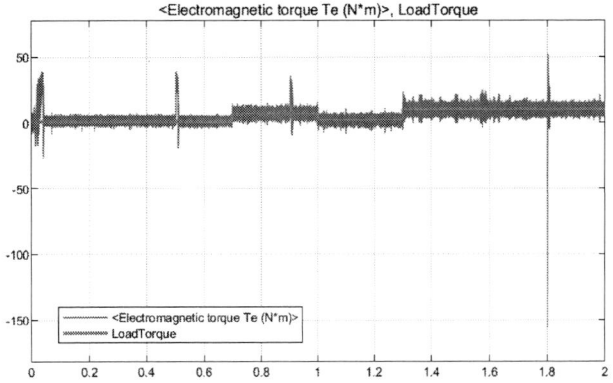

Fig. 11 Torque Response of the third implementation

V. CONCLUSIONS

This paper presents three implementations of a vector control block for a permanent magnet synchronous machine. Starting with an open loop control for both components of the stator current. Next an open loop control for i_{ds} and a close loop control for i_{qs}, and the last with close loop control for both currents. Based on the results obtained from the three simulations, a good compromise between responses and resources is in the second implementation. It requires medium computation resources, the speed respond is tracking the reference value with high accuracy and, also, the torque response has low ripple.

REFERENCES

[1] B.K. Bose, "Modern Power Electronics and AC Drives", Prentice Hall PTR, USA, 2002

[2] T. Ishikawa, Y. Seki, N. Kurita, "Analysis for Fault Detection of Vector-Controlled Permanent Magnet Synchronous Motor With Permanent Magnet Defect", IEEE Transactions on magnetics, Vol. 49, No. 5, pp. 2331-2334, May, 2013

[3] R. Souad, H. Zeroug, "Comparison Between Direct Torque Control and Vector Control of Permanent Magnet Synchronous Motor Drive", 13th International Power Electronics and Motion Control Coference, Poland, 2008

[4] Shaoshen Xue, H. Xu, C. Fang, W. Niu and S. Xue, "Study on the High Accuracy Vector Control System of Direct Drive Permanent-Magnet Synchronous Machine", 15th International Conference on Electrical Machines and Systems, Sapporo, Japan, 2012

[5] S. N. Vukosavic, "Power Electronics and Power Systems – Electrical Machines", Springer Science Business Media, New York, USA, 2013

Gap in pagination due to formatting issues.

Pages 394-397

Comparison between LLC and Phase-Shift converter with synchronous rectification for high power, high current applications

T. M. Patarau, D. M. Petreus, I. Ferencz
Applied Electronics, Technical University of Cluj-Napoca
Cluj-Napoca, Romania
toma.patarau@ael.utcluj.ro

Z. Orban
SC Datronix Computer SRL
Cluj-Napoca, Romania

Abstract —The paper presents a comparison between LLC and Phase-Shift converter with synchronous rectification for converters that need low voltage and high current at the output. The analysis is aimed to help engineers to choose properly the high power, high current topology providing a comparison based on design and simulation of both the above-mentioned topologies. The comparison shows that the best suited topology for this type of applications is the phase shift converter with current doubler rectifier and synchronous rectification.

Keywords— phase shift converter, LLC converter, high output current.

I. INTRODUCTION

There is an increasing demand in industry for high power converters that require at the output low voltage levels and very high currents. Applications like solar battery inverters (48V,104A), chemical electrolysers (10V, 300A), DC arc welding (60V, 200A), copper refining, telecommunication equipment, plasma torch, magnetic confinement are only some of the main areas where high output current converters are needed [1].

There are many papers that analyze the advantages and disadvantages of different converter topologies for low voltage and very high output current. P. Alou et al. makes a comparison between current doubler rectifier and center tap rectifier for a phase shift converter operating at high output current. The article concludes that in applications with an output current lower than 15A the center taped rectifier performs better than the current doubler rectifier. For high output current values (higher than 40A) the current doubler is better. In the range of (20-30A) the choice depends on the allowed output current ripple [2].

The authors in [3] make a comparison between a LLC converter and a phase shift converter for battery chargers used in electrical vehicles. 1kW converters where compared with 400V output voltage. For these converters conventional rectifiers with diodes where used because of the high output voltage and small currents. Both converters have merits and can be used in battery charger applications at this power levels. The advantages of LLC converter include: ZVS for the full load, low EMI, lower cost compared to the advantages presented by the phase shift converter: low conduction losses at high loads and easier to control.

Authors in [4] make a comparison between three soft switching topologies with high frequency transformer isolation used in high output current applications: 1) LCL SRC with capacitive output filter; 2) LCL SRC with inductive output filter; and 3) phase-shifted ZVS PWM full-bridge converter. The authors showed that the LCL SRC with capacitive output filter is best suited for the presented application. None of the above topologies except the LCL SRC maintained zero voltage switching (ZVS) for the complete load range without overcoming the drawbacks of inductive output load. It is also demonstrated that the LCL SRC needed low resonant inductor for proper operation with low input voltages and its better to boost firstly the input voltage when using this converter. Usually adding another stage ad additional costs and complexity to the converter making this topology less preferred.

The LLC converter with high output current and fast dynamic response is analyzed in detail in [5] for server and data center, telecom, PV and battery charging. It was shown that for high current applications the design of the LLC converter becomes challenging. High output currents lead to high ripple current through the output capacitor causing high losses for this capacitor. There are also problems with the synchronous rectifier. The package inductance of the rectifier induces parasitic voltages that can cause improper timing for the gate drivers.

The authors in [6] developed a medium power push-pull DC/DC converter for hydrogen generation from photovoltaic sources. The topology performs: high step-down ratio, galvanic isolation, good efficiency and low part count. The push-pull converter uses high-voltage IGBT transistors running at 50kHz switching frequency with a peak efficiency of 90%. The main disadvantage of this converter is its low power density caused by the low switching frequency used.

The operation of the two topologies is well described in literature and will not be resumed in this paper. The reminder of the paper is organized as follows: Section II presents the design of the two power topologies; Section III presents the simulation results validating the calculated values from the previous section; Section IV is a discussion that compares the results and highlights the bottlenecks encountered for the two topologies for high current applications and the last section draws the conclusions.

II. DESIGN OF THE TWO POWER TOPOLOGIES

The specifications for the design of the two converters are presented in the following table:

TABLE I. DESIGN SPECIFICATIONS

Input voltage (V_{in})	Output characteristics (V_{out}, I_{out})	Output power and efficiency (P_{out}, η_{min})	Switching frequency (f_{sw})
$V_{inmin} = 350V$ $V_{in} = 400V$ $V_{inmax} = 450V$	$V_{out} = 10V$ $I_{out} = 300A$	$P_{out} = 3kW$ $\eta_{min} = 96\%$	$f_{sw} = 250kHz$

A. The phase shift converter with current doubler rectifier

The schematic of the power stage of the phase-shift converter with current doubler rectifier and the waveforms are represented in Fig.1 and Fig.2.

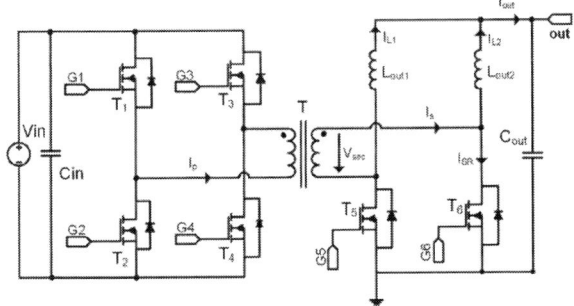

Fig. 1. Phase shift converter schematic

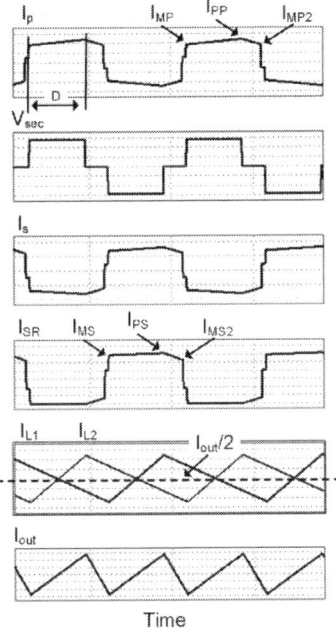

Fig. 2. Phase shift converter waveforms

The design procedure of the phase-shift converter power stage starts with calculating the transformer turns ratio, n. A maximum duty cycle (calculated at minimum input voltage) $D_{max} = 80\%$ is considered. This condition leaves room for the duty cycle loss and for the dead times required for zero-voltage switching of the primary MOSFET transistors [7]. In (1) $V_{RDSON\_prim} = 10mV$ is the maximum allowed voltage drop on the primary side MOSFET transistor and $V_{RDSON\_sec} = 500mV$ is the maximum allowed voltage drop on the synchronous rectifier MOSFET transistor.

$$n = \frac{D_{max}\left(V_{in\,min} - 2V_{RDSON\_prim}\right)}{2(V_{out} + _{RDSON\_sec})} \quad 14 \tag{1}$$

The nominal value of the duty cycle is than found to be $D_{typ} = 70\%$. For low core losses of the output inductor the output current ripple, ΔI_{Lout}, is chosen to be 10% of the output current. The minimum magnetizing inductance can be calculated as in (2).

$$L_{mag} = \frac{V_{in}(1 - _{typ})}{\frac{\Delta I_{Lout}}{2n}} f_{sw} \quad 500uH \tag{2}$$

The maximum primary winding current is:

$$I_{PP} = \left(\frac{P_{out}}{2V_{out}\eta} + \frac{\Delta I_{Lout}}{)}\right)\frac{1}{n} + \frac{V_{in\,min}D_{max}}{L_{mag}f_{sw}} = 14.5 \tag{3}$$

The RMS value of the primary current was calculated as follows:

$$I_{prms} = \sqrt{\frac{1}{T_{sw}}\int_0^{T_{sw}} i_p(t)dt} \quad 13.5A \tag{4}$$

In order to have small power losses in the primary bridge transistors with low R_{DSON} should be selected. If a maximum dissipation of 8W is allowed on one transistor than a transistor with $R_{DSONmax} = 40m\Omega$ must be selected.

The maxim value of the secondary current is calculated in (4). The RMS value of this current is $I_{srms} = 151A$.

$$I_{PS} = \frac{P_{out}}{2V_{out}} + \frac{\Delta_{Lout}}{} = 165 \tag{5}$$

It can be observed that, in the power transformer, the secondary winding currents are high. In high current applications, it is recommended to have in secondary only one winding to minimize the losses [2]. In the case of the current doubler rectifier the current passing through the secondary winding is only half of the load current, the other half being supplied by one of the inductors. This constitute an important advantage for this topology.

The maximum voltage on the synchronous rectifier in the secondary is $V_{SR} = 2V_{out}/D_{typ} = 28.5V$. In order to calculate the conduction losses of the synchronous rectifier MOSFET transistors, the RMS value of current through each transistor is needed.

$$I_{SRrms} = \frac{P_{out}}{V_{out}} \sqrt{\frac{D_{typ} + 1}{4}} \quad 195 \tag{6}$$

It can be observe that a transistor having only a R_{DSON} of 7.5mΩ would dissipate 28.7W. Two or more transistors in parallel will be needed.

The resonant inductor to achieve zero voltage switching for output currents smaller than $I_{PP}/7$ is calculated in (7). $C_{oss} = 70pF$ is the average output capacitance of a MOSFET from the primary bridge.

$$L = 2C_{oss} \frac{V_{in}^{2}}{\left(\dfrac{I_{PP}}{7} - \dfrac{\Delta I_{Lout}}{n} \right)^{2}} \quad 22uH \tag{7}$$

The output inductor is calculated as follows:

$$L_{out} = \frac{V_{out} \left(1 - 0.5 \; _{typ} \right)}{\Delta I_{Lout} \left(\dfrac{f_{sw}}{2} \right)} \quad 2uH \tag{8}$$

The RMS value of the current through each output inductor is:

$$I_{Lout\,RMS} = \sqrt{\left(\frac{P_{out}}{2V_{out}} \right)^{2} + \left(\frac{\Delta_{Lout}}{\sqrt{3}} \right)^{2}} = 150 \tag{9}$$

The minimum value for the output capacitor, considering a maximum output voltage ripple of $\Delta V_{out} = 2mV$, is:

$$_{out} = \frac{V_{out} \left(1 - \; _{typ} \right)}{16 L_{out} \Delta V_{out} f_{sw}^{2}} \quad 750 \mu F \tag{10}$$

The RMS value of the current through this capacitor is 4A.

$$I_{Cout\_rms} = \frac{\Delta I_{out}}{2\sqrt{}} \quad 4 \tag{11}$$

B. The LLC rezonant converter

The power stage of the LLC converter will be designed in the same conditions as for the phase-shift converter. The schematic and the waveforms representing the operation of the converter at resonance are represented in the figure below. The converter is designed to operate near the resonant frequency (f_0) in order to minimize switching frequency variations. The design procedure is adopted from [8].

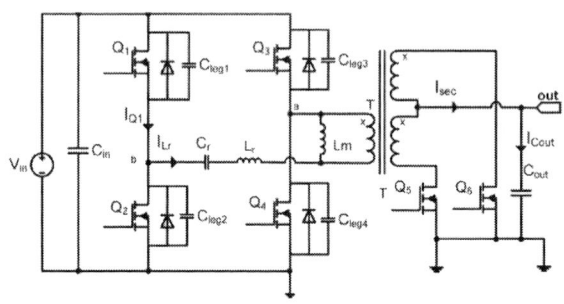

Fig. 3. LLC converter schematic

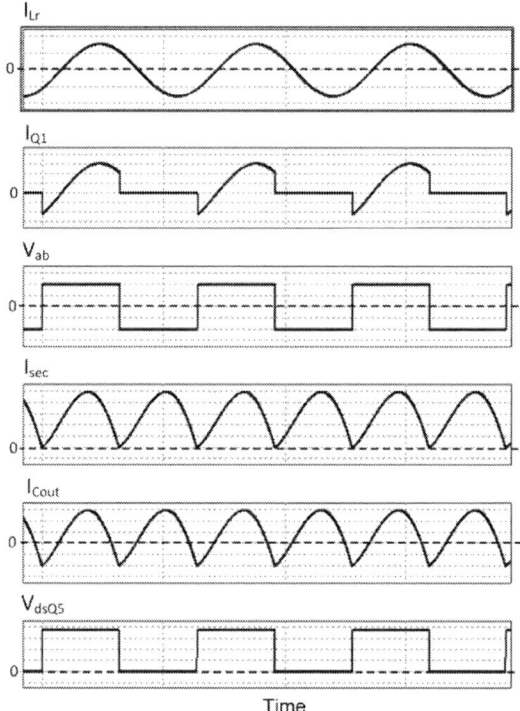

Fig. 4. LLC waveforms

The gain at f_0 is a function of m ($m=L_p/L_r$). A higher peak gain is obtained for smaller m value however too small value of m leads to poor coupling of the transformer. Typically, m is chosen between 3 and 7. In this paper a value of 4 was chosen. In this case the minimum gain M_{min} can be calculated as:

$$M_{min} = \sqrt{\frac{m}{m-1}} \quad 1.10 \tag{12}$$

For minimum input voltage the maximum gain obtained is:

$$M_{max} = \frac{V_{in\,max}}{V_{in\,min}} \quad _{min} \quad 1.47 \tag{13}$$

The turns ratio can be now calculated:

$$n = \frac{N_p}{N_s} = \frac{V_{in\,max}}{2\left(V_{out} + V_F\right)}\bigg|_{min} = 43 \tag{14}$$

where V_F is the voltage drop on the synchronous rectifier. The equivalent load resistance is then calculated as:

$$R_{ac} = \frac{8n^2}{\pi^2}\frac{V_{out}}{P_{out}} = 49\Omega \tag{15}$$

The value of the quality factor is chosen from Fig. 5.

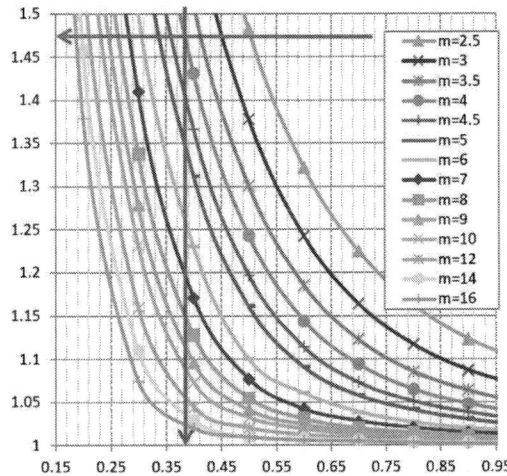

Fig. 5. Peak fain vs Q for different m values

$Q = 0.38$ gives enough margin for load transients and stable zero voltage switching. With the value of Q chosen the values for the components of the resonant network can be calculated.

$$C_r = \frac{1}{2\pi Q f_0 R_{ac}} \quad 85nF$$

$$L_r = \frac{1}{\left(2\pi f_0\right)^2 C_r} \quad 30uH \tag{16}$$

$$L_p = mL \quad 120uH$$

The RMS value of the primary current is $I_{pRMS}=10.82A$.

$$I_{pRMS} = \sqrt{\frac{\pi I_{out}}{2\sqrt{2}n} + \frac{n\left(V_o + V\right)}{4\sqrt{2}f_0 M(L - L_r)}} = 10.82 \tag{17}$$

This value is the same as for the inductor and resonant capacitor current. The maximum voltage drop on C_r can be calculated as:

$$V_{cr\,max} = V_{in\,max} + \frac{I_{overcurent}}{2\pi f_{sw\,min}\,r} = 878V \tag{18}$$

This value is calculated for $I_{overcurent} = 320A$. The maximum voltage drop on the synchronous rectifier is $V_{dsQ5}=20.6V$. The RMS value for the current passing through $Q4$ and $Q5$ can be calculated as $I_{Q4RMS}=\pi/4I_o=235.5A$. A capacitor with a value of $1000\mu F$ was chosen with a ESR of maximum $1m\Omega$ was chosen. The RMS current through the output capacitor can be calculated as follows:

$$I_{CORMS} = \sqrt{\frac{\pi^2 - 8}{8}}I_o \quad 144.64A \tag{19}$$

The calculated output voltage ripple is 502mV.

III. SIMULATION MODELS

For each converter a simulation model was created to validate the calculation results.

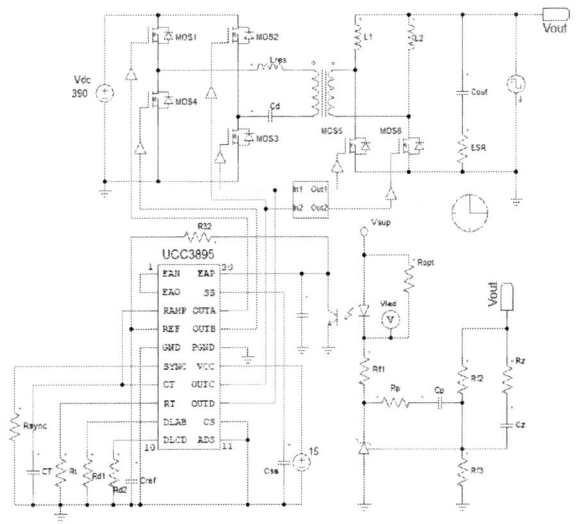

Fig. 6. Simulation model for the phase-shift converter

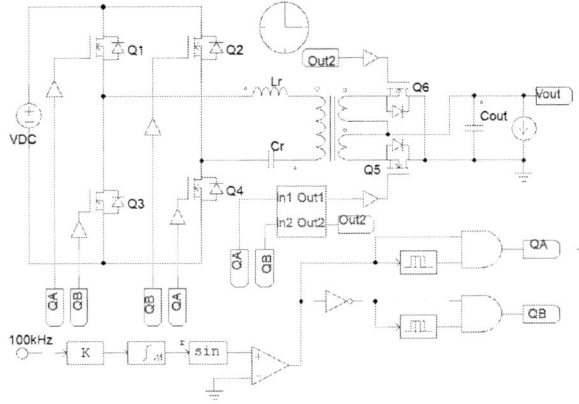

Fig. 7. Simulation model for the LLC converter

The results for each simulation model is represented in the figures below

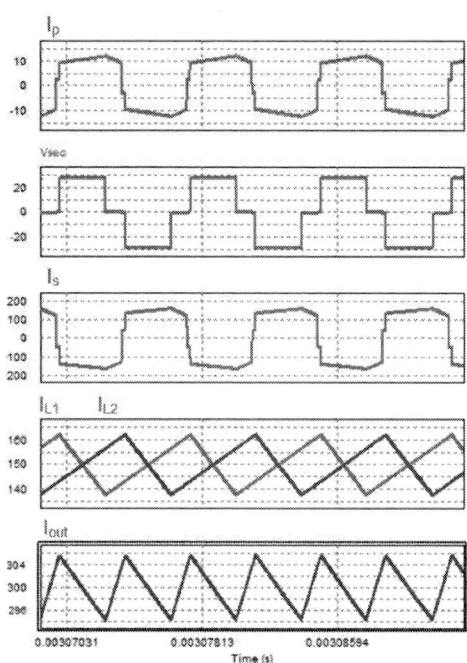

Fig. 8.   Simulation results for the PHS converter

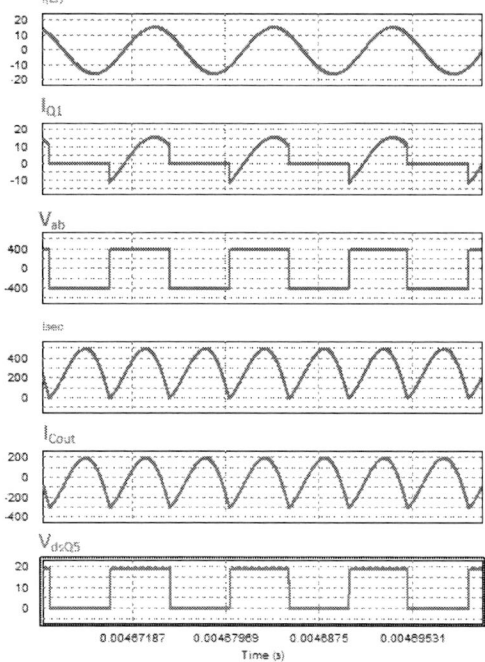

Fig. 9.   Simulation results for the LLC converter

IV. RESULTS

Comparison results are presented in the tables below.

TABLE II.   RESULTS FOR THE PHS CONVERTER

PHS	
Primary side	
- ZVS for primary transistors at heavy loads	
- Fixed frequency	
- I_{prim_rms} = 13.52A	
- I_{prim_pk} = 14.5A	
- no need for resonant capacitor	
Secondary side	
- Hard switching for secondary rectifier	
- Needs two high current secondary inductors (I_{Lout_rms} = 150A)	
- low current ripple through output capacitor filter	
- current ripple is cancelled similar to interleaved 2-phase buck. - I_{cap_rms} = 4A	
- small capacitor is needed to meet the voltage ripple requirement (1 000uF 1mΩ needed for 40mV ripple)	
- easy to control because the power stage has a second order transfer function for the full range	
- easy to obtain wide voltage range (large gain)	
- fixed frequency	
- good regulation for full operation range	
- simpler transformer with only one secondary winding - I_{sec_rms} = 151A	
- Synchronous rectifier rms current: 195A	
- may need active snubber circuits	

TABLE III.   RESULTS FOR THE LLC CONVERTER

LLC	
Primary side	
- ZVS for primary transistors full range	
- Variable frequency	
- I_{prim_rms} = 10.8A	
- I_{prim_pk} = 15A	
- high current and voltage resonant capacitor I_{cr_rms} = 21.17A, V_{crmax} 307V	
Secondary side	
- ZCS for secondary rectifier	
- without secondary inductor	
- high current ripple causes high loss on the capacitor filter	
- current ripple is NOT cancelled - I_{cap_rms} = 144A	
- large capacitor is needed to meet the voltage ripple requirement (10 000uF 1mΩ needed for 500mV ripple)	
- difficult control because at different operating points the open loop transfer function varies between a first order system and a second order system	
- difficult to obtain wide voltage range (large gain)	
- variable frequency	
- at light load the output voltage regulation is poor.	
- doble winding on the secondary side of the transformer - I_{sec_rms} = 235A	
- Synchronous rectifier rms current: 235A	
- No need for snubber circuits	

V. Conclusions

The paper presented a comparison between the PHS and the LLC converter used in high power applications that require low output voltage and high output power. It was found that even though both converters have merits, the PHS converter is better than the LLC converter in respect to efficiency, cost, PCB size, control, line regulation and output voltage range. Advantages and disadvantages for both converters were provided in order to help engineers to choose the proper topology for their applications.

Acknowledgment

This research was supported by the project "High power density and high efficiency micro-inverters for renewable energy sources" - MICROINV - Contract no. 16/1.09.2016, project co-funded from the European Regional Development Fund through the Competitiveness Operational Program 2014-2020, Romania.

References

[1] Rakesh Maurya, S.P. Srivastava, Pramod Agarwal, "Symmetrical and asymmetrical controlled three-phase high frequency isolated DC–DC converter", Electrical Power and Energy Systems, Vol. 52, pp. 132–142, November 2013;

[2] P. Alou, J. A. Oliver, O. Garcia, R. Prieto and J. A. Cobos, "Comparison of current doubler rectifier and center tapped rectifier for low voltage applications," Twenty-First Annual IEEE Applied Power Electronics Conference and Exposition, 2006. APEC '06., Dallas, TX, 2006, doi: 10.1109/APEC.2006.1620622.

[3] T. Gherman, D. Petreus, T. Patarau and A. Ignat, "A study of an Electrical Vehicle Battery Charger's DC-DC Stage," 2018 41st International Spring Seminar on Electronics Technology (ISSE), Zlatibor, 2018, pp. 1-6, doi: 10.1109/ISSE.2018.8443658.

[4] D. S. Gautam and A. K. S. Bhat, "A Comparison of Soft-Switched DC-to-DC Converters for Electrolyzer Application," in IEEE Transactions on Power Electronics, vol. 28, no. 1, pp. 54-63, Jan. 2013, doi: 10.1109/TPEL.2012.2195682.

[5] Y. Chen and Y. Liu, "Latest advances of LLC converters in high current, fast dynamic response, and wide voltage range applications," in CPSS Transactions on Power Electronics and Applications, vol. 2, no. 1, pp. 59-67, 2017, doi: 10.24295/CPSSTPEA.2017.00007.

[6] A. Garrigo´s, J. M. Blanes, J. A. Carrasco, J. L. Liza´n, R. Beneito, J.A. Molina, " 5 kW DC/DC converter for hydrogen generation from photovoltaic sources" International journal of hydrogen energy 35, vol 35, no. 12, pp 6123-6130, https://doi.org/10.1016/j.ijhydene.2010.03.131

[7] UCCx895 BiCMOS Advanced Phase-Shift PWM Controller Data Sheet.

[8] Half-Bridge LLC Resonant Converter Design Using FSFR-Series Fairchild Power Switch (FPS™). Application Note AN-4151.

Impact Protection of Vehicles by Automatic Cutting of General Power Supply with GTO

Alexandru Vasile,
Electronics and Telecommunications Faculty
Politehnica" University of Bucharest,
Bucharest, Romania,
alexandru.vasile@cetti.ro

Irina Bristena Bacis,
Electronics and Telecommunications Faculty
Politehnica" University of Bucharest,
Bucharest, Romania,
irina.bacis@cetti.ro

Ciprian Ionescu
Electronics and Telecommunications Faculty
Politehnica" University of Bucharest,
Bucharest, Romania,
ciprian.ionescu@cetti.ro

Summary: In the field of car electronics, in addition to the harsh conditions of temperature (-40°C - +130°C) and wide-spectrum vibrations, there is a danger of the vehicle igniting in the event of impacts producing a short circuit of the electrical conductors, which remain under voltage. The paper describes an automatic disconnection system of the car general power supply made with a modern device produced by Infineon - Gate Turn-Off Thyristor (GTO) - automatically operated by impact-sensitive electronic systems. At the end of the paper are mentioned the problems that occur in the modern supply of vehicles with DC voltage of 42V and the results obtained by the authors in the design of electronic circuits.

Keywords: impact protection, GTO, automotive electronics, car power supply

I INTRODUCTION

Any circuit or electronic protection device in the field of automobiles must meet two requirements:
- Correct starting of cars without additional manoeuvres in the whole range of vehicle operating temperatures [1];
- The protection circuit must not affect the normal operation of the electrical and electronic circuits of the vehicle.

This second point mainly refers to the voltage drop caused by the protection circuit. They could be significant and difficult to control. Also, the operation of the electric battery charger should not be influenced by these protection systems [2].

In the conditions of the increase of the car traffic and of the traffic speed on the roads, the number of vehicles that are buffered is increasing, and the number of vehicles that catch on fire after the impact is increasing. The ignition of the body of a crashed vehicle can be prevented by the impossibility of producing electrical sparks produced by various electric power cables existing throughout the body. Deformation of the body and crushing of the cables is usually done in a few tens of milliseconds.

It is known that in the event of an impact, the first action taken by the intervention teams is to disconnect the battery to prevent the production of sparks, an action that in many cases is late. The system proposed in this paper and on which various tests have been made has a response time of 10 milliseconds, a coverage time to eliminate the possibility of short circuit event, spark development and fire starting. The paper illustrates the electrical interconnection of the system with the rest of the electrical installation and of the electrical power subassemblies (alternator, starter, headlights, fan motor group, etc.). The impact moment is detected with the accelerometer system for deploying the airbag. A special problem of the proposed system is that of ensuring the starting conditions in a reasonable time. The paper illustrates three variants of starting circuits depending on the different energy consumption on three types of cars.

II CONSTRUCTIVE DESCRIPTION OF THE SYSTEM

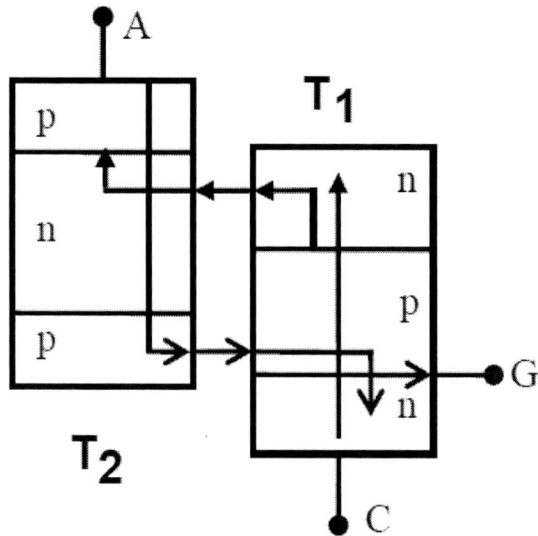

Fig.1. Current distribution in GTO at shutdown

The thyristor invented more than 50 years ago is a semi-controlled device that mainly once primed (entered into conduction) cannot be turned off (stopped conduction) except by interrupting the conduction current by various methods, very inconvenient procedure in many power applications, more chosen in DC-AC and DC to DC converters. In the case of these converters, the conduction current does not become zero naturally, but only with the help of sophisticated and expensive circuits (see Fig. 1).

The development of technology has made it possible to invent a Gate Turn Off (GTO) device which by injecting a reverse current into the gate which causes a considerable decrease in gate current and this in turn by the typical amplification of the main thyristor produces a very low conduction although in the ON state. and the device stops driving.

A GTO is a four-layer p-n-p-n device. To obtain a high efficiency of the emitter at the cathode end, the n + cathode layer is highly doped. in order to maintain a good efficiency of the emitter, the doping level of this layer is gradually reduced. In order to optimize the current-stopping capacity, the gate cathode junction must be interwoven from individual segments that are accessed through a common contact. The most popular device has several segments arranged in concentric rings around the center of the device.

The GTO device is started in conduction by increasing the anodic voltage until ICBO1 and ICBO2 (the respective base-collector currents) increase by the avalanche multiplication process or by injecting a gate current. The current gain of the transistors increases rapidly as the emitter current increases. Any mechanism to increase the emitter current can be used to start the device in conduction. Normally,

this is done by injecting a current pulse into the base region of the outer gate. Physically, as the sum of the carriers approaches the unit, the device begins to regenerate and each transistor conducts its saturation area. Once saturated, all junctions involve direct polarization and the total potential drop of the device becomes approximately equal to that of a single p-n diode. The anodic current is restricted only by the external circuit. Once the device has been switched on in this way, the external current on the grid is no longer needed to keep it running, as the regeneration process is self-sustaining. Returning to blocking mode only takes place when the anodic current is brought below the "holding current" level.

To stop a GTO from driving, a negative current is introduced on the grid regarding the cathode. The holes in the anode are extracted from the base of T2, and an inverse phenomenon begins. This is the most critical phase of the GTO shutdown process, as highly localized high temperature regions can cause the device to malfunction unless the segments shut down quickly. Once the anodic current has completely disappeared, the device regains its steady state blocking characteristics.

A picture of the device is illustrated in Fig. 1. The picture shows the mechanical-electrical system for interconnecting the GTO in the general electrical circuit of the vehicle. The GTO used is an Infineon device type T901SJBTW8. An interconnection problem is the electrical connection and heat transfer. The mechanical assembly is mounted on the positive terminal of the battery. It is provided with screws connecting the high current cable (hundreds of Amps) to operate the starter.

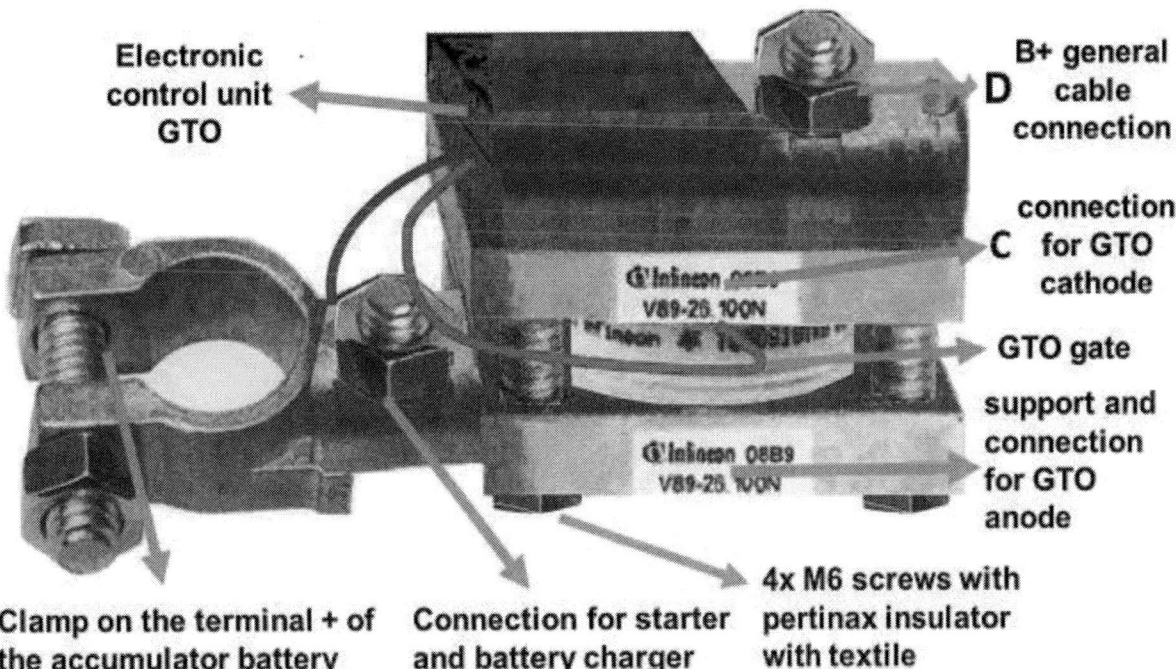

Fig. 2.The system for automatic disconnection of the general power supply with GTO

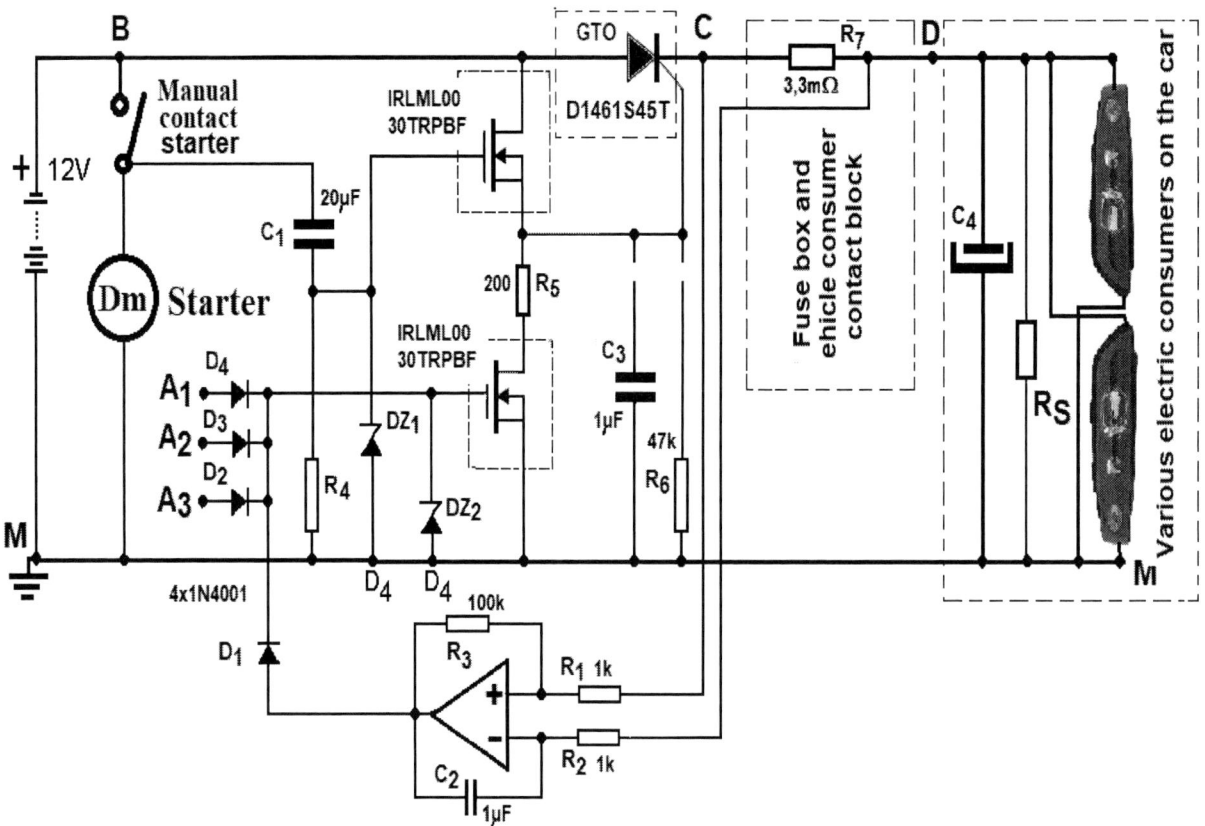

Fig. 3. Electronic diagram of the vehicle impact protection device

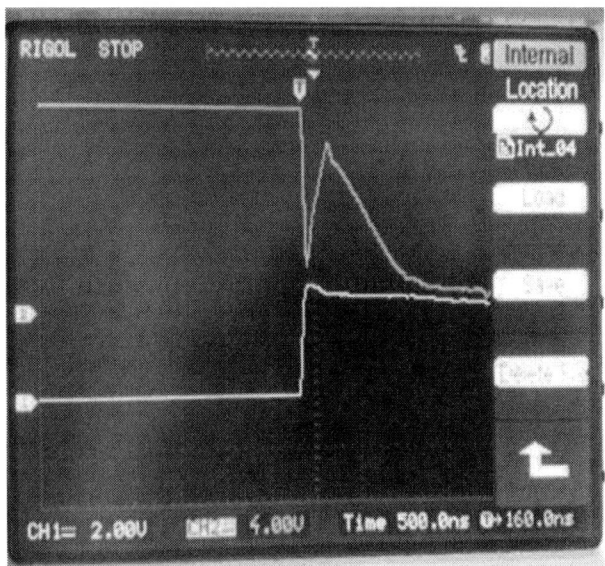

Fig. 4. Response of the GTO trigger system to impact

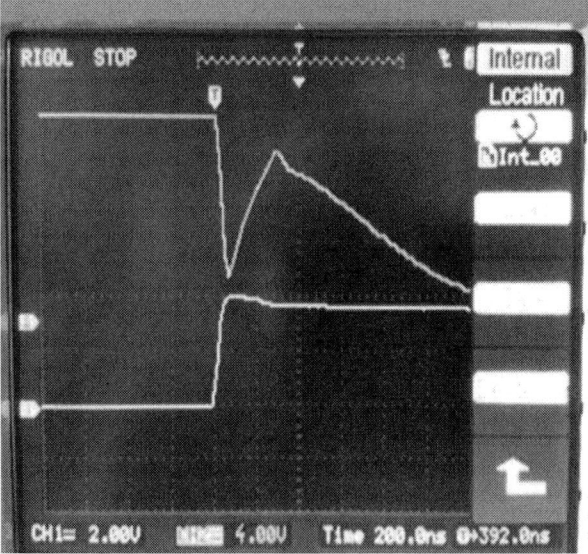

Fig. 5. The response of the GTO trigger system to the impact on the extended time base

II. FUNCTIONAL ILLUSTRATION OF THE SYSTEM

Figure 2 shows the electronic control circuit of the GTO (Gate Turn-Off Thyristor) device that meets all the requirements mentioned in paragraph 2. The "Manual contact starter" command activates the power supply of all electrical consumers on the vehicle in the following Way

The current pulse through the capacitor C1 and R4 in the first moment a positive voltage is applied on the grid of the CMOS device CM1, it enters the conduction and the GRTO gate receives a current so that the GTO enters the conduit. In a relatively short time of 40 ms the capacitor is charged and the voltage at point F decreases considerably so that the Device CM1 freezes and is taken out of operation. However, after 1 second the CMS is interrupted because the engine has started and the bendix on the starter interrupts its supply. In addition, the gate of the CM1 transistor is protected by a 16V Zenner DZ1 diode. After 40 ms, the entire electrical installation of the car is supplied with a nominal voltage of 12V in the case of cars and 24 V in the case of vehicles (see Figure 3 and Figure 4). In case of an event, the positive amplitude signal that triggers the explosion in the airbag is also applied in the CM2 grid CM2 by means of a Diode A1, A2, or / and A3. In this case, it is provided to receive signals from 3 airbags (front, left side and right side). Many vehicles have several impact protection devices and consequently the SI circuit consisting of diodes D2, D3 and D3 has an appropriate number of inputs [3]. The signal captured from the control circuit of the AIRBAG (which can come from the specialized computer) and applied on the grid of the CM2 device leads it to conduction and through the resistor R5 the GTO is closed in a relatively short time THX. Dida Dz 2 additionally provides the grid of the CM2 device for parasitic pulses [8] At the first impact, a deformation of the body materials begins, which is usually greater than 10 ms. In this way, the cables laid through the entire body can be crushed and short-circuited, leading to the ignition of the fuel scattered by the fuel system at a pressure of 2 bar.[11], [12],

The device described above in a much shorter time interrupts the electric current through the entire installation and the electrical handling no longer occurs, so the vehicle no longer catches fire [5]. The resistor R5 is a safety element in the operation of the circuit for limiting a direct current through CM1 and CM2, and the C3-R6 is an integration circuit for parasites captured from high voltage discharges produced by spark plugs [6]. In many cases it has been found that certain vehicles ignite when parked. This phenomenon occurs due to the aging of some insulators and the existence of accidental mechanical stresses that produce short circuits. To eliminate this unwanted event, a current detector element has been provided, dimensioned so that at a consumption exceeding 100 A, a comparator circuit triggers a positive signal that is applied through A1 and the GTO is extinguished in the same way as in the paragraph above.[9],[13]

It should be noted that if the starter is actuated within 50 ms, the supply is stopped and must be intervened by a specialized service to remove the cause that prevents the engine from starting [14].

Fig. 4 shows the time response of the GTO circuit (blue curve) to actuating the airbag deployment pulse (yellow curve). It is observed that a total turning off of the GTO device, connected in series with the general power supply circuit, occurs in a very short time 1.5 ms [10].

REFERENCES

[1.] B. Mihăilescu, I. Plotog, P. Svasta, M. Vladescu, "EM simulation and experimental assessment of VPS process control module", SIITME 2010, Romania, Pitesti, 23-26 Sept. 2010, DOI: 10.1109/ SIITME. 2010. 5650919, Page(s): 233-236;

[2.] R. K. Jurgen, "Automotive electronics Handbook", McGraw-Hill, Inc. New York; 2002;

[3.] Drumea, A. , "Low power aspects of a microcontroller-based module with wireless communication", Proc. of the 23rd International Symposium for Design and Technology in Electronic Packaging (SIITME2017), pp.134-137, October 2017.

[4.] C. Marghescu, and A. Drumea, "Embedded systems for controlling LED matrix displays," Proc. SPIE 10010 Advanced Topics in Optoelectronics Microelectronics and Nanotechnologies VIII (ATOM-N2016), vol. 10010, no. 1, pp. 100101E-100101E-6, 2016.

[5.] E. Marcu, R. A. Dobre, and M. Vlădescu, 'Key Aspects of Infrastructure-to-Vehicle Signaling Using Visible Light Communications', in Future Access Enablers for Ubiquitous and Intelligent Infrastructures, Cham, 2018, pp. 212–217, doi: 10.1007/978-3-319-92213-3_31.

[6.] R.-A. Dobre, R.-O. Preda, and M. Vlădescu, 'Sonic Watermarking Method for Ensuring the Integrity of Audio Recordings', Applied Sciences, vol. 10, no. 10, Art. no. 10, May 2020, doi: 10.3390/app10103367.

[7.] IB Brezeanu, PA Paraschivoiu, R Negroiu, LA Chiva, Applications of Kramers-Kronig relations, Design and Technology in Electronic Packaging (SIITME), 2017 IEEE 23rd International Symposium for, Page(s): 82-85, Location: Constanta, Romania, WOS:000428032300013, ISBN:978-1-5386-1626-0, Publisher: IEEE

[8.] P Svasta, R Negroiu, Al Vasile, Supercapacitors—An alternative electrical energy storage device, Electrical and Electronics Engineering (ISEEE), 2017 5th International Symposium on, Page(s): 1-5, Conference Location: Galati, Romania, WOS:000428234400002, ISBN:978-1-5386-2059-5,

[9.] L. A. Perişoară, "BER Analysis of STBC Codes for MIMO Rayleigh Flat Fading Channels", TELFOR Journal, Vol. 4, No. 2, pp. 78-82, 2012

[10.] Ş. G. Roşu, M. Ş. Teodorescu, Adriana Florescu, L. A. Perişoară, "Study of Operating Conditions Impact on Wireless Power Transfer Systems Performance", 11th International Symposium on Advanced Topics in Electrical Engineering (ATEE), Bucharest, Romania, March 28-30, 2019

[11.] Ciprian Ionescu, A. Drumea, N. Codreanu, A. Vasile, "Thermal investigations on high power LEDs", 6th Advanced Topics in Optoelectronics, Microelectronics and Nanotechnologies Conference, ATOM 2012, Constanţa, România, 23 - 26 august 2012, p. 139 - 141, ISBN:978-0-8194-9089-6, ISSN: 0277-786XWOS:000327457500085

[12.] Andrei Drumea, A. Vasile, "Aspects of serial communication in a network of medical devices", ISSE'06, 29th International Spring Seminar on Electronics Technology, 2006, ISBN 978-1-4244-0550-3, pp. 364-368, Accession Number: WOS:000246825800071.

[13.] L. A. Perişoară, A. Vasile, D. I. Săcăleanu, „Vehicles Diagnosis based on LabVIEW and CAN interfaces", 23rd International Symposium for Design and Technology in Electronic Packaging (SIITME 2017), Constanţa, Romania, pp. 383-386, Oct. 26-29, 2017

[14.] R Negroiu, P Svasta, C Pirvu, Al Vasile, C Marghescu, Electrochemical impedance spectroscopy for different types of supercapacitors, 2017/5/10, Electronics Technology (ISSE), 2017, Location: Sofia, Bulgaria, WOS:000426973000012, ISBN:978-1-5386-0582-0, Publisher: IEEE

2020 IEEE 26th International Symposium for Design and Technology in Electronic Packaging (SIITME)

Two-Stage Converter for Piezoelectric Energy Harvesting using Buck Configuration

C. Covaci and A. Gontean
Applied Electronics Department
"Politehnica" University
Timisoara, Romania
corina.covaci@student.upt.ro

Abstract—**Piezoelectric energy harvesting is considered a feasible solution for small scale energy harvesting, although it requires rectification, maximum power extraction, and output voltage regulation. For this purpose, several electronic circuits are used in literature, these being chosen depending on the application. In this paper, we simulate a piezoelectric energy harvesting circuit, using a two-stage converter based on a diode bridge rectifier and a buck configuration. The simulation is optimized for the SMD10T2R111WL piezoelectric transducer, whose SPICE model was developed in our previous work.**

Keywords—piezoelectric; energy harvesting; two-stage converter; buck configuration

I. INTRODUCTION

Piezoelectric energy harvesting is receiving more attention from the scientific world due to its advantages as [1-5]:

- high energy;
- power density;
- simple structure;
- low cost;
- it does not need an external voltage source;
- good scalability;
- ease of application.

Because of the direct piezoelectric effect, when an external mechanical force is applied to a piezoelectric transducer, it will generate an electric field, whose amplitude is highly dependent on the speed and magnitude of the input force. Since in the real world it is unlikely for the input force to be constant, without an external source, the piezoelectric energy harvesting solutions need to rectify and store the generated energy [6]. Among the best-known electronic circuits for piezoelectric energy harvesting are: diode bridges, two-stage converter circuits, Synchronized Switch Harvesting on Inductor (SSHI), and Synchronous Electric Charge Extraction (SECE).

This paper is focusing on a two-stage converter circuit using a buck configuration.

II. THE STANDARD SOLUTION FOR PIEZOELECTRIC ENERGY HARVESTING

The most common solution used to rectify the AC power generated by the piezoelectric transducers is the bridge rectifier electronic circuit. This solution uses a diode bridge to convert the piezoelectric transducer's output voltage of both polarities to a single polarity and charge a storage capacitor to a required voltage [7].

In Fig. 1, we use the SMD10T2R111WL piezoelectric transducer's SPICE model [8] to simulate the piezoelectric element in a standard AC-DC piezoelectric energy harvesting circuit.

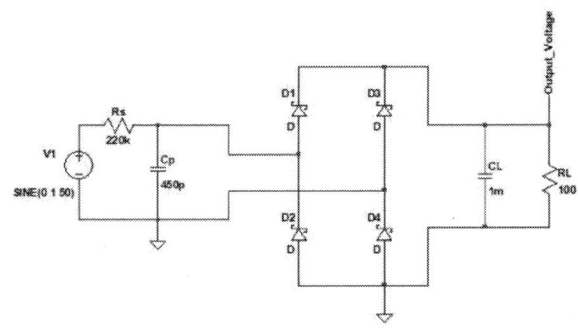

Fig. 1. Standard AC-DC energy harvesting circuit with SMD10T2R111WL piezoelectric model

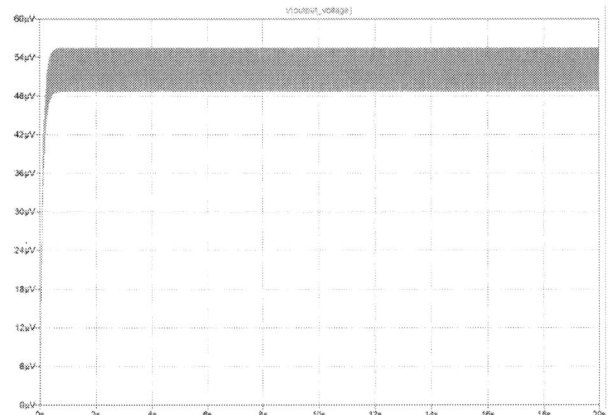

Fig. 2. The output voltage of the standard AC-DC energy harvesting circuit with SMD10T2R111WL piezoelectric model

978-1-7281-7507-2/20 $31.00 © 2020 IEEE

408

21-24 October, Pitesti, Romania

As you can see in Fig. 2, the output voltage of the standard AC-DC energy harvesting circuit is low (~50 µV). Therefore, the rectification of the generated voltage is not enough, it is also needed a DC-DC converter to maximize the power transferred to the storage device.

III. TWO-STAGE CONVERTOR USING A BUCK CONFIGURATION

A. General information

In Fig. 3 is represented the block diagram of a two-stage converter used to maximize the rectified power.

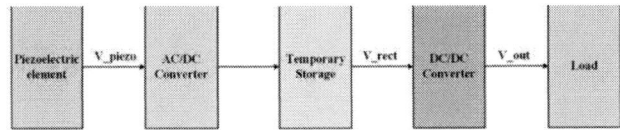

Fig. 3. Two-stage converter's block diagram

As in the previous chapter, we use the SMD10T2R111WL piezoelectric transducer's SPICE model to simulate the piezoelectric element. The AC voltage generated by the transducer is rectified using a standard diode bridge circuit and after that, the voltage is stored in a temporary storage device (we use a 1 mF capacitor). A Buck converter is used to maximize the power transfer from the temporary storage and the load. The diode bridge contains Schottky diodes to increase the circuit's efficiency.

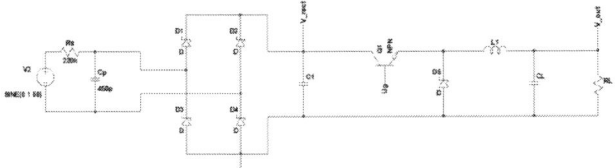

Fig. 4. Two-stage converter using buck configuration

The converter's efficiency η_C is calculated in (1):

$$\eta_C = \frac{V_{rect} + V_D - V_{ces}}{V_{rect}} \cdot \frac{V_{out}}{V_{out} + V_D} \quad (1)$$

where V_{rect} is the rectified voltage, V_D is the forward bias of the diode $D5$, V_{ces} is the voltage drop of the internal switch of $Q1$ and V_{out} is the circuit's output voltage.

B. Converter's dimensioning

The starting point of our simulation was based on Guan and Liao [9] work. We simulated their two-stage converter, but we used the SMD10T2R111WL piezoelectric transducer's SPICE model. The simulation's results showed that $V_{rect} = 2.6$ V and $V_{out} = 2.1$ V.

After the first simulation, we dimensioned the converter to obtain 1.8 V (standard supply voltage for several devices) at the circuit's output. As we know that the buck converter's input voltage is 2.6 V, we can calculate the control signal's duty cycle D using (2):

$$D = \frac{V_{rect}}{V_{out}} = 75\% \quad (2)$$

To calculate the inductance and capacitance of the buck converter, we used (3) and (4), found in the literature [10-12]:

$$L = D \cdot \frac{V_{rect} - V_{out}}{f_{SW} \cdot \Delta I_L} \quad (3)$$

$$C_{min} = \frac{\Delta I_L}{8 \cdot f_{SW} \cdot \Delta V_C} \quad (4)$$

Knowing that ΔI_L's formula for Buck converter is calculated in (5), the resulted value for $L1$ is 425 mH and the minimum value of CL is 4.41 nF.

$$\Delta I_L = 2 \cdot \frac{V_{out}}{R} \quad (5)$$

For inductance and capacitance calculation, we considered the switching frequency f_{SW} 20 kHz and the desired output voltage ripple ΔV_C 100 mV.

The results of the simulation are represented in Fig. 5.

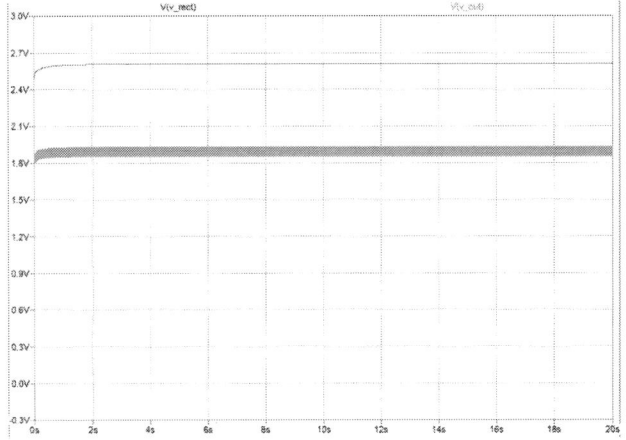

Fig. 5. The simulation's results

After the circuit's stabilization, V_{rect} mean value is approximative 2.6 V and V_{out} is around 1.8 V.

Although the simulation's results are as desired, the values of the components obtained using the standard Buck converter formulas are unrealistic. Therefore, we did another simulation using more realistic values for the inductance and capacitance.

C. Simulation using realistic components

This time, we used not only realistic components, but we used also SPICE models for real diodes and transducer. We chose the MBRS340 Schottky diode model and we change the NPN transistor with IRF7309P MOS transistor.

Moreover, the SPICE model of the piezoelectric transducer simulates this time the behavior of several SMD10T2R111WL transducers connected in series.

The new simulation and its results are presented in Figure 6 and Figure 7.

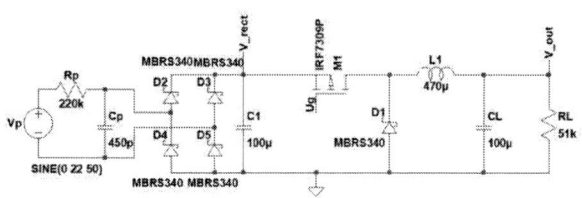

Fig. 6. The simulation's schematic

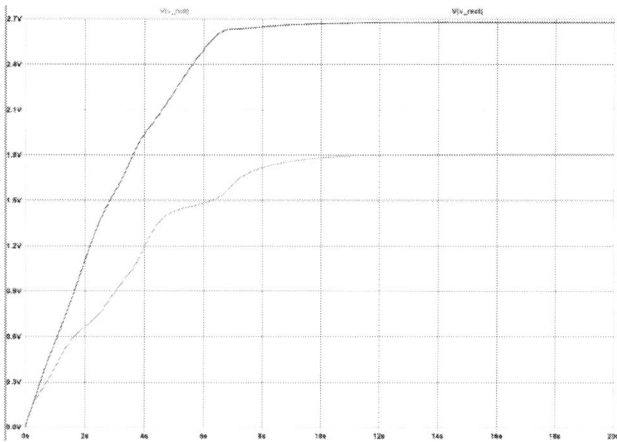

Fig. 7. The simulation's results

We used the same 20 kHz PWM with 75% duty cycle to control the transistor. We set the inductance to 470 µH, the capacitance to 100 µF, and the load resistance to 51 kΩ.

As observed in Figure 7, the output voltage of the two-stage converter using the buck configuration is 1.8V.

D. Converter's efficiency

The converter's efficiency can be determined using (1). V_{rect} and V_{out} are known from the previous sub-chapter, while V_D and V_{ces} values depend on the components used. In the simulation described above, we used ideal models of the bipolar transistor $Q1$ and diode $D5$.

The options for bipolar transistors and diodes on the market are numerous and hence the values of V_D and V_{ces} are various. The most common values are:

- For $Q1$, the internal switch usually has a voltage drop of 0.1V - 0.3V;

- If $D5$ is a rectifier diode, its forward bias is $0.6V - 0.7V$;

- If $D5$ is a Schottky, the typical forward bias is 0.1V - 0.3V.

Due to the numerous possible values for V_D and V_{ces}, the converter's efficiency cannot be generalized to a single value. Therefore, the efficiency of the converter using different diodes and bipolar transistors is represented in Fig. 8.

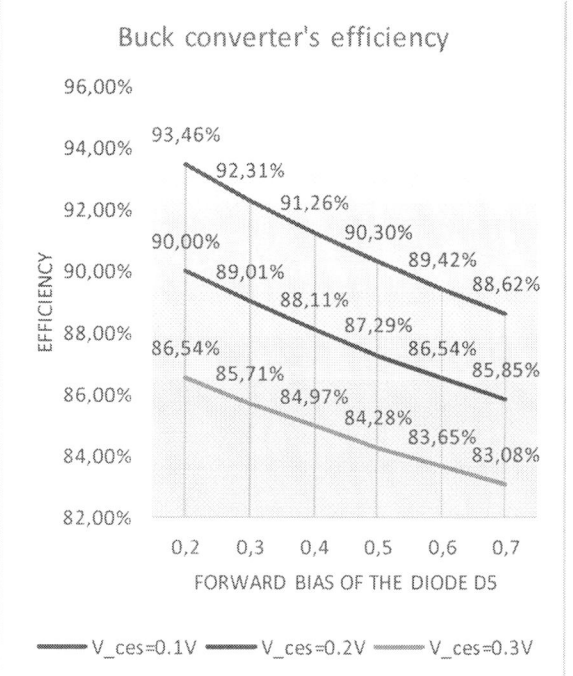

Fig. 8. Buck converter's efficiency for different values of V_D and V_{ces}

Naturally, the best efficiency of 93.46% is obtained when we use a bipolar transistor with the voltage drop of the internal switch of 0.1V and a Schottky diode with the forward bias voltage of 0.2V.

On the other hand, the worst efficiency of 83.08% is obtained when the voltage drop of the internal switch of $Q1$ is 0.3 V and $D1$ is a rectifier diode with the forward bias voltage is 0.7V.

To increase the converter's efficiency, one option is to use a synchronous configuration. In the synchronous alternative, the low side switch is composed of a transistor, instead of the diode present in Figure 4 and Figure 7. The two transistors are controlled so that they are never turned on simultaneously.

When a synchronous buck configuration is used, the V_D and V_{ces} values are equal. Taking this into consideration in equation (1), for synchronous configuration, the converter' efficiency is calculated using (2):

$$\eta_C = \frac{V_{out}}{V_{out} + V_D} \tag{2}$$

The efficiency of the two-stage converter using a synchronous buck configuration is presented in Figure 9.

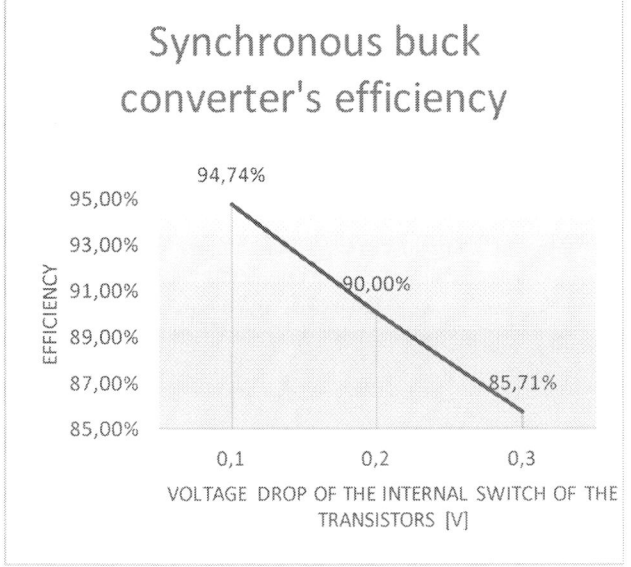

Fig. 9. Buck converter's efficiency for different values of V_D and V_{ces}

In this case, when we use two transistors with the voltage drop of the internal switch of 0.1V each, the efficiency of the converter is 94.74%. In the worst case, when the voltage of the internal switch is 0.3V, the converter's efficiency is 85.71%. Therefore, an efficiency improvement can be observed, indeed.

CONCLUSIONS

In conclusion, the standard rectifier using a diode bridge is not a feasible standalone solution for piezoelectric energy harvesting systems due to its low output.

The two stage-converter for piezoelectric energy harvesting presented in this paper contains beside the bridge rectifier a DC-DC buck converter. The converter stabilizes the generated output voltage to 1.8 V, which is the standard supply voltage for several devices.

Because of the low power of the piezoelectric transducer, when the standard buck's formulas are used, the values of the components are unrealistic.

The efficiency of the converter is highly dependent on V_D (the forward bias of the diode) and V_{ces} (the voltage drop of the internal switch of the transistor). Considering that the typical forward bias of the diode is between 0.1 V and 0.7 V (both rectifier and Schottky diodes are taken in consideration), while the transistor's voltage drop of the internal switch is usually between 01 V and 0.3 V, the calculated efficiency of the converter is 93.46% when precise components are used or 83.08% in the worst-case scenario.

The limits of the efficiency interval can be increased by using a synchronous configuration. This means that the buck's diode is replaced with a transistor controlled to be turned on only when the other transistor is turned off.

In this case, the efficiency varies from 85.71% in the worst-case scenario to 94.74% in best case scenario. The cost for this efficiency increasing is the complex circuit that has to control the two transistors so that they are never turned on simultaneously.

REFERENCES

[1] Maghsoudi Nia, E., Wan Abdullah Zawawi, N., & Mahinder Singh, B. (2019). Design of a pavement using piezoelectric materials. Materialwissenschaft Und Werkstofftechnik, 50(3), 320-328. doi:10.1002/mawe.201900002

[2] Gareh, S., Kok, B. C., Yee, M. H., Borhana, A. A., & Alswed, S. K. Optimization of the Compression-Based Piezoelectric Traffic Model (CPTM) for Road Energy Harvesting Application. Int. J. Renew. Energy Res. 2019, 9(3).

[3] Wu, L., Do, X.-D., Lee, S.-G., & Ha, D. S. A Self-Powered and Optimal SSHI Circuit Integrated With an Active Rectifier for Piezoelectric Energy Harvesting. IEEE Trans. Circuits Syst. I Regul. Pap. 2017, 64(3), 537–549. doi: 10.1109/tcsi.2016.2608999

[4] Erturk, A.; Inman, D. J. 1.1 Vibration-Based Energy Harvesting Using Piezoelectric Transduction. In Piezoelectric Energy Harvesting, 1st ed.; John Wiley & Sons, 2011; pp. 1–4.

[5] Piliposian, G., Hasanyan, A., & Piliposyan, D. The effect of the location of piezoelectric patches on the sensing, actuating and energy harvesting properties of a composite plate. J. Phys. D: Appl. Phys. 2019, 52(44), 445501. doi: 10.1088/1361-6463/ab37be

[6] Briscoe, J.; Dunn, S. 3.2.6 Applications. In Nanostructured Piezoelectric Energy Harvesters, 1st ed., Springer International Publishing, 2014; pp. 39–42.

[7] Covaci, C., & Gontean, A. (2020). Piezoelectric Energy Harvesting Solutions: A Review. Sensors, 20(12), 3512. doi:10.3390/s20123512

[8] Covaci, C., & Gontean, A. (2018). SPICE Model of a Piezoelectric Transducer. 2018 IEEE 24th International Symposium for Design and Technology in Electronic Packaging (SIITME). doi:10.1109/siitme.2018.8599212

[9] Guan, M. J., & Liao, W. H. (2007). On the efficiencies of piezoelectric energy harvesting circuits towards storage device voltages. Smart Materials and Structures, 16(2), 498-505. doi:10.1088/0964-1726/16/2/031

[10] Discontinuous Conduction Mode of Simple Converters - Technical Articles. (n.d.). Retrieved August 20, 2020, from https://www.allaboutcircuits.com/technical-articles/discontinuous-conduction-mode-of-simple-converters/

[11] Basic Calculation of a Buck Converter's Power Stage. (n.d.). Retrieved August 20, 2020, from https://www.ti.com/lit/an/slva477b/slva477b.pdf?ts=1597824376893&ref_url=https://www.google.com/

[12] Basic Calculation of a Buck Converter's Power Stage. (n.d.). Retrieved August 20, 2020, from https://www.richtek.com/Design Support/Technical Document/~/media/AN_PDF/AN041_EN.ashx

2020 IEEE 26th International Symposium for Design and Technology in Electronic Packaging (SIITME)

The Energy Efficiency of a Prosumer in a Photovoltaic System

Marius-Alexandru Dobrea
Dept. of Automatic Control and Industrial Informatics
Faculty of Automatic Control and Computer Science, UPB
Bucharest, Romania
marius.alexandru.dobrea@gmail.com

Stefan Bichiu, Ioana Opris
Dept. of Power Generation and Use
Faculty of Power Engineering, UPB
Bucharest Romania
b_stefan94@yahoo.ro , ilopris@gmail.com

Mihaela Vasluianu
Collective of Automation and Applied Informatics
Faculty of Hidrotehnics, UTCB
Bucharest Romania
mihaelavasluianu89@gmail.com

Abstract—This article aims to calculate the net income achieved by a prosumer for an existing On-Grid system of photovoltaic panels consisting of a set of 12 polycrystalline panels with a nominal power of 250 W and 60 cells and an On-Grid inverter (StecaGrid model) with a maximum power of 3000 W. The prosumer is located in southern Romania, Ilfov County. Modeling of the photovoltaic panel in Matlab / SIMULINK was performed using a mathematical model, and it was compared with a real data set. Next, an analysis was made of the energy produced by the photovoltaic system, and the energy consumed by the prosumer, the excess energy being introduced in the national energy system. We have achieved a net income saving for March, so the analyzed photovoltaic system has energy efficiency that can significantly reduce electricity costs over time. Thus, using the Matlab / Simulink model, photovoltaic systems can be designed to satisfy the needs of prosumers.

Keywords—Prosumer, photovoltaic system, mathematical model, energy efficiency

I. INTRODUCTION

Solar energy conversion technologies (which can be used in the form of heat or electricity) are increasingly used in homes, commercial, administrative and industrial units. Photovoltaic (PV) solar cells directly convert solar energy into electricity. They consist of semiconductor materials that absorb sunlight and activate the detachment of electrons from the atoms of the semiconductor material, allowing the formation of a flux that constitutes the electric current[1]. These cells are contained in modules of about 40-60 units, and the sunlit silicon layer behaves like a battery (which has a negative pole and a positive pole)[2].

As with batteries, the voltage can be increased by connecting photovoltaic cells in series, and the capacity by connecting them in parallel. From a constructive point of view, silicon is used in various forms: single crystals, multi-crystals and thin-layer amorphous cells[2].

In Romania, the photovoltaic field is active in renewable energies. This is revealed by the massive number of connection contracts signed by distribution companies compared to the small number of projects developed so far[3]. Significant changes have taken place in the energy sector in most European countries due to the need to increase security in the supply of energy to consumers, and within this requirement, renewable energy sources offer a viable solution, including environmental protection[4].

Prosumers are those who consume energy from the electricity grid and also produce and deliver energy from renewable sources. They produce for their own consumption and inject the excess into the network, for a fee, based on a sale-purchase contract with their electricity supplier. At the end of January, there were 160 officially registered prosumers in Romania[5].

In the last ten years, the price of photovoltaic systems has fallen steadily amid increasing use. Today we find photovoltaic cells on the garden lighting system, on external batteries for mobile phones, boats, even on trains or planes[6].

The increasing promotion of photovoltaic systems, the advancement of research in the field and the fact that more and more states have already implemented or are implementing policies to subsidize solar energy systems, contribute to an even greater accessibility of this technology. Besides, investment in such technology recovers in about five years by lowering electricity bills and increasing the market value of the building where they are located. The closer the energy production is to the place of consumption, the lower its transmission/distribution losses are and, implicitly, money is saved[1].

II. RELATED WORK

An on-grid photovoltaic system is a system connected in parallel to the public grid, and when the energy produced by the photovoltaic panels is insufficient, the necessary surplus is provided by the electricity distribution network[7]. On grid photovoltaic systems may not have batteries, and the electricity produced during the day is used for own consumption or injected into the network to be used by other consumers. Because the ON-GRID inverter is connected in parallel to the grid, consumers can take some of the energy produced from the panels and the rest from the national grid[7].

If desired, the ON-GRID inverter can limit energy production to the value of consumption, and thus the contract with the energy supplier is not necessarily necessary. It is the ideal solution for companies that have significant consumption during the day, thus having the possibility to reduce the electricity bill by over 75%[2].

In the paper, [8] the authors present a review of the grid-connected solar photovoltaic system. The complete grid-connected photovoltaic system architecture includes the construction of the photovoltaic matrix, MPPT methods, DC-DC converters, inverters and control algorithms. Different topologies and control methods of the grid-connected photovoltaic system are studied to justify the potential of the grid-connected photovoltaic system.

The authors of the paper [9] developed a hybrid generation system. The purpose of this system is to take minimum energy consumption from the national distribution network. The tests were performed over a period of one day, using real experimental data.

In the paper [10], an overview was made on the integration of solar energy in electricity networks. Integration technology has become essential due to the energy requirements of the world which have imposed a significant need for different methods by which energy can be produced or integrated. In addition to the fact that the integration of solar energy into non-renewable sources is important because it reduces the rates of consumption of non-renewable resources thus minimises the dependence on fossil fuels. The photovoltaic or photovoltaic system drives this revolution by using the available power of the sun and transforming it from direct current to alternating current.

III. THE STRUCTURE OF THE ON-GRID PHOTOVOLTAIC SYSTEM

A. Principle of operation

Photovoltaic systems are generally composed of six individual components: solar photovoltaic panels, charge regulator, inverter, intelligent electricity meter and electricity grid[1].

The correct installation of these components determines how efficient the solar panels are. However, a charge controller and battery bank are optional. Even if these two components help you store and, as such, make better use of the electricity generated, they could also increase the total price of the photovoltaic installation. Even if photovoltaic solar installations produce power when exposed to the sun, the other components are necessary for the correct conversion, distribution and storage of energy produced by solar panels[1].

The specific equipment of a prosumer is a bidirectional power meter (import/export) that allows the monitoring of imported and exported energy from the national energy system.

As can be seen in the figure, the photovoltaic system used is composed of the following elements:

1. Photovoltaic panels
2. DC Isolator
3. Inverter
4. Fuse Board
5. Import / Export power meter
6. National Energy System
7. Load

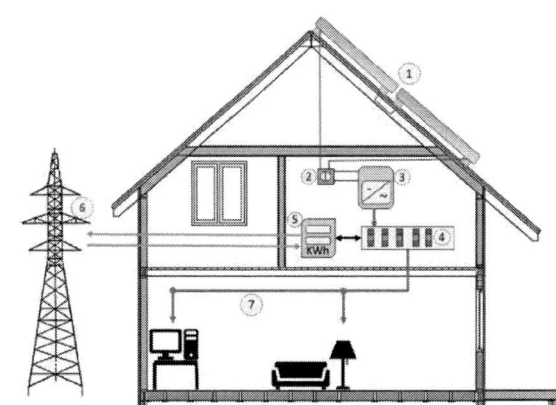

Fig. 1. On Grid Photovoltaic System

B. Modeling the photovoltaic system

The article aims to compare the energy produced by the photovoltaic system with the energy consumed by the prosumer. The consumer is located in southern Romania, Ilfov County.

The photovoltaic system used by the prosumer is an On-Grid system of photovoltaic panels consisting of a set of 12 panels (polycrystalline model) with a rated power of 250 W and 60 cells and an On-grid inverter (StecaGrid model) with a maximum power of 3000 W.

To calculate the potential income, which a prosumer can achieve by providing energy in the energy system, it is necessary to model and simulate the photovoltaic system.

The mathematical model described in the article [6] is implemented in Matlab / Simulink, as follows:

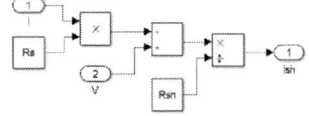

Fig. 2. Shunt Current model in Simulink

978-1-7281-7507-2/20 $31.00 © 2020 IEEE

The model presented in figure 2 shows the linear behaviour of the incident solar radiation and, at the same time, it is temperature-dependent. As can be seen in the model, the subsystem is affected by saturation current and short circuit current.

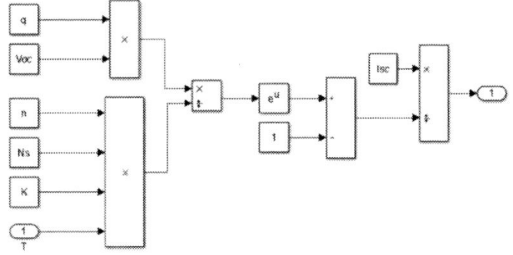

Fig. 3. Reverse saturation current model in Simulink

Figure 3 is a subsystem of the photovoltaic system and uses bandgap, reference temperature and electron charge.

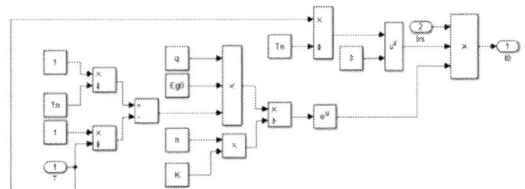

Fig. 4. Saturation current model for simulation in Simulink

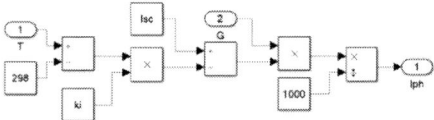

Fig. 5. Photocurrent model for simulation in Simulink

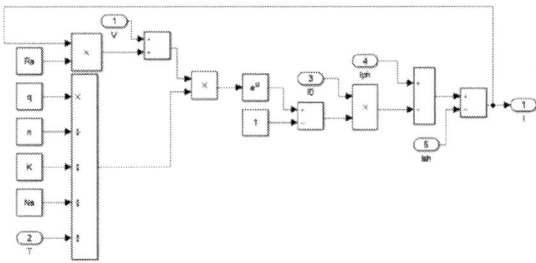

Fig. 6. Photovoltaic current model in Simulink

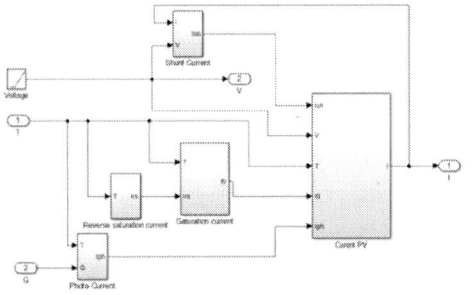

Fig. 7. Modelling the photovoltaic system in Simulink

Figure 7 shows the photovoltaic system, which is the addition of simulation models of Shunt Current, Photocurrent, Saturation Current, and Reverse Saturation Current. This complete system use temperature and irradiation as input and results in current and voltage as output.

IV. RESULTS

After the realization of the mathematical model in Simulink, the characteristics of the photovoltaic system are realized.

Figure 8 shows the performance of the photovoltaic panel through the four types of characteristics:

- Current-Voltage characteristic, where the temperature is constant and the solar radiation is variable (8.a)
- Current-Voltage characteristic, where the temperature is variable and the solar radiation is constant (8.b)
- Power-Voltage characteristic, where the temperature is constant and the solar radiation is variable (8.c)
- Power-Voltage characteristic, where the temperature is variable and the solar radiation is constant (8.d)

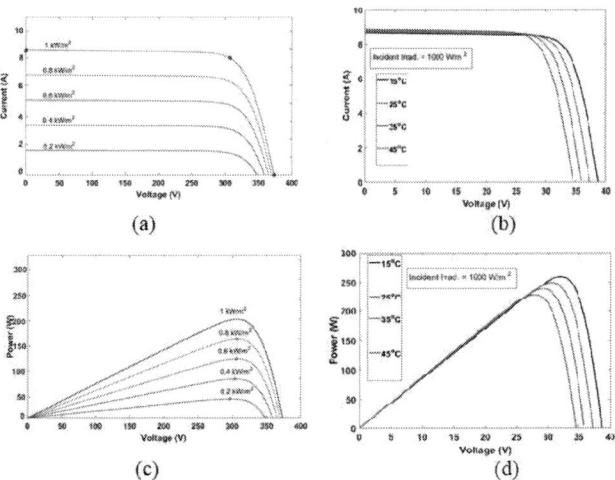

Fig. 8. Characteristics of the photovoltaic system

In March 2020, values of Solar Temperature and Radiation were registered and using a mathematical model, we performed the simulation of the output power [W] at the inverter where the input quantities are represented by the temperature and the solar radiation measured every 30 minutes.

At the same time, we monitored the real power obtained from the photovoltaic panel system with the help of a measuring device (SMA Energy Meter). In the figure below, you can see the variation of electricity obtained from simulation and measurement. From the comparison of the two sets of values, a correlation coefficient of approximately 95% was obtained, the main disturbances coming from the partial shading phenomenon.

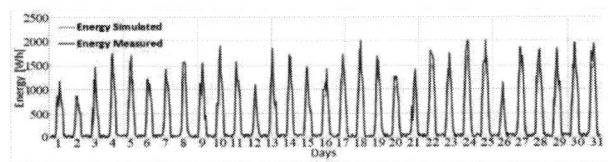

Fig. 9. Energy produced by the PV system during month March 2020

The correlation coefficient represents the degree of accuracy that describes the relationship between two variables using a monotone function, as follows[12]:

$$Coef(X,Y) = \frac{\sum(x-\bar{x})(y-\bar{y})}{\sqrt{\sum(x-\bar{x})^2 \ \sum(y-\bar{y})^2}} \qquad (1)$$

where, $\bar{x}$ is the average measured data, and $\bar{y}$ is the average simulated data.

The following figure graphically represents a comparison between the energy produced by the photovoltaic panel system and the energy consumed in the building for the average hourly value. For March 2020, consumption of 136 kWh was achieved, and the energy production was 331 kWh. The surplus energy was used to deliver energy to the national energy system and bring in a foreign exchange benefit.

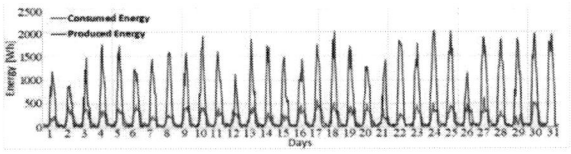

Fig. 10. The energy produced and consumed during month March 2020

The figure below represents the amount of energy delivered to the grid when there is a excess or taken from the grid when there is a power deficit. The energy deficit that appears in March comes from the fact that the professional activity takes place at home and the work schedule is until 19:00. Thus, in the time interval 17:00 - 08:00, where the production of electricity from the system with photovoltaic panels is relatively negligible, it is necessary to supply the building with energy taken from the local supplier. Between 08:00 and 17:00, a surplus of electricity is generated and it can be delivered to the network. The revenue from the supply of electricity in the system is 0.251 Ron / kwh, and the cost for taking over the energy from the network is 0.524 Ron / kwh.

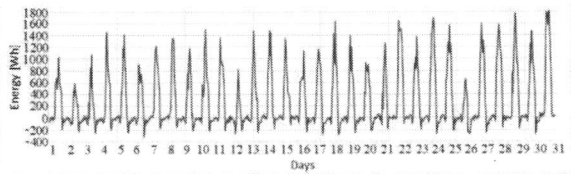

Fig. 11. Excess energy achieved during month March 2020

The income from the electricity delivery in the grid is 0.251 Ron / kWh, and the cost for taking over the energy from the system is 0.524 Ron / kWh [5].

To calculate the net benefit, we applied the following equation [5]:

Net benefit = Net Value Export Income + Net Value Electricity Bill Savings – Net Value Costs (2)

Where, Net Value Export Income represents the value of energy sold to the grid, at the prosumer price, Net Value Electricity Bill Savings represents the value of the energy consumed from the own generation of electricity (from the PV system), at the price of household consumers, Net Value Costs represents the amount of energy consumed bought from the energy supplier at the price of the household consumer.

Net benefit = 229.34 kWh * 0.251 Ron + 101.44 kWh * 0.524 Ron - 34.68 kWh * 0.524 Ron (3)

The net income achieved for the analyzed period was 76.56 Ron.

V. Conclusions. Perspectives. Requirements

The main purpose of this paper is to calculate the annual revenue generated by a prosumer for On-Grid existing system that contains a set of twelve polycrystalline solar panels with a rated output of 250W and an inverter with a maximum power of 3000 W.

To make the set of photovoltaic panels a mathematical model was developed, following the data obtained from this mathematical model, a comparison was made between these data and real data and a modeling and simulation was done in Matlab / Simulink, and the real data were taken by a prosumer from Romania, Ilfov County.

After modeling was performed an analysis of energy consumption and energy prosumer total proposed system and the excess was introduced into the national power grid.

The real power obtained from the photovoltaic panel system is monitored with the help of a measuring device (SMA Energy Meter) to see the variation of electricity obtained from simulation and measurement. The comparison of the sets of values, a correlation coefficient of approximately 95% was obtained, the primary disturbances coming from the partial shading phenomenon.

Following the analysis, an energy-saving was achieved, and an income of 0.251 RON / kWh was delivered in the national network.

References

[1] M. Mathew and J. Hossain, "Analysis of a grid connected solar photovoltaic system with different PV technologies," 2017 IEEE International Conference on Circuits and Systems (ICCS), Thiruvananthapuram, 2017, pp. 264-269, doi: 10.1109/ICCS1.2017.8326002.

[2] K.N. Nwaigwe, P. Mutabilwa, E. Dintwa , An overview of solar power (PV systems) integration into electricity grids, Materials Science for Energy Technologies 2 (2019) 629–633

[3] Mihaela Vasluianu, Oana Carmen Niculesc Faida, Ramona-Oana Flangea, Neculoiu Giorgian, Mariana Marinescu, Microgrid System for a Residential Ensamble, 2019 22nd International Conference on Control Systems and Computer Science (CSCS), May 2019, Bucharest, ISBN: 978-1-7281-2331-8

[4] Mihaela Puianu, Ramona – Oana Flangea, Mariana Marinescu, Viorel Marinescu, Cloud Computing for a Hybrid System, 2017 16th RoEduNet Conference: Networking in Education and Research (RoEduNet), Targu Mures, Romania, Decembrie 2017, ISBN: 978-1-5386-3411-0

[5] GfK Belgium consortium, Study on "Residential Prosumers in the European Energy Union" JUST/2015/CONS/FW/C006/0127, Framework Contract EAHC/2013/CP/04, pp. 11, 182-183, 2017.

[6] Manoharan Premkumar, Chandrasekaran Kumar, Ravichandran Sowmya, Mathematical Modelling of Solar Photovoltaic Cell/Panel/Array Based on the Physical Parameters from the Manufacturer's Datasheet, Int. Journal of Renewable Energy Development 9 (1) 2020: 7-22.

[7] Michał Jasiński, Jacek Rezmer, Tomasz Sikorski, Jarosław Szymańda, Integration Monitoring of On-grid Photovoltaic System:Case Study, Periodica Polytechnica Electrical Engineering and Computer Science, 63(2), pp. 99–105, 2019

[8] Ajay Kumar, Dr Nitin Gupta, Vikas Gupta, A Comprehensive Review on Grid-Tied Solar Photovoltaic System, Journal of Green Engineering 7(1):213-254, January 2017, DOI: 10.13052/jge1904-4720.71210

[9] Eric O'Shaughnessy, Dylan Cutler, Kristen Ardani, Robert Margolis, Solar plus: A review of the end-user economics of solar PV integration with storage and load control in residential buildings, Applied Energy, Volume 228, 2018, Pages 2165-2175, ISSN 0306-2619, https://doi.org/10.1016/j.apenergy.2018.07.048.

[10] Rezvani, A., Esmaeily, A., Etaati, H. et al. Intelligent hybrid power generation system using new hybrid fuzzy-neural for photovoltaic system and RBFNSM for wind turbine in the grid connected mode. Front. Energy 13, 131–148 (2019). https://doi.org/10.1007/s11708-017-0446-x

3D tracking system at maximum solar emissivity with microcontroller

B. Dumitrascu, L. Baicu, A. Culea Florescu and N. Nistor

Department of Electronics and Telecommunications
University "Dunarea de Jos" of Galati
Galati, Romania
bogdan.dumitrascu@ugal.ro

Abstract—In this paper a smart real-time tracking and positioning for a solar cell or solar panel in order to maximize the solar energy recovered from the electromagnetic radiation from the Sun. The idea of the paper is to use a mathematical transform from the Cartesian coordinates to spherical coordinates, with the mechanical part implementation based on two changes (the azimuth change and angular change) using two actuators, made of precision stepper motors. The system will position and correct the positioning of the solar panel by reading in real-time the instantaneous current on a test load (different from the real charging circuit load), the current is digitized using the internal analog to digital convertor of a microcontroller and stored in a memory buffer. The position for the maximum solar electromagnetic radiation that can to reach the solar panel is determined by periodic scanning for the maximum generated current. In order to scan for the maximum current, the solar cell's orientation is changed for the horizontal angle and azimuth. The solar panel will be rotated in small increments a few degrees in both directions and the instantaneous current will be recorded along with the orientation. The rotations will be for the horizontal angle and azimuth (up and down). A two steps algorithm is used to determine each local maximum. The first step determines the average component using the analog to digital convertor with lower resolution. The second step involves a higher resolution search in the vicinity of the maximum from step one. The higher resolution current data will be stored in a memory buffer along with the corresponding horizontal position and azimuth. With the latest current data and the values stored in the memory buffer, the microcontroller will perform corrections to the position of the solar panel.

Keywords—Current monitoring, microcontroller process, energy efficiency, optimum position

I. INTRODUCTION

According to [1] the solar radiation that reaches the ground is influenced by several parameters, such as longitude and latitude, which influences the distance travelled by the solar radiation in the atmosphere before reaching the solar panel. The energy quantity that can reach a solar panel (after passing through the atmosphere at a certain angle) depends in principle by the orientation of the solar cell in reference to the solar ray. The maximum efficiency is when the angle between the solar panel and the solar rays is 90 degrees. The efficiency of the panels can be improved by changing the orientation of the panels [2] and [3].

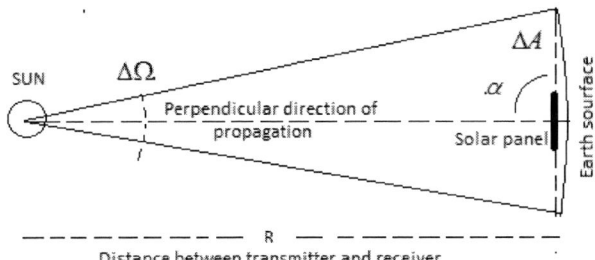

Fig. 1. Representation of the relation between the sun and solar panel.

In this paper a study regarding the orientation of a mini solar panel, based on an algorithm of dynamic and periodic testing of the position of the solar panel compared to the solar rays, using an electronic circuit that measures the current generated by the panel on a constant load (during the test that takes a few seconds). During large time intervals (comparable with the time constants of the sun) the circuit disconnects the main load and connects a constant known load to the solar panel for testing. During testing the panel is rotated in small increments to determine the optimum horizontal and azimuth angles. The currents measured during the test movements are recorded in a matrix along with the position of the solar panel during the measurement. At the end of the test the optimum angle and azimuth is determined and panel is rotated to the optimal orientation. After the positioning the main load is reconnected.

The maximum efficiency is obtained when the angle between the panel and solar rays is 90 degrees see fig. 1. The radiant energy flow is determined by the energy passing through a surface in a given time, t, and has the following expression:

$$\phi = \frac{W}{t}[Watt] \tag{1}$$

The solid angle, measured in steradians, under which a spherical surface ∆A is seen from a distance R is by definition:

$$\Delta\Omega = \frac{\Delta A}{R^2}[sr] \tag{2}$$

The intensity of the solar radiant energy is defined by the energy flow on the solid angle:

$$I = \frac{\Delta\phi}{\Delta\Omega}\,[W/sr] \tag{3}$$

Finally the value we are interested in is the energy density defined as:

$$E = \frac{\Delta\phi}{A} = \frac{I}{R^2}\sin\alpha, \tag{4}$$

where α is the angle of the solar rays and the parallel through the point of contact with the earth surface. The energy intensity is important for this study because the power of the recovered energy using a solar panel depends directly on this value, which influences the number of photons taking part in the energy conversion.

It can be seen that the energy intensity is inversely proportional by the distance to the emitter and receiver, a value that cannot be changed and directly proportional to sin(α), which has the maximum value when the solar rays are perpendicular to the solar panel. The article proposes to use this property of the energy efficiency of the solar panel by positioning the solar panel in a way that insures the orientation of the panel is perpendicular to the solar rays.

According to [4] the following theoretical relation between the intensity of the solar radiation and the angle of travel through the atmosphere exists:

$$I_{a,n} = I_0\left(a_0 + a_1 e^{\frac{-k}{\cos(\theta)}}\right)\left[W/m^2\right], \tag{5}$$

where:Io is the extraterrestrial irradiance; θ is the solar zenith angle. For the clear 2-3 km visibility haze model, the three constants in equation (5) are [4]:

$$\begin{cases} a_0 = 0.423 - 0.00821\cdot(6-A)^2 \\ a_1 = 0.505 - 0.00595\cdot(6.5-A)^2, \\ k = 0.2711 - 0.01858\cdot(2.5-A)^2 \end{cases} \tag{6}$$

where A is the local elevation in kilometers.

II. EXPERIMENTAL PLATFORM

The experimental platform is presented in fig. 2. It consists of a solar panel, microcontroller, driver circuit for stepper motors, two stepper motors, a constant load, a current sensor and a light source. A stepper motor is used to set the panel's horizontal angle and the second motor is used to set the azimuth. The microcontroller is used to control the 2 motors and to read the current generated by the panel. An algorithm is implemented to search for the maximum efficiency by changing the panel's orientation and measuring the current and

a two steps algorithm to determine the position where the generated current is the highest and record the position after this the panel is moved to the recorded position of maximum efficiency.

A block schematic of the algorithm is presented in fig.3. The panel is set to the start position and checks if current is being generated (a light source exists) and if not it will remain in a loop until a current is being generated. If a current is generated the maximum current is scanned using the two step algorithm. If a higher current is detected the position is saved and the panel is moved to this position until the new scan for a new higher current. If the maximum current found after the scanning algorithm is lower than the minimum set current, the panel is moved back to the starting position. A delay between maximum current scans is implemented to take into account for the time constants for the changes in position of the light source.

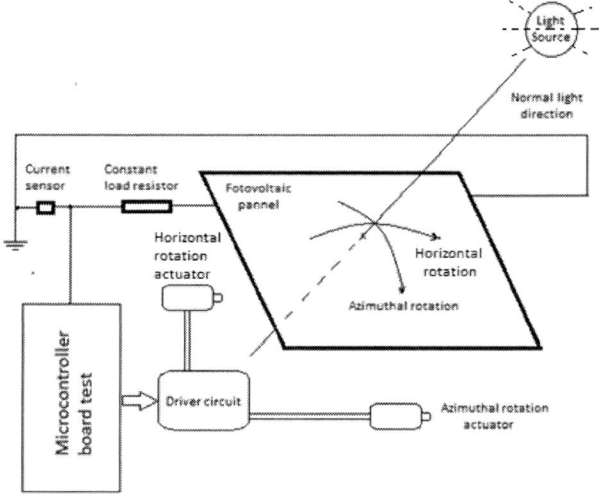

Fig. 2. Experimental platform

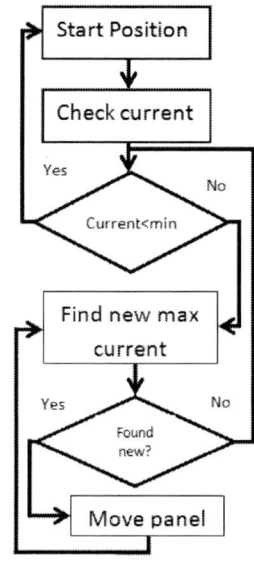

Fig. 3. Schematic block of the positioning algorithm

III. PRECISION CURRENT SCANNING USING TWO STEP ALGORITHM

The current scanning algorithm receives the initial position and measures the current in the vicinity of the given position and returns the position of the maximum current. The algorithm for precision scanning consists of two steps and is presented in fig.4. In the first step the voltage domain for the ADC is set to the maximum (0-5 V) and the voltage from the current sensing resistor is measured while the solar panel is rotated with small angle (2-4 degrees).

In step two the measurements are resumed for the previous measurements but the voltage references are changed according to the average current from the first step. The current variation data is stored in a memory buffer. The maxim current is determined from the available data and the position is stored. After the maximum current is determined the position of the maximum current is used for the position of the panel.

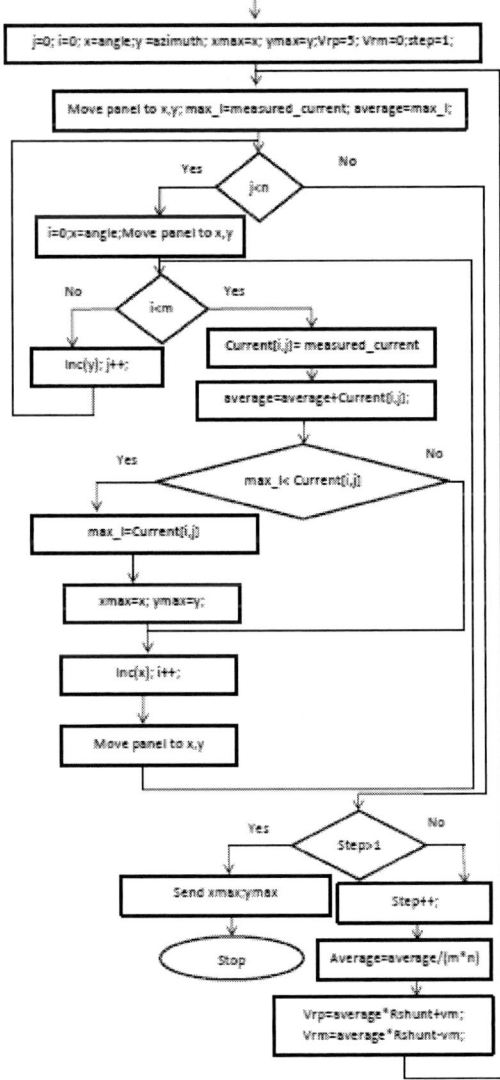

Fig. 4.   Current scanning using two steps algorithm

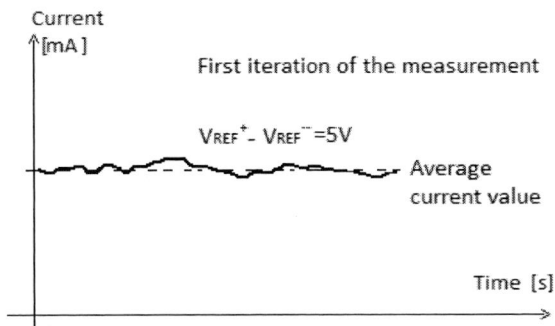

Small current variations measured with the analog port of the microcontroller for large voltage reference values

Fig. 5.   First step current scanning

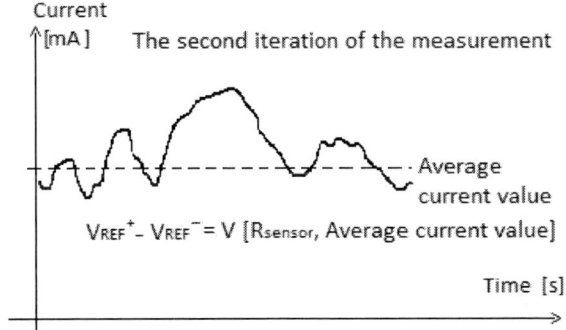

Large current variations measured with the analog port of the microcontroller for small voltage reference values

Fig. 6.   Second step current scanning

IV. RESULTS

Preliminary results obtained, using laboratory equipment,have revealed the existence of an easy detectable variation of the generated current for small variations of the horizontal angle and azimuth. These results are presented in table 1 and fig. 7 and 8. The experiment contains a small set of measurements that were manually recorded.

TABLE I.   PRELIMINARY EXPERIMENTAL MEASUREMENTS

Angle variation	Azimuthal variation	Horizonthal variation
Degree	μA	μA
-2	0.09	0.2075
-1.5	0.142	0.2275
-1	0.166	0.24125
-0.5	0.18	0.24875
0	0.19	0.25375
0.5	0.181	0.25
1	0.175	0.24625
1.5	0.168	0.235
2	0.114	0.205

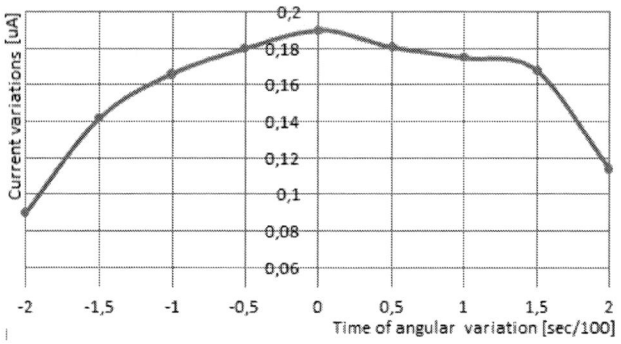

Fig. 7. Current variation during horizontal angle change

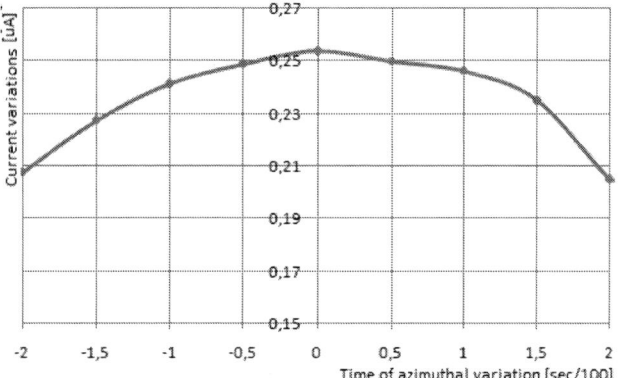

Fig. 8. Current variation during azimuthal variation

After implementing the proposed method in this paper, and following the two steps of initial scaling and rescaling with improved precision, using a microcontroller with a 10 bit resolution DAC and variable Vref+ and Vref-, we obtained a set of data. Vref+ and Vref- are determined by the algorithm during the first step and used in the second step of the search to improve the precision of the measured current. The set of data was obtained using the proposed algorithm and was sufficient in order to plot the correction characteristics.

The experimental data were obtained with high resolution and at a rate of around 500 samples per second. The results are presented in fig.9 and fig.10.

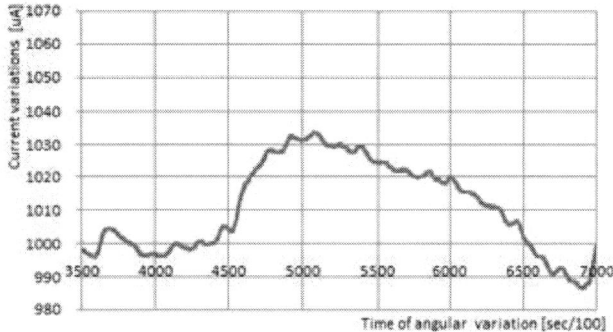

Fig. 9. Experimental data for angular change from -4 to 4 degrees

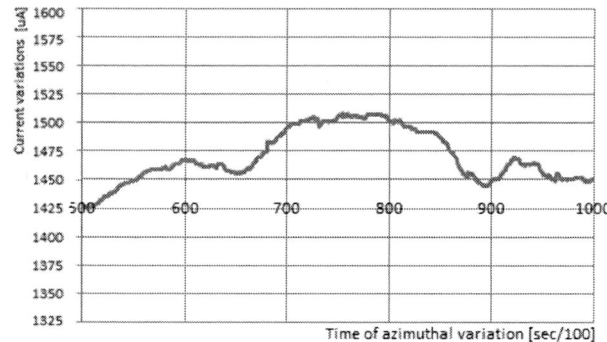

Fig. 10. Experimental data for azimuth change from -2 to 2 degrees

V. CONCLUSIONS

The research shows that high precision measurement of the current generated at constant illumination of a solar cell shows small variations that can be detected using dynamic rescaling of the microcontroller's voltage reference, resulting in a new voltage domain based on previous determination.

Using the proposed method very small azimuth and angular movements of the solar panel can be achieved at low rotation speeds, almost unnoticeable both in a visual, mechanical and extra power consumption point of view.

The method is based on the dynamic correction of the position of the solar cell so that the solar rays will always reach in the optimal position.

Using this method local maximum efficiency generation can be obtained.

The method requires the sampling storing and processing of the data about the currents corresponding to different special positions and periodic correction of the position with a refresh rate of 15 minutes.

Further studies involve a more detailed analysis of stepper motors used to correct the two angles and the adaptation of the method for a variable load.

REFERENCES

[1] W. B. Stine and M. Geyer, "Power From The Sun", Power From The Sun.net;

[2] F. I. Mustafa, S. Shakir, F. F. Mustafa, A. T. Naiyf ,"Simple design and implementation of solar tracking system two axis with four sensors for Baghdad city", 2018 9th International Renewable Energy Congress (IREC), Hammamet, 2018, pp. 1-5, doi: 10.1109/IREC.2018.8362577

[3] A. Masih and I. Odinaev, "Performance Comparison of Dual Axis Solar Tracker with Static Solar System in Ural Region of Russia," 2019 Ural Symposium on Biomedical Engineering, Radioelectronics and Information Technology (USBEREIT), Yekaterinburg, Russia, 2019, pp. 375-378, doi: 10.1109/USBEREIT.2019.8736642;

[4] https://www.powerfromthesun.net/Book/chapter02/chapter02.html

[5] Bendt, P. and A. Rabl (1980), "Effect of Circumsolar Radiation on Performance of Focusing Collectors; "SERI Report TR-34 -093, April.

[6] Bird, R. E., R. L. Hurlstrom, and J. L. Lewis (1983), "Terrestrial Solar Spectral Data Sets," Solar Energy 30(6), 563.

Virtual Investigations of a Stand-Alone Photovoltaic System with Supercapacitor Bank Used to Power an Irrigation System

B. I. Evstatiev
Department of Electronics
University of Ruse Angel Kanchev
Ruse, Bulgaria
bevstatiev@uni-ruse.bg

N. D. Codreanu
Center for Technological Electronics and
Interconnection Techniques
University POLITEHNICA of Bucharest
Bucharest, Romania
norocel.codreanu@cetti.ro

K. G. Gabrovska-Evstatieva
Department of Computer Science
University of Ruse Angel Kanchev
Ruse, Bulgaria
kgg@ami.uni-ruse.bg

Abstract—**The paper presents virtual investigation studies of a stand-alone Photovoltaic (PV) system with a supercapacitor bank, used for powering an irrigation system. It powers two water pumps of 200W each, with input power varying from 50W to 350W. An algorithm was developed for controlling the charging and discharging of supercapacitors, in order to optimize the operation of the water pumps. Next, an equivalent electric circuit is created, and SPICE simulations are performed in the Micro-Cap environment. Two scenarios are investigated, using 300F and 250F supercapacitor banks, rated at 27V and 32V, respectively. Furthermore, for each scenario, the capacitors charging/discharging currents during the different operation regimes are assessed. The results showed that the maximal currents during the worst-case scenarios are within (12.3 - 16.6) A. Another important parameter that was investigated is the minimum operating time of the pumps during "sudden absence of power" for the two scenarios - 59s and 141s, respectively.**

Keywords—PV energy; irrigation pump; supercapacitor

I. INTRODUCTION

The recent national and EU documents regulate the necessity for further expansions of renewable energy sources (RES). On the other hand, the use of decentralized energy sources in agricultural areas has high potential, which remains unrealized. Numerous previous studies exist on the topic. Xiang et al. compared the feasibility of sole solar and hybrid wind-solar generators for powering of water pumps in China [6]. The obtained results showed that in terms of reliability and price, it would be better to rely on PV energy, rather than hybrid wind-solar energy. In another study, Wazed et al. investigated various PV and solar thermal technologies for powering an irrigation system [4]. Their conclusions were that the most effective solution is CdTe PV module running a pump based on permanent magnet DC motor. In general, the studies agree that PV generators are the preferred renewable technology for such applications.

The PV energy generation is a random process, which depends on numerous factors, such as month of the year, time of the day, cloudiness, temperature, etc. This means that the power provided by PV installations rarely matches the rated power of the pumps. Furthermore, it is known that the dependency of pumps' flow rate on the input power is nonlinear, which has a significant impact on their efficiency [1]. Therefore, different approaches were investigated aimed at optimizing the water pumping process.

In [3] was presented a methodology for sizing of PV powered irrigation systems using a water tank storage. The idea is that the pumped water is stored in a reservoir and is later used for irrigation when necessary. The pumping process is controlled by a pump converter, which is selected depending on the parameters of the system and underground water. In [2] it was presented that the energy utilization of pumps, directly connected to PV sources, is quite low. However, the study also showed that the system efficiency could be improved significantly with the use of DC-DC controllers, which maintain a constant voltage. In [5] was investigated the influence of different pumping heads and different PV array configurations on the radiation threshold and the quantity of pumped water. The results showed that the total efficiency of a pumping system powered directly by PV modules could vary significantly, for the different scenarios but does not go above 60%.

In order to improve the efficiency of PV powered pumping systems, some studies have considered hybrid scenarios. In [7] and [8] were investigated systems with batteries and supercapacitors, where the role of the capacitor is to provide the surge currents. In another study, Das and Mandal (2018) compared the efficiency of a small water pump powered by sole PV (1), PV with batteries and charge controller (2), PV with supercapacitor without charge controller (3) and PV with a combination of supercapacitor and batteries (4) [9]. The highest system efficiencies were obtained for the second and the third scenario, depending on the water head. In a similar study, Das et al. (2017) showed that the inclusion of a supercapacitor in parallel with the DC pump could increase significantly the efficiency of the system, compared to direct powering [10]. To the best of our knowledge, no study has previously investigated the application of charge controllers with supercapacitors in irrigation systems.

The goal of the paper is to identify and investigate the working scenarios of a PV energy source with a supercapacitor bank, used for powering irrigation pumps in the agriculture.

II. MATERIALS AND METHODS

A. Architecture and algorithms

The scientific work investigates a situation where two water pumps are powered by a stand-alone PV system. A block diagram of the control system is presented in Fig. 1. The DC/DC converter maximizes the power output of the PV generator, which is then stored in a supercapacitor bank. The system controls the two pumps independently using a switching regulator. Furthermore, each pump has its own voltage regulator, responsible for providing the required power.

Next, an operation algorithm for the control system is developed, which is presented in Fig. 2. Initially both pumps are turned off. Depending on the energy stored in the supercapacitor bank, two possibilities exist:

- If the capacitor voltage U_C is above the higher critical value $U_C \geq U_{CRIT\_HI}$ (block 2), this means there is enough energy stored to begin irrigation. If the instantaneous power $P_{PV}(t)$ provided by the PV modules (through the DC/DC converter) is lower than the power of the first pump (block 3), only the first pump (P_{Pump1}) is turned on (block 5); otherwise, both pumps are turned on (blocks 4 and 5);

- Once the capacitor voltage U_C goes below the lower critical value $U_C \leq U_{CRIT\_LOW}$ (block 6), the control system can no longer provide enough power. If $P_{PV}(t)$ is lower than the power of the first pump, both pumps are turned off (block 8 and 9). Otherwise, only the second pump is stopped.

B. Methodology of the simulation

In the present study, it is assumed that the supercapacitor bank is implemented with capacitors of 3000F and 2.7V, manufactured by Maxwell Technologies. Furthermore, it is assumed that the used DC pumps are with the following characteristics: power - 200W, voltage - 24V, maximum water flow - 4m³/h and maximum head - 35m. When sizing the PV system, it is also assumed that the supercapacitor bank should be able to provide at least 1 - 2 minutes autonomous running time for the pumps in order to minimize the surge losses.

Two scenarios (**Scn** in table 1) are investigated:

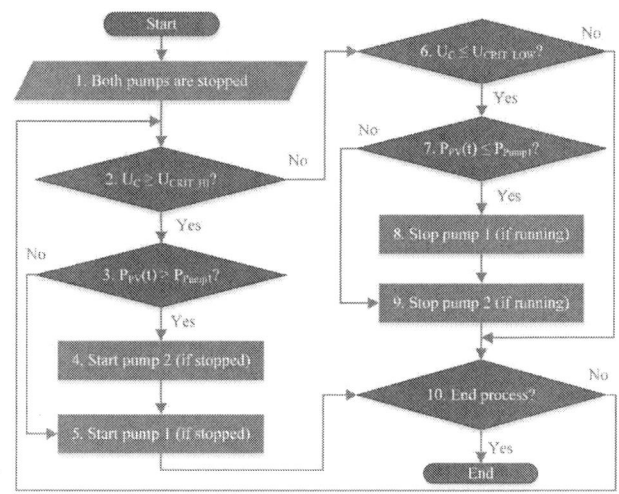

Fig. 2. Algorithm of the control system.

1. 10 supercapacitors are connected in series, therefore the bank capacity is $C_{BANK} = 300F$ and the rated voltage is $U_{C.R} = 27V$;

2. 12 supercapacitors are connected in series, therefore the bank capacity is $C_{BANK} = 250F$ and the rated voltage is $U_{C.R} = 32.4V$.

The available capacitor energy for the two scenarios can be calculated in Joules, according to formula 1:

$$E_{AVL} = \frac{1}{2} C \cdot \left(U_{CRIT\_HI}^2 - U_{CRIT\_LOW}^2 \right), \quad J \qquad (1)$$

In both cases it is assumed that the higher critical voltage is equal to the rated voltage ($U_{CRIT\_HI} = U_{C.R}$) and the lower critical voltage is the voltage of the DC pumps, i.e. 24V. The parameters of each scenario are summarized in Table 1.

TABLE I. USAGE SCENARIOS FOR THE PV POWERED IRRIGATION SYSTEM

Scn	C_{BANK}, F	$U_{C.R}, V$	U_{CRIT_HI}, V	U_{CRIT_LO}, V	E_{AVL}, Wh
1	300	27	27	24	6.4
2	250	32.4	32.4	24	16.45

The investigation of the current flow through the capacitors is especially important for this study for the following reasons:

1. The efficiency of the battery bank depends on the energy losses, i.e. on the magnitude of the current;

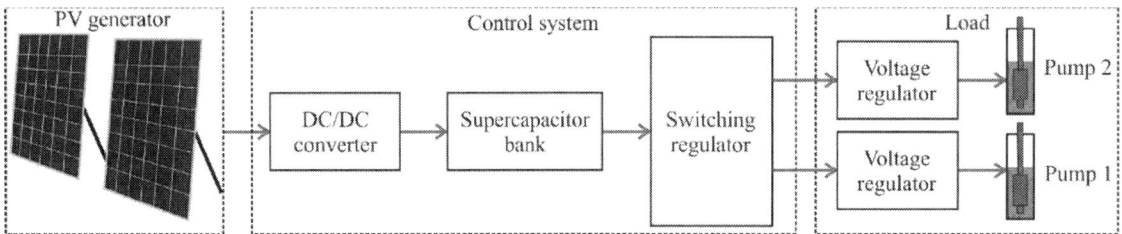

Fig. 1. Block diagram of the investigated stand-alone photovoltaic (PV) system.

2. The energy losses leave the system in the form of heat. Considering an irrigation system is commonly exposed to harsh environmental conditions with temperatures reaching up to $40°C$, the system's ability to release heat might be seriously limited. This will result in significant increase of the temperature of the supercapacitors and, as a result, in reduced lifetime expectancy.

Therefore, when investigating the described system, the following parameters should be obtained:

1. The capacitor's current during the initial charging time, when both pumps are turned off (I_{INIT});

2. The capacitor's current during discharging (I_{DIS});

3. The capacitor's current during charging (I_{CH});

4. The capacitor's current during "sudden absence of power" (SAP) when the pumps are powered entirely by the Supercapacitor bank (I_{SAP});

5. The minimum working time of the pump(s) during "sudden absence of power" (t_{MIN}).

III. RESULTS AND DISCUSSION

In order to model the autonomous PV system, the following basic approximations were accepted:

1. The surge currents are not considered as they are specific for each pump;

2. The voltage of the capacitor never gets below the lower critical value U_{CRIT_LOW}.

Next, an equivalent electric circuit was created in SPICE, using the Micro-Cap environment (Fig. 3). The combination of V1, E1 and Rv simulates a constant power source, therefore they implement the behavior of the PV generator and the DC/DC converter, together. To achieve this, the dependent voltage source E1 is implemented with the following equation:

$$E1 = \frac{P_{PV}}{\frac{v(5,2)}{r(Rv)}} - v(V1) \quad (2)$$

where P_{PV} is the supplied constant power, $v(5,2)$ is the voltage drop on Rv, $r(Rv)$ is the resistance of Rv and $v(V1)$ is the voltage of the independent source V1.

The supercapacitor bank is implemented with the capacitor C1 and the resistor R1, whose value is chosen according to the

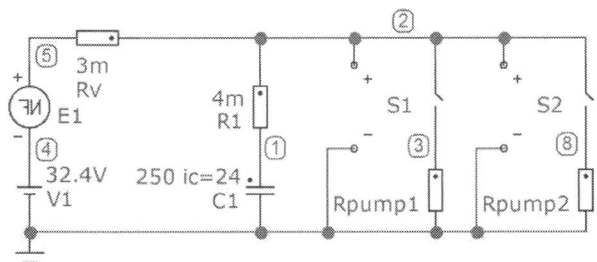

Fig. 3. The equivalent electric circuit developed for investigating the autonomous PV system.

manufacturer specification. Next is implemented the "Switching regulator" using the switches S1 and S2, running in hysteresis mode. Finally, the variable resistors Rpump1 and Rpump2 implement the behavior of the voltage regulators and the pumps, together. To achieve this, they are defined as nonlinear resistors. For example, Rpump1 is defined with:

$$Rpump1 = v(3) > 23.9 \, ? \, v(3) * \frac{v(3)}{P_{Pump1}} : 100k \quad (3)$$

where v(3) is the voltage of node 3 in V and P_{Pump1} is the power of pump 1 in W.

In equation 3, the question mark (?) allows to use the SPICE simulator's conditional logic. Its behavior could be explained as follows: if the applied voltage is less than 23.9V, the resistance of the pump is $Rpump = 100k\Omega$, i.e. no power is consumed; otherwise, the resistor's value is calculated according to $Rpump = v(3) * \frac{v(3)}{P_{Pump1}}$.

Using the developed equivalent circuit, simulations were prepared and performed for the two investigated scenarios. The following input powers P_{IN} were used: 50W, 100W, 250W, 300W and 350W.

Considering the complexity of the nonlinear circuit, the simulations were implemented in three steps, with appropriate initial conditions:

1. Initial charging;

2. Charging/discharging cycle (Fig. 4);

3. Sudden absence of power situation (Fig. 5).

Fig. 4 presents the transient response for two input powers ($P_{IN} = 50W$ and $P_{IN} = 350W$) of the following quantities:

1. The capacitor voltage: $v(c1)$;

2. The pump(s) voltage: $v(Rpump1)$ and $v(Rpump2)$;

3. The capacitor's current: $i(c1)$;

4. The supplied power: $-v(2) * i(v1)$;

5. The pump(s) power: $PD(Rpump1)$ and $PD(Rpump2)$.

As can be seen, for $50W$ input power only the first pump starts periodically, depending on the available energy. On the other hand, when $350W$ power is provided, the first pump is running continuously, while the second one starts when enough energy is stored in the battery bank.

In Fig. 5 is presented the transient response in the worst-case scenario when the input PV power suddenly becomes zero with one and two pumps running. The simulation results for the two scenarios are summarized in Table 2 and Table 3. As expected, the highest discharges currents occur during "sudden absence of power". For Scenario 1, they are in the interval (14.7…16.6)A and for Scenario 2 - (12.3…16.4)A. The highest charging currents occur when both pumps are turned off and increase with the input power.

2020 IEEE 26ᵗʰ International Symposium for Design and Technology in Electronic Packaging (SIITME)

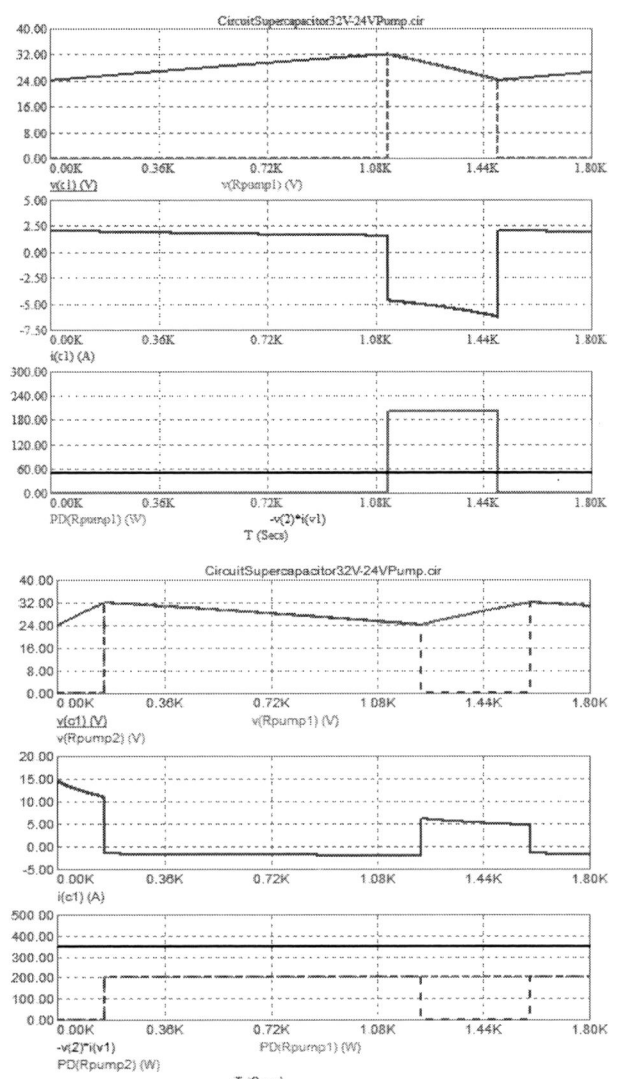

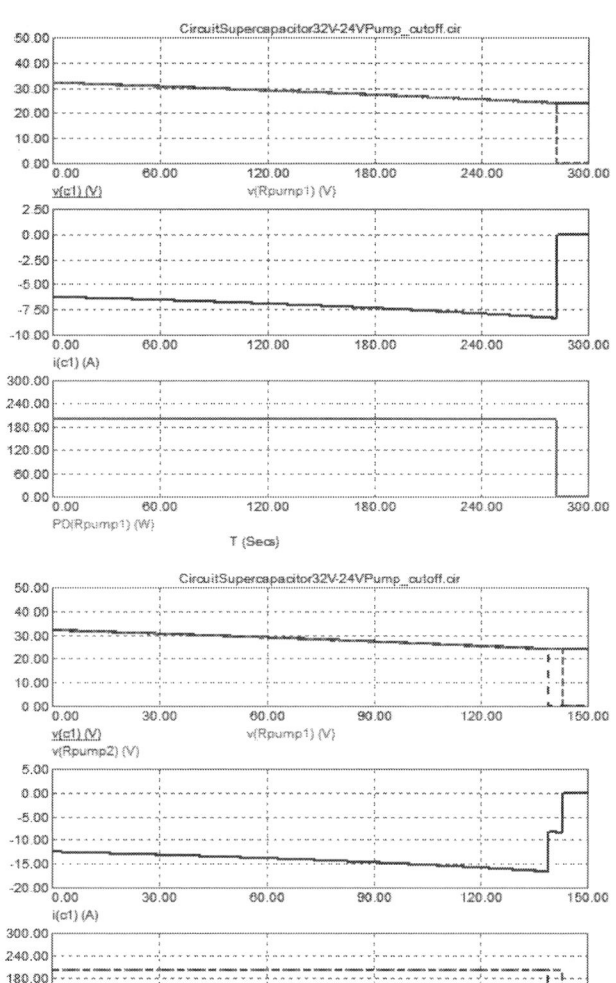

Fig. 4. Simulation results for Scenario 2 with provided PV power 50W (top) and 350W (bottom).

Fig. 5. Simulation results during "sudden absence PV power" when one pump (top) and two pumps (bottom) are running.

The simulation results showed that the expected minimum working time of the pumps for Scenario 1 is 117s if one pump is running and 59s if both pumps are running. Similarly, for Scenario 2 the values are 282s and 141s, respectively.

IV. CONCLUSIONS

The study investigates, based on simulations, the working regimes of a stand-alone PV system with supercapacitor bank, used to power an irrigation system. A control system is responsible for regulating the voltage of the supercapacitor bank. Its charge is controlled by the switching regulator, which is responsible for turning on/off the loads. The decision is taken according to an algorithm, which is aimed at optimizing the process and minimizing the energy losses due to surge currents.

Next, the study presents an equivalent circuit for simulation, in order to investigate the behavior of the system for different scenarios and working regimes. The simulation is aimed at

TABLE II. SUMMARY OF THE SIMULATION RESULTS FOR SCENARIO 1

P_{IN}, W	I_{INIT}, A	I_{DIS}, A	I_{CH}, A	I_{SAP}, A	t_{MIN}, s
50	~1.9	5.6 – 6.3	~1.9	7.3 – 8.3	117
100	4.2 – 3.8	3.7 – 4.2	4.2 – 3.8	7.3 – 8.3	117
250	10.3 – 9.2	5.6 – 6.3	~1.9	14.7 – 16.6	59
300	12.5 – 10.9	3.75 – 4.1	4.2 – 3.8	14.7 – 16.6	59
350	14.7 – 13.0	1.9 – 2.0	6.1 – 5.5	14.7 – 16.6	59

TABLE III. SUMMARY OF THE SIMULATION RESULTS FOR SCENARIO 2

P_{IN}, W	I_{INIT}, A	I_{DIS}, A	I_{CH}, A	I_{SAP}, A	t_{MIN}, s
50	2.1 – 1.6	4.6 – 6.1	2.1 – 1.6	6.2 – 8.2	282
100	4.4 – 3.1	3.1 – 4.2	4.1 – 3.1	6.2 – 8.2	282
250	10.3 – 7.9	4.7 – 6.4	2.1 – 1.6	12.3 – 16.4	141
300	12.5 – 9.4	3.1 – 4.2	4.1 – 3.1	12.3 – 16.4	141
350	14.7 – 11.1	1.4 – 2.0	6.3 – 4.7	12.3 – 16.4	141

obtaining several important parameters of the system, such as the capacitor's charging/discharging currents and the minimum working time of the pumps during sudden absence of power from the PV source. For the investigated scenarios, the minimal working time of the pumps varies from 59s to 282s and the currents in the worst-case scenarios vary from 12.3V to 16.6V.

Considering such installations would typically be used in remote rural areas, it could be exposed to harsh environmental conditions, with temperatures as high as 40°C. Therefore, it is important to investigate the temperature response and reliability of the supercapacitors under the expected currents and environment parameters, which is a task for future investigations.

ACKNOWLEDGMENT

This work was supported by the University of Ruse Research Fund under contract no 2020-FNI-FEEA-02.

REFERENCES

[1] M. Benghanem, K.O. Daffallah, A. Almohammedi, "Estimation of daily flow rate of photovoltaic water pumping systems using solar radiation data", Results in Physics, Vol. 8, pp. 949-954, 2018.

[2] M. A. Elgendy, B. Zahawi, D. J. Atkinson, "Comparison of Directly Connected and Constant Voltage Controlled Photovoltaic Pumping Systems", IEEE Transactions on Sustainable Energy, Vol. 1, No. 3, pp. 184-192, October 2010.

[3] L. Stoyanov, I. Govedarski and V. Lazarov, "Sizing of PV Based Power Supply for Irrigation System – Application in Sandanski, Bulgaria", XVI-th International Conference on Electrical Machines, Drives and Power Systems ELMA 2019, 6-8 June 2019, Varna, Bulgaria, pp. 579-585, 2019.

[4] S. M. Wazed, B. R. Hughes, D. O`Connor and J. K. Calautit, "Solar Driven Irrigation Systems for Remote Rural Farms", Energy Procedia, Vol. 142, pp. 184-191, 2017.

[5] V. Ch. Sontake, A. K. Tiwari and V. R. Kalamkar, "Performance investigations of solar photovoltaic water pumping system using centrifugal deep well pump", Thermal Science, Vol. 24, No. 5A, pp. 2915-2927, 2020.

[6] Ch. Xiang, J. Liu, Y. Yu, W. Shao, Ch. Mei and L. Xia, "Feasibility assessment of renewable energies for cassava irrigation in China", Energy Procedia, Vol. 142, pp. 17-22, 2017.

[7] M.E. Glavin and W.G. Hurley, "Optimisation of a photovoltaic battery ultracapacitor hybrid energy storage system", Solar Energy, Vol. 86, pp. 3009–3020, 2012.

[8] M. A. Camara, A. Djellad, P. O. Logerais, O. Riou and J. F. Durastanti, "Modeling of a hybrid energy storage system supplied by a photovoltaic source to feed a DC motor", International Journal of Renewable and Sustainable Energy, Vol. 2, No. 6, pp. 222-228, 2013.

[9] M. Das and P. Mandal, "A comparative performance analysis of direct, with battery, supercapacitor, and battery-supercapacitor enabled photovoltaic water pumping systems using centrifugal pump", Solar Energy, Vol. 171, pp. 302–309, 2018.

[10] M. Das, D. Mukherjee and S. R. B. Chaudhuri, "An approach to study the performance of photovoltaic water pumping using supercapacitor", Materials Today: Proceedings, Vol. 4, pp. 10400–10406, 2017.

Islanded Microgrid Simulation and Cost Optimisation

A. Ignat, E. Szilagyi, and D. Petreuș
Department of Applied Electronics
Technical University of Cluj-Napoca
Cluj-Napoca, Romania
Andreea.Ignat@ael.utcluj.ro

Abstract—**The operation of a microgrid implies multiple levels of control, ranging from energy production to optimisation methods. Testing these control methods on the actual microgrid can be time consuming and expensive. The simulated model of a microgrid can be adapted to suit the tested control method in an expedited and cost-effective manner. Furthermore, the simulation offers proof of concept for the microgrid architecture as well as its operation. This paper proposes a microgrid model designed for testing cost optimisation algorithms before using them in real life. By running the simulation of the proposed microgrid together with the cost optimisation algorithms the two validate each other and provide significant data that can be used for improvements and fine tuning before real life testing, thus eliminating most of the risks. This can be valuable for other researchers in the field.**

Keywords—microgrid; renewable energy; cost optimization.

I. INTRODUCTION

A microgrid is defined as a low voltage grid composed of interconnected power sources, controllable loads and critical loads. It can operate in either isolated or grid-connected, being subject to the operational tasks characteristic of the main network [1,2]. Microgrids offer several advantages, such as reducing greenhouse gas emissions, improving the voltage profile, and decentralizing energy supply [3].

Due to the increased focus on climate change, socio-economic development and the need to reduce greenhouse gas emissions, microgrids include, in particular, sustainable sources, such as renewable energy sources and energy-efficient systems [4]. The optimization of these systems is achieved with the help of energy management systems, which implement decision-making strategies. These strategies consider increasing the energy efficiency of the system, increasing reliability, reducing energy consumption, reducing operating costs, reducing losses and reducing greenhouse gas emissions.

Although they have the role of covering local energy needs, the structure and operation of microgrids are usually quite complex. Complexity arises due to several factors: in the first instance, a variety of operating modes - among these we can mention the autonomous operation whenever the electricity distribution network is not available; in addition, the variety of types of energy in a microgrid - not only electricity, but also heat, for example; also the various functions that an energy management system must perform - such as coordination of multiple generation sources, but also energy transfer, transformation and storage; finally, external and internal random factors affecting operations. All these aspects make the control and planning of a microgrid quite difficult. On the other hand, this widespread complexity leaves much room for improvement in the current state of the art.

Energy management means the development of a high-level algorithm that determines the amount of energy generated and its division, aiming for the lowest possible cost of operating the system. This algorithm must ensure an energy balance between the system consumers and the generated powers [5].

The energy management of a microgrid is subject to several constraints: the limits of energy generation, the limits of energy consumption, the limits of the energy storage system. Renewable energy sources, such as photovoltaic panels and wind turbines, are increasingly being used to reduce carbon emissions. These resources bring additional limitations to the system because they depend on the weather, and the programming of their use must be made according to the weather forecast.

Various algorithms have been used for the energy management problem, which presents as an optimisation problems with several constraints. The methods of solving this optimisation problem can be divided into classical (mathematical) ones [6,7] metaheuristics [8,9].

Testing these various management methods using a simulated model of the microgrids is a widely used technique [10,11] due to its versatility, fast response time, and advantage of spotting and correcting errors before deploying the management system. The MATLB/Simulink environment is by far the most popular for simulating the energy management of microgrids, both islanded and grid-connected.

This paper aims to test a cost optimisation and day-ahead scheduling algorithm developed by the authors in [12] using a Simulink model of the microgrid based on [13] and developed in [14].

The paper has five main sections. The first provides a description of the microgrid model and its parameters. The second describes the optimisation algorithm and shows its results. Section 3 shows how the algorithm and the simulation work together and discusses the obtained results, while section 5 outlines the conclusions.

II. MICROGRID MODEL

The microgrid model is illustrated in Fig. 1. Developed in Simulink, it is used for testing a cost optimisation

2020 IEEE 26ᵗʰ International Symposium for Design and Technology in Electronic Packaging (SIITME)

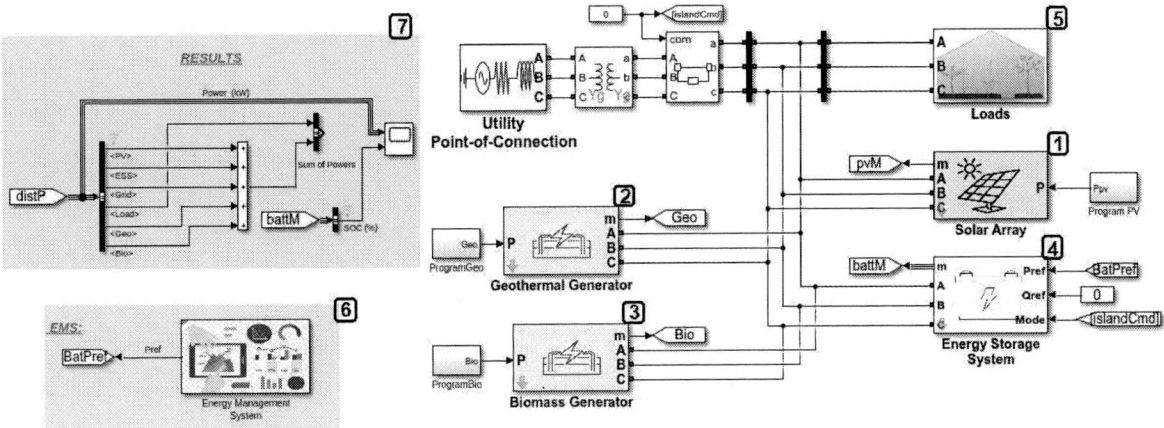

Fig. 1. Microgrid Simulink model.

algorithm, Harmony Search Algorithm (HSA), implemented for the microgrid.

The proposed microgrid model includes three renewable energy sources (1, 2, and 3), an energy storage system (4), loads (5), an energy management system (6), and a block for displaying the results (7). Although the microgrid that it simulates only functions in isolated mode, the microgrid model can function either in isolated or grid-connected mode for potential future developments.

The microgrid model is aimed at testing optimisation algorithms and energy management schemes in as little time as possible. Therefore, the primary control level is not simulated. What is relevant for the purpose of the simulation are the amounts of energy generated and consumed during certain periods of time. This is achieved by simulating all three renewable energy sources with ideal components. This ensures the minimum possible simulation time for a Software in the Loop scenario.

The simulation is run for an entire day. This translates to 86400 s in the Simulink environment. The algorithm used for cost optimisation, Harmony Search Algorithm, is included in the initialization file of the model and, therefore, performs the necessary calculations before the simulation starts.

III. MICROGRID COST OPTIMISATION

Based in the improvisation process of musicians, the Harmony Search Algorithm starts with an initial population of randomly generated possible solutions, called the Harmony Memory. During runtime, this population is constantly updated with new and better possible solutions. These are obtained either by generating random new ones or by adapting existing ones from the Harmony Memory. When a newly generated solution is better than the worst one in the Harmony Memory, the latter is replaced. This process makes full use of the initial population, while providing a balance between the exploitation and exploration characteristics, that are common for all metaheuristics. The combination makes the algorithm suitable for global optimality.

For the proposed microgrid the optimisation problem consists in finding the appropriate amounts of energy that must be generated hourly by each renewable energy source so that, firstly, the load demand is covered and, secondly, the minimum operating cost is obtained. Covering the load demand is the main purpose of the microgrid and therefore, takes precedence. Finding a generation scenario that satisfies the aforementioned purpose and results in a minimum operating cost is the purpose of the optimisation algorithm.

The input parameters of the algorithm are the generation scenario of the photovoltaic panels, the load demand scenario, and the initial state of charge of the energy management system. The output consists of a generation scenario for the geothermal generator, one for the biomass generator, and a charge/discharge scenario for the energy management system. Satisfying the load demand is one of the constraints. The other two constraints concern the lower and upper limits of the energy storage system, which must be followed.

The generation scenario produced by the Harmony Search Algorithm for a typical December day can be observed in Table 1. The table data can be interpreted as follows:

- $E_{PV}(t)$ [Wh] – the amount of energy generated using the photovoltaic panels in the [t-1,t] time slot;

- $E_{Geo}(t)$ [Wh] – the amount of energy generated using the geothermal generator in the [t-1,t] time slot;

- $E_{Bio}(t)$ [Wh] – the amount of energy generated using the biomass generator in the [t-1,t] time slot;

- $E_{Bio}(t)$ [Wh] – the amount of energy generated using the biomass generator in the [t-1,t] time slot;

- $E_{Bat}(t)$ [Wh] – the amount of energy charged in or discharged from the energy storage system in the [t-1,t) time slot (a positive value represents discharge, while a negative value represents charge);

- $SoC(t)$ [%] – the state of charge of the energy storage system at the end of the [t-1,t) time slot.

TABLE I.   GENERATION SCENARIO PROVIDED BY THE HARMONY SEARCH ALGORITHM

t	$E_{PV}(t)$ [Wh]	$E_{Geo}(t)$ [Wh]	$E_{Bio}(t)$ [Wh]	$E_{Bat}(t)$ [Wh]	$E_{Load}(t)$ [Wh]	SoC(t) [%]	t	$E_{PV}(t)$ [Wh]	$E_{Geo}(t)$ [Wh]	$E_{Bio}(t)$ [Wh]	$E_{Bat}(t)$ [Wh]	$E_{Load}(t)$ [Wh]	SoC(t) [%]
1	0	0	0	0	0	50	13	1038	0	0	-838	200	92.046
2	0	0	0	250	250	48.697	14	859	0	0	-859	0	96.520
3	0	1500	1500	-200	2800	49.739	15	562	0	0	-562	0	99.447
4	0	1500	1500	-300	2700	51.302	16	217	0	0	83	300	99.015
5	0	0	0	0	0	51.302	17	0	0	0	0	0	99.015
6	0	0	0	300	300	49.739	18	0	200	0	0	200	99.015
7	0	0	0	600	600	46.614	19	0	1259	77	2164	3500	87.744
8	0	1500	1500	400	3400	44.531	20	0	1500	1500	600	3600	84.619
9	300	1500	1500	200	3500	43.489	21	0	1500	1500	600	3600	81.494
10	643	1500	1500	-3043	600	59.338	22	0	1500	1500	700	3700	77.848
11	917	1500	1500	-3317	600	76.614	23	0	1500	1500	500	3500	75.244
12	1059	1062	104	-2125	100	87.682	24	0	0	0	0	0	75.244

The SoC value for a time slot depends on the corresponding value for the previous time slot and E_{Bat} value for the current one. Therefore, the initial state of charge, SoC(0) = 50%, must be specified.

IV. MICROGRID SIMULATION RESULTS

The solution generated by the HSA is an energy generation scenario that takes into account the different renewable energy sources (photovoltaic panels, geothermal generator, and biomass generator) and their operating cost, the initial State of Charge (SoC) of the energy storage system and its operating cost, as well as the load profile. The energy generation scenario obtained in this manner is used for controlling the model of the microgrid by correlating the desired power output of the energy sources and the load demand with the simulated time. The SoC of the batteries is calculated during the simulation and compared to the expected one, generated by the optimisation algorithm for validating the results.

The scenario with the initial SoC of 50% was simulated and its results are illustrated in Fig. 2.

The first step in simulating the generation scenario is to transform the output of the algorithm into a suitable input for the microgrid model. Assuming that the generators operate at maximum power (1500 W for each), the amount of energy for each time slot in the generation scenario from Table 1 is transformed in operation time for each time slot. This means that for time slots [2,3) and [3,4) both geothermal and biomass generators will operate continuously for 120 minutes, 60 for each time slot. Similarly, for the [11,12) time slot the geothermal generator will operate for 42 minutes and the biomass generator for 4 minutes. The generation profile of the photovoltaic panels and the load profile are transformed into inputs for the simulation model in the same manner.

The state of charge of the energy storage system is used as control data. If the calculated results and the simulated ones correspond, than the algorithm and simulation validate each other.

The main advantage of using an optimisation algorithm, apart from achieving an optimum operation cost, is the efficient use of available resources. To illustrate this, some sections in Fig. 2 are highlighted.

Section 1 refers to time slots [2,3) and [3,4). As one can see, the output of the photovoltaic panels is 0 W and the load demand is 2800W and 2700 W respectively. This means that the load demand must be covered using the geothermal and biomass generators. This charges the energy storage system slightly, as visible in Table 1, as well as Fig. 2. In order to cover the load demand for time slots [4,5) – [6,7) only the energy storage system is used.

Section 2 refers to time slots [7,8) and [8,9). Here the output of the photovoltaic panels is minimal and the load demand must, again, be covered by the two generators. The generators continue to function further, coming to a stop in time slot [11,12).

Section 3 refers to time slots [12,13) – [16,17), where both generators are turned off and the load demand is close to 0 W. This charges the energy storage system. Because the two generators were stopped in time slot [11,12), the upper limit of the energy storage system is respected and no excess energy is produced. This ensures that the equipment is not damaged.

Section 4 refers to time slots [18,19) – [22,23). Here the load demand is covered again by the two generators since the output of the photovoltaic panels is again 0 W.

Looking at the variation of the State of charge through the entire 24-hour simulation one can observe that the optimisation algorithm creates a generation scenario that maintains the SoC value around 50% until the photovoltaic panels begin to produce energy. This allows the charging of the batteries when the load demand is lowest and the output of the photovoltaic panels is highest. This eliminates the need to limit the panels' output, need that would arise without the scheduled use of the microgrid resources.

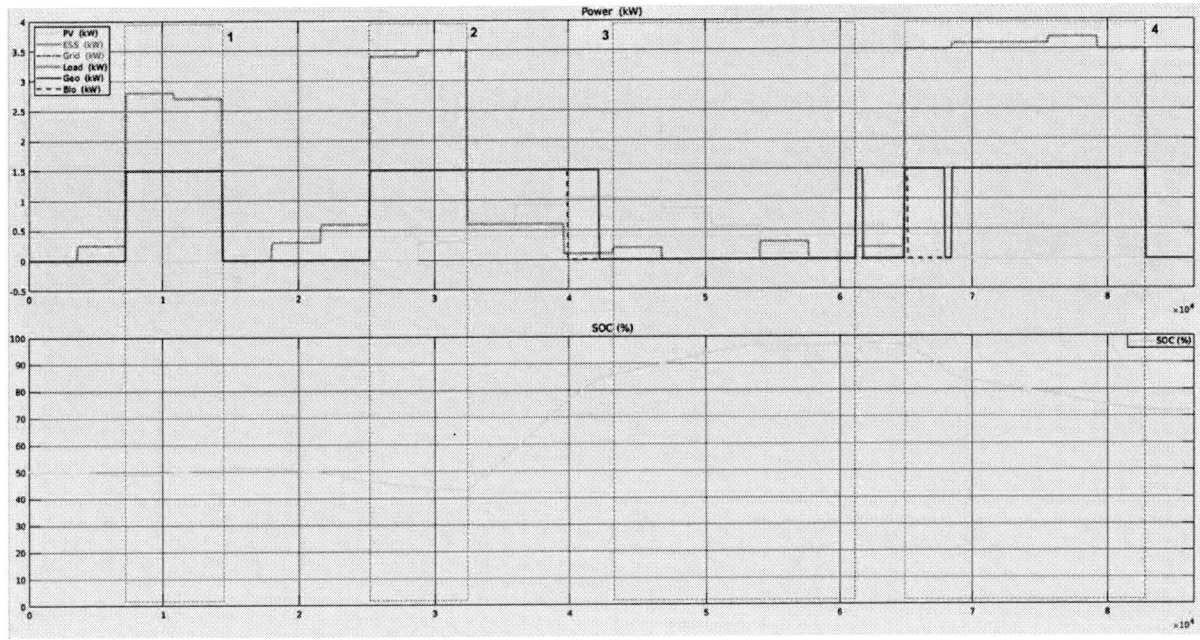

Fig. 2. Simulation results.

The operating cost obtaine with the calculated and simulated generation scenarion is 0.34353€ for the entire 24-hour period.

The combination of optimisation algorithm and microgrid simulation takes a total of 16 seconds to run on an Intel Core i7 computer with 16 GB of RAM memory.

V. CONCLUSIONS

A Simulink model of the microgrid, based on [13] and developed in [14], was used to test and validate the Harmony Search Algorithm.

The algorithm is used for optimising the operation cost of the microgrid while preserving its fundamental purpose of covering the load demand.

The advantages of using an optimisation algorithm for scheduling the microgrid operation include the efficient use of resources, the guaranteed coverage of load demand, using all microgrid resources within their parameters (thus, avoiding equipment damage) and eliminating the need of output power limiting or load cut-off.

The calculated and simulated results were compared and a suitable correspondence between them was found.

REFERENCES

[1] Hatziargyriou N, Asano H, Iravani R, Marnay C. Microgrids 2007;5(4):78–94.

[2] DOE. Summary report: 2012 DOE microgrid workshop, Chicago, Illinois; 2012.

[3] Lasseter R, Akhil A, Marnay C, Stevens J, Dagle J, Guttromson R, et al. The certs microgrid concept, white paper on integration of distributed energy resources. California Energy Commission, Office of Power Technologies-US Department of Energy, LBNL-50829; 2002.

[4] Li M, Zhang X, Li G, Jiang C. A feasibility study of microgrids for reducing energy use and GHG emissions in an industrial application. Appl Energy 2016;176:138–48.

[5] M. Elsied, A. Oukaour, T. Youssef, H. Gualous, O. Mohammed, "An advanced real time energy management system for microgrids", Energy, vol. 114, pp. 742-752, 2016.

[6] Quiggin D, Cornell S, Tierney M, Buswell R. A simulation and optimisation study: towards a decentralised microgrid, using real world fluctuation data. Energy 2012;41(1):549–59.

[7] E. Lazar, D. Petreus, R. Etz, T. Patarau, "Optimal Scheduling of an Islanded Microgrid Based on Minimum Cost", 39th International Spring Seminar on Electronics Technology (ISSE2016), Pilsen, Czech Republic, pp. 155-156, May 2016.

[8] Riva Sanseverino E, Di Silvestre M, Graditi G. A generalized framework for optimal sizing of distributed energy resources in micro-grids using an indicator-based swarm approach; 2013.

[9] Logenthiran T, Srinivasan D, Khambadkone AM, Sundar Raj T. Optimal sizing of distributed energy resources for integrated microgrids using evolutionary strategy. In: 2012 IEEE congress on evolutionary computation (CEC). IEEE; 2012. p. 1–8.

[10] Enrique Kremers, Jose Gonzalez de Durana, Oscar Barambones, Multi-agent modeling for the simulation of a simple smart microgrid, Energy Conversion and Management, Volume 75, 2013, Pages 643-650.

[11] O. Nzimako and A. Rajapakse, "Real time simulation of a microgrid with multiple distributed energy resources," 2016 International Conference on Cogeneration, Small Power Plants and District Energy (ICUE), Bangkok, 2016, pp. 1-6.

[12] E. Lazar, A. Ignat, D. Petreus and R. Etz, "Energy Management for an Islanded Microgrid Based on Harmony Search Algorithm", 2018 41st International Spring Seminar on Electronics Technology (ISSE), Zlatibor, Serbia, 2018, pp. 1-6.

[13] J. LeSage, "Microgrid Energy Management System (EMS) using Optimization", MathWorks File Exchange, https://www.mathworks.com/matlabcentral/fileexchange/73139-microgrid-energy-management-system-ems-using-optimization?s_tid=prof_contriblnk [accessed on 30.01.2020].

[14] A. Ignat, E. Szilagyi and D. Petreuş, "Renewable Energy Microgrid Model using MATLAB — Simulink," *2020 43rd International Spring Seminar on Electronics Technology (ISSE)*, Demanovska Valley, Slovakia, 2020, pp. 1-6, doi: 10.1109/ISSE49702.2020.9120923

2020 IEEE 26th International Symposium for Design and Technology in Electronic Packaging (SIITME)

Theoretical and Numerical Aspects Concerning the Stress in a Superconducting Solenoid

Radu Jubleanu
University POLITEHNICA of
Bucharest
Bucharest, Romania
radu_17jub@yahoo.com

Dumitru Cazacu
Faculty of Electronics,
Communication and Computers
University of Pitesti
Pitesti, Romania
dumitru.cazacu@upit.ro

Nicu Bizon
Faculty of Electronics,
Communication and Computers
University of Pitesti
Pitesti, Romania
nicu.bizon@upit.ro

Abstract— **This paper deals with the numerical modeling of stress in superconducting coils made of three materials.**
Mechanical aspects must be carefully analyzed in the case of these coils due to the high values of stresses that tend to deform the winding. The von Misses stress is obtained by a coupled magneto-structural finite element analysis. Numerical and analytical results were compared for free normal density and adaptive mesh. The geometry of the coils was designed so that each of them should store 11 kJ of energy. The performances of the materials were compared.

Keywords—solenoid; stress; magnetic energy; superconductor; finite element

I. INTRODUCTION

A Superconducting Magnetic Energy Storage (SMES) system has as central component the superconducting coil [1], [2]. This coil is subjected to high magnetic field and stress [3]. In literature, for the studying the stress, numerous numerical models have been used [4], [5]. These numerical models are useful in problems with a high degree of complexity and where the realization of experimental models is difficult as in the case of superconducting systems [6], [7]. In terms of geometry, modeling of solenoids is mainly used due to the fact that they are the most common in practical applications [8].

This paper presents a comparative study using finite element numerical models for three superconducting coils, made of three different materials: Niobium-Titanium (Nb-Ti), Yttrium Barium Copper Oxide (YBCO) and Bismuth Strontium Calcium Copper Oxide (BSCCO). Each of them has different mechanical and electrical properties. To perform the comparative study the same value of the stored magnetic energy was imposed.

II. THEORETICAL BACKGROUND

In Fig.1 we consider a cylindrical coil that was designed so that it can store a given amount of magnetic energy W_m. For the design of the coil, the length of coil L and its inner radius R_i are also given.

The expression of the stored magnetic energy is described by (1). For a cylindrical coil it becomes relation (2):

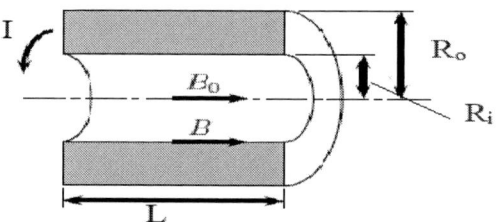

Fig. 1. Geometric parameters of solenoid

$$W_m = \iiint \frac{B^2}{2\mu_0} dV \qquad (1)$$

$$W_m = \frac{B^2}{2\mu_0} \pi R^2 L \qquad (2)$$

Using (2) the magnetic flux density can be obtained:

$$B = \sqrt{\frac{W_m \cdot 2\mu_0}{\pi \cdot R^2 \cdot L}} \qquad (3)$$

After obtaining the value of the magnetic induction the number of turns is computed, using a known electric supply current I. From relation (4) we obtain the number of turns of the coil N.

$$B = \mu_0 \frac{N \cdot I}{L} \rightarrow N = \frac{B \cdot L}{\mu_0 \cdot I} \qquad (4)$$

To find the coil thickness (outer radius) the current density in a conductor is used, more precisely the conductor section is calculated, the number of conductors being also known.
It is known that in a coil the magnetic induction is limited by the saturation of the magnetic core, reaching the value of about 2T. This leads to the storage of a low value magnetic energy. For the storage of a large amount of magnetic energy the superconducting coil with energy values of hundreds of MJ is used. This amount of energy also entails a rather serious problem of a mechanical nature.
Specifically, the magnetic field created in the superconducting coil behaves like a magnetic pressure that tends to expand the coil. This problem was presented in paper [1]. The radial

978-1-7281-7507-2/20 $31.00 © 2020 IEEE   430   21-24 October, Pitesti, Romania

stress, for a certain value of the magnetic induction B, is as follows:

$$\sigma_\theta = \frac{B^2}{2\mu_n}\frac{\alpha^2+1}{\alpha^2-1} \qquad (5)$$

where: α-ratio of outer to inner radius of solenoid, σ_θ-radial stress. For a good operation of the superconducting device, the value of the radial stress must be as small as possible, and imperiously it must not exceed the breaking value of the material from which the coil is built.

III. NUMERICAL MODELING

Numerical models were considered for three coils made of different superconducting materials. Table I shows the respective electrical and mechanical properties for the three types of superconducting materials: Nb-Ti, YBCO, BSCCO [9], [10], [11], [12].

TABLE I. ELECTRICAL AND MECHANICHAL PROPERTIES OF SUPERCONDUCTIND MATERIALS

Material	Electrical proprieties		Mechanical proprieties		
	Critical current density [A/mm^2]	Current used [A/mm^2]	Young's modulus [GPa]	Poisson's ratio	Density [Kg/m^3]
Nb-Ti	~200	126	100	0.32	6350
YBCO	~400	200	110	0.33	6300
BSCCO	~700	400	80	0.34	6400

The determination of the geometric parameters for each coil is made based on the respective known parameters: the value of stored energy $W_m = 11$ kJ, the length of the coils (same length for all coils) L = 126 mm and the inner radius R_i = 56mm [13]. Using the relations (2) - (4) the values for the respective outer radius the number of turns for each coil presented in Table I were obtained.

TABLE II

	Nb-Ti	YBCO	BSCC0
I[A]	126	200	400
N[turns]	4207	2650	1325
R$_0$[mm]	90	78	67

Due to the differences among the critical currents of the materials, different thicknesses have been obtained.

The superconductive coils carry high currents (hundreds of amps) and for this reasons it is necessary a mechanical analysis to determine the value of the stress produced by the Lorentz force. Thus there is a need for a coupled analysis that determines on the one hand the value of the Lorentz force and then the value of the stress.

A. Coupled field analysis

Generally there are two methods used for coupling the fields: serial coupling (weak coupling) and direct (strong or simultaneously coupling).

In the last approach the coupled analysis are described by a single system of equations. Solving it allows to obtain all the unknowns or degrees of freedom. It is used for highly nonlinear coupled problems. In the serial approach the stress analysis is performed after the magnetic analysis, which is used as an input for the stress one (Fig.1). The serial analysis is generally used for liniar problems, as in our paper.

In order to model and simulate this coupled problem the finite element method software Comsol Multiphysics was used.

Due to the symmetry properties of the solenoid geometries and in order to reduce the calculation times, axially symmetrical 2D models were created.

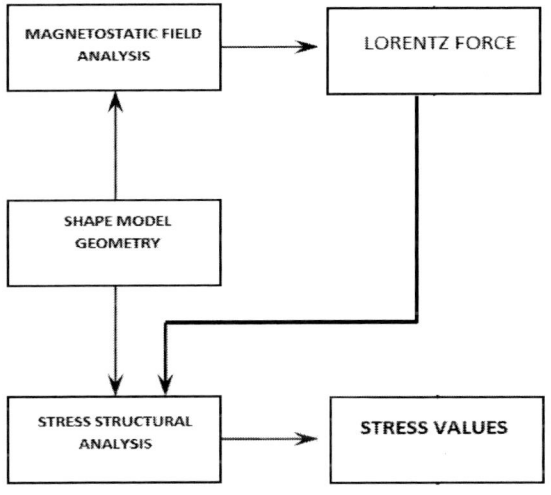

Fig.1. Serial coupling

B. Coupled magnetostatic structural matrix formulation

The steady state magnetic field is described by the following equation

$$\nabla \times \left(\frac{1}{\mu}\nabla \times \overline{A}\right) = \overline{J} \qquad (6)$$

Where $\overline{A}$ is the magnetic vector potential (MVP) and J is the source current density. In the 2D problems the MVP has only one component. For the axial symmetric models is the modified MVP ρA_θ. Solving the algebraic systems of equations obtained by the application of finite element method (FEM) the values of MVP in the mesh nodes are determinate. Then the magnetic flux density B is obtained. For this type of model equation (5) becomes:

$$\frac{\partial}{\partial\rho}\left[\frac{1}{\rho}\frac{\partial(\rho A_\varphi)}{\partial\rho}\right]+\frac{\partial}{\partial z}\left[\frac{1}{\rho}\frac{\partial(\rho A_\varphi)}{\partial z}\right]=-\mu J_\varphi \qquad (7)$$

By applying the Galerkin finite element method to (7) the global system of equation is obtained.

$$[S]\left[A_\varphi\right]=-[J] \qquad (8)$$

where [S] and [J] have the following expressions [5]:

$$[S]=\iint\frac{1}{\mu\rho}\left[\frac{\partial}{\partial\rho}\rho[N^T]\frac{\partial}{\partial\rho}(\rho[N])+\frac{\partial}{\partial z}[\rho N^T]\frac{\partial}{\partial z}(\rho[N])\right]d\rho dz$$
(9)
and

$$[J]=\iint\rho[N]^T J_\varphi\,d\rho dz \qquad (10)$$

where N- is the elemental shape function
The mechanical equation has the following matrix form

$$[K][x]=[F] \qquad (11)$$
$$K=\iint 2\pi\rho[C]^T[D][C]d\rho dz \qquad (12)$$
$$F=\iint 2\pi\rho\{[N]^T[\vec{J}\times\vec{B}]\}d\rho dz \qquad (13)$$

K-structural stiffness matrix
C-displacement stress matrix
D-elasticity matrix
F-Lorentz force
Equation (7) and (8) are solved serial to determinate the MVP, the magnetic flux density, the Lorentz force and the mechanical stress.

C. Numerical models

The numerical models presented in this paper are linear with the coil metal having a static elastic behavior. In order to create a numerical model with a finite element software the next stages have to be followed: preprocessing, processing and post processing.
The preprocessing stage includes the next steps:
- creating the geometry.
An axial-symmetric model is used to reduce the calculation time.
- setting the material properties according to the data in tables I. Materials used for the coils are considered to be homogeneous and isotropic, being already in the superconductive state.
-setting the conditions on the boundaries, for both the steady state magnetic and mechanical problems. The magnetic insulation boundary condition is set for the magnetostatics problem and the zero axial displacement for the mechanics one, as in Fig.2.
-generating the mesh. A free unstructured normal density mesh with triangular elements was created.
In the processing stage the problems are solved, using direct or iterative solvers.
For this problem second order finite elements of quadratic Lagrange type were used and a PARDISO direct solver. In the postprocessing step different quantities are obtain and graphically processed in 2D or 3D.

IV. RESULTS

In Fig. 3-a, b, c, the maps for the numerical values of the radial stress for each type of material used are described.

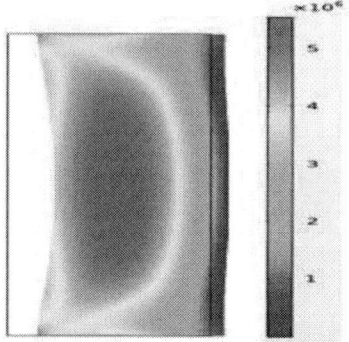

a. Nb-Ti material thickness-34 mm

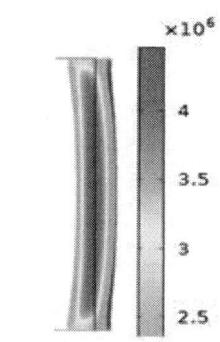

b) BSCCO material thickness-11mm

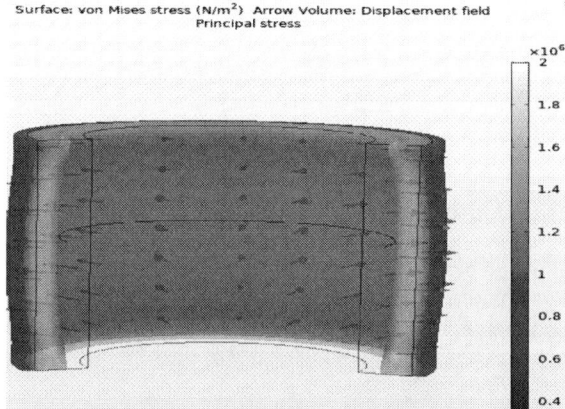

c) YBCO material- thickness-22mm
Fig.3 Stress map for each material

The numerical results obtained for the magnetic induction, stress and stored energy are presented in table III. Also, the analytical values of stress are calculated using relation (5) to achieve comparisons with numerical values. In order to increase the accuracy of the numerical solution the p-version finite element approach was used.

Table. III

Coil material	Nb-Ti	YBCO	BSCCO
FE	434	432	498
DOFs	4977	4955	5681
Magnetic flux density values [T]	3.5	3.6	4
Numerical values of stress [N/m²]	$8.2 \cdot 10^6$	$8.9 \cdot 10^6$	$3.3 \cdot 10^7$
Analytical values of stress [N/m²]	$2.3 \cdot 10^7$	$3.2 \cdot 10^7$	$5.8 \cdot 10^7$
Computed stored energy [J]	11255	10637	10040

FE-finite elements; DOFs-degrees of freedom

This means using second degree polynomial approximation and increasing the mesh density. With this approach the approximation error decreases algebraically versus the number of the unknowns. An adaptive mesh implementation was applied.

It is important that the numerical model accurately captures local variations in the solution, such as stress concentrations. Adaptive mesh refinement adds mesh elements based on an error criterion, to resolve those areas where the error is large. A three level adaptive mesh is used, i.e. the refining of the mesh is performed three times in a row. Each finite element is divided in four elements at each degree. This implies greater computational resources (time and memory). The obtained results are presented in table IV. There is a decrease in numerical errors in all three cases.

Table IV. Numerical results using initial and adaptive mesh

Material	Nb-Ti		YBCO		BSCCO	
	Stress [N/m²]	Storage energy [J]	Stress [N/m²]	Storage energy [J]	Stress [N/m²]	Storage energy [J]
Free mesh	$8.2 \cdot 10^6$	11255	$8.9 \cdot 10^6$	10637	$3.3 \cdot 10^7$	10040
Adaptive mesh	$2.17 \cdot 10^7$	11257	$3.12 \cdot 10^7$	10642	$4.8 \cdot 10^7$	10042
Analytical values	$2.3 \cdot 10^7$	11000	$3.2 \cdot 10^7$	11000	$5.8 \cdot 10^7$	11000
Relative error free mesh[%]	64.2	2.3	72.1	3.3	43.10	8.7
Relative error adaptive mesh[%]	5.652	2.29	2.500	3.25	17.2	8.68

V. CONCLUSIONS

Analyzing the numerical and analytical results we can conclude the following:
- to store 11 kJ of energy, for the BSCCO superconducting material the coil has a thickness three times smaller than for the Nb-Ti.
– the values of von Misses stress has the lowest value for Nb-Ti.

– Lorentz forces exerted are radial, and they tend to expand the coils and the highest values are recorded in the center of the coils
– the values obtained for stresses are lower than the stress failure stress, in all three cases.
-the use of an adaptive mesh involves higher computational resources and longer resolution times, but numerical errors decrease significantly.
– in the case of the numerical determination for the stress values, a dependence on the mesh density was noticed.
In the future work nonlinear aspects will be considered.

ACKNOWLEDGMENT

This study is realized under Ph.D. stage at the Doctoral School of Electronics, Telecommunications and Information Technology, University Polytechnic of Bucharest (Romania), contract number SD04/35/2018.

REFERENCES

[1] R.Jubleanu, D.Cazacu, N.Bizon "Stress in cylindrical and toroidal superconducting coils" –paper accepted and presented at ", The 12th International Conference on Electronics, Computers and Artificial Intelligence (ECAI 2020)

[2] E. Euxibie, J. L. Coulomb, G. Meunier, and J. C. Sabonnadiere, "Mechanical deformation of a conductor under the electromagnetic stresses," IEEE Trans. Magn., vol. 22, no. 5, pp. 828-830, 1986.

[3] Y. Iwasa, Case Studies in Superconducting Magnets: Design and Operational Issues, Springer-Nb-ti Berlin : Springer, 2009

[4] L. Hirsinger and R. Billardon, "Magneto-elatic finite element analysis including magnetic forces and magnetostriction effects," IEEE. Trans. Magn., vol. 31, no. 3, pp. 1877-1880, May 1995.

[5] A. P. S. Baghel, R. V. Thakar, S. V. Kulkarni, S. Chauhan "Linear Magneto-Structural Analysis by Sequential Coupling Method" 28 February 2014.

[6] Schneider-Muntau, H. J. "Magnet Technology Beyond 50T." IEEE Transactions On Applied Superconductivity, 2006: 926-933.

[7] F. Toral "Mechanical Design of Superconducting Accelerator Magnets" CIEMAT, Madrid, Spain, Superconductivity for Accelerators, Erice, Italy, 24 April - 4 May 2013

[8] Szabolcs Rembeczki " Design and Optimization of Force-Reduced High Field Magnets" Melbourne, Florida May, 2009

[9] B. ten Haken, L.J.M van de Klundert, V.S. Vysotsky, V.R. Karasik "The critical current in a NbTi tape measured in different directions of magnetic field and the current reduction due to the self field" IEEE TRANSACTIONS ON MAGNETICS, VOL. 28, NO. 1, JANUARY 1992

[10] Celentano, A. Augieri, A. Mauretti, A. Vannozzi, A. Angrisani Armenio, V. Galluzzi, S. Gaudio, A. Mancini, A. Rufoloni, I. Davoli, C. Del Gaudio, and F. Nann "Electrical and Mechanical Characterization of Coated Conductors Lap Joints" IEEE TRANSACTIONS ON APPLIED SUPERCONDUCTIVITY, VOL. 20, NO. 3, JUNE 2010

[11] Frederic Trillaud, Kevin Berger, Bruno Douine and Jean Levque "Distribution of current density, temperature and mechanical deformation in YBCO bulks under Field-Cooling magnetization" IEEE Transactions on Applied Superconductivity, Institute of Electrical and Electronics Engineers, 2018

[12] Ugur Kölemen, Orhan Uzun Cem Emeksiz, Fikret Yılmaz, Atilla Coskun, Ahmet Ekicibil, Bekir Özçelik "Mechanical Properties of BSCCO Superconductor by Oliver–Pharr Method and Work of Indentation Approach"J Supercond Nov Magn (2013) 26:3215–3219

[13] Antonio Bartalesi "Design of High Field Solenoids made of High Temperature Superconductors"May 19, 2010

2020 IEEE 26th International Symposium for Design and Technology in Electronic Packaging (SIITME)

Comparative Analysis of Pad Geometries Used for Multi-Layer Ceramic Capacitors in Power Distribution Networks

Adrian-Razvan Petre
TENSOR Romania
Bucharest, Romania
razvan.petre@tensor.ro

A.Drumea, M.Pantazica, C.I.Marghescu
Dept. of Electronics Technology
Politehnica University Bucharest
Bucharest, Romania
cristina.marghescu@cetti.ro

Abstract—Decoupling capacitors have already been mandatory for circuit board designs for decades, but lower supply voltages combined with higher currents required by the integrated circuits (ICs) made these passive components of higher importance. Capacitors provide a temporary source of energy from a time perspective and also a low-impedance path from a frequency perspective being essential components in Power Distribution Networks (PDNs). To achieve a certain target impedance, it is important to accurately characterize the Equivalent Series Inductance (ESL) of ceramic decoupling capacitors. Even if manufacturers give highly detailed simulation models for their parts, the ESL can easily be increased by the mounting inductance. This addition results from the VIA pair required to connect the component to the power planes of the board. In this paper, we investigate different pad geometries used for ceramic capacitors in PDNs which can increase or decrease the mounting inductance. A comparative analysis is performed based on the results and best design practices are listed in the end.

Index Terms—mounting inductance, loop inductance, reverse-aspect-ratio capacitors, PDN

I. INTRODUCTION

In the last years system complexity has increased along with operation frequencies for modern circuit boards requiring more power in a smaller area resulted from miniaturization processes. Moreover, supply voltages have been lower continuously to enhance better power consumption performances. A typical Power Distribution Networks (PDN) includes besides the voltage reference module and the Integrated Circuits (ICs) decoupling capacitors. These passive components ensure a local energy source next to the ICs and also a small impedance from the IC to the voltage reference module [10], [11].

Mounting structure has also a high influence in the performances of capacitors by adding an additional inductance in series with the existing Equivalent Series Inductance (ESL). This mounting inductance is caused by the VIA structure connecting the passive part to the power and ground plane structure and is highly influenced by the geometry of the capacitor footprint geometry.

In this paper we investigate different pad geometries used for Multi-Layer Ceramic Capacitors (MLCC) in Power Distribution Networks (PDN) by both a simulation-based approach

and also by measurements. A 4-layered circuit board including all the geometries of interest was designed and measurements were performed in a laboratory environment. The same circuit board was also simulated using ANSYS HFSS 3D full-wave solver. To accurately extract and separate mounting inductance from internal ESL, a simple RLC circuit is used to characterize the investigated structure. Results are presented in a tabular and graphical form and conclusions are drawn on what are the best geometries to be used, each with its own advantages and disadvantages.

II. CIRCUIT BOARD DESIGN

In this paper we study several pad geometries used to connect decoupling capacitors to internal power and ground planes in a 4-layered circuit board. The VIA structure is the one leading to an additional mounting inductance in series with the ESL of the capacitor. As seen in Figure 1, the source of this inductance are the two parallel VIAs crossed by opposite currents. If each VIA is considered to have a self inductance L_{self} and a mutual one L_{mutual} with regard to the other one, the total loop inductance is described by Equation 1. A few design guidelines can therefore be drawn: as short as possible and as close as possible VIAs are desired, since closer VIAs will result in a higher L_{mutual} which reduces the total inductance. Moreover, multiple VIA pairs in parallel for each power and ground connection will dramatically reduce the total loop inductance.

$$L_{loop} = 2 \cdot L_{self} - 2 \cdot L_{mutual} \tag{1}$$

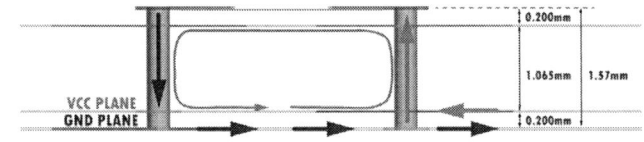

Fig. 1. Source of loop inductance added by the VIA structure interconnecting it to the power and ground plane pair. Dimensions for the stack-up used in this investigation are also presented here.

978-1-7281-7507-2/20 $31.00 © 2020 IEEE   434   21-24 October, Pitesti, Romania

As previously mentioned, placing the VIAs connecting decoupling capacitor to power and ground planes as close as possible is mandatory to reduce loop inductance. In order to enhance this feature, reverse aspect ratio capacitors are used. This type of components have the electrical pads situated on the length of the case and not on the width as regular aspect ratios have, thus allowing contact VIAs to be placed even closer. Considered of interest to investigate were pad geometries corresponding to case sizes of 0805 and 0603 with their reverse aspect ratios correspondents, 0508 and 0306.

In Figure 2 the pad geometries investigated for 0805 cases are presented for further discussions. Each structure has VIA holes spaced from the capacitor pads at specific values varying from 0.9mm for geometries no. 1, 2 or 3, 0.5mm for no. 4 or 5 and to 0mm for no. 6 and 7, in the last case the VIA annulus being tangent tot the capacitor pad. For the reverse aspect ratio geometries was considered of interest to investigate only the case of VIAs placed tangent to the capacitor pad, structures no. 9, 10 or 11. Cases no. 8 and 12 are two extreme ones where the VIA was place directly in pad, a design that could pose issues in the assembly process but which however has very good performances from a power integrity perspective. For each spacing value, the case of multiple VIA pairs was also investigated, as is the case of geometry no. 3, 5, 7, 10 or 11. This structures significantly reduce loop inductance but subsequently take more board area which could become an issue for highly populated boards. All the geometries from Figure 2 were also investigated for 0306 cases, with spacing from VIA to capacitor pad scaled to their corresponding aspect ratio.

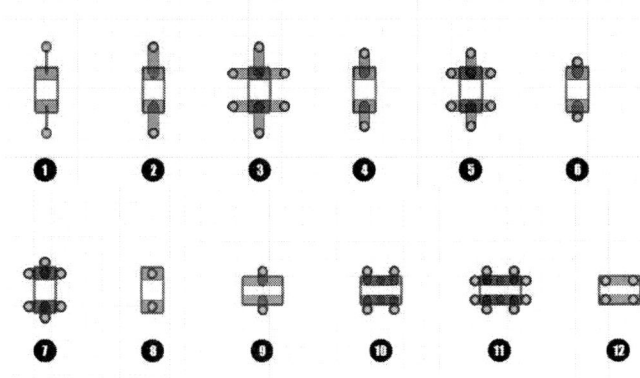

Fig. 2. Different pad geometries used for Multi-Layer Ceramic Capacitors investigated in this paper. The 12 geometries presented here were used for both 0805 and 0508 capacitors respectively 0603 and 0306.

In Figure 3 the 4-layered circuit board developed to measure and simulate the 12 different pad geometries is displayed for analysis. The four delay tune tracks visible on the upper layer are not part of this investigation. The 12 pad geometries are separated in four columns, one for each case style arranged from left to right: 0805, 0603, 0508 and 0306. The stack-up of this circuit board is the one presented in Figure 1 with layers 3 and 4 used as power and ground planes. Layer 1 is where

the investigated pad geometries are situated and layer 2 is not used. One female SMA connector connected to the power and ground plane structure is used to interface the circuit board with the measurement tool.

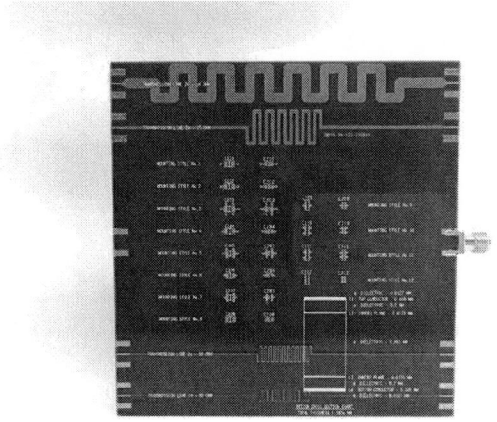

Fig. 3. 4-layered circuit board designed for this study. One SMA edge connector connected to the power and ground plane pair is visible in the right part. The four delay tune tracks of different thickness are not part of this investigation.

III. MEASUREMENT AND SIMULATION SETUP

This section describes the measurement and simulation setup, methodologies and materials used. Calibration of the measurement device and also characterization of the circuit under test are described.

A. Measurement Setup

In order to accurately measure the mounting loop inductance an Agilent 4396B Impedance Analyzer was used in conjunction with Agilent 43961A Impedance Test Adapter. Various methods to investigate impedance variation over frequency exist such as measuring the reflection coefficient and determining the value of the DUT from its definition [10], [13]. However, all these methods are limited to either measurements of large or small values of impedance. The U-I method of measurement in which the voltage and current at the contact points of the DUT are measured is the most precise over a large interval of values. Agilent 43961A Impedance Test Adapter also uses this method, this being also the reason for opting for this measurement setup displayed in Figure 4.

The measurement range is limited inferior by the 100kHz lower limit of the Agilent 4396B Impedance Analyzer and superior by the upper limit of the Impedance Test Adapter of 1MHz. However, this was not an issue for this investigation since the effects of interest were in the tenths to hundreds of MHz range as further discussed in the following paragraphs.

In order to calibrate the measurement device, short, load and open terminations from an Agilent 85032B Type N Calibration Kit were used to bring the calibration plane up to the output connector of the Impedance Test Adapter. These terminations

2020 IEEE 26th International Symposium for Design and Technology in Electronic Packaging (SIITME)

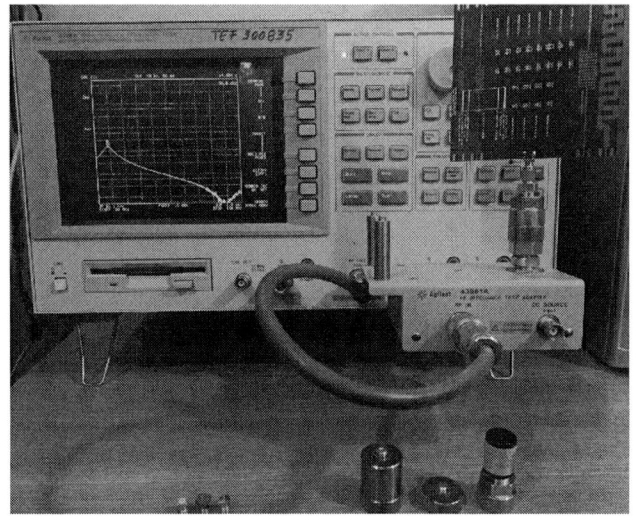

Fig. 4. Laboratory measurement setup with an Agilent 4396B Impedance Analyzer used with Agilent 43961A Impedance Test Adapter.

are seen in the bottom side of Figure 4. Since a female SMA connector was used to interface the circuit board with the impedance analyzer, this fixture also had to be calibrated. By proper terminating the SMA male connector from the Impedance Test Adapter with some custom-made terminations displayed in Figure 5, the calibration plane was brought up to the connector side.

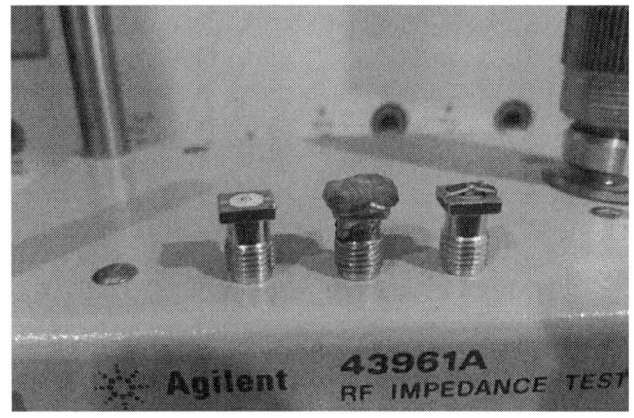

Fig. 5. Custom-made terminations of open, short and load (from left to right) used to terminate SMA male connector from the Impedance Test Adapter.

Various methodologies for measuring the loop inductance for each pad geometry exist. One method could be to short the two pads for each geometry with a $0\,\Omega$ resistor but this would also include the parasitic elements of the resistor and the ones of the power and ground planes. If the latter can be easily measured, characterizing a zero ohm resistor is difficult since the very small values of resistance and inductance are involved. If however a capacitor is soldered on the landing pad, a different behavior with three resonant peaks appears, each influenced by certain figures of interest of the structure. This was also the method used in this investigation. By soldering only one capacitor at a time

on the specific geometry and leaving all the others not connected, the loop inductance was accurately measured. Capacitors X7R all of 100nF nominal value from Murata were used, 0805 case style (Mfr. No. GCD21BR71H104KA01L), 0508 (Mfr. No. LLL216R71E104MA01L), 0603 (Mfr. No. GCJ188R71E104KA12D) and 0306 (Mfr. No. LLL185R71A104MA11L). By also using the online tool SimSurfing [13] from Murata, accurate S-parameters of these capacitors were downloaded and inserted into the profile simulation as presented in the next subsection.

With only one capacitor connected at a time using the specific pad geometry under test, the circuit formed by the power and ground planes, loop inductance of VIAs and mounted capacitor was characterized as the simple lumped RLC circuit from Figure 6. In this circuit the capacitor was described as a series circuit, elements C300, R300 and L300 from the presented figure where a series mounting inductance L301 also adds. Besides the L301 inductance which had to be determined, the other three values are known from Murata's SimSurfing online tool. The capacitance of the power and ground planes, C200, comes in parallel with the existing capacitor and L100 and R100 add in series with the circuit, each corresponding to spreading inductance and resistance of the planes.

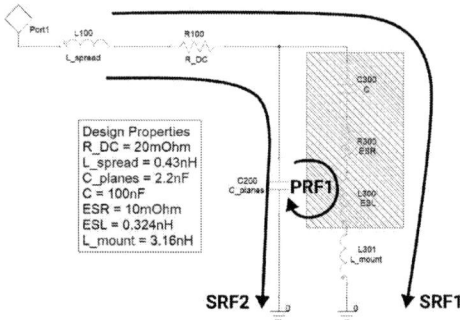

Fig. 6. Equivalent series and parallel RLC circuit developed to accurate characterize the interactions between capacitors and board inductance and capacitance.

The circuit from Figure 6 will have an impedance profile with three resonances points, two maxims and a minimum. The minimum ones are caused by the two series circuits called "SRF1" and "SRF2" and the parallel one by the parallel circuit pointed as "PRF1". In Table I the figures of interest for each of these peaks are displayed both in a literal way and also in a numerical one. The numerical values are calculated for values displayed in Figure 6. Resonance peak "PRF1" has its frequency determined by elements C_{planes} in series with C with ESL in series with L_{mount}, all the values except L_{mount} being known. This also points out the measurement techniques used in this paper: by measuring the resonance frequency of this peak, the mounting inductance was obtained.

Using the resonance method to measure the loop inductance for each pad geometry was the best approach for this specific task since resonance points can easily be measured due to

TABLE I

RESONANT PARALLEL OR SERIES POINTS FOR THE CIRCUIT SCHEMATIC IN FIGURE 6 WHICH ACCURATELY DESCRIBES THE POWER PLANES AND CAPACITOR STRUCTURE MEASURED IN THIS INVESTIGATION.

		SRF1	PRF1	SRF2
Resonant loop R [Ω]	Formula	$ESR + R_{DC}$	$ESR + R_{DC}$	R_{DC}
	Value	30	10	20
Resonant loop L [nH]	Formula	$ESL + L_{mount} + L_{spread}$	$ESL + L_{mount}$	$(ESL + L_{mount})\|\|L_{spread}$
	Value	3.91	3.48	0.38
Resonant loop C [nF]	Formula	C	$\frac{C_{planes} \cdot C}{C_{planes} + C}$	C_{planes}
	Value	100	2.15	2.2
Z_0 [mΩ]	Formula	$\sqrt{\frac{L}{C}}$	$\sqrt{\frac{L}{C}}$	$\sqrt{\frac{L}{C}}$
	Value	197.84	1272.19	417.11
$P_{resonance}$ [MHz]	Formula	$\frac{1}{2\pi \cdot L \cdot C}$	$\frac{1}{2\pi \cdot L \cdot C}$	$\frac{1}{2\pi \cdot L \cdot C}$
	Value	7.89	59.96	170.00
Z_{peak} [Ω]	Formula	$ESR + R_{DC}$	$\frac{Z_0^2}{ESR}$	R_{DC}
	Value	0.030	161	0.020

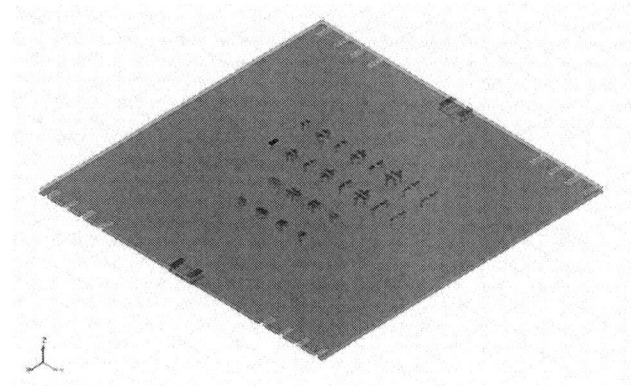

Fig. 7. Simulation setup in ANSYS HFSS full-wave 3D solver. The design from Figure 3 was simulated using S parameter models for each capacitor provided by the manufacturer.

their peak or valley characteristic. Even if the other two series resonance points could have been used to determine mounting inductance, they both also include the spreading inductance, a quantity that is highly dependent of the contact point diameter and length of the current path. Moreover, measuring the resonance frequency of a series circuit could result in large errors due to the low value of the inductance at the resonance point which could be close to or below the noise level on the measurement device.

B. Simulation Setup

ANSYS HFSS was used to determine via a simulation-based approach the mounting inductance for each pad geometry. ECAD design introduced in the simulation tool had a port configured in the place where the SMA connector was designed to be attached and the impedance profile from that point was investigated. For each simulation, the specific pad geometry under investigation was populated with a corresponding capacitor Touchstone model downloaded from Murata's tool, SimSurfing.

The stack-up of the board is the same already presented in Figure 1, the outer copper layers with thickness of 36um and the inner ones with 18um. The dielectric materials have a relative electrical constant of 4.6.

The impedance profile viewed from the port was investigated for each of the 12*2 pad geometries from 100kHz to 1GHz. As in the case of measurements, the loop inductance value was extracted indirectly by measuring the frequency point "PRF1" and extracting its value using the same equations from Table I.

IV. DISSEMINATION AND RESULTS

This section presents the results obtained from both simulation and measurement which correlated to a better than 10% error. These errors are attributed to the measurement process since this one is more prone to errors. Even if the impedance analyzer was proper calibrated as described in the previous section, only a 0.1GHz deviation in the measured PRF resonance frequency could have added 0.01nH to the mounting inductance.

The results of this investigation are synthesized in Table II. The pad geometries numbered from 1 to 12 for each capacitor case are listed in the first two columns. Internal capacitance and ESL were extracted from the online tool SimSurfing. By measuring the bare board with no capacitor attached to it the capacitance of planes, C_{planes}, was extracted. The last quantity measured is the frequency "PRF1" point when a single capacitor mounted on a specific pad geometry is soldered to the board. The frequency of the same "PRF1" frequency point was extracted by simulation in Ansys HFSS. Using the equations from I, the figure of interest of this study, the mounting inductance resulted.

In Figure 8 the board impedance profile is displayed in various situations. Firstly, no capacitor is placed on the board and its impedance is measured. This is how the power and ground planes capacitance was obtained. Secondly, a 0805 100nF capacitor was placed using pad geometry 1 and the board impedance profile was both measured using Agilent 4396B and also simulated using ANSYS HFSS. The last trace is the same board impedance profile estimated by a simple RLC circuit from Figure 6. The very good agreement between the impedance predicted by this circuit and the measured profile verified the accuracy of the RLC model created.

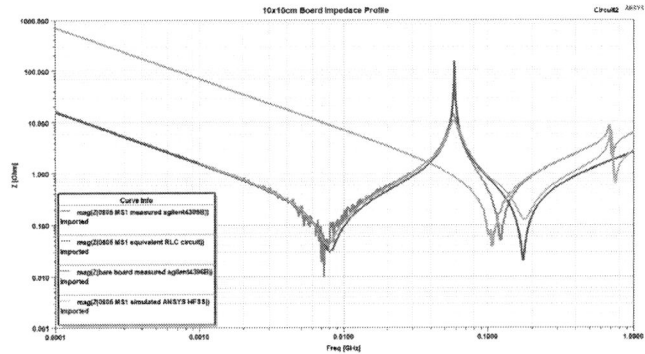

Fig. 8. Board impedance profile resulted from the full-wave 3D simulation, measurement and circuit characterization for a 0805 capacitor mounted with pad geometry No.1 to the board.

TABLE II

SYNTHESIZED RESULTS OF THIS INVESTIGATION WITH INTERNAL TO CAPACITORS VALUES, MEASURED AND SIMULATED ONES. ERROR IS CALCULATED FOR MEASUREMENT WITH REGARD TO ANSYS HFSS 3D FULL-WAVE SIMULATION.

CASE SIZE	Pad Geometry	INTERNAL		C_{planes} [nF]	MEASURED			SIMULATED			Error
		C [nF]	ESL [nH]		F_{PRF1} [MHz]	L_{mount} + ESL [nH]	L_{mount} [nH]	F_{PRF1} [MHz]	L_{mount} + ESL [nH]	L_{mount} [nH]	
0805	1	100	0.324	2.2	56.8	3.64	3.32	57.42	3.56	3.24	2.41%
	2				61.9	3.07	2.74	62.94	2.96	2.64	3.80%
	3				93.5	1.34	1.02	90.98	1.42	1.09	6.89%
	4				65.2	2.76	2.44	65.91	2.70	2.38	2.49%
	5				94.2	1.32	1.00	91.68	1.40	1.07	6.87%
	6				69.6	2.42	2.10	70.09	2.39	2.07	1.63%
	7				93.5	1.34	1.02	90.98	1.42	1.09	6.89%
	8				76.6	2.00	1.68	76.85	1.99	1.66	0.78%
0508	9	90	0.120		78.5	1.91	1.79	78.64	1.90	1.78	0.38%
	10				96.1	1.27	1.15	98.24	1.22	1.10	4.99%
	11				115.5	0.88	0.76	117.21	0.86	0.74	3.47%
	12				111.2	0.95	0.83	114.60	0.90	0.78	7.17%
0603	1	100	0.320		57.9	3.50	3.18	58.30	3.45	3.13	1.53%
	2				62.1	3.04	2.72	63.40	2.92	2.60	4.75%
	3				95.8	1.28	0.96	96.23	1.26	0.94	1.20%
	4				65.5	2.73	2.41	66.70	2.64	2.32	4.21%
	5				96.3	1.27	0.94	97.80	1.23	0.91	4.25%
	6				70.1	2.39	2.07	71.23	2.32	1.99	3.77%
	7				90.1	1.45	1.13	92.34	1.38	1.06	6.56%
	8				79.3	1.87	1.54	81.22	1.78	1.46	5.97%
0306	9	90	0.090		80.1	1.83	1.74	81.09	1.79	1.70	2.62%
	10				99.9	1.18	1.09	99.76	1.18	1.09	0.30%
	11				126.4	0.74	0.65	127.53	0.72	0.63	2.05%
	12				122.2	0.79	0.70	123.68	0.77	0.68	2.76%

The numerical result from Table II for the mounting inductance are displayed in a graphical form in Figure 9 for all the geometries used in this investigation. Mounting inductance is very low for pad geometries such as No. 3, 5 and 7 where three VIA pairs in parallel were used to connect the capacitor to the power and ground planes. As expected, the reverse aspect ratio of 0508 and 0306 give the best results in terms of mounting inductance since these cases allow VIA holes to be placed closer one to another. Moreover, this type of capacitors also come with a lower internal equivalent series inductance as their homologous regular aspect ratio capacitors.

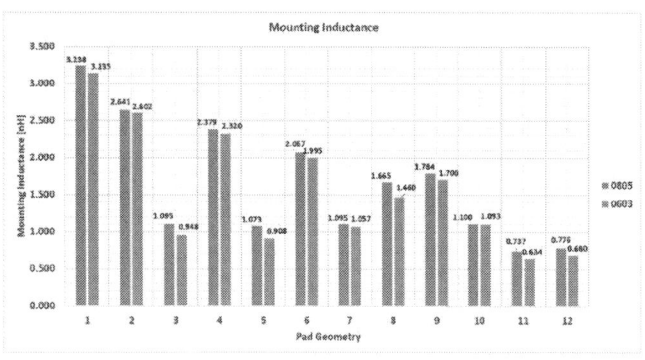

Fig. 9. Mounting inductance for each pad geometry investigated in this paper, both for 0805 and 0603 capacitor sizes. Results from ANSYS HFSS 3D full-wave simulation were used.

In Figure 10 the total inductance of the capacitors is displayed from the smallest to the largest only for 0805 pad geometries. As expected, the best geometries to be used are the ones corresponding to reverse aspect ratio capacitors or to multiple VIA pairs, each of them with a certain increase in cost: the latter occupying a larger board area and the fist with a higher price. As expected, pad geometries such as no. 1, 2 or 4 with long traces connecting a capacitor to its corresponding VIA pair have the worst performances and should never be used when connecting capacitor to power and ground planes.

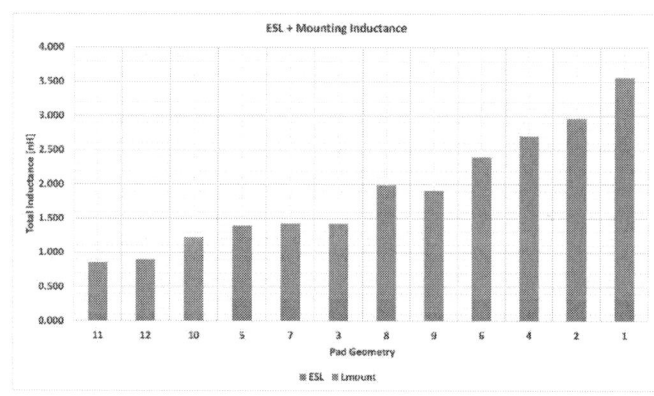

Fig. 10. Contribution of mounting inductance to the total inductance of a capacitor for each pad geometry investigated in this paper, only for 0805 case sizes. Results from ANSYS HFSS 3D full-wave simulation were used.

V. CONCLUSIONS

A variety of pad geometries was investigated in this paper for both 0805 and 0603 capacitors. Reverse aspect ratio cases such as 0508 and 0306 were also studied. Results validated initial assumptions that as closely as possible VIAs should be used to connect capacitors to the power and ground planes and that the VIA loop should be minimized as far as is possible. Multiple VIA pairs used in parallel gave the best results but unfortunately these structures come with the cost of a larger occupied board area. Reverse aspect ratio capacitors are also of interest since they allow power and ground VIAs to be placed closer on the board.

In conclusion, pad geometries with VIAs spaced as close as possible and with multiple VIA pairs placed in parallel should be used if mounting inductance is desired to be reduced. Alternatively, if this figure of interest is desired to be reduced even more, reverse aspect ratio capacitors should be used.

ACKNOWLEDGMENT

Special thanks to TENSOR Company, Ansys Channel Partner in Romania, for supporting the authors with the necessary software tools to simulate some of the works done for this research.

REFERENCES

[1] Marcel Manofu, Radu Voina and Cătălin Negrea, "Analysis of Multi-Layer Ceramic Capacitors used in Power Distribution Networks" presented at the 2nd PCNS 10-13th September 2019, Bucharest, Romania.

[2] T. Roy, L. Smith and J. Prymak, "ESR and ESL of ceramic capacitor applied to decoupling applications," IEEE 7th Topical Meeting on Electrical Performance of Electronic Packaging (Cat. No.98TH8370), West Point, NY, USA, 1998, pp. 213-216, doi: 10.1109/EPEP.1998.733985.

[3] R. Fizeşan and D. Pitică, "Power integrity design tips to minimize the effects of mounting inductance of decoupling capacitors," 2012 13th International Conference on Optimization of Electrical and Electronic Equipment (OPTIM), Brasov, 2012, pp. 36-41, doi: 10.1109/OPTIM.2012.6231953.

[4] M. Pantazica, A. Drumea, and C. Marghescu, "Analysis of self discharge characteristics of electric double layer capacitors", Proc. of 2017 IEEE 23st International Symposium for Design and Technology in Electronic Packaging (SIITME2017), pp. 90-93, October 2017.

[5] A. Drumea, and M. Blejan, "Design, implementation and testing of an electrohydraulic system for automated winding machine for aluminum wire rods," Proc. of the International Conference on Electronics, Computers and Artificial Intelligence (ECAI2013), pp.1-4, June 2013.

[6] A. Drumea, and M. Pantazica, "Aspects of using low layer count PCBs for embedded systems with FPGA devices in BGA packages", Proc. of 2016 IEEE 22nd International Symposium for Design and Technology in Electronic Packaging (SIITME2016), pp. 74-77, October 2016.

[7] A. Drumea, and R. Dobre, "Analysis of power supply circuits for electroluminescent panels," Proc. SPIE 10010 Advanced Topics in Optoelectronics Microelectronics and Nanotechnologies VIII (ATOM-N2016), vol. 10010, no. 1, pp. 100101D-100101D-6, 2016.

[8] Caramaliu, R.V., Vasile, Al., and Bacis, I.B. , "Wearable Vital Parameters Monitoring System", Proc. SPIE 9258, Advanced Topics in Optoelectronics, Microelectronics, and Nanotechnologies VII (ATOM-N2015), 92580R, 2015.

[9] Vasile, Al. and Bacis, I.B.V., "Intrusions and their lock in data communications through Internet", Proc. of the 25th International Symposium for Design and Technology in Electronic Packaging (SIITME2019), pp. 399-403, October 2019.

[10] I. Nowak and J. R. Miller, Frequency-Domain Characterization of Power Distribution Networks. USA: Artech House, Inc., 2007. Simulation Methods and Tools. pp. 32–54.

[11] Larry D. Smith, Eric Bogatin. Principles of Power Integrity for PDN Design - SIMPLIFIED. Prentice Hall, 2017. ISBN 978-0-13-273555-1.

[12] Murata SimSurfing https://ds.murata.co.jp/simsurfing/mlcc.html?lcid=en-us accessed at 10th of October 2020

[13] Agilent Technologies, USA. Agilent Impedance Measurement Handbook, 4 ed., 2009.

EMC Simulation of Conducted Emissions Produced by a DC-DC converter

Andrei-Marius Silaghi
Dept. of Measurements and Optical Electronics
University Politehnica Timisoara
Timisoara, Romania
andrei.silaghi@upt.ro

Florin Berinde
Dept. of Power Engine Systems
Vitesco Technologies
Timisoara, Romania
florin.berinde@continental-corporation.com

Ciprian Bleoju
Dept. of Mechanics / Qualification Laboratories
University Politehnica Timisoara / Continental Automotive
Timisoara, Romania
ciprian.bleoju@continental-corporation.com

Aldo De Sabata
Dept. of Measurements and Optical Electronics
University Politehnica Timisoara
Timisoara, Romania
aldo.de-sabata@upt.ro

Abstract—**The Conducted Emissions (CE) measurement is mandatory for a product to enter mass production. This paper presents LTspice and Matlab tools application for EMC Simulation, namely of CE – LISN (Line Impedance Stabilization Network) method described in CISPR 25 Standard. Based on analog simulation (without any layout information or mechanical models) optimizations are made for a DC-DC converter, to reduce the levels of CE.**

Keywords—*LTspice; CE-LISN; EMC Simulation; Common mode choke*

I. INTRODUCTION

Modern cars are equipped with complex electric and electronic systems that control the operation of the engine, brakes, and other elements and assist the driver to ensure a convenient and safe driving of the vehicle. Communication facilities are provided to allow for exchange of information between electronic units of the car and between the car system and driver and the exterior. The ever-increasing complexity of car electronics made electromagnetic compatibility an important issue in the car industry. Research in this field, standards conception and industrial application have gain momentum in the recent period [1].

Improving CE measurement results according to CISPR 25 [2] has been approached by some of the authors in the past [3]. In this paper, a case study concerning on-spot devised methods for reducing CE levels are presented, so that the client demands are satisfied and smallest levels for conducted emissions obtained [3].

The measurement in a real-life EMC Laboratory is a cost-intensive task so, in this context, application of simulation for improving test results has been investigated by the authors [4], [5]. For example, in [5] PSpice is used to provide a circuit simulation tool that can be used to verify Automotive products.

In the literature, various authors tackle the problem of conducted emissions simulation [6]-[8]. PSpice has been used in the past to model a DC-DC converter and to simulate the levels of CE it generates between 100 and 500 kHz. A comparison is made between simulated noise and measured noise according to the CISPR 25 Standard [6].

Afterwards, PSpice is used to predict the conducted emissions of a switch-mode power converter [7]. The time domain simulation takes into consideration discrete element parasitics and also the ones from the printed circuit board. This model based on CAD PCB layout data can be used to study voltage waveforms and spectra [7].

CST Microwave Studio is also used for modelling the CE of a Switched-Mode Power Supply (SMPS) [8]. The noise voltage source is calculated by relying on measurement data and the coupling path between aggressor and victim is characterized by simulation and measurement. Measurements are made to validate the simulation model and show that a good prediction is obtained using numerical simulation [8].

In this paper we present the use of LTspice tool for EMC Simulation purpose: namely Conducted Emissions -LISN method described in the CISPR 25 Standard. Based on analog simulation (without any layout information or mechanical models), optimizations are made to reduce the levels of common mode currents, which are the basis for calculating conducted emissions.

The paper is structured as follows. Section II presents the configuration setup used for initial simulations and also results obtained with LTspice and Matlab. In Section III, more simulation scenarios are reported. Conclusions are drawn in the last Section.

II. INITIAL SIMULATIONS

EMC Simulation has been widely used in past years to reduce the time-to-market of products. By using analog and 3D

2020 IEEE 26th International Symposium for Design and Technology in Electronic Packaging (SIITME)

simulation, a project team can skip expensive investigations in a laboratory, for its Automotive DUT (Device Under Test) [4].

For example, a type of DC-DC converter has been modelled in PSpice in [6], but no solutions have been proposed to reduce CE levels. In our case study, we are concerned with design modifications to improve CE test results, according to CISPR 25 standard [2], for a DC-DC converter. Some of our proposed tests and improvements consist of: placement of common mode chokes (CMC) (with different values), adding an extra low voltage ground, simulation with grounding or no grounding from the DUT to test table, placing 3 LISNs.

Firstly, in Fig. 1 the EMC Setup of the EUT (Equipment under test) is represented. The EUT is a DC-DC converter used in automotive area. It has a load of 0.25 Ω. The setup uses two LISNs (Line Impedance Stabilization Network) and the EUT is supplied with 48 V. Also, the parasitic coupling between the load and the chassis is simulated by a 1nF capacitor.

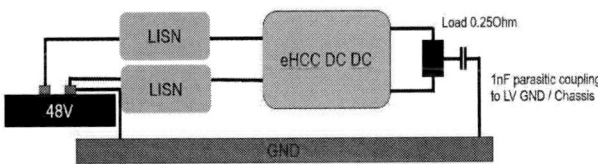

Fig. 1. EMC Setup Conducted Emissions Simulation

This setup has been built in LTspice [9]. LTspice is a high-performance SPICE simulation software, schematic capture and waveform viewer with enhancements and models for easing the simulation of analog circuits [9]. For example, in Fig. 2 the supply part (48 V), the 2 LISN models and the EMC Filter at the input of the DC-DC converter schematic are reported. All the values are parametrized, so that a convenient adjustment can be applied to different component values. Also, the 470 μH CMC is depicted in this schematic, although at the beginning it was excluded from simulation.

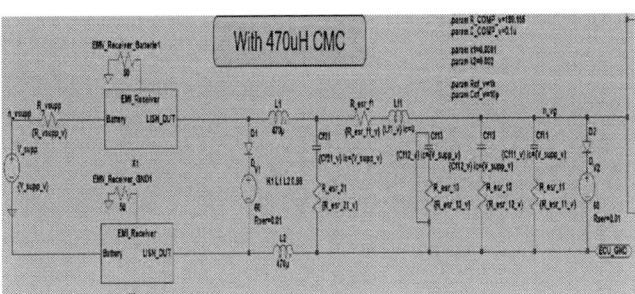

Fig. 2. LISN + EMC Filter part from schematic

The EMC filter at the input of the DC-DC converter has the purpose to ensure a damping effect to maintain the stability of the converter. For example, the capacitor C_{fl3} had this role, but it has not been connected in the considered situation, because it did not resist through the test of "alternative voltage superimposed". This role of maintaining the stability has thus been realized by adjusting the parameters of the feedback loop of the DC-DC. Also, the sources V1 and V2 together with diodes D1 and D2 have the role of simulating the effect of limiting

diodes that fade the voltage pulses that appear at sudden interruptions of current flow through the filter coil.

For the initial schematic, without modifications (and without CMC), the voltage as a function of time measured at the "+" LISN has been saved and the result has been plotted by means of a Matlab script (created to simulate the functioning of a measuring receiver with Peak and Average detectors).

In Fig. 3 the interference voltage level (expressed in dBμV, on Y axis) versus frequency (between 9kHz and 10 MHz, on X axis) is represented. The peak and average detectors outputs are compared with the corresponding limits from CISPR 25 Standard (CE-LISN method). The result is a fail because the peak and average detectors cross the limits between 500 kHz – 2 MHz.

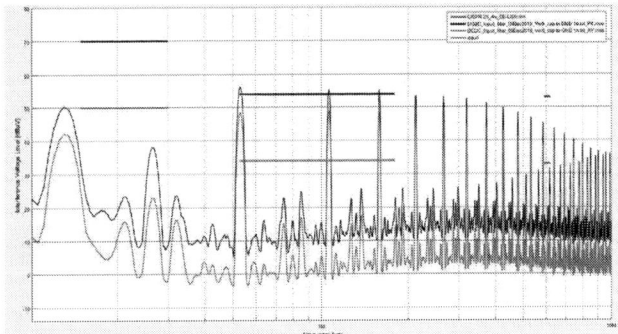

Fig. 3. Initial fail simulation

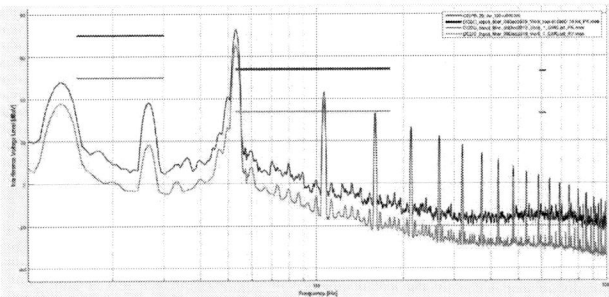

Fig. 4 Fail result with CMC of 100μH

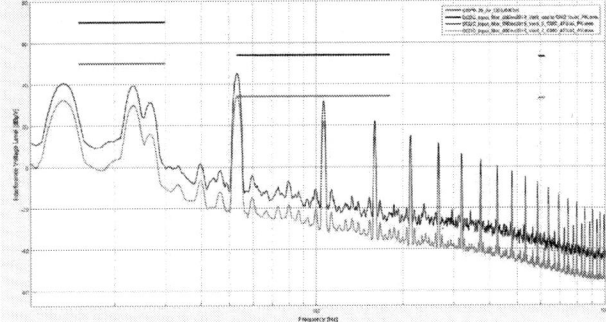

Fig. 5. Pass result with CMC of 470 uH

Thus, solutions to improve CE simulation results had to be tested. Addition of a CMC is reported in this paper. Common mode choke coils are used to suppress common mode noise. This type of coil is produced by winding the signal or supply wires on a ferrite core. Since magnetic flux flows inside the

978-1-7281-7507-2/20 $31.00 © 2020 IEEE   441   21-24 October, Pitesti, Romania

ferrite core, common mode choke coils work as an inductor against common mode current [10]. The CMC is placed between D1 and V1 and Cf21 on the other side. With the use of a 100 µH CMC, the results are improved but still a fail results (see Fig. 4). If a 470 µH CMC is used, a pass is obtained (Fig. 5), due to the fact that the peak and average detectors are below their corresponding limits from the CISPR 25 Standard. In this way, a good solution to reduce CE levels is obtained.

III. OTHER SIMULATIONS

Although the solution with CMC reduces significantly the CE levels, this solution is an expensive one, and also the CMC occupies a large area on the PCB, so other variants must be found.

For example, in Fig. 6 a new concept is presented. From CST Microwave Studio tool, a wiring harness model is exported for use in LTspice. The load from the right side is simulated now with the following values for the components: R_{10}=272 mΩ, L_7=0.8 µH and C_2=C_3=100 pF. If we assume the use of a loadbox in an EMC Laboratory test, C_2 and C_3 become 3.4nF (both cases were simulated).

Also, to improve CE test, an extra low voltage (LV) ground has been added to the schematic. To simulate the connection between ECU GND and LV GND, the following components have been used: C_1=60µF in series with L_2=0.166nH and R_5=0.75mΩ. In laboratory testing, we could encounter two scenarios regarding the grounding of the DUT: with and without a connection between the chassis of the DUT and the metallic test table. The first scenario is simulated by L_5=3.17µH and R_8=22.4mΩ. The result for the modified setup from Fig. 6 is visible in Fig. 7, where another pass result has been obtained.

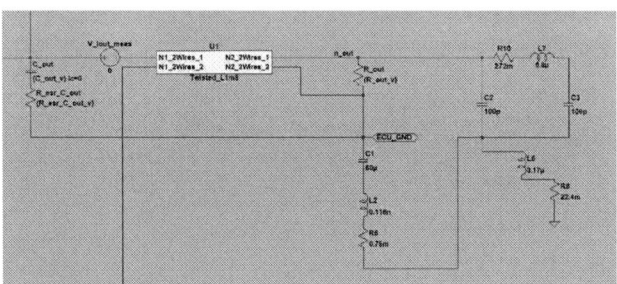

Fig. 6. Improved schematic with GND LV (and no ground from DUT to test table)

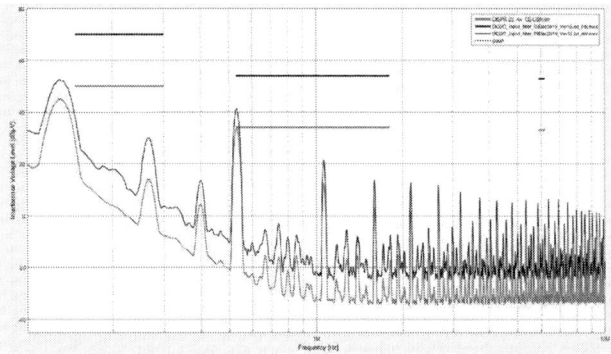

Fig. 7. Pass result for Fig. 6

The second scenario has been simulated as represented in Fig. 8. This time we did not have in the schematic the combination of components: L_5=3.17 µH and R_8=22.4 mΩ (which represented the ground connection between the DUT and the test table). Again, in this case a pass result has been obtained (Fig. 9).

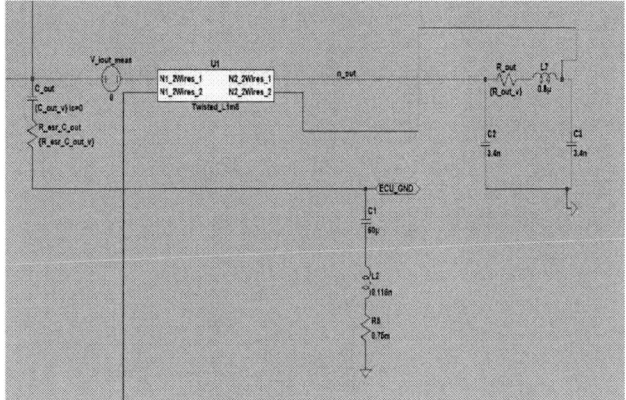

Fig. 8. Ground from DUT to test table – 2 LISN

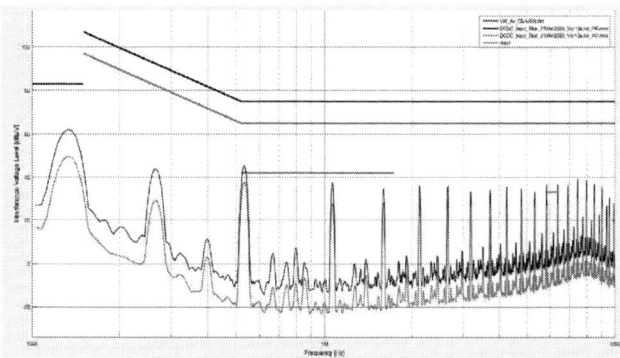

Fig. 9. Pass result for the setup from Fig. 8

Afterwards, an extra LISN has been placed at the LV GND, to study its influence at this new ground connection (Fig. 10). Here, we considered again the scenario with no ground from DUT to test table (with L_5=3.17 µH and R_8=22.4 mΩ). The final result is visible in Fig. 11, which is again a pass one.

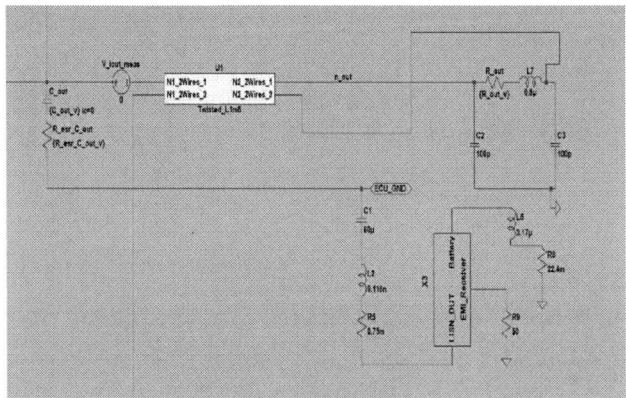

Fig. 10. No ground from DUT to test table – 3 LISN

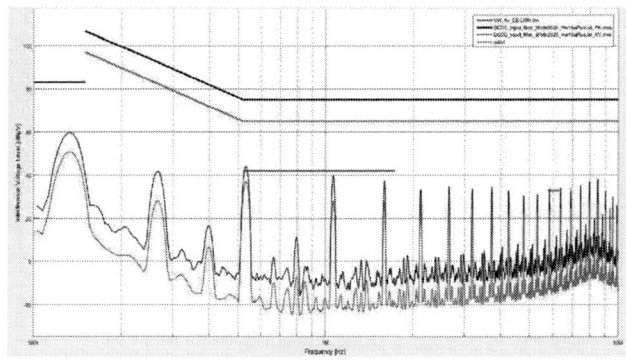

Fig. 11. Pass result for final trial (Fig. 10)

IV. CONCLUSIONS

In this paper we reported the use of a combination of simulation tools, namely LTspice and Matlab with the intent to lower the CE levels of an automotive product. These requirements are stated in CISPR 25 Standard (CE-LISN testing method).

Firstly, we built the initial setup of the DC-DC converter in LTspice with 2 LISNs. Analog simulation has been used, without information regarding layout. The interference voltage level versus frequency was plotted in a special Matlab tool, to simulate the behavior of a measuring receiver from EMC laboratory testing.

The initial result was a fail, so a common mode choke was added to the EMC filter of the product. This solution lead to a pass result, but the component being expensive, alternative solutions had to be considered. We can mention the following proposed solutions: adding an extra low voltage ground, simulation with grounding or no grounding from the DUT to test table, placing 3 LISNs.

As future work, these solutions will be implemented in the product, and tested in an EMC laboratory, to prove the existence of a match between simulation and measurement results.

ACKNOWLEDGMENT

The authors wish to acknowledge the support given by Continental Automotive Timisoara (Qualification Laboratory) and Continental Automotive Regensburg (Design and Consultancy Group – Mr. Felix Mueller).

The research was funded by the Ministry of Education and Research of Romania through UEFISCDI, project code PN-III-P1-1.2-PCCDI-2017-0917.

REFERENCES

[1] T. Rybak, M. Steffka, Automotive Electromagnetic Compatibility, Kluwer Academic Publishers, USA, 2004.

[2] International Standard CISPR 25, "Vehicles, boats and internal combustion engines - Radio disturbance characteristics - Limits and methods of measurement for the protection of on-board receivers", Ed. 4, 2016.

[3] A. Silaghi, A. Buta, S. Baderca, A. De Sabata, "Methods for reducing Conducted Emissions levels", IMEKO 2017, 14-15 Sept 2017, Iasi (Romania), 2017.

[4] A. Silaghi, F. Mueller, A. De Sabata, A.-P. Buta, P.-M. Nicolae, "Analysis of Shielding Effectiveness of an Automotive Display through Simulation and Testing", EMC Europe, 2020, accepted for publication.

[5] A. Silaghi, A. De Sabata, A. Graur, R. Fechet, "Simulation of Surge Pulse Generator and Applications in Automotive Immunity Testing", 2020 International Conference on Development and Application Systems (DAS), 21-23 May 2020, Suceava (Romania), 2020.

[6] A. Durier, C. Marot, O. Crepel, "Using the EM Simulation tools to predict Conducted Emissions level of a DC/DC boost converter: Introducing EBEM-CE model", 2013 9th International Workshop on Electromagnetic Compatibility of Integrated Circuits (EMC Compo), 15-18 Dec. 2013, Nara (Japan).

[7] S. Wessling, S. Dickmann, "Predicting the Conducted Emissions of Switched-Mode Power Converters Including Component and Printed Circuit Board Parasitics", 2015 IEEE International Symposium on Electromagnetic Compatibility (EMC), 16-22 August 2015, Dresden (Germany), 2015.

[8] Y. Wang, S. Bai, X. Guo, et. al, "Conducted Emission Modelling for a Switched-Mode Power Supply (SPMS)", 2015 IEEE Symposium on Electromagnetic Compatibility and Signal Integrity, 15-21 March 2015, Santa Clara (USA), 2015.

[9] LTspice v.2020, www.analog.com

[10] www.murata.com

IEEE
445 Hoes Lane
Piscataway, NJ 08854-4141

ISBN 978-1-7281-7507-2